Measurement Systems

Application and Design

McGraw-Hill Series in Mechanical Engineering

Measurement Systems
Application and Design

Fifth Edition

Ernest O. Doebelin
Department of Mechanical Engineering
The Ohio State University

Mc Graw Hill **Higher Education**

Boston   Burr Ridge, IL   Dubuque, IA   Madison, WI   New York   San Francisco   St. Louis
Bangkok   Bogotá   Caracas   Kuala Lumpur   Lisbon   London   Madrid   Mexico City
Milan   Montreal   New Delhi   Santiago   Seoul   Singapore   Sydney   Taipei   Toronto

The McGraw·Hill Companies

MEASUREMENT SYSTEMS: APPLICATION AND DESIGN, FIFTH EDITION
International Edition 2003

Published by McGraw-Hill, a business unit of The McGraw-Hill Companies, Inc., 1221 Avenue of the Americas, New York, NY 10020.

10 09 08 07 06 05 04 03 02 01
20 09 08 07 06 05 04 03
CTF SLP

Cover concept: *Ernest O. Doebelin*; computer image: *©Photodisc, Global Communications, Vol. 64*

Library of Congress Cataloging-in-Publication Data
Doebelin, Ernest O.
  Measurement systems : application and design / Ernest O. Doebelin.—5th ed.
      p.   cm. – (McGraw-Hill series in mechanical and industrial engineering)
  Includes index.
  ISBN 0-07-243886-X
  1. Measuring instruments. 2. Physical measurements. I. Title. II. Series.
  QC100.5.D63   2004
  681'.2—dc21   2003044176
                                                                 CIP

When ordering this title, use ISBN 0-07-119465-7

Printed in Singapore

www.mhhe.com

Ernest O. Doebelin has received his B.S., M.S., and Ph.D. degrees in Mechanical Engineering from Case Institute of Technology and Ohio State University, respectively. While working on his Ph.D. at Ohio State University, he started teaching as a full-time instructor, continuing this activity for four years. Upon completion of his Ph.D., he continued teaching as Assistant Professor. At this time (1958), required courses in control were essentially unheard of in mechanical engineering, but the department chair encouraged Dr. Doebelin to pursue this development. Over the years, he initiated, taught, and wrote texts for eight courses in system dynamics, measurement, and control, ranging from sophomore level to Ph.D. level courses. Of these courses, seven had laboratories, which Dr. Doebelin designed, supervised the construction of, and taught. Throughout his career, he continued to actually teach in all the laboratories in addition to training graduate-student assistants. In an era when one could opt for an emphasis on teaching, rather than contract research, and with a love of writing, he published 11 textbooks: Dynamic Analysis and Feedback Control (1962); Measurement Systems (1966); System Dynamics: Modeling and Response (1972); Measurement Systems, Revised Edition (1975); System Modeling and Response: Theoretical and Experimental Approaches (1980); Measurement Systems, 3rd edition (1983); Control System Principles and Design (1985); Measurement Systems, 4th edition (1990); Engineering Experimentation (1995); System Dynamics: Modeling Analysis, Simulation, Design (1998); and Measurement Systems, 5th edition (2004). Student manuals for all the laboratories, plus condensed, user-friendly software manuals were also produced.

The use of computer technology for system analysis and design, and as embedded hardware/software in operating control and measurement systems, has been a feature of all his texts, beginning with the first analog computers in the 1950s and continuing to today's ubiquitous PC. Particularly emphasized was the use of dynamic system simulation software as a powerful teaching/learning tool in addition to its obvious number-crunching power in practical design work. This started with the use of IBM's CSMP, and gradually transitioned into the PC versions of MATLAB/SIMULINK. All the texts tried to strike the best balance between theoretical concepts and practical implementation, using myriad examples to familiarize readers with the "building blocks" of actual systems, vitally important in an era when many engineering students are "computer savvy" but often unaware of the available control and measurement hardware.

In a career which emphasized teaching, Dr. Doebelin was fortunate to win many awards. These included several departmental, college, and alumni recognitions, and the university-wide distinguished teaching award (five selectees yearly from the entire university faculty). The ASEE also presented him with the Excellence in Laboratory Instruction Award. After his retirement in 1990, he continued to maintain a full-time teaching schedule of lectures and laboratories, but only for one quarter each year. He also worked on a volunteer basis at Otterbein College, a local liberal arts school, developing and teaching a course on Understanding Technology. This was an effort to address the nationwide problem of technology illiteracy within the general population. As a further "hobby" of retirement, he has become a politics/economics junkie, focusing particularly on alternative views of globalization.

CONTENTS

PREFACE TO THE FIFTH EDITION

This book first came out in 1966; it might be useful to quickly review how it has changed (and in some ways stayed the same) over the span of some 38 years. Its original premise was that measurement science and technology was a significant field of engineering interest in its own right, rather than an adjunct to various specialty areas such as fluid mechanics or vibration. Thus, it warranted its own courses and labs that emphasized this general viewpoint. This does not mean that specialty courses in, say, vibration measurement or heat transfer measurement are not appropriate in a curriculum, but that preceding such courses (or at least at *some* point), students should encounter measurement as a basic method for studying and solving engineering problems of all types. The background needed to appreciate this generalist view has two major components: the hardware and software of measurement systems, and the methodology of experimental analysis. *Measurement Systems* has focused on the first of these, and in 1995, I addressed the second in a new text.[1] This viewpoint continues in this fifth edition.

In 1966 personal computers were still far in the future, but mainframe machines used in a "batch mode" were already having major impacts on engineering and engineering education. As computer technology became more and more pervasive, the text recognized this trend and gradually added those computer-related topics that were relevant to the measurement process. These included computer simulation of measurement-system dynamic response, convenient statistical software, and the *vital role played by sensors in computer-aided machines and processes.* This latter application area is today a major justification for the general view of measurement espoused above. Almost every machine and process being designed today by engineers uses some form of feedback control implemented by digital hardware and software. *Every* such system includes one or more sensors that are absolutely vital to proper system functioning. A designer who has not been exposed to the "generalist" view of measurement and thus made aware of the devices and analysis methods available is at a distinct disadvantage in "inventing" a new process or machine. Since the needed computer technology is so powerful and cost/effective, the major roadblocks to implementing a new design concept are often not there but rather in the sensors and actuators. While this text is certainly not a controls book, the use of simple control concepts was always included because *feedback-control systems use sensors* and *many sensors use feedback principles* (hot-wire anemometers, servo accelerometers, chilled-mirror hygrometers, etc.). Since the book does not presume a previous course on control, these applications are presented so they

---

[1] E. O. Doebelin, "Engineering Experimentation: Planning, Execution, Reporting," McGraw-Hill, New York, 1995.

are understandable to such readers. It is perhaps surprising to some that a good understanding of such dynamic systems can be achieved by simple descriptions augmented by powerful and easy-to-use simulation software. In the current edition, major use of MATLAB/SIMULINK simulation provides this effective learning tool.

From the 1966 beginnings, the text devoted considerable space to the *system-dynamics viewpoint* of measurement-system dynamic response. This was originally influenced by the author's teaching of system-dynamics courses at various levels and the writing of several texts focused on this area.[2] (The 1972 text was revised and expanded in 1998.[3]) When a system-dynamics course is included early in the curriculum, this general background can then be applied and reinforced in later application courses such as control, vibration, measurement systems, vehicle dynamics, acoustics, etc. This curricular design is efficient and effective since the basic system dynamics need be presented only once, while the later application courses can penetrate more deeply into their specialty focus, while at the same time reinforcing student understanding of earlier material. While I believe that required system-dynamics courses serve this valuable function, some readers of *Measurement Systems* will certainly not have this preparation. Thus, this and earlier editions provide the needed background material in condensed, but effective, form. The current edition continues the heavy emphasis on *frequency-spectrum methods,* utilizing MATLAB (e.g., FFT) software wherever applicable.

The original organization into three major parts is retained in this new edition:

1. General concepts
2. Measuring devices
3. Manipulation, transmission, and recording of data

Within this framework, the Table of Contents gives a more detailed breakdown, which is useful in selecting the parts of the text that might be appropriate for a particular course and instructor. While the length of the text may at first seem daunting to a prospective user (instructor or student), it is not difficult to browse the content and pick out a coherent set of topics that suits the needs of a specific course. We face a similar situation at Ohio State where this text is used in three courses, two required and one elective. The first required course has a 4-hour lab and 3 hours of separate lecture for a total of 5 credit hours for one quarter. The lecture component is perhaps stronger than in a typical measurement course because we have chosen to include a "minicourse" in applied statistics and considerable material on technical communication (written and oral). These two topics are taught from my *Engineering Experimentation* text, which has a detailed coverage. The statistics material is intended for *general* applicability, not just for measurement situations, since statistics is not taught elsewhere in the curriculum. Requiring two textbooks

---

[2]E. O. Doebelin, "System Dynamics: Modeling and Response," Merrill, Columbus, OH, 1972; "System Modeling and Response: Theoretical and Experimental Approaches," Wiley, New York, 1980.

[3]E. O. Doebelin, "System Dynamics: Modeling, Analysis, Simulation, Design," Marcel Dekker, New York, 1998.

(*Measurement Systems* and *Engineering Experimentation*) for a single course seems prohibitively expensive, but the same two texts are also used in a required "project lab" course that follows on the heels of this course so the total expense is not unreasonable. The third course, which uses only *Measurement Systems,* is an elective for seniors and graduate students, and extends in breadth and depth from the first required course. If *Measurement Systems* seems to be too lengthy for a single course, consider that most students after graduation will likely encounter the need for this kind of information either for the design of computer-aided systems, which always require sensors and associated signal processing, or for experimental design/ development projects. If they have become familiar with the text by using parts of it in a course, it will become a valuable resource for their engineering practice, a feature not shared by texts that are less comprehensive.

An important part of many measurement systems is the data-acquisition and -processing software, usually implemented in a personal computer (desktop or laptop). When the previous edition was being written (late 1980s), personal computers were just arriving on the scene, and data-acquisition software for them was not widely available. Chapter 14 of that fourth edition was a brief presentation of a personal computer/software system (MACSYM) that had been designed, built, and marketed by Analog Devices specifically for data-acquisition and control applications, an unserved niche market that the company hoped to capitalize on. We acquired several of these systems for student and research use, and at that time, they met this need very well. Unfortunately for Analog Devices (which was highly successful, and continues to be with other product lines), personal computers shortly became a mass market with plummeting prices, making the MACSYM system, while technically excellent, economically unviable. Since then, many software products for personal computer data acquisition and control have appeared and today compete in this important field. Certainly the best known and most widely used is LABVIEW from National Instruments, and many engineering educators use this product for teaching/research, especially since the company offers very good educational discounts. It is not possible for a single individual to comprehensively exercise and then evaluate all the software of this class that is available, so judgments as to suitability for undergraduate teaching purposes are likely to be colored by personal experience and preferences. Based on my own surveys and hands-on experience with students in our labs, I have concluded that the DASYLAB software offers significant advantages for both teaching and many industrial applications. Perhaps National Instruments also recognized this potential since they recently bought the German software company that produces DASYLAB.

Chapter 13 of this edition is devoted to an introduction to DASYLAB, and a version of the software is provided with each copy of the book. This version does, of course, *not* allow its use with actual sensors, but one of the useful features of all DASYLAB versions is a *simulation* mode of operation, where one can easily and quickly build the entire software portion of the data-acquisition system and try it out with simulated sensor signals of any desired kind. Thus, we can develop and "debug" the software before connecting the external sensors, amplifiers, etc. This feature also makes DASYLAB an unsurpassed *teaching tool* since each student can quickly try out any ideas for a particular application before committing to specific measurement hardware for the system. I have found the learning process for DASYLAB to be *much* quicker than for LABVIEW so you do not have to commit an entire course to learning the system; it can be easily integrated into any existing measurement lab. Also, while LABVIEW is sometimes used in a "black box" mode (where the instructor or graduate students do the programming and undergraduate students just use the resulting system to gather data), with DASYLAB, even sophisticated systems can be put together by undergraduate students themselves with just a few hours of exposure. In Chapter 13, I have tried to make this initial experience even quicker, easier, and more illuminating for the reader. I have heard from industry contacts that many companies are also finding DASYLAB to be very cost/effective, even for rather complex applications. I believe that LABVIEW is often used by applications programmers who *do nothing else,* that is, they spend *all their time* developing sophisticated software for some complex measurement/control system or for automating some commercial instrument (like a rheometer). Each rheometer sold then includes this same software; thus, the programming cost (time and money) is amortized over many instruments. When one is using the same (LABVIEW) software over and over, one can justify a long learning curve, and since it is used daily, we *do not forget* how to use it. Also, LABVIEW's versatility allows it to deal with situations that might frustrate a less comprehensive software package. Of course, as is usual with any class of software, this versatility comes at the price of complexity. Most mechanical engineers, however, are *not* programming specialists, but rather they need to develop a data-acquisition system *occasionally,* on a "one-shot" basis, which means that the learning curve has to be short and the recall after having not used the software for a few months must be quick. I believe DASYLAB meets this sort of need in an optimum way. I hope you will at least try it to reach your own judgment.

Details of the text's topical coverage can be quickly surveyed from the Table of Contents. Also, I have taken pains to develop a very comprehensive *index,* so try that when looking for a specific item. For users of previous editions, it might be useful to here mention some of the more significant changes (such as Chapter 13 just discussed) found in the new edition. Chapter 14 also is new; there, I decided to *focus on a particular industry* and show how measurement systems apply. Of the many possibilities, I chose integrated circuit and MEMS manufacturing. These depend heavily on micro- and nanotechnology, which use:

Scanning probe microscopes

Partial-pressure analyzers for vacuum systems

Micromotion measurement and control

Contaminant particle measurement systems and clean rooms

Magnetic-levitation conveyers

to manufacture microcircuits and microscale sensors and actuators. Each of these listed topic areas is examined in some detail, and the contributions of measurement technology identified. [MEMS-type sensors (pressure transducers, accelerometers, infrared imagers, mass flow sensors, etc.) are also discussed elsewhere in the text where appropriate.]

In addition to Chapters 13 and 14, there are a number of significant changes and additions in the fifth edition, plus many minor ones too numerous to list here. The more significant changes include:

1. The material on *calibration* and *uncertainty calculations* has been thoroughly updated to reflect the latest positions of ISO and NIST.

2. Simulation examples have been updated to replace the obsolete CSMP with MATLAB/SIMULINK, and the use of *apparatus simulation* as an aid to sensor selection has been added.

3. *Sensor fusion* ("complementary filtering") with examples from aircraft altitude and attitude sensing is covered, as is the use of *observers* for the measurement of inaccessible variables.

4. Footnotes on reference material and hardware manufacturers have been augmented with *Internet* addresses.

5. The relation between *calibration accuracy* and *installed accuracy* is explained.

6. The use of *overlap graphs* to decide whether an experiment verifies or contradicts a theory is explained.

7. The effect of measurement-system errors on *quality-control decisions* is covered.

8. *MINITAB statistics software* is used wherever it is applicable and illuminating.

9. *Multiple regression* in computer-aided calibration and measurement is covered.

10. The concept of a *noise floor* caused by intrinsic random fluctuations in all physical variables is discussed.

11. Classical frequency response graphs of amplitude ratio and phase angle are augmented with *time-delay graphs,* which makes judgment of accurate frequency range much easier.

12. *Magnetoresistance* and *Hall effect* motion sensors are discussed.

13. The treatment of *capacitance motion sensors* has been expanded.

14. The use of *motion-control systems* for positioning sensors or other components has been added.

15. The use of *high-speed film* and *video cameras* for motion study has been expanded.

16. Velocity sensing using *tachometer encoders, lasers,* and *microwave* ("radar") methods has been added.

17. The treatment of "nonclassical" gyros such as the *GyroChip* and *fiber-optic* types, has been expanded.

18. The use of the *Global Positioning System* in measurement applications has been added.

19. Detailed *strength-of-materials* analysis of a load cell, augmented with a *finite-element* study and experimental verification, is included.

20. Methods for measuring *pressure distribution,* using Fuji pressure film, photoluminescent paint, and "crossbar" type electrical piezoresistance sensor arrays are covered.

21. Addition of *particle-image-velocimetry* (PIV) for fluid flow analysis is covered.

22. The treatment of orifice flowmetering for *compressible flow* has been revised.

23. Flow measurement with *turbine flowmeters* has been updated and revised.

24. A conceptual error in the basic *thermocouple principle* has been corrected.

25. Thermal radiation *detectors* are covered in more detail, and *uncooled microbolometer imaging systems* have been added.

26. The material on *heat flux sensors* has been updated.

27. The design example on *analog electrical differentiation* has been thoroughly revised.

28. *Digital offline dynamic compensation* using MATLAB FFT methods has been added.

29. Galvanometers used in *optical oscillographs* has been eliminated, but the use of galvanometers in motion-control systems, such as *laser scanners,* has been added.

30. A discussion of the popular *sigma-delta analog/digital converters* has been added.

31. The radio telemetry section has been thoroughly revised, and more current wireless technologies, such as *Bluetooth,* have been added.

32. A new section on *instrument connectivity* has been added.

33. The section on *strip-chart, x/y,* and *galvanometer recorders* has been revised.

34. The concept of *virtual instruments* is now included.

35. A section on *electrical current* and *power* measurement has been added.

A final comment on changes must be made on the subject of *solutions manuals.* This is my eleventh engineering textbook, and for the first ten, I consistently declined to produce a solutions manual. This peculiarity is not due to laziness on my part but relates rather to some "philosophical" positions that I, rightly or wrongly, hold dear. (I will not here burden you with these but have always been happy to discuss them with anyone who would listen.) My various publishers have always explained, and I agreed, that the lack of a solutions manual will surely lose some adoptions. For the present book, the publisher made clear that this time *there would be* a solutions manual, whether I, or someone else, did it. Faced with this situation, I decided that if there was to be a solutions manual, I wanted it to be a good one and thus determined to do it myself. No graduate or other students were used, and I personally produced "camera ready" copy, including all equations and illustrations. I hope it will be found useful, but since it is my first endeavor along these lines, I will welcome any comments or criticisms.

By judicious selection of topics, the two texts, *Measurement Systems* and *Engineering Experimentation,* can be used effectively, singly or together, in a wide variety of contexts. For a freshman course that introduces students to engineering and uses a hands-on lab, perhaps including "reverse engineering" of some device, to demonstrate the two major solution paths (theory and experimentation) for engineering problems, *Engineering Experimentation* could supply many useful reading assignments. These include an easily understandable and practically useful introduction to statistical viewpoints and methods, the role of experimentation in design and development, and guidance for written and oral communication. Later in the curriculum, we often find labs tied to some theory course or stand-alone labs that come after certain theory courses have been completed. When a lab is focused on a specific area such as, say, vibration, *Measurement Systems* can supply the needed background on the pertinent sensors, signal conditioning, and data-acquisition and -processing software. Such use, of course, only employs a fraction of the material available in the text, so the expense becomes an issue. There may or may not exist a suitable measurement text devoted only to vibration, but this book will likely be just as expensive. If a curriculum has a number of such specialty labs, *Measurement Systems* will likely have the material needed in all of them. In such a case, one would hope that textbook requirements would be coordinated so that students would purchase only one text for use in all these labs. If statistical methods, experiment design, and technical communication are included in some or all of these labs, the cost of *Engineering Experimentation* might be "amortized" over the several courses. If, as at Ohio State, you find it difficult to "squeeze in" a statistics course taught in your mathematics or statistics department, the "minicourse" provided by *Engineering Experimentation* can be embedded in one or more labs and may provide a practical viewpoint often lacking in mathematics department presentations.

Many curricula now include one or more "capstone" courses that emphasize design and give students practice in applying the specialty courses encountered earlier in their studies. At Ohio State, we have traditionally had two such required senior courses, one focused on design and another devoted to experimental methods. At present, we are trying out another approach, which uses a sequence of courses/labs that allow students to design, build, and experimentally test a machine or process. These projects are often suggested by industrial sponsors who interact with the students and instructors to provide an experience more typical of actual engineering practice. These sponsors provide some equipment or apparatus, and lend some financial support. For courses devoted specifically to experimentation or for sequences that include it as an important component, *Engineering Experimentation,* possibly augmented by *Measurement Systems,* can provide useful content.

As mentioned earlier, I believe the optimum organization is to provide, somewhere in the curriculum, a *general* measurement lab/course where the science and technology of measurement is presented as an important engineering field in its own right. For such a course, *Measurement Systems* could be a good choice, perhaps augmented by *Engineering Experimentation,* depending on the course's intended focus and coverage. Even for such a course, it will be necessary, due to the breadth and depth of the book, to carefully select the student assignments, but this is actually made *easier* because there is so much to choose from that most needs can be satisfied. If, as at Ohio State, there is a more advanced measurement-systems course (probably elective, for seniors and/or graduate students), then *Measurement Systems* will again provide the needed material for a wide variety of needs. For this advanced course, I have over the years developed some homework problems and projects that, due to their length, were not included in any of my books but rather were provided in a locally printed manual. In teaching this course, in addition to weekly homework assignments (some from *Measurement Systems,* some from the manual), I assign a "project" that runs for most of the quarter. The manual provides extensive background notes in addition to the requested student homework. Three such projects currently are in the manual:

1. Preliminary design of a viscosimeter
2. Vibration isolation methods for sensitive instruments and machines
3. Design of a vibrating-cylinder ultra-precision pressure transducer

Some of the "weekly" homework problems in the manual are in the following areas:

1. Theory and simulation study of a carrier-amplifier system
2. Accelerometer selection for a drop-test shock machine
3. Dynamic compensation for a thermocouple
4. Use of the correlation function in pipeline leak detection
5. Sensor fusion ("complementary filtering")
6. Frequency-modulated (FM) sensors and digital integration
7. FFT methods for sensor dynamic compensation
8. Use of FFT analysis to document pressure transducer dynamics based on shock tube testing

If any instructor wants a copy of this manual or a "Xeroxable" master for printing copies for students, please contact me at 614-882-2670 to make arrangements to get the material, "at cost." I do not have an electronic copy.

General Concepts

Types of Applications of Measurement Instrumentation

1.1 WHY STUDY MEASUREMENT SYSTEMS?

The study of any subject matter in engineering should be motivated by an appreciation of the uses to which the material might be put in the everyday practice of the profession. Measurement systems are used for many detailed purposes in a wide variety of application areas. Our approach will be to start with some *specific applications* in a *specific industry* and then *generalize* this picture by developing *classification schemes* that apply to all possible situations.

While measurement is used in many contexts, I want to introduce some basic ideas using the automotive industry as an example. This industry employs measurement in many ways and is thus a good choice for exploring the various uses of measurement tools. In the text title, the term "measurement system" is meant to include all components in a chain of hardware and software that leads from the measured variable to processed data. Let us start examining the use of measurement in the automotive industry "at the beginning," that is, with the *conceptual design process,* where a new automobile or truck is first conceived and the basic configuration developed. Because a modern automobile uses as many as 40 or 50 sensors (measuring devices) in implementing various functions necessary to the operation of the car, an automobile designer must be aware of the instruments available for the various measurements and how they operate and interface with other parts of the system. As new sensors are invented, designers must keep up with such developments since they may allow improvements in car design and operation. Lack of such sensor knowledge can severely restrict the range of designs that one can conceive, thus limiting improvements in overall car performance. While sensor specialists will at later stages of design consider the measuring devices in great detail, the conceptual designer must have a basic appreciation of their capabilities, so that the initial design does not neglect any useful possibilities.

Once the conceptual stage of design is well underway, measurement system considerations arise in new contexts. Many engineered products are nowadays designed using the methods of *concurrent engineering* where design and manufacturing are *integrated,* rather than being considered *sequentially,* as was often the case in earlier times. Before concurrent engineering became common, design was generally completed *first,* manufacturing considerations addressed only *later,* and costly revisions and delays (or poor designs) were often the result. With concurrent engineering, product design concepts are not "frozen" until both function and manufacturability have been reconciled. That is, the design and manufacturing engineers work in coordinated teams, blending their expertise right from the beginning of the design process. Both functionality and manufacturability considerations often require the design process to include *laboratory testing* of one kind or another. For example, if a new material is being considered, we may need to run strength tests to develop data needed by the design engineers. Or, a new or revised manufacturing process may require statistical response surface experiments[1] to find the effects of process variables on performance and/or cost. Finally, availability from suppliers of new components, such as improved shock absorbers, may require performance testing to decide whether their use is warranted in the new design. We see that laboratory testing and the associated measurement systems are thus a vital part of the design process.

As design and development proceed, prototype subsystems and finally entire vehicles will be produced. These are used as "test beds" to evaluate performance and then feed back information to the design/manufacturing teams. That is, initial designs usually have unsuspected flaws, which are revealed by building and testing the prototypes. Also, "pencil and paper" or computer-aided designs *always* are based on theories that are never exactly correct, so experimental testing is needed to verify, or improve, theoretical calculations. We begin to appreciate that design relies heavily on experimental testing at every stage of the process.

We have seen that experimentation is often needed during the design phase to help in the development of the manufacturing processes for the product. Once the design has been finalized, then manufacture of the product in quantity, rather than the "one of a kind" mode used during development, can commence. When we examine actual production machinery and processes, we often find that these manufacturing tools are controlled by a so-called *feedback mechanism.* In such a scheme, some quality parameter of the part produced is *measured* with appropriate sensors. This measured value is compared with a desired value of the parameter, and if the desired and measured values do not agree within some allowable tolerance, a controller adjusts the machine or process until the product is "on specification." Perhaps the most obvious example of this general situation is the machining of parts to specific dimensions. Here the measuring devices are precision gages that measure shaft diameters, hole sizes, lengths, etc. Robots used to weld, spray paint, or assemble parts are also usually feedback devices that use motion and force sensors

---

[1]E. O. Doebelin, "Engineering Experimentation," McGraw-Hill, New York, 1995, p. 273.

to control the robots' operation. Again it is clear that measurement plays a significant role in almost every manufacturing enterprise.

Turning now to the final product, a modern automobile, as mentioned earlier, relies on a multitude of sensors for its optimum operation. Some of these play essentially a "monitoring" role, that is, they measure and display to the driver, information useful for safe and efficient operation of the car. Speedometers tell us the vehicle's speed, while tachometers display engine RPM. Fuel gages keep track of the gas supply, and temperature sensors warn of overheating. Recent developments include use of the Global Positioning System (based on satellites) to locate the car on an electronic map and guide the driver to a desired destination. Many other sensors are part of feedback controls that optimize engine operation by measuring such variables as atmospheric pressure, air flow rate, fuel/air ratio, engine temperatures, etc. Acceleration sensors (*accelerometers*) measure vehicle motion during a crash and signal air bags to deploy if the crash is sufficiently severe. Brake-cylinder pressure and wheel-speed sensors control the antilock braking system to give better driver control on slippery surfaces. To keep costs down, many automotive sensors use micro-electro-mechanical systems (MEMS). Using manufacturing techniques borrowed from integrated-circuit technology, miniature sensors are mass produced at low cost from materials such as silicon. A recent example is the GyroChip, a replacement for the classical gyroscopic instrument used to measure angular velocity. This sensor is being used in cars to augment vehicle stability during severe or emergency maneuvers.

1.2 CLASSIFICATION OF TYPES OF MEASUREMENT APPLICATIONS

I used the automotive industry as a familiar example to introduce you to the varied applications of measurement in engineering. To help you organize your thinking on this subject I now want to *generalize* the topic of measurement applications. Fortunately, all the specific examples I gave from the auto industry, and in fact, examples from *any industry,* can be classified into only three major categories:

1. Monitoring of processes and operations.
2. Control of processes and operations.
3. Experimental engineering analysis.

That is, I suggest that *every* application of measurement, including those not yet "invented," can be put into one of the three groups just listed or some combination of them. Let us now explore this scheme of classification in general terms and also relate it to our earlier automotive examples.

Monitoring of processes and operations refers to situations where the measuring device is being used to keep track of some quantity. The thermometers, barometers, radars, and anemometers used by the weather bureau fit this definition. They simply indicate the condition of the environment, and their readings do not serve any control functions in the ordinary sense. Similarly, water, gas, and electric

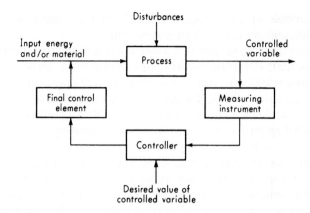

Figure 1.1 Feedback-control system.

meters in the home keep track of the quantity of the commodity used so that the cost to the user can be computed. In our automotive illustration, the speedometer, fuel gage, outdoor temperature sensor, and compass would belong to this monitoring class of applications.

Control[2] *of processes and operations* is one of the most important classes of measurement application. This usually refers to an automatic feedback control system, as diagramed in generic terms in Fig. 1.1. This type of application is sufficiently important that most undergraduate curricula in mechanical, aerospace, electrical, chemical, and industrial engineering will include a required course (and several electives) in control systems.

The subject of feedback control is pertinent to this text on measurement systems in two basic ways, one of which is the use of sensors in feedback control systems, as just mentioned. The other relates to the fact that many measurement systems *themselves* use feedback principles in their operation. One could in fact say that sensors are used in feedback systems and feedback systems are used in sensors. Of the many possible examples of the latter, we mention the hot-wire anemometer, a device for measuring rapidly varying fluid velocity. Without feedback, the hot wire used in the instrument is accurate only for velocity fluctuations of frequency less than about 100 Hz. By redesigning the instrument to use feedback, this limit is extended to about 30,000 Hz, making the instrument much more useful.

The operation of systems such as that of Fig. 1.1 is briefly described as follows: We want to control some "process," such as the heating of our house; to be specific, we want to keep the temperature near some desired value, such as 70° F. The process is influenced by various "disturbances" (such as the outdoor temperature) that we can *not* control and also by an input of energy and/or material that we *are* able to manipulate, using some "final control element" (the gas valve in our furnace). The design principle of all feedback control systems says that we should

[2]E. O. Doebelin, "Control System Principles and Design," Wiley, New York, 1985.

measure the variable which we want to control, compare it (in a "controller") with its desired value, and then, based on the "error" between the two, manipulate the final control element in such a way as to drive the controlled variable closer to its desired value. We see that this basic design concept means that *every* feedback control system will have at least one measuring device as a vital component. Since feedback systems are used in literally millions of applications for controlling temperature, pressure, shaft speed, fluid flow, robot arm position, aircraft speed and altitude, etc., these control applications are one of the most important uses of measurement systems.

Returning to our earlier automotive examples, feedback control applications are found in the car's speed control system, the antilock braking system, the coolant temperature regulating system, the air-conditioning system, the engine pollution controls, and many more. Also, the majority of the manufacturing tools and processes used to produce the car are under feedback control.

This text is not one on feedback control; however, when feedback is used in *the measurement system itself,* we will not avoid discussion of its implications. Fortunately, this can usually be done without requiring that the reader have taken a controls course or be expert in this technology.

Experimental engineering analysis is that part of engineering design, development, and research that relies on laboratory testing of one kind or another to answer questions. That is, as engineers, we have only two basic ways of solving engineering problems: theory and experimentation.[3] Some (usually simple) problems can be adequately solved using theory alone. Most problems require a judiciously selected blend of theory and experiment. It is not unusual for the "lab testing" portion of an engineering project to consume more than half of the total resources. As a result, most engineers need to be proficient in planning and conducting this phase of the effort. The text just referenced addresses the entire process; the current text concentrates on that portion intimately related to the measurement system itself.

Since the choice of how much theory and how much experiment to use in a particular application is difficult and important, we want to provide some guidelines to help organize your thinking, at least in a general way. Figures 1.2 and 1.3 compare and contrast the features of these two problem-solving methods. If we decide to use experimentation, it is helpful to realize that *all* engineering experiments can be put into a relatively small number of classes. This classification can be accomplished in several ways, but one which I have found meaningful is given in Fig. 1.4.

1.3 COMPUTER-AIDED MACHINES AND PROCESSES

In constructing useful machines and processes for society, it is now extremely common for engineers to include in the design, as dedicated components of an overall system, computers of various sizes. Inexpensive, compact, and powerful computer hardware and software can make possible significant advances in productivity,

---

[3]E. O. Doebelin, "Engineering Experimentation," McGraw-Hill, New York, 1995.

1. Often give results that are of general use rather than for restricted application.
2. Invariably require the application of simplifying assumptions. Thus, not the actual physical system but rather a simplified "mathematical model" of the system is studied. This means the theoretically predicted behavior is *always* different from the real behavior.
3. In some cases, may lead to complicated mathematical problems. This has blocked theoretical treatment of many problems in the past. Today, increasing availability of high-speed computing machines allows theoretical treatment of many problems that could not be so treated in the past.
4. Require only pencil, paper, computing machines, etc. Extensive laboratory facilities are not required. (Some computers are very complex and expensive, but they can be used for solving all kinds of problems. Much laboratory equipment, on the other hand, is special-purpose and suited only to a limited variety of tasks.)
5. No time delay engendered in building models, assembling and checking instrumentation, and gathering data.

Figure 1.2 Features of theoretical methods.

1. Often give results that apply only to the specific system being tests. However, techniques such as dimensional analysis may allow some generalization.
2. No simplifying assumptions necessary if tests are run on an actual system. The true behavior of the system is revealed.
3. *Accurate* measurements necessary to give a true picture. This may require expensive and complicated equipment. *The characteristics of all the measuring and recording equipment must be thoroughly understood.*
4. Actual system or a scale model required. If a scale model is used, similarity of all significant features must be preserved.
5. Considerable time required for design, construction, and debugging of apparatus.

Figure 1.3 Features of experimental methods.

product quality, efficiency, flexibility, and safety. While the nontechnical public often (wrongly) views the *entire system* as a "computer," it is important that we not encourage this misconception. The computer is *helpless* to control any machine or process without the sensors that measure critical process variables or the actuators

1. Testing the validity of theoretical predictions based on simplifying assumptions; improvement of theory, based on measured behavior.
   Example: frequency-response testing of mechanical linkage for resonant frequencies.

2. Formulation of generalized empirical relationships in situations where no adequate theory exists.
   Example: determination of friction factor for turbulent pipe flow.

3. Determination of material, component, and system parameters, variables, and performance indices.
   Example: determination of yield point of a certain alloy steel, speed-torque curves for an electric motor, thermal efficiency of a steam turbine.

4. Study of phenomena with hopes of developing a theory.
   Example: electron microscopy of metal fatigue cracks.

5. Solution of mathematical equations by means of analogies
   Example: solution of shaft torsion problems by measurements on soap bubbles.

Figure 1.4 Types of experimental-analysis problems.

("final control elements") that manipulate process inputs and thus affect the process controlled variables. Thus, many of the amazing feats of engineering accomplished by computer-aided devices depend heavily on the availability and proper operation of associated measurement systems.

1.4 CONCLUSION

Whatever the nature of the application, intelligent selection and use of measurement instrumentation depend on a broad knowledge of what is available and how the performance of the equipment may be best described in terms of the job to be done. New equipment is continuously being developed, but certain basic devices have proved their usefulness in broad areas and undoubtedly will be widely used for many years. A representative cross section of such devices is discussed in this text. These devices are of great interest in themselves; they also serve as the vehicle for the presentation and development of general techniques and principles needed in handling problems in measurement instrumentation. In addition, these general concepts are useful in treating any devices that may be developed in the future.

The treatment is also intended to be on a level that will be of service to not only the user, but also the *designer* of measurement instrumentation equipment. There are two main reasons for this emphasis. First, much experimental equipment (including measurement instruments) is often "homemade," especially in smaller companies where the high cost of specialized gear cannot always be justified. Second, the instrument industry is a large and growing one which utilizes many engineers in a design capacity. While the general techniques of mechanical and electrical design as applied to *machines* are also applicable to instruments, in many cases a rather different point of view is necessary in instrument design. This is due, in part, to the fact that the design of machines is mainly concerned with considerations of *power* and *efficiency,* whereas instrument design almost completely neglects these areas and concerns itself with the acquisition and manipulation of *information.* Since a considerable number of engineering graduates will work in the instrument industry, their education should include treatment of the most significant aspects of this area.

The third class of applications listed earlier, experimental engineering analysis, requires not only familiarity with measurement systems, but also some understanding of the planning, execution, and evaluation of experiments. While all these aspects of experimental work might be treated in a single text or course, I have chosen in the present text to concentrate on a thorough exposition of the measurement system itself. A comprehensive treatment of the *overall* problems and methods of engineering experimentation is presented in my companion text.[4] There, a major emphasis is on statistical methods, especially some simplified and practical approaches to statistical design of experiments. The two books together give a complete and in-depth coverage of all aspects of engineering experimental work.

PROBLEMS

1.1 By consulting various technical journals in the library, find accounts of experimental studies carried out by engineers or scientists. Find three such articles, reference them completely, explain briefly what was accomplished, and attempt to classify them according to one or more categories of Fig. 1.4.

1.2 Give three specific examples of measuring-instrument applications in each of the following areas: (*a*) monitoring of processes and operations, (*b*) control of processes and operations, (*c*) experimental engineering analysis.

1.3 Compare and contrast the experimental and the theoretical approaches to the following problems:

   (*a*) What is the tolerable vibration level to which astronauts may safely be exposed in launch vehicles?

   (*b*) Find the relationship between applied force F and resulting friction torque T_f in the simple brake of Fig. P1.1.

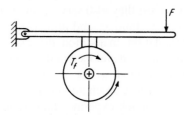

Figure P1.1

[4]E. O. Doebelin, "Engineering Experimentation: Planning, Execution, Reporting," McGraw-Hill, New York, 1995.

(*c*) Find the location of the center of mass of the rocket shown in Fig. P1.2 if the shapes, sizes, and materials of all the component parts are known.

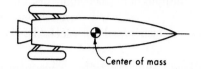
Center of mass

Figure P1.2

(*d*) At what angle with the horizontal should a projectile be launched to achieve the greatest horizontal range?

BIBLIOGRAPHY

Books

1. C. P. Wright, "Applied Measurement Engineering," Prentice-Hall, Englewood Cliffs, 1995.
2. R. S. Figliola and D. E. Beasley, "Theory and Design for Mechanical Measurements," 3rd ed., Wiley, New York, 2000.
3. R. B. Northrop, "Introduction to Instrumentation and Measurements," CRC Press, New York, 1997.
4. J. P. Holman, "Experimental Methods for Engineers," 7th ed., McGraw-Hill, New York, 2001.
5. M. S. Ray, "Engineering Experimentation," McGraw-Hill, New York, 1988.
6. K. S. Lion, "Instrumentation in Scientific Research," McGraw-Hill, New York, 1959.
7. C. F. Hix and R. P. Alley, "Physical Laws and Effects," Wiley, New York, 1958.
8. I. J. Busch-Vishniac, "Electromechanical Sensors and Actuators," Springer, New York, 1999.
9. R. V. Jones, "Instruments and Experiences," Wiley, New York, 1988.
10. A. H. Slocum, "Precision Machine Design," Prentice-Hall, Engelwood Cliffs, 1992.
11. E. O. Doebelin, "Engineering Experimentation," McGraw-Hill, New York, 1995.
12. J. C. Gibbings, "The Systematic Experiment," Cambridge Univ. Press, New York, 1986.
13. H. Schenck, Jr., "Theories of Engineering Experimentation," 3rd ed., McGraw-Hill, New York, 1979.
14. C. Lipson and J. Sheth, "Statistical Design and Analysis of Engineering Experiments," McGraw-Hill, New York, 1973.
15. F. B. Wilson, "An Introduction to Scientific Research," McGraw-Hill, New York, 1952.
16. H. K. P. Neubert, "Instrument Transducers," Oxford Univ. Press, London, 1963.
17. P. H. Sydenham (ed.), "Handbook of Measurement Science," vol. 1: "Theoretical Fundamentals," Wiley, New York, 1982.
18. P. H. Sydenham, "Mechanical Design of Instruments," Instrument Society of America, Research Triangle Park, NC, 1986.

19. T. G. Beckwith and R. D. Marangoni, "Mechanical Measurements," 5th ed., Addison-Wesley, Reading, 1993.

20. C. S. Draper, W. McKay, and S. Lees, "Instrument Engineering," vols. 1 to 3, McGraw-Hill, New York, 1955.

21. J. G. Webster, "The Measurement, Instrumentation, and Sensors Handbook," CRC Press, Boca Raton, FL, 1999.

22. S. Soloman, "Sensors Handbook," McGraw-Hill, New York, 1998.

23. J. Fraden, "Handbook of Modern Sensors," AIP, New York, 1997.

Periodicals

1. *Sensors (A searchable index of all articles is available at www.sensorsmag.com/articles/article_index/.)*

2. *The Review of Scientific Instruments*

3. *Journal of Physics E: Scientific Instruments*

4. *Proc. of Society for Experimental Mechanics*

5. *Journal of Instrument Society of America*

6. *Instruments and Control Systems*

7. *ASME Journal of Dynamic Systems, Measurement, and Control*

8. *Transactions of the Institute of Measurement and Control*

9. *IEEE Transactions in Instruments and Measurements*

10. *Precision Engineering*

11. *Measurement Science and Technology*

12. *CalLab*

CHAPTER 2

Generalized Configurations and Functional Descriptions of Measuring Instruments

2.1 FUNCTIONAL ELEMENTS OF AN INSTRUMENT

It is possible and desirable to describe both the operation and the performance (degree of approach to perfection) of measuring instruments and associated equipment in a generalized way without recourse to specific physical hardware. The operation can be described in terms of the functional elements of instrument systems, and the performance is defined in terms of the static and dynamic performance characteristics. This section develops the concept of the functional elements of an instrument or measurement system.

If you examine diverse physical instruments with a view toward generalization, soon you recognize in the elements of the instruments a recurring pattern of similarity with regard to function. This leads to the concept of breaking down instruments into a limited number of types of elements according to the generalized function performed by the element. This breakdown can be made in a number of ways, and no standardized, universally accepted scheme is used at present. We now give one such scheme which may help you to understand the operation of any new instrument with which you may come in contact and to plan the design of a new instrument.

Consider Fig. 2.1, which represents a possible arrangement of functional elements in an instrument and includes *all* the basic functions considered necessary for a description of any instrument. The *primary sensing element* is that which first receives energy from the measured medium and produces an output depending in some way on the measured quantity ("measurand"). It is important to note that an instrument *always* extracts some energy from the measured medium. Thus the measured quantity is *always* disturbed by the act of measurement, which makes a

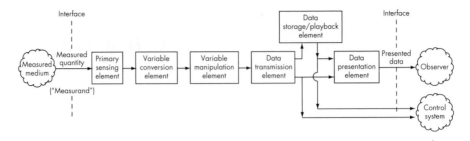

Figure 2.1
Functional elements of an instrument or a measurement system.

perfect measurement theoretically impossible. Good instruments are designed to minimize this "loading effect," but it is always present to some degree.

The output signal of the primary sensing element is some physical variable, such as displacement or voltage. For the instrument to perform the desired function, it may be necessary to convert this variable to another more suitable variable while preserving the information content of the original signal. An element that performs such a function is called a *variable-conversion element.* It should be noted that not every instrument includes a variable-conversion element, but some require several. Also, the "elements" we speak of are *functional* elements, not physical elements. That is, Fig. 2.1 shows an instrument neatly separated into blocks, which may lead you to think of the physical apparatus as being precisely separable into subassemblies performing the specific functions shown. That is, in general, not the case; a specific piece of hardware may perform *several* of the basic functions, for instance.

In performing its intended task, an instrument may require that a signal represented by some physical variable be manipulated in some way. By "manipulation," we mean specifically a change in numerical value according to some definite rule but a preservation of the physical nature of the variable. Thus an electronic amplifier accepts a small voltage signal as input and produces an output signal that is also a voltage but is some constant times the input. An element that performs such a function is called a *variable-manipulation element.* Again, you should not be misled by Fig. 2.1. A variable-manipulation element does not necessarily *follow* a variable-conversion element, but may precede it, appear elsewhere in the chain, or not appear at all.

When the functional elements of an instrument are actually physically separated, it becomes necessary to transmit the data from one to another. An element performing this function is called a *data-transmission element.* It may be as simple as a shaft and bearing assembly or as complicated as a telemetry system for transmitting signals from satellites to ground equipment by radio.

If the information about the measured quantity is to be communicated to a human being for monitoring, control, or analysis purposes, it must be put into a form recognizable by one of the human senses. An element that performs this "translation" function is called a *data-presentation element.* This function includes the simple *indication* of a pointer moving over a scale and the *recording* of a pen moving

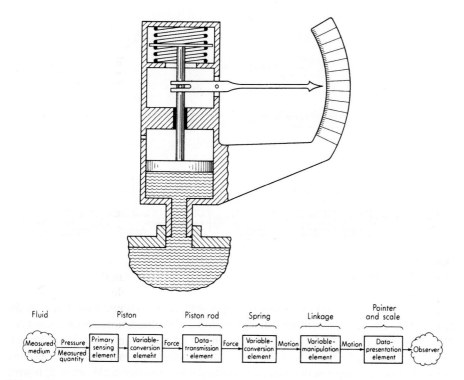

Figure 2.2
Pressure gage.

over a chart. Indication and recording also may be performed in discrete increments (rather than smoothly), as exemplified by a digital voltmeter or printer. While the majority of instruments communicate with people through the visual sense, the use of other senses such as hearing and touch is certainly conceivable.

Although data storage in the form of pen/ink recording is often employed, some applications require a distinct *data storage/playback* function which can easily re-create the stored data upon command. The magnetic tape recorder/reproducer is the classical example here. However, many recent instruments digitize the electric signals and store them in a computerlike digital memory (RAM, hard drive, floppy disk, etc.).

Before we go on to some illustrative examples, let us emphasize again that Fig. 2.1 is intended as a vehicle for presenting the concept of functional elements, and not as a physical schematic of a generalized instrument. A given instrument may involve the basic functions in any number and combination; they need not appear in the order of Fig. 2.1. A given physical component may serve several of the basic functions.

As an example of the above concepts, consider the rudimentary pressure gage of Fig. 2.2. One of several possible valid interpretations is as follows: The primary sensing element is the piston, which also serves the function of variable conversion

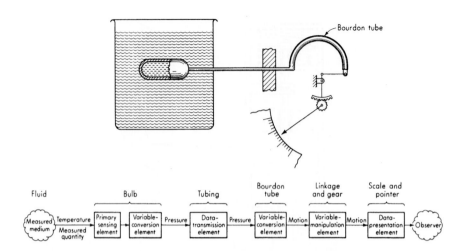

Figure 2.3
Pressure thermometer.

since it converts the fluid pressure (force per unit area) to a resultant force on the piston face. Force is transmitted by the piston rod to the spring, which converts force to a proportional displacement. This displacement of the piston rod is magnified (manipulated) by the linkage to give a larger pointer displacement. The pointer and scale indicate the pressure, thus serving as data-presentation elements. If it were necessary to locate the gage at some distance from the source of pressure, a small tube could serve as a data-transmission element.

Figure 2.3 depicts a pressure-type thermometer. The liquid-filled bulb acts as a primary sensor and variable-conversion element since a temperature change results in a pressure buildup within the bulb, because of the constrained thermal expansion of the filling fluid. This pressure is transmitted through the tube to a Bourdon-type pressure gage, which converts pressure to displacement. This displacement is manipulated by the linkage and gearing to give a larger pointer motion. A scale and pointer again serve for data presentation.

A remote-reading shaft-revolution counter is shown in Fig. 2.4. The microswitch sensing arm and the camlike projection on the rotating shaft serve both a primary sensing and a variable-conversion function since rotary displacement is converted to linear displacement. The microswitch contacts also serve for variable conversion, changing a mechanical to an electrical oscillation (a sequence of voltage pulses). These voltage pulses may be transmitted relatively long distances over wires to a solenoid. The solenoid reconverts the electrical pulses to mechanical reciprocation of the solenoid plunger, which serves as input to a mechanical counter. The counter itself involves variable conversion (reciprocating to rotary motion), variable manipulation (rotary motion to decimalized rotary motion), and data presentation.

As a final example, let us examine Fig. 2.5, which illustrates schematically a D'Arsonval galvanometer as used in oscillographs and optical scanning systems.

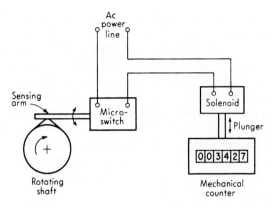

Figure 2.4
Digital revolution counter.

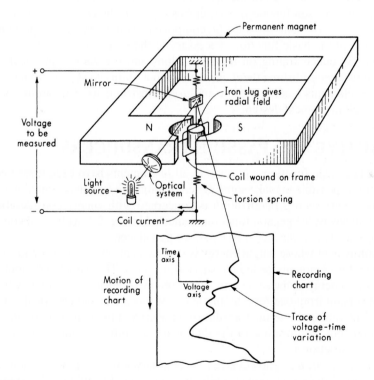

Figure 2.5
D'Arsonval galvanometer.

A time-varying voltage to be recorded is applied to the ends of the two wires which transmit the voltage to a coil made up of a number of turns wound on a rigid frame. This coil is suspended in the field of a permanent magnet. The resistance of the coil converts the applied voltage to a proportional current (ideally). The interaction between the current and the magnetic field produces a torque on the coil, which gives another variable conversion. This torque is converted to an angular deflection by the torsion springs. A mirror rigidly attached to the coil frame converts the frame rotation to the rotation of a light beam which the mirror reflects. The light-beam rotation is twice the mirror rotation, which gives a motion magnification. The reflected beam intercepts a recording chart made of photosensitive material which is moved at a fixed and known rate, to give a time base. The combined horizontal motion of the light spot and vertical motion of the recording chart generates a graph of voltage versus time. The "optical lever arm" (the distance from the mirror to the recording chart) has a motion-magnifying effect, since the spot displacement per unit mirror rotation is directly proportional to it.

In this instrument, the coil and magnet assembly probably would be considered as the primary sensing element since the lead wires (which serve a transmission function) are not really part of the instrument, and the coil resistance (which acts in a variable-conversion function) is an intrinsic part of the coil. In any case, the assignment of precise names to specific components is not nearly as important as the recognition of the basic functions necessary to the successful operation of the instrument. By concentrating on these functions and the various physical devices available for accomplishing them, we develop our ability to synthesize new combinations of elements leading to new and useful instruments. This ability is fundamental to all instrument design.

2.2 ACTIVE AND PASSIVE TRANSDUCERS

Once certain basic functions common to all instruments have been identified, then we see if it is possible to make some generalizations on *how* these functions may be performed. One such generalization is concerned with energy considerations. In performing any of the general functions indicated in Fig. 2.1, a physical component may act as an active transducer or a passive transducer.

A component whose output energy is supplied entirely or almost entirely by its input signal is commonly called a *passive transducer.* The output and input signals may involve energy of the same form (say, both mechanical), or there may be an energy conversion from one form to another (say, mechanical to electrical). (In much technical literature, the term "transducer" is restricted to devices involving energy *conversion;* but, conforming to the dictionary definition of the term, we do not make this restriction.)

An *active transducer,* however, has an auxiliary source of power which supplies a major part of the output power while the input signal supplies only an insignificant portion. Again, there may or may not be a conversion of energy from one form to another.

In all the examples of Sec. 2.1, there is only one active transducer—the microswitch of Fig. 2.4; all other components are passive transducers. The power to drive the solenoid comes not from the rotating shaft, but from the ac power line, an auxiliary source of power. Some further examples of active transducers may be in order. The electronic amplifier shown in Fig. 2.6 is a good one. The element supplying the input-signal voltage, e_i need supply only a negligible amount of power since almost no current is drawn, owing to negligible gate current and a high R_g. However, the output element (the load resistance R_L) receives significant current and voltage and thus power. This power must be supplied by the battery E_{bb}, the auxiliary power source. Thus the input *controls* the output, but does not actually supply the output power.

Another active transducer of great practical importance, the *instrument servomechanism,* is shown in simplified form in Fig. 2.7. This is actually an instrument *system* made up of components, some of which are passive transducers and others active transducers. When it is considered as an entity, however, with input voltage e_i and output displacement x_o, it meets the definition of an active transducer and is profitably thought of as such. The purpose of this device is to cause the motion x_o to follow the variations of the voltage e_i in a proportional manner. Since the motor torque is proportional to the error voltage e_e, it is clear that the system can be at rest only if e_e is zero. This occurs only when $e_i = e_{sl}$; since e_{sl} is proportional to x_o, this means that x_o must be proportional to e_i in the static case. If e_i varies, x_o will tend to follow it, and by proper design, accurate "tracking" of e_i by x_o should be possible. You should recognize this device as an instrument which uses the feedback principle of Fig. 1.1.

2.3 ANALOG AND DIGITAL MODES OF OPERATION

It is possible further to classify how the basic functions may be performed by turning attention to the analog or digital nature of the signals that represent the information.

For analog signals, the precise value of the quantity (voltage, rotation angle, etc.) carrying the information is significant. However, digital signals are basically of a binary (on/off) nature, and variations in numerical value are associated with changes in the logical state ("true/false") of some combination of "switches." In a

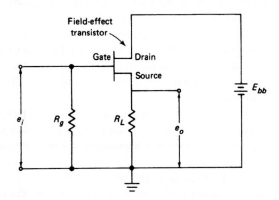

Figure 2.6
Electronic amplifier.

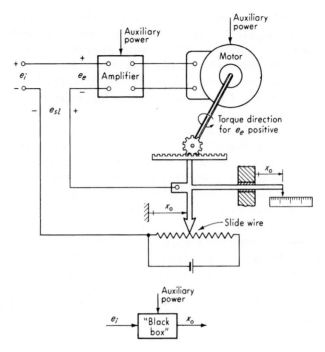

Figure 2.7
Instrument servomechanism.

typical digital electronic system, *any* voltage in the range of +2 to +5 V produces the on state, while signals of 0 to +0.8 V correspond to off. Thus whether the voltage is 3 or 4 V is of *no* consequence. The same result is produced, and so the system is quite tolerant of spurious "noise" voltages which might contaminate the information signal. In a digitally represented value of, say, 5.763, the least significant digit (3) is carried by on/off signals of the same (large) size as for the most significant digit (5). Thus in an all-digital device such as a digital computer, there is no limit to the number of digits which can be accurately carried; we use whatever can be justified by the particular application. When *combined* analog/digital systems are used (often the case in measurement systems), the digital portions need not limit system accuracy. These limitations generally are associated with the analog portions and/or the analog/digital conversion devices.

The majority of primary sensing elements are of the analog type. The only digital device illustrated in this text up to this point is the revolution counter of Fig. 2.4. This is clearly a digital device since it is impossible for this instrument to indicate, say, 0.79; it measures only in steps of 1. The importance of digital instruments is increasing, perhaps mainly because of the widespread use of digital computers in both data-reduction and automatic control systems. Since the digital computer works only with digital signals, any information supplied to it must be in digital form. The computer's output is also in digital form. Thus any communication with the computer at either the input or the output end must be in terms of digital signals. Since most measurement and control apparatus is of an analog nature, it is necessary to have both *analog-to-digital converters* (at the input to the computer) and *digital-to-analog converters* (at the output of the computer). These devices (which are discussed in greater detail in a later chapter) serve as "translators" that enable the computer to communicate with the outside world, which is largely of an analog nature.

2.4 NULL AND DEFLECTION METHODS

Another useful classification separates devices by their operation on a null or a deflection principle. In a *deflection-type* device, the measured quantity produces some physical effect that engenders a similar but opposing effect in some part of the instrument. The opposing effect is closely related to some variable (usually a mechanical displacement or deflection) that can be directly observed by some human sense. The opposing effect increases until a balance is achieved, at which point the "deflection" is measured and the value of the measured quantity inferred from this. The pressure gage of Fig. 2.2 exemplifies this type of device, since the pressure force engenders an opposing spring force as a result of an unbalance of forces on the piston rod (called the force-summing link), which causes a deflection of the spring. As the spring deflects, its force increases; thus a balance will be achieved at some deflection if the pressure is within the design range of the instrument.

In contrast to the deflection-type device, a *null-type* device attempts to maintain deflection at zero by suitable application of an effect opposing that generated by the measured quantity. Necessary to such an operation are a detector of unbalance and a means (manual or automatic) of restoring the balance. Since deflection is kept at zero (ideally), determination of numerical values requires accurate knowledge of the magnitude of the opposing effect. A pressure gage operating on a null principle is depicted in simplified form in Fig. 2.8. By adding the proper standard weights to the platform of known weight, the pressure force on the face of the piston may be balanced by gravitational force. The condition of force balance is indicated by the platform remaining at rest between the upper and lower stops. Since the weights and the piston area are all known, the unknown pressure may be computed.

Upon comparing the null and deflection methods of measurement exemplified by the pressure gages described above, we note that, in the deflection instrument, accuracy depends on the calibration of the spring, whereas in the null instrument it depends on the accuracy of the standard weights. In this particular case (and for most measurements in general), the accuracy attainable by the null method is of a higher level than that by the deflection method. One reason is that the spring is not in itself a primary standard of force, but must be calibrated by standard weights, whereas in the null instrument a *direct* comparison of the unknown force with the standard is achieved. Another advantage of null methods is the fact that, since the measured quantity is balanced out, the detector of unbalance can be made very sensitive, because it need cover only a small range around zero. Also the detector need not be calibrated since it must detect only the presence and direction of unbalance,

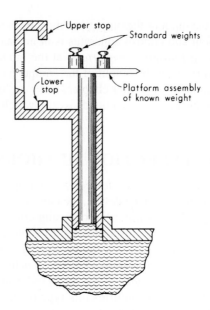

Figure 2.8
Deadweight pressure gage.

but not the amount. However, a deflection instrument must be larger, more rugged, and thus less sensitive if it is to measure large magnitudes.

The disadvantages of null methods appear mainly in dynamic measurements. Let us consider the pressure gages again. The difficulty in keeping the platform balanced for a fluctuating pressure should be apparent. The spring-type gage suffers not nearly so much in this respect. By use of automatic balancing devices (such as the instrument servomechanism of Fig. 2.7) the speed of null methods may be improved considerably, and instruments of this type are of great importance.

2.5 INPUT-OUTPUT CONFIGURATION OF INSTRUMENTS AND MEASUREMENT SYSTEMS

Before we discuss instrument performance characteristics, it is desirable to develop a generalized configuration that brings out the significant input-output relationships present in all measuring apparatus. A scheme suggested by Draper, McKay, and Lees[1] is presented in somewhat modified form in Fig. 2.9. Input quantities are classified into three categories: desired inputs, interfering inputs, and modifying inputs. *Desired inputs* represent the quantities that the instrument is specifically intended to measure. *Interfering inputs* represent quantities to which the instrument is unintentionally sensitive. A desired input produces a component of output

[1]C. S. Draper, W. McKay, and S. Lees, "Instrument Engineering," vol. 3, p. 58, McGraw-Hill, New York, 1955.

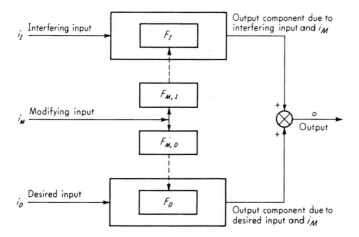

Figure 2.9
Generalized input-output configuration.

according to an input-output relation symbolized by F_D, where F_D denotes the mathematical operations necessary to obtain the output from the input. The symbol F_D may represent different concepts, depending on the particular input-output characteristic being described. Thus F_D might be a constant number K that gives the proportionality constant relating a constant static input to the corresponding static output for a linear instrument. For a nonlinear instrument, a simple constant is not adequate to relate static inputs and outputs; a mathematical *function* is required. To relate dynamic inputs and outputs, differential equations are necessary. If a description of the output "scatter," or dispersion, for repeated equal static inputs is desired, a statistical distribution function of some kind is needed. The symbol F_D encompasses all such concepts. The symbol F_I serves a similar function for an interfering input.

The third class of inputs might be thought of as being included among the interfering inputs, but a separate classification is actually more significant. This is the class of modifying inputs. *Modifying inputs* are the quantities that cause a change in the input-output relations for the desired and interfering inputs; that is, they cause a change in F_D and/or F_I. The symbols $F_{M,I}$ and $F_{M,D}$ represent (in the appropriate form) the specific manner in which i_M affects F_I and F_D, respectively. These symbols, $F_{M,I}$ and $F_{M,D}$ are interpreted in the same general way as F_I and F_D.

The block diagram of Fig. 2.9 illustrates the above concepts. The circle with a cross in it is a conventional symbol for a *summing device*. The two plus signs as shown indicate that the output of the summing device is the instantaneous algebraic sum of its two inputs. Since an instrument system may have several inputs of each of the three types as well as several outputs, it may be necessary to draw more complex block diagrams than in Fig. 2.9. This extension is, however, straightforward.

The above concepts can be clarified by means of specific examples. Consider the mercury manometer used for differential-pressure measurement as shown in

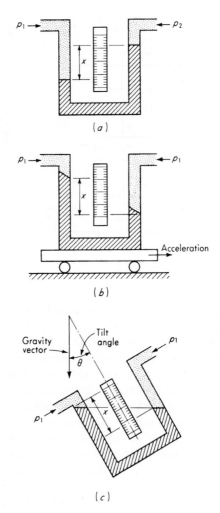

Figure 2.10
Spurious inputs for manometer.

Fig. 2.10*a.* The desired inputs are the pressures p_1 and p_2 whose difference causes the output displacement x, which can be read off the calibrated scale. Figures 2.10*b* and *c* show the action of two possible interfering inputs. In Fig. 2.10*b* the manometer is mounted on some vehicle that is accelerating. A simple analysis shows that there will be an output x even though the differential pressure might be zero. Thus if you are trying to measure pressures under such circumstances, an error will be engendered because of the interfering acceleration input. Similarly, in Fig. 2.10*c,* if the manometer is not properly aligned with the gravity vector, it may give an output signal x even though no pressure difference exists. Thus the tilt angle θ is an interfering input. (It is also a modifying input.)

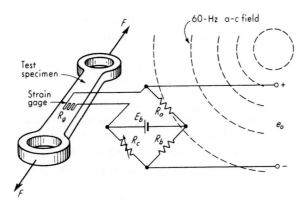

Figure 2.11
Interfering input for strain-gage circuit.

Modifying inputs for the manometer include ambient temperature and gravitational force. Ambient temperature manifests its influence in a number of ways. First, the calibrated scale changes length with temperature; thus the proportionality factor relating $p_1 - p_2$ to x is modified whenever temperature varies from its basic calibration value. Also, the density of mercury varies with temperature, which again leads to a change in the proportionality factor. A change in gravitational force resulting from changes in location of the manometer, such as moving it to another country or putting it aboard a spaceship, leads to a similar modification in the scale factor. Note that the effects of *both* the desired and the interfering inputs may be altered by the modifying inputs.

As another example, consider the electric-resistance strain-gage setup shown in Fig. 2.11. The gage consists of a fine-wire grid of resistance R_g firmly cemented to the specimen whose unit strain ϵ at a certain point is to be measured. When strained, the gage's resistance changes according to the relation

$$\Delta R_g = (GF)R_g\epsilon \tag{2.1}$$

where
$$\Delta R_g \triangleq \text{change in gage resistance, } \Omega^* \tag{2.2}$$
$$GF \triangleq \text{gage factor, dimensionless} \tag{2.3}$$
$$R_g \triangleq \text{gage resistance when unstrained, } \Omega \tag{2.4}$$
$$\epsilon \triangleq \text{unit strain, cm/cm} \tag{2.5}$$

The resistance change is proportional to the strain. Thus if we could measure the resistance, we could compute the strain. The resistance is measured by using the Wheatstone-bridge arrangement shown. When no load F is present, the bridge is balanced (e_o set to zero) by adjusting R_c. Application of load causes a strain, a ΔR_g,

---

*The symbol $\triangleq$ means "equal by definition."

and thus unbalances the bridge, causing an output voltage e_o which is proportional to ϵ and can be measured on a meter or an oscilloscope. The voltage e_o is given by

$$e_o = -(GF)R_g \epsilon E_b \frac{R_a}{(R_g + R_a)^2} \tag{2.6}$$

The desired input here is clearly the strain ϵ which causes a proportional output voltage e_o. One interfering input which often causes trouble in such apparatus is the 60-Hz magnetic field caused by nearby power lines, electric motors, etc. This field induces voltages in the strain-gage circuit, causing output voltages e_o even when the strain is zero. Another interfering input is the gage temperature. If this varies, it causes a change in gage resistance that will cause a voltage output even if there is no strain. Temperature has another interfering effect since it causes a differential expansion of the gage and the specimen, which gives rise to a strain ϵ and a voltage e_o even though no force F has been applied. Temperature also acts as a modifying input since the gage factor is sensitive to temperature. The battery voltage E_b is another modifying input. Both these are modifying inputs since they tend to change the proportionality factor between the desired input ϵ and the output e_o or between an interfering input (gage temperature) and output e_o.

Methods of Correction for Interfering and Modifying Inputs

In the design and/or use of measuring instruments, a number of methods for nullifying or reducing the effects of spurious inputs are available. We briefly describe some of the most widely used.

The *method of inherent insensitivity* proposes the obviously sound design philosophy that the elements of the instrument should *inherently* be sensitive to only the desired inputs. While usually this is not entirely possible, the simplicity of this approach encourages one to consider its application wherever feasible. In terms of the general configuration of Fig. 2.9, this approach requires that somehow F_I and/or $F_{M,D}$ be made as nearly equal to zero as possible. Thus, even though i_I and/or i_M may exist, they cannot affect the output. As an example of the application of this concept to the strain gage of Fig. 2.11, we might try to find some gage material that exhibits an extremely low temperature coefficient of resistance while retaining its sensitivity to strain. If such a material can be found, the problem of interfering temperature inputs is at least partially solved. Similarly, in mechanical apparatus that must maintain accurate dimensions in the face of ambient-temperature changes, the use of a material[2] of very small temperature coefficient of expansion may be helpful. Two such materials are the metal alloy Invar and the glass/ceramic Zerodur. If stiffness (such as in a spring) must be temperature insensitive, consider the metal alloy Ni-Span C.

---

[2]D. G. Chetwynd, "Selection of Structural Materials for Precision Devices," *Precision Eng.,* vol. 9, no. 1, pp. 3–6, January 1987.

The *method of high-gain feedback* is exemplified by the system shown in Fig. 2.12*b*. Suppose we wish to measure a voltage e_i by applying it to a motor whose torque acts on a spring, causing a displacement x_o, which may be measured on a calibrated scale. By proper design, the displacement x_o might be made proportional to the voltage e_i according to

$$x_o = (K_{Mo}K_{SP})e_i \qquad (2.7)$$

where K_{Mo} and K_{SP} are appropriate constants. This arrangement, shown in Fig. 2.12*a*, is called an open-loop system. If modifying inputs i_{M1} *and* i_{M2} exist, they cause changes in K_{Mo} and K_{SP} that lead to errors in the relation between e_i and x_o. These errors are in *direct proportion* to the changes in K_{Mo} and K_{SP}. Suppose, instead, we construct a system as in Fig. 2.12*b*. Here the output x_o is measured by

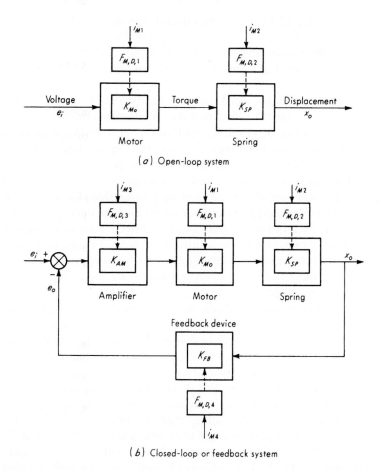

(*a*) Open-loop system

(*b*) Closed-loop or feedback system

Figure 2.12
Use of feedback to reduce effect of spurious inputs.

the feedback device, which produces a voltage e_o proportional to x_o. This voltage is subtracted from the input voltage e_i, and the difference is applied to an amplifier which drives the motor and thereby the spring to produce x_o. We may write

$$(e_i - e_o)K_{AM}K_{Mo}K_{SP} = (e_i - K_{FB}x_o)K_{AM}K_{Mo}K_{SP} = x_o \qquad (2.8)$$

$$e_i K_{AM}K_{Mo}K_{SP} = (1 + K_{AM}K_{Mo}K_{SP}K_{FB})x_o \qquad (2.9)$$

$$x_o = \frac{K_{AM}K_{Mo}K_{SP}}{1 + K_{AM}K_{Mo}K_{SP}K_{FB}} e_i \qquad (2.10)$$

Suppose, now, that we design K_{AM} to be very large (a "high-gain" system), so that $K_{AM}K_{Mo}K_{SP}K_{FB} \gg 1$. Then

$$x_o \approx \frac{1}{K_{FB}} e_i \qquad (2.11)$$

The significance of Eq. (2.11) is that the effect of variations in K_{Mo}, K_{SP}, and K_{AM} (as a result of modifying inputs i_{M1}, i_{M2}, and i_{M3}) on the relation between input e_i and output x_o has been made negligible. *We now require only that K_{FB} stay constant (unaffected by i_{M4}) in order to maintain constant input-output calibration as shown by Eq. (2.11).*

You may question whether much really has been gained by this somewhat elaborate scheme, since we merely transferred the requirements for stability from K_{Mo} and K_{SP} to K_{FB}. In practice, however, this method often leads to great improvements in accuracy. One reason is that, since the amplifier supplies most of the power needed, the feedback device can be designed with low power-handling capacity. In general, this leads to greater accuracy and linearity in the feedback-device characteristics. Also, the input signal e_i need carry only negligible power; thus the feedback system extracts less energy from the measured medium than the corresponding open-loop system. This, of course, results in less distortion of the measured quantity because of the presence of the measuring instrument. Finally, if the open-loop chain consists of several (perhaps many) devices, each susceptible to its own spurious inputs, then *all* these bad effects can be negated by the use of high amplification and a stable, accurate feedback device.

Before we pass on to other methods, we should mention that application of the feedback principle is not without its own peculiar problems. The main one is dynamic instability, wherein excessively high amplification leads to destructive oscillations. The study of the design of feedback systems is a whole field in itself, and many texts treating this subject are available.[3]

The *method of calculated output corrections* requires one to measure or estimate the magnitudes of the interfering and/or modifying inputs and to know quantitatively how they affect the output. With this information, it is possible to calculate corrections which may be added to or subtracted from the indicated output so as to leave (ideally) only that component associated with the desired input. Thus, in the manometer of Fig. 2.10, the effects of temperature on both the calibrated scale's length and the density of mercury may be quite accurately computed if the

---

[3]E. O. Doebelin, "Control System Principles and Design," Wiley, New York, 1985.

temperature is known. The local gravitational acceleration is also known for a given elevation and latitude, so that this effect may be corrected by calculation. Since many measurement systems today can afford to include a microcomputer to carry out various functions, if we also provide sensors for the spurious inputs, the microcomputer can implement the method of calculated output corrections on an automatic basis, giving a so-called *smart sensor.*

The *method of signal filtering* is based on the possibility of introducing certain elements ("filters") into the instrument which in some fashion block the spurious signals, so that their effects on the output are removed or reduced. The filter may be applied to any suitable signal in the instrument, be it input, output, or intermediate signal. The concept of signal filtering is shown schematically in Fig. 2.13 for the

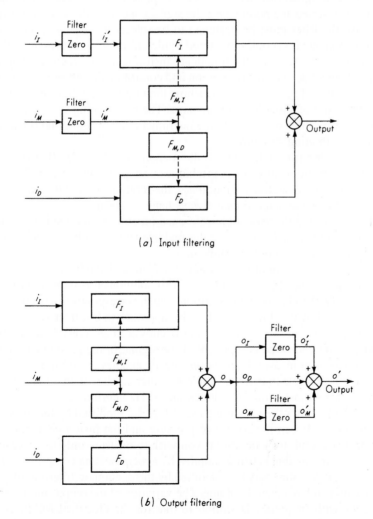

(*a*) Input filtering

(*b*) Output filtering

Figure 2.13
General principle of filtering.

cases of input and output filtering. The application to intermediate signals should be obvious. In Fig. 2.13a the inputs i_I and i_M are caused to pass through filters whose input-output relation is (ideally) zero. Thus i_I' and i_M' are zero even if i_I and i_M are not zero. The concept of output filtering is illustrated in Fig. 2.13b. Here the output o, though really one signal, is thought of as a superposition of o_I (output due to interfering input), o_D (output due to desired input), and o_M (output due to modifying input). If it is possible to construct filters that selectively block o_I and o_M but allow o_D to pass through, this may be symbolized as in Fig. 2.13b and results in o' consisting entirely of o_D.

The filters necessary in the application of this method may take several forms; they are best illustrated by examples. If put directly in the path of a spurious input, a filter can be designed (ideally) to block completely the passage of the signal. If, however, it is inserted at a point where the signal contains both desired and spurious components, the filter must be designed to be selective. That is, it must pass the desired components essentially unaltered while effectively suppressing all others.

Often it is necessary to attach delicate instruments to structures that vibrate. Electromechanical devices for navigation and control of aircraft or missiles are outstanding examples. Figure 2.14a shows how the interfering vibration input may be filtered out by use of suitable spring mounts. The mass-spring system is actually a mechanical filter which passes on to the instrument only a negligible fraction of the motion of the vibrating structure.

The interfering tilt-angle input to the manometer of Fig. 2.10c may be effectively filtered out by means of the gimbal-mounting scheme of Fig. 2.14b. If the gimbal bearings are essentially frictionless, the rotations θ_1 and θ_2 cannot be communicated to the manometer; thus it always hangs vertical.

In Fig. 2.14c the thermocouple reference junction is shielded from ambient-temperature fluctuations by means of thermal insulation. Such an arrangement acts as a filter for temperature or heat-flow inputs.

The strain-gage circuit of Fig. 2.14d is shielded from the interfering 60-Hz field by enclosing it in a metal box of some sort. This solution corresponds to filtering the interfering *input*. Another possible solution, which corresponds to selective filtering of the *output*, is shown in Fig. 2.14e. For this approach to be effective, it is essential that the frequencies in the desired signal occupy a range considerably separated from those in the undesired component of the signal. In the present example, suppose the strains to be measured are mainly steady and never vary more rapidly than 2 Hz. Then it is possible to insert a simple *RC* filter, as shown, that will pass the desired signals but almost completely block the 60-Hz interference.

Figure 2.14f shows the pressure gage of Fig. 2.2 modified by the insertion of a flow restriction between the source of pressure and the piston chamber. Such an arrangement is useful, for example, if you wish to measure only the average pressure in a large air tank that is being supplied by a reciprocating compressor. The pulsations in the air pressure may be smoothed by the pneumatic filtering effect of the flow restriction and associated volume. The variation of the output-input amplitude ratio $|p_o/p_i|$ with frequency is similar to that for the electrical *RC* filter of Fig. 2.14e. Thus steady or slowly varying input pressures are accurately measured while

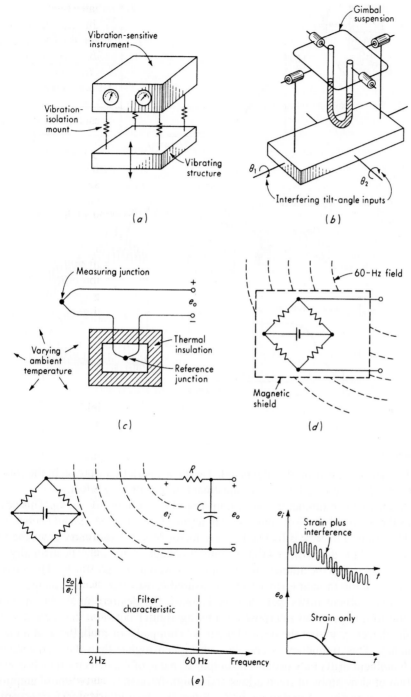

Figure 2.14
Examples of filtering.

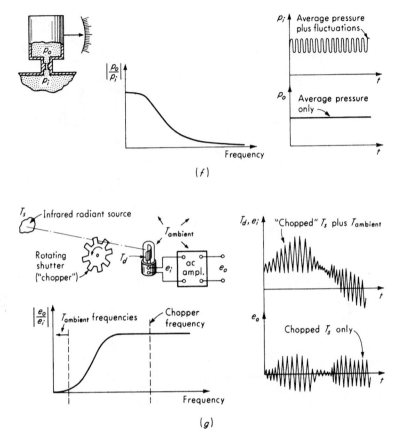

Figure 2.14
(*Continued*)

rapid variations are strongly attenuated. The flow restriction may be in the form of a needle valve, which allows easy adjustment of the filtering effect.

A "chopped" radiometer is shown in simplified form in Fig. 2.14g. This device senses the temperature T_s of some body in terms of the infrared radiant energy emitted. The emitted energy is focused on a detector of some sort and causes the temperature T_d of the detector, and thus its output voltage e_i, to vary. The difficulty with such devices is that the ambient temperature, as well as T_s, affects T_d. This effect is serious since the radiant energy to be measured causes very small changes in T_d; thus small ambient drifts can completely mask the desired input. An ingenious solution to this problem interposes a rotating shutter between the radiant source and the detector, so that the desired input is "chopped," or modulated, at a known frequency. This frequency is chosen to be much higher than the frequencies at which ambient drifts may occur. The output signal e_i of the detector thus is a superposition of slow ambient fluctuations and a high-frequency wave whose amplitude varies in proportion to variations in T_s. Since the desired and interfering components are thus widely separated in frequency, they may be selectively filtered. In this case, we desire a filter that rejects constant and slowly varying signals, but faithfully

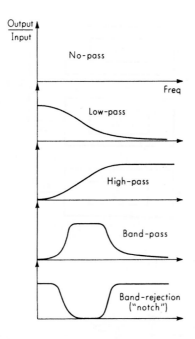

Figure 2.15
Basic filter types.

reproduces rapid variations. Such a characteristic is typical of an ordinary ac amplifier, and since amplification is necessary in such instruments in any case, the use of an ac amplifier as shown solves two problems at once.

In summing up the method of signal filtering, it may be said that, in general, it is usually possible to design filters of mechanical, electrical, thermal, pneumatic, etc., nature which separate signals according to their frequency content in some specific manner. Figure 2.15 summarizes the most common useful forms of such devices.

The *method of opposing inputs* consists of intentionally introducing into the instrument interfering and/or modifying inputs that tend to cancel the bad effects of the unavoidable spurious inputs. Figure 2.16 shows schematically the concept for interfering inputs. The extension to modifying inputs should be obvious. The intentionally introduced input is designed so that the signal o_{I1} and o_{I2} are essentially equal but act in the opposite sense; thus the net contribution $o_{I1} - o_{I2}$ to the output is nearly zero. This method actually might be considered as a variation on the method of calculated output corrections. However, the "calculation" and application of the correction are achieved automatically owing to the structure of the system, rather than by numerical calculation by a human operator. Thus the two methods are similar; however, the distinction between them is a worthwhile one since it helps to organize your thinking in inventing new applications of these generalized correction concepts.

Some examples of the method of opposing inputs are shown in Fig. 2.17. A millivoltmeter, shown in Fig. 2.17*a*, is basically a *current*-sensitive device.

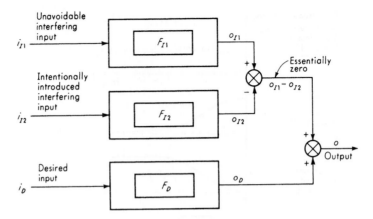

Figure 2.16
Method of opposing inputs.

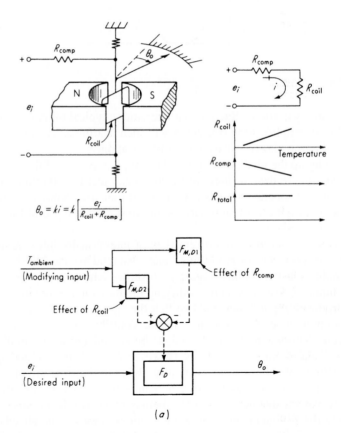

$$\theta_o = ki = k\left[\frac{e_i}{R_{coil} + R_{comp}}\right]$$

(a)

Figure 2.17
Examples of method of opposing inputs.

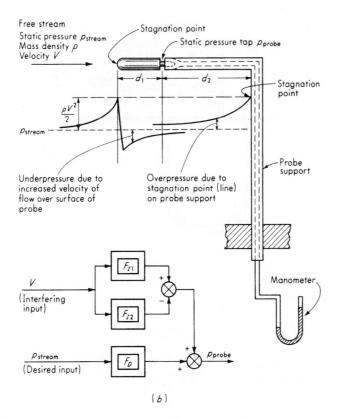

(b)

Figure 2.17
(*Continued*)

However, as long as the total circuit resistance is constant, its scale can be calibrated in voltage, since voltage and current are proportional. A modifying input here is the ambient temperature, since it causes the coil resistance R_{coil} to change, thereby altering the proportionality factor between current and voltage. To correct for this error, the compensating resistance R_{comp} is introduced into the circuit, and its material is carefully chosen to have a temperature coefficient of resistance *opposite* to that of R_{coil}. Thus when the temperature changes, the total resistance of the circuit is unaffected and the calibration of the meter remains accurate.

Figure 2.17*b* shows a static-pressure-probe design due to L. Prandtl. As the fluid flows over the surface of the probe, the velocity of the fluid must increase since these streamlines are longer than those in the undisturbed flow. This velocity increase causes a drop in static pressure, so that a tap in the surface of the probe gives an incorrect reading. This underpressure error varies with the distance d_1 of the tap from the probe tip. Prandtl recognized that the probe support will have a stagnation point (line) along its front edge and that this overpressure will be felt upstream, the effect decreasing as the distance d_2 increases. By properly choosing

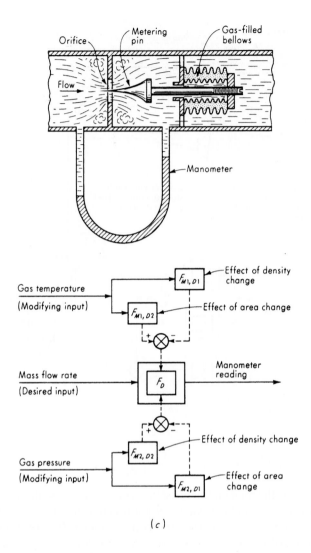

Figure 2.17
(*Continued*)

distances d_1 and d_2 (by experimental test), these two effects can be made exactly to cancel, giving a true static-pressure value at the tap.

A device for the measurement of the mass flow rate of gases is shown in Fig. 2.17c. The mass flow rate of gas through an orifice may be found by measuring the pressure drop across the orifice, perhaps by means of a U-tube manometer. Unfortunately, the mass flow rate also depends on the density of the gas, which varies with pressure and temperature. Thus the pressure-drop measuring device usually cannot be calibrated to give the mass flow rate, since variations in gas temperature and pressure yield different mass flow rates for the same orifice pressure drop. The instrument of Fig. 2.17c overcomes this problem in an ingenious fashion. The flow

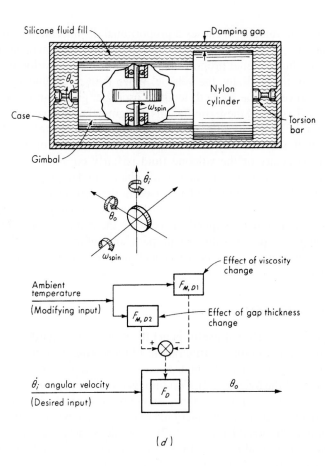

Figure 2.17
(*Concluded*)

rate through the orifice also depends on its flow area. Thus if the flow area could be varied in just the right way, this variation could compensate for pressure and temperature changes so that a given orifice pressure drop would *always* correspond to the same mass flow rate. This is accomplished by attaching the specially shaped metering pin to a gas-filled bellows as shown. When the temperature drops (causing an increase in density and therefore in mass flow rate), the gas in the bellows contracts, which moves the metering pin into the orifice and thereby reduces the flow area. This returns the mass flow rate to its proper value. Similarly, should the pressure of the flowing gas increase, causing an increase in density and mass flow rate, the gas-filled bellows would be compressed again, reducing the flow area and correcting the mass flow rate. The proper shape for the metering pin is revealed by a detailed analysis of the system.

A final example of the method of opposing inputs is the rate gyroscope of Fig. 2.17*d*. Such devices are widely used in aerospace vehicles for the generation of stabilization signals in the control system. The action of the device is that a vehicle rotation at angular velocity θ_i causes a proportional displacement θ_o of the gimbal relative to the case. This rotation θ_o is measured by some motion pickup (not shown in Fig. 2.17d). Thus a signal proportional to vehicle angular velocity is available, and this is useful in stabilizing the vehicle. When the vehicle undergoes rapid motion changes, however, the angle θ_o tends to oscillate, giving an incorrect angular-velocity signal. To control these oscillations, the gimbal rotation θ_o is damped by the shearing action of a viscous silicone fluid in a narrow damping gap. The damping effect varies with the viscosity of the fluid and the thickness of the damping gap. Although the viscosity of the silicone fluid is fairly constant, it does vary with ambient temperature, causing an undesirable change in damping characteristics. To compensate for this, a nylon cylinder is used in the gyro of Fig. 2.17d. When the temperature increases, viscosity drops, causing a loss of damping. Simultaneously, however, the nylon cylinder expands, narrowing the damping gap and thus restoring the damping to its proper value. By proper choice of materials and geometry, the two effects can be made to very nearly cancel over the operating temperature range of the equipment.

2.6 CONCLUSION

In this chapter we developed useful generalizations with regard to the functional elements and the input-output configurations of measuring instruments and systems. In the analysis of a given instrument or in the design of a new one, the starting point is the separation of the overall operation into its functional elements. Here you must take a broad view of *what* must be done, but not be concerned with *how* it is actually accomplished. Once the general functional concepts have been clarified, the details of operation may be considered fruitfully. The ideas of active and passive transducers, analog and digital modes of operation, and null versus deflection methods give a systematic approach for either analysis or design.

Finally, compensation of spurious inputs and detailed evaluation of performance are facilitated by application of input-output block diagrams. These configuration diagrams show clearly which physical analyses must be made to evaluate performance with respect to accurate measurement of the desired inputs and rejection of spurious inputs. The evaluation of the relative quality of different instruments (or the same instrument with different numerical parameter values) requires the definition of performance criteria against which competitive designs may be compared. This is the subject of Chapter 3.

PROBLEMS

2.1 Make block diagrams such as Fig. 2.1, showing the functional elements of the instruments depicted in the following:

 (*a*) Fig. 2.7.
 (*b*) Fig. 2.8.
 (*c*) Fig. 2.10*a*.
 (*d*) Fig. 2.11. Take F as input and e_o as output.
 (*e*) Fig. 2.14*g*. Take T_s as input and e_o as output.
 (*f*) Fig. 2.17*b*. Take V as input and manometer Δh as output.
 (*g*) Fig. 2.17*d*. Take θ_i as input and θ_o as output.

2.2 Identify the active transducers, if any, in the instruments of (*a*) Fig. 2.8, (*b*) Fig. 2.10*a*, (*c*) Fig. 2.11, (*d*) Fig. 2.17*b*, (*e*) Fig. 2.17*c*.

2.3 Consider a man, driving a car along a road, who sees the opportunity to pass and decides to accelerate.

 (*a*) If the light waves entering his eyes are considered input and accelerator-pedal travel is taken as output, is the man functioning as an active or a passive transducer?

 (*b*) If the accelerator-pedal travel is considered input and car velocity as output, is the automobile engine an active or a passive transducer?

2.4 Give an example of a null method of force measurement.

2.5 Give an example of a null method of voltage measurement.

2.6 Sketch and explain two possible modifications of the system of Fig. 2.4 that will allow measurement to 1/10 revolution.

2.7 Identify desired, interfering, and modifying inputs for the systems of (*a*) Fig. 2.2, (*b*) Fig. 2.3, (*c*) Fig. 2.4, (*d*) Fig. 2.5.

2.8 Why is tilt angle in Fig. 2.10*c* a modifying input?

2.9 Suppose in Eq. (2.7) that $K_{Mo} = K_{SP} = e_i = 1.0$. Now let K_{Mo} change by 10 percent to 1.1. What is the change in x_o? In Eq. (2.10), let $K_{Mo} = K_{SP} = K_{FB} = e_i = 1.0$, and $K_{AM} = 100$. Now let K_{Mo} change by 10 percent to 1.1. What is the change in x_o? Investigate the effect of similar changes in K_{AM}, K_{SP}, and K_{FB}.

2.10 The natural frequency of oscillation of the balance wheel in a watch depends on the moment of inertia of the wheel and the spring constant of the (torsional) hairspring. A temperature rise results in a reduced spring constant, which lowers the oscillation frequency. Propose a compensating means for this effect. Non-temperature-sensitive hairspring material is not an acceptable solution.

3
CHAPTER

Generalized Performance Characteristics of Instruments

3.1 INTRODUCTION

If you are trying to choose, from commercially available instruments, the one most suitable for a proposed measurement, or, alternatively, if you are engaged in the design of instruments for specific measuring tasks, then the subject of performance criteria assumes major proportions. That is, to make intelligent decisions, there must be some quantitative bases for comparing one instrument (or proposed design) with the possible alternatives. Chapter 2 has served as a useful preliminary to these considerations since there we developed systematic methods for breaking down the overall problem into its component parts. Now we propose to study in considerable detail the performance of measuring instruments and systems with regard to how well they measure the desired inputs and how thoroughly they reject the spurious inputs.

The treatment of instrument performance characteristics generally has been broken down into the subareas of *static characteristics* and *dynamic characteristics,* and this plan is followed here. The reasons for such a classification are several. First, some applications involve the measurement of quantities that are constant or vary only quite slowly. Under these conditions, it is possible to define a set of performance criteria that give a meaningful description of the quality of measurement without becoming concerned with dynamic descriptions involving differential equations. These criteria are called the *static* characteristics. Many other measurement problems involve rapidly varying quantities. Here the dynamic relations between the instrument input and output must be examined, generally by the use of differential equations. Performance criteria based on these dynamic relations constitute the *dynamic* characteristics.

Actually, static characteristics also influence the quality of measurement under dynamic conditions, but the static characteristics generally show up as nonlinear or statistical effects in the otherwise linear differential equations giving the dynamic characteristics. These effects would make the differential equations analytically unmanageable, and so the conventional approach is to treat the two aspects of the problem separately. Thus the differential equations of dynamic performance generally neglect the effects of dry friction, backlash, hysteresis, statistical scatter, etc., even though these effects influence the dynamic behavior. These phenomena are more conveniently studied as static characteristics, and the overall performance of an instrument is then judged by a semiquantitative superposition of the static and dynamic characteristics. This approach is, of course, approximate but a necessary expedient for convenient mathematical study. Once tentative designs and numerical values are available, we can of course use *simulation* to investigate the nonlinear and statistical effects.

3.2 STATIC CHARACTERISTICS AND STATIC CALIBRATION

Meaning of Static Calibration

We begin our study by considering the process of static calibration since all the static performance characteristics are obtained by one form or another of this process. In general, *static calibration* refers to a situation in which all inputs (desired, interfering, modifying) except one are kept at some constant values. Then the one input under study is varied over some range of constant values, which causes the output(s) to vary over some range of constant values. The input-output relations developed in this way comprise a static calibration *valid under the stated constant conditions of all the other inputs.* This procedure may be repeated, by varying in turn each input considered to be of interest and thus developing a family of static input-output relations. Then we might hope to describe the overall instrument static behavior by some suitable form of superposition of these individual effects. In some cases, if overall rather than individual effects were desired, the calibration procedure would specify the variation of several inputs simultaneously. Also if you examine any practical instrument critically, you will find many modifying and/or interfering inputs, each of which might have quite small effects and which would be impractical to control. Thus the statement "*all* other inputs are held constant" refers to an ideal situation which can be only approached, but never reached, in practice. *Measurement method* describes the ideal situation while *measurement process* describes the (imperfect) physical realization of the measurement method.

The statement that one input is varied and all others are held constant implies that all these inputs are determined (measured) independently of the instrument being calibrated. For interfering or modifying inputs (whose effects on the output should be relatively small in a good instrument), the measurement of these inputs usually need not be at an extremely high accuracy level. For example, suppose a pressure gage has temperature as an interfering input to the extent that a temperature change of $100°$ causes a pressure error of 0.100 percent. Now, if we had measured the $100°$ interfering input with a thermometer which itself had an error of 2.0 percent, the pressure error actually would have been 0.102 percent. It should be clear that the difference between an error of 0.100 and 0.102 percent is entirely negligible in most engineering situations. However, when calibrating the response of the instrument to its *desired* inputs, you must exercise considerable care in choosing the means of determining the numerical values of these inputs. That is, if a pressure gage is inherently capable of an accuracy of 0.1 percent, you must certainly be able to determine its input pressure during calibration with an accuracy somewhat greater than this. In other words, it is impossible to calibrate an instrument to an accuracy greater than that of the standard with which it is compared. A rule often followed is that the calibration system (the standard and any auxiliary apparatus used with it) have a total uncertainty four times better than the unit under test.[1]

Details about standards for specific physical variables will be given at the beginning of the chapter devoted to that variable. Here we want to give some general information. A *standard* for a certain physical variable is often "just another" measuring device for that variable. However, to be called a standard, its accuracy must be at a higher level than the instrument to be calibrated, the 4-to-1 ratio mentioned above being a common requirement. There is actually a *hierarchy* of standards which arranges them in order of decreasing accuracy, with *primary* standards being the most accurate (at the top of the hierarchy). Primary standards are considered the "state of the art," that is, the most accurate way known to measure the quantity of interest. Such standards are developed, maintained, and improved by national laboratories such as the National Institute for Standards and Technology (NIST)[2] in the United States. These labs also may provide calibration services to industry or other customers. Large, "high-tech" companies sometimes maintain calibration laboratories which also are capable of calibration at this highest level.

As you might guess, primary standards tend to be complex and expensive and are needed only for the most critical situations. Thus we need *lower level* (secondary, tertiary, etc.) standards which are simpler and cheaper to use for most engineering calibration work. Such standards are available for calibration service at the national laboratories, commercial calibration laboratories, and in-house calibration laboratories associated with industrial companies, universities, etc. When planning a specific experimental project, we need to decide how accurate our measurements need to be, and then arrange to calibrate each instrument against a standard that is about four times more accurate, if possible. Thus, if we need a 1 percent accuracy in a pressure gage, we need to calibrate it against a standard accurate to about 0.25 percent or better. Of course, the gage must be *capable* of 1 percent accuracy. If it has random errors of, say, 3 percent, calibrating it against a 0.25 percent standard will *not* make it a "1 percent gage." This, of course, would be discovered during calibration, but we do not want to waste our time, so our initial selection of instruments must be carefully made.

In performing a calibration, the following steps are necessary:

---

[1]ISO Guide 25, ANSI/NCSL Z540-1.

[2]Formerly (before 1989) called the National Bureau of Standards (NBS).

1. Examine the construction of the instrument, and identify and list all the possible inputs.
2. Decide, as best you can, which of the inputs will be significant in the application for which the instrument is to be calibrated.
3. Procure apparatus that will allow you to vary all the significant inputs over the ranges considered necessary. Procure standards to measure each input.
4. By holding some inputs constant, varying others, and recording the output(s), develop the desired static input-output relations.

Now we are ready for a more detailed discussion of specific static characteristics. These characteristics may be classified as either general or special. General static characteristics are of interest in *every* instrument. Special static characteristics are of interest in only a particular instrument. We concentrate mainly on general characteristics, leaving the treatment of special characteristics to later sections of the text in which specific instruments are discussed.

Measured Value versus True Value

When we measure some physical quantity with an instrument and obtain a numerical value, usually we are concerned with how close this value may be to the "true" value. It is first necessary to understand that this so-called true value is, in general, unknown and unknowable, since perfectly exact definitions of the physical quantities to be measured are impossible. This can be illustrated by specific example, for instance, the length of a cylindrical rod. When we ask ourselves what we *really* mean by the length of this rod, we must consider such questions as these:

1. Are the two ends of the rod planes?
2. If they are planes, are they parallel?
3. If they are not planes, what sort of surfaces are they?
4. What about surface roughness?

We see that complex problems are introduced when we deal with a real object rather than an abstract, geometric solid. The term "true value," then, really refers to a value called the *reference value* that would be obtained if the quantity under consideration were measured by an *exemplar method,*[3] that is, a method agreed on by experts as being sufficiently accurate for the purposes to which the data ultimately will be put. The measurement process consists of actually carrying out, as well as possible, the instructions for performing the measurement, which are the measurement method. (Since calibration is essentially a refined form of measurement, these remarks apply equally to the process of calibration.) If this process is repeated over and over under *assumed* identical conditions, we get a large number of readings from the instrument. Usually these readings will not all be the same, and so we note immediately that we may *try* to ensure identical conditions for each trial, but it is

---

[3]Churchill Eisenhart, "Realistic Evaluation of the Precision and Accuracy of Instrument Calibration Systems," *J. Res. Natl. Bur. Std., C,* vol. 67C, no. 2, April–June 1963.

never exactly possible. The data generated in this fashion may be used to describe the measurement process so that, if it is used in the future, we may be able to attach some numerical estimates of error to its outputs.

If the output data are to give a meaningful description of the measurement process, the data must form what is called a *random sequence.* Another way of saying this is that the process must be in a state of *statistical control.*[4] The concept of the state of statistical control is not a particularly simple one, but we try to explain its essence briefly. First we note that it is meaningless to speak of the accuracy of an instrument as an isolated device. We must always consider the instrument plus its environment and method of use, that is, the instrument plus its inputs. This aggregate constitutes the measurement process. Every instrument has an infinite number of inputs; that is, the causes that can conceivably affect the output, if only very slightly, are limitless. Such effects as atmospheric pressure, temperature, and humidity are among the more obvious. But if we are willing to "split hairs," we can uncover a multitude of other physical causes that could affect the instrument with varying degrees of severity. In defining a calibration procedure for a specific instrument, we specify that certain inputs must be held "constant" within certain limits. These inputs, it is hoped, are the ones that contribute the largest components to the overall error of the instrument. The remaining infinite number of inputs is left uncontrolled, and it is hoped that each of these individually contributes only a very small effect and that in the aggregate their effect on the instrument output will be of a random nature. If this is indeed the case, the process is said to be in statistical control. Experimental proof that a process is in statistical control is not easy to come by; in fact, achieving *strict* statistical control is unlikely. Thus we can only approximate this situation.

Lack of control is sometimes obvious, however, if we repeat a measurement and plot the result (output) versus the trial number. Figure 3.1*a* shows such a graph for the calibration of a particular instrument. In this instance, it was ascertained after some study that the instrument actually was much more sensitive to temperature than had been thought. The original calibration was carried out in a room without temperature control. Thus the room temperature varied from a low in the morning to a peak in the early afternoon and then dropped again in the late afternoon. Since the 10 trials covered a period of about one day, the trend of the curve is understandable. By performing the calibration in a temperature-controlled room, the graph of Fig. 3.1*b* was obtained. For the detection of more subtle deviations from statistical control, the methods of statistical quality-control charts are useful.[5]

If the measurement process is in reasonably good statistical control and if we repeat a given measurement (or calibration point) over and over, we will generate a set of data exhibiting random scatter. As an example, consider the pressure gage of Fig. 3.2. Suppose we wish to determine the relationship between the desired input (pressure) and the output (scale reading). Other inputs which could be significant

[4]Ibid.

[5]E. B. Wilson, Jr., "An Introduction to Scientific Research," Chap 9, McGraw-Hill, New York, 1952.

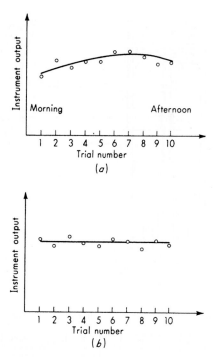

Figure 3.1
Effect of uncontrolled input on calibration.

and which might have to be controlled during the pressure calibration include temperature, acceleration, and vibration. Temperature can cause expansion and contraction of instrument parts in such a way that the scale reading will change even though the pressure has remained constant. An instrument acceleration along the axis of the piston rod will cause a scale reading even though pressure again has remained unchanged. This input is significant if the pressure gage is to be used aboard a vehicle of some kind. A small amount of vibration actually may be helpful to the operation of an instrument, since vibration may reduce the effects of static friction. Thus if the pressure gage is to be attached to a reciprocating air compressor (which always has some vibration), it may be more accurate under these conditions than it would be under calibration conditions where no vibration was provided. These examples illustrate the general importance of carefully considering the relationship between the calibration conditions and the actual application conditions.

Some Basic Statistics

Suppose, now, that we have procured a sufficiently accurate pressure standard and have arranged to maintain the other inputs reasonably close to the actual application conditions. Repeated calibration at a given pressure (say, 10 kPa) might give the data of Fig. 3.3. Suppose we now order the readings from the lowest (9.81) to the

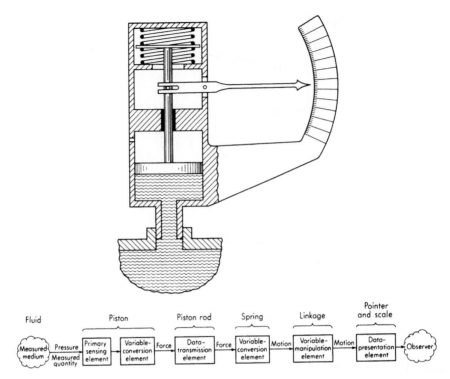

Figure 3.2
Pressure gage.

highest (10.42) and see how many readings fall in each interval of, say, 0.05 kPa, starting at 9.80. The result can be represented graphically as in Fig. 3.4*a*. Suppose we now define the quantity *Z* by

$$Z \triangleq \frac{(\text{number of readings in an interval})/(\text{total number of readings})}{\text{width of interval}} \quad \textbf{(3.1)}$$

and we plot a "bar graph" with height *Z* for each interval. Such a *"histogram"* is shown in Fig. 3.4*b*. Statistical software, such as the *MINITAB* used in this text, will of course plot histograms for us. It should be clear from Eq. (3.1) that the area of a particular "bar" is numerically equal to the *probability* that a specific reading will fall in the associated interval. The area of the entire histogram must then be 1.0 (100 percent = 1.0), since there is 100 percent probability that the reading will fall somewhere between the lowest and highest values, at least based on the data available. If it were now possible to take an infinite number of readings, each with an infinite number of significant digits, we could make the chosen intervals as small as we pleased and still have each interval contain a finite number of readings. Thus the steps in the graph of Fig. 3.4*b* would become smaller and smaller, with the graph approaching a smooth curve in the limit. If we take this limiting abstract case as a mathematical model for the real physical situation, the function *Z* = *f(x)* is called

True pressure (reference value) = 10,000 ± .001 kPa
Acceleration = 0
Vibration level = 0
Ambient temperature = 20 ± 1°C

Trial number	Scale reading, kPa
1	10.02
2	10.20
3	10.26
4	10.20
5	10.22
6	10.13
7	9.97
8	10.12
9	10.09
10	9.90
11	10.05
12	10.17
13	10.42
14	10.21
15	10.23
16	10.11
17	9.98
18	10.10
19	10.04
20	9.81

(a)

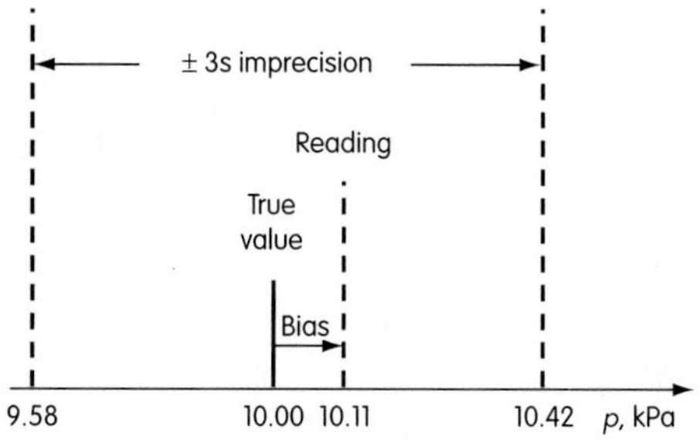

(b)

Figure 3.3
Pressure-gage calibration data.

the *probability density function* for the mathematical model of the real physical process (see Fig. 3.5a). From the basic definition of Z, clearly

$$\text{Probability of reading lying between } a \text{ and } b \triangleq P(a < x < b) = \int_a^b f(x)\, dx \tag{3.2}$$

The probability information sometimes is given in terms of the *cumulative distribution function F(x)*, which is defined by

$$F(x) \triangleq \text{probability that reading is less than any chosen value of } x$$

$$F(x) = \int_{-\infty}^x f(x)\, dx \tag{3.3}$$

and is shown in Fig 3.5b.

From the infinite number of forms possible for probability density functions, a relatively small number are useful mathematical models for practical applications; in fact, *one* particular form is quite dominant. The most useful density function or distribution is the normal or *Gaussian* function, which is given by

$$f(x) = \frac{1}{\sqrt{2\pi}\,\sigma} e^{-(x-\mu)^2/(2\sigma^2)} \qquad -\infty < x < +\infty \tag{3.4}$$

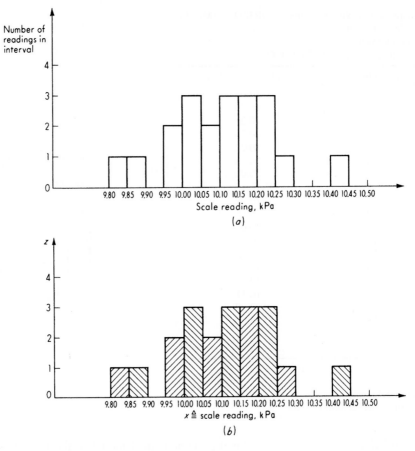

Figure 3.4
Distribution of data.

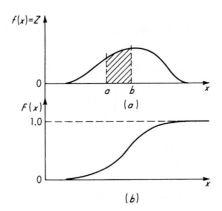

Figure 3.5
Probability distribution function.

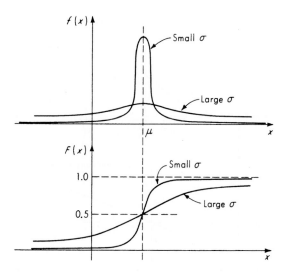

Figure 3.6
Gaussian distribution

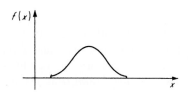

Figure 3.7
Non-Gaussian distribution.

Equation (3.4) defines a whole family of curves depending on the particular numerical values of μ (the mean value) and σ (the standard deviation). The shape of the curve is determined entirely by σ, with μ serving only to locate the position of the curve along the x axis. The cumulative distribution function $F(x)$ cannot be written explicitly in this case because the integral of Eq. (3.3) cannot be carried out; however, the function has been tabulated by performing the integration by numerical means. Figure 3.6 shows that a small value of σ indicates a high probability that a "reading" will be found close to μ. Equation (3.4) also shows that there is a small probability that very large ($\rightarrow \pm \infty$) readings will occur. This is one of the reasons why a true Gaussian distribution can never occur in the real world; physical variables are always limited to finite values. There is *zero* probability, for example, that the pointer on a pressure gage will read 100 kPa when the range of the gage is only 20 kPa. Real distributions must thus, in general, have their "tails" cut off, as in Fig. 3.7.

Although actual data will not conform *exactly* to the Gaussian distribution, very often they are sufficiently close to allow use of the Gaussian model in engineering

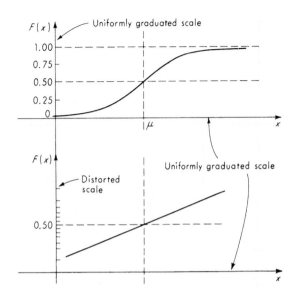

Figure 3.8
Rectification of Gaussian curve.

work. It would be desirable to have available tests that would indicate whether the data were "reasonably" close to Gaussian. We must admit however, that in much practical work the time and effort necessary for such tests cannot be justified, and the Gaussian model is simply *assumed* until troubles arise which justify a closer study of the particular situation.

One method of testing for an approximate Gaussian distribution involves the use of probability graph paper. If we take the cumulative distribution function for a Gaussian distribution and suitably distort the vertical scale of the graph, the curve can be made to plot as a straight line, as shown in Fig. 3.8. (This, of course, can be done with any curvilinear relation, not just probability curves.) Such graph paper is commercially available and may be used to give a rough, but useful qualitative test for conformity to the Gaussian distribution. (If you have access to statistical software, special graph paper is not needed; the software plots your points to the proper scales.) For example, consider the data of Fig. 3.3. These data may be plotted on Gaussian probability graph paper as follows: First lay out on the uniformly graduated horizontal axis a numerical scale that includes all the pressure readings. Now the probability graph paper represents the cumulative distribution, so that the ordinate of any point represents the probability that a reading will be less than the abscissa of that particular point. This probability, in terms of the sample data available, is simply the percentage (in decimal form) of points that fell at or below that particular value. Figure 3.9a shows the resulting plot. Note that the highest point (10.42) cannot be plotted since 100 percent cannot appear on the ordinate scale. Also shown in Fig 3.9a is the "perfect Gaussian line," the straight line that would be perfectly followed by data from an infinitely large sample of Gaussian data which had the

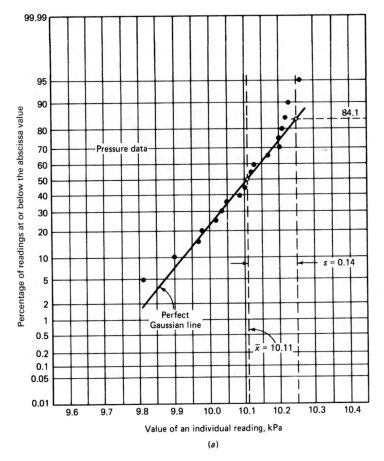

Figure 3.9
Graphical check of Gaussian distribution.

same μ and σ values as our actual data sample. To plot this line, we must estimate μ from the *sample mean value* $\overline{X}$, using

$$\overline{X} \triangleq \frac{\sum\limits_{i=1}^{N} X_i}{N} \tag{3.5}$$

where $\qquad\qquad X_i \triangleq$ individual reading $\tag{3.6}$

$\qquad\qquad\qquad N \triangleq$ total number of readings $\tag{3.7}$

and σ from the *sample standard deviation s*, using

$$s \triangleq \sqrt{\frac{\sum\limits_{i=1}^{N} (X_i - \overline{X})^2}{N - 1}} \tag{3.8}$$

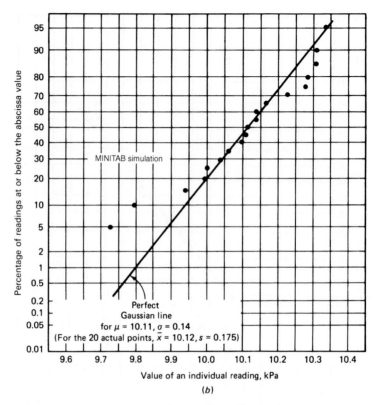

Percentage of readings at or below the abscissa value

MINITAB simulation

Perfect
Gaussian line
for $\mu = 10.11$, $\sigma = 0.14$
(For the 20 actual points, $\bar{x} = 10.12$, $s = 0.175$)

Value of an individual reading, kPa

(*b*)

Figure 3.9

(*Continued*)

The data of Fig. 3.3 give $\bar{X} = 10.11$ and $s = 0.14$ kPa. Two points that may be used to plot any perfect Gaussian line are $(\bar{X}, 50\%)$ and $(\bar{X} + s, 84.1\%)$, which yield the line of Fig. 3.9*a* for our data. Superimposing this line on our actual data, we may judge visually and qualitatively whether our data are "close to" or "far from" Gaussian. Note that there is *no hope* of ever *proving* real-world data to be Gaussian. There are always physical constraints which require real data to be at least somewhat non-Gaussian.

Since our graphical test is clearly subjective, it may be useful to provide some additional data to help reach the required decision. Figure 3.9*b* and *c* show "data" generated by a digital-computer Gaussian random-number generator which is part of the statistical software called MINITAB. Such random-number algorithms are excellent simulations of perfect Gaussian distributions. Figure 3.9*b* graphs data generated when the program was asked to produce a sample of 20 readings with $\mu = 10.11$ and $\sigma = 0.14$. Note that even though the sample was drawn from a perfect Gaussian distribution, the points do not fall on the perfect Gaussian line. This does not mean that the computer algorithm is faulty. It simply shows that a sample of size 20 is too small to give a statistically reliable prediction. In Fig. 3.9*c*, the sample size was increased to 100, which gives a clear improvement but still not

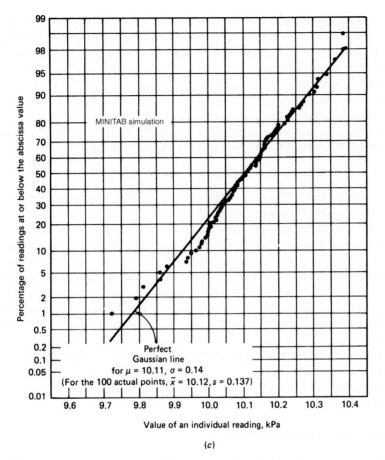

Figure 3.9
(Concluded)

perfection. The results of Fig. 3.9*a* would be considered by most people to indicate a reasonable approximation to a Gaussian distribution. However, this test is clearly only qualitative, and its main usefulness perhaps is in revealing *gross* departures from the theoretical distribution. Such deviations would lead us to examine the instrument and measurement process more closely before attempting to make any statistical statements about accuracy. For a perfect Gaussian distribution, it can be shown[6] that

$$68\% \text{ of the readings lie within } \pm 1\sigma \text{ of } \mu$$
$$95\% \text{ of the readings lie within } \pm 2\sigma \text{ of } \mu \tag{3.9}$$
$$99.7\% \text{ of the readings lie within } \pm 3\sigma \text{ of } \mu$$

---

[6]E. O. Doebelin, "Engineering Experimentation," McGraw-Hill, New York, 1995, p. 42.

Thus if we assume that our real distribution is nearly Gaussian, we might predict, for instance, that if more readings were taken, 99.7 percent would fall within ± 0.42 kPa of 10.11. Since the purpose of calibration is to convert instrument readings into estimates of the true value, let us suppose we now use the calibrated gage to measure an unknown pressure and the gage happens to read 10.11. Based on the calibration we would say our best estimate of the true pressure is 10.00, but this value is uncertain. We cannot *remove* this uncertainty, but we can quantify it. We are 68 percent sure the true value is somewhere within ± 0.14 of 10.00, 95 percent sure it is within ± 0.28, and 99.7 percent sure it is within ± 0.42.

$$\text{instrument bias} \triangleq \text{reading} - \text{reference (“true”) value} \qquad \textbf{(3.10)}$$

The instrument *bias* is $(10.11 - 10.00) = +0.11$. The basic *precision index* is s (0.14), but it has been common practice to quote the imprecision as $\pm\ 3s$. Figure 3.3*b* displays these definitions.

When cumulative distribution graphs such as those in Fig. 3.9 are plotted, it can be shown[7] that it is more correct to plot the *j*th point of an *n*-point sample, not as $(j/n)100$ percent, but rather as $[(j - 0.3)/(n + 0.4)]100$ percent. This also allows the plotting of the "last" point, which would otherwise always be at 100 percent and thus not plottable. While this procedure gives a useful improvement for samples of 15 or less, it has almost no effect when, as in Fig. 3.9, samples of 20 or more are used.

A somewhat more quantitative method for deciding whether data are close to Gaussian involves the chi-square (χ^2) goodness-of-fit test.[8] The reference explains this test but concludes it is inferior to our "Gaussian graph paper" method; thus we pursue it no further.

Least-Squares Calibration Curves

Up to now, we have been examining the situation (Fig. 3.3) in which a single true value is applied repeatedly and the resulting measured values are recorded and analyzed. In an actual instrument calibration, the true value is varied, in increments, over some range, causing the measured value also to vary over a range. Very often there is no multiple repetition of a given true value. The procedure is merely to cover the desired range in both the increasing and the decreasing directions. Thus a given true value is applied, at most, twice if we choose to use the same set of true values for both increasing and decreasing readings.

As an example, suppose we wish to calibrate the pressure gage of Fig. 3.2 for the relation between the desired input (pressure) and the output (scale reading). Figure 3.10*a* gives the data for such a calibration over the range 0 to 10 kPa. In this instrument (as in most but not all), the input-output relation is ideally a straight line. The *average calibration curve* for such an instrument generally is taken as a straight line which fits the scattered data points best as defined by some chosen criterion.

[7]C. Lipson and N. J. Sheth, "Statistical Design and Analysis of Engineering Experiments," McGraw-Hill, New York, 1973, p. 18.

[8]E. O. Doebelin, "Engineering Experimentation," McGraw-Hill, New York, 1995, p. 55.

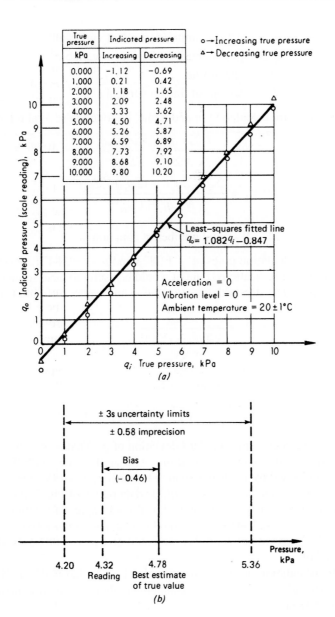

Figure 3.10
Pressure-gage calibration.

The most common is the least-squares criterion, which minimizes the sum of the squares of the vertical deviations of the data points from the fitted line. (The least-squares procedure also can be used to fit curves other than straight lines to scattered data.) The equation for the straight line is taken as

$$q_o = mq_i + b \qquad (3.11)$$

where $\qquad$ $q_o \triangleq$ output quantity (dependent variable) $\qquad$ **(3.12)**

$q_i \triangleq$ input quantity (independent variable) $\qquad$ **(3.13)**

$m \triangleq$ slope of line $\qquad$ **(3.14)**

$b \triangleq$ intercept of line on vertical axis $\qquad$ **(3.15)**

The equations for calculating m and b may be found in several references:[9]

$$m = \frac{N\Sigma q_i q_o - (\Sigma q_i)(\Sigma q_o)}{N\Sigma q_i^2 - (\Sigma q_i)^2} \qquad \textbf{(3.16)}$$

$$b = \frac{(\Sigma q_o)(\Sigma q_i^2) - (\Sigma q_i q_o)(\Sigma q_i)}{N\Sigma q_i^2 - (\Sigma q_i)^2} \qquad \textbf{(3.17)}$$

where $\qquad$ $N \triangleq$ total number of data points $\qquad$ **(3.18)**

In this example, calculation gives $m = 1.0823$ and $b = -0.8471$ kPa. Since these values are derived from scattered data, it would be useful to have some idea of their possible variation. The standard deviations of m and b may be found from

$$s_m^2 = \frac{N s_{q_o}^2}{N\Sigma q_i^2 - (\Sigma q_i)^2} \qquad \textbf{(3.19)}$$

$$s_b^2 = \frac{s_{q_o}^2 \Sigma q_i^2}{N\Sigma q_i^2 - (\Sigma q_i)^2} \qquad \textbf{(3.20)}$$

where $\qquad$ $s_{q_o}^2 = \frac{1}{N-2} \Sigma (mq_i + b - q_o)^2 \qquad \textbf{(3.21)}$

The symbol s_{q_o} represents the standard deviation of q_o. That is, if q_i were fixed and then repeated over and over, q_o would give scattered values, with the amount of scatter being indicated by s_{q_o}. If we assume that this s_{q_o} would be the same for *any* value of q_i, we can calculate s_{q_o}, using *all* the data points of Fig. 3.10a and *without* having to repeat any one q_i many times. For this example, calculation gives $s_{q_o} = 0.208$ kPa. Then $s_m = 0.0140$ and $s_b = 0.0830$ kPa. Assuming a Gaussian distribution and the 99.7 percent limits ($\pm 3s$), we could give m as 1.082 ± 0.042 and b as -0.847 ± 0.249.

In *using* the calibration results, the situation is such that q_o (the indicated pressure) is known and we wish to make a statement about q_i (the true pressure). The least-squares line gives

$$q_i = \frac{q_o + 0.847}{1.082} \qquad \textbf{(3.22)}$$

However, the q_i value computed in this way must have some plus-or-minus error limits put on it. These can be obtained since s_{q_i} can be computed from

---

[9]H. D. Young "Statistical Treatment of Experimental Data," McGraw-Hill, New York, 1962, p. 121.

$$s_{qi}^2 = \frac{1}{N-2} \sum \left(\frac{q_o - b}{m} - q_i \right)^2 = \frac{s_{q_o}^2}{m^2} \tag{3.23}$$

which in this example gives $s_{q_i} = 0.192$ kPa. Thus if we were using this gage to measure an unknown pressure and got a reading of 4.32 kPa, our estimate of the true pressure would be 4.78 ± 0.58 kPa if we wished to use the $\pm 3s$ limits.

We should note that in computing s_{q_o} either of two approaches could be used. We might use data such as in Fig. 3.10a and apply Eq. (3.21) or, alternatively, repeat a given q_i many times and compute s_{q_o} from Eq. (3.8). If s_{q_o} is actually the same for all values of q_i (as assumed above), these two methods should give the same answer for large samples. In computing s_{q_i}, however, the second method is not feasible because we cannot, in general, fix q_o in a calibration and then repeat that point over and over to get scattered values of q_i. This is because q_i is truly an independent variable (subject to choice), whereas q_o is dependent (not subject to choice). Thus, in computing s_{q_i}, an approach such as Eq. (3.23) is necessary.

A calibration such as that of Fig. 3.10a allows decomposition of the total error of a measurement process into two parts, the *bias* and the *imprecision* (Fig. 3.10b). That is, if we get a reading of 4.32 kPa, the true value is given as 4.78 ± 0.58, the bias would be -0.46 kPa, and the imprecision ± 0.58. Of course, once the instrument has been calibrated, the bias can be removed, and the only remaining error is that due to imprecision. The bias is also called the *systematic error* (since it is the same for each reading and thus can be removed by calibration). The error due to imprecision is called the *random error* since it is, in general, different for every reading and we can only put bounds on it, but cannot remove it. Thus calibration is the process of removing bias and defining imprecision numerically.

While the method just described was used for many years and does not cause any major problems, a more refined technique is available and is now recommended by national and international standards groups.[10] The improvements have two major features. First, our simple method considers a standard deviation calculated from a small number of points to be as accurate as one gotten from a large number of points. Intuition alone tells us that this is not true. Statistical theory (confidence intervals)[11] allows us to adjust the uncertainty to suit the number of points. The second improvement substitutes for our $3s$ limits (99.7 percent level) a limit analogous to $2s$ (95 percent level). This change can be justified without complex statistical arguments as follows. When we use $\pm 3s$ limits, we imply that our uncertainty band is 99.7 percent sure to include the correct value. For Gaussian and near-Gaussian distributions, 99.7 percent puts us well into the "tails" of the distribution. In the tail

---

[10]B. N. Taylor and C. E. Kuyatt, "Guidelines for Evaluating and Expressing the Uncertainty of NIST Measurement Results," NIST Tech. Note 1297, 1993, sec. 6.5; "Guide to the Expression of Uncertainty in Measurement," 1st ed., ISO, 1993, sec. G1.2; "Measurement Uncertainty," ANSI/ASME PTC 19.1-1986 Part 1, 1986; R. B. Abernethy, "Measurement Uncertainty Handbook," Instrument Society of America, 1980; H. W. Colemen, "Experimentation and Uncertainty Analysis for Engineers," 2nd ed., Wiley, New York, 1999; "ASTM Standards on Precision and Bias for Various Applications," 5th ed., ASTM, 1997.

[11]E. O. Doebelin, "Engineering Experimentation," p. 520.

Degrees of freedom $(N-2)$	t_{95} value	Degrees of freedom $(N-2)$	t_{95} value
1	12.706	14	2.145
2	4.303	15	2.131
3	3.182	16	2.120
4	2.776	17	2.110
5	2.571	18	2.101
6	2.447	19	2.093
7	2.365	20	2.086
8	2.306	25	2.060
9	2.262	30	2.042
10	2.228	40	2.02
11	2.201	60	1.980
12	2.179	infinity	1.960
13	2.160		

Figure 3.11
t distribution values for uncertainty calculation.

regions, it takes *very* large samples to get reliable results for probabilities. In most engineering experiments (including calibration) the sample sizes are *much* too small to make any reliable predictions about the tail regions. For example, to actually compute a 99.7 percent probability from a sample of real data would require about 1000 points (997 within the band and 3 outside). If we had only, say, 100 points, we might get a 99 percent probability (99 in, 1 out) but we *could not* compute a 99.1 percent limit. You just cannot compute valid probabilities resolved to tenths of 1 percent unless you have about 1000 or more points. With samples of size about 20, if we get 1 of the 20 outside the band, that would be 5 percent outside and 95 percent inside. Since most engineering samples are relatively small, it was decided that quoting uncertainty bands at the 95 percent level of confidence was more realistic than using the older 99.7 percent criterion.

Since our present interest is in specifying uncertainty for readings taken from calibration curves, let us now define a recommended procedure which follows the spirit of the "new" methods. We will still find the best fit calibration line by the standard least-squares method given earlier. When we *use* this calibration line to get a true value from a measured value, we will still use Eq. (3.22). However, we need a new method for calculating the uncertainty band that surrounds our nominal value. Mandel[12] provides a method which meets our needs. To make our uncertainty band ("confidence interval") sensitive to sample size, we will need to use another of the common probability distributions, the so-called *t distribution.* Since we have decided that we want a 95 percent level of confidence, we will need only an abbreviated table (Fig. 3.11) of this distribution. Mandel's Eq. (12.35) gives a $\pm$ 95 percent confidence interval, defined by two hyperbolas on either side of the

---

[12]J. Mandel, "The Statistical Analysis of Experimental Data," Wiley, NY, 1964, p. 286. A paperback reprint (Dover, 1984) was still available in 1999.

least-squares lines. When we read the instrument (q_o), we draw a horizontal line through that value. This line intersects the two hyperbolas, and the q_i values at these two intersections define the ends of a 95 percent confidence interval for the true value. That is, we are 95 percent sure that the true value lies between these two values of q_i.

The "vertical" location of the two hyperbolas as a function of q_i is computed from

$$\Delta q_o = \pm t_{95, N-2} \, s_{q_o} \sqrt{\frac{1}{n} + \frac{1}{N} + \frac{N(q_i - \bar{q}_i)^2}{N \Sigma q_i^2 - (\Sigma q_i)^2}} \tag{3.24}$$

That is, the upper hyperbola is plotted by adding the positive Δq_0 to the best-fit line, and the lower one is plotted by adding the negative one. The numerical value of $t_{95, N-2}$ is taken from the t table, using N as the total number of points on the calibration graph (22 points in Fig. 3.10a makes $N - 2 = 20$ and the t value $= 2.086$). The standard deviation s_{q_o} is computed from Eq. (3.21), and q_i has the same meaning as in Fig. 3.10a. In some applications, we are able to repeat the measurement n times ($n = 2$ if two q_o readings are taken) and use for the q_o value the *average* of the n readings. If this is done, the confidence interval for q_i will be smaller (better), as the formula shows. We now want to use the new method to compute the uncertainty in q_i for the same q_o value as used earlier (4.32) and then compare this with an "old method" $2s_{q_i}$ value of (2) (0.192) $= 0.384$.

Using Eq. (3.24) (with $n = 1$), we can compute the values needed to plot the graph of Fig. 3.12. Visually, the two "hyperbolas" seem to be straight lines, but inspection of the tabular results shows that Δq_o does vary with q_i, the largest value being 0.376 (at the left and right ends of the curves) and the smallest being 0.350 at

Draw a horizontal line at the measured value of q_o. Its intersections with the prediction bands define a 95% confidence interval for true pressure q_i.

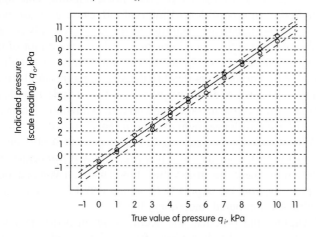

Figure 3.12
Calibration curve with 95% confidence interval.

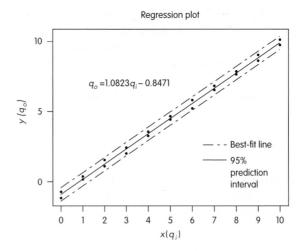

Figure 3.13
MINITAB calibration curve.

the center. For a q_o reading of 4.32, the 95 percent confidence band for q_i is found to be 4.78 ± 0.392. Comparing the "old" and "new" methods, we note that the old method gives a *fixed* size to the confidence interval while the new will give a slightly different value for different q_o readings. Also, the numerical values are somewhat different; 0.384 versus 0.392 in our example.

While we have given various computing formulas for these curve-fitting and uncertainty calculations, many engineers nowadays have on their personal computers statistical software which makes the computing and graphing quick and easy. In MINITAB, the fitting and plotting of the line, with the 95 percent uncertainty bands included, gives Fig. 3.13. Using MINITAB's nomenclature, you should request 95 percent *prediction* (not confidence) intervals. Because a major assumption of the analysis is that the statistical variability of the measurements is the same over the entire calibration range, it is good practice to also make a plot of the residuals versus q_i. Such a graph can show whether this assumption is reasonable. For our example, Fig. 3.14 shows no obvious trend in the size of the residuals; the variability seems to be about the same over the whole range. When there is a clear indication that the variability is not constant, other methods are available.[13] However, these are rarely needed, and we thus leave them to the reference.

Be sure you are clear on the *meaning* of our 95 percent confidence intervals. If you use these methods in your day-to-day work, you will be correct 95 percent of the time in predicting where the "true value" lies. That is, if you use the method for 100 different projects, you can expect to be *wrong* about five times; the true value will *not* lie within the computed uncertainty band.

In actual engineering practice, the accuracy of an instrument sometimes is given by a single numerical value; very often it is not made clear just what the

[13]J. Mandel, "Evaluation and Control of Measurements," Marcel Dekker, New York, 1991, sect. 5.4.

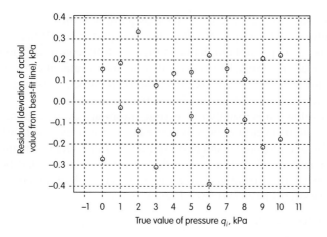

Figure 3.14
Check for size of variability over calibration range.

precise meaning of this number is meant to be. Even though a calibration, as in Fig. 3.10, has been carried out, s_{q_i} is not calculated. Instead, the error is taken as the largest horizontal deviation of any data point from the fitted line. In Fig. 3.10 this occurs at $q_i = 0$ and amounts to 0.25 kPa. The inaccuracy in this case thus might be quoted as ± 2.5 percent of full scale. This practice is no doubt due to the practical viewpoint that when a measurement is taken, all we really want is to say that it cannot be incorrect by more than some specific value; thus the "easy way out" is simply to give a single number.

Irrespective of the precise *meaning* to be attached to accuracy figures provided, say, by instrument manufacturers, the *form* of such specifications is fairly uniform. More often than not, accuracy is quoted as a percentage figure based on the full-scale reading of the instrument. Thus if a pressure gage has a range from 0 to 10 kPa and a quoted inaccuracy of ± 1.0 percent of full scale, this is to be interpreted as meaning that no error greater than ± 0.1 kPa can be expected for any reading that might be taken on this gage, provided it is "properly" used. The manufacturer may or may not be explicit about the conditions required for "proper use." Note that for an actual reading of 1 kPa, a 0.1-kPa error is 10 percent of the *reading*.

Another method sometimes utilized gives the error as a percentage of the particular reading with a qualifying statement to apply to the low end of the scale. For example, a spring scale might be described as having an inaccuracy of ± 0.5 percent of reading or ± 0.1 N, whichever is greater. Thus for readings less than 20 N, the error is constant at ± 0.1 N, while for larger readings the error is proportional to the reading.

Calibration Accuracy versus Installed Accuracy

Because calibration is a vital function in every measurement application, we have devoted considerable effort to explaining the procedure. We must, however, remember that it is not an end in itself; our *real* purpose is to *measure* something. When we earlier stated that calibration *removes* the bias portion of the error, this is true *only* for the conditions under which the calibration was performed. When we use the instrument, it is rarely possible to maintain that carefully controlled environment in our experimental apparatus. This means that the measurement error (bias and imprecision) must be re-evaluated, taking into account, as best possible, the deviation of the measurement conditions from the calibration conditions.

This re-evaluation is usually not as straightforward as the calibration was because the measurement environment is rarely as controlled as a standards laboratory calibration. In fact, judgement and past experience (rather than "scientific" calculation) is often necessary.[14] One aspect of this situation is that the bias portion of the error is now *not* zero. (Recall that we earlier said that calibration *removes* the bias.) Abernethy classifies biases into five different types:

1. Large known biases.
2. Small known biases.
3. Large unknown biases.
4. Small unknown biases with unknown algebraic sign.
5. Small unknown biases with known algebraic sign.

Large known biases *are* eliminated by calibration. Small known biases may or may not be corrected, depending on the difficulty of correction and the magnitude of the bias. Large unknown biases are not correctable; they may exist, but the magnitude and perhaps even the sign is not known. They usually come from human errors in data processing, incorrect installation and/or handling of instrumentation, and unexpected environmental disturbances. In a well-controlled measurement process, the assumption is that there are *no* large unknown biases. The reference gives extensive detail on methods for ensuring this.

Small, unknown biases then remain as a contribution to the measurement error. Quoting from Abernethy:

> *"It is both difficult and frustrating to estimate the limit of an unknown bias. To determine the exact bias in a measurement, it would be necessary to compare the true value and the measurements. This is almost always impossible . . . If there is no source of data for the bias, the judgement of the most knowledgeable instrumentation expert on the measurement must be used."*

The reference does give some hints on how to deal with this situation, but we need to admit that past experience and judgement are generally necessary.

A result of this is that the bias in the measurement situation (as contrasted with calibration) is treated as a *random* effect rather than as systematic. The number that we estimate for this bias is called the *bias limit*. It is defined as the range of values within which we feel that the actual bias will be found 95 percent of the time. The 95 percent value was chosen to make it consistent with our 95 percent confidence limit for the imprecision. Using this scheme, the "total error" in the measurement is

---

[14]R. B. Abernethy, "Measurement Uncertainty Handbook," Instrument Society of America, 1980, Research Triangle Park, NC, p. 3.

defined as the sum of the bias limit and the imprecision, and is given the name *uncertainty.*

$$\text{Uncertainty} \triangleq U \triangleq \pm (B + t_{95,\,N-1}\,s) \tag{3.25}$$

where $\qquad B \triangleq$ bias limit $\tag{3.26}$

$t_{95,\,N-1} \triangleq$ value from t table (Fig. 3.11) and N is the number of readings used to compute s $\tag{3.27}$

$s \triangleq$ sample standard deviation, computed from the usual formula $\tag{3.28}$

The meaning attached to the uncertainty U is that we are 95 percent sure that the "true" value of the measured quantity lies within $\pm U$ of the measured value. While the term "true value" is in common use, it is preferred to call this the *reference value* since, as we argued earlier, the true value of a measured quantity can never be found. Rather, we can only talk about the "best available" measurement or calibration technique for the quantity. The value given by this method is called a reference value.

The basic measure of imprecision is the standard deviation s, and it is called the *precision index.* To attach a certain confidence level to uncertainties, s is multiplied by some constant, called the *coverage factor.* It is most common, and our suggestion, that the coverage factor be $t_{95,\,N-1}$, giving 95 percent confidence. If we could enforce at measurement the same conditions as at calibration, then we could use our calibration uncertainty number as the measurement uncertainty. Otherwise, we must add to the calibration uncertainty some adjustment to account for, as best we can, the increased uncertainty of the measurement conditions. Let us briefly consider some examples to clarify this concept. In Fig. 3.15a we show a simple spring-type

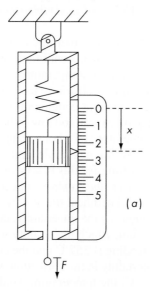

Figure 3.15
Differences between calibration and measurement situations.

force measuring scale. We could easily calibrate this with standard masses and find a best-fit line and uncertainty, and remove any scale bias present. If we then move this scale to our experimental apparatus to measure an unknown force, the uncertainty will increase for a number of possible reasons. For example, if the temperatures at calibration and at measurement are not the same, the scale will exhibit an uncorrected bias. This bias has two sources; thermal expansion (which shifts the zero point) and temperature sensitivity of the spring's elastic modulus (which changes the spring stiffness). Unless we measure temperature and correct for these effects, a bias of unknown sign and magnitude will be present and increase the measurement uncertainty in an unknown (and thus "random") way. Other possible effects include angular misalignment of the unknown force with the scale's sensitive axis, giving a further increase in uncertainty. To compute how much we need to increase the uncertainty, we need to estimate these temperature, misalignment, and any other effects felt to be significant. Note that we do not measure the temperature and misalignment and then *correct* for these effects, rather we estimate some limits on how large we think these effects might be and then add this to the uncertainty.

Figure 3.15*b* shows a displacement-measuring dial indicator, for both calibration and measurement situations. The instrument is first calibrated against a more accurate micrometer and then used to measure deflection of a beam, as part of an experiment to find the beam's spring constant, F/δ. Such dial indicators have an internal spring to maintain contact between the indicator stylus and the surface being displaced. When we measure the beam displacement, a bias error will be introduced because the indicator spring force acts against the applied force F, causing the measured deflection to be less than it should be. If the force F is always downward, this bias error would be treated as having an unknown magnitude, but a known sign. That is, the deflection is always measured too low. If we estimate an upper limit for its magnitude, this bias would give an *unsymmetrical* uncertainty; for example, -0.003 in. to $+0.001$ in.

A final example, Fig. 3.15*c*, shows a thermocouple (electrical temperature sensor) for both calibration and measurement conditions. Such devices are calibrated in an accurately controlled and measured temperature environment. Specifically, the wires are immersed in a liquid-filled well whose temperature is uniform at T_{hot} over a long distance. This is done to prevent conduction heat transfer along the wires, which would cause the sensing tip to read too low. When the calibrated thermocouple is used to measure the temperature of a hot gas, the wires are in contact with a cool duct wall, conduction is now not negligible, and the sensing tip will read low. We can again estimate an upper limit for this bias error and add it to the calibration uncertainty. If such an error causes *unacceptable* uncertainty, we may measure the wall temperature, estimate the needed heat transfer parameter, and compute a correction. This correction will *improve* the uncertainty, but not eliminate it, since the correction will itself be uncertain, which uncertainty we will have to estimate and include. For example, if our reading is 357°C and the correction is $+8$°C, the nominal value is 365°C. If the uncertainty in the correction is ± 2°C, and the uncertainty due to other sources was ± 5°C, the temperature would be quoted as 365 ± 7°C.

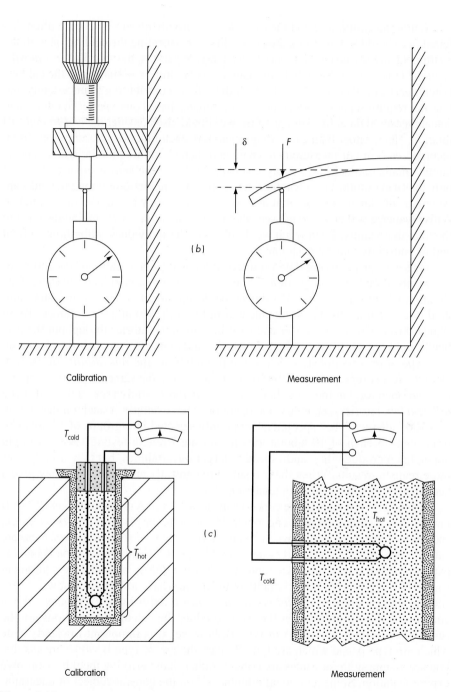

(b)

Calibration

Measurement

(c)

Calibration

Measurement

Figure 3.15
(Concluded)

While the calibration and measurement environments are by choice often different, we should not overlook the possibility of performing the calibration with the instrument already installed in our experimental apparatus, the so-called *in situ calibration*. This, if possible, would in many cases be preferred since then the calibration numerical results would include all the effects contributing to uncertainty and thus not require separate judgements and estimates based on experience rather than actual measured data. In a similar spirit, we should also consider the *end-to-end* calibration. Here, rather than calibrating separately each link (sensor, amplifier, filter, recorder, etc.) in our measurement chain and then combining the individual uncertainties mathematically, we apply a standard to only the sensor input and record only the final output. The advantage here is that all interactions among the links are automatically taken into account and the procedure may be considerably quicker. A disadvantage is that we do not see which components are contributing the most to the total uncertainty. Even when we do perform the individual calibrations, a final end-to-end study may be desirable.

One way of gathering data on uncertainty is to *repeat* calibrations and measurements, though this may often not be technically or economically feasible. One situation where it is regularly used is in the development and use of industry standard tests, such as the many documented in the ASTM standards volumes. These standard tests will be used over and over in many laboratories throughout the particular industry, so it is possible to devote considerable effort to their development. One type of uncertainty study leads to information on the *repeatability* of the test. Here the test is repeated several or many times, using the same laboratory, apparatus, and operator. On the other hand, a *reproducibility* study repeats the test using *different* laboratories, apparatus, and operators. (For example, a steel rod may be cut into 100 "identical" test specimens, randomly selected into groups of 10, and one group sent to each of 10 laboratories for tensile strength testing.) As you might guess, the reproducibility studies lead to larger uncertainties, but they usually give more realistic results. When these results are carefully analyzed, they may give a keen insight into the sources of uncertainty.

A final bit of nomenclature concerns the terms "type A uncertainty" and "type B uncertainty," which you may encounter. If you actually *perform experimentation,* such as we have described above, to get numerical values for uncertainties, these are called "type A"; all other sources are called "type B." Examples of type B include expert opinion, past experience, and specification sheets provided by equipment suppliers. Such sources of uncertainty numbers may have to be used, but we always want to make clear that *we did not actually run a calibration experiment.* When combining uncertainties to get the overall value, the calculation methods are the same whether they are type A or type B, however it is important to clearly state which are type A and which are type B since the use of type B values means that you are assuming these values are correct, rather than verifying them by your own experimentation. Some additional references[15] on the general subject of uncertainties may be helpful.

---

[15]S. J. Kline and F. A. McClintock, "Describing Uncertainties in Single Sample Experiments," *Mech. Eng.,* vol. 75, January 1953, p. 3; L. W. Thrasher and R. C. Binder, "A Practical Application of Uncertainty Calculations to Measured Data," *Trans. ASME,* February 1957, p. 373.

Combination of Component Errors in Overall System-Accuracy Calculations

A measurement system is often made up of a chain of components, each of which is subject to individual inaccuracy. If the individual inaccuracies are known, how is the overall inaccuracy computed? A similar problem occurs in experiments that use the results (measurements) from several different instruments to compute some quantity. If the inaccuracy of each instrument is known, how is the inaccuracy of the computed result estimated? Or, inversely, if there must be a certain accuracy in a computed result, what errors are allowable in the individual instruments?

To answer the above questions, consider the problem of computing a quantity y which is a known function of n independent variables, $x_1, x_2, x_3, \ldots x_n$, that is,

$$y = f(x_1, x_2, x_3, \ldots x_n) \tag{3.29}$$

For small changes in the independent variables from given "operating points," a Taylor series expansion gives a good approximation for the corresponding change in y.

$$\Delta y \approx \frac{\partial f}{\partial x_1} \cdot \Delta x_1 + \frac{\partial f}{\partial x_2} \cdot \Delta x_2 + \frac{\partial f}{\partial x_3} \cdot \Delta x_3 + \cdots + \frac{\partial f}{\partial x_n} \cdot \Delta x_n \tag{3.30}$$

Think of the partial derivative as the *sensitivity* of y to changes in the particular x. When a partial derivative has a large numerical value, y is very sensitive to that particular x. Since the partial derivatives are numerically evaluated at the operating point, they are constants (not functions) in Eq. (3.30), so this equation defines y as a *linear* function of the x's, even though the original function (f) may be nonlinear.

If the Δx's are now considered to be the uncertainties u_{xi} in each measured value x_i, then the corresponding uncertainty U_y in y is given by

$$U_y \approx \sqrt{\left(\frac{\partial f}{\partial x_1} \cdot u_{x1}\right)^2 + \left(\frac{\partial f}{\partial x_2} \cdot u_{x2}\right)^2 + \left(\frac{\partial f}{\partial x_3} \cdot u_{x3}\right)^2 + \cdots + \left(\frac{\partial f}{\partial x_n} \cdot u_{xn}\right)^2} \tag{3.31}$$

This relation is called the *root-sum-square* (rss) formula. We have not here proven its validity, but this rests on the fact that the standard deviation of any linear function of Gaussian independent variables is given by the square root of the sum of squares of the individual standard deviations. It is an approximate result because y is not really a linear function of the x's; it is close to linear only for small changes. Since U_y is computed from individual uncertainties that are 95 percent confidence intervals, it has the same meaning, that is, it is a 95 percent confidence interval for the dependent variable y. Equation (3.31) is the basis for computing the error of complete measurement systems from the errors in individual components.

When we originally plan an experiment, we must decide how accurate the final results must be to meet the goals of the study. Once this decision has been made, then we can explore the accuracy needed in each individual instrument. It should be clear that this kind of problem does not have a unique solution; there will be *many* combinations of individual errors that give the same overall error. To get started on such problems, we can use the *method of equal effects*. If we have nothing else to go on, it seems reasonable to, at least initially, force all the instruments to contribute *equally* to the overall error. Using Eq. (3.31), this leads to

$$\mu_{xi} \approx \frac{U_y}{\sqrt{n} \cdot \dfrac{\partial f}{\partial x_i}} \tag{3.32}$$

This equation defines initial values for the allowable error in each measurement. We must then compare these *requirements* with the *capabilities* of the actual instruments available to us. If we can find instruments which meet all these needs, we have at least one solution to our problem. If one or more requirements *cannot* be met, it does not mean we are defeated. We then need to see if some of our instruments are *better* than Eq. (3.32) requires. If so, we can use them to relax the requirements that we were not able to meet. A careful evaluation of all such possibilities may lead to a combination of instruments that meets the overall specification. It of course can also happen that, if the overall accuracy is set too high, we can find *no* set of real instruments that meets our needs. If so, the systematic approach used here allows a convincing argument when we need to discuss the dilemma with superiors or colleagues.

As an example of the above procedures, consider an experiment for measuring, by means of a dynamometer, the average power transmitted by a rotating shaft. The formula for power can be written as

$$\text{Watts} = \frac{2\pi RFL}{t} \qquad \text{or} \qquad \text{hp} = \frac{2\pi RFL}{550t} \tag{3.33}$$

where

$R \triangleq$ revolutions of shaft during time t
$F \triangleq$ force at end of torque arm, lbf (N)
$L \triangleq$ length of torque arm, ft (m)
$t \triangleq$ time length of run, s

$$(3.34)$$

A sketch of the experimental setup is shown in Fig. 3.16*a*. The revolution counter is of the type shown in Fig. 2.4 and can be turned on and off with an electric switch. The instants of turning on and off are recorded by a stopwatch. If it is assumed the counter does not miss any counts, the maximum error in R is ± 1, because of the digital nature of the device (see Fig. 3.16*b*).

There is a related error, however, in determining the time t since perfect synchronization of the starting and stopping of watch and counter is not possible. The stopwatch might be known to be a quite accurate time-measuring instrument, but this does not guarantee that it will always measure the time interval intended. In assigning an error to t, then, we are not helped much by the watch manufacturer's guarantee of 0.10 percent inaccuracy if our synchronization error is much larger than this. This synchronization error is certainly not precisely known since it involves human factors. An experiment to determine its statistical characteristics would be a more expensive and involved undertaking than the power measurement of which it is a part. So we are in the rather common position of having to rely on experience and judgement in arriving at an estimate of the proper numerical value,

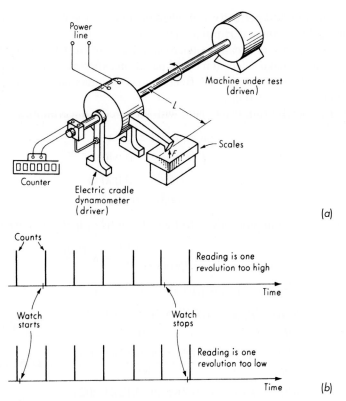

Figure 3.16
(*a*) Dynamometer test setup. (*b*) Revolution-counting error.

and we begin to appreciate that some of the statistical niceties and fine points of theory considered earlier may appear somewhat academic in such a situation. They are always useful in terms of the understanding of basic concepts that they develop; however, they cannot be relied on to give clear-cut answers in situations where the basic data are ill-defined. In the present case, suppose it is decided that a total starting and stopping error is taken as ± 0.50 s. Whether this is to be considered as an absolute limit or as, say, a 95 percent uncertainty, becomes a matter of judgement and preference. *Our* preference, as stated earlier, is to treat this number as an uncertainty at the 95 percent level of confidence. The measurement of the torque arm length L is also subject to similar vagaries, depending on the care taken in its measurement. Suppose we use a fairly rough procedure and decide on a 95 percent uncertainty of ± 0.05 in.

The scales used to measure the force F can be calibrated with deadweights, yielding a set of data analogous to that of Fig. 3.10*a*, which can be processed to calculate a 95 percent uncertainty as done there. Suppose this turns out to be ± 0.040 lbf. This *calibration uncertainty* must however be translated into a corresponding *measurement uncertainty*, as we discussed above. When actually used, the scales will be subject to vibration (not present at calibration), which may reduce frictional effects and decrease the error (uncertainty). At the same time, the pointer on the scale will not stand perfectly still when the dynamometer is running; thus, in reading the scale, we must perform a mental averaging process, which may introduce a new error not present at calibration. Such effects are clearly difficult to quantify, and again we must make a decision based partly on experience and judgement. Suppose we assume the two mentioned effects cancel and thus take the force measurement uncertainty as $\pm$ 0.040 lbf.

For a specific run, if the data, with 95 percent uncertainties for each item are

$$R = 1{,}202 \pm 1.0 \, \text{r}$$
$$F = 10.12 \pm 0.040 \, \text{lbf}$$
$$L = 15.63 \pm 0.050 \, \text{in} \tag{3.35}$$
$$t = 60.0 \pm 0.50 \, \text{s}$$

then the calculation proceeds as follows: In terms of inch units, we have

$$\text{hp} = \frac{2\pi}{(550)(12)} \frac{FLR}{t} = K \frac{FLR}{t} \tag{3.36}$$

Then, computing the various partial derivatives to three significant figures gives

$$\frac{\partial(\text{hp})}{\partial F} = \frac{KLR}{t} = \frac{(0.000952)(15.63)(1{,}202)}{60} = 0.298 \, \text{hp/lbf} \tag{3.37}$$

$$\frac{\partial(\text{hp})}{\partial R} = \frac{KFL}{t} = \frac{(0.000952)(10.12)(15.63)}{60} = 0.00251 \, \text{hp/r} \tag{3.38}$$

$$\frac{\partial(\text{hp})}{\partial L} = \frac{KFR}{t} = \frac{(0.000952)(10.12)(1{,}202)}{60} = 0.193 \, \text{hp/in} \tag{3.39}$$

$$\frac{\partial(\text{hp})}{\partial t} = \frac{KFLR}{t^2} = -\frac{(0.000952)(10.12)(15.63)(1{,}202)}{3{,}600}$$
$$= -0.0500 \, \text{hp/s} \tag{3.40}$$

While we have chosen to use the statistical uncertainty concept as our preferred measure of error, other viewpoints are possible. One such treats the individual errors as *absolute limits;* the measured quantity is assumed to never fall outside these bounds. When we combine such errors, we want the combined error to have this same meaning. Since the individual errors are as likely to be negative as positive (we usually have no information on the sign), to get the *worst case* we would use Eq. (3.30), but with absolute values for each term, so that they add up in the worst possible way. If we call this absolute error E_a, our example calculation goes as follows:

$$\pm E_a = (0.298)(0.040) + (0.00251)(1.0) + (0.193)(0.050) + (0.05)(0.50) \tag{3.41}$$
$$\pm E_a = 0.0119 + 0.00251 + 0.00965 + 0.025 = 0.049 \, \text{hp} \tag{3.42}$$

We now compute horsepower as

$$\text{hp} = \frac{(2.0000)(3.1416)(10.120)(15.630)(1{,}202.0)}{(550.00)(12.000)(60.000)} = 3.0167 \tag{3.43}$$

which we round off to 3.017. Then the result may be quoted as hp = 3.017 ± 0.049 hp or hp = 3.017 ± 1.6 percent.

We prefer to treat the individual errors as 95 percent uncertainties, giving the total uncertainty U as

$$U = \pm\sqrt{0.0119^2 + 0.00251^2 + 0.00965^2 + 0.025^2} = \pm0.029 \text{ hp} \quad \textbf{(3.44)}$$

The uncertainty U will of course always be less than the corresponding absolute error E_a, as the two formulas show. We choose to use uncertainties because, due to the random nature of the errors, it is *not* likely that the worst possible combination will actually occur, making the absolute error method too conservative. As engineers, we *do* need to be appropriately conservative, but not when this leads to unrealistic and uneconomic decisions. Finally, we must again admit that, since the individual errors are often *estimates* rather than "scientific" calculations with hard data, the distinction between absolute errors and uncertainties may be subject to some challenge from "practical" critics. Nevertheless, the uncertainty viewpoint is a reasonable one, is widely accepted, and we continue to recommend it.

When computing values for the partial derivatives, Eqs. (3.37) to (3.40), an approximate numerical method (rather than the exact analytical one used there) may be useful when programming data acquisition software. We simply compute the change in horsepower Δhp caused by a small (typically 1 percent) change in each of the independent variables, say ΔF. The ratio (Δhp/ΔF), for example, will be very close to the analytically computed value ∂(hp)/∂F.

Finally, suppose we wish to measure hp to 0.5 percent accuracy in the previous example. What accuracies are needed in the individual measurements? Using Eq. (3.32), we get

$$\Delta F = \frac{(3.02)(0.005)}{\sqrt{4}\,(0.298)} = 0.025 \text{ lbf}$$

$$\Delta R = \frac{(3.02)(0.005)}{\sqrt{4}\,(0.0025)} = 3.0 \text{ r}$$

$$\Delta L = \frac{(3.02)(0.005)}{\sqrt{4}\,(0.193)} = 0.039 \text{ in} \qquad \textbf{(3.45)}$$

$$\Delta t = \frac{(3.02)(0.005)}{\sqrt{4}\,(0.05)} = 0.15 \text{ s}$$

If it is found that the best instrument and technique available for measuring, say, F are good to only 0.04 lbf rather than the 0.025 lbf called for by Eq. (3.45), this does not necessarily mean that horsepower cannot be measured to 0.5 percent. However, it does mean that one or more of the other quantities, R, L, and t *must* be measured *more* accurately than required by Eq. (3.45). Making one or more of these measurements more accurately may offset the excessive error in the F measurement. The given formulas allow calculation of whether this will be true.

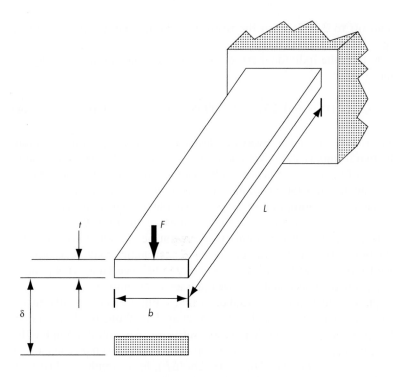

Figure 3.17
Example of theory/experiment comparison.

Theory Validation by Experimental Testing

One of the most common and important application areas for experimental work is the testing of new *theoretical* relations. When a new theory is developed, we always question its validity until we get experimental confirmation of its predictions. Such confirmation (or refutation) is usually of a statistical nature since both the theoretical prediction and the experimental measurement have some uncertainty attached to them. A useful way of visualizing this situation is an "overlap graph." Here we plot the theoretical result and the measured result on the same line, each with its own uncertainty band. If the two uncertainty bands do not overlap, then it is unlikely that the theory is correct. If there *is* some overlap, then the theory gets more and more likely to be correct as the overlap region increases. A simple example will illustrate this concept.

The cantilever beam of Fig. 3.17 is a common structural element that has been theoretically analyzed in several ways. If the beam is sufficiently "long and thin"

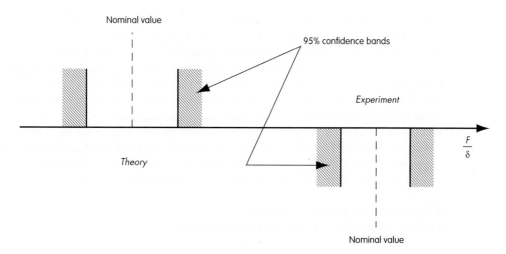

Figure 3.18
Overlap graph compares theoretical prediction with experimental measurement.

and if several other assumptions are also reasonable, then the end deflection δ due to the force F is given by

$$\delta = \frac{4FL^3}{Ebt^3} \qquad (3.46)$$

To check this theory we might build a beam with certain dimensions (L, b, t) from a material with a certain modulus of elasticity E. We then apply a force F and measure the resulting end deflection δ. Suppose our interest is in the "spring constant" F/δ. We have a measurement of the applied force F and an estimate of its 95 percent confidence interval and similar data for the deflection. The uncertainty in the ratio F/δ can be computed in the usual way. We now have numbers for the best estimate of the spring constant and a 95 percent confidence interval for this value.

Turning now to theory, note that *every* "theoretical" result relies on one or more parameter values that can only be found by "experiment." That is, to compute the theoretical value of the spring constant for a specific beam, we need numbers for the dimensions and the material property E. *These can only be found by experiment!* Thus we must measure b, t, and L and "look up" a value for E in some reference. When we measure the dimensions, we can as usual attach an uncertainty to each. If you look up an E value for, say, some aluminum alloy, you often do not find any data on its uncertainty, though the quoted value surely has uncertainty. Let us assume that a phone call to the material supplier gets us at least an estimate of E's uncertainty. We now have enough data to calculate, in the usual way, the uncertainty in the theoretical value of F/δ.

We now can plot the "overlap graph" of Fig. 3.18 which shows a situation where the theory is *not* validated. That is, at the 95 percent level of confidence, there is *no* overlap between the two confidence bands, making it highly unlikely that the theory is accurate for the particular beam tested. (This might be the case if the beam is very short. For short beams, the theory behind Eq. (3.46), which is based on pure bending, would be inaccurate since now shearing effects become important.) Note that in this example, as in the general case, the question "Is the theory correct?" cannot be answered by a simple yes or no. As the confidence bands exhibit more and more overlap, the theory becomes more and more acceptable. Once more we see that statistical viewpoints give us a more realistic picture than would the more naive direct comparison of only the "base values" of measured and theoretical spring constants without considering the uncertainties involved in both. While the overlap graphs are often sufficient, a more numerical comparison is available.[16]

Effect of Measurement Error on Quality-Control Decisions in Manufacturing

When a measurement system is used for quality control of a manufacturing process, measurement errors can result in the acceptance of "bad" product and the rejection of "good," both undesirable. Manufacturing processes have their own statistical variability, and it is usually uneconomical to "tune" them to produce 100 percent good product. Rather, we allow a small percentage (say, 5 percent) of bad product and use a measurement system to detect these and divert them for rework or scrap so they do not get shipped to customers. Since all measurement systems exhibit statistical scatter, some good product may be *measured* as bad and some bad product may be *measured* as good, partially defeating our quality-control goals. If both the process and measurement variabilities are assumed to be Gaussian, statistical analysis[17] can be used to quantify these effects.

In Fig. 3.19 we show a process variable (solid curve) whose "target" value is 10 but which is acceptable in the range of 4 to 16. We have set our "gage" (dashed curves) to reject product which *measures* below 4 and above 16. Due to measurement errors, some percentage of bad product will be measured as good and vice versa, as shown by the shaded squares. If our main concern is that bad product should *not* be measured as good and shipped to customers, we can *bias* our gage limits, that is, we choose to reject units which *measure* as large as say, 6, and as small as 14. Then the chance of accepting bad product is nearly zero. Of course the price we pay is that now we will be rejecting more good product. Quality-control managers have to decide where to strike a balance between these conflicting goals. Their judgement is made more rational when they use the statistical tools referenced above.

Juran states that the effect of measurement errors can usually be ignored if the measurement system standard deviation is less than about 4 percent of the product acceptance range. In Fig. 3.19 the percentage is $0.5/(16 - 4) = 4.2$ percent, so this

---

[16]E. O. Doebelin, "Engineering Experimentation," p. 73.

[17]A. R. Eagle, "A Method for Handling Errors in Testing and Measuring," *Industrial Quality Control,* March 1954, pp. 10–15; J. M. Juran, "Quality Planning and Analysis," McGraw-Hill, New York, 1970, pp. 377–380; D. R. McCarville and D. C. Montgomery, "Optimal Guard Bands for Gauges in Series," *Quality Engineering,* vol. 9(2), 1996–97, pp. 167–177.

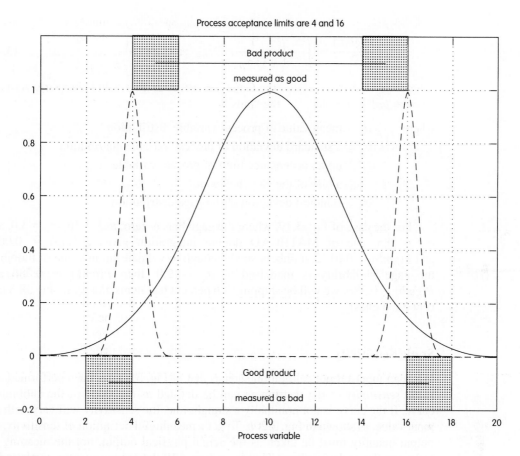

Figure 3.19
Effect of measurement errors in manufacturing quality control.

example almost meets Juran's criterion. When these errors *cannot* be ignored, McCarville and Montgomery give formulas to compute the two percentages needed to make decisions. These formulas are double integrals which must be *numerically* evaluated because the Gaussian distribution function has never been analytically integrated. Software such as MATHCAD[18] provides a convenient tool for the integration. The probability *P(bmg)* of a bad item being measured as good is

$$\frac{P(bmg)}{2} = \int_{ual}^{+\infty} \int_{z}^{+\infty} \frac{1}{\sqrt{2\pi}} e^{\frac{-y-(ual-b))^2}{2\sigma_g^2}} dy \frac{1}{\sqrt{2\pi}} e^{\frac{-(z-\mu_p)^2}{2\sigma_p^2}} dz \qquad \textbf{(3.47)}$$

while the probability *P(gmb)* of a good item being measured as bad is

---

[18]Mathsoft Inc., 201 Broadway, Cambridge, MA 02139, phone 1-800-mathcad.

$$\frac{P(gmb)}{2} = \int_{-\infty}^{ual} \int_{-\infty}^{z} \frac{1}{\sqrt{2\pi}} e^{\frac{-(y-(ual-b))^2}{2\sigma_g^2}} \, dy \, \frac{1}{\sqrt{2\pi}} e^{\frac{-(z-\mu_p)^2}{2\sigma_p^2}} \, dz \qquad \textbf{(3.48)}$$

$+\infty \approx \mu_p + 4\sigma_p$ (the Gaussian curves go to $\pm\infty$ but are nearly zero at
$-\infty \approx ual - b - 4\sigma_g$ the "tails")

where $\mu_p \triangleq$ mean value of process variable distribution

$\sigma_p \triangleq$ a standard deviation of process variable distribution

$ual \triangleq$ upper acceptance limit of process variable

$b \triangleq$ a bias of the gage below ual

$\sigma_g \triangleq$ standard deviation of the gage distribution

For the case of Fig. 3.19, where the gage bias is zero, $ual = 16$, $\sigma_p = 3.0$, and $\sigma_g = 0.5$, use of MATHCAD double integration gives $P(bmg) = 0.0088$ and $P(gmb) = 0.0132$, numbers small enough to validate Juran's rule of thumb. If the gage variability is increased to $\sigma_g = 2.0$, then $P(bmg) = 0.086$ and $P(gmb) = 0.388$; we will be shipping 8.6 percent bad items and scrapping 38.8 percent good ones!

Static Sensitivity

When an input-output calibration such as that of Fig. 3.10 has been performed, the *static sensitivity* of the instrument can be defined as the slope of the calibration curve. If the curve is not nominally a straight line, the sensitivity will vary with the input value, as shown in Fig. 3.20b. To get a meaningful definition of sensitivity, the output quantity must be taken as the actual physical output, not the meaning attached to the scale numbers. That is, in Fig. 3.10 the output quantity is plotted as kilopascals; however, the actual physical output is an angular rotation of the pointer. Thus to define sensitivity properly, we must know the angular spacing of the kilopascal marks on the scale of the pressure gage. Suppose this is 5 angular degrees/kPa. Since we already calculated the slope in kilopascals per kilopascal as 1.082 in Fig. 3.10, we get the instrument static sensitivity as (5)(1.082) = 5.41 angular degrees/kPa. In this form the sensitivity allows comparison of this pressure gage with others as regards its ability to detect pressure changes.

While the instrument's sensitivity to its desired input is of primary concern, its sensitivity to interfering and/or modifying inputs also may be of interest. As an example, consider temperature as an input to the pressure gage mentioned above. Temperature can cause a relative expansion and contraction that will result in a change in output reading even though the pressure has not changed. In this sense, it is an interfering input. Also, temperature can alter the modulus of elasticity of the pressure-gage spring, thereby affecting the pressure sensitivity. In this sense, it is a modifying input. The first effect is often called a *zero drift* while the second is a *sensitivity drift* or *scale-factor drift*. Figure 3.20c shows how the superposition of these two effects determines the total error due to temperature. If the instrument is used

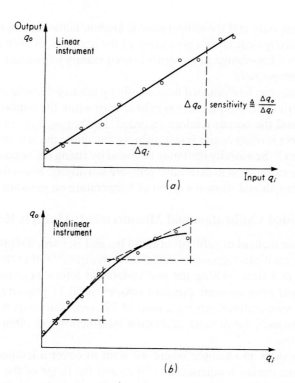

Figure 3.20 (a) and (b)
Definition of sensitivity.

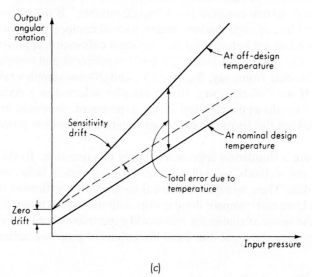

Figure 3.20 (c)
Zero and sensitivity drift.

for measurement only and the temperature is known, numerical knowledge of zero drift and sensitivity drift allows correction of the readings. If such corrections are not feasible, then knowledge of the drifts is used mainly to estimate overall system errors due to temperature.

These effects can be evaluated numerically by running suitable calibration tests. To evaluate zero drift, the pressure is held at zero while the temperature is varied over a range and the output reading recorded. For reasonably small temperature ranges, the effect is often nearly linear; then we can quote the zero drift as, say, 0.01 angular degree/C°. Sensitivity drift may be found by fixing the temperature and running a pressure calibration to determine pressure sensitivity. Repeating this for various temperatures should show the effect of temperature on pressure sensitivity.

Computer-Aided Calibration and Measurement: Multiple Regression

While the above method of calibration could be, and is, used, a more cost/effective approach uses multiple-regression statistical techniques[19] rather than the classical "one variable at a time." Using the one variable at a time approach, the calibration of an actual ultra-accurate pressure sensor[20] used 11 pressure points at each of 7 different temperatures, giving a total of 77 runs, quite expensive. Using the regression approach, the desired calibration information can often be obtained in many fewer runs.

Let us consider an example where we want to cover a temperature range of 40 to 100°F (the design temperature is 70°F) and the range of the pressure transducer is from 0 to 100 psig. In laying out the calibration plan, we must choose the total number of runs and also the specific combinations of pressure and temperature to actually use since there are 77 possibilities, but we do not want to use them all. The choice of number of runs usually involves some experience with the equipment being calibrated; no one can give you a "magic number." If the system behavior is quite linear and free of large random effects, a small number of runs will give accurate data. If we had not before used the statistical calibration approach, we would probably initially run the classical ("77-run") experiment and compare its results with those obtained from, say, 5-, 10-, 15-, and 20-run samples taken from the 77-run data. If we find that, say, 10-run samples selected in a certain way give nearly as good results as the classical 77-run experiment, we would in the future be justified in making the 10-run example our standard calibration procedure for this system.

We will use a simulation approach to study this question. To allow a realistic evaluation of our methods, we will postulate a *known* system behavior and use it to generate our data. Then, when the statistical methods fit coefficients to an assumed form of model, we can compare these results with the *known* true values. This luxury is of course never available for real-world experiments but is a valuable analysis tool for learning the capabilities and limitations of proposed techniques. Let us

---

[19]E. O. Doebelin, "Engineering Experimentation," p. 213.

[20]E. O. Doebelin, "Measurement Systems," 4th ed., pp. 89–94.

assume that the pressure transducer output voltage e_o is given in terms of pressure p and temperature T by

$$e_o = (0.1000 + 0.00006667 \, (T - 70.00))p + 0.01000 \, (T - 70.00) \quad \textbf{(3.49)}$$

Using this assumed model, temperature values of 40, 50, 60, 70, 80, 90, and 100, and pressure values of 0, 10, 20, 30, 40, 50, 60, 70, 80, 90, and 100, we can calculate the output voltage for each of these 77 points. We then pretend that we have *measured* these values and submit them to our regression software to see if it can find the correct coefficients in our model. In a real-world situation we would of course know neither the form of Eq. (3.49) nor the numerical coefficients in it. We must *assume* some form and then the software will find the best set of coefficients. We can then check whether the predicted values adequately match the measured data. If not, we study the nature of the errors to see how to modify our assumed model form for a second try. In many instrument calibration situations, the simple model of Eq. (3.49) is actually close to reality and also has theoretical foundation, so choosing this form would be reasonable.

We have suggested that 77 runs are not really necessary, and we now want to explore this possibility. Statistical experiment design[21] suggests optimum ways of selecting the combinations of variables to actually be run, but in our simple case, common sense can also suggest choices which give good results. We can display the two parameters p and T as in Fig. 3.21*a* and reason that we need to choose combinations which "cover" the ranges in some reasonable way. If we decide to try a total of nine runs, the selection shown does not seem foolish. (If we decided to try 13 runs, what "common sense" combinations would *you* suggest to add to the 9 runs shown? Why?) Equation 3.49 can be rearranged to give

$$e_o = -0.7000 + 0.01000T + 0.09533p + 0.00006667pT \quad \textbf{(3.50)}$$

Since this equation has four numerical parameters, if our model and data were perfect, it would require *only four* sets of p, T, and e_o measurements to find the parameter values (four equations in four unknowns). Since our models are never perfect and some random effects always exist, we generally need more than the minimum number of runs to get good results. *How many* more than this number depends on the individual situation.

Since the 9 runs we have decided to try are not all found in the 77 runs mentioned earlier, we now use Eq. (3.50) to compute the needed voltage values. We also use our statistical software[22] to generate Gaussian random "noise" values to add to our computed e_o numbers to simulate real-world random effects. If we keep these random effects small, nine-run experiments will give good results. If the random effects are large, more runs will be needed to "average out" their influence and give good coefficient values. Because, in calibration, we carefully try to hold constant all inputs other than those being calibrated, any random effects observed in the data are

---

[21]E. O. Doebelin, "Engineering Experimentation," pp. 224–254.

[22]STATGRAPHICS, MINITAB, SYSTAT, STATISTICA, SAS, etc.

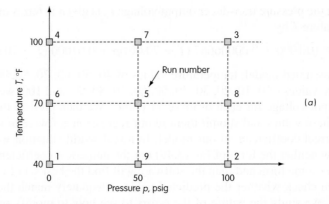

Run Number	P (psig)	T (°F)	e_o (v)	Noise to be added to e_o (V)
1	0.0	40.0	−0.30	−0.02316
2	100.0	40.0	9.50	−0.08362
3	100.0	100.0	10.50	0.03864
4	0.0	100.0	0.30	−0.04066
5	50.0	70.0	5.00	0.004274
6	0.0	70.0	0.00	0.10358
7	50.0	100.0	5.40	−0.04730
8	100.0	70.0	10.00	0.04298
9	50.0	40.0	4.60	0.06667
10	25.0	55.0		
11	25.0	85.0		(b)
12	75.0	55.0		
13	75.0	85.0		

Figure 3.21
Multiple regression used to cut cost of calibration.

usually characteristic of the instrument itself, not its inputs. In high-quality devices (accuracy ≥ 99 percent), random effects in calibration data are thus expected to be small (≤ 1 percent of full scale). This gives us an estimate of how large to make the random "noise" in our simulation experiments.

All these questions can be conveniently studied with our simulation approach and personal computer statistical software. If you have available convenient statistical software, it is easy to show that *any* four runs will give the exact coefficients in Eq. (3.50) if no random noise is assumed. When the small random noise of Fig. 3.21b is added, and we try runs of size four and nine, the results are

$$e_o = -0.7115 + 0.009708T + 0.09380p + 0.00008996pT \quad \text{1-4 runs} \quad \textbf{(3.51a)}$$

$$e_o = -0.5968 + 0.008784T + 0.09356p + 0.00008996pT \quad \text{1-9 runs} \quad \textbf{(3.51b)}$$

Note that neither of these results shows the correct values for the coefficients, though we would of course *not* know this in a real-world situation. The result for four runs gives an *exact* fit to the four points used since we have four equations in four unknowns. The *model,* however, is not perfect because we have fitted it exactly to four "noisy" points, *not* the true values. When we use nine points to estimate four coefficients in our model, we have nine equations in four unknowns and least-squares methods of finding the coefficients are used. This method tends to "average out" the noise in our data, giving a better estimate of the underlying relation among p, T, and e_o. In fact, if we used a very large number of points, we would converge on the exact values of the coefficients.

To see that a 9-run calibration improves on a 4-run, we actually have to *use* the two models found, to predict p values from measured values of T and e_o and then compare the errors. This could of course be done for any or all of our original nine points, but a more critical test, called a *validation experiment,* is available. (This test is recommended for *any* model-building regression studies, not just instrument calibrations.) Here we check the predictions of our model, not for the points used to derive it, but for some *other* points within the range of use of the model. In a real-world situation we never know the "true values," and the validation experiment must compare measured values of e_o with model-predicted ones. In our present simulation study we can do better. We can compare model-predicted values with *true* values. This could be done for the points used to derive the model or for any other points that might be of interest. In Fig. 3.21*b* the entries numbered 10-13 give 4 combinations of T and p that were not used to derive either of the models. I included them in this data table because my statistical software provides for these "validation" calculations automatically, so I routinely enter such "missing values" when I do a regression. If your software does not provide such a capability, it is of course easy to do it as a separate calculation. When the differences between the model-predicted and true pressure values are calculated for the four "validation" points and we add up the absolute values of the four errors to get an overall measure of error, we get 1.09 psig for the four-run model and 0.42 psig for the nine-run model, showing the superiority of the nine-run model. If our simulation study included more than nine runs, we would find that the number 0.42 would gradually approach zero as we used larger and larger numbers of runs. (This does *not* mean that in *real-world* calibrations, we can make errors as small as we please by simply using more runs. Why?)

The nine runs shown in the table of Fig. 3.21*b* need not be performed in the top-to-bottom sequence shown there. Depending on circumstances, the runs of *any* experiment may need to be performed in a random sequence.[23] When performing calibrations, however, we often want to be conservative and give effects such as hysteresis and friction every chance to occur in their worst forms. This may lead us to sequence our chosen runs in the nonrandom pattern low → medium → high → medium → low → . . . for the desired input (pressure in our case). This still leaves unspecified the sequencing of spurious inputs such as our temperature, but this sequence is often not critical and can be decided by convenience. That is, it may be more convenient to change temperature from 40 to 70 and then 100°F, rather than from 100 to 40 and then 70°F, and this convenient sequence usually causes no

---

[23]E. O. Doebelin, "Engineering Experimentation," pp. 193–196 and 202, 215, 287.

trouble. Applying this philosophy we might sequence our runs as 1, 9, 2, 8, 5, 6, 4, 7, and 3. Depending on specific company practices, calibration equipment characteristics, and past experience, other sequences might also be justifiable.

Since, during calibration, we consider pressure and temperature to be independent variables set at chosen values and output voltage as a dependent variable *caused* by them, the form of Eq. (3.50) is natural. However, when we use our calibration results to make a measurement, our "inputs" are a voltage and a temperature reading that we use to calculate the corresponding pressure. Now the dependent variable is p, and we would like to have a formula which gives p as a known function of T and e_o. Equation (3.50) is easily rearranged to give this, and the bottom diagram in Fig. 3.22 gives a graphic representation of this formula. When our calibration equation includes higher power terms (as often needed to get the highest accuracy out of precise instruments), this "inversion" of the formula is *not* possible. As a simple example, if y is related to x by $y = 0.34 + 1.56x + 0.12x^2 + 0.0034x^3$, then y is easily computed for any given x. However, if we are given y, then getting x is *not* as simple. In fact, one must solve for the three roots of a cubic polynomial and then decide which of the three we really want. If our calibration equation includes terms that are not simple powers of x, then analytical solutions may not even exist and approximate numerical methods must be used. These complications

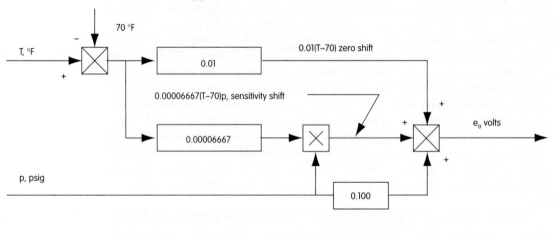

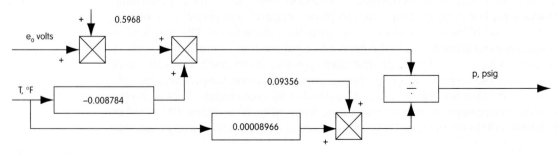

Figure 3.22
Results of regression calibration displayed in block diagram form.

are unwanted in computerized data-processing software since they use more memory and take more time. Fortunately, a relatively simple solution is available.

When we do a calibration and formulate an equation that models the relation between the instrument output and the desired and spurious inputs, we generate a table with columns (for the output and each input) and rows (for the individual calibration points). In a physical sense, the input variables really do *cause* the output variable; temperature and pressure in our pressure transducer *cause* the output voltage. When we decide on the form of the calibration equation, we often have in mind known physical effects that act in a certain way. For example, mechanical springs will change their lengths in proportion to temperature and their stiffness will also decrease as temperature increases. These *linear* effects are often sufficient, but for large temperature changes, higher powers may be necessary for the highest accuracy. We usually do not have a "scientific" basis (such as detailed atomic or molecular properties) for choosing polynomial models for various physical effects; they are simply good and simple "curve fits" for measured material behaviors.

When we want to process our table of calibration data so that we can easily use it to compute, say, pressure values from measured output voltage and instrument temperature, we can simply now *define* pressure to be the "dependent" variable and use multiple-regression software (least-squares curve fitting) to find the best coefficients in an equation that models the relation among the variables. What *form* should we assume for the "inverse" model? If our "direct" model had been a polynomial, then the "inverse" would require the mathematical operations we earlier stated as undesirable. At this point we simply decide that, mathematically, our table of data is "neutral" with respect to which variables are called dependent and which independent. The regression software doesn't really care. Thinking graphically, if we have a smooth curve for y versus x, if the axis labels were left off, would you really know which variable was dependent and which independent?

A numerical example will be helpful in better understanding this situation and validating the methods used. Let us again use a pressure transducer as our illustration but now let it have a built-in temperature sensor that puts out a voltage e_T proportional to temperature. (A more complex example might let the temperature sensor be nonlinear, for example, $e_T = a_0 + a_1T + a_2T^2$.) A calibration might produce a table of 33 entries; 11 pressure values from 0 to 100 psig, at each of three temperatures, 0, 85, and 170°F. To do a simulation, let us again assume that we *know* the relation among the variables, something we *never* know in practice. This artifice will allow us to compare our curve-fitting results with the known perfect values. Suppose the *correct* equation relating pressure transducer output voltage e_p to true pressure p and transducer temperature T is

$$e_p = -0.1080 + 0.09930p + 0.00001000pT \qquad (3.52a)$$
$$+ 0.003000T + 0.00004000p^2 + (2.000 \times 10^{-7})p^3$$

We now use this equation to calculate e_p values for 33 "calibration runs" where temperature is set at 0, 85, and 170°F, and pressure goes from 0 to 100 psig in steps of 10 psig, for each temperature. Treating this data as if it were the result of a real calibration, we submit it to a multiple regression program (I used MINITAB) and ask the program to find the "best" coefficients in a model equation. In the real world, we of course *would not know* the correct equation and would be forced to conjure up a reasonable one. We would thus never know whether we had the "right answer," although we could of course tell whether it was a good fit. With simulation we can tell whether the method is capable of finding the correct coefficients *when we do know the correct equation form.* As you might expect from our earlier regression study, when the data is uncorrupted by noise, the least-squares method does find the correct coefficient values (in this example, correct to about 12 decimal places).

Let us now do the more practically useful exercise in which we want to find an equation from which we can calculate values of pressure from measured values of output voltage and transducer temperature. Let us also deal with the common situation where the pressure sensor has a built-in temperature sensor that outputs voltage values, not temperature values. This requires that we have a calibration equation for the temperature sensor, relating temperature to voltage. Our method can deal with any form of equation, but let us assume a simple linear one, that is, $e_T = 0.1051 + 0.019517T$. Using this equation we can "add a column" to our data table for the temperature sensor voltage e_T. We now want the regression method to find the best coefficients in an equation that gives p as a function of e_p and e_T. Note that *theoretically,* this would be an "inverse" relation to that of Eq. (3.52*a*), would *not* be a simple polynomial, and would require complex and time-consuming calculations. Instead, we assume a simple polynomial relation similar to Eq. (3.52*a*), accept that this cannot give the perfect results we just saw in the direct relation, but hope that the fit will be close enough for our practical use.

The form of equation that we submit to the software is

$$p = C_0 + C_1 e_p + C_2 e_T + C_3 e_p \cdot e_T + C_4 \cdot e_p^2 + C_5 \cdot e_p^3 \qquad \textbf{(3.52\textit{b})}$$

Remember, this *cannot* be the theoretically correct equation, but it *may* be a good fit. If not, we can adjust the form of the equation, using more and/or different terms. The values that MINITAB came up with are

$$p = 1.270 + 10.06 e_p - 1.555 e_T - 0.02428 e_p \cdot e_T - 0.04733 e_p^2 - 0.0006953 e_p^3$$

$$\textbf{(3.52\textit{c})}$$

To check the quality of the fit, we substitute e_p and e_T values and compare these p values with the known 0, 10, 20, . . . , 100 psig values. The fit turns out to be very good, and acceptable for almost any engineering purpose. Typical values are, for example, 60.01, 59.97, and 60.01 (should be 60) and 9.985, 10.01, and 10.01 (should be 10).

We suggested earlier that good results can often be obtained with fewer calibration points if they are carefully chosen. Taking 0°F-temperature points at 0, 30, 60, and 90 psig, 85°F-temperature points at 10, 40, and 70 psig, and 170°F-temperature points at 20, 50, 80, and 100 psig, the fitting equation was

$$p = 1.255 + 10.09 \, e_p - 1.567 e_T - 0.02290 e_p \cdot e_T - 0.05241 e_p^2 - 0.000399 e_p^3$$

$$\textbf{(3.52\textit{d})}$$

Even though these coefficients are somewhat different than those in Eq. (3.52*c*), the *fit* is still very good, and we have cut our calibration time and cost significantly

(11 points rather than 33). Our simulation studies here have used "perfect" data, but it is not difficult to add random noise, as we have in our other study, to appraise the behavior in more realistic circumstances. As usual, the fit cannot be as good as with perfect data, but our methods are still valid.

In the polynomial models we have suggested, it is usually best to keep the highest power lower than five. If we break the calibration curve into contiguous *segments,* each with its own equation, a high degree of accuracy can be achieved with quite low powers, often just the first power. This approach has some advantages and is described in detail in several references.[24]

Linearity

If an instrument's calibration curve for desired input is not a straight line, the instrument may still be highly accurate. In many applications, however, linear behavior is most desirable. The conversion from a scale reading to the corresponding measured value of input quantity is most convenient if we merely have to multiply by a fixed constant rather than consult a nonlinear calibration curve or compute from a nonlinear calibration equation. Also, when the instrument is part of a larger data or control system, linear behavior of the parts often simplifies design and analysis of the whole. Thus specifications relating to the degree of conformity to straight-line behavior are common.

Several definitions[25] of linearity are possible. However, *independent linearity* seems to be preferable in many cases. Here the reference straight line is the least-squares fit, as in Fig. 3.10. Thus the linearity is simply a measure of the maximum deviation of any calibration points from this straight line. This may be expressed as a percentage of the actual reading, a percentage of full-scale reading, or a combination of the two. The last method is probably the most realistic and leads to the following type of specification:

Independent nonlinearity $= \pm A$ percent of reading or

$$\pm B \text{ percent of full scale, whichever is greater} \quad \textbf{(3.53)}$$

The first part ($\pm A$ percent of reading) of the specification recognizes the desirability of a constant-percentage nonlinearity, while the second ($\pm B$ percent of full scale) recognizes the impossibility of testing for extremely small deviations near zero. That is, if a fixed percentage of reading is specified, the absolute deviations approach zero as the readings approach zero. Since the test equipment should be about 4 times as accurate as the instrument under test, this leads to impossible requirements on the test equipment. Figure 3.23 shows the type of tolerance band allowed by specifications of the form (3.53).

---

[24]L. H. Eccles, "The Representation of Physical Units in IEEE 1451.2," *Sensors,* April 1999, pp. 28–35; D. L. Ersland, "A Consistent Mathematical Approach for Multiple Input Calibration," *Sensors,* May 1999, pp. 104–106; L. H. Eccles, "IEEE-1451.2 Engineering Units Conversion Algorithm," *Sensors,* May 1999, pp. 107–112.

[25]J. G. Webster, ed., *The Measurement, Instrumentation and Sensors Handbook,* CRC Press, Boca Raton, 1999, pp. 3–12.

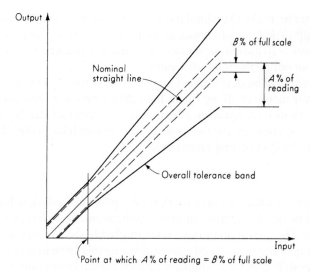

Figure 3.23
Linearity specification.

Note that in instruments considered essentially linear, the specification of non-linearity is equivalent to a specification of overall inaccuracy when the common (nonstatistical) definition of inaccuracy is used. Thus in many commercial linear instruments, only a linearity specification (and not an accuracy specification) may be given. The reverse (an accuracy specification but not a linearity specification) may be true if nominally linear behavior is implied by the quotation of a fixed sensitivity figure.

In addition to overall accuracy requirements, linearity specifications often are useful in dividing the total error into its component parts. Such a division is sometimes advantageous in choosing and/or applying measuring systems for a particular application in which, perhaps, one type of error is more important than another. In such cases, different definitions of linearity may be especially suitable for certain types of systems. The Scientific Apparatus Makers Association standard load-cell (force-measuring device) terminology,[26] for instance, defines linearity as follows: "The maximum deviation of the calibration curve from a straight line drawn between no-load and full-scale load outputs, expressed as a percentage of the full-scale output and measured on increasing load only." The breakdown of total inaccuracy into its component parts is carried further in the next few sections, where hysteresis, resolution, etc., are considered.

Threshold, Noise Floor, Resolution, Hysteresis, and Dead Space

Consider a situation in which the pressure gage of Fig. 3.2 has the input pressure slowly and smoothly varied from zero to full scale and then back to zero. If there

---

[26]"Standard Load Cell Terminology and Definitions: II," Scientific Apparatus Makers Association, Chicago, Jan. 11, 1962.

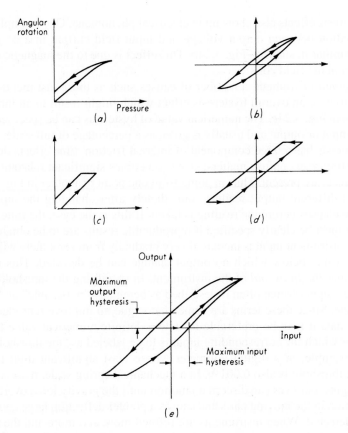

Figure 3.24
Hysteresis effects. (Magnitude exaggerated for graphical clarity.)

were no friction due to sliding of moving parts, the input-output graph might appear as in Fig. 3.24a. The noncoincidence of loading and unloading curves is due to the internal friction or hysteretic damping of the stressed parts (mainly the spring). That is, not all the energy put into the stressed parts upon loading is recoverable upon unloading, because of the second law of thermodynamics, which rules out perfectly reversible processes in the real world. Certain materials[27] exhibit a minimum of internal friction, and they should be given consideration in designing highly stressed instrument parts, provided that their other properties are suitable for the specific application. For instruments with a usable range on both sides of zero, the behavior is as shown in Fig. 3.24b.

If it were possible to reduce internal friction to zero but external sliding friction were still present, the results might be as in Fig. 3.24c and d, where a constant coulomb (dry) friction force is assumed. If there is any free play or looseness in the mechanism of an instrument, a curve of similar shape will result.

---

[27]Iso-Elastic, John Chatillon Co., Greensboro, North Carolina.

Hysteresis effects also show up in electrical phenomena. One example is found in the relation between output voltage and input field current in a dc generator, which is similar in shape to Fig. 3.24b. This effect is due to the magnetic hysteresis of the iron in the field coils.

In a given instrument, a number of causes such as those just mentioned may combine to give an overall hysteresis effect which might result in an input-output relation as in Fig. 3.24e. The numerical value of hysteresis can be specified in terms of either input or output and usually is given as a percentage of full scale. When the total hysteresis has a large component of internal friction, time effects during hysteresis testing may confuse matters, since sometimes significant relaxation and recovery effects are present. Thus in going from one point to another in Fig. 3.24e, we may get a different output reading immediately after changing the input than if some time elapses before the reading is taken. If this is the case, the time sequence of the test must be clearly specified if reproducible results are to be obtained.

If the instrument input is increased very gradually from zero, there will be some minimum value below which no output change can be detected. This minimum value defines the *threshold* of the instrument. In specifying the threshold, the first detectable output change often is described as being any "noticeable" or "measurable" change. Since these terms are somewhat vague, to improve reproducibility of threshold data it may be preferable to state a definite numerical value for output change for which the corresponding input is to be labeled as "the threshold."

For example, in a digital voltmeter whose least significant digit represents 1 mV, the threshold is also 0.001 V. In a mechanical spring scale, if we apply very small weights, our eyes can discern no motion until the gravity force overcomes the static friction in the moving parts and causes a pointer deflection large enough to be visually detected. When instruments are refined more and more and the threshold gets closer and closer to zero, we inevitably encounter tiny *random fluctuations* that put a lower bound on what can be measured. This limit is sometimes called the *noise floor* of the instrument.

That is, when the input, say, voltage, force, or pressure is truly zero, the instrument output does not "stand still" but rather fluctuates in a random fashion. If we then apply a small unknown voltage, force, or pressure to the instrument, we are not able to distinguish the signal from the noise, making any measurement impossible. Many routine engineering measurements do not require consideration of this phenomenon since they are usually at a much higher level. However, some do, and in any case, we should be aware of such fundamental limits on measurement. A classic paper[28] analyzes various mechanical and electrical instruments to establish numerical relations for the random fluctuations. For example, a thin quartz fiber can be used in torsion as an elastic element to transduce tiny torques into angular deflections measured optically by reflecting a light beam from an attached mirror, to a

---

[28]R. B. Barnes and S. Silverman, "Brownian Motion as a Natural Limit to all Measuring Processes," *Reviews of Modern Physics,* Vol. 6, July 1934, pp. 162–192.

scale. With no external torques applied, the mirror exhibits random angular deflections (ϕ radians) with a mean-squared amplitude $\overline{\phi^2}$ given by

$$\overline{\phi^2} = \frac{kT}{A} \tag{3.54}$$

where $A \triangleq$ torsional spring constant, $(N - m)/radian$
$k \triangleq$ Boltzmann's constant, 1.38×10^{-23} J/K
$T \triangleq$ absolute temperature, K

Recall that the mean-squared value is the time-average of the square of the angular deflection and is a measure of the "size" of the random fluctuations. Formulas similar to Eq. (3.54) can be derived (see Ref. 28) for many measurement situations and always show that the fluctuations go to zero as temperature T goes to absolute zero. Thus, very sensitive instruments may employ cooling (as with liquid-nitrogen-cooled infrared detectors) to achieve the lowest possible thresholds.

The *rapidity of variation* of random fluctuations is characterized by their *power spectral density,* which shows the relative strengths of different frequency regions, high-frequency strength corresponding to rapid variation. If we need to measure very small signals immersed in the random noise, we can use filtering to help us, with, however, certain trade-offs. The simplest situation is for a desired signal that is *constant.* Here the "filtering" consists of simple time averaging:

$$\frac{1}{T} \cdot \int_0^T (x_{\text{signal}} + x_{\text{noise}})(dt) = \frac{1}{T} \cdot \int_0^T x_{\text{signal}} \, dt + \frac{1}{T} \cdot \int_0^T x_{\text{noise}} \, dt \tag{3.55}$$

Since the noise has an average value of zero, as time goes by, the second integral will gradually go to zero. If the noise has strong low-frequency content, we will have to average (integrate) for a long time before we can accurately read the desired signal. Thus, the improved threshold is paid for with a slower response time. When our desired signal is not a simple constant, but rather has *its own* frequency range (say f_1 to f_2 Hz) then a bandpass filter can be used to advantage. For example, a sound-measuring microphone[29] may need to measure low-level sound pressure in a range from, say, 100 to 1000 Hz. For microphones, the random fluctuation in pressure has a "flat" spectral density called *white noise;* all frequencies are present in equal strength. The formula for computing the mean-square pressure P_N^2 (N^2/m^4) of the random noise found between frequencies f_1 and f_2 is

$$P_{N,\,(f_2-f_1)}^2 = 4kTR_a \cdot (f_2 - f_1) \tag{3.56}$$

where $R_a \triangleq$ microphone acoustic resistance, pressure/volume flow rate
$k \triangleq$ Boltzmann's constant, 1.38×10^{-23} J/K

---

[29]V. Tarnow, "Thermal Noise in Microphones and Preamplifiers," *Bruel and Kaer Technical Review,* #3, 1972, pp. 3–14.

R_a is typically 3×10^7 (N/m^2)/(m^3/s). Using the numbers given, the mean-square noise pressure is calculated as 4.44×10^{-10} N^2/m^4. We usually prefer to work with the root-mean-square (rms) value, which is easily obtained as the square root of the mean-square, in this case 2.11×10^{-5} N/m^2. The rms values are preferred because they have the same dimensions as the physical quantity, in our case, pressure. If the sound pressure we wish to measure has an rms value much larger than that of the noise, we should have no trouble measuring it accurately with the given microphone, assuming no other significant noise sources are present. As a matter of interest, the accepted threshold of *human* hearing is about 2×10^{-5} N/m^2, so this microphone is quite similar to the human ear, at least in this respect.

Equation (3.56) clearly shows that the random noise gets smaller as we restrict the measurement bandwidth $(f_2 - f_1)$. Thus, if our desired signal frequency spectrum is, say, from 100 to 200 Hz, the noise will be much less. The price paid for reduced noise is again *slower response* since bandpass filters used to restrict the frequency range have response times of about $4/(f_2 - f_1)$, 0.04 s in our example.

Since microphone signals are usually electronically amplified, the random noise of the amplifier adds to that produced by the microphone to give the total system noise floor. Tarnow gives a formula for estimating the amplifier noise voltage, which has four sources: the FET gate resistor, "shot noise" from the gate current, noise from the FET channel, and "1/f" noise from the transistor. Figure 3.25 shows the frequency spectrum of the total noise and that of the amplifier alone. These amplifiers have been carefully designed for low noise, so we see that the microphone noise is the limiting factor for most of its frequency range.

If the input is increased slowly from some arbitrary (nonzero) input value, again the output does not change at all until a certain input increment is exceeded. This increment is called the *resolution;* again, to reduce ambiguity, it is defined as the input increment that gives some small but definite numerical change in the output. Thus resolution defines the smallest measurable input *change* while threshold

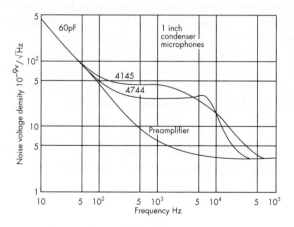

Figure 3.25
Noise floors for microphones and amplifiers.

defines the smallest measurable *input.* Both threshold and resolution may be given either in absolute terms or as a percentage of full-scale reading. An instrument with large hysteresis does not necessarily have poor resolution. Internal friction in a spring can give a large hysteresis, but even small changes in input (force) cause corresponding changes in deflection, giving high resolution.

The terms "dead space," "dead band," and "dead zone" sometimes are used interchangeably with the term "hysteresis." However, they may be defined as the total range of input values possible for a given output and thus may be numerically twice the hysteresis as defined in Fig. 3.24*e*. Since none of these terms is completely standardized, you should always be sure which definition is meant.

Scale Readability

Since the majority of instruments that have analog (rather than digital) output are read by a human observer noting the position of a "pointer" on a calibrated scale, usually it is desirable for data takers to state their opinions as to how closely they believe they can read this scale. This characteristic, *which depends on both the instrument and the observer,* is called the *scale readability.* While this characteristic logically should be *implied* by the number of significant figures recorded in the data, it is probably good practice for the observer to stop and think about this before taking data and to then *record* the scale readability. It may also be appropriate at this point to suggest that all data, including scale readabilities, be given in decimal rather than fractional form. Since some instrument scales are calibrated in $\frac{1}{4}$'s, $\frac{1}{2}$'s, etc., this requires data takers to convert to decimal form before recording data. This procedure is considered preferable to recording a piece of data as, say, $21\frac{1}{4}$ and then *later* trying to decide whether 21.250 or 21.3 was meant.

Span

The range of variable that an instrument is designed to measure is sometimes called the *span.* Equivalent terminology also in use states the "low operating limit" and "high operating limit." For essentially linear instruments, the term "linear operating range" is also common. A related term, which, however, implies dynamic fidelity also, is the *dynamic range.* This is the ratio of the largest to the smallest dynamic input that the instrument will faithfully measure. The number representing the dynamic range often is given in decibels, where the decibel (dB) value of a number N is defined as dB $\triangleq$ 20 log N. Thus a dynamic range of 60 dB indicates the instrument can handle a range of input sizes of 1,000 to 1.

Generalized Static Stiffness and Input Impedance: Loading Effects

We mentioned very early that the introduction of any measuring instrument into a measured medium always results in the extraction of some energy from the medium, thereby changing the value of the measured quantity from its undisturbed state and thus making perfect measurements theoretically impossible. Since the instrument designer wishes to approach perfection as nearly as practicable, some numerical means of characterizing this "loading" effect of the instrument on the measured medium would be helpful in comparing competitive instrument designs. The concepts of *stiffness* and *input impedance*[30] serve such a function. While both terms are useful for both static and dynamic conditions, here we consider their static aspects only.

In Fig. 2.1 and subsequent schematic and block diagrams, the connection of functional elements by single lines perhaps gives the impression that the transfer of information and energy is described by a single variable only. Closer examination reveals that energy transfers require the specification of two variable quantities for their description. The definitions of stiffness and generalized input impedance are in terms of two such variables. At the input of each component in a measuring system, there exists a variable q_{i1} of primary interest insofar as information transmission is concerned. At the same point, however, there is associated with q_{i1} another variable q_{i2} such that the product $q_{i1}q_{i2}$ has the dimensions of power and represents an instantaneous rate of energy withdrawal from the preceding element. When these two signals are identified, we can define the generalized input impedance Z_{gi} by

$$Z_{gi} \triangleq \frac{q_{i1}}{q_{i2}} \tag{3.57}$$

if q_{i1} is an "effort variable." (Effort variable is defined shortly.) [At this point we consider only systems where Eq. (3.57) is an ordinary algebraic equation. However, the concept of impedance can be extended easily to dynamic situations, and then (3.57) must be given a more general interpretation.] Using (3.57), we see that the power drain is $P = q_{i1}^2/Z_{gi}$ and that a *large input impedance is needed to keep the power drain small.* The concept of generalized impedance (and, of course, the terminology itself) is a generalization of electric impedance, and we first give some examples from this perhaps somewhat more familiar field.

Consider a voltmeter of the common type shown in Fig. 2.17a. Suppose this meter is to be applied to a circuit in order to measure an unknown voltage *E*, as in Fig. 3.26a. As soon as the meter is attached to terminals *a* and *b*, the circuit is changed and the value of *E* is no longer the same. For the meter alone, the input variable of direct interest (q_{i1}) is the terminal voltage E_m. If we look for an associated variable (q_{i2}) which, when multiplied by q_{i1}, gives the power withdrawal, we find the meter current i_m meets these requirements. In this example, then, $Z_{gi} = e_m/i_m = R_m$, the meter resistance.

To further illustrate the significance of input impedance, let us determine just how much error is caused when the meter is connected to the circuit. To facilitate this, first we cite without proof a very useful network theorem called Thévenin's theorem.[31] Consider any network made up of linear, bilateral impedances and power sources. A linear impedance is one whose elements (*R, L, C*) do not change value with the magnitude of the current or voltage. Most resistances, capacitances, and air-core inductances are linear; iron-core inductances are nonlinear. A bilateral

---

[30]R. G. Boiten, "The Mechanics of Instrumentation," *Proc. Inst. Mech. Engrs.* (*London*), vol. 177, no. 10, 1963.

[31]G. Rizzoni, "Principles and Applications of Electrical Engineering," 2nd ed., Irwin, 1996, p. 82.

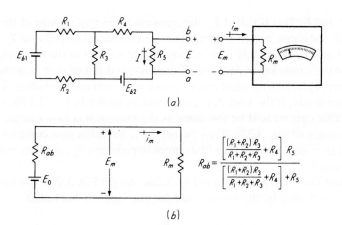

(a)

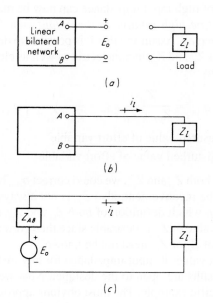

(b)

Figure 3.26
Voltmeter loading effect.

(a)

(b)

Load

(c)

Figure 3.27
Thévenin's theorem.

impedance transmits energy equally well in either direction. Resistances, capacitances, and inductances are essentially bilateral. Field-effect transistors (FETs) are unilateral since they effectively transmit energy in only one direction (from gate to source, *not* the reverse).

A linear bilateral network is shown in Fig. 3.27*a* as a "black box" with terminals *A* and *B*. A load of impedance Z_l may be connected across the terminals *A* and *B*. When the load Z_l is *not* connected, a voltage will, in general, exist at terminals A and B. This is called E_o, the open-circuit output voltage of the network. Also with Z_l *not* connected, it is possible to determine the impedance Z_{AB} between terminals A and B. When this is done, any power sources in the network are to be replaced by their internal impedance. If their internal impedance is assumed to be zero, they are replaced by a short circuit (just a wire with no resistance). Thévenin's theorem then states: If the load Z_l is connected as shown in Fig. 3.27b, a current i_l will flow. This current will be the *same* as the current that flows in the fictitious equivalent circuit of Fig. 3.27c. Thus the network, no matter how complex it is, may be replaced by a single impedance (the output impedance) Z_{AB} in series with a single voltage source E_o.

Applying Thévenin's theorem to Fig. 3.26a, we get Fig. 3.26b. We see here that the value E_m indicated by the meter is *not* the true value E, but rather

$$E_m = \frac{R_m}{R_{ab} + R_m} E \tag{3.58}$$

and that if E_m is to approach E, we must have $R_m \gg R_{ab}$. Thus our earlier statement about the desirability of high input impedance can now be made more specific. *The input impedance must be high relative to the output impedance of the system to which the load is connected.* Assuming that it is possible to define generalized input and output impedances Z_{gi} and Z_{go} in nonelectric as well as electric systems, we may generalize Eq. (3.58) to

$$q_{i1m} = \frac{Z_{gi}}{Z_{go} + Z_{gi}} q_{i1u} = \frac{1}{Z_{go}/Z_{gi} + 1} q_{i1u} \tag{3.59}$$

where $q_{i1m} \triangleq$ measured value of effort variable

$q_{i1u} \triangleq$ undisturbed value of effort variable

Of course, if we knew both Z_{gi} and Z_{go}, we could correct q_{i1m} by means of Eq. (3.59). However, this would be inconvenient; also Z_{go} is not always known, especially in nonelectric systems, in which definition of *both* Z_{gi} and Z_{go} is not always straightforward. Thus a high value of Z_{gi} is desirable since then corrections are unnecessary and the actual values of Z_{gi} or Z_{go} need not be known.

To achieve a high value of input impedance for any instrument, not just voltmeters, a number of paths are open to the designer. Now we describe three, using the voltmeter as a specific example. The most obvious approach is to leave the configuration of the instrument unchanged, but to alter the numerical values of physical parameters so that the input impedance is increased. In the voltmeter of Fig. 3.26, this is accomplished simply by winding the coil in such a way (higher resistance material and/or more turns) that R_m is increased. While this accomplishes the desired result, certain undesirable effects also appear. Since this type of voltmeter is basically a *current*-sensitive rather than a voltage-sensitive device, an increase in R_m will *reduce* the magnetic torque available from a given impressed voltage. Thus if the spring constant of the restraining springs is not changed, the angular deflection for a given voltage (the sensitivity) is reduced. To bring the sensitivity back to its former value, we must reduce the spring constant. Also, because of lower torque levels, pivot bearings with less friction must be employed. These design changes generally result in a less rugged and less reliable instrument, so that this method of increasing input impedance is limited in the degree of improvement possible before other performance features are compromised. This situation occurs in most instruments, not just in this specific example.

If input impedance is to be increased without compromising other characteristics, different approaches are needed. One of general utility employs a change in configuration of the instrument so as to include an auxiliary power source. The concept is that a rugged instrument requires a fair amount of power to actuate its output elements, but that this power need *not* necessarily be taken from the measured medium. Rather, the low power signal from the primary sensing element may *control* the output of the auxiliary power source so as to realize a power-amplifying effect.

To continue our voltmeter example, this approach is exemplified by a transistor version of the classic vacuum-tube voltmeter (VTVM) shown in rudimentary form in Fig. 3.28a. When the input voltage is zero (short circuit across the E_m terminals), adjustment of R_a allows setting of the meter voltage E'_m to zero. If an input E_m is

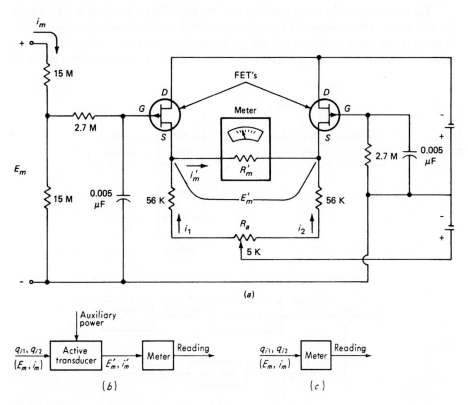

(a)

(b)

(c)

Figure 3.28
Field-effect transistor (FET) voltmeter.

applied, the gate bias is no longer the same, the currents i_1 and i_2 are no longer equal, and a meter voltage E'_m exists. While the meter current i'_m may still be as large as in a conventional meter, the current that determines the input impedance is i_m, which will be very small. Thus a very high input impedance can be realized while a rugged meter element is used. A block diagram for the FET voltmeter is seen in Fig. 3.28b, which may be compared with that for an ordinary meter in Fig. 3.28c.

Still another approach to the problem of increasing input impedance uses the principle of feedback or null balance. For the specific area of voltage measurements, this technique is exemplified by the potentiometer. The simplest form of this instrument is shown in Fig. 3.29. Clearly, in Fig. 3.29a each position of the sliding contact corresponds to a definite voltage between terminals a and b. Thus the scale can be calibrated, and any voltage between zero and the battery voltage can be obtained by properly positioning the slider. If we now connect an unknown voltage E_m and a galvanometer (current detector), as in Fig. 3.29b, then if E_m is less than the battery voltage, there will be some point on the slider scale at which the voltage picked off the slide wire just equals the unknown E_m. This point of null balance can be detected by a zero deflection of the galvanometer since the net loop voltage a, c, d, b, a will be zero. Then the unknown voltage can be read from the calibrated scale.

We should note that under conditions of perfect balance the current drawn from the unknown voltage source is exactly zero, which yields an infinite input impedance. In actual practice, there must always remain some unknown unbalance current since a galvanometer always has a threshold below which currents cannot be

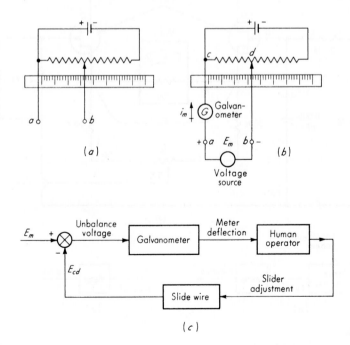

Figure 3.29
Potentiometer voltage measurement.

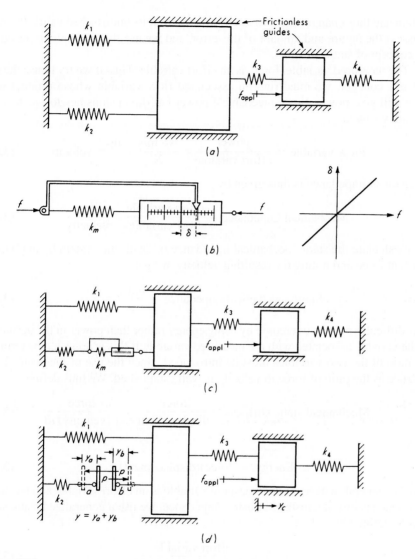

Figure 3.31
Force-gage loading effect.

Consider the system of Fig. 3.31a as an idealized model of some elastic structure under applied load f_{appl}. This load will cause forces in the various structural
members. Suppose we wish to measure the force in the member represented by the
spring k_2. A common method of force measurement employs a calibrated elastic link
whose deflection is proportional to force; thus a deflection measurement allows a
force measurement. Such a device, with spring constant k_m, is shown in Fig. 3.31b.
To measure the force in link k_2, we insert the force-measuring device "in series"
with the link k_2, as shown in Fig. 3.31c. The usual difficulty is encountered here in
that the insertion of the measuring instrument alters the condition of the measured

system and thus changes the measured variable from its undisturbed value. We wish to assess the nature and amount of this error, and we use this example to introduce the concept of static stiffness.

The measured variable, force, is an effort variable. Thus if we try to use the impedance concept, we must find an associated flow variable whose product with force will give power. Since mechanical power has dimensions newton-meters per second, we have

$$\text{Flow variable} = \frac{\text{power}}{\text{effort variable}} = \frac{\text{N} \cdot \text{m/s}}{\text{N}} = \frac{\text{m}}{\text{s}} = \text{velocity} \qquad \textbf{(3.66)}$$

Mechanical impedance is thus given by

$$\text{Mechanical impedance} = \frac{\text{effort variable}}{\text{flow variable}} = \frac{\text{force}}{\text{velocity}} \qquad \textbf{(3.67)}$$

If we calculate the static mechanical impedance of an elastic system by applying a constant force and noting the resulting velocity, we get

$$\text{Static mechanical impedance} = \frac{\text{force}}{0} = \infty \qquad \textbf{(3.68)}$$

This difficulty may be overcome by using energy rather than power in the definition of the variable associated with the measured variable. If this is done, a new term for the ratio of the two variables must be introduced, since the use of mechanical impedance as the ratio of force to velocity is well established. We thus define

$$\text{Mechanical static stiffness} \triangleq \frac{\text{force}}{\text{displacement}} = \frac{\text{force}}{\int (\text{velocity})\, dt} \qquad \textbf{(3.69)}$$

since

$$\text{Energy} = (\text{force})(\text{displacement}) \qquad \textbf{(3.70)}$$

Thus, in general, whenever the measured variable is an effort variable and the static impedance is infinite, instead of using impedance, we use a generalized static stiffness S_g defined by

$$S_g \triangleq \frac{\text{effort variable}}{\int (\text{flow variable})\, dt} \qquad \textbf{(3.71)}$$

If this is done, it can be shown that the same formulas can be used for calculating the error due to inserting the measuring instrument as were utilized for impedance, except S is employed instead of Z. Thus, Eq. (3.59) becomes

$$q_{i1m} = \frac{S_{gi}}{S_{go} + S_{gi}}\, q_{i1u} = \frac{1}{S_{go}/S_{gi} + 1}\, q_{i1u} \qquad \textbf{(3.72)}$$

where $q_{i1m} \triangleq$ measured value of effort variable

$q_{i1u} \triangleq$ undisturbed value of effort variable

$S_{gi} \triangleq$ generalized static input stiffness of measuring instrument

$S_{go} \triangleq$ generalized static output stiffness of measured system

Let us apply these general concepts to the specific case at hand. The output stiffness of the system of Fig. 3.31a at the point of insertion of the measuring device is simply the ratio of force p to deflection y at terminals a and b in Fig. 3.31d. This stiffness can be found theoretically by applying a fictitious load p and calculating the resulting y; or, if the structure (or a scale model) has been constructed, we can obtain the stiffness experimentally by applying known loads and measuring the resulting deflections. A theoretical analysis might proceed as below:

$$\Sigma \text{ forces} = 0$$

$$p - y_b k_1 + k_3(y_c - y_b) = 0 \tag{3.73}$$

$$f_{\text{appl}} - k_3(y_c - y_b) - k_4 y_c = 0 \tag{3.74}$$

$$(-k_1 - k_3)y_b + k_3 y_c = -p \tag{3.75}$$

$$k_3 y_b + (-k_3 - k_4)y_c = -f_{\text{appl}} \tag{3.76}$$

Using determinants yields

$$y_b = \frac{\begin{vmatrix} -p & k_3 \\ -f_{\text{appl}} & -(k_3 + k_4) \end{vmatrix}}{\begin{vmatrix} -(k_1 + k_3) & k_3 \\ k_3 & -(k_3 + k_4) \end{vmatrix}} = \frac{p(k_3 + k_4) + f_{\text{appl}}(k_3)}{(k_3 + k_4)(k_1 + k_3) - k_3^2} \tag{3.77}$$

The output stiffness is now obtained from Eq. (3.77) by letting f_{appl} be zero:

$$S_{go} = \frac{p}{y} = \frac{p}{y_a + y_b} = \frac{p}{p/k_2 + p(k_3 + k_4)/[(k_3 + k_4)(k_1 + k_3) - k_3^2]} \tag{3.78}$$

$$S_{go} = \frac{1}{1/k_2 + (k_3 + k_4)/[(k_3 + k_4)(k_1 + k_3) - k_3^2]} \tag{3.79}$$

The input stiffness of the measuring instrument is given by

$$S_{gi} = \frac{\text{force}}{\text{displacement}} = k_m \tag{3.80}$$

We may now apply Eq. (3.72) to get

$$\frac{\text{Measured value of force}}{\text{True value of force}} = \frac{k_m}{1/[1/k_2 + (k_3 + k_4)/(k_1 k_3 + k_1 k_4 + k_3 k_4)] + k_m} \tag{3.81}$$

From Eq. (3.72) it is apparent that in general we should like to have $S_{gi} \gg S_{go}$ in order to have the measured value close to the true value. In this example, this requirement corresponds to

$$k_m \gg \frac{1}{1/k_2 + (k_3 + k_4)/(k_1 k_3 + k_1 k_4 + k_3 k_4)} \tag{3.82}$$

Thus the measuring device must have a sufficiently stiff spring.

We saw earlier that when the measured variable is not an effort variable, admittance (rather than impedance) is a more convenient tool. Again, however, under static conditions it may happen that admittance is infinite; thus a concept analogous

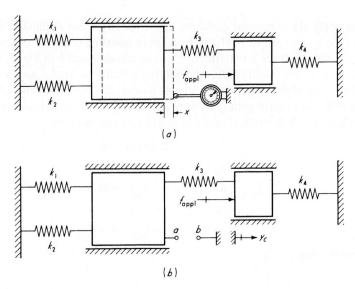

Figure 3.32
Displacement-gage loading effect.

to stiffness is needed to facilitate the treatment of such situations. For such cases, the generalized compliance C_g is defined by

$$C_g \triangleq \frac{\text{flow variable}}{\int (\text{effort variable})\, dt} \tag{3.83}$$

As a mechanical example, suppose we wish to measure the displacement x in Fig. 3.32a by means of a dial indicator. Such indicators generally have a spring load (stiffness k_m) to ensure positive contact with the body whose motion is being measured. This spring load adds a force to the measured system, thereby causing error in the motion measurement. It is clear that the indicator spring load should be as light as possible, but we wish to make more quantitative statements about measurement accuracy. Analysis[33] shows that:

$$\frac{\text{Measured value of deflection}}{\text{True value of deflection}} = \frac{1}{k_m/[k_1 + k_2 + k_3 k_4/(k_3 + k_4)] + 1} \tag{3.84}$$

and k_m must be sufficiently small to get accurate displacement measurement.

Although we have used some specific electrical and mechanical measurement examples to develop the basic concepts of impedance, admittance, stiffness, and compliance as they apply to loading effects in measurement systems, these concepts can be easily generalized to other measured variables. This generalization and proofs of formulas, such as Eq. (3.72) are left for the reference.[34]

In concluding this section on loading effects, the following comments are appropriate. In every measurement, the instrument input causes a load on the

[33]E. O. Doebelin, "Measurement Systems," 4th ed., p. 85.

[34]E. O. Doebelin, "Measurement Systems," 3d ed., pp. 87–97, McGraw-Hill, New York, 1983.

measured-medium output. If the instrument system consists of several inter-connected stages (the "elements" of Fig. 2.1), there may be significant loading effects *between* stages. This is often the case when the "elements" are general-purpose devices such as amplifiers and recorders which are connected in various ways at different times to create a measurement system suited to a particular problem. In using such a "building block" approach, loading problems must be carefully considered, and methods such as have been outlined above must be used if satisfactory and predictable results are to be obtained. If, however, the elements are merely functional components (permanently connected) of a specific instrument, then loading between elements may be a necessary consideration for the instrument *designer* but not for the user. The user need be concerned only with the loading situation at the interface between measured medium and primary sensing element.

Concluding Remarks on Static Characteristics

In this section we presented the most significant static characteristics, commonly used in describing instrument performance. It is not possible to list *all* the specific static characteristics that might be pertinent in a particular instrument; rather, those of general interest were considered. It should also be emphasized that the terminology of the measurement field has not been thoroughly standardized. Therefore, you should be careful to determine the precise definition intended when an apparently familiar term is encountered.

Also, errors that are not intrinsically associated with the instrument itself, but are due to human factors involved in taking readings or incurred by incorrect installation of the instrument, were not considered in detail. However, such errors certainly could be included in our general framework merely by extending the concept of inputs to include human factors and installation effects.

3.3 DYNAMIC CHARACTERISTICS

The study of measurement system dynamics is just one of many practical applications of the more general area, often called *system dynamics,* which is treated in depth in other texts[35] and is in some curricula a prerequisite for courses in measurement, control, vibration, etc. Since not all readers have had a system dynamics course, we include those fundamentals necessary for our purposes.

Generalized Mathematical Model of Measurement System

As in so many other areas of engineering application (vibration theory, circuit theory, automatic-control theory, aircraft stability and control theory, etc.), the most widely useful mathematical model for the study of measurement-system dynamic response is the ordinary linear differential equation with constant coefficients. We assume that the relation between any particular input (desired, interfering, or

---

[35]E. O. Doebelin, "System Dynamics: "Modeling, Analysis, Simulation Design," Marcel Dekker, New York, 1998.

modifying) and the output can, by application of suitable simplifying assumptions, be put in the form

$$a_n \frac{d^n q_o}{dt^n} + a_{n-1} \frac{d^{n-1} q_o}{dt^{n-1}} + \cdots + a_1 \frac{dq_o}{dt} + a_0 q_o = b_m \frac{d^m q_i}{dt^m}$$

$$+ b_{m-1} \frac{d^{m-1} q_i}{d^{m-1}} + \cdots + b_1 \frac{dq_i}{dt} + b_0 q_i \quad (3.85)$$

where $q_o \triangleq$ output quantity
    $q_i \triangleq$ input quantity
    $t \triangleq$ time
    a's, b's $\triangleq$ combinations of system physical parameters, assumed constant

If we define the differential operator $D \triangleq d/dt$, then Eq. (3.85) can be written as

$$(a_n D^n + a_{n-1} D^{n-1} + \cdots + a_1 D + a_0)q_o$$
$$= (b_m D^m + b_{m-1} D^{m-1} + \cdots + b_1 D + b_0)q_i \quad (3.86)$$

The solution of equations of this type has been put on a systematic basis by using either the "classical" method of D operators or the Laplace-transform[36] method. With the D-operator method, the complete solution q_o is obtained in two separate parts as

$$q_o = q_{ocf} + q_{opi} \qquad (3.87)$$

where $q_{ocf} \triangleq$ complementary-function part of solution
    $q_{opi} \triangleq$ particular-integral part of solution

The solution q_{ocf} has n arbitrary constants; q_{opi} has none. These n arbitrary constants may be evaluated numerically by imposing n initial conditions on Eq. (3.87). The solution q_{ocf} is obtained by calculating the n roots of the algebraic *characteristic equation*

$$a_n D^n + a_{n-1} D^{n-1} + \cdots + a_1 D + a_0 = 0 \qquad (3.88)$$

where the operator D is treated as if it were an algebraic unknown. Once these roots $s_1, s_2, \ldots, s_n$ have been found, the complementary-function solution is immediately written by following the rules stated below:

1. *Real roots, unrepeated.* For each real unrepeated root s one term of the solution is written as Ce^{st}, where C is an arbitrary constant. Thus, for example, roots -1.7, $+3.2$, and 0 give a solution $C_1 e^{-1.7t} + C_2 e^{3.2t} + C_3$.
2. *Real roots, repeated.* For each root s which appears p times, the solution is written as $(C_0 + C_1 t + C_2 t^2 + \cdots + C_{p-1} t^{p-1})e^{st}$. Thus, if there are roots $-1, -1, +2, +2, 0, 0$, the solution is written as $(C_0 + C_1 t)e^{-t} + (C_2 + C_3 t + C_4 t^2)e^{2t} + C_5 + C_6 t$.

---

[36]E. O. Doebelin, "System Dynamics," 1998.

3. *Complex roots, unrepeated.* A complex root has the general form $a + ib$. It can be shown that if the a's of Eq. (3.88) are themselves real numbers (which they generally will be, since they are physical quantities such as mass, spring rate, etc.), then when any complex roots occur, they will always occur in pairs of the form $a \pm ib$. For each such root pair, the corresponding solution is

$$Ce^{at} \sin (bt + \phi)$$

where C and ϕ are the two arbitrary constants. Thus roots $-3 \pm i4$, $2 \pm i5$, and $0 \pm i7$ give a solution $C_0 e^{-3t} \sin (4t + \phi_0) + C_1 e^{2t} \sin (5t + \phi_1) + C_2 \sin (7t + \phi_2)$.

4. *Complex roots, repeated.* For each pair of complex roots $a \pm ib$ which appears p times the solution is $C_0 e^{at} \sin (bt + \phi_0) + C_1 t e^{at} \sin (bt + \phi_1) + \cdots + C_{p-1} t^{p-1} e^{at} \sin (bt + \phi_{p-1})$. Roots $-3 \pm i2$, $-3 \pm i2$, and $+3 \pm i2$ thus give a solution $C_0 e^{-3t} \sin (2t + \phi_0) + C_1 t e^{-3t} \sin (2t + \phi_1) + C_2 t^2 e^{-3t} \sin (2t + \phi_2)$.

The complete complementary-function solution is simply the algebraic sum of the individual parts found from the four rules. Whereas the above method for finding q_{ocf} *always* works, no universal method for finding the particular solution q_{opi} exists. This is because q_{opi} depends on the form of q_i, and we can always define a sufficiently "pathological" form of q_i to prevent solution for q_{opi}. However, if q_i is restricted to functions of prime engineering interest, a relatively simple method for finding q_{opi} is available. This is the *method of undetermined coefficients,* which is briefly reviewed. Since the method does not work for all q_i, the first question to be answered is whether it will work for the q_i of interest. For a given q_i, the right side of Eq. (3.85) is some known function of time $f(t)$. To test whether this method can be applied, we repeatedly differentiate $f(t)$ and then examine the functions created by these differentiations. There are three possibilities:

1. After a certain-order derivative, all higher derivatives are zero.
2. After a certain-order derivative, all higher derivatives have the same functional form as some lower-order derivative.
3. Upon repeated differentiation, new functional forms continue to arise.

If case 1 or 2 occurs, the method will work. If case 3 occurs, this method will not work, and others must be tried. If the method is applicable, the solution q_{opi} is immediately written as

$$q_{opi} = Af(t) + Bf'(t) + Cf''(t) + \cdots \tag{3.89}$$

where the right side includes one term for each functionally different form found by examining $f(t)$ and all its derivatives. The constants, A, B, C, etc., can be found *immediately* (they do *not* depend on the initial conditions) by substituting q_{opi} [as given in Eq. (3.89)] into Eq. (3.85) and requiring (3.85) to be an *identity.* This procedure always generates as many simultaneous algebraic equations in the unknowns A, B, C, etc., as there are unknowns; thus the equations can be solved for A, B, C, etc.

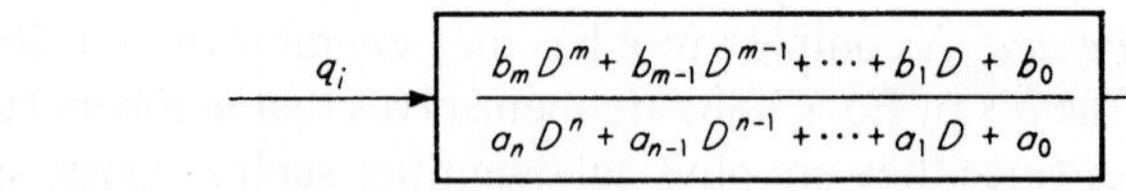

Figure 3.33
General operational transfer function.

Digital Simulation Methods for Dynamic Response Analysis

In studying the dynamic performance of measurement systems, the classical operator and Laplace-transform methods outlined or referenced earlier are capable of providing analytical solutions even for high-order differential equations. However, digital simulation becomes an increasingly attractive alternative as system complexity increases, and becomes a necessity if nonlinear and/or time-variant effects are to be examined. Since a detailed treatment of digital simulation application is not within the scope of this text, we use it only as a working tool whenever appropriate. Fortunately, the basic methods are rather easy to comprehend, so the reader with little or no previous experience should have no difficulty following the examples. We use the MATLAB/SIMULINK language, which is well documented[37] for the reader who desires greater detail.

Operational Transfer Function

In the analysis, design, and application of measurement systems, the concept of the operational transfer function is very useful. The operational transfer function relating output q_o to input q_i is defined by treating Eq. (3.86) as if it were an algebraic relation and forming the ratio of output to input:

$$\text{Operational transfer function} \triangleq \frac{q_o}{q_i}(D)$$

$$\triangleq \frac{b_m D^m + b_{m-1} D^{m-1} + \cdots + b_1 D + b_0}{a_n D^n + a_{n-1} D^{n-1} + \cdots + a_1 D + a_0} \tag{3.90}$$

In writing transfer functions, we always write $(q_o/q_i)(D)$, not just q_o/q_i, to emphasize that the transfer function is a *general* relation between q_o and q_i and very definitely *not* the instantaneous ratio of the time-varying quantities q_o and q_i.

One of the several useful features of transfer functions is their utility for graphic symbolic depiction of system dynamic characteristics by means of block diagrams. That is, if we wish to depict graphically a device with transfer function (3.90), we can draw a block diagram as in Fig. 3.33. Furthermore, the transfer function is helpful in determining the overall characteristics of a system made up of components whose individual transfer functions are known. This combination is most simply achieved when there is negligible loading[38] (the input impedance of the second

---

[37]E. O. Doebelin, "System Dynamics," 1998.

[38]E. O. Doebelin, "System Dynamics," 1998, pp. 522–527.

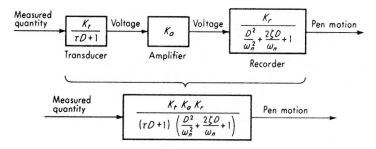

Figure 3.34
Combination of individual transfer functions.

device is much higher than the output impedance of the first, etc.) between the connected devices. For this case, the overall transfer function is simply the product of the individual ones, since the output of the preceding device becomes the input of the following one. Figure 3.34 illustrates this procedure. When significant loading *is* present, we may apply the impedance concepts of Sec. 3.2 (extended to the dynamic case) or simply analyze the complete system "from scratch" without using the individual transfer functions.

In the technical literature, the Laplace-transform method is in common use for the study of linear systems. When such methods are employed, the *Laplace transfer function* is defined as the ratio of the Laplace transform of the output quantity to the Laplace transform of the input quantity when all initial conditions are zero. Thus, analogous to Eq. (3.90), the Laplace transfer function would be written as

$$\frac{q_o(s)}{q_i(s)} \triangleq \frac{q_o}{q_i}(s) \triangleq \frac{b_m s^m + b_{m-1}s^{m-1} + \cdots + b_1 s + b_0}{a_n s^n + a_{n-1}s^{n-1} + \cdots + a_1 s + a_0} \tag{3.91}$$

where $s \triangleq \sigma + i\omega$ is the complex variable of the Laplace transform. We note that, as far as the *form* of the transfer function is concerned, we can shift from the Laplace form to the D-operator form (or vice versa) simply by interchanging s and D. Thus, if we encounter a block diagram using the Laplace notation, we can *always* convert to the D notation by a simple substitution. Then all the methods we subsequently develop may be applied to the operational transfer function.

Sinusoidal Transfer Function

In studying the quality of measurement under dynamic conditions, we analyze the response of measurement systems to certain "standard" inputs. One of the most important of such responses is the steady-state response to a sinusoidal input. Here the input q_i is of the form $A_i \sin \omega t$. If we wait for all transient effects to die out (the complementary-function solution of a stable linear system always dies out eventually), we see that the output quantity q_o will be a sine wave of exactly the same frequency (ω) as the input. However, the amplitude of the output may differ from that of the input, and a phase shift may be present. These results are easily shown by

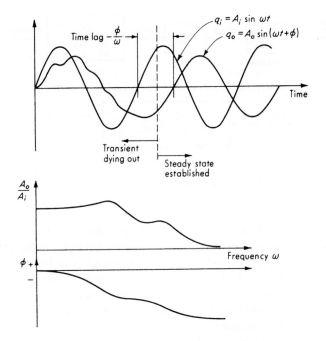

Figure 3.35
Frequency-response terminology.

obtaining the particular (steady-state) solution by the method of undetermined coefficients. Since the frequency is the same, the relation between the input and output sine waves is completely specified by giving their amplitude ratio and phase shift. Both quantities, in general, change when the driving frequency ω changes. Thus the *frequency response* of a system consists of curves of amplitude ratio and phase shift as a function of frequency. Figure 3.35 illustrates these concepts.

While the frequency response of any linear system may be obtained by getting the particular solution of its differential equation with

$$q_i = A_i \sin \omega t$$

much quicker and easier methods are available. These methods depend on the concept of the sinusoidal transfer function. The sinusoidal transfer function of a system is obtained by substituting $i\omega$ for D in the operational transfer function:

Sinusoidal transfer function $\triangleq \dfrac{q_o}{q_i}(i\omega)$

$$\triangleq \frac{b_m(i\omega)^m + b_{m-1}(i\omega)^{m-1} + \cdots + b_1 i\omega + b_0}{a_n(i\omega)^n + a_{n-1}(i\omega)^{n-1} + \cdots + a_1 i\omega + a_0} \qquad (3.92)$$

where $i \triangleq \sqrt{-1}$ and $\omega \triangleq$ frequency in radians per unit time. For any given frequency ω, Eq. (3.92) shows that $(q_o/q_i)(i\omega)$ is a complex number, which can always be put in the polar form $M \angle \phi$. It can be shown[39] that *the magnitude M of*

---

[39]E. O. Doebelin, "System Dynamics," 1998, pp. 631–633.

the complex number is the amplitude ratio A_o/A_i while the angle ϕ is the phase angle by which the output q_o leads the input q_i. (If the output *lags* the input, ϕ is negative.) The sinusoidal transfer function is of great practical importance and is widely used in many applications of system dynamics, such as vibration, acoustics, circuit analysis, control systems, and measurement systems. When we need to compute and graph the frequency response curves, most mathematical software (such as the MATLAB, which is used in this text) provides easy and rapid methods, which we will shortly be illustrating. If you prefer the Laplace transfer function to the operational, you can of course get the sinusoidal by substituting $i\omega$ for s, rather than for D.

Zero-Order Instrument

While the general mathematical model of Eq. (3.85) is adequate for handling any linear measurement system, certain special cases occur so frequently in practice that they warrant separate consideration. Furthermore, more complicated systems can be studied profitably as combinations of these simple special cases.

The simplest possible special case of Eq. (3.85) occurs when all the a's and b's other than a_0 and b_0 are assumed to be zero. The differential equation then degenerates into the simple algebraic equation

$$a_0 q_o = b_0 q_i \tag{3.93}$$

Any instrument or system that closely obeys Eq. (3.93) over its intended range of operating conditions is defined to be a *zero-order instrument.* Actually, two constants a_0 and b_0 are not necessary, and so we define the static sensitivity (or steady-state "gain") as follows:

$$q_o = \frac{b_0}{a_0} q_i = K q_i \tag{3.94}$$

$$K \triangleq \frac{b_0}{a_0} \triangleq \text{static sensitivity} \tag{3.95}$$

Since the equation $q_o = K q_i$ is algebraic, it is clear that, no matter how q_i might vary with time, the instrument output (reading) follows it *perfectly* with no distortion or time lag of any sort. *Thus, the zero-order instrument represents ideal or perfect dynamic performance* and is thus a standard against which less perfect instruments may be compared.

A practical example of a zero-order instrument is the displacement-measuring potentiometer. Here (see Fig. 3.36) a strip of resistance material is excited with a voltage and provided with a sliding contact. If the resistance is distributed linearly along length L, we may write

$$e_o = \frac{x_i}{L} E_b = K x_i \tag{3.96}$$

where $K \triangleq E_b/L$ volts per inch.

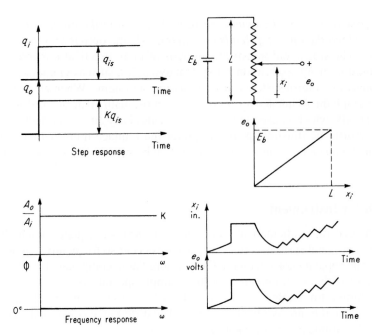

Figure 3.36
Zero-order instrument.

If you examine this measuring device more critically, you will find that it is not *exactly* a zero-order instrument. This is simply a manifestation of the *universal* rule that no mathematical model can *exactly* represent *any* physical system. In our present example, we would find that, if we wish to *use* a potentiometer for motion measurements, we must attach to the output terminals some voltage-measuring device (such as an oscilloscope). Such a device will always draw some current (however small) from the potentiometer. Thus, when x_i changes, the potentiometer winding current will also change. This in itself would cause no dynamic distortion or lag *if the potentiometer were a pure resistance.* However, the idea of a pure resistance is a *mathematical model,* not a real system; thus the potentiometer will have some (however small) inductance and capacitance. If x_i is varied relatively slowly, these parasitic inductance and capacitance effects will not be apparent. However, for sufficiently fast variation of x_i, these effects are no longer negligible and cause dynamic errors between x_i and e_o. The reasons why a potentiometer is normally called a zero-order instrument are as follows:

1. The parasitic inductance and capacitance can be made very small by design.
2. The speeds ("frequencies") of motion to be measured are not high enough to make the inductive or capacitive effects noticeable.

Another aspect of nonideal behavior in a real potentiometer comes to light when we realize that the sliding contact must be attached to the body whose motion is to be measured. Thus, there is a mechanical loading effect, due to the inertia of the sliding contact and its friction, which will cause the measured motion x_i to be different from that which would occur if the potentiometer were not present. Thus this effect is different *in kind* from the inductive and capacitive phenomena mentioned earlier, since they affected the relation [Eq. (3.96)] between e_o and x_i whereas the mechanical loading has no effect on this relation but, rather, makes x_i different from the undisturbed case.

First-Order Instrument

If in Eq. (3.85) all a's and b's other than a_1, a_0, and b_0 are taken as zero, we get

$$a_1 \frac{dq_o}{dt} + a_0 \, q_o = b_0 \, q_i \tag{3.97}$$

Any instrument that follows this equation is, by definition, a first-order instrument. There may be some conflict here between mathematical terminology and common engineering usage. In mathematics, a first-order *equation* has the general form

$$a_1 \frac{dq_o}{dt} + a_0 \, q_o = (b_m D^m + b_{m-1} D^{m-1} + \cdots + b_1 D + b_0) q_i \tag{3.98}$$

where m could have any numerical value. However, through long usage, in engineering we commonly understand a first-order *instrument* to be defined by Eq. (3.97). Since in technical presentations both words *and equations* generally are employed, confusion on this point is rarely a problem.

While Eq. (3.97) has three parameters a_1, a_0, and b_0, only two are really essential since the whole equation could always be divided through by a_1, a_0, or b_0, thus making the coefficient of one of the terms numerically equal to 1. The conventional procedure is to divide through by a_0, which gives

$$\frac{a_1}{a_0} \frac{dq_o}{dt} + q_o = \frac{b_0}{a_0} q_i \tag{3.99}$$

which becomes

$$(\tau D + 1) q_o = K q_i \tag{3.100}$$

when we define

$$K \triangleq \frac{b_0}{a_0} \triangleq \text{static sensitivity} \tag{3.101}$$

$$\tau \triangleq \frac{a_1}{a_0} \triangleq \text{time constant} \tag{3.102}$$

The time constant τ always has the dimensions of time, while the static sensitivity K has the dimensions of output divided by input. For *any*-order instrument, K always is defined as b_0/a_0 and always has the same physical meaning, that is, the amount of output per unit input when the input is static (constant), because under such conditions all the derivative terms in the differential equation are zero. The operational transfer function of any first-order instrument is

$$\frac{q_o}{q_i}(D) = \frac{K}{\tau D + 1} \tag{3.103}$$

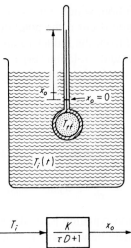

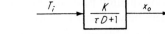

Figure 3.37
First-order instrument.

    As an example of a first-order instrument, let us consider the liquid-in-glass thermometer of Fig. 3.37. The input (measured) quantity here is the temperature $T_i(t)$ of the fluid surrounding the bulb of the thermometer, and the output is the displacement x_o of the thermometer fluid in the capillary tube. We assume the temperature $T_i(t)$ is uniform throughout the fluid at any given time, but may vary with time in an arbitrary fashion. The principle of operation of such a thermometer is the thermal expansion of the filling fluid which drives the liquid column up or down in response to temperature changes. Since this liquid column has inertia, mechanical lags will be involved in moving the liquid from one level to another. However, we assume that this lag is negligible compared with the thermal lag involved in transferring heat from the surrounding fluid through the bulb wall and into the thermometer fluid. This assumption rests (as all such assumptions necessarily must) on experience, judgement, order-of-magnitude calculations, and, ultimately, experimental verification (or refutation) of the results predicted by the analysis. Assumption of negligible mechanical lag allows us to relate the temperature of the fluid in the bulb to the reading x_o by the instantaneous (algebraic) equation

$$x_o = \frac{K_{ex}V_b}{A_c} T_{tf} \tag{3.104}$$

where $x_o \triangleq$ displacement from reference mark, m

        $T_{tf} \triangleq$ temperature of fluid in bulb (assumed uniform throughout bulb volume), $T_{tf} = 0$ when $x_o = 0$, °C

        $K_{ex} \triangleq$ differential expansion coefficient of thermometer fluid and bulb glass, $m^3/(m^3 \cdot C°)$

        $V_b \triangleq$ volume of bulb, m^3

        $A_c \triangleq$ cross-sectional area of capillary tube, m^2

To get a differential equation relating input and output in this thermometer, we consider conservation of energy over an infinitesimal time dt for the thermometer bulb:

$$\text{Heat in} - \text{heat out} = \text{energy stored}$$

$$U A_b(T_i - T_{tf}) \, dt - 0(\text{assume no heat loss}) = V_b \, \rho C \, dT_{tf} \qquad \textbf{(3.105)}$$

where $U \triangleq$ overall heat-transfer coefficient across bulb wall, W/(m² · °C)
$A_b \triangleq$ heat-transfer area of bulb wall, m²
$\rho \triangleq$ mass density of thermometer fluid, kg/m³
$C \triangleq$ specific heat of thermometer fluid, J/(kg · °C)

Equation (3.105) involves many assumptions:

1. The bulb wall and fluid films on each side are pure resistance to heat transfer with no heat-storage capacity. This will be a good assumption if the heat-storage capacity (mass) (specific heat) of the bulb wall and fluid films is small *compared* with $C\rho V_b$ for the bulb.

2. The overall coefficient U is constant. Actually, film coefficients and bulb-wall conductivity all change with temperature, but these changes are quite small as long as the temperature does not vary over wide ranges.

3. The heat-transfer area A_b is constant. Actually, expansion and contraction would cause this to vary, but this effect should be quite small.

4. No heat is lost from the thermometer bulb by conduction up the stem. Heat loss will be small if the stem is of small diameter, made of a poor conductor, and immersed in the fluid over a great length and if the exposed end is subjected to an air temperature not much different from T_i and T_{tf}.

5. The mass of fluid in the bulb is constant. Actually mass must enter or leave the bulb whenever the level in the capillary tube changes. For a fine capillary and a large bulb, this effect should be small.

6. The specific heat C is constant. Again, this fluid property varies with temperature, but the variation is slight except for large temperature changes.

The above list of assumptions is not complete, but should give some appreciation of the discrepancies between a mathematical model and the real system it represents. Many of these assumptions could be relaxed to get a more accurate model, but we would pay a heavy price in increased mathematical complexity. The choice of assumptions that are *just good enough* for the needs of the job at hand is one of the most difficult and important tasks of the engineer.

Returning to Eq. (3.105), we may write it as

$$V_b \, \rho C \frac{dT_{tf}}{dt} + U A_b T_{tf} = U A_b T_i \qquad \textbf{(3.106)}$$

Using Eq. (3.104), we get

$$\frac{\rho C A_c}{K_{ex}} \frac{dx_o}{dt} + \frac{U A_b A_c}{K_{ex} V_b} x_o = U A_b T_i \qquad \textbf{(3.107)}$$

which we recognize to be the form of Eq. (3.99), and so we immediately define

$$K \triangleq \frac{A_{ex}V_b}{A_c} \qquad \text{m/C}° \tag{3.108}$$

$$\tau \triangleq \frac{\rho C V_b}{U A_b} \qquad \text{s} \tag{3.109}$$

Having shown a concrete example of a first-order instrument, let us return to the problem of examining the dynamic response of first-order instruments in general. Once you have obtained the differential equation relating the input and output of an instrument, you can study its dynamic performance by taking the input (quantity to be measured) to be some known function of time and then solving the differential equation for the output as a function of time. If the output is closely proportional to the input at all times, the dynamic accuracy is good. The fundamental difficulty in this approach lies in the fact that, in actual practice, the quantities to be measured usually do not follow some simple mathematical function, but rather are of a random nature. Fortunately, however, much can be learned about instrument performance by examining the response to certain, rather simple "standard" input functions. That is, just as you are able to analyze not the real *system,* but rather an idealized model of it, so also you can work not with the real *inputs* to a system, but rather with simplified representations of them. This simplification of inputs (just as that of systems) can be carried out at several different levels, which leads to either simple, rather unrealistic input functions that are readily handled mathematically or complex, more realistic representations that lead to mathematical difficulties.

We commence our study by considering several quite simple standard inputs that are in wide use. Although these inputs are, in general, only crude approximations to the actual inputs, they are extremely useful for studying the effects of parameter changes in a given instrument or for comparing the *relative* performance of two competitive measurement systems.

Step Response of First-Order Instruments

To apply a step input to a system, we assume that initially it is in equilibrium, with $q_i = q_o = 0$, when at time $t = 0$ the input quantity increases instantly by an amount q_{is} (see Fig. 3.38). For $t > 0$, Eq. (3.100) becomes

$$(\tau D + 1)q_o = Kq_{is} \tag{3.110}$$

It can be shown generally (by mathematical reasoning) or in any specific physical problem, such as the thermometer (by physical reasoning), that the initial condition for this situation is $q_o = 0$ for $t = 0^+$ ($t = 0^+$ means an infinitesimal time after $t = 0$). The complementary-function solution is

$$q_{ocf} = Ce^{-t/\tau} \tag{3.111}$$

while the particular solution is

$$q_{opi} = Kq_{is} \tag{3.112}$$

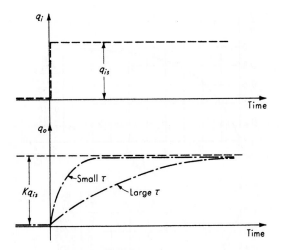

Figure 3.38
Step-function response of first-order instrument.

giving the complete solution as

$$q_o = Ce^{-t/\tau} + Kq_{is}$$ (3.113)

Applying the initial condition, we get

$$0 = C + Kq_{is}$$
$$C = -Kq_{is}$$

which gives finally

$$q_o = Kq_{is}(1 - e^{-t/\tau})$$ (3.114)

Examination of Eq. (3.114) shows that the speed of response depends on *only* the value of τ and is faster if τ is smaller. Thus in first-order instruments we strive to minimize τ for faithful dynamic measurements.

These results may be nondimensionalized by writing

$$\frac{q_o}{Kq_{is}} = 1 - e^{-t/\tau}$$ (3.115)

and then plotting $q_o/(Kq_{is})$ versus t/τ, as in Fig. 3.39a. This curve is then universal for any value of K, q_{is}, or τ that might be encountered. We could also define the measurement error e_m as

$$e_m \triangleq q_i - \frac{q_o}{K}$$ (3.116)

$$e_m = q_{is} - q_{is}(1 - e^{-t/\tau})$$

and nondimensionalize for plotting in Fig. 3.39b as

$$\frac{e_m}{q_{is}} = e^{-t/\tau}$$ (3.117)

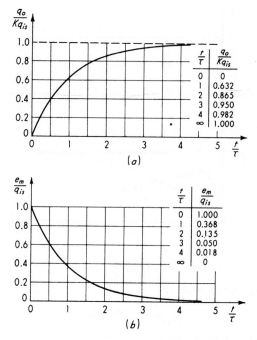

(a)

(b)

Figure 3.39
Nondimensional step-function response of first-order instrument.

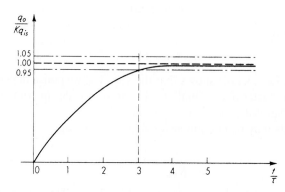

Figure 3.40
Settling-time definition.

A dynamic characteristic useful in characterizing the speed of response of any instrument is the *settling time*. This is the time (after application of a step input) for the instrument to reach and stay within a stated plus-and-minus tolerance band around its final value. A small settling time thus indicates fast response. It is obviously that the numerical value of a settling time depends on the percentage tolerance band used; you must always state this. Thus you speak of, say, a 5 percent settling time. For a first-order instrument a 5 percent settling time is equal to three time constants (see Fig. 3.40). Other percentages may be and are used when appropriate.

Knowing now that fast response requires a small value of τ, we can examine any specific first-order instrument to see what physical changes would be needed to reduce τ. If we use our thermometer example, Eq. (3.109) shows that τ may be reduced by

1. Reducing ρ, C, and V_b
2. Increasing U and A_b

Since ρ and C are properties of the fluid filling the thermometer, they cannot be varied independently of each other, and so for small τ we search for fluids with a small ρC product. The bulb volume V_b may be reduced, but this will also reduce A_b unless some extended-surface heat-transfer augmentation (such as fins on the bulb) is introduced. Even more significant is the effect of reduced V_b on the static sensitivity K, as given by Eq. (3.108). We see that attempts to reduce τ by decreasing V_b will result in reductions in K. Thus increased speed of response is traded off for lower sensitivity. This trade-off is not unusual and will be observed in many other instruments.

The fact that τ depends on U means that we cannot state that a certain *thermometer* has a certain time constant, but only that a specific thermometer *used in a certain fluid under certain heat-transfer conditions* (say, free or forced convection) has a certain time constant. This is because U depends partly on the value of the film coefficient of heat transfer at the outside of the bulb, which varies greatly with changes in fluid (liquid or gas), flow velocity, etc. For example, a thermometer in stirred oil might have a time constant of 5 s while the same thermometer in stagnant air would have a τ of perhaps 100 s. Thus you must always be careful in giving (or using) performance data to be sure that the conditions of use correspond to those in force during calibration or that proper corrections are applied.

To illustrate the nature of nonlinear instrument responses, linearization techniques available for approximate analysis, and the utility of the digital simulation methods mentioned above, consider the vacuum furnace of Fig. 3.41a. The wall temperature T_i is steady at 400 K when a thermometer at 300 K is suddenly inserted, subjecting it to a 100 K step change. Because of the vacuum environment, heat transfer from the furnace to the thermometer is assumed to be strictly by radiation, and Eq. (3.105) assumes the form

$$MC\, dT_o = E(T_i^4 - T_o^4)dt \qquad \frac{dT_o}{dt} = \frac{E}{MC}(T_i^4 - T_o^4) = 10^{-8}(T_i^4 - T_o^4) \quad \textbf{(3.118)}$$

where some typical numerical values have been inserted for M, C, and E. While nonlinear Eq. (3.118) presents difficulties for analytical solution, digital simulation obtains a near-perfect numerical solution with very little effort, and approximate analytical linearizations are also available. The most common linearizing approximation is the Taylor-series method.[40] If Eq. (3.118) is considered as an isolated model relating T_o to T_i, then T_i plays the role of a given input (rather than an unknown) and

---

[40]E. O. Doebelin, "System Dynamics," 1998, pp. 43–47.

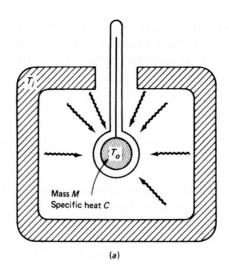

(a)

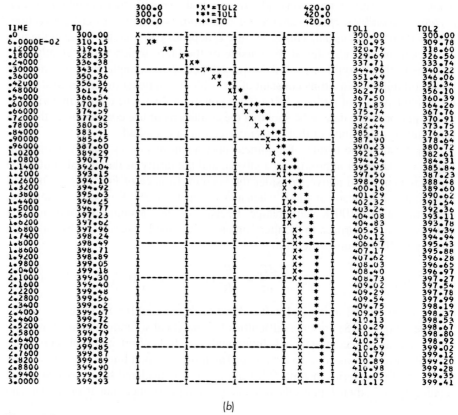

(b)

Figure 3.41

Digital simulation of nonlinear instrument.

the term T_i^4 need not be linearized to allow analytical solution. The Taylor-series approach then gives

$$T_o^4 \approx T_{o0}^4 - 4T_{o0}^3(T_o - T_{o0}) = -4.50 \times 10^{10} + (1.715 \times 10^8)T_o \quad \textbf{(3.119)}$$

where the operating point T_{o0} has been chosen as 350 K, midway between 300 and 400 K. The linearized approximate version of Eq. (3.118) is then

$$0.583\frac{dT_o}{dt} + T_o = 411.8 \qquad \textbf{(3.120)}$$

This can easily be solved analytically and clearly has a time constant $\tau = 0.583$ s.

If the system under study included not only the thermometer response model but also, say, a temperature control system for the furnace, then T_i would now play the role of an *unknown* and T_i^4 would also need to be linearized:

$$\frac{dT_o}{dt} = 10^{-8}[T_{i0}^4 + 4T_{i0}^3(T_i - T_{i0}) - T_{o0}^4 - 4T_{o0}^3(T_o - T_{o0})] \qquad \textbf{(3.121)}$$

$$0.583\frac{dT_o}{dt} + T_o = T_i = 400 \qquad \textbf{(3.122)}$$

A final linearizing scheme, which does not employ Taylor series, expands

$$T_i^4 - T_o^4 = (T_i^2 + T_o^2)(T_i + T_o)(T_i - T_o) \qquad \textbf{(3.123)}$$

and then takes $(T_i^2 + T_o^2)(T_i + T_o)$ constant at the operating point (350 K) values to give

$$0.583\frac{dT_o}{dt} + T_o = T_i = 400$$

which is the same result as Eq. (3.122).

Of the three different models just explained, the two linear ones could easily be solved analytically, but the nonlinear one could not; it requires digital simulation. Since we want to compare the results from the three models, we will solve all three with digital simulation, in one run, and graph the three temperature readings superimposed on a single sheet. In the 1960s and 1970s, electronic analog computers were widely used for simulation studies, but digital computers with special simulation software gradually took over this task. Originally, digital simulation software used a command line approach, where statements expressing the equations and operations had to be written out in some engineering language such as FORTRAN, augmented by a set of special simulation statements that actuated operations such as numerical integration with a single command. Today, this approach is still available but has been expanded to include a graphical method that uses block diagrams, very similar to the old analog computer diagrams. In these graphical user interfaces, one simply connects functional blocks selected from a menu into a simulation diagram. "Behind the scenes," the software actually writes the equations which the block diagram represents and then integrates them to get the desired solutions; the software never actually displays the set of simultaneous equations being solved. While the older command line approach might still be preferred for very complex simulations, the relatively simple ones used in this text are most conveniently implemented using the graphical approach. In this first example, we will show *both* methods so that you can easily compare them, but for subsequent applications, we will usually show only the graphical technique.

For the thermometer example, we will use the command-line language CSMP and the graphical software SIMULINK. Although CSMP is no longer supported by a software company, we prefer it for this illustration since it is easy for one slightly familiar with FORTRAN to understand and it is very similar to current command-line languages such as ACSL. If one were today purchasing software, languages such as ACSL or MATLAB/SIMULINK, which provide both command-line and graphical programming, would be considered. A CSMP program for our three-thermometer models would go as follows:

```
PARAM K = 1.E - 08,TI = 400., TOI = 300.      gives
                                              parameter
                                              values

TODOT = K*(TI**4 - TO**4)       computes highest derivative
                                of unknown

TO = INTGRL(TOI,TODOT)       numerically integrates TODOT to get TO,
                             starting from initial value TOI

TODTL1 = 706.2 - 1.715*TOL1      same as above for the
TOL1 = INTGRL(TOI,TODTL1)        TO linearized equation

TODTL2 = 686. - 1.715*TOL2       same as above for the
TOL2 = INTGRL(TOI,TODTL2)        TI and TO linearized equation

TIMER FINTIM = 3.0,DELT = .001,OUTDEL = .06      gives
                                                 finish time,
                                                 computing
                                                 increment,
                                                 and plotting
                                                 increment

OUTPUT TO,TOL1,TOL2      requests graphs of TO, TOL1, TOL2 versus time

PAGE GROUP,WIDTH = 50       requests all three curves to same scale,
                            50 characters wide
END
```

Because CSMP (and other similar languages) have many "built-in" convenience features, we are able to solve the above linear and nonlinear equations with very little time and effort. Most of the statements are explained above. In the TIMER statement, we chose FINTIM = 3.0 s since our linearized models have $\tau = 0.583$ s and a first-order step response is nearly complete in 5τ. Computing increment DELT usually is taken near FINTIM/2000, but its value is not critical since

CSMP self-adjusts to optimize accuracy and speed. Plotting increment OUTDEL generally is taken as FINTIM/50 to just fill the printer page unless this obscures details, in which case the curves are spread over several pages by using a smaller OUTDEL.

Figure 3.41b shows that the two approximate linearizations which follow Eq. (3.122) give very good results, while that of Eq. (3.120) suffers from an incorrect final value. It appears, then, that a linearized first-order model with $\tau = 0.583$ s would be acceptable for many purposes under the given conditions. Changing conditions (larger step inputs, input forms other than steps, etc.) may decrease the accuracy of the linearized model; however, all such situations can be easily studied by using simulation and appropriate decisions can be made.

Figure 3.42a shows the SIMULINK block diagram that generates the same solutions as our CSMP program. In the blocks called **Fcn**, the symbol **u** stands for *whatever* variable is connected to the block's input. The integrator blocks (transfer function 1/s) use Laplace-transform symbology; we could substitute 1/D if we preferred the operator notation. The **To Workspace** blocks are used to select and name variables that we wish to tabulate and/or graph. Figure 3.42b shows the graphs produced, which of course agree with the CSMP results.

Ramp Response of First-Order Instruments

To apply a ramp input to a system, we assume that initially the system is in equilibrium, with $q_i = q_o = 0$, when at $t = 0$ the input q_i suddenly starts to change at a constant rate $\dot{q}_{is}$. We thus have

$$q_i = \begin{cases} q_0 = 0 & t \le 0 \\ \dot{q}_{is}t & t \ge 0 \end{cases} \qquad (3.124)$$

and therefore $\qquad (\tau D + 1)q_o = K\dot{q}_{is}\, t$

The necessary initial condition again can be shown to be $q_o = 0$ for $t = 0^+$. Solution of Eq. (3.124) gives

$$q_{ocf} = Ce^{-t/\tau}$$
$$q_{opi} = K\dot{q}_{is}(t - \tau)$$
$$q_o = Ce^{-t/\tau} + K\dot{q}_{is}(t - \tau)$$

and applying the initial condition gives

$$q_o = K\dot{q}_{is}(\tau e^{-t/\tau} + t - \tau) \qquad (3.125)$$

We again define measurement error e_m by

$$e_m \triangleq q_i - \frac{q_o}{K} = \dot{q}_{is}t - \dot{q}_{is}\tau e^{-t/\tau} - \dot{q}_{is}t + \dot{q}_{is}\tau \qquad (3.126)$$

$$e_m = \underbrace{- \dot{q}_{is}\tau e^{-t/\tau}}_{\substack{\text{transient error} \\ e_{m,\,t}}} + \underbrace{\dot{q}_{is}\tau}_{\substack{\text{steady-} \\ \text{state} \\ \text{error} \\ e_{m,\,ss}}} \qquad (3.127)$$

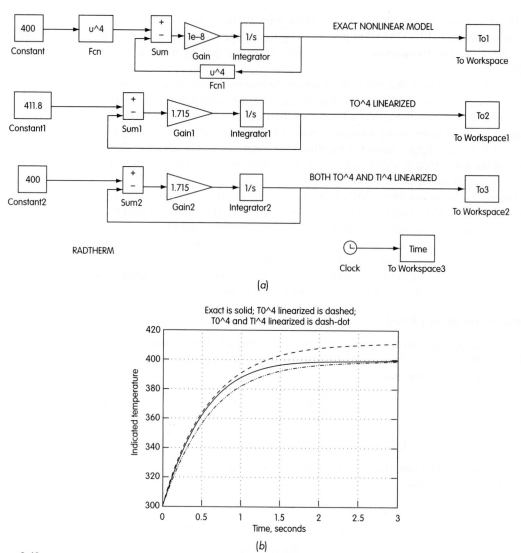

Figure 3.42

SIMULINK block diagram and results for nonlinear thermometer model.

We note that the first term of e_m gradually will disappear as time goes by, and so it is called the *transient error*. The second term, however, persists forever and is thus called the *steady-state error*. The transient error disappears more quickly if τ is small. The steady-state error is directly proportional to τ; thus small τ is desirable here also. Steady-state error also increases directly with $\dot{q}_{is}$, the rate of change of the measured quantity. In steady state, the horizontal (time) displacement between input and output curves is seen to be τ, and so we may make the interpretation that the instrument is reading what the input *was* τ seconds ago. The above results together with a nondimensionalized representation, are given graphically in Fig. 3.43.

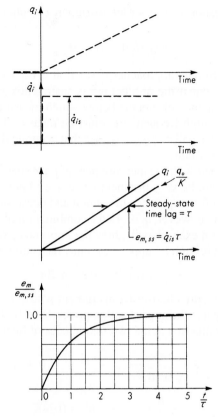

Figure 3.43
Ramp response of first-order instrument.

Frequency Response of First-Order Instruments

Equation (3.92) may be applied directly to the problem of finding the response of first-order systems to sinusoidal inputs. We have

$$\frac{q_o}{q_i}(i\omega) = \frac{K}{i\omega\tau + 1} = \frac{K}{\sqrt{\omega^2\tau^2 + 1}} \angle [\tan^{-1}(-\omega\tau)] \qquad (3.128)$$

Thus the amplitude ratio is

$$\frac{A_o}{A_i} = \left|\frac{q_o}{q_i}(i\omega)\right| = \frac{K}{\sqrt{\omega^2\tau^2 + 1}} \qquad (3.129)$$

and the phase angle is

$$\phi = \angle \frac{q_o}{q_i}(i\omega) = \tan^{-1}(-\omega\tau) \qquad (3.130)$$

The ideal frequency response (zero-order instrument) would have

$$\frac{q_o}{q_i}(i\omega) = K \angle 0° \qquad (3.131)$$

Thus a first-order instrument approaches perfection if Eq. (3.128) approaches Eq. (3.131). We see this occur if the product $\omega\tau$ is sufficiently small. Thus for *any* τ there will be some frequency of input ω below which measurement is accurate; or, alternatively, if a q_i of high frequency ω must be measured, the instrument used must have a sufficiently small τ. Again, we see that accurate dynamic measurement requires a small time constant.

If we were concerned with the measurement of *pure* sine waves only, the above considerations would not be very pertinent since if we knew the frequency and τ, we could easily correct for amplitude attenuation and phase shift by simple calculations. In actual practice, however, q_i is often a combination of several sine waves of different frequencies. An example will show the importance of adequate frequency response under such conditions. Suppose we must measure a q_i given by

$$q_i = 1 \sin 2t + 0.3 \sin 20t \qquad (3.132)$$

(where t is in seconds) with a first-order instrument whose τ is 0.2 s. Since this is a linear system, we may use the superposition principle to find q_o. We first evaluate the sinusoidal transfer function at the two frequencies of interest:

$$\left.\frac{q_o}{q_i}(i\omega)\right|_{\omega=2} = \frac{K}{\sqrt{0.16+1}} \angle -21.8° = 0.93K \angle -21.8° \qquad (3.133)$$

$$\left.\frac{q_o}{q_i}(i\omega)\right|_{\omega=20} = \frac{K}{\sqrt{16+1}} \angle -76° = 0.24K \angle -76° \qquad (3.134)$$

We can then write q_o as

$$q_o = (1)(0.93K) \sin (2t - 21.8°) + (0.3)(0.24K) \sin (20t - 76°) \qquad (3.135)$$

$$\frac{q_o}{K} = 0.93 \sin (2t - 21.8°) + 0.072 \sin (20t - 76°) \qquad (3.136)$$

Since ideally $q_o/K = q_i$, comparison of Eq. (3.136) with (3.132) shows the presence of considerable measurement error. A graph of these two equations in Fig. 3.44b shows that the instrument gives a severely distorted measurement of the input. Furthermore, the high-frequency (20 rad/s) component present in the instrument output is now so small relative to the low-frequency component that any attempts at correction are not only inconvenient, but also inaccurate.

Suppose we consider use of an instrument with $\tau = 0.002$ s. Then we have

$$\left.\frac{q_o}{q_i}(i\omega)\right|_{\omega=2} = \frac{K}{\sqrt{1.6 \times 10^{-5}+1}} \angle -0.23° = 1.00K \angle -0.23° \qquad (3.137)$$

$$\left.\frac{q_o}{q_i}(i\omega)\right|_{\omega=20} = \frac{K}{\sqrt{1.6 \times 10^{-3}+1}} \angle -2.3° = 1.00K \angle -2.3° \qquad (3.138)$$

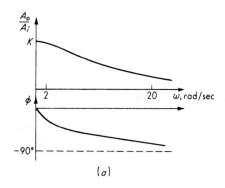

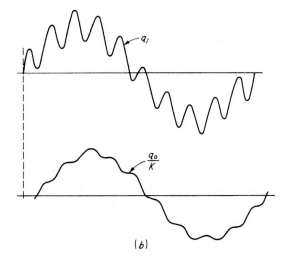

Figure 3.44
Example of inadequate frequency response.

which yields

$$\frac{q_o}{K} = 1.00 \sin (2t - 0.23°) + 0.3 \sin (20t - 2.3°) \qquad \textbf{(3.139)}$$

Comparison of Eq. (3.132) and (3.139) shows clearly that this instrument faithfully
measures the given q_i.

A nondimensional representation of the frequency response of any first-order
system may be obtained by writing Eq. (3.128) as

$$\frac{q_o/K}{q_i} (i\omega) = \frac{1}{\sqrt{(\omega\tau)^2 + 1}} \angle [\tan^{-1} (-\omega\tau)] \qquad \textbf{(3.140)}$$

and plotting as in Fig. 3.45.

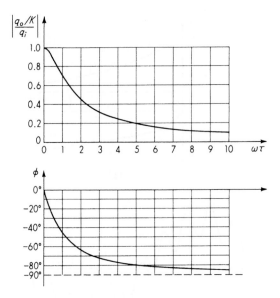

Figure 3.45
Frequency response of first-order instrument.

While the frequency response calculations just completed are not difficult, simulation methods may sometimes be preferred. Figure 3.46a shows a SIMULINK diagram useful in the present problem. SIMULINK uses the Laplace transfer function notation, so D's are replaced by s's. Icons for many dynamic system operations are available from menus and are "dragged" onto a blank worksheet to construct the simulation diagram for any system. Interconnecting "wires" are quickly "connected" with the mouse. Variables to be recorded are sent to **To Workspace** icons that gather the time-varying values for later MATLAB plotting. Our simple example uses two sine wave icons (amplitude, frequency, and phase angle are selectable), two **Transfer fcn** icons (we enter coefficients for the numerator and denominator polynomials), a **Sum** icon to add various inputs, and a **Clock** icon (always needed) to record time.

Running this simulation and plotting results yields Fig. 3.46b. A nice feature here is that the simulation obtains the *complete* solution of the differential equation, thus showing both the starting transient and also the sinusoidal steady state. Our hand calculations gave only the sinusoidal steady state. The accurate plot shows the fast first-order system to be almost perfect; its plotted curve lies so close to the perfect response that the two curves plot as one. (We could of course "zoom" the graphs to show the small error, if we wished, or make a separate plot of the error itself.) While you may not be a SIMULINK user, you should be able to follow the diagrams and comprehend the results. We will be using simulation widely as we progress through the book. It is particularly helpful for studying nonlinearities since analytical solutions there are usually not possible.

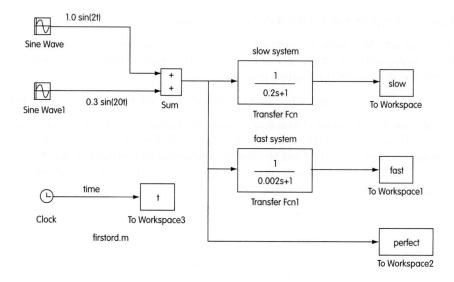

(a)

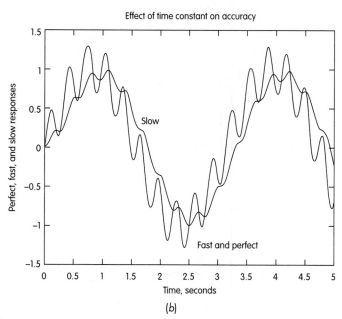

(b)

Figure 3.46

Effect of first-order system time constant on measurement accuracy.

Impulse Response of First-Order Instruments

The final standard input we consider is the *impulse function.* Consider the pulse function $p(t)$ defined graphically in Fig. 3.47a. The impulse function of "strength" (area) A is defined by the limiting process

$$\text{Impulse function of strength } A \triangleq \lim_{T \to 0} p(t) \tag{3.141}$$

We see that this "function" has rather peculiar properties. Its time duration is infinitesimal, its peak is infinitely high, and its area is A. If A is taken as 1, it is called the *unit* impulse function $\delta(t)$. Thus an impulse function of any strength A may be written as $A\delta(t)$. This rather peculiar function plays an important role in system dynamic analysis, as we see in greater detail later.

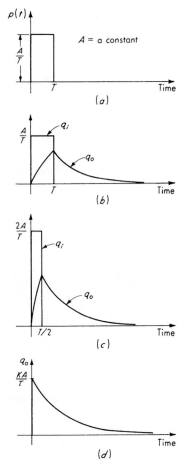

Figure 3.47
Impulse response of first-order instrument.

We now find the response of a first-order instrument to an impulse input. We do this by finding the response to the pulse $p(t)$ and then applying the limiting process to the result. For $0 < t < T$ we have

$$(\tau D + 1)q_o = Kq_i = \frac{KA}{T} \tag{3.142}$$

Since, up until time T, this is no different from a *step* input of size A/T, our initial condition is $q_o = 0$ at $t = 0^+$, and the complete solution is

$$q_o = \frac{KA}{T}(1 - e^{-t/\tau}) \tag{3.143}$$

However, this solution is valid only up to time T. At this time we have

$$q_o\bigg|_{t=T} = \frac{KA}{T}(1 - e^{-T/\tau}) \tag{3.144}$$

Now for $t > T$, our differential equation is

$$(\tau D + 1)q_o = Kq_i = 0 \tag{3.145}$$

which gives
$$q_o = Ce^{-t/\tau} \tag{3.146}$$

The constant C is found by imposing initial condition (3.144),

$$\frac{KA}{T}(1 - e^{-T/\tau}) = Ce^{-T/\tau} \tag{3.147}$$

$$C = \frac{KA(1 - e^{-T/\tau})}{Te^{-T/\tau}} \tag{3.148}$$

giving finally
$$q_o = \frac{KA(1 - e^{-T/\tau})e^{-t/\tau}}{Te^{-T/\tau}} \tag{3.149}$$

Figure 3.47b shows a typical response, and Fig. 3.47c shows the effect of cutting T in half. As T is made shorter and shorter, the first part ($t < T$) of the response becomes of negligible consequence, so that we can get an expression for q_o by taking the limit of Eq. (3.149) as $T \rightarrow 0$.

$$\lim_{T \to 0}\left[\frac{KA(1 - e^{-T/\tau})}{Te^{-T/\tau}}\right]e^{-t/\tau} = KAe^{-t/\tau}\lim_{T \to 0}\frac{1 - e^{-T/\tau}}{Te^{-T/\tau}} \tag{3.150}$$

$$\lim_{T \to 0}\frac{1 - e^{-T/\tau}}{T} = \frac{0}{0} \qquad \text{an indeterminate form}$$

Applying L'Hospital's rule yields

$$\lim_{T \to 0}\frac{1 - e^{-T/\tau}}{T} = \lim_{T \to 0}\frac{(1/\tau)e^{-T/\tau}}{1} = \frac{1}{\tau} \tag{3.151}$$

Thus we have finally for the impulse response of a first-order instrument

$$q_o = \frac{KA}{\tau} e^{-t/\tau} \tag{3.152}$$

which is plotted in Fig. 3.47*d*.

We note that the output q_o is also "peculiar" in that it has an infinite (vertical) slope at $t = 0$ and thus goes from zero to a finite value in infinitesimal time. Such behavior is clearly impossible for a physical system since it requires energy transfer at an infinite rate. In our thermometer example, for instance, to cause the temperature of the fluid in the bulb *suddenly* to rise a finite amount requires an infinite rate of heat transfer. Mathematically, this infinite rate of heat transfer is provided by having the input $T_i(t)$ be infinite, i.e., an impulse function. In actuality, of course, T_i cannot go to infinity; however, if it is large enough and of sufficiently short duration (relative to the response speed of the system), the system may respond very nearly as it would for a perfect impulse.

To illustrate this, suppose in Fig. 3.47*a* we take $A = 1$ and $T = 0.01\tau$. The response to this approximate unit impulse is

$$q_o = \begin{cases} \dfrac{100K}{\tau}(1 - e^{-t/\tau}) & 0 \le t \le T \tag{3.153} \\[2mm] \dfrac{100K(1 - e^{-0.01})e^{-t/\tau}}{\tau e^{-0.01}} & T \le t \le \infty \tag{3.154} \end{cases}$$

Figure 3.48 gives a tabular and graphical comparison of the exact and approximate response, showing excellent agreement. The agreement is quite acceptable in most

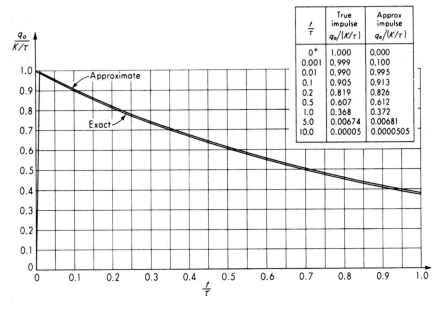

$\dfrac{t}{\tau}$	True impulse $q_o/(K/\tau)$	Approx impulse $q_o/(K/\tau)$
0^+	1.000	0.000
0.001	0.999	0.100
0.01	0.990	0.995
0.1	0.905	0.913
0.2	0.819	0.826
0.5	0.607	0.612
1.0	0.368	0.372
5.0	0.00674	0.00681
10.0	0.00005	0.0000505

Figure 3.48
Exact and approximate impulse response.

cases if T/τ is even as large as 0.1. It can also be shown that *the shape of the pulse is immaterial;* as long as its duration is sufficiently short, only its *area* matters. The plausibility of this statement may be shown by integrating the terms in the differential equation as follows:

$$\tau \frac{dq_o}{dt} + q_o = Kq_i \tag{3.155}$$

$$\int_0^{0+} \tau \, dq_0 + \int_0^{0+} q_o \, dt = \int_0^{0+} Kq_i \, dt \tag{3.156}$$

$$\tau(q_o\big|_{0+} - q_o\big|_0) + 0 = K \,(\text{area under } q_i \text{ curve from } t = 0 \text{ to } t = 0^+) \tag{3.157}$$

$$q_o\big|_{0+} = \frac{K}{\tau} \,(\text{area of impulse}) \tag{3.158}$$

This analysis holds strictly for an exact impulse and is a good approximation for a pulse of arbitrary shape if its duration is sufficiently short. It should be noted that, since the right side of the differential equation is zero for $t > 0^+$, an impulse (or a short pulse) is equivalent to a zero forcing function and a nonzero initial ($t = 0^+$) condition. That is, the solution of

$$(\tau D + 1)q_o = 0$$

$$q_o = \frac{K}{\tau} \quad \text{at } t = 0^+ \tag{3.159}$$

is exactly the same as the unit impulse response.

Another interesting aspect of the impulse function is its relation to the step function. Since a perfect step function is also physically unrealizable because it changes from one level to another in infinitesimal time, consider an approximation such as in Fig. 3.49. If this approximate step function is fed into a differentiating device, the output will be a pulse-type function. As the approximate step function is made to approach the mathematical ideal more and more closely, the output of the differentiating device will approach a perfect impulse function. *In this sense, the impulse function may be thought of as the derivative of the step function,* even though the discontinuities in the step function preclude the rigorous application of the basic definition of the derivative. In Fig. 3.44, the truth of these assertions is demonstrated by passing the output of the differentiating device through an integrating device ($1/D$).

Second-Order Instrument

A *second-order instrument* is one that follows the equation

$$a_2 \frac{d^2 q_o}{dt^2} + a_1 \frac{dq_o}{dt} + a_0 q_o = b_0 q_i \tag{3.160}$$

Again, a second-order *equation* could have more terms on the right-hand side, but in common engineering usage, Eq. (3.160) is generally accepted as defining a second-order *instrument*.

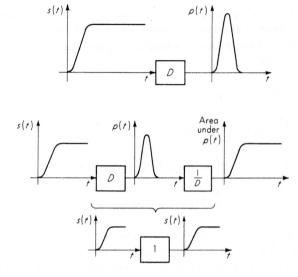

Figure 3.49
Approximate step and impulse functions.

The essential parameters in Eq. (3.160) can be reduced to three:

$$K \triangleq \frac{b_0}{a_0} \triangleq \text{static sensitivity} \tag{3.161}$$

$$\omega_n \triangleq \sqrt{\frac{a_0}{a_2}} \triangleq \text{undamped natural frequency, rad/time} \tag{3.162}$$

$$\zeta \triangleq \frac{a_1}{2\sqrt{a_0 a_2}} \triangleq \text{damping ratio, dimensionless} \tag{3.163}$$

which gives
$$\left(\frac{D^2}{\omega_n^2} + \frac{2\zeta D}{\omega_n} + 1 \right) q_o = K q_i \tag{3.164}$$

The operational transfer function is thus

$$\frac{q_o}{q_i}(D) = \frac{K}{D^2/\omega_n^2 + 2\zeta D/\omega_n + 1} \tag{3.165}$$

A good example of a second-order instrument is the force-measuring spring scale of Fig. 3.50. We assume the applied force f_i has frequency components only well below the natural frequency of the spring itself. Then the main dynamic effect of the spring may be taken into account by adding one-third of the spring's mass to the main moving mass. This total mass we call M. The spring is assumed linear with spring constant K_s newtons per meter. Although in a real scale there might be considerable dry friction, we assume perfect film lubrication and therefore a viscous damping effect with constant B (in newtons per meter per second).

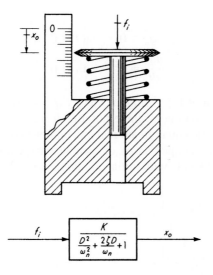

Figure 3.50
Second-order instrument.

The scale can be adjusted so that $x_o = 0$ when $f_i = 0$ (gravity force will then drop out of the equation), which yields

$$\Sigma \text{ forces} = (\text{mass})(\text{acceleration})$$

$$f_i - B\frac{dx_o}{dt} - K_s x_o = M\frac{d^2 x_o}{dt^2} \tag{3.166}$$

$$(MD^2 + BD + K_s)x_o = f_i \tag{3.167}$$

Noting this to fit the second-order model, we immediately define

$$K \triangleq \frac{1}{K_s} \qquad \text{m/N} \tag{3.168}$$

$$\omega_n \triangleq \sqrt{\frac{K_s}{M}} \qquad \text{rad/s} \tag{3.169}$$

$$\zeta \triangleq \frac{B}{2\sqrt{K_s M}} \tag{3.170}$$

Step Response of Second-Order Instruments

For a step input of size q_{is} we get

$$\left(\frac{D^2}{\omega_n^2} + \frac{2\zeta D}{\omega_n} + 1\right)q_o = Kq_{is} \tag{3.171}$$

with initial conditions

$$q_o = 0 \qquad \text{at } t = 0^+$$

$$\frac{dq_o}{dt} = 0 \qquad \text{at } t = 0^+ \tag{3.172}$$

The particular solution of Eq. (3.171) is clearly $q_{opi} = Kq_{is}$. The complementary-function solution takes on one of three possible forms, depending on whether the roots of the characteristic equation are real and unrepeated (overdamped case), real and repeated (critically damped case), or complex (underdamped case). The complete solutions of Eq. (3.171) with initial conditions (3.172) are, in nondimensional form,

$$\frac{q_o}{Kq_{is}} = -\frac{\zeta + \sqrt{\zeta^2 - 1}}{2\sqrt{\zeta^2 - 1}} e^{(-\zeta + \sqrt{\zeta^2 - 1})\,\omega_n t}$$

$$+ \frac{\zeta - \sqrt{\zeta^2 - 1}}{2\sqrt{\zeta^2 - 1}} e^{(-\zeta - \sqrt{\zeta^2 - 1})\,\omega_n t} + 1 \qquad \text{overdamped} \tag{3.173}$$

$$\frac{q_o}{Kq_{is}} = -(1 + \omega_n t)e^{-\omega_n t} + 1 \qquad \text{critically damped} \tag{3.174}$$

$$\frac{q_o}{Kq_{is}} = -\frac{e^{-\zeta \omega_n t}}{\sqrt{1 - \zeta^2}} \sin\left(\sqrt{1 - \zeta^2}\,\omega_n t + \phi\right) + 1 \qquad \text{underdamped} \tag{3.175}$$

$$\phi \triangleq \sin^{-1}\sqrt{1 - \zeta^2}$$

Since t and ω_n always appear as the product $\omega_n t$, the curves of $q_o/(Kq_{is})$ may be plotted against $\omega_n t$, which makes them universal for any ω_n, as in Fig. 3.51. This fact also shows that ω_n is *a direct indication of speed of response*. For a given ζ,

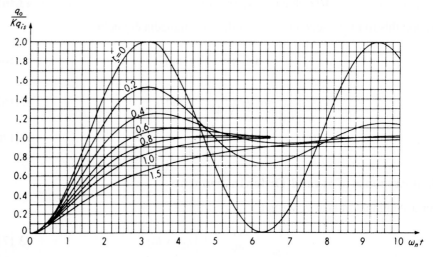

Figure 3.51
Nondimensional step-function response of second-order instrument.

doubling ω_n will halve the response time since $\omega_n t$ [and thus $q_o/(Kq_{is})$] achieves the same value at one-half the time. The effect of ζ is not clearly perceived from the equations, but is evident from the graphs. An increase in ζ reduces oscillation, but also slows the response in the sense that the first crossing of the final value is retarded. A settling time actually may be a better indication of response speed; however, then the optimum value of ζ will vary with the chosen tolerance band. For example, if we choose a 10 percent settling time, the curve for $\zeta = 0.6$ gives a settling time of about $2.4/\omega_n$, and this is optimum, since either larger or smaller ζ gives a longer settling time. However, if we had chosen a 5 percent settling time, a ζ between 0.7 and 0.8 would give the shortest value. In choosing a proper ζ value for a practical application, the situation is further complicated by the fact that the real inputs will not be step functions and their *actual* form influences what will be the best ζ value. If the actual inputs are quite variable in form, some compromise must be struck. Many commercial instruments use $\zeta = 0.6$ or 0.7. We show shortly that this range of ζ gives good frequency response over the widest frequency range.

Terminated-Ramp Response of Second-Order Instruments

Under certain circumstances, the response of second-order instruments to perfect step inputs is misleading. The best example is perhaps found in piezoelectric pressure pickups, accelerometers, and load cells. While these devices are discussed in detail later, at present it is sufficient to state that usually they have an extremely high natural frequency and very little damping ($\zeta < 0.01$, often). Based on a perfect step input, such an instrument appears highly undesirable because of its large overshoot and strong oscillation (Fig. 3.52). Actually, these instruments may give excellent response. The explanation of this apparent inconsistency lies in the fact that *perfect* step inputs do not occur in nature, since a macroscopic quantity cannot change a finite amount in an infinitesimal time. Thus a more realistic input than the step is the *terminated-ramp input,* defined in Fig. 3.53. This input has a *finite* slope equal to

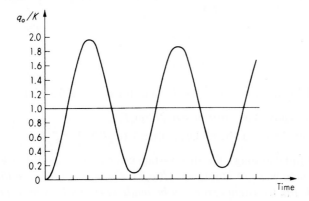

Figure 3.52
Step response of lightly damped system.

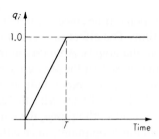

Figure 3.53
Terminated-ramp input.

$1/T$, whereas a step input has an infinite slope. By letting T get smaller and smaller, we can approach the perfect step input. For a second-order system, we would have

$$\left(\frac{D^2}{\omega_n^2} + \frac{2\zeta D}{\omega_n} + 1\right) q_o = K q_i \tag{3.176}$$

$$q = \begin{cases} \dfrac{t}{T} & 0 \le t \le T \\ 1.0 & T \le t < \infty \end{cases} \tag{3.177}$$

$$q_o = \frac{dq_o}{dt} = 0 \qquad \text{at } t = 0^+ \tag{3.178}$$

Since we are concerned here with lightly damped systems, we obtain the solution for only the underdamped case:

$$\frac{q_o}{K} = \frac{t}{T} - \frac{2\zeta}{\omega_n T} + \frac{1}{\omega_n T \sqrt{1 - \zeta^2}} e^{-\zeta \omega_n t} \sin\left(\sqrt{1 - \zeta^2}\,\omega_n t + \phi\right) \quad 0 \le t \le T \tag{3.179}$$

$$\frac{q_o}{K} = \frac{t}{T} - \frac{2\zeta}{\omega_n T} + \frac{1}{\omega_n T \sqrt{1 - \zeta^2}} e^{-\zeta \omega_n t} \sin\left(\sqrt{1 - \zeta^2}\,\omega_n t + \phi\right)$$

$$- \left\{ \frac{t}{T} - 1 - \frac{2\zeta}{\omega_n T} + \frac{1}{\omega_n T \sqrt{1 - \zeta^2}} e^{-\zeta \omega_n (t - T)} \sin\left[\sqrt{1 - \zeta^2}\,\omega_n (t - T) + \phi\right] \right\}$$

$$T \le t < \infty \tag{3.180}$$

$$\phi \triangleq 2 \tan^{-1} \frac{\sqrt{1 - \zeta^2}}{\zeta} \tag{3.181}$$

From Eq. (3.179) we note immediately that, for $0 \le t \le T$, the following is true:

1. There is a steady-state error of size $2\zeta/(\omega_n T)$.
2. The transient error can be no larger than $1/(\omega_n T \sqrt{1 - \zeta^2})$.

Thus if $\zeta = 0$ (no damping), the steady-state error is zero and the "transient" error is a sustained sine wave of amplitude $1/(\omega_n T)$. *Therefore, if ω_n is sufficiently large relative to $1/T$, the transient error can be made very small even if the damping is practically nonexistent.* This result is based on Eq. (3.179), but similar results are obtained from (3.180) for $T \le t < \infty$, since the transient induced at $t = 0$ by the

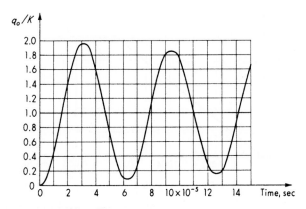

Figure 3.54
Step response of lightly damped system.

increasing ramp is essentially the same as that induced at $t = T$ by a decreasing ramp. That is, the q_i of Fig. 3.53 is really a superposition of an increasing ramp starting at $t = 0$ and a decreasing ramp starting at $t = T$.

As a numerical example, suppose a pressure pickup with $\zeta = 0.01$ and $\omega_n = 100,000$ rad/s is subjected to terminated-ramp-type inputs with $T = 0.00628$ s. The step response of such an instrument is shown in Fig. 3.54 and indicates the severe overshooting and oscillation which could lead us to reject the instrument. Figure 3.55, however, shows the terminated-ramp response corresponding to the *actual* input. It is clear that the response is almost perfect. In fact, if we allow a transient error $1/(\omega_n T)$ of 1 percent, T can be as short as 0.001.

Ramp Response of Second-Order Instruments

The differential equation here is

$$\left(\frac{D^2}{\omega_n^2} + \frac{2\zeta D}{\omega_n} + 1\right) q_o = K\dot{q}_{is}\, t \qquad (3.182)$$

$$q_o = \frac{dq_o}{dt} = 0 \qquad \text{at } t = 0^+$$

Figure 3.56 shows the general character of the response. There is a steady-state error $2\zeta \dot{q}_{is}/\omega_n$. Since the value of $\dot{q}_{is}$ is set by the measured quantity, the steady-state error can be reduced only by reducing ζ and increasing ω_n. For a given ω_n, reduction in ζ results in larger oscillations. There is also a steady-state time lag $2\zeta/\omega_n$. Figure 3.57 gives a set of nondimensionalized curves that summarize system behavior.

Frequency Response of Second-Order Instruments

The sinusoidal transfer function is

$$\frac{q_o}{q_i}(i\omega) = \frac{K}{(i\omega/\omega_n)^2 + 2\zeta i\omega/\omega_n + 1} \qquad (3.183)$$

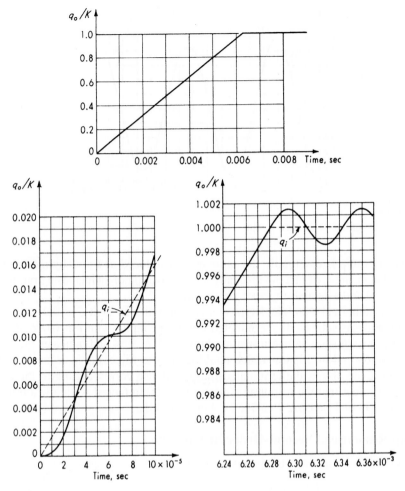

Figure 3.55
Terminated-ramp response of lightly damped system.

which can be put in the form

$$\frac{q_o/K}{q_i}(i\omega) = \frac{1}{\sqrt{[1 - (\omega/\omega_n)^2]^2 + 4\zeta^2\omega^2/\omega_n^2}} \angle \phi \qquad (3.184)$$

$$\phi \triangleq \tan^{-1}\frac{2\zeta}{\omega/\omega_n - \omega_n/\omega} \qquad (3.185)$$

Figure 3.58 gives the nondimensionalized frequency-response curves. Clearly, increasing ω_n will increase the range of frequencies for which the amplitude-ratio curve is relatively flat; thus a high ω_n is needed to measure accurately high-frequency q_i's. An optimum range of values for ζ is indicated by both amplitude-ratio and phase-angle curves. The widest flat amplitude ratio exists for ζ of about 0.6 to 0.7. While zero phase angle would be ideal, it is rarely possible to realize this

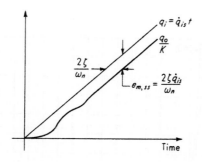

Figure 3.56
Ramp response of second-order instrument.

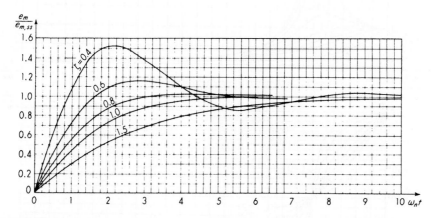

Figure 3.57
Nondimensional ramp response.

even approximately. Actually, if the main interest is in q_o reproducing the correct *shape* of q_i and if a time delay is acceptable, we show shortly that ϕ need not be zero; rather, it should vary *linearly* with frequency ω. Examining the phase curves of Fig. 3.58, we note that the curves for $\zeta = 0.6$ to 0.7 are nearly straight for the widest frequency range. These considerations lead to the widely accepted choice of $\zeta = 0.6$ to 0.7 as the optimum value of damping for second-order instruments. There are exceptions, however, as noted in the section on terminated-ramp response.

Impulse Response of Second-Order Instruments

In the section on first-order instruments we showed that the impulse response is equivalent to the free (unforced) response if the initial $(t = 0^+)$ conditions produced by the impulse are taken into account. To find the initial conditions produced by applying an impulse of area A to a second-order instrument, redraw the block diagram of Fig. 3.59a as in Fig. 3.59b. (The equivalence of the two diagrams is easily demonstrated by tracing through the signals in Fig. 3.59b to get the differential

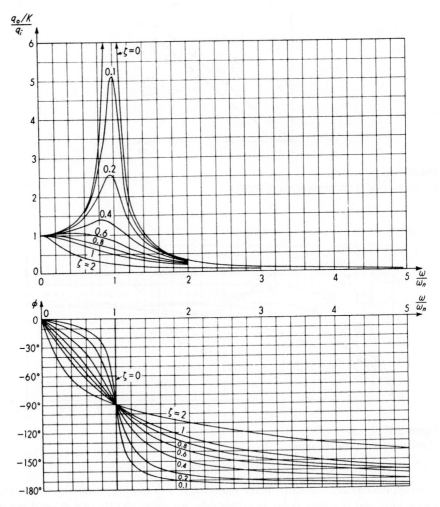

Figure 3.58
Frequency response of second-order instrument.

equation relating q_o to q_i.) In Fig. 3.59c the impulse is applied at q_i, and the "propagation" of this input signal is traced through the rest of the diagram. This analysis shows that at $t = 0^+$ we have $q_o = 0$ and $\dot{q}_o = KA\omega_n^2$. The differential equation to be solved is then

$$\left(\frac{D^2}{\omega_n^2} + \frac{2\zeta D}{\omega_n} + 1\right) q_o = 0$$

$$q_o = 0 \qquad \frac{dq_o}{dt} = KA\omega_n^2 \qquad \text{at } t = 0^+ \qquad \textbf{(3.186)}$$

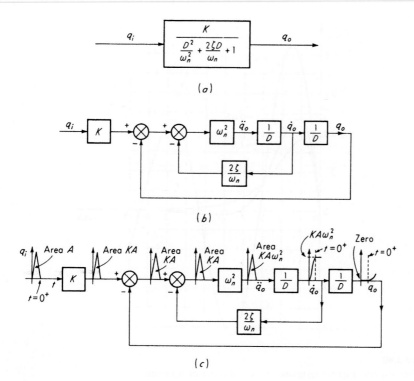

Figure 3.59
Block-diagram analysis of impulse response.

The solutions are found to be

$$\frac{q_o}{K A \omega_n} = \frac{1}{2\sqrt{\zeta^2 - 1}} \left(e^{(-\zeta + \sqrt{\zeta^2 - 1})\omega_n t} - e^{(-\zeta - \sqrt{\zeta^2 - 1})\omega_n t} \right) \qquad \text{overdamped} \qquad \textbf{(3.187)}$$

$$\frac{q_o}{K A \omega_n} = \omega_n t e^{-\omega_n t} \qquad \text{critically damped} \qquad \textbf{(3.188)}$$

$$\frac{q_o}{K A \omega_n} = \frac{1}{\sqrt{1 - \zeta^2}} e^{-\zeta \omega_n t} \sin\left(\sqrt{1 - \zeta^2}\,\omega_n t\right) \qquad \text{underdamped} \qquad \textbf{(3.189)}$$

Figure 3.60 displays these results graphically.

Dead-Time Elements

Some components of measuring systems are adequately represented as dead-time elements. A *dead-time element* is defined as a system in which the output is exactly the same form as the input, but occurs τ_{dt} seconds (the dead time) later. Mathematically,

$$q_o(t) = K q_i(t - \tau_{dt}) \qquad t \geq \tau_{dt} \qquad \textbf{(3.190)}$$

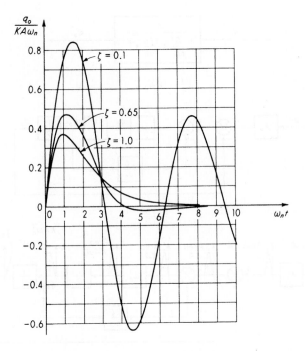

Figure 3.60
Nondimensional impulse response of second-order instrument.

This type of element is also called a *pure delay* or *transport lag.* An example of such an effect is found in pneumatic signal-transmission systems. A pressure signal at one end of a length of pneumatic tubing will cause no response at all at the other end until the pressure wave has had time to propagate the distance between them. Because this speed of propagation is the same as the speed of sound, a 350-m length of tubing will have a dead time of about 1 s, since the speed of sound in standard air is 345 m/s.

The response of dead-time elements to the standard inputs is easily found. For steps, ramps, and impulses the results are given in Fig. 3.61. For sinusoidal input we have

$$q_i = A_i \sin \omega t$$

and from Eq. (3.190)

$$q_o = KA_i \sin \omega(t - \tau_{dt}) \tag{3.191}$$

$$q_o = KA_i \sin (\omega t - \omega \tau_{dt}) = KA_i \sin (\omega t + \phi) \tag{3.192}$$

Thus

$$\frac{q_o/K}{q_i} (i\omega) = 1 \angle \phi = e^{-i\omega \tau_{dt}} \tag{3.193}$$

The frequency-response curves for a dead-time element are shown in Fig. 3.62.

The next type of term is $(i\omega)^n$, where $n = \pm 1, \pm 2, \ldots$. The phase angle of such terms is constant with frequency and is given by $90n°$. The amplitude ratio is ω^n, so that the decibel value is

$$20 \log \omega^n = 20n \log \omega$$

Since we plot against $\log \omega$, the decibel curves become straight lines of slope $20n$ dB/decade (see Fig. 3.63). A *decade* is defined as any 10:1 frequency range; an *octave* is any 2:1 range.

Terms of the form $i\omega\tau + 1$ and $1/(i\omega\tau + 1)$ give, respectively,

$$dB = 20 \log \sqrt{(\omega\tau)^2 + 1} \qquad\qquad \textbf{(3.200)}$$

and $\qquad\qquad dB = -20 \log \sqrt{(\omega\tau)^2 + 1} \qquad\qquad \textbf{(3.201)}$

When $\omega\tau \gg 1$, these become

$$dB \approx 20 \log \omega\tau = 20 \log \tau + 20 \log \omega \qquad\qquad \textbf{(3.202)}$$

and $\qquad\qquad dB \approx -20 \log \omega\tau = -20 \log \tau - 20 \log \omega \qquad\qquad \textbf{(3.203)}$

We see that both represent straight lines of slope ± 20 dB/decade, and these straight lines will be the high-frequency asymptotes of the actual amplitude-ratio curves. Similarly, for $\omega\tau \ll 1$,

$$dB \approx 20 \log 1 = 0 \qquad\qquad \textbf{(3.204)}$$

and $\qquad\qquad dB \approx -20 \log 1 = 0 \qquad\qquad \textbf{(3.205)}$

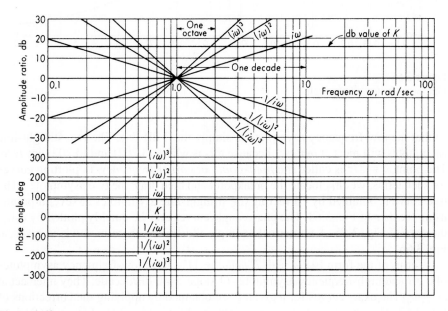

Figure 3.63
Integrator and differentiator frequency response.

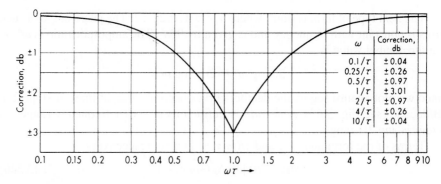

Figure 3.64
First-order-system amplitude-ratio corrections.

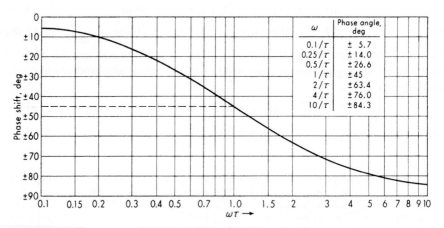

Figure 3.65
First-order-system phase angle.

so that the low-frequency asymptote is simply the 0-dB line. The two straight-line asymptotes will meet at $\omega\tau = 1$ because this is where (3.202) and (3.203) are zero. The point $\omega = 1/\tau$ is called the *breakpoint,* or *corner frequency.* In plotting curves for such terms, we first locate the breakpoint and then draw the two asymptotes. The true curve is obtained by correcting the straight-line asymptotes at several points, using the data of Fig. 3.64. The phase-angle curves may be quickly plotted by using the data of Fig. 3.65. A numerical example illustrating these methods is given in Fig. 3.66.

Terms of the form $[(i\omega/\omega_n)^2 + 2\zeta i\omega/\omega_n + 1]^{\pm 1}$ have low-frequency asymptotes of 0 dB and high-frequency asymptotes of slope $\pm$ 40 dB/decade. They intersect at $\omega = \omega_n$. The exact curves for a given ζ are obtained by applying the corrections of Fig. 3.67. The phase-angle curves are obtained from Fig. 3.68. Figure 3.69 gives numerical examples.

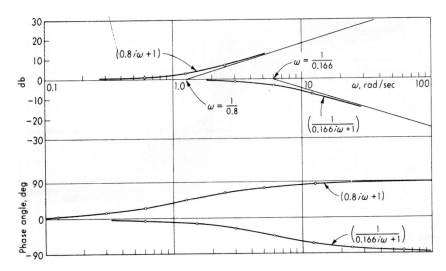

Figure 3.66
Example of first-order terms.

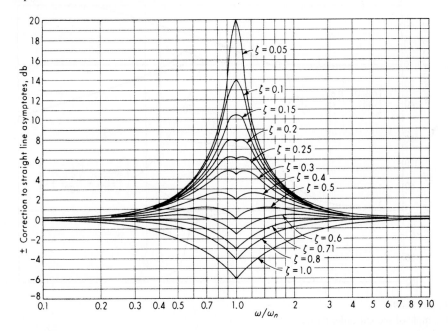

Figure 3.67
Second-order-system amplitude-ratio corrections.

The final type of term considered is the dead-time term $e^{-i\omega\tau_{dt}}$. Since the amplitude ratio is 1.0 for all frequencies, the decibel curve is simply the 0-dB line. The phase-angle curve is easily plotted from $\phi = -\omega\tau_{dt}$ for any given dead time.

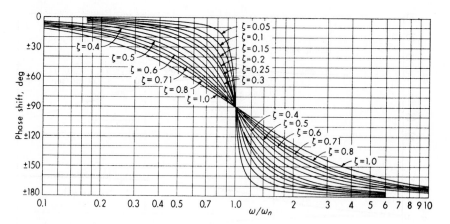

Figure 3.68
Second-order-system phase angle.

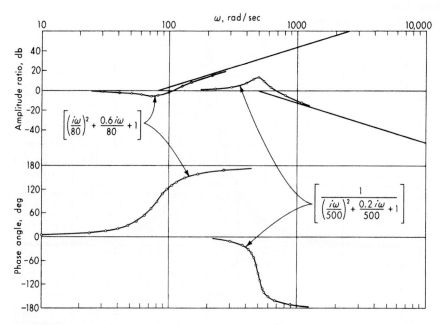

Figure 3.69
Example of second-order terms.

To illustrate the procedure for combining the individual terms to obtain the overall frequency-response curves, we consider the following example:

$$\frac{q_o}{q_i}(i\omega) = \frac{4.4(i\omega)}{(i\omega + 1)(0.2i\omega + 1)} \qquad \textbf{3.206}$$

Figure 3.70 shows the procedure and results.

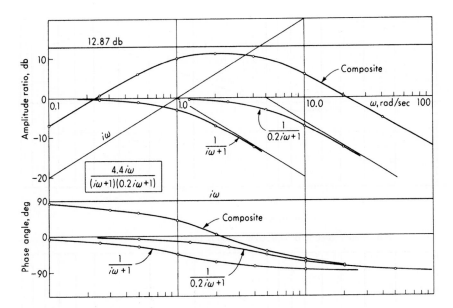

Figure 3.70
Example of frequency-response plot.

The "manual" graphing technique just explained is today mainly used as a quick and rough *sketching* tool; we use computer software when we need *accurate* graphs. Manual sketching will *always* remain a useful skill at the system *design* stage since it gives clear insight into the effect of individual system components on the overall system behavior. When we do finally use computer aids to plot accurate graphs, the manual sketches allow us to also catch computer data-entry errors since we know roughly what the curves *should* look like. Most mathematical software provides for easy frequency response plotting. For the MATLAB used in this book, we enter,

$$\text{den} = [a_n \, a_{n-1} \ldots a_1 \, a_0]$$
$$\text{num} = [b_m \, b_{m-1} \ldots b_1 \, b_0]$$
$$\text{bode (num, den)}$$

using the a's and b's from Eq. (3.90). Figures 3.71 and 3.72 give the results, corresponding to Figs. 3.69 and 3.70, respectively.

Response of a General Form of Instrument to a Periodic Input

Our approach to the dynamic response of measurement systems has, to this point, been limited in two ways. First, we considered only rather simple types of instruments (zero-order, first-order, and second-order). Second, we subjected these instruments to rather simple inputs (steps, ramps, sine waves, and impulses). At this point, by applying more advanced mathematical tools, we begin to remove both limitations. We see that the concept of frequency response plays a central role in

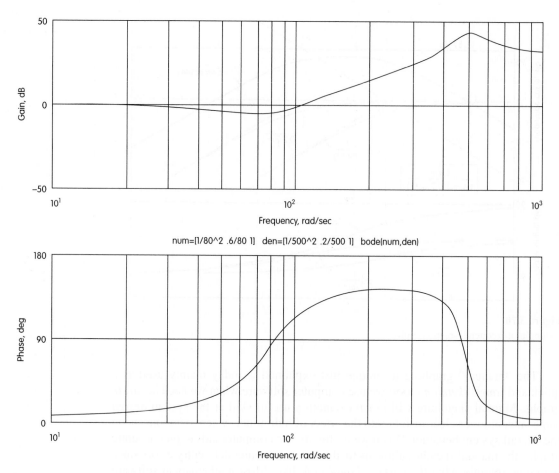

Figure 3.71
MATLAB Bode plot.

these developments. The first step involves the study of the response of a general (linear, time-invariant) instrument to periodic inputs.

By a periodic function we mean one that repeats itself cyclically over and over, as in Fig. 3.73. If this function meets the *Dirichlet conditions* (it must be single-valued, be finite, and have a finite number of discontinuities and maxima and minima in one cycle), it may be represented by a *Fourier series*.[41] That is,

$$q_i(t) = q_{i,\,av} + \frac{1}{L}\left(\sum_{n=1}^{\infty} a_n \cos\frac{n\pi t}{L} + \sum_{n=1}^{\infty} b_n \sin\frac{n\pi t}{L}\right) \qquad (3.207)$$

[41]E. Kreyszig, "Advanced Engineering Mathematics," 8th ed., Wiley, New York, 1999, Chap. 10.

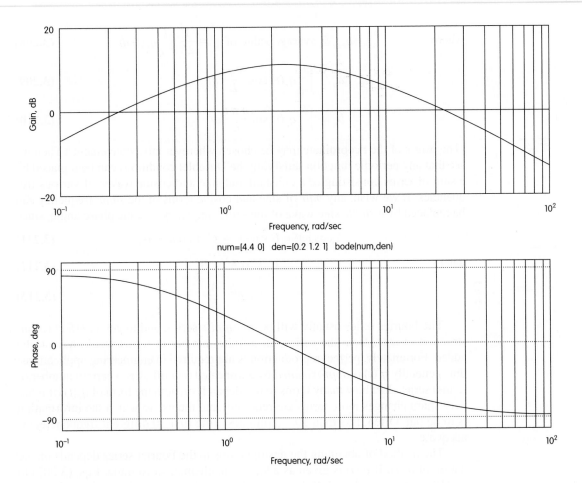

num=[4.4 0]   den=[0.2 1.2 1]   bode(num,den)

Figure 3.72
MATLAB Bode plot.

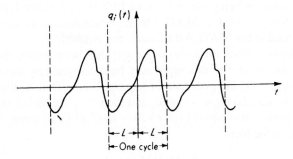

Figure 3.73
General periodic function.

where
$$q_{i\,av} \triangleq \text{average value of } q_i = \frac{1}{2L} \int_{-L}^{L} q_i(t)\, dt \qquad \text{(3.208)}$$

$$a_n \triangleq \int_{-L}^{L} q_i(t) \cos \frac{n\pi t}{L}\, dt \qquad \text{(3.209)}$$

$$b_n \triangleq \int_{-L}^{L} q_i(t) \sin \frac{n\pi t}{L}\, dt \qquad \text{(3.210)}$$

(The origin of the t coordinate may be chosen wherever most convenient.) Then we see that any periodic function satisfying the Dirichlet conditions can be replaced by a sum of terms consisting of a constant and sine and cosine waves of various frequencies. If we wish, any *pair* of sine and cosine terms of the *same* frequency can be replaced by a *single* sine wave of the same frequency at some phase angle, since

$$A \cos \omega t + B \sin \omega t \equiv C \sin (\omega t + \alpha) \qquad \text{(3.211)}$$

where
$$C \triangleq \sqrt{A^2 + B^2} \qquad \text{(3.212)}$$

$$\alpha \triangleq \tan^{-1} \frac{A}{B} \qquad \text{(3.213)}$$

The Fourier series usually will be an *infinite* series, and to get a *perfect* reconstruction of $q_i(t)$ from the series, an infinite number of terms would have to be added. Fortunately, perfect reproduction is not required in engineering applications; thus generally $q_i(t)$ is *approximated* by a truncated (cut off after a certain number of terms) series. Just how many terms to use depends on both the form of $q_i(t)$ (if it has very sharp changes more terms are required) and the use to which the information is to be put. Often, less than 10 "harmonics" (the first 10 different frequencies) is adequate.

The method of obtaining the desired terms in the Fourier series depends on the nature of $q_i(t)$. If $q_i(t)$ is given as a known mathematical formula, Eqs. (3.207) to (3.210) may be employed. If the required integrations cannot be performed analytically because of the complexity of $q_i(t)$ or because it is given by an experimental graph rather than a formula, then various approximate numerical methods are available, the most common being some form of FFT (Fast Fourier Transform) algorithm, such as that provided in MATLAB. Explanations on its use, giving more detail than provided in the MATLAB manual, are available.[42]

Once the Fourier series for a particular $q_i(t)$ has been found, the *steady-state* response of any instrument to this input may be determined by use of frequency-response techniques and the principle of superposition. That is, the response for each individual sinusoidal term is found, and then they are added algebraically to get the total response. By use of Eqs. (3.211) to (3.213) all terms in the Fourier series can be put in the form

$$A_{ik} \sin (\omega_k t + \alpha_k)$$

---

[42]E. O. Doebelin, "System Dynamics," 1998, pp. 644–662.

We now define the complex number $Q_i(i\omega_k)$ by

$$Q_i(i\omega_k) \triangleq A_{ik} \angle \alpha_k \qquad (3.214)$$

For example, the constant term -7.2 becomes $7.2 \angle 180°$, and the term $9.3 \sin(20t + 37°)$ becomes $9.3 \angle 37°$. When the Fourier series representing $q_i(t)$ is expressed in this form, it is called $Q_i(i\omega)$, the *input-frequency spectrum*. Thus, if

$$q_i(t) = A_{i0} + A_{i1} \sin(\omega_1 t + \alpha_1) + A_{i2} \sin(\omega_2 t + \alpha_2) + \cdots \qquad (3.215)$$

then
$$Q_i(i\omega) = |A_{i0}| \angle (0° \text{ or } 180°) + A_{i1} \angle \alpha_1 + A_{i2} \angle \alpha_2 + \cdots \qquad (3.216)$$

Such a spectrum, which exists only at isolated frequencies, is called a *discrete spectrum*.

Now, we recall that the sinusoidal transfer function of the system, $(q_o/q_i)(i\omega)$, is also a complex number for any given frequency. An alternative name for the sinusoidal transfer function is the *system frequency response*. If we pick any frequency ω_k and multiply the complex numbers $Q_i(i\omega_k)$ and $(q_o/q_i)(i\omega_k)$, we get another complex number that we define as $Q_o(i\omega_k)$. If we do this for *all* frequencies and add all the $Q_o(i\omega_k)$, the sum is called $Q_o(i\omega)$, the *output frequency spectrum,* and it has the form

$$Q_o(i\omega) = A_{o0} \angle (0° \text{ or } 180°) + A_{o1} \angle \beta_1 + A_{o2} \angle \beta_2 + \cdots \qquad (3.217)$$

This is now interpreted as in Eqs. (3.215) and (3.216) to give

$$q_o(t) = \pm A_{o0} + A_{o1} \sin(\omega_1 t + \beta_1) + A_{o2} \sin(\omega_2 t + \beta_2) + \cdots \qquad (3.218)$$

Since $q_i(t)$ is periodic, the frequencies ω_2, ω_3, etc., are all integer multiples of ω_1, and thus $q_o(t)$ also will be a periodic function. For accurate measurement, $q_i(t)$ and $q_o(t)$ must have nearly identical waveforms, of course. The validity of the statement

$$Q_o(i\omega) = Q_i(i\omega) \left[\frac{q_o}{q_i}(i\omega) \right] \qquad (3.219)$$

used in the above manipulations follows easily from the basic definition of sinusoidal transfer function, the superposition theorem, and the rules for multiplying complex numbers. Figure 3.74 illustrates the method graphically.

The above method is extremely useful for understanding the basic relationships governing the choice of a suitable measuring instrument for a periodic signal. However, other approaches may be more useful for actual numerical calculation of specific cases. Digital simulation can yield very complete information with little effort, as our next example shows. Figure 3.75a represents the periodic volume flow rate of a simple reciprocating pump running at constant speed:

$$\text{Flow rate} = Q_p |\sin \omega_p t| \qquad \text{m}^3/\text{s} \qquad (3.220)$$

Application of Eqs. (3.207) to (3.213) gives the frequency spectrum of this signal for $\omega_k = 0, 2\omega_p, 4\omega_p, \ldots, 2k\omega_p$ ($k = 1, 2, 3, \ldots$):

$$Q_i(0) = \frac{2Q_p}{\pi} \qquad Q_i(i\omega_k) = \left| \frac{4Q_p}{\pi(1 - 4k^2)} \right| \angle -90° \qquad (3.221)$$

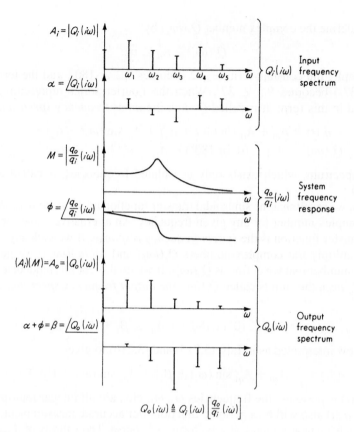

Figure 3.74
System response to periodic input.

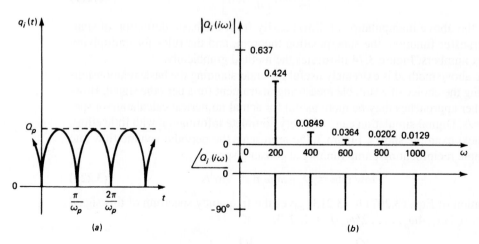

Figure 3.75
Reciprocating-pump flow rate and Fourier series.

Figure 3.75*b* shows this spectrum for $Q_p = 1.0$ and $\omega_p = 100$ rad/s. We now use SIMULINK digital simulation to compare the exact Q_i with a five-term truncated Fourier series, examine four candidate second-order flowmeter instruments (with various values of ω_n and ζ) for measurement quality, and observe both "starting transients" and "periodic steady-state" behavior of the instruments. (The calculation method of Fig. 3.74 gives *no* clue as to transient behavior.)

Figure 3.76*a* shows the SIMULINK simulation diagram. Recall that function blocks, (**Fcn** and **Fcn1**) use the symbol **u** for *whatever* variable is the input to the block (time in our present example). We use two such blocks to form the exact $q_i(t)$ and also a five-term Fourier series approximation as given by Eq. (3.221), which we could use in an analytical (rather than simulation) approach to the problem. Four second-order transfer function blocks, **Transfer Fcn** to **Transfer Fcn3**, are used to model four candidate flowmeters with (ω_n, ζ) values of (1000, 0.65), (100, 0.65), (5000, 0.05), and (200, 0.10), respectively, where the natural frequencies are in rad/sec. The responses in Fig. 3.76*b* show that two of the flowmeters might be acceptable, but the other two are clearly inadequate. The (1000, 0.65) meter seems to follow the input accurately except for a delay (which is usually acceptable in practice and explained later in Fig. 3.106). It also is not able to perfectly reproduce the sharp "points" at zero flow rate. Both these defects are largely corrected by the (5000, 0.05) meter, which visually appears to be perfect in this graph. We usually recommend damping ratios near 0.65, but the low (0.05) value is acceptable when the natural frequency is sufficiently high. The (100, 0.65) meter is clearly unacceptable since it totally distorts the waveform and also attenuates the amplitude of the

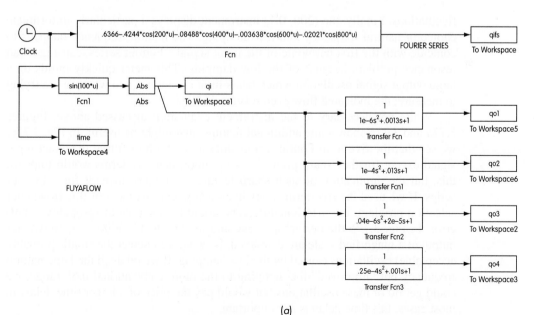

(a)

Figure 3.76
Simulation diagram and results for flowmeter accuracy study.

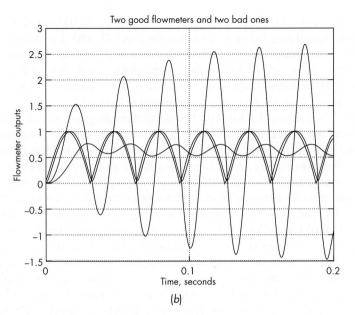

Figure 3.76
(Concluded)

fluctuations. Finally, the (200, 0.1) instrument displays a particularly unfortunate choice of natural frequency and low damping. The natural frequency happens to coincide with the first harmonic of the input signal's Fourier series, causing a bad resonance problem because of the low damping. This meter quickly builds up a large output signal oscillation which bears little resemblance to what we are trying to measure; the indicated flow even reverses!

While not necessary to the instrument evaluation discussed above, Figures 3.77*a* and 3.77*b* show some additional features that might be useful. In Fig. 3.77*a*, we see that the five-term Fourier series does a good job of fitting the exact input signal except for the "sharp points." Using more than five terms would improve this, but the duplication of such sharp features is always difficult for a Fourier series. If we used the five-term series in an analytical treatment of this flowmeter selection problem, the mentioned defects would be a source of (probably small) error. In Fig. 3.77*b* the perfect q_i curve and the "perfect" (5000, 0.5) flowmeter output at a magnified scale are displayed. Now we can detect the small (probably acceptable) oscillations caused by the low damping. If we retained the large natural frequency but "improved" the damping to the more conventional 0.65 range, we could get rid of these oscillations but would pay the price of a larger time delay. In most cases, this time delay is *not* important.

For readers new to simulation software, this example illustrates the ease with which we can investigate useful details of dynamic behavior.

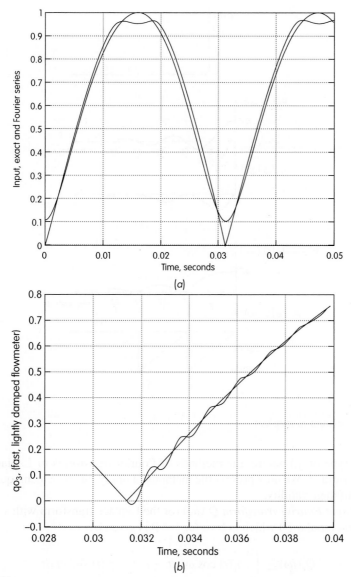

Figure 3.77

a. Accuracy of Fourier series fit. *b.* Flowmeter oscillations.

Response of a General Form of Instrument to a Transient Input

By a transient input, we mean a $q_i(t)$ that is identically zero for all values of time greater than some finite value t_o, that is, an input that eventually dies out. For transient inputs of specific mathematical form, usually we can solve the differential equation and get $q_o(t)$ directly. For q_i's given by experimental data or, more important, if we wish to bring out certain important results of a *general* (not restricted to

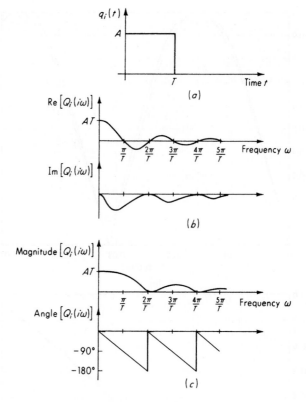

Figure 3.78
Frequency spectrum of transient.

a specific type of q_i) nature, the methods of Fourier transforms or Laplace transforms are useful.[43] We now present the methods of applying these techniques, without proof of their validity.

The *direct Fourier transform* $Q_i(i\omega)$ (or the Laplace transform with $s = i\omega$) of the transient input $q_i(t)$ which is zero for $t < 0$ is given by

$$Q_i(i\omega) \triangleq \int_0^\infty q_i(t) \cos \omega t \, dt - i \int_0^\infty q_i(t) \sin \omega t \, dt \qquad (3.222)$$

where ω can take all values from $-\infty$ to $+\infty$. Equation (3.222) is said to transform the input function from the time domain $[q_i(t)]$ to the frequency domain $[Q_i(i\omega)]$. The function $Q_i(i\omega)$ is also called the *frequency spectrum* of the input and plays the same role for transient inputs as Eq. (3.216) does for periodic inputs. However,

[43]E. O. Doebelin, "System Modeling and Response," Wiley, New York, 1980; M. F. Gardner and J. L. Barnes, "Transients in Linear Systems," Wiley, New York, 1942; A. Papoulis, "The Fourier Integral and Its Applications," McGraw-Hill, New York, 1962.

whereas $Q_i(i\omega)$ is a *discrete* spectrum for $q_i(t)$ periodic, it is a *continuous* spectrum for $q_i(t)$ transient. That is, if you carry out Eq. (3.222) for a given $q_i(t)$, you will find $Q_i(i\omega)$ to be a complex number that varies with (is a function of) frequency ω and exists for *all* ω, not just at isolated points. As an example, consider the transient input of Fig. 3.78a. Applying Eq. (3.222), we get

$$Q_i(i\omega) = \int_0^T A \cos \omega t \, dt - i \int_0^T A \sin \omega t \, dt \qquad (3.223)$$

$$Q_i(i\omega) = \underbrace{\frac{A \sin \omega T}{\omega}}_{\text{real part}} + i \underbrace{\frac{A}{\omega}(-1 + \cos \omega T)}_{\text{imaginary part}} \qquad (3.224)$$

or, alternatively,

$$Q_i(i\omega) = \underbrace{\frac{\sqrt{2}A}{\omega}\sqrt{1 - \cos \omega T}}_{\text{magnitude}} \underbrace{\angle \, \alpha}_{\text{angle}} \qquad (3.225)$$

where

$$\tan \alpha = \frac{\cos \omega T - 1}{\sin \omega T} \qquad (3.226)$$

Plots of this frequency spectrum are given in Figs. 3.78b and c. [While $Q_i(i\omega)$ exists for both positive and negative values of ω, because of symmetry, often we can get the desired results by considering only the range $0 \leq \omega < +\infty$. The stated symmetry consists of the following:

1. Re $Q_i(-i\omega) = $ Re $Q_i(+i\omega)$
2. Im $Q_i(-i\omega) = -$ Im $Q_i(+i\omega)$
3. Magnitude $Q_i(-i\omega) = $ magnitude $Q_i(+i\omega)$
4. Angle $Q_i(-i\omega) = -$ angle $Q_i(+i\omega)$

In most of our graphs and calculation methods, we employ the range $0 \leq \omega < +\infty$, but $Q_i(-i\omega)$ always exists and can be found from a given $Q_i(+i\omega)$ by application of the above symmetry rules.] These graphs indicate the "frequency content" of the transient input, just as the Fourier series indicates the frequency content of a periodic input. Thus we see that if T is small, large values of $Q_i(i\omega)$ persist out to higher frequencies than if T is large. Therefore a short-duration pulse is said to have greater high-frequency content than a long one. It is important to point out that the concept of frequency content for transients is not as clear-cut as for periodic functions; $q_i(t)$ for a transient *cannot* be built up by simply adding distinct sine waves, because $Q_i(i\omega)$ is now a *continuous function and no distinct frequencies exist.*

A further illustration of this distinction may be found by examining the dimensions of $Q_i(i\omega)$ in both cases. As an example, consider a $q_i(t)$ which is a pressure in (kPa). If this pressure is periodic, Eq. (3.212), etc., shows that $Q_i(i\omega)$ has the same dimensions as $q_i(t)$, that is, (kPa). Now, however, if the pressure is a transient, Eq. (3.222) gives

$$Q_i(i\omega) = \int_0^\infty (\text{kPa}) \underbrace{\cos \omega t}_{\text{dimensionless}} (\text{seconds}) - i \, (\text{same dimensions}) \qquad (3.227)$$

We see that $Q_i(i\omega)$ now has dimensions of N · s/m², or, reinterpreting this, kPa/(rad/s). That is, $Q_i(i\omega)$ is thought of as the amount of signal *per unit frequency increment,* rather than as the actual amount of signal at a discrete frequency. This is analogous to the concept of distributed (rather than concentrated) loads in strength of materials. When a beam has sand (or water, etc.) piled on it, the applied load at any particular *point* is zero, but over an *area* the load is the force density times the area. Similarly, a transient signal has no discrete frequencies, but does contain a certain amount of signal within any frequency *band.* Thus, for a transient, $Q_i(i\omega)$ may be thought of as the *density* of signal per frequency bandwidth rather than as the signal itself.

The main purpose of using Eq. (3.222) is to convert functions from the time domain to the frequency domain, perform certain desired operations (which are *easier* or more *revealing* in the frequency domain than in the time domain), and then convert the information back to the time domain, since this is the more familiar and directly applicable (in an engineering sense) form. The conversion from frequency domain to time domain is accomplished by the *inverse-Fourier-* (or Laplace-) *transform formula* given by

$$q_i(t) = \frac{2}{\pi} \int_0^\infty \mathrm{Re}\,[Q_i(i\omega)] \cos \omega t\, d\omega \qquad t > 0$$

$$q_i(t) \equiv 0 \qquad t < 0$$

$$(3.228)$$

Since these transformations are unique, if a $Q_i(i\omega)$ for a given $q_i(t)$ is found from Eq. (3.222), it should be possible to reconstruct the original $q_i(t)$ from the $Q_i(i\omega)$ by Eq. (3.228). Carrying this out for our example, we get

$$q_i(t) = \frac{2}{\pi} \int_0^\infty \frac{A \sin \omega T}{\omega} \cos \omega t\, d\omega \qquad t > 0$$

$$q_i(t) \equiv 0 \qquad t < 0$$

$$(3.229)$$

After some transformation, this can be put in a standard form found in integral tables and gives

$$q_i(t) = \begin{cases} 0 & t > 0 \\ A & 0 < t < T \\ A/2 & t = T \\ 0 & T < t \end{cases}$$

$$(3.230)$$

This function is shown in Fig. 3.79, and we see that it is practically identical to Fig. 3.78a. Actually, in Fig. 3.78a, we were not mathematically precise in defining $q_i(t)$ at $t = T$ since the graph shows $q_i(t)$ taking on *all* values between 0 and A. The usual practice for such discontinuities is to define the function as single-valued and equal to the midpoint. The Fourier transform is set up on this basis and thus always gives results similar to (3.230). The Fourier *series* for a periodic function with step discontinuities also behaves in this fashion; that is, it converges to the midpoint. Thus

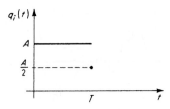

Figure 3.79
Single-valued definition of transient.

in using numerical schemes, if an ordinate falls right on a discontinuity, the mid-point should be used as the numerical entry in the computation schedule.

To get a better feeling for the above methods and to show how they are graphically or numerically applied to functions (data) for which mathematical formulas are not available, let us consider the following development. Figure 3.80*a* shows a typical transient $q_i(t)$ as might be experimentally recorded in, say, a shock test. The first thing we note is that, for all practical purposes, the transient is ended in a finite time t_0. Thus, since $q_i(t)$ is a multiplying factor in the integrand and becomes zero for $t > t_0$, we may write Eq. (3.222) as

$$Q_i(i\omega) = \int_0^{t_0} q_i(t) \cos \omega t \, dt - i \int_0^{t_0} q_i(t) \sin \omega t \, dt \qquad \textbf{(3.231)}$$

We obtain $Q_i(i\omega)$ one point (frequency) at a time as follows:

1. Choose a numerical value of ω, say ω_1.
2. Now, $\cos \omega_1 t$ is a perfectly definite curve and may be plotted against t.
3. Multiply $q_i(t)$ and $\cos \omega_1 t$ point by point to get the curve $q_i(t) \cos \omega_1 t$.
4. Integrate, by any suitable numerical, graphical, or mechanical means, the curve $q_i(t) \cos \omega_1 t$ from $t = 0$ to $t = t_0$. Call the integral (area under curve) a_1.
5. Repeat the above procedure for $q_i(t) \sin \omega_1 t$ and call the integral b_1.
6. Then $Q_i(i\omega_1)$ is $a_1 - ib_1$.
7. Repeat for as many ω's as desired to generate the curves for $Q_i(i\omega)$ versus ω.

Figure 3.80 illustrates these procedures.

To appreciate the difference in "frequency content" between a "slow" transient and a "fast" one, we consider Fig. 3.81. Here $Q_i(i\omega)$ is found (for a high value of ω) for both a slow transient (Fig. 3.81*a*) and a fast one (Fig. 3.81*b*). It is clear that $Q_i(i\omega)$ will be nearly zero for ω's at or above the chosen ω_1 for the slow transient; thus its frequency content is limited to lower frequencies. The fast transient, however, has a nonzero value for $Q_i(i\omega_1)$ and thus "contains" frequencies at and somewhat above this value. For any real-world transient, we can always find *some* ω_1 high enough to make $Q_i(i\omega_1) \approx 0$; that is, all *real* transients are limited in frequency

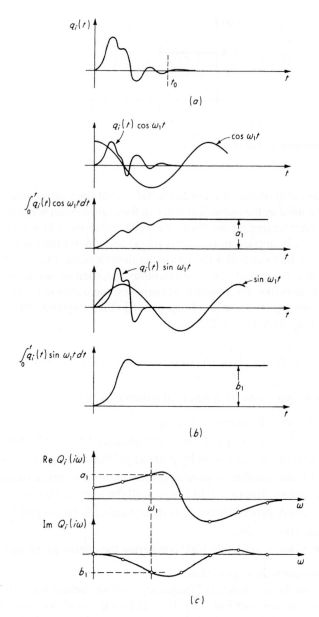

Figure 3.80
Graphical interpretation of direct Fourier transform.

content at the high end. An "unreal" (only mathematically possible) transient is the impulse function, which we can easily show to contain *all* frequencies from 0 to ∞ and all in equal "strength." For an impulse of area A

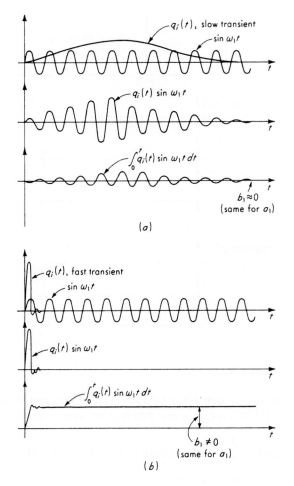

Figure 3.81
Frequency content of fast and slow transients.

$$Q_i(i\omega) = \int_0^\infty A\delta(t) \cos \omega t \, dt - i \int_0^\infty A\delta(t) \sin \omega t \, dt \qquad \textbf{(3.232)}$$

$$Q_i(i\omega) = A - i0 = A \qquad \text{for any finite } \omega \qquad \textbf{(3.233)}$$

Thus Fig. 3.82 shows the frequency content of an impulse. This property of an impulse makes it most useful as a "test signal" for investigating unknown systems, since all frequencies will be excited equally and the true nature of the system will be revealed by its response. We develop this important concept in greater detail later.

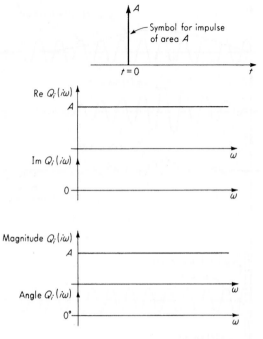

Figure 3.82
Frequency spectrum of impulse.

The process of *inverse* transformation also can be interpreted graphically. The defining equation (3.228) may, in actual practice, be written as

$$q_i(t) = \frac{2}{\pi} \int_0^{\omega_0} \text{Re}\,[Q_i(i\omega)] \cos \omega t \, d\omega \qquad t > 0$$

$$q_i(t) \equiv 0 \qquad t < 0$$

(3.234)

since all $Q_i(i\omega)$ representing physical quantities become approximately equal to zero for ω greater than some finite value ω_0. This follows directly from the fact that the $Q_i(i\omega)$ come from the $q_i(t)$ and they cannot contain infinitely high frequencies. A step-by-step procedure for finding $q_i(t)$ one point at a time from a given $Q_i(i\omega)$ is as follows:

1. Choose a numerical value of t, say t_1.
2. Now $\cos \omega t_1$ is a perfectly definite curve and may be plotted against ω.
3. Multiply $\text{Re}[Q_i(i\omega)]$ and $\cos \omega t_1$ point by point to get the curve $\text{Re}[Q_i(i\omega)]$ $\cos \omega t_1$.
4. Integrate, by any suitable numerical, graphical, or mechanical means, the curve $\text{Re}[Q_i(i\omega)] \cos \omega t_1$ from $\omega = 0$ to $\omega = \omega_0$. The integral (area under curve) is $(\pi/2)q_i(t_1)$, from Eq. (3.234). Plot $q_i(t_1)$ versus t.
5. Repeat for as many t's as desired to generate the curve $q_i(t)$ versus t.

Figure 3.83 illustrates this procedure.

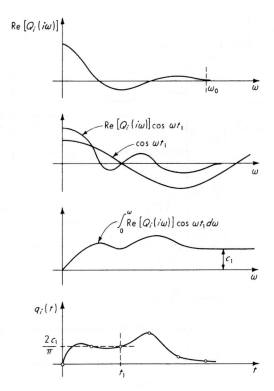

Figure 3.83
Graphical interpretation of inverse Fourier transform.

The main usefulness of the above transform methods is based on the important result[44] relating the Fourier transform of the input signal $Q_i(i\omega)$, the system frequency response $(q_0/q_i)(i\omega)$, and the Fourier transform of the output signal $Q_o(i\omega)$:

$$Q_0(i\omega) = Q_i(i\omega)\left[\frac{q_o}{q_i}(i\omega)\right] \tag{3.235}$$

The restriction on this result is that the initial conditions must be zero; the system starts from rest. Many important practical problems meet this requirement. (Also, for nonzero initial conditions, the method may still be applied if the input is suitably modified.[45])

The meaning of Eq. (3.235) is that when the frequency-response curves for a system are known and a transient input $q_i(t)$ is applied, we can transform $q_i(t)$ to $Q_i(i\omega)$, multiply $Q_i(i\omega)$ and $(q_o/q_i)(i\omega)$ point by point to get $Q_o(i\omega)$, and then inverse-transform $Q_0(i\omega)$ to $q_o(t)$ to get the output or response of the system. Figure 3.84 illustrates the procedure.

---

[44]J. A. Aseltine, "Transform Method in Linear System Analysis," McGraw-Hill, New York, 1958.
[45]Ibid.

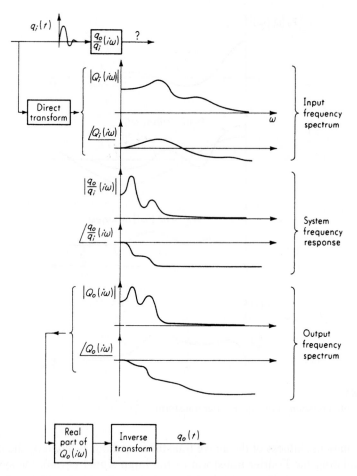

Figure 3.84
System response to transient input.

From a measurement-system point of view, Eq. (3.235) has the following important interpretation. For accurate measurement, $q_o(t) \approx Kq_i(t)$, and since the Fourier transforms are unique [only one possible $F(i\omega)$ for each $f(t)$ and vice versa], this requires $Q_o i\omega) \approx KQ_i(i\omega)$. Since $Q_o(i\omega)$ is obtained by multiplying $Q_i(i\omega)$ and $(q_o/q_i)(i\omega)$, this means $(q_o/q_i)(i\omega)$ must be $K \angle 0°$ over the entire range of frequencies for which $Q_i(i\omega)$ is not practically zero *but can be anything elsewhere*. The requirement that $(q_o/q_i)(i\omega)$ be $K \angle 0°$ for *all* frequencies for perfect measurement is obvious without the use of transform methods. The condition that this need be so only for a definite, finite *range* of frequencies (corresponding to the frequency content of the input) is the contribution of the transform methods and is of great practical significance since it puts much more realistic demands on the measurement system. A further relaxation of these requirements (which allows phase shift) is developed later in this chapter.

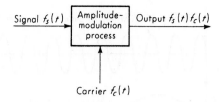

Figure 3.85
Amplitude modulation.

While Eq. (3.222) was useful for defining and calculating $Q_i(i\omega)$ for the purposes of this chapter, an alternative approach that uses a Laplace transform as an intermediate step may expedite numerical calculations for transients expressible as mathematical formulas. In this method, $q_i(t)$ is analytically Laplace-transformed to $Q_i(s)$ by using available transform theorems and tables. Then, by simply substituting $s = i\omega$, $Q_i(i\omega)$ can be calculated numerically for a desired range of frequency.

Frequency Spectra of Amplitude-Modulated Signals

Interest in amplitude-modulated signals stems mainly from two considerations:

1. Physical data that are to be measured and interpreted sometimes are amplitude-modulated.
2. Certain types of measurement systems intentionally introduce amplitude modulation for one or more benefits.

While, in general, the signal that modulates the amplitude of a carrier wave may be of any form (single sine wave, general periodic function, random wave, transient, etc.) and the carrier may be given different forms (sine wave, square wave, etc.), perhaps the process is most easily understood for a single sine wave modulating a sinusoidal carrier. The modulation process is basically one of multiplying the signal carrying the information by a carrier wave of constant frequency and amplitude (see Fig. 3.85). For our simple example, we have

$$\text{Output} = (A_s \sin \omega_s t)(A_c \sin \omega_c t) \tag{3.236}$$

where $A_s \triangleq$ amplitude of signal
$\omega_s \triangleq$ frequency of signal
$A_c \triangleq$ amplitude of carrier
$\omega_c \triangleq$ frequency of carrier

The frequency ω_c is greater (usually considerably greater) than ω_s. For such a situation, the output has the shape shown in Fig. 3.86a. The frequency spectrum of such a signal is obtained easily from the following trigonometric identity:

$$\sin \alpha \sin \beta \equiv \tfrac{1}{2} \cos(\alpha - \beta) - \tfrac{1}{2} \cos(\alpha + \beta) \tag{3.237}$$

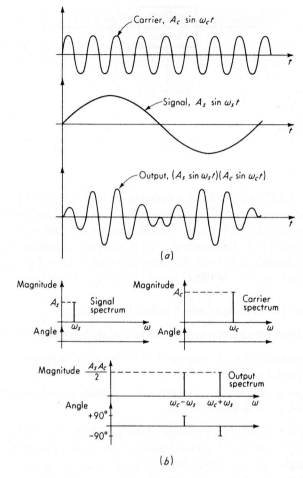

Figure 3.86
Frequency spectrum of amplitude-modulated signals.

Applying this to Eq. (3.268), we get

$$\text{Output} = \frac{A_s A_c}{2} [\cos(\omega_c - \omega_s)t - \cos(\omega_c + \omega_s)t] \tag{3.238}$$

$$\text{Output} = \frac{A_s A_c}{2} \sin[(\omega_c - \omega_s)t + 90°] + \frac{A_s A_c}{2} \sin[(\omega_c + \omega_s)t - 90°] \tag{3.239}$$

We see that the frequency spectrum of this signal is a discrete spectrum existing only at the frequencies $\omega_c - \omega_s$ and $\omega_c + \omega_s$, the so-called *side frequencies*. If such a signal is the input $q_i(t)$ to a measurement system, we can find the steady-state output easily by the methods of Fig. 3.74.

Some applications of these concepts may be appropriate at this point. In a first, rough consideration of the vibration and noise of shafts with gears, perhaps we would expect the important frequencies to correspond to the rotational speeds of the shafts and to the tooth-meshing frequencies. For example, a shaft with a 20-tooth gear running at 200 r/s would be expected to generate noise at 200 and 4,000 Hz. However, actual noise measurements in such situations may show the peak noise to occur at frequencies different from those expected from the above crude analysis.[46] These discrepancies often may be resolved by the application of amplitude-modulation concepts as follows:

For a pair of absolutely true-running gears, the tooth forces (which cause vibration and thus noise) would have a fundamental frequency equal to the tooth-meshing frequency. These forces would not be pure sine waves, but *would be periodic*. Thus we could get a Fourier series for them. For simplicity, let us assume these forces to be pure sine waves of fixed amplitude. In an actual set of gears, there is always some eccentricity or "runout"; that is, the gears are closer together at some points in their rotation than at others. It is postulated that this runout leads to a force amplitude that *varies* as the gear rotates; that is, the tooth-force amplitude is *modulated* as a function of rotational position. If this is so, the frequencies of generated noise (corresponding to tooth-force frequencies) would be expected to be the side frequencies generated by modulating the 4,000-Hz tooth-meshing frequency with the 200-Hz (once per rotation) runout frequency. These frequencies (3,800 and 4,200 Hz in our example) actually have been measured, which confirms the original conjecture. Without the amplitude-modulation concepts, the engineer would have been hard pressed to explain these frequencies in the measured data.

Another interesting example is found in the carrier amplifier. To measure easily and record very small voltages coming from transducers (such as strain gages) requires a very-high-gain amplifier. Because of drift problems, a high-gain amplifier is easier to build as an ac rather than a dc unit. An ac amplifier, however, does not amplify constant or slowly varying voltages and so would appear to be unsuitable for measuring static strains. This problem is overcome by exciting the strain-gage bridge with ac voltage (say 5 V at 3,000 Hz) rather than dc. Thus when the bridge is unbalanced by strain-induced resistance changes, the output voltage will be a 3,000-Hz ac voltage whose amplitude will be modulated by the strain changes. Thus if we are measuring strains which vary from, say, 0 Hz (static) to 10 Hz, the amplifier will have an input-frequency spectrum bounded by 2,990 and 3,010 Hz. This range of frequencies is easily handled by an ac amplifier. Figure 3.87 illustrates these concepts. This "shifting" of the information frequencies from one part of the frequency range to another is the basis of many useful applications of amplitude modulation.

As a final example, suppose that the wires leading from the bridge to the amplifier in Fig. 3.87 are subjected to a stray 60-Hz field from surrounding ac machinery and a 60-Hz noise, or "hum," is superimposed (additively) on the desired signals. This 60-Hz noise easily could be larger than the desired strain signals. With a carrier system, however, this noise may be eliminated merely by designing the ac amplifier so that it does not respond to 60 Hz. Since the desired band of frequencies

---

[46]P. K. Stein, "Measurement Engineering," vol. 1, sec. 17, Stein Engineering Services, Phoenix, 1962; A. L. Gu and R. H. Badgley, "Prediction of Vibration Sidebands in Gear Meshes," *ASME Paper* 74-DET-95, 1974.

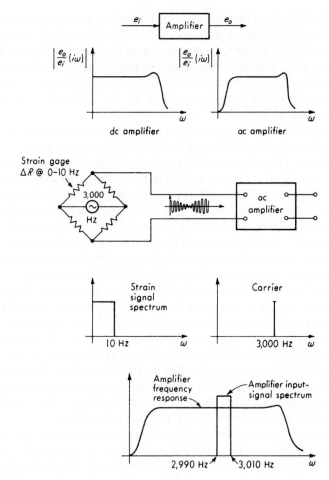

Figure 3.87
Application of amplitude modulation.

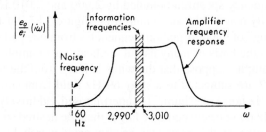

Figure 3.88
Noise rejection through amplitude modulation.

is 2,990 to 3,010 Hz, making the low-frequency cutoff of the amplifier greater than 60 Hz is not difficult. Figure 3.88 illustrates this situation.

We now extend the amplitude-modulation concept to signals other than just a single sine wave. If the modulating signal is a periodic function $f_i(t)$, it may be expanded in a Fourier series to get the output of the modulator as [see Eq. (3.207)]

$$\text{Output} = \left[f_{i,\text{av}} + \frac{1}{L} \left(\sum_{n=1}^{\infty} a_n \cos \frac{n\pi t}{L} + \sum_{n=1}^{\infty} b_n \sin \frac{n\pi t}{L} \right) \right] A_c \sin \omega_c t \quad \textbf{(3.240)}$$

which can be written as

$$\text{Output} = A_0 A_c \sin \omega_c t + (A_1 A_c \cos \omega_1 t \sin \omega_c t$$
$$+ A_2 A_c \cos \omega_2 t \sin \omega_c t + \cdots) + (B_1 A_c \sin \omega_1 t \sin \omega_c t$$
$$+ B_2 A_c \sin \omega_2 t \sin \omega_c t + \cdots) \quad \textbf{(3.241)}$$

Now, $\qquad \sin \alpha \sin \beta \equiv \frac{1}{2} \cos (\alpha - \beta) - \frac{1}{2} \cos (\alpha + \beta)$

and $\qquad \sin \alpha \cos \beta \equiv \frac{1}{2} \sin (\alpha + \beta) + \frac{1}{2} \sin (\alpha - \beta)$

and so

$$\text{Output} = A_0 A_c \sin \omega_c t + C_1 \{ \sin [(\omega_c + \omega_1)t - \alpha_1]$$
$$+ \sin [\omega_c - \omega_1) t + \alpha_1] \} + \cdots \quad \textbf{(3.242)}$$

where $\qquad C_1 \triangleq \dfrac{A_c}{2} \sqrt{A_1^2 + B_1^2}$

$$\alpha_1 \triangleq \tan^{-1} \frac{B_1}{A_1} \quad \textbf{(3.243)}$$

We see that the spectrum of the output signal is a discrete spectrum containing the frequencies ω_c, $\omega_c \pm \omega_1$, $\omega_c \pm \omega_2$, $\omega_c \pm \omega_3$, etc. That is, each frequency component of the modulating signal produces one pair of side frequencies (see Fig. 3.89). If the output of the modulator is applied to the input of a system with known frequency response, the methods of Fig. 3.74 can be used again to find the steady-state output.

If the modulating signal is a transient, the spectrum of the modulator output may be determined with the help of the modulation theorem[47] for Fourier (or Laplace) transforms. If the modulating signal is a transient $f_i(t)$, it will have a Fourier transform $F_i(i\omega)$, which can be obtained in the usual ways. The modulation theorem leads to the following result if $f_i(t)$ is multiplied by the carrier $A_c \sin \omega_c t$ to produce the modulated output:

$$\text{Fourier transform of modulated output} \triangleq |Q_i(i\omega)| \angle Q_i(i\omega) \quad \textbf{(3.244)}$$

where $\qquad |Q_i(i\omega)| = \dfrac{A_c}{2} \text{magnitude}\{ F_i[i(\omega - \omega_c)] \} \quad \textbf{(3.245)}$

$$\angle Q_i(i\omega) = \text{angle } \{ F_i[i(\omega - \omega_c)] \} - 90° \quad \textbf{(3.246)}$$

and $0 \leq \omega < \infty$. Note that the argument (independent variable) of the F_i function is now $i(\omega - \omega_c)$, so that if we want to find $Q_i(i4{,}000)$ and $\omega_c = 3{,}000$, we must

---

[47]G. A. Korn and T. M. Korn, "Mathematical Handbook for Scientists and Engineers," p. 219, McGraw-Hill, New York, 1961; Papoulis, "Fourier Integral and Its Applications," p. 15.

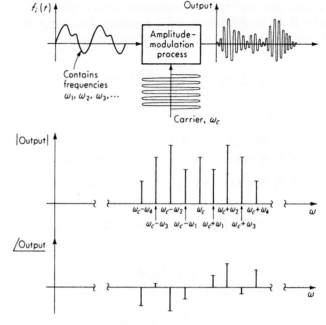

Figure 3.89
Frequency spectrum when modulating signal is periodic.

evaluate $F_i(i1,000)$. Also note that, to get $Q_i(i\omega)$ for $0 < \omega < \infty$, we must know $F_i(i\omega)$ for *negative* ω's, since any $\omega < \omega_c$ gives $F_i[i(\omega - \omega_c)]$ a negative argument. While generally we have worked with positive ω's, the transform for negative ω's always exists and is easily found from the previously given symmetry rules. The spectrum given by Eq. (3.244) will be a continuous one, and if the modulated output is applied as an input to some system with known frequency response, then the corresponding output can be obtained by the methods of Fig. 3.84.

For measurement systems in which amplitude modulation is intentionally introduced to allow the use of carrier-amplifier techniques, the carrier frequency must be considerably greater (usually 5 to 10 times) than any significant frequencies present in the modulating signal. For such a situation, the pertinent frequency spectra are as shown in Fig. 3.90. We see again that the amplitude-modulation process shifts the frequency spectrum by the amount ω_c.

When amplitude modulation is intentionally introduced to facilitate data handling in one way or another, it generally plays the role of an intermediate step, and the amplitude-modulated signal usually is not considered a suitable final readout. Rather, the original form of the modulating signal (the basic measured data from, say, a transducer) should be recovered. The process for accomplishing this involves *demodulation* (or *detection,* as it is sometimes called) and filtering. Demodulation may be full-wave, half-wave, phase-sensitive, or non-phase-sensitive (Fig. 3.91). Here we treat the form giving the best reproduction of the original data, full-wave phase-sensitive demodulation, and consider only the process, not the hardware for

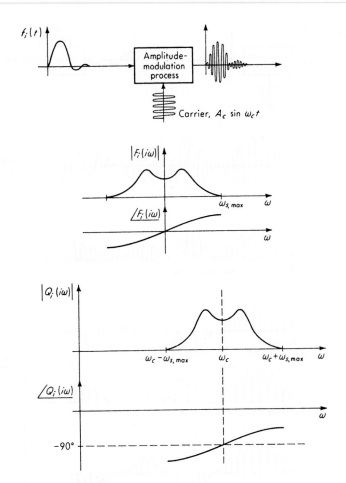

Figure 3.90
Frequency spectrum when modulating signal is a transient.

accomplishing it. Again, it is necessary to consider whether the form of the original signal was a single sine wave, periodic function, or transient. For a single sine wave $A_s \sin \omega_s t$ which is modulating a carrier $A_c \sin \omega_c t$, the expression for the full-wave phase-sensitive demodulated signal is

$$\text{Demodulator output} = (A_s \sin \omega_s t)|A_c \sin \omega_c t| \qquad \textbf{(3.247)}$$

as seen from Fig. 3.92a. Now $|A_c \sin \omega_c t|$ is a periodic function and may be expanded in a Fourier series by application of Eq. (3.207). The results are

$$|A_c \sin \omega_c t| = \frac{2A_c}{\pi}\left(1 - \frac{2}{3}\cos 2\omega_c t - \frac{2}{15}\cos 4\omega_c t + \cdots\right.$$

$$\left. + \frac{2}{1-4n^2}\cos 2n\omega_c t + \cdots\right) \qquad n = 3, 4, 5.\ldots \qquad \textbf{(3.248)}$$

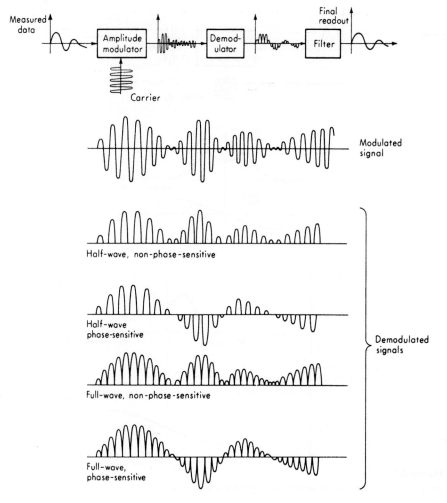

Figure 3.91
Types of demodulation.

Equation (3.247) can then be written as

Demodulator output $= (A_s \sin \omega_s t)$

$$\left[\frac{2A_c}{\pi}\left(1 - \frac{2}{3}\cos 2\omega_c t - \frac{2}{15}\cos 4\omega_c t + \cdots\right)\right] \quad \textbf{(3.249)}$$

which, when multiplied, gives

$$\text{Demodulator output} = \frac{2A_c A_s}{\pi}\sin \omega_s t - \frac{4A_c A_s}{3\pi}\sin \omega_s t \cos 2\omega_c t$$

$$- \frac{4A_c A_s}{15\pi}\sin \omega_s t \cos 4\omega_c t + \cdots \quad \textbf{(3.250)}$$

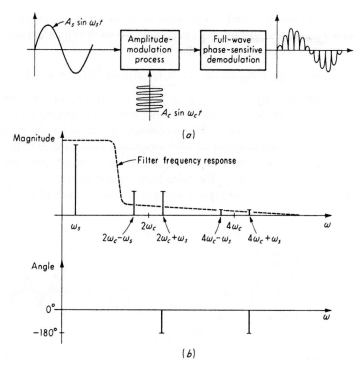

Figure 3.92
Frequency spectrum of full-wave phase-sensitive demodulation.

Now, by a trigonometric identity, terms of the form $(\sin \omega_s t)(\cos 2n\omega_c t)$ can be written as $[\sin (2n\omega_c + \omega_s)t - \sin (2n\omega_c - \omega_s)t]/2$. Thus we can write Eq. (3.250) as

$$\text{Demodulator output} = \frac{2A_c A_s}{\pi} \sin \omega_s t - \frac{2A_c A_s}{3\pi} [\sin (2\omega_c + \omega_s)t$$

$$- \sin (2\omega_c - \omega_s)t] - \frac{2A_c A_s}{15\pi} [\sin (4\omega_c + \omega_s)t - \sin (4\omega_c - \omega_s)t] + \cdots \quad \textbf{(3.251)}$$

From this we see that the frequency spectrum of the demodulator output signal is a discrete spectrum with frequency content at ω_s, $2\omega_c \pm \omega_s$, $4\omega_c \pm \omega_s$, etc., as shown in Fig. 3.92b. If this signal were an input to a system of known frequency response (such as the filter of Fig. 3.91), the output of this system would be found by the methods of Fig. 3.74. If the output of the filter is to look like the original data, the filter must be designed to *reject* frequencies $2\omega_c \pm \omega_s$, $4\omega_c \pm \omega_s$, etc., while *passing* with a minimum of distortion the signal frequency ω_s. The design of such a low-pass filter is made simpler if the passband and the rejection band are more widely separated. This is the basis of our earlier statement that carrier frequencies usually are chosen to be 5 to 10 times the highest expected signal frequency.

When the modulating signal is a periodic wave rather than a single sine wave, a procedure similar to that just used is employed, except now the modulating signal is *also* expressed as a Fourier series of the form

$$\text{Modulating signal} = A_{s0} + A_{s1} \sin (\omega_s t + \alpha_1) + A_{s2} \sin (2\omega_s t + \alpha_2) + \cdots \quad \textbf{(3.252)}$$

When this is multiplied by $|A_c \sin \omega_c t|$ as given by Eq. (3.248), we find exactly the same situation as for a single sine wave, but it must be applied for *each* signal frequency $(0, \omega_s, 2\omega_s, 3\omega_s, \text{etc.})$. The frequency spectrum of the demodulated signal thus will be a discrete spectrum with frequency content at $\omega = 0$, $(\omega_s, 2\omega_c, 2\omega_c \pm \omega_s, 4\omega_c, 4\omega_c \pm \omega_s, 6\omega_c, 6\omega_c \pm \omega_s, \text{etc.})$, $(2\omega_s, 2\omega_c \pm 2\omega_s, 4\omega_c \pm 2\omega_s, 6\omega_c \pm 2\omega_s, \text{etc.})$, etc. Figure 3.93 illustrates these concepts. Again, if such a signal is applied to a system of known frequency response, the methods of Fig. 3.74 allow calculation of the output.

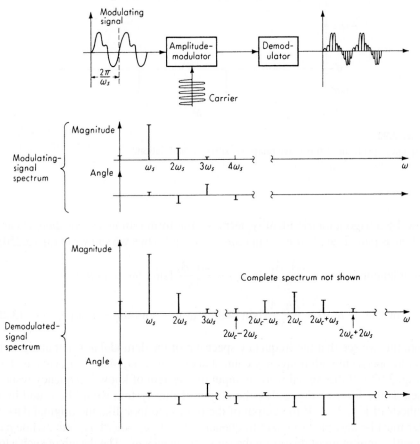

Figure 3.93
Demodulation spectrum for periodic input.

When the modulating signal is a transient $f_i(t)$, the demodulated signal will be $f_i(t)|A_c \sin \omega_c t|$, which can be written as

$$\text{Demodulated signal} = f_i(t)\left[\frac{2A_c}{\pi}\left(1 - \frac{2}{3}\cos 2\omega_c t - \frac{2}{15}\cos 4\omega_c t + \cdots\right)\right] \quad \textbf{(3.253)}$$

or, after multiplying,

$$\text{Demodulated signal} = \frac{2A_c}{\pi}f_i(t) - \frac{4A_c}{3\pi}f_i(t)\cos 2\omega_c t$$

$$- \frac{4A_c}{15\pi}f_i(t)\cos 4\omega_c t + \cdots \quad \textbf{(3.254)}$$

Application of the modulation theorem to each of the modulated terms of Eq. (3.254) leads to the result

$$\text{Fourier transform of demodulated signal} \triangleq |Q_i(i\omega)| \angle Q_i(i\omega)$$

where

$$|Q_i(i\omega)| = \frac{2A_c}{\pi}|F_i(i\omega)| + \frac{2A_c}{3\pi}|F_i[(i(\omega - 2\omega_c)]|$$

$$+ \frac{2A_c}{15\pi}|F_i[(i(\omega - 4\omega_c)]| + \cdots \quad \textbf{(3.255)}$$

and $0 \le \omega \le \infty$. The expression for $\angle Q_i(i\omega)$ is not easily given since in Fourier-transforming Eq. (3.254) (each term may be treated separately since superposition is allowed), we find $Q_i(i\omega)$ as a *sum* of complex numbers rather than a product. To find the overall angle of a sum of complex numbers, we must express *each* number as $a_i + ib_i$ and then add the a's and b's to get the overall $a + ib$. The angle is then $\tan^{-1}(b/a)$. The angle due to transforming $f_i(t)$ is $\angle F_i(i\omega)$, the angle due to transforming $-f_i(t)\cos 2\omega_c t$ is $-180° + \angle F_i[i(\omega - 2\omega_c)]$, the angle due to transforming $-f_i(t)\cos 4\omega_c t$ is $-180° + \angle F_i[i(\omega - 4\omega_c)]$, etc. These results allow calculation of $\angle Q_i(i\omega)$. The frequency spectrum of such a signal is shown in Fig. 3.94. Since it is a continuous spectrum, the response of a system to such an input may be found by the methods of Fig. 3.84. In Fig. 3.94 it is assumed that the $F_i(i\omega)$ is practically zero for $\omega > \omega_{s,\,max}$ and that $\omega_c > \omega_{s,\,max}$. For such a situation, the phase-angle calculation is greatly simplified, since the individual terms of Eq. (3.255) do not coexist over any frequency range. That is, for $0 < \omega < \omega_{s,\,max}$, only the first term is nonzero; for $(2\omega_c - \omega_{s,\,max}) < \omega < (2\omega_c + \omega_{s,\,max})$, only the second term is nonzero; etc. Thus the overall phase angle is determined by one term only within each of the specified frequency bands.

This concludes our treatment of amplitude-modulated and -demodulated signals. The methods developed can be readily applied to the other common variations such as half-wave demodulation, non-phase-sensitive demodulation, non-sinusoidal-carrier waveforms (square wave, for instance), etc. Non-phase-sensitive demodulation cannot detect a sign change in the modulating signal. Half-wave systems shift the side frequencies of demodulated signals to $\omega_c \pm \omega_s$ rather than

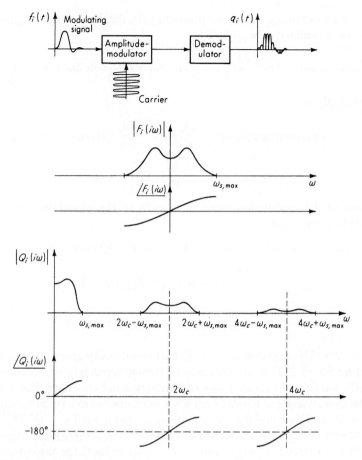

Figure 3.94
Demodulation spectrum for transient input.

$2\omega_c \pm \omega_s$, thus making the filtering problem more difficult. They also have less amplitude than full-wave systems. Nonsinusoidal carriers may be useful in reducing heating effects in resistive transducers such as strain gages. The carrier waveform is such as to give a high ratio of peak value to rms (effective heating) value. The output signal is related to peak value while the power dissipated in the strain gage is related to rms value; thus a high peak/rms ratio increases output for a given allowable heating level. A carrier in the form of a train of high, narrow pulses satisfies this sort of criterion.

Characteristics of Random Signals

The final class of signals we consider is the so-called random or stochastic type. Such signals are of considerable importance since they serve as more realistic mathematical models of many physical processes than do deterministic signals. By

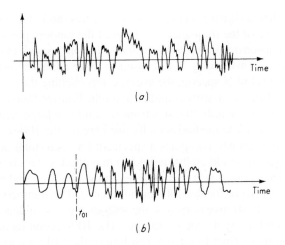

Figure 3.95
Random signals.

a random signal we mean one that can be described only statistically before it actually occurs; that is, it cannot be described by a specific function of time prior to its occurrence. Of course, *all* signals in the real world have some degree of randomness, so that we should make it clear that we are *always* dealing with random signals, even though we may take a specific time function (periodic, transient, etc.) as a *model* of what is really going on to simplify analysis in some types of problems.

Figure 3.95*a* shows a time record of a typical random signal such as might be measured by a pressure pickup mounted in a booster-rocket nozzle subjected to acoustic pressures generated by the rocket exhaust. These pressures are strongly random in that no specific frequencies are apparent in a time record. The stresses caused by such pressures can lead to fatigue failure of the structure. Thus engineers are concerned with means to analyze the effects of such random forcing functions on structures and machinery. Since experimental methods are needed in much of this work, accurate measurement of random signals is important. We will see that frequency-response techniques are most useful for this type of problem.

First, we should note that, as far as deciding on the requirements for a measuring system is concerned, *after* a random process has occurred and there is a time record of it, we can define the function as zero for some $t > t_0$, treat it as a transient, and calculate its Fourier transform to determine its frequency content and thus the required frequency response of the instrumentation. The only problem here is the selection of the cutoff time t_0 so as to have an adequate statistical "sample" of the random function. In Fig. 3.95*b*, for example, if we based our calculation on the record for $0 < t < t_{01}$, we would miss completely the high-frequency character apparent in the record for $t > t_{01}$. The existence of some minimum valid cutoff time t_0 implies that the random process is *stationary,* that is, its statistical properties (such as average value, mean-square value, etc.) do not change with time. When this is true, there will exist some t_0 corresponding to a chosen level of confidence in the results.

Actually, when a signal is entirely random or has random content, one can *not* get a good estimate of the frequency spectrum of the random portion by computing a *single* Fourier spectrum, *no matter how long the time record.* This important fact, which intuitively seems wrong, is discussed in more advanced texts.[48] It is necessary to *average* a set of N spectra, the average approaching the correct spectrum as N is increased. That is, if we computed a single Fourier transform from a time record of 100 seconds length, this spectrum would have large errors and present a "jagged" appearance. The method actually used breaks the 100-second record into, say, ten 10-second records, computes a spectrum for each (total of 10 spectra) and then averages these (just add up the individual magnitudes at each frequency and divide by 10). This averaged spectrum will be "smoother" and will be more accurate than the single spectrum for the 100-second record. The numbers in this example are given only for illustrative purposes; the reference gives guidance on how proper values are chosen for any practical example. The 100-second record *does* have the advantage of 10 times finer frequency resolution; the "10-second" records chosen for averaging would have to meet the user's specification for this criterion. That is, we might need 10 (or more) records of, say, 33-second length to meet *all* our needs.

We should make it clear that, in dealing with random processes, theoretically an *infinite* record length is needed to give precise results, and results based on finite-length records must always be qualified by statistical statements referring to the *probability* of the result's being correct within a certain percentage. This situation leads to an engineering trade-off since long records are desirable from accuracy considerations, but are undesirable in terms of the cost involved in obtaining and analyzing them. Also, in many situations the maximum available length of record is limited by the lifetime of the device under study, which may be quite short in the case of a missile or rocket. If a long enough record is available, the minimum allowable t_0 can be found by choosing a small t_0 and calculating the Fourier transform. Then a larger t_0 is chosen and the Fourier transform again evaluated. If the second transform differs significantly from the first, the first t_0 was not long enough. By choosing successively longer t_0's, you will find some range of t_0 beyond which further increases in t_0 cause no significant change in the transform. Any t_0 beyond this range thus would be considered acceptably long. The averaging procedure explained above might not be necessary if our interest is *only* in finding the highest frequencies present in our signal, to choose a measurement system with sufficient bandwidth. This is so because an (inaccurate) spectrum computed from a single record, *will* often give a usable indication of where the signal's frequency content "dies out." These questions of record length and number of spectra to average have been studied theoretically but one can learn a lot by simulating these situations using the DASYLAB software provided with this book. (In actual practice, more refined statistical procedures may be required, and are available,[49] to resolve questions of this sort.)

---

[48]E. O. Doebelin, "System Modeling and Response," Wiley, New York, 1980, p. 268.

[49]E. O. Doebelin, "System Modeling and Response," pp. 111, 267.

While the above concepts are adequate for understanding the requirements put on measurement systems for random variables, they do not cover the means available for statistically *describing* the signals. That is, while the exact form of a random function cannot be predicted ahead of time, certain of its statistical characteristics *can* be predicted; these may be useful in predicting (in a statistical way) the output of some physical system that has the random variable as an input. Now we develop some of the more common methods of statistically describing random signals. These methods fall mainly into two groups: those concerned with describing the magnitude of the variable and those concerned with describing the rapidity of change (frequency content) of the variable.

If we call a random variable $q_i'(t)$, the *average* or *mean value* $\overline{q_i'(t)}$ is defined by

$$\overline{q_i'(t)} \triangleq \lim_{T \to \infty} \left(\frac{1}{T}\right) \int_0^T q_i'(t) \, dt \tag{3.256}$$

Since this is a constant component of the total signal and we are more concerned with the *fluctuations,* it usually is subtracted from the total signal, to give a signal with zero mean value. Also, many real random processes inherently have zero mean value. For these reasons, from here on we consider only signals $q_i(t)$ with zero mean value. An indication of the magnitude of the random variable is the *mean-squared value* $\overline{q_i^2(t)}$ given by

$$\overline{q_i^2(t)} \triangleq \lim_{T \to \infty} \left(\frac{1}{T}\right) \int_0^T q_i^2(t) \, dt \tag{3.257}$$

The mean-squared value has dimensions of $[q_i(t)]^2$. Thus to get a measure of the size of $q_i(t)$ itself, the *root-mean-square* (rms) value $q_i(t)_{\text{rms}}$ is defined by

$$q_i(t)_{\text{rms}} \triangleq \sqrt{\overline{q_i^2(t)}} \tag{3.258}$$

The above definitions are illustrated in Fig. 3.96. The quantities $\overline{q_i^2(t)}$ and $q_i(t)_{\text{rms}}$ give an indication of the overall size of $q_i(t)$, but no clue as to the distribution of "amplitude," that is, the probability of occurrence of large or small values of $q_i(t)$. The specification of this important information is provided by the *amplitude-distribution function* (probability density function) $W_1(q_i)$. To define this function, we consider Fig. 3.97. We define the probability P that $q_i(t)$ will be found between some specific value q_{i1} and $q_{i1} + \Delta q_i$ by

$$\text{Probability } [q_{i1} < q_i < (q_{i1} + \Delta q_i)] \triangleq P[q_{i1}, (q_{i1} + \Delta q_i)] = \lim_{T \to \infty} \frac{\sum \Delta t_i}{T} \tag{3.259}$$

where $\sum \Delta t_i$ represents the total time spent by $q_i(t)$ within the band Δq_i during time interval T. We now define $W_1(q_i)$ by

$$\text{Amplitude-distribution function} \triangleq W_1(q_i) \triangleq \lim_{\Delta q_i \to 0} \frac{P[q_i, (q_i + \Delta q_i)]}{\Delta q_i} \tag{3.260}$$

From this definition it should be clear that

$$W_1(q_i) \, dq_i = \text{probability that } q_i \text{ lies in } dq_i \tag{3.261}$$

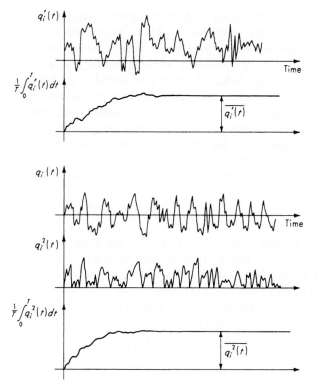

Figure 3.96
Average and mean-square values.

and thus

$$\int_{q_{i1}}^{q_{i2}} W_1(q_i)\, dq_i = \text{probability that } q_i \text{ lies between } q_{i1} \text{ and } q_{i2} \qquad \textbf{(3.262)}$$

The function $W_1(q_i)$ theoretically can take on an infinite number of different forms; however, certain forms have been found to be adequate mathematical models for real physical processes. The most common is the Gaussian or normal distribution given by

$$W_1(q_i) = \frac{1}{\sqrt{2\pi}\,\sigma}\, e^{-q_i^2/(2\sigma^2)} \qquad \textbf{(3.263)}$$

where $\sigma \triangleq$ standard deviation. We encountered the general (average value $\neq 0$) form earlier in Eq. (3.4). While much real-world data is close to Gaussian, we should check each application, using the method of Fig. 3.9.

While the quantities $q_i(t)_{\text{rms}}$ and $W_1(q_i)$ usually are sufficient to describe the magnitude of a random variable, they give no indication as to the *rapidity* of variation in time. That is, two random processes could both be Gaussian with the same

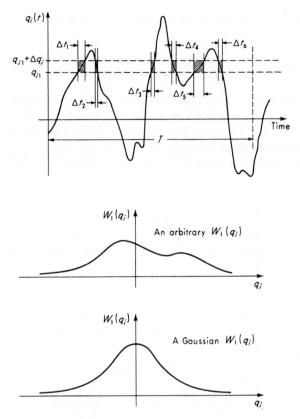

Figure 3.97
Definition of probability density function.

numerical value of σ, but one could vary much more rapidly than the other. To describe the time aspect of random variables, the concepts of *autocorrelation function* and *mean-square spectral density* (power spectral density) are employed.[50] The autocorrelation function $R(\tau)$ of a random variable $q_i(t)$ is given by

$$R(\tau) \triangleq \lim_{T \to \infty} \left(\frac{1}{T}\right) \int_0^T q_i(t) q_i(t + \tau)\, dt \qquad (3.264)$$

The function $q_i(t + \tau)$ is simply $q_i(t)$ shifted in time by τ seconds. Thus, to find $R(\tau)$, we select a value of τ, say 2 s, plot against t the functions $q_i(t)$ and $q_i(t + 2)$, multiply them together point by point, integrate the product curve from 0 to T, and divide by T. Then this procedure is repeated for other values of τ to generate the curve $R(\tau)$ versus τ. The shifting of $q_i(t)$ by τ sec could be accomplished by writing

[50]S. H. Crandall, "Random Vibration," Wiley, New York, 1958; E. O. Doebelin, "System Modeling and Response," p. 120.

$q_i(t)$ on magnetic tape at one point and reading it off at another. The time delay τ then is simply

$$\tau = \frac{\text{distance between read and write heads}}{\text{tape velocity}}$$

The multiplication of $q_i(t)$ and $q_i(t + \tau)$, the integration, and the division by T were in earlier times accomplished by standard electronic analog-computer components. However, today the entire $R(\tau)$ calculation would usually be done digitally. Figure 3.98 illustrates these concepts. For $\tau = 0$, Eq. (3.264) gives $R(0) = \overline{q_i^2(t)}$; that is, the autocorrelation function is numerically equal to the mean-square value for $\tau = 0$.

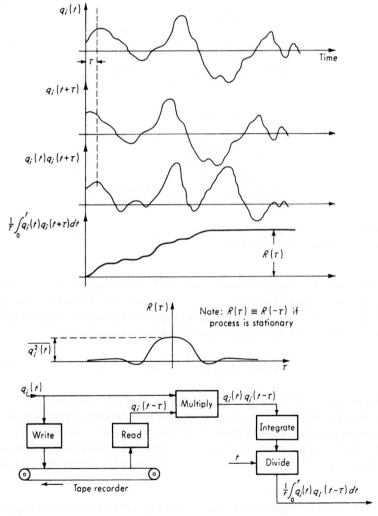

Figure 3.98
Definition of autocorrelation function.

To appreciate the relation between $R(\tau)$ and the rapidity of variation of $q_i(t)$, consider Fig. 3.99, where both a slowly varying and a rapidly varying $q_i(t)$ are shown. For *any* $q_i(t)$, fast or slow, when $\tau = 0$, $q_i(t)q_i(t + \tau)$ is *positive* for all t. This integration gives the largest possible value, $\overline{q_i^2(t)}$. Any shift ($\tau \neq 0$) will "misalign" the positive and negative parts of $q_i(t)$ and $q_i(t + \tau)$, causing the product curve to be sometimes positive, sometimes negative. Thus integration of this curve gives a smaller value than for $\tau = 0$. If $q_i(t)$ is rapidly varying, only a small shift (small τ) is needed to cause this misalignment, whereas a slowly varying $q_i(t)$ requires a larger shift before $R(\tau)$ drops off significantly. Thus a sharp peak in $R(\tau)$ at $\tau = 0$ indicates the presence of rapid variation (strong high-frequency content) in $q_i(t)$.

The mean-square spectral density (power spectral density) is another method of determining the frequency content of a random signal. The mean-square spectral density is proportional to the Fourier transform of $R(\tau)$ and conveys in the frequency domain exactly the same information as $R(\tau)$ conveys in the time domain. While they are mathematically related, in actual practice where they must be determined experimentally, one or the other may be preferable. It appears that in random-vibration work the mean-square spectral-density approach is largely preferred. To develop this concept, let us consider first a periodic function expanded into a Fourier series to give

$$q_i(t) = A_{i1} \sin (\omega_1 t + \alpha_1) + A_{i2} \sin (2\omega_1 t + \alpha_2) + \cdots \qquad (3.265)$$

It is easy to show that the total mean-square value of $q_i(t)$ is equal to the sum of the individual mean-square values for each of the harmonic terms:

$$\overline{q_i^2(t)} = \overline{q_{i1}^2(t)} + \overline{q_{i2}^2(t)} + \cdots \qquad (3.266)$$

Thus the contribution of each frequency to the overall mean-square value is easily found.

We now develop, for a *random* function, a related technique that will show how the total mean-square value is "distributed" over the frequency range. First we note that for a random function, no isolated, discrete frequencies exist; thus the frequency spectrum is a continuous one. The concept of mean-square spectral density

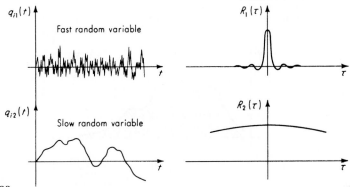

Figure 3.99
Frequency significance of autocorrelation function.

perhaps is most clearly visualized in terms of the analog instrumentation which could be used to measure it experimentally. Figure 3.100a shows the arrangement necessary to measure the *overall* mean-square value of a signal $q_i(t)$. To find out how much each part of the frequency range contributes to this overall value, we simply filter out (with a narrow-bandpass filter of bandwidth $\Delta\omega$) all frequencies other than the narrow band of interest and then perform the squaring and averaging operations on what remains, as in Fig. 3.100b. The results of this operation are thus the mean-square value of that part of the signal $q_i(t)$ lying within the chosen frequency band. We call this value $\overline{q_{i,\,\omega}^2}$. The filter is adjustable in the sense that we can shift its passband anywhere along the frequency axis, and so we can obtain $\overline{q_{i,\,\omega}^2}$ for any chosen center frequency ω. A narrow passband is desirable for resolving closely spaced peaks in the spectrum, but is undesirable in terms of the increased time required to cover a given frequency range in small steps rather than large. Thus a compromise is needed. The mean-square spectral density $\phi(\omega)$ is defined by

$$\phi(\omega) \triangleq \frac{\overline{q_{i,\,\omega}^2}}{\Delta\omega} \tag{3.267}$$

and represents the "density" (amount per unit frequency bandwidth) of the mean-square value, since $\phi(\omega)\Delta\omega = \overline{q_{i,\,\omega}^2}$. [Theoretically, the filter bandwidth should be

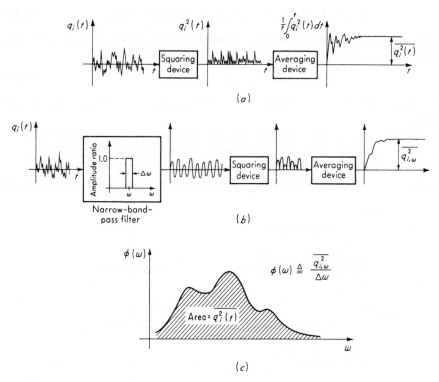

Figure 3.100
Definition of mean-square spectral density.

shrunk toward zero and $\phi(\omega)$ defined as the *limit* of $\overline{q_{i,\omega}^2}/\Delta\omega$. In practice, $\Delta\omega$ has been made "small enough" when further reductions cause no significant change in the numerical value of $\phi(\omega)$. Where $\phi(\omega)$ changes little with ω, a wider $\Delta\omega$ will suffice than where $\phi(\omega)$ has "sharp spikes."] If we evaluate $\phi(\omega)$ for a whole range of frequencies, we can plot it as a curve versus ω, as in Fig. 3.100c. Note that the dimensions of $\phi(\omega)$ are those of $q_i^2/(\text{rad}/\text{s})$. The total area under the $\phi(\omega)$-versus-ω curve will be the total mean-square value $\overline{q_i^2(t)}$.

You will find that in the literature the term "power spectral density" (sometimes "auto-spectral density") is used almost exclusively for the quantity we call mean-square spectral density. This is a carry-over from communications engineering where the concept was originally developed and where the signal $q_i(t)$ is a voltage applied to a 1-Ω resistor. Under these conditions, $\phi(\omega)$ would have dimensions of power/(rad/s). Actually, however, the concept of $\phi(\omega)$ is a *mathematical* one related to the mean-square value, and *not* a physical one related to electrical engineering. In physical applications the dimensions of $\phi(\omega)$ would be $(^\circ\text{C})^2/(\text{rad}/\text{s})$ if $q_i(t)$ were temperature, $g^2/(\text{rad}/\text{s})$ if $q_i(t)$ were acceleration, etc., which are in general *not* power/(rad/s). However, the term "power spectral density" and its symbol PSD seem to be firmly entrenched and probably will continue to prevail. Here we merely suggest that mean-square spectral density might be more appropriate terminology. Thus when dealing with, say, random pressures, you would refer to the mean-square spectral density of pressure.

Perhaps the main interest in the mean-square spectral density lies in the fact that if a $q_i(t)$ with a known $\phi_i(\omega)$ is applied as input to a linear system of known frequency response, the mean-square spectral density $\phi_o(\omega)$ of the output is easily computed from the relation.[51]

$$\phi_o(\omega) = \phi_i(\omega)\left|\frac{q_o}{q_i}(i\omega)\right|^2 \tag{3.268}$$

We can thus compute the $\phi_o(\omega)$ curve point by point (see Fig. 3.101). The area under this curve will be the total mean-square value of $q_o(t)$. Furthermore, if the input is Gaussian, the output will also be Gaussian. It is then possible to make statements such as

$$|q_o(t)| > \sqrt{\overline{q_o^2(t)}} \qquad 31.7\% \text{ of the time}$$

$$|q_o(t)| > 2\sqrt{\overline{q_o^2(t)}} \qquad 4.6\% \text{ of the time} \tag{3.269}$$

$$|q_o(t)| > 3\sqrt{\overline{q_o^2(t)}} \qquad 0.3\% \text{ of the time}$$

Other useful results of this nature can be found in the lilterature.[52]

A particular form of $\phi(\omega)$ is of great utility in practice. This is the *white noise*. For a mathematically perfect white noise, $\phi(\omega)$ is equal to a constant for all frequencies; that is, it contains all frequencies in equal amounts (see Fig. 3.102). This

---

[51]J. S. Bendat, "Principles and Application of Random Noise Theory," p. 199, Wiley, New York, 1958.

[52]J. S. Bendat, L. D. Enochson, and A. G. Piersol, "Analytical Study of Vibration Data Reduction Methods," *NASA N*64-15529, 1963.

makes it useful as a test signal, just as the impulse is useful as a transient test signal because of its uniform frequency content. From Eq. (3.268), if $\phi_i(\omega) = C$, then

$$\phi_o(\omega) = C \left| \frac{q_o}{q_i}(i\omega) \right|^2 \tag{3.270}$$

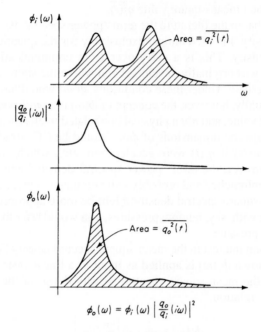

$$\phi_o(\omega) = \phi_i(\omega) \left| \frac{q_o}{q_i}(i\omega) \right|^2$$

Figure 3.101
System response to random input.

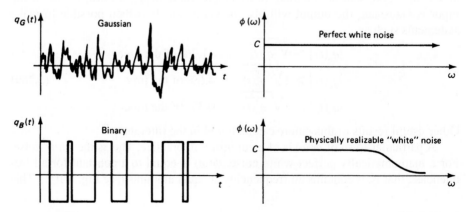

Figure 3.102
White noise.

and thus

$$\left| \frac{q_o}{q_i}(i\omega) \right| = \sqrt{\frac{\phi_o(\omega)}{C}} \tag{3.271}$$

While it is not possible to build a generator of perfect white noise, the flat range of $\phi(\omega)$ of a practical generator can be quite large, and as long as its flat range extends beyond the frequency response of the system being tested, the "nonwhiteness" will not present any difficulty. If a white-noise generator is available, we can "construct" almost any $\phi(\omega)$ that we wish simply by passing the white noise through a suitably designed filter according to Eq. (3.270). That is, $|(q_o/q_i)(i\omega)|^2$ for the filter must have the shape desired in $\phi(\omega)$. This technique is widely used in computer simulation studies of aircraft gust response, control-system evaluation, etc., where the frequency characteristics of some random-input quantity are known and it is desired to study their effect on some system.

In addition to the Gaussian (continuous) white noise we emphasize here, a flat mean-square spectral density also can be achieved with *binary random noise,* a signal that switches between two fixed levels at time intervals which vary randomly. A further variation is pseudorandom binary noise; details on the application of these types of signals are available in the literature.[53]

Most of the previous material has referred to statistical properties of a *single* random variable. Useful practical results may be derived by consideration of two random variables. The *joint amplitude-distribution function* (joint probability density function) of two random variables $q_1(t)$ and $q_2(t)$ is given by

$$W_1(q_1, q_2) = \lim_{T \to \infty} \lim_{\substack{\Delta_{q1} \to 0 \\ \Delta_{q2} \to 0}} \frac{1}{T \, \Delta q_i \, \Delta q_2} \Sigma \, \Delta t_i \tag{3.272}$$

where $\Sigma \, \Delta t_i$ represents the total time (during time T) that $q_1(t)$ and $q_2(t)$ spent *simultaneously* in the bands $q_1 + \Delta q_1$ and $q_2 + \Delta q_2$. Figure 3.103 illustrates these concepts. From the basic definition, it should be clear that

$$\text{Probability } (q_{1a} < q_1 < q_{1b}, q_{2a} < q_2 < q_{2b}) = \int_{q_{2a}}^{q_{2b}} \int_{q_{1a}}^{q_{1b}} W_1(q_1, q_2) \, dq_1 \, dq_2 \tag{3.273}$$

Again, $W_1(q_1, q_2)$ can take an infinite variety of forms. The most useful is probably the bivariate Gaussian (normal) distribution.[54] The main purpose in experimentally measuring $W_1(q_1, q_2)$ is, just as for $W_1(q_1)$, to determine whether the physical data follow approximately some simple mathematical form such as the Gaussian. If this can be proved, many useful theoretical results can be applied. Also, certain calculations can be made directly from $W_1(q_1, q_2)$. For example, if q_1 and q_2 represent the random vibratory motions of two adjacent machine parts, knowledge of $W_1(q_1, q_2)$ allows calculation of the probability that the two parts will strike each other. In actual practice, engineering applications of $W_1(q_1, q_2)$ are somewhat limited because of the difficulty of measuring this function.

---

[53]E. O. Doebelin, "System Modeling and Response," p. 273.

[54]A. M. Mood, "Introduction to the Theory of Statistics," p. 165, McGraw-Hill, New York, 1950.

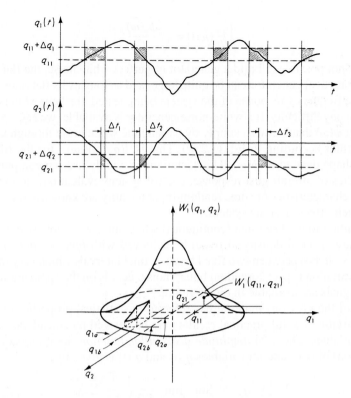

Figure 3.103
Bivariate probability density function.

The *cross-correlation function* $R_{q1q2}(\tau)$ for two random variables $q_1(t)$ and $q_2(t)$ is defined by

$$R_{q1q2}(\tau) = \lim_{T \to \infty} \left(\frac{1}{T}\right) \int_0^T q_1(t)q_2(t + \tau) \, dt \qquad (3.274)$$

An example[55] of its application might be as follows: Suppose a source of vibratory motion $q_1(t)$ exists at one point in a structure and causes a vibratory response motion $q_2(t)$ at another point in the structure. Suppose also that the transmission of vibration from the first point to the second could occur by either (or both) of two mechanisms. The first mechanism is by acoustic (air-pressure) wave propagation through the air separating the two points; the second is by elastic wave propagation through the metallic structure connecting the two points. Since the propagation velocities of waves in air and metal are greatly different, we would expect that an input at q_1 would cause an output at q_2 that would be delayed by different times, depending on whether the transmission was mainly through the air or mainly through the metal. If we know the transmission path length and the wave velocity, these delays can be calculated. Suppose the air-path delay is 0.01 s and the structure-path delay is

[55]Bendat, Enochson, and Piersol, op. cit.

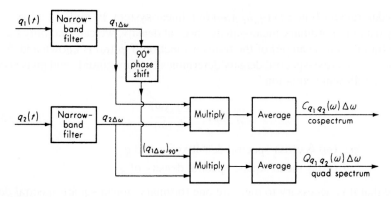

Figure 3.104
Cross-spectral density.

0.002 s. If we now experimentally measure $R_{q_1q_2}(\tau)$ and find a large peak at $\tau = 0.01$ s and a smaller one at $\tau = 0.002$ s, we conclude that the acoustic transmission is responsible for most of the vibration at q_2. Thus, since $R_{q_1q_2}$ is a measure of the correlation ("relatedness") between two signals delayed by various amounts, it can be used as a diagnostic tool for investigating the presence and/or nature of the relation, as in the above example. Developments in LSI (large-scale integration) electronics and microprocessors provide low-cost correlators which can serve as the basis for various useful instrument systems.[56]

Information equivalent to that contained in the cross-correlation function but in a different (and often more practically useful) form is found in the *cross-spectral density* (cross-power spectral density) $\phi_{q_1q_2}(\omega)$ given by

$$\phi_{q_1q_2}(\omega) = C_{q1q2}(\omega) - iQ_{q_1q_2}(\omega) \tag{3.275}$$

where $\quad C_{q_1q_2}(\omega) \triangleq$ cospectrum

$$\triangleq \lim_{T \to \infty} \lim_{\Delta\omega \to 0} \frac{1}{T\,\Delta\omega} \int_0^T (q_{1\Delta\omega})(q_{2\Delta\omega})\,dt \tag{3.276}$$

$Q_{q_1q_2}(\omega) \triangleq$ quad spectrum

$$\triangleq \lim_{T \to \infty} \lim_{\Delta\omega \to 0} \frac{1}{T\,\Delta\omega} \int_0^T (q_{1\Delta\omega})_{90°}(q_{2\Delta\omega})\,dt \tag{3.277}$$

$q_{1\Delta\omega} \triangleq$ output of narrow- ($\Delta\omega$) band filter whose input is $q_1(t)$ $\quad$ **(3.278)**

$q_{2\Delta\omega} \triangleq$ output of narrow- ($\Delta\omega$) band filter whose input is $q_2(t)$

$(q_{1\Delta\omega})_{90} \triangleq$ signal $q_{1\Delta\omega}$ with phase shift of $90°$ $\quad$ **(3.279)**

A block diagram illustrating the definition of $\phi_{q_1q_2}(\omega)$ is given in Fig. 3.104. Note that $\phi_{q_1q_2}(\omega)$ is a *complex* quantity whereas $\phi(\omega)$ is real. Perhaps the main application of the cross-spectral density is in the experimental determination of the

[56]M. S. Beck, "Correlation in Instruments: Cross Correlation Flowmeters," *J. Phys. Eng.: Sci. Instrum.*, vol. 14, pp. 7–19, 1981.

sinusoidal transfer function $(q_o/q_i)(i\omega)$ of a linear system. From Eq. (3.268) we see that by using the ordinary mean-square spectral density, we can find only the magnitude (not the phase angle) of the transfer function if $\phi_i(\omega)$ is known and $\phi_o(\omega)$ is measured. The cross-spectral density determines both magnitude and phase according to the following equation:

$$\frac{q_o}{q_i}(i\omega) = \frac{\phi_{q_i q_o}(\omega)}{\phi_{q_i}(\omega)} \tag{3.280}$$

where $\phi_{q_i q_o}(\omega) \triangleq$ cross-spectral density of q_i and q_o

$\phi_{q_i}(\omega) \triangleq$ mean-square spectral density of q_i

We see that it is necessary to measure one (ordinary) mean-square spectral density and one cross-spectral density.

While some operations with random numbers are *analytically* challenging, simulation approaches are often easy and quick. In SIMULINK, we can generate Gaussian random signals using an icon called **Band-Limited White Noise,** which has three adjustable parameters. The **seed** value (use any five-digit integer) is used to start the random number algorithm. If you want two random signals to be independent (not related), give them *different* seed numbers. If you want to *repeat* a simulation, use the *same* seed you originally used, to get the same set of random numbers. The **noise power** sets the size of the fluctuations; larger values give larger fluctuations. Finally, the frequency content is set by **sample time.** If you set it at, say, 0.001 s, you will generate a new Gaussian random value every millisecond. The values are held fixed between samples, so you get a "square-cornered" waveform. This "random-amplitude square wave" is *not* white noise, since its frequency spectrum rolls off smoothly from its zero-frequency value to exactly 0.0 at the frequency $1/T$, where T is the sample time. Thus the spectrum of our example with $T = 0.001$ s rolls off to zero at 1000 Hz. This spectrum is such that for the range $0-0.1/T$ (0-100 Hz for our example), the spectrum is nearly flat, like white noise (it *does* extend beyond $1/T$).

Since real-world random signals are usually "smooth," rather than "square-cornered," we often follow the noise generator with a low-pass filter to "round the corners" and further control the frequency spectrum. In our example, we might use an 8th-order Butterworth filter with cutoff frequency set at 100 Hz. This "discards" the non-white part of the square-wave signal's spectrum, which is beyond 100 Hz.

Figure 3.105 demonstrates some of these concepts. In Fig. 3.105*a* we see the block diagram, where we pass the original "square-wave" noise through SIMULINK's Butterworth low-pass filter, set to "cut off" at 100 Hz. The diagram also shows how we can compute the mean-square and rms values of any signal, using the basic definition and the appropriate SIMULINK modules. Note that the division $1/T$ goes to infinity at $t = 0$, so we add a small constant to avoid an error message. Figure 3.105*b* shows the original noise and the filtered signal, which makes clear that the filter produces a smooth signal which is still random and Gaussian. Spectrum analysis of this signal would show that it has the desired flat spectrum from 0 to about 100 Hz, rolling off smoothly to zero at higher frequencies.

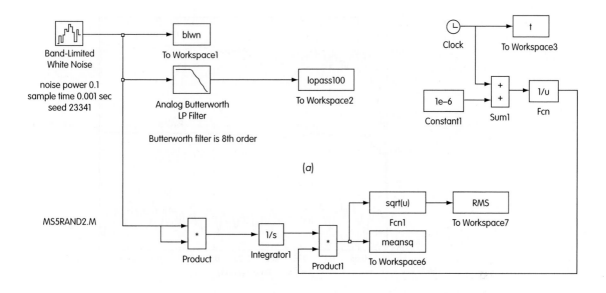

(a)

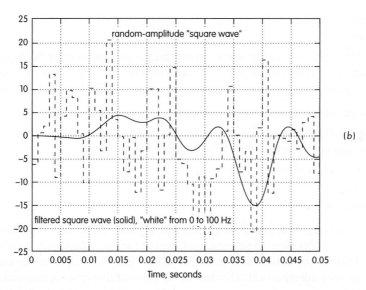

(b)

Figure 3.105
SIMULINK simulation for random signals.

By the use of such methods, it is possible to closely model the behavior of real-world random signals. The "convergence" of the mean-square and rms values is shown in Fig. 3.105c which also shows a longer record of the filtered signal.

When we run actual experiments (rather than simulations), we often use data acquisition/processing software, such as the DASYLAB explained in Chap. 13 and provided on the CD-ROM that comes with this book. DASYLAB and SIMULINK

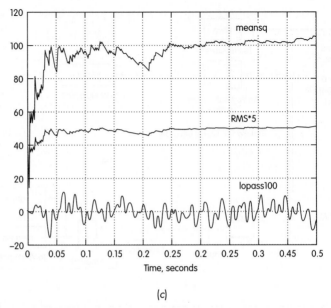

(c)

Figure 3.105
(Concluded)

both use "icons connected by wires" to represent systems, making both quite easy to learn and use. DASYLAB provides modules that perform (on actual data from sensors) most of the random-signal processing that we might need in practice:

1. Average, mean-square, and rms values.
2. Correlation functions.
3. Auto- and cross-spectral densities, transfer function.
4. Fast Fourier Transforms, direct and inverse.
5. Filtering.
6. Histograms.

This concludes our treatment of random signals. The most important, commonly useful results have been presented, and the main terminology has been developed. Further theoretical and practical details may be found in the literature. Today frequency spectrum calculations for all types of dynamic signals and statistical calculations for random signals most often are done digitally by using fast Fourier transform (FFT) methods. These are implemented either on general-purpose digital computers using ready-made programs or on special-purpose computers marketed as signal and/or system analyzers. We discuss these latter instruments in greater detail in a later chapter.

Requirements on Instrument Transfer Function to Ensure Accurate Measurement

Having expressed up to this point many different forms of signals in terms of the common denominator of frequency content, we can now give a general statement of the requirements that a measurement system must meet in order to measure accurately a given form of input.

1. For perfect shape reproduction with no time delay between q_i and q_o and with:
   a. Periodic inputs. $(q_o/q_i)(i\omega)$ must equal $K \angle 0°$ for all frequencies contained in q_i with significant amplitude.
   b. Transient inputs. $(q_o/q_i)(i\omega)$ must equal $K \angle 0°$ for the entire frequency range in which the Fourier transform of $q_i(t)$ has significant magnitude.
   c. Amplitude-modulated signals. Same criteria as in parts a and b, depending on whether the modulating signal is periodic or transient.
   d. Demodulated signals. $(q_o/q_i)(i\omega)$ for everything following the demodulator should be $K \angle 0°$ for all significant frequency bands of the modulating signal and should be zero for all carrier and side frequency bands produced by the modulation process.

While proper choice of parameters allows many measurement systems to meet the requirement of flat amplitude ratio, the *simultaneous* achievement of near-zero phase angle over the same frequency range is rarely possible (second-order instruments with small ζ and large ω_n, such as piezoelectric devices, are an exception). A relaxed criterion which *can* be met by most practical systems allows the phase angle to be nonzero, but requires that it vary *linearly* with frequency over the range of flat amplitude ratio. This requirement results in q_o being a "perfect" reproduction of q_i; however, q_o will be delayed in time by a specific amount, as if the system contained a dead-time effect τ_{dt}. For many applications, the fact that q_o appears on our oscilloscope screen or recorder chart τ_{dt} seconds "late" is of no importance whatever. Thus requiring a system frequency response of $K \angle (-\omega \tau_{dt})$ is widely acceptable. There are, however, two situations in which such time (phase) shift might cause difficulty. In *multichannel* systems, unless each channel has the *same* τ_{dt} (not likely), the following happens: If you pick, say a gas-temperature value from channel 1 and a gas-pressure value from channel 4, by using the same chart-paper time value, any computed gas density value would be *wrong* because the temperature and pressure values, while aligned on the chart, were not simultaneous in actual time occurrence (this is called *time-skew*). Of course, if you know the τ_{dt} values for each channel, suitable corrections can be applied, perhaps including those in a computer program used for data reduction. In fact, by using microprocessor technology, these corrections could be made part of the multichannel recorder itself, with each channel being properly shifted in time *before* it is written on the chart.

The second type of application where $K \angle (-\omega \tau_{dt})$ suffers relative to $K \angle 0°$ occurs when the measuring system is embedded in a feedback control loop. Here phase lag detracts from system stability and should be minimized.[57]

Revising our earlier accuracy statement, we now say:

2. For perfect shape reproduction with time delay τ_{dt} between q_i and q_o and with:
   a. Periodic inputs.
   b. Transient inputs.

---

[57]E. O. Doebelin, "Control System Principles and Design," Wiley, New York, 1985, Chapter 6.

  c. Amplitude-modulated signals.
  d. Demodulated signals.
  e. Random signals.

Same as 1 except that wherever $(q_o/q_i)(i\omega) = K\angle\ 0°$ was required, now (q_o/q_i) $(i\omega) = K\angle\ (-\omega\tau_{dt})$ is required. *That is, amplitude ratio is constant but phase lag increases linearly with frequency* ω.

  The validity of the above statements is readily perceived by consideration of Fig. 3.106. To get the output in the frequency domain, the frequency-domain input is always multiplied by the sinusoidal transfer function. If this transfer function is $K\angle\ (-\omega\tau_{dt})$ the output will have a magnitude equal to K times the input magnitude and an angle equal to the angle of the input minus $\omega\tau_{dt}$. This is exactly what we would get if we passed $q_i(t)$ through a pure gain K followed by a dead time τ_{dt}. Thus when we inverse-transform to get from $Q_o(i\omega)$ to $q_o(t)$, we are *bound* to get $Kq_i(t)$ delayed by τ_{dt} seconds. Thus, while the *actual* instrument transfer function will be of the form of Eq. (3.194), over the pertinent range of frequencies it must effectively amount to $K\angle\ (-\omega\tau_{dt})$ if accurate waveform reproduction is to be

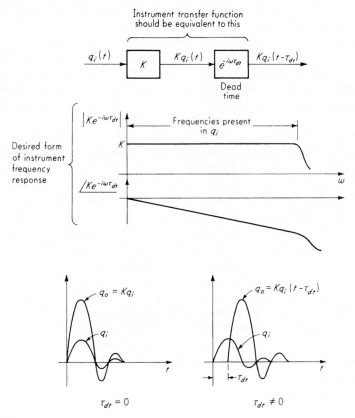

Figure 3.106
Requirements for accurate measurement.

expected. This is the basis for the selection of $\zeta \approx 0.6$ to 0.7 in second-order instruments since this range of ζ values makes the amplitude ratio most nearly constant and the phase-angle-versus-frequency curve most nearly linear over the widest possible frequency range for a given ω_n.

While the conventional frequency-response curves display amplitude ratio and phase-angle, for *measurement* systems, a graph of the time-delay, $-\phi/\omega$, is really more useful than the phase angle. Our accuracy criterion requires a straight-line phase angle, which means a constant time delay. The "flatness" of a time-delay graph is much easier to visually judge than is the "straightness" of a slightly curved and sloping phase-angle plot. Also, for the popular logarithmic frequency axes, the desired time-delay curve is still flat, but the desired phase-angle curve is now *not* straight and thus really unusable. For these reasons, we recommend that *all* measurement system frequency-response graphs show amplitude ratio and time-delay $-\phi/\omega$. If there are good reasons to show the phase angle in a specific application, all three curves can be displayed.

Figure 3.107a shows a nondimensional delay graph useful for all first-order systems (compare it with the phase-angle display of Fig. 3.65 to see the advantage). If we require a delay constant within, say, 5 percent, then $\omega\tau$ must be less than 0.4 (if $\tau = 2.0$ s, $\omega < 0.2$ rad/s). If we relax our requirement to 10 percent, then $\omega\tau < 0.6$. For second-order systems, Fig. 3.107b shows that $\zeta = 0.80$ has a very constant delay. However, we earlier chose $\zeta \approx 0.65$ as optimum because $\zeta = 0.80$ unduly sacrifices a flat *amplitude ratio,* that is, we desire *both* the amplitude ratio and delay to be flat.

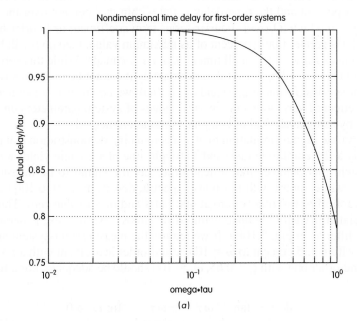

Figure 3.107
Time-delay curves for first- and second-order systems.

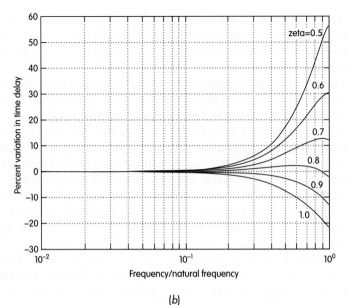

Figure 3.107
(Concluded)

It should be noted that the above accuracy criteria do not actually state any *numerical* results. That is, no statement of the form "If the amplitude ratio is flat within $\pm x$ percent and the time-delay is flat within $\pm y$ percent over the range of frequencies in which $|Q_i(i\omega)|$ (Fourier-series term coefficient or Fourier-transform magnitude) is greater than z percent of its maximum value, then $q_o(t)$ will be within $\pm w$ percent of $Kq_i(t - \tau_{dt})$ at all times" is or *can be* made. While this sort of statement would be exceedingly useful, unfortunately it cannot be made in any *general* sense. For specific forms of $q_i(t)$ and $(q_o/q_i)(i\omega)$, we could investigate mathematically the effect of variations in numerical values of system parameters on dynamic accuracy by using analytical or computer simulation methods.

We can use digital simulation to give a convincing demonstration that requiring essentially a flat amplitude ratio and a straight-line phase angle yields a measurement system that behaves like a dead time. Let the signal to be measured be a Gaussian "white" noise with flat power spectral density from 0 to 100 rad/s with a smooth rolloff to essentially zero at 400 rad/s and mean value zero. This type of random signal is easily obtained using SIMULINK's white noise generator and a low-pass filter as in Fig. 3.108a. If we require the measurement-system amplitude ratio flat within ± 5 percent to $\omega = 100$, a first-order instrument with a $\tau \approx 0.005$ s or a second-order one with $\zeta = 0.65$, $\omega_n \approx 100$ should be adequate. For a first-order system,

$$\phi = -\tan^{-1}\omega\tau \approx -\omega\tau \qquad \text{for } \omega \to 0$$

$$\tau_{dt} \triangleq \tau = 0.005 \text{ s} \tag{3.281}$$

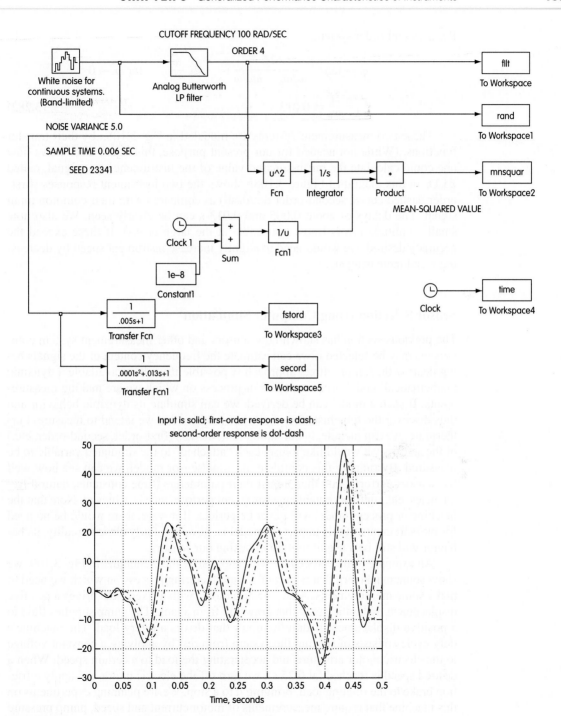

Figure 3.108
Good waveform accuracy is preserved when time delay is nearly constant.

For a second-order system,

$$\phi = \tan^{-1} \frac{2\zeta}{\omega/\omega_n - \omega_n/\omega} \approx - 2\zeta \frac{\omega}{\omega_n} \qquad \text{for } \omega \to 0$$

$$\tau_{dt} = \frac{2\zeta}{\omega_n} = 0.013 \text{ s} \tag{3.282}$$

These two measurement systems are modeled in Fig. 3.108*a* by their transfer functions. (While not needed for our present purpose, this figure also shows how one could calculate the mean-squared value of the instrument input signal, called `filt` in this diagram.) Figure 3.108*b* shows the two instrument responses (first-order dashed curve, second-order dot/dash) as compared with their common input signal. The delays of about 0.005 and 0.013 s can be clearly seen. We also note small amplitude errors (most noticeable at the large peaks). If these exceed the accuracy desired, we would need to slightly increase instrument speed by decreasing τ and increasing ω_n.

Sensor Selection Using Computer Simulation

The previous section has shown how sensors and other measurement system components may be selected if we can estimate the frequency content of the signals being dealt with. A more direct approach is possible if we have available a dynamic mathematical model of the machine or process on which we are making measurements. If such a model can be derived, we can simulate its dynamic behavior and thus discover the time histories of all the variables that we intend to measure. Furthermore, we can include, in our simulation, models (first-order, second-order, etc.) of the sensors we would like to use, each "attached" to the simulated variable to be measured. By running this complete machine/sensor model, we can see how well the sensors perform, and then adjust their parameters (time constants, natural frequencies, etc.) until acceptable measurement performance is achieved. Note that the machine or process model will never be perfect; if it were, there would be no need for measurements. If this imperfect model is, however, fairly close to reality, its behavior will still be a useful tool for choosing our sensors.

An example of this procedure will help clarify the technique. In Fig. 3.109, we show some machinery that is part of a manufacturing process on which we need to make some measurements. A DC electric motor with a fixed field drives a positive displacement hydraulic pump that draws oil from a tank. The pump supplies fluid to a positive displacement hydraulic motor that drives a rotary load. The machine's duty cycle consists of starting the system from rest by applying a constant voltage to the electric motor armature and accelerating the load to a certain speed. When a desired speed is reached (at 0.25 s), we turn off the electric motor and apply a friction brake to the rotating load, bringing it to a stop. We are planning experiments on this machine that require measurements of motor current and speed, pump pressure and flow rate, hydraulic motor pressure and speed, and temperatures of the brake friction material.

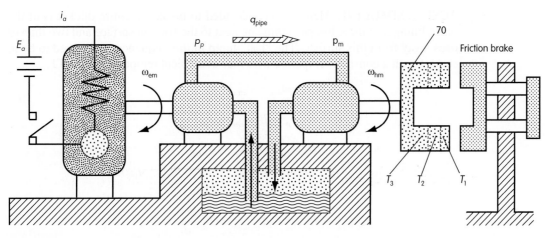

Figure 3.109
Machinery system to be instrumented.

Deriving a math model of this machine requires some experience in system dynamics, but our interest here is not in developing this expertise. We will thus just state without derivation a set of simultaneous equations that might be used for this model and then show how it could be simulated using SIMULINK. Those unfamiliar with the machinery modeling are referred to references.[58] We will first state the equations with all parameters given as letters. In the simulation, we must of course assign numerical values to all parameters and inputs. One of the advantages of this simulation approach lies in the ease with which we may change parameters of both the machine model and the sensors to explore all the possibilities of interest.

The first equation is a Kirchoff law for the armature circuit. The second and sixth are Newton's laws for the electric motor/pump shaft and the hydraulic motor/load shaft. We assume the brake, when applied, produces a constant friction torque. The third and fifth equations are flow budgets for pump and motor, both of which have leakage. The fourth equation is a Newton's law for the liquid in the pipe connecting the pump to motor. This liquid has inertia and friction, but is assumed incompressible. The thermal model of the brake lining uses a number of assumptions. The total frictional heat generated is given by the product of friction torque and instantaneous load speed. We assume that this total heat splits into two parts; one going into the brake-lining material and the other into the steel that rubs against it. This split is apportioned according to the geometry and thermal properties of the two materials.[59] (Three equations similar to the last three could be written for the steel member; however, we have no interest in those temperatures.) In estimating temperatures, we must use a lumped model for the thermal resistances and capacitances if we want to use standard dynamic system simulation software such as

---

[58]E. O. Doebelin, "System Dynamics," 1998.

[59]E. O. Doebelin, "System Dynamics," 1998, p. 497.

ACSL or SIMULINK. Here we have decided to break the entire thickness of the brake lining into three lumps; a thin one next to the friction surface and two thicker ones away from this surface. The "last" lump in this sequence is assumed to be in contact with a constant-temperature surface at ambient temperature of 70°F.

$$R_a i_a + K_{be}\omega_{em} - e_{\text{arm}} = 0$$

$$K_{te} i_a - K_{tp} p_p - B_p \omega_{em} = J_{pm}\frac{d\omega_{em}}{dt}$$

$$D_p \omega_{em} - K_{lp} p_p = q_{\text{pipe}}$$

$$I_f q_{\text{pipe}} = (p_p - p_m) - q_{\text{pipe}} R_f$$

$$q_{\text{pipe}} - p_m K_{lm} = D_m \omega_m$$

$$p_m K_{tm} - B_m \omega_m - T_{\text{friction}} = J_{ml}\frac{d\omega_m}{dt}$$

$$(T_{\text{friction}}\,\omega_m)\left(R_{12}\frac{K_f}{K_f + K_s}\right) - (T_1 - T_2) = R_{12}C_1\frac{dT_1}{dt}$$

$$(T_1 - T_2) - \frac{R_{12}}{R_{23}}(T_2 - T_3) = R_{12}C_2\frac{dT_2}{dt}$$

$$\frac{R_{33}}{R_{23}}(T_2 - T_3) - (T_3 - 70) = R_{33}C_3\frac{dT_3}{dt}$$

The SIMULINK block diagram for the first six equations is shown in Fig. 3.110a. There was not room on the computer screen for the last three equations, so this first diagram was "compacted" into a single block (called **unNamed**), using SIMULINK's "group" command. The three thermal equations were then added to get the diagram of Fig. 3.110b. The signal coming out of **unNamed,** the frictional heating rate called **htpwr** in Fig. 3.110a, is the only one we need from the first six equations. Even though the details of the first six equations are not apparent on this screen, we can still run the complete simulation from it and request plots of any of the variables. Once we have the simulation of our machine running properly, we can "attach" sensors of any desired type (first-order, second-order, etc.) to any variable of interest and "design" that sensor to give the accuracy we want. In Fig. 3.111, this has been done with temperature T_1, which appears to be accurately measured with a first-order sensor with a 0.01-s time constant. Fig. 3.112 shows some other variables of interest, each of which could be similarly studied to find the sensor dynamics needed for accurate measurement. Because machinery and processes that have significant dynamic actions are often dynamically modeled at the design stage, we can often use the approach explained here to choose any sensors that might be needed in the experimental testing stage of process development.

Numerical Correction of Dynamic Data

Theoretically, if $q_o(t)$ (the actual measured data) is known and if $(q_o/q_i)(i\omega)$ for the measurement system is known, we can always reconstruct a *perfect* record of $q_i(t)$ by the following process:

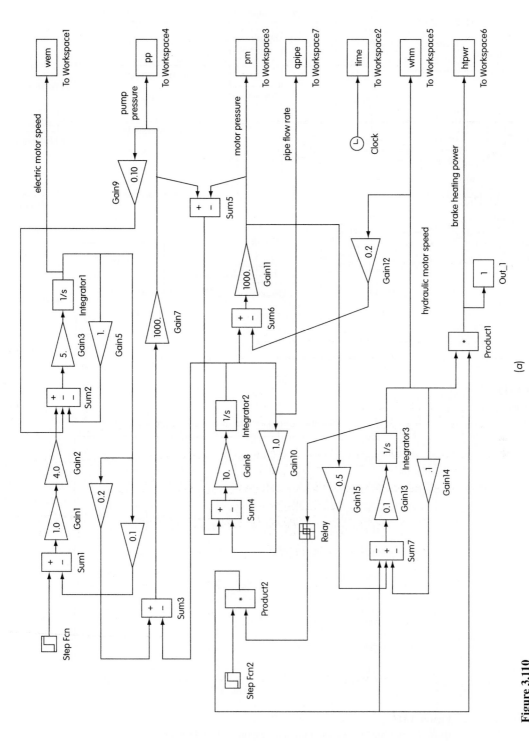

Figure 3.110
SIMULINK simulation of machine system.

(a)

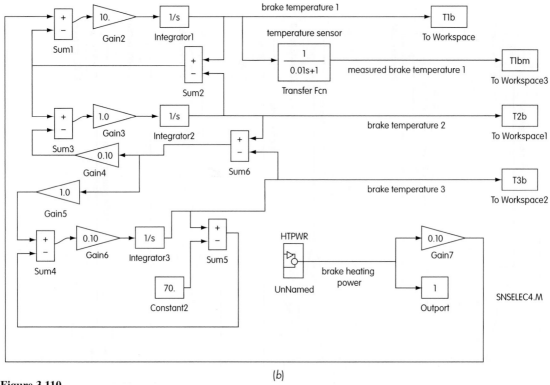

Figure 3.110
(Concluded)

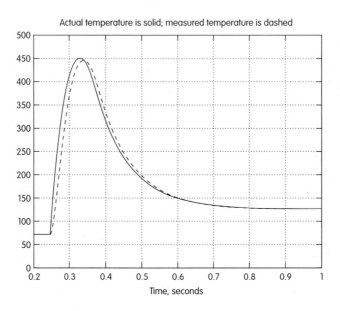

Figure 3.111
Brake friction surface temperature and response of trial sensor.

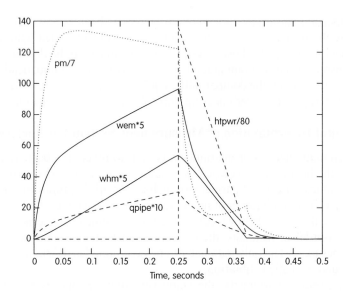

Figure 3.112
Some other machine variables to be sensed.

1. Transform $q_o(t)$ to $Q_o(i\omega)$.
2. Apply the formula

$$Q_o(i\omega) = Q_i(i\omega)\frac{q_o}{q_i}(i\omega)$$

in the inverse sense as

$$Q_i(i\omega) = \frac{Q_o(i\omega)}{(q_o/q_i)(i\omega)}$$

to find $Q_i(i\omega)$.
3. Inverse-transform $Q_i(i\omega)$ to $q_i(t)$.

This procedure theoretically will give the exact $q_i(t)$ whether the measurement system meets the $K\angle\ (-\omega\tau_{dt})$ requirements or not. In actual practice, of course, while the measurement system does not have to meet $K\angle\ (-\omega\tau_{dt})$, it *does* have to respond fairly strongly to all frequencies present in q_i; otherwise, some parts of the q_o frequency spectrum will be so small as to be submerged in the unavoidable "noise" present in all systems and thus be unrecoverable by the above mathematical process. As general-purpose digital computers are used more in data processing and as computing power is built into more "instruments," such dynamic correction becomes increasingly practical and is a usable alternative in those situations where measurement systems meeting $K\angle\ (-\omega\tau_{dt})$ cannot be constructed with the present state of the art. Many FFT signal/system analyzers have this capability built into their software.

An important variation of the above process has been successfully applied in cases where the primary sensor is inadequate but can be cascaded with frequency-sensitive analog elements whose transfer functions make up the deficiencies in the primary sensor. The above computations are then, in a sense, automatically and continuously carried out by the compensating equipment to reconstruct $q_i(t)$. This subject is discussed in detail later under the topic Dynamic Compensation.

Experimental Determination of Measurement-System Parameters

While theoretical analysis of instruments is vital to reveal the basic relationships involved in the operation of a device, it is rarely accurate enough to provide usable numerical values for critical parameters such as sensitivity, time constant, natural frequency, etc. Thus calibration of instrument systems is a necessity. We earlier discussed static calibration; here we concentrate on dynamic characteristics.

For zero-order instruments, the response is instantaneous and so no dynamic characteristics exist. The only parameter to be determined is the static sensitivity K, which is found by static calibration.

For first-order instruments, the static sensitivity K also is found by static calibration. There is only one parameter pertinent to dynamic response, the time constant τ, and this may be found by a variety of methods. One common method applies a step input and measures τ as the time to achieve 63.2 percent of the final value. This method is influenced by inaccuracies in the determination of the $t = 0$ point and also gives no check as to whether the instrument is really first-order. A preferred method uses the data from a step-function test replotted semilogarithmically to get a better estimate of τ and to check conformity to true first-order response. This method goes as follows: From Eq. (3114) we can write

$$\frac{q_o - Kq_{is}}{Kq_{is}} = -e^{-t/\tau} \tag{3.283}$$

$$1 - \frac{q_o}{Kq_{is}} = e^{-t/\tau} \tag{3.284}$$

Now we define

$$Z \triangleq \log_e\left(1 - \frac{q_o}{Kq_{is}}\right) \tag{3.285}$$

and then

$$Z = \frac{-t}{\tau} \qquad \frac{dZ}{dt} = \frac{-1}{\tau} \tag{3.286}$$

Thus if we plot Z versus t, we get a straight line whose slope is numerically $-1/\tau$. Figure 3.113 illustrates the procedure. This gives a more accurate value of τ since the best line through *all* the data points is used rather than just two points, as in the 63.2 percent method. Furthermore, if the data points fall nearly on a straight line, we are assured that the instrument is behaving as a first-order type. If the data deviate considerably from a straight line, we know the instrument is not truly first-order and a τ value obtained by the 63.2 percent method would be quite misleading.

An even stronger verification (or refutation) of first-order dynamic characteristics is available from frequency-response testing, although at considerable cost of

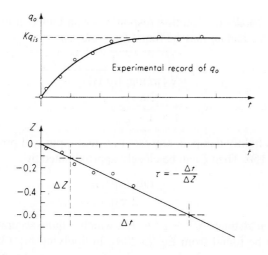

Figure 3.113
Step-function test of first-order system.

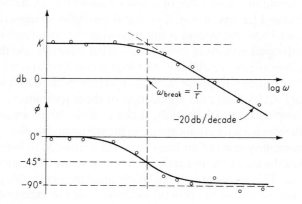

Figure 3.114
Frequency-response test of first-order system.

time and money if the system is not completely electrical, since nonelectrical sine-wave generators are neither common nor necessarily cheap. If the equipment is available, the system is subjected to sinusoidal inputs over a wide frequency range, and the input and output are recorded. Amplitude ratio and phase angle are plotted on the logarithmic scales. If the system is truly first-order, the amplitude ratio follows the typical low- and high-frequency asymptotes (slope 0 and -20 dB/decade) and the phase angle approaches $-90°$ asymptotically. If these characteristics are present, the numerical value of τ is found by determining ω at the breakpoint and using $\tau = 1/\omega_{break}$ (see Fig. 3.114). Deviations from the above amplitude and/or phase characteristics indicate non-first-order behavior.

For second-order systems, K is found from static calibration, and ζ and ω_n can be obtained in a number of ways from step or frequency-response tests. Figure

3.115a shows a typical step-function response for an underdamped second-order system. The values ζ and ω_n may be found from the relations

$$\zeta = \sqrt{\frac{1}{[\pi/\log_e{(a/A)}]^2 + 1}} \qquad (3.287)$$

$$\omega_n = \frac{2\pi}{T\sqrt{1 - \zeta^2}} \qquad (3.288)$$

When a system is lightly damped, any fast transient input will produce a response similar to Fig. 3.115b. Then ζ can be closely approximated by

$$\zeta \approx \frac{\log_e{(x_1/x_n)}}{2\pi n} \qquad (3.289)$$

This approximation assumes $\sqrt{1 - \zeta^2} \approx 1.0$, which is quite accurate when $\zeta < 0.1$, and again ω_n can be found from Eq. (3.288). In applying Eq. (3.288), if several cycles of oscillation appear in the record, it is more accurate to determine the period T as the average of as many distinct cycles as are available rather than from a single cycle. If a system is strictly linear and second-order, the value of n in Eq. (3.289) is immaterial; the same value of ζ will be found for any number of cycles. Thus if ζ is calculated for, say, $n = 1, 2, 4$, and 6 and *different* numerical values of ζ are obtained, we know the system is not following the postulated mathematical model. For overdamped systems ($\zeta > 1.0$), no oscillations exist, and the determination of ζ and ω_n becomes more difficult. Methods for dealing with this case are left to a reference.[60] Frequency-response methods also may be used to find ζ and ω_n or τ_1 and τ_2. Figure 3.116 shows the application of these techniques. The methods shown use the amplitude-ratio curve only. If phase-angle curves are available, they constitute a valuable check on conformance to the postulated model.

For measurement systems of arbitrary form (as contrasted to first- and second-order types), description of the dynamic behavior in terms of frequency response usually is desired. This information may be obtained by sinusoidal, pulse, or random signal testing, following the general methods[61] used to experimentally determine mathematical models for physical systems. When the physical system being studied is a *measurement* system, the output signal q_o is itself generally useful and no separate output sensor is required. However, we do usually need to measure the input signal q_i with a separate sensor, which serves as the calibration standard and whose accuracy is known, and, if possible, is about 10 times better than that of the system being calibrated. If we can obtain $(q_o/q_i)(i\omega)$ thus for the measurement system, this defines the frequency range over which corrections are negligible and provides the data needed to make dynamic corrections (using the transform methods of the previous section) if we wish to use the instrument in its nonflat range of frequency response.

Sometimes we can *assume* with good accuracy the form of the input signal, and we then need *measure* only the instrument output signal. An example of this is

---

[60]E. O. Doebelin, *Measurement Systems,* 4th ed., p. 192.

[61]E. O. Doebelin, "System Modeling and Response," 1980, chap. 6.

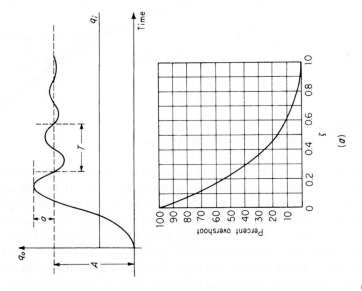

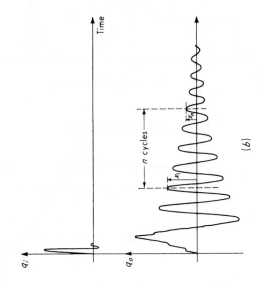

Figure 3.115
Step and pulse tests for second-order system.

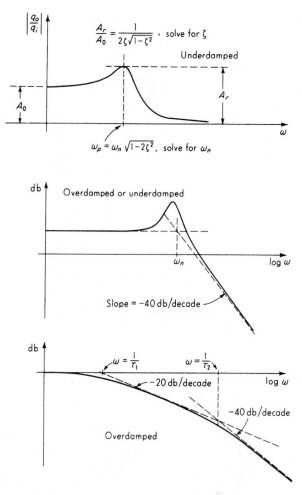

Figure 3.116
Frequency-response test of second-order system.

found in shock-tube testing of pressure transducers. Here, theory and previous experience guarantees a nearly perfect step input. If the transducer frequency response is wanted, one approach[62] differentiates the measured step response signal, giving the *impulse response.* Then a Fourier transform using standard FFT software gives the amplitude ratio and phase angle of the sinusoidal transfer function. In shock tubes the step response may not decay completely to zero before reflected waves deform the perfect step input. Using an exponential window on the response data, convergence to zero can be *forced,* making the response truly a transient and expediting the subsequent FFT operation. This makes the apparent transducer damping larger than reality, but this error is often acceptable. The differentiation technique can of course be applied to step input tests in general, no just for shock

---

[62]Personal communication, Dr. Jeff Dosch, PCB Piezotronics, Depew, NY, 888-684-0011

tubes. If the initial step response dies out properly, similar methods, without the need to differentiate, have been successful.[63]

Loading Effects under Dynamic Conditions

The treatment of loading effects by means of impedance, admittance, etc., was discussed in Sec. 3.2 for static conditions. All these results can be immediately transferred to the case of dynamic operation by generalizing the definitions in terms of transfer functions. The basic equations relating the undisturbed value q_{i1u} and the actual measured value q_{i1m} at the input of a device are

$$q_{i1m} = \frac{1}{Z_{go}/Z_{gi} + 1} q_{i1u} \tag{3.59}$$

$$q_{i1m} = \frac{1}{Y_{go}/Y_{gi} + 1} q_{i1u} \tag{3.65}$$

$$q_{i1m} = \frac{1}{S_{go}/S_{gi} + 1} q_{i1u} \tag{3.72}$$

The quantities Z, Y, and S previously were considered to be the ratios of small changes in two related system variables under stated conditions. To generalize these concepts, we now define the quantities Z, Y, and S as *transfer functions* relating the same two variables under the same conditions except that now dynamic operation is considered. That is, we must get (theoretically or experimentally) $Z(D)$, $Y(D)$, and $S(D)$ if we wish to use operational transfer functions and $Z(i\omega)$, $Y(i\omega)$, and $S(i\omega)$ if we wish to use frequency-response methods.

Usually the frequency-response form is most useful if these quantities must be found experimentally. This means, then, that in finding, say, $Z(i\omega)$, one of the two variables involved in the definition of Z plays the role of an "input" quantity which we vary sinusoidally at different frequencies. This causes a sinusoidal change in the other ("output") variable, and thus we can speak of an amplitude ratio and phase angle between these two quantities, making $Z(i\omega)$ now a complex number that varies with frequency. (If the system is somewhat nonlinear, the effective approximate Z becomes a function also of input amplitude. This situation was adequately described under static conditions in Sec. 3.2.) In Eq. (3.59), for example, both Z_{go} and Z_{gi} would now be complex numbers; if these were known, we could calculate the amplitude and phase of $q_{i1\mu}$ if the amplitude, phase, and frequency of a sinusoidal q_{i1m} were given. The quantity q_{i1m} then would be the *actual* input (q_i) to the measuring device, and we could calculate q_o if the transfer function $(q_o/q_i)(i\omega)$ were known. That is,

$$Q_o(i\omega) = \frac{1}{Z_{go}(i\omega)/Z_{gi}(i\omega) + 1} \left[\frac{q_o}{q_i}(i\omega) \right] Q_{i1u}(i\omega) \tag{3.290}$$

---

[63]J. Weiss, H. Knauss, and S. Wagner, "Method for the determination of frequency response and signal to noise ratio for constant-temperature hot-wire anemometers," *Review of Scientific Instruments,* vol. 72, #3, March 2001, pp. 1904–1909.

Thus we could define a *loaded transfer function* $(q_o/q_{i1u})(i\omega)$ as

$$\frac{q_o}{q_{i1u}}(i\omega) \triangleq \frac{1}{Z_{go}(i\omega)/Z_{gi}(i\omega) + 1} \frac{q_o}{q_i}(i\omega) \qquad (3.291)$$

where $q_o \triangleq$ actual output of measuring device that has no load at *its* output

$q_{i1u} \triangleq$ measured variable value that would exist if measuring device caused *no* loading on measured medium

Equations (3.65) and (3.72) may also be modified in similar fashion. Also, if differential equations relating $q_o(t)$ are desired, we may write

$$\frac{q_o}{q_{i1u}}(D) = \frac{1}{Z_{go}(D)/Z_{gi}(D) + 1} \frac{q_o}{q_i}(D) \qquad (3.292)$$

and then obtain the differential equation in the usual way by "cross-multiplying":

$$[Z_{go}(D) + Z_{gi}(D)](a_n D^n + a_{n-1} D^{n-1} + \cdots + a_1 D + a_0)q_o$$
$$= [Z_{gi}(D)](b_m D^m + b_{m-1} D^{m-1} + \cdots + b_1 D + b_0)q_{i1u} \qquad (3.293)$$

An example of the above methods will be helpful. Consider a device for measuring translational velocity, as in Fig. 3.117a. The unloaded transfer function relating the output displacement x_o and the input (measured) velocity v_i is obtained as follows:

$$B_i(\dot{x}_i - \dot{x}_o) - K_{is}x_o = M_i \ddot{x}_o \qquad (3.294)$$

$$\frac{x_o}{v_i}(D) = \frac{K_i}{D^2/\omega_{ni}^2 + 2\zeta_i D/\omega_{ni} + 1} \qquad (3.295)$$

where $K_i \triangleq$ instrument static sensitivity $\triangleq \dfrac{B_i}{K_{is}}$ m/(m/s)   (3.296)

$\zeta_i \triangleq$ instrument damping ratio $\triangleq \dfrac{B_i}{2\sqrt{K_{is} M_i}}$ (3.297)

$\omega_{ni} \triangleq$ instrument undamped natural frequency $\triangleq \sqrt{\dfrac{K_{is}}{M_i}}$ rad/s   (3.298)

We see that the instrument is second-order and thus will measure v_i accurately for frequencies sufficiently low relative to ω_{ni}. Suppose we now attach this instrument to a vibrating system whose velocity we wish to measure, as in Fig. 3.117b. The presence of the measuring instrument will distort the velocity we are trying to measure. The character of this distortion may be assessed by application of Eq. (3.65), since the measured quantity is velocity (a flow variable), and thus admittance is the appropriate quantity to use. We determine the input admittance $Y_{gi}(D) = (v/f)(D)$ from Fig. 3.117c as follows:

$$f - K_{is}x_o = M_i \ddot{x}_o \qquad (3.299)$$

Also $$f = B_i(v - \dot{x}_o) \qquad (3.300)$$

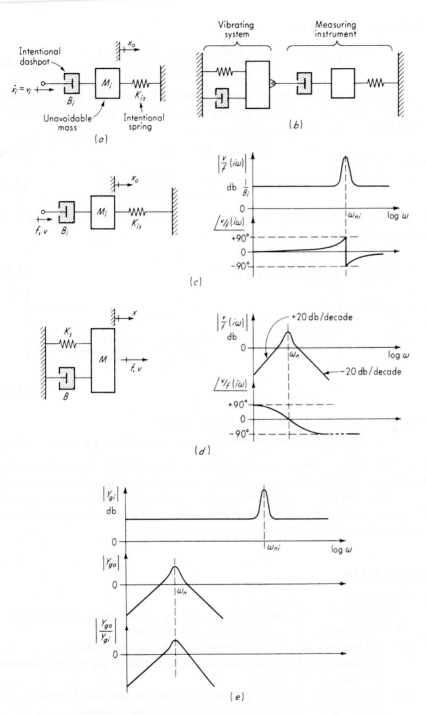

Figure 3.117
Example of dynamic loading analysis.

and, eliminating x_o, we get

$$Y_{gi}(D) = \frac{v}{f}(D) = \frac{(1/B_i)(D^2/\omega_{ni}^2 + 2\zeta_i D/\omega_{ni} + 1)}{D^2/\omega_{ni}^2 + 1} \quad \textbf{(3.301)}$$

Figure 3.117c also shows the frequency characteristics of this input admittance. The output admittance $Y_{gi}(D) = (v/f)(D)$ of the measured system is obtained from Fig. 3.117d:

$$f - B\dot{x} - K_s x = M\ddot{x} \quad \textbf{(3.302)}$$

$$Y_{go}(D) = \frac{v}{f}(D) = \frac{(1/K_s)D}{D^2/\omega_n^2 + 2\zeta D/\omega_n + 1} \quad \textbf{(3.303)}$$

The frequency characteristic of this output admittance is shown in Fig. 3.117d. We may now write

$$\frac{x_o}{v_{i1u}}(D) = \frac{1}{Y_{go}(D)/Y_{gi}(D) + 1} \frac{x_o}{v_i}(D) \quad \textbf{(3.304)}$$

$$\frac{x_o}{v_{i1u}}(D) = \left[\cfrac{1}{\underbrace{\cfrac{(1/K_s)D}{D^2/\omega_n^2 + 2\zeta D/\omega_n + 1} \cfrac{(D^2/\omega_{ni}^2) + 1}{(1/B_i)(D^2/\omega_{ni}^2 + 2\zeta_i D/\omega_{ni} + 1)} + 1}_{\text{loading effect}}} \right]$$

$$\times \left[\frac{K_i}{D^2/\omega_{ni}^2 + 2\zeta_i D/\omega_{ni} + 1} \right] \quad \textbf{(3.305)}$$

where $x_o \triangleq$ actual output of measuring device

$v_{i1u} \triangleq$ velocity that would exist if measuring device caused no loading

Figure 3.117e shows that in this example the loading effect is most serious for frequencies near the natural frequency of the measured system, but approaches zero for both very low and very high frequencies. Since the loading effects can be expressed in frequency terms, they can be handled for all kinds of inputs by using appropriate Fourier series, transform, or mean-square spectral density.

PROBLEMS

3.1 Consider the system of Fig. 2.3.

(a) Explain how you would carry out a static calibration to determine the relation between the desired input and the output.

(b) The temperature of the air surrounding the capillary tube is an interfering input. Explain how you would calibrate the relation between this input and the output.

(c) The elevation difference between the Bourdon tube and the bulb is another interfering input. Discuss means for its calibration.

3.2 Does the system of Fig. 2.4 require calibration? Explain.

3.3 What fundamental difficulties arise in trying to define the true temperature of a physical body?

3.4 Slide a coin along a smooth surface, trying to make it come to rest at a drawn line. Measure the distance of the coin from the line. Repeat 100 times and check the resulting data for conformance to a Gaussian distribution, using probability graph paper.

3.5 In the section "Numerical Correction of Dynamic Data," a method, using frequency-response techniques is suggested for correcting dynamic measurements from instruments which do *not* meet our usual requirements of flat amplitude ratio and constant time delay. Figure P3.1 shows a SIMULINK diagram for such a situation. If you run it, you will see that the measured value q_o is quite different from the actual q_i. Using the technique described in the text, apply MATLAB (or other available) software to compute the system sinusoidal transfer function and the direct and inverse FFTs needed to implement the method. Show that the corrected data is very close to the ideal.

3.6 In Eq. (2.6), solve for the strain ϵ in terms of the other parameters; $\epsilon = f(GF, R_g, E_b, R_a, e_o)$. Then take the natural log of both sides; $\ln \epsilon = \ln f$. Now take the differential of both sides so that terms such as $d\epsilon/\epsilon$, de_o/e_o, dR_a/R_a, etc., are formed. This will give the percentage error $d\epsilon/\epsilon$ in ϵ as a function of the percentage errors in the other parameters. If GF, R_g, E_b, R_a, and e_o are all measured to ± 1 percent error, what is the possible error in the computed value of ϵ?

3.7 Is the logarithmic differentiation method of Prob. 3.6 applicable to all forms of functional relations? Explain. *Hint:* Apply it to the relation $w = \sin x + 5y^3 - 6e^z$.

3.8 The discharge coefficient C_q of an orifice can be found by collecting the water that flows through during a timed interval when it is under a constant head h. The formula is

$$C_q = \frac{W}{t\rho A \sqrt{2gh}}$$

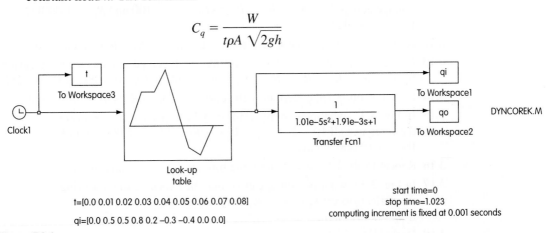

t=[0.0 0.01 0.02 0.03 0.04 0.05 0.06 0.07 0.08]

qi=[0.0 0.5 0.5 0.8 0.2 −0.3 −0.4 0.0 0.0]

start time=0
stop time=1.023
computing increment is fixed at 0.001 seconds

Figure P3.1

Find C_q and its possible error if:

$W = 865 \pm 0.5$ lbm	$A = \pi d^2/4$	$d = 0.500 \pm 0.001$ in
$t = 600.0 \pm 2$s	$g = 32.17 \pm 0.1\%$ ft/s^2	
$\rho = 62.36 \pm 0.1\%$ lbm/ft^3	$h = 12.02 \pm 0.01$ ft	

considering both the following:
(a) The errors are the absolute limits.
(b) The errors are 95 percent uncertainties.

3.9   In Prob. 3.8 if C_q must be measured within ± 0.5 percent for the numerical mean values given, what errors are allowable in the measured data? Use the method of equal effects.

3.10 Static calibration of an instrument gives the data of Fig. P3.2. Calculate (a) the best-fit straight line and (b) q_i and its error limits if the instrument is used after calibration and reads $q_o = 5.72$.

q_i	q_o Increasing values	q_o Decreasing values
0	−0.07	+0.01
5	1.08	1.16
10	2.05	2.10
15	3.27	3.29
20	4.28	4.36
25	5.41	5.45
30	6.43	6.53
35	7.57	7.61
40	8.66	8.75

Figure P3.2

3.11 In Fig. 3.26, what percentage error may be expected in measuring the voltage across R_5 if $R_1 = R_2 = R_3 = R_4 = R_5 = 100$ and $R_m = 1,000\Omega$? If R_m 10,000Ω?

3.12 Repeat Prob. 3.11, except now the voltage across R_3 is to be measured.

3.13 In Fig. 3.30, what percentage error may be expected in measuring the current through R_5 if $R_1 = R_2 = R_3 = R_4 = R_5 = 100$ and $R_m = 10\Omega$? If $R_m = 1\Omega$?

3.14 Repeat Prob. 3.13, except now the current through R_3 is to be measured.

3.15 In Fig. 3.31, what percentage error may be expected in measuring the force in k_2 if $k_1 = k_2 = k_3 = k_4 = 100$ and $k_m = 1,000$ N/cm? If $k_m = 10,000$ N/cm?

3.16 Repeat Prob. 3.15, except now the force in k_3 is to be measured.

3.17 In Fig. 3.32, what percentage error may be expected in measuring the deflection x if $k_1 = k_2 = k_3 = k_4 = 1$ and $k_m = 0.1$ N/cm? If $k_m = 0.01$ N/cm?

3.18 Derive Eq. (3.84).

3.19 A first-order pressure sensor must meet the following dynamic response specifications:
(*a*) At least 95 percent accuracy within 0.05 s after a step input.
(*b*) Steady-state error of no more than 2 psi for a ramp input of 100.0 psi/s.
(*c*) Amplitude accuracy no worse than 90 percent for a sine wave input of frequency 20.0 Hz.
Find the largest allowable time constant for this sensor.

3.20 A mercury thermometer has a capillary tube of 0.010-in diameter. If the bulb is made of a zero-expansion material, what volume must it have if a sensitivity of 0.10 in/F° is desired? Assume operation near 70°F. If the bulb is spherical and is immersed in stationary air, estimate the time constant.

3.21 A balloon carrying a first-order thermometer with a 15-s time constant rises through the atmosphere at 6 m/s. Assume temperature varies with altitude at 0.15°C/30m. The balloon radios temperature and altitude readings back to the ground. At 3000 m the balloon says the temperature is 0°C. What is the true altitude at which 0°C occurs?

3.22 A first-order instrument must measure signals with frequency content up to 100 Hz with an amplitude inaccuracy of 5 percent. What is the maximum allowable time constant? What will be the phase shift at 50 and 100 Hz?

3.23 For the spring scale of Fig. 3.50, discuss the trade-off between sensitivity and speed of response resulting from changes in K_s.

3.24 Derive Eqs. (3.179) to (3.181).

3.25 Find the transfer function of a spring scale (Fig. 3.50) whose mass is negligible. Show that the steady-state time lag for a ramp input is the same whether mass is zero or not.

3.26 We wish to design (choose τ_1, τ_2 ζ, ω_n) the measurement system of Fig. P3.3*a* so as to achieve an amplitude ratio which is flat $\pm$ 5 percent for

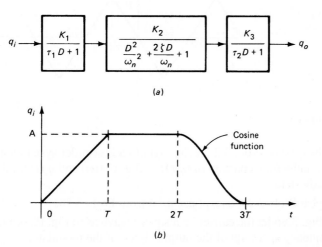

(*a*)

(*b*)

Figure P3.3

the frequency range 0 to 100 Hz. (Note that such a problem does not have a single unique answer.) Strive for the *largest* τ's and *smallest* ω_n which will meet the specifications. (Why?) Note that the rising amplitude ratio associated with a small ζ can compensate for the dropoff due to the τ's and that $\zeta = 0.65$ is not necessarily optimum in this case. When you find a set of values which is satisfactory:

(*a*) Check the linearity of phase angle with frequency, and estimate the effective time delay between q_o and q_i.

(*b*) If $q_i(t)$ is a transient of the form shown in Fig. P3.3*b*, use SIMULINK (or other available digital simulation) to find the smallest T for which the instantaneous error between q_o and q_i does not exceed 5 percent of the full-scale q_i value.

3.27 Find $Q_i(i\omega)$ for the $q_i(t)$ of Fig. P3.4 by the exact analytical method.

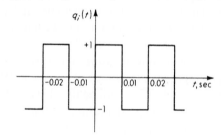

Figure P3.4

3.28 Repeat Prob. 3.27 for Fig. P3.5.

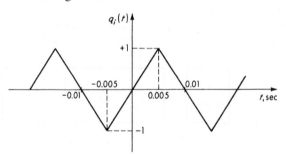

Figure P3.5

3.29 If the $q_i(t)$ of Prob. 3.27 is the input to a first-order system with a gain of 1 and a time constant of 0.001 s, find $Q_o(i\omega)$ and $q_o(t)$ for the periodic steady state.

3.30 Repeat Prob. 3.29, except use $q_i(t)$ from Prob. 3.28.

3.31 In Fig. 3.86 let the carrier be a square wave as in Fig. P3.6. Find the frequency spectrum of the output signal of the modulator.

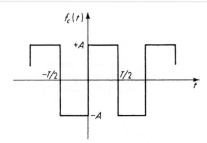

Figure P3.6

3.32 Repeat Prob. 3.31 if the carrier is a square wave as in Fig. P3.7.

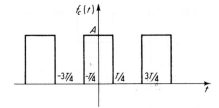

Figure P3.7

3.33 It is desired to simulate a random atmospheric turbulence whose mean-square spectral density $\phi_t(\omega)$ is adequately represented as $10/(1 + 0.0001\omega^2)$, where ω is in radians per second. A white-noise generator having $\phi_{wn}(\omega) = 10$ is available. Select a suitable filter configuration and numerical values to follow the generator and produce the desired $\phi_t(\omega)$. The output of the noise generator should "see" a filter input resistance of 10,000Ω.

3.34 Tests on a gyroscope show that it can withstand any random vibration along a given axis if the frequency content is between 0 and 1,000 rad/s and the rms acceleration is less than 80 in/s^2. This gyro is to be mounted in a rocket where it will be subjected to acoustic-pressure-induced vibration. The transfer function between pressure and acceleration and the mean-square spectral density of pressure are as given in Fig. P3.8. Will this gyro withstand the vibration?

3.35 In Fig. 3.36, a potentiometer displacement transducer is treated as a zero-order instrument. When the output of this device is connected to, say, an oscilloscope, the circuit is modified and the zero-order model may no longer be valid for this complete measurement system. Note that *some kind* of voltage-measuring device *must* be used to actually measure displacement with the potentiometer. Most instrumentation oscilloscopes

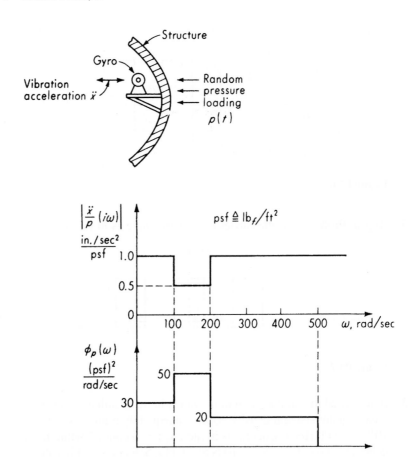

Figure P3.8

have an input circuit consisting of a 1 MΩ (intentional) resistance in parallel with about 50 pF of (parasitic) capacitance. Let the potentiometer have a total resistance of 1 kΩ and be excited with 10 V DC. The motion to be measured starts at time zero with the wiper at 50 percent of full stroke (say, 0.5 cm) and moves to 40 percent of stroke in 0.001 s. Derive the differential equation which describes this situation, using letter values for all the parameters. Then solve it for the numerical values of this example. (You will need to use SIMULINK (or other available) simulation because the differential equation will have time-varying coefficients. This class of equation generally does *not* have analytical solutions.)

3.36 Explain how the sinusoidal transfer function of a system may be obtained from measured records of $q_i(t)$ and $q_o(t)$ if q_i is a transient of any shape whatever.

3.37 Reanalyze the force-measuring problem of Fig. 3.31 for *dynamic* operation, assuming the two blocks have masses M_1 and M_2. That is, get the operational transfer function analogous to Eq. (3.81).

3.38 Reanalyze the voltage-measuring problem of Fig. 3.26 for dynamic operations, i.e., replace the batteries with sources of time-varying voltage. Also, let the voltage-measuring device be an oscilloscope with R_m shunted by a capacitor C_m.

3.39 Reanalyze the current-measuring problem of Fig. 3.30 for dynamic operation; i.e., replace the batteries with sources of time-varying voltage. Also, let the current-measuring device have an inductance L_m in series with R_m.

BIBLIOGRAPHY

1. H. E. Koenig and W. A. Blackwell, "Electromechanical System Theory," McGraw-Hill, New York, 1961.

2. C. L. Cuccia, "Harmonics, Sidebands, and Transients in Communication Engineering," McGraw-Hill, New York, 1952.

3. T. N. Whitehead, "The Design and Use of Instruments and Accurate Mechanisms," Dover, New York, 1954.

4. "ASTM Standards on Precision and Bias for Various Applications," 5th ed., ASTM, Philadelphia, 1997.

5. W. A. Wildhack et al., "Accuracy in Measurements and Calibrations," *NBS, Tech. Notes* 262, 1965.

6. J. Mandel, "The Statistical Analysis of Experimental Data," Dover, New York, 1984.

7. Special Issue on the Fast Fourier Transform. *IEEE Trans. Audio Electroacoustics,* vol. AV-17, June 1969.

8. J. S. Bendat and A. G. Piersol, "Measurement and Analysis of Random Data," Wiley, New York, 1966.

9. G. A. Korn, "Random Process Simulation and Measurement," McGraw-Hill, New York, 1966.

10. E. O. Doebelin, "System Modeling and Response," Wiley, New York, 1980.

11. J. S. Bendat and A. G. Piersol, "Engineering Applications of Correlation and Spectral Analysis," Wiley, New York, 1980.

12. E. O. Doebelin, "System Dynamics," Marcel Dekker, New York, 1998.

13. E. O. Doebelin, "Control System Principles and Design," Wiley, New York, 1985.

14. H. H. Ku, "Precision Measurement and Calibration: Statistical Concepts and Procedures," *NBS* SP300, 1969.

15. M. G. Natrella, "Experimental Statistics," *NBS Handbook,* vol. 91, 1966.

16. J. G. Proakis and D. G. Manolakis, "Digital Signal Processing: Principals, Algorithms, and Applications," Prentice Hall, Upper Saddle Park, NJ, 1996.

17. S. M. Kay, "Modern Spectral Estimation," Prentice Hall, Englewood Cliffs, NJ, 1988.

18. E. O. Doebelin, "Engineering Experimentation," McGraw-Hill, New York, 1995.

19. C. P. Wright, "Applied Measurement Engineering," Prentice Hall PTR, Englewood Cliffs, NJ, 1995.

20. G. Rizzoni, "Principles and Applications of Electrical Engineering," 2d ed., Irwin, Chicago, 1996.

21. H. W. Coleman and W. G. Steele, Jr., "Experimentation and Uncertainty Analysis for Engineers," 2d ed., Wiley, New York, 1999.

22. B. N. Taylor and C. E. Kuyutt, "Guidelines for Evaluating and Expressing the Uncertainty of NIST Measurement Results," *NIST Tech.* Note 1297, 1993.

23. "Guide to the Expression of Uncertainty in Measurement," 1st ed., ISO, Geneva, Switzerland, 1993.

24. "Measurement Uncertainty," *ANSI/ASME PTC* 19.1-1986 Part 1, 1986.

25. R. B. Abernethy, "Measurement Uncertainty Handbook," Instrument Society of America, Research Triangle Park, NC, 1980.

26. P. H. Sydenham, "Handbook of Measurement Science," Wiley, New York, 1982.

PART 2

Measuring Devices

CHAPTER 4

Motion and Dimensional Measurement

4.1 INTRODUCTION

We commence our study of specific measuring devices with motion and dimensional measurements because they are based on two of the fundamental quantities in nature (length and time) and because so many other quantities (such as force, pressure, temperature, etc.) are often measured by transducing them to motion and then measuring this resulting motion. Our main interest is motion (a *changing* displacement); however, many of the sensors described are used in manufacturing processes where gaging of part dimensions is required.

We are also mainly (though not exclusively) concerned with electromechanical transducers which convert motion quantities into electrical quantities. The intent is not to present a catalog listing of the myriad physical effects which have been, or might be, used as the basis of a motion transducer, but rather to provide sufficient detail for practical application of the relatively small number of transducer types which form the basis of the majority of practical measurements. The above-mentioned catalog-listing type of information is extremely useful to one who has a measurement problem not solvable by standard techniques and who must therefore invent and/or develop a new instrument. Material of this type is available in several references.[1]

4.2 FUNDAMENTAL STANDARDS

The four fundamental quantities of the International Measuring System, for which independent standards have been defined, are length, time, mass, and temperature. Units and standards for all other quantities are *derived* from these. In motion measurement, the fundamental quantities are length and time. Prior to 1960 the

---

[1]K. S. Lion, "Instrumentation in Scientific Research," McGraw-Hill, New York, 1959; C. F. Hix, Jr., and R. P. Alley, "Physical Laws and Effects," Wiley, New York, 1958.

standard of length was the carefully preserved platinum-iridium International Meter Bar at Sèvres, France. In 1960 the meter was redefined in terms of the wavelength of a krypton-86 lamp as the length equal to 1,650,763.73 wavelengths in vacuum corresponding to the transition between two energy levels of the atom krypton 86. This standard is believed to be reproducible to about 2 parts in 10^8 and can be applied at this precision level to measurements of length in the range of about 10^{-8} to 40 in.[2]

The above National Prototype Standard is not available for routine calibration work. Rather, to protect such top-level standards from deterioration, the National Institute of Standards and Technology (NIST) has set up National Reference Standards and, below these, Working Standards. Further down the line in accuracy are the so-called Interlaboratory Standards, which are standards sent in to NIST for calibration and certification by factories and laboratories all over the country. These last-mentioned standards are usually readily available to the working engineer for calibration of motion transducers.

The fundamental unit of time is the second, which was redefined for scientific use as 1/31,556,925.9747 of the tropical year at 12^h ephemeris time, 0 January 1900, by the International Committee on Weights and Measures in 1956. A serious fault in this definition is that no one can measure an interval of time by direct comparison with the interval of time defining the second. Rather, lengthy astronomical measurements over several years are necessary to relate the current value of the mean solar second to the basic standard. These measurements and calculations result in an estimated probable error of about 1 part in 10^9, which is quite poor compared with the precision implied in the basic definition of the second. To remedy this difficulty, metrologists in 1964 again redefined the second in terms of the frequencies of atomic resonators.[3] Now the second is defined as the interval of time corresponding to 9,192,631,770 cycles of the atomic resonant frequency of cesium 133. Figure 4.1*a* and *b* (taken from *NBS Spec. Publ.* 250, NBS Calibration Services, 1987) gives data on the accuracy of the length and angle standards, while Fig. 4.2 (*NBS Spec. Publ.* 445-1, The National Measurement System for Time and Frequency, A. S. Risley, 1976) gives an interesting historical perspective on time-keeping accuracy.

The above short discussion is concerned with the *fundamental* standards of length and time, rather than the practical working standards with which most engineers will be concerned. These practical standards and associated calibration procedures are discussed in each specific section, such as relative displacement, acceleration, etc.

---

[2]J. S. Beers, "A Gage Block Measurement Process Using Single Wavelength Interferometry," *NBS Monograph* 152, December 1975.

[3]H. Hellwig, "Frequency Standards and Clocks: A Tutorial Introduction," *NBS Tech. Note* 616, 1972.

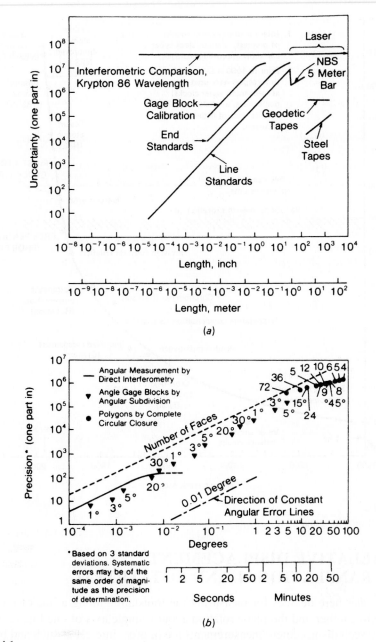

Figure 4.1
Length and angle standards.

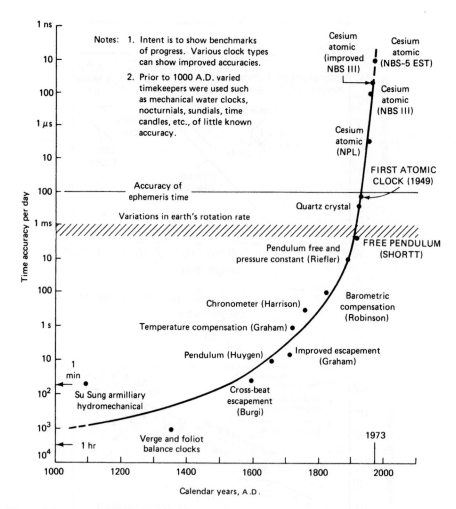

Figure 4.2
History of time keeping.

4.3 RELATIVE DISPLACEMENT: TRANSLATIONAL AND ROTATIONAL

We consider here devices for measuring the translation along a line of one point relative to another and the plane rotation about a single axis of one line relative to another. Such displacement measurements are of great interest as such and because they form the basis of many transducers for measuring pressure, force, acceleration, temperature, etc., as shown in Fig. 4.3.

Calibration

Static calibration of translational devices often can be satisfactorily accomplished by using ordinary micrometers as the standard. When used directly to measure the

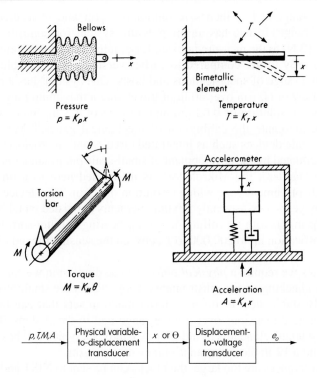

Figure 4.3
Transducer applications of displacement measurement.

displacement of the transducer, these devices usually are suitable to read to the nearest 0.0001 in or 0.01 mm. If smaller increments are necessary, lever arrangements (about a 10:1 ratio is fairly easy to achieve) or wedge-type mechanisms (about 100:1) can be employed for motion reduction. Micrometer-based calibrators are available from various gage manufacturers and provide higher accuracy and resolution than the micrometers commonly used in machine shops. One such unit[4] has a minimum graduation of 10 μin and an accuracy of $\pm$ 10 μin. Another[5] uses a differential micrometer to achieve a resolution of 0.07 μm and minimum graduations of 0.5 μm. Levers (10 to 1 range) or wedges (100 to 1) can be used to reduce motion. The use of levers based on elastic deflection rather than pivot bearings allows "ordinary" micrometers to control extremely small motions. A family of such units is available;[6] a typical device providing a range of 1000 nm has a resolution of 1 nm and a thermal sensitivity of 0.00017 nm/K. More complex (and expensive) calibrators use linear encoder scales or laser interferometers to realize greater accuracy, longer ranges, and digital electronic readouts. One instrument[7] using a linear

[4]Universal Calibrator 400B, Federal Products, Providence, RI, 800-343-2050.

[5]Newport Corp., Irvine, CA, 800-222-6440.

[6]Precision Instrument Laboratory, Alson E. Hatheway, Inc., Pasadena, CA, 818-795-0514.

[7]Trimos Universal Length Measuring Instrument TULM 210, Fowler Tools and Instruments, Boston, MA, 800-788-2353.

encoder has a range of 17.7 inches, resolution of 1 μin, and an accuracy of 28 μin over the full range. It also has an air pressure system for controlling the force applied (1 to 12 N) to the measured object. Laser interferometer systems, now quite common and easy to use compared to earlier versions, are available from many sources, with a range of specifications and costs. One system[8] has a resolution of 25 μin, accuracy of 60 μin over 40 in of travel, and a maximum range of 2000 in. Another[9] has a resolution of 0.62 nm and a range of 1.3 m. Such systems have been used to automate the calibration of mechanical gages.[10] Measurement of microscopic-scale devices such as integrated circuits and microelectromechanical systems is facilitated by the development of small-scale pitch standards.[11] While not considered a standard, the Mikrokator[12] is mentioned here as a unique, *totally mechanical* displacement gage with a resolution of 1 μin. This device, invented in Sweden many years ago, uses only flexural elements (a twisted metal band and flat diaphragm springs), with no rolling or sliding bearings of any sort, to produce a motion magnification up to 200,000 to 1 between the sensing tip and the indicating needle!

Sometimes we require a *physical object* whose dimension we can rely on with certainty, for checking various instruments. *Gage blocks* are small blocks of hard, dimensionally stable steel or other material made in sets that can be stacked to provide accurate dimensions over a wide range and in small steps. They are the basic working length standards of industry. A 1-in block typically[13] has an accuracy of $\pm$ 2 μin and a 12-in block has an accuracy of $\pm$ 12 μin.

If these tolerances are too large, the blocks can be sent to NIST and calibrated[14] against light wavelengths to the nearest 10^{-7} in. Some precision-manufacturing operations currently require and use the latter calibration service. When transducers are calibrated to very high accuracies, it is extremely important to control all interfering and/or modifying inputs such as ambient temperature,[15] electrical excitation to the transducer, etc.

Rotational or angular displacement is not itself a fundamental quantity since it is based on length, and so a fundamental standard is not necessary. However, reference and working standards for angles (and thus angular displacement) are desirable and available. The basic standards (against which other standards or instruments may be calibrated) are called *angle blocks*.[16] These are carefully made steel blocks

[8]Optodyne, Inc., Compton, CA, 800-766-3920.

[9]ZMI1000, Zygo Corp., Middlefield, CT, 203-347-8506.

[10]X. Gan and K. K. Kang, "Automatic Calibration of Height Setting Micrometers Using a Laser Interferometer," *Cal Lab,* July–August 1995, pp. 14–17.

[11]J. Potzick, "A New NIST-Certified Small Scale Pitch Standard," *Cal Lab,* May–June 1997, pp. 18–24.

[12]Federal Products Co., Providence, RI, 401-784-3375.

[13]Ibid.

[14]P. E. Pontius, "Measurement Assurance Program, Long Gage Blocks," *NBS Monograph* 149, November 1975.

[15]J. B. Bryan et al., "Thermal Effects in Dimensional Metrology," *ASME Paper* 65-*Prod*—13, 1965.

[16]C. E. Haven and A. G. Strong, "Assembled Polygon for the Calibration of Angle Blocks," *Natl. Bur. Std. (U.S.), Handbook 77,* vol. 3, p. 318, 1961.

about $\frac{5}{8}$ in wide and 3 in long, with a specified angle between the two contact surfaces. Just as for length gage blocks, these angle blocks can be stacked to "build up" any desired angle accurately and in small increments. The blocks can be calibrated to an accuracy of 0.1 second of arc by NIST.[17]

Rotational transducers rarely require such accuracy for calibration, nor can the laborious and expensive techniques necessary to realize these limits be economically justified. Thus most static calibration of angular-displacement transducers can be adequately carried out by using more convenient and readily available equipment.

A precision dividing head may be used to both produce the angular motion and measure it. One such unit[18] has a resolution of 0.0001° and repeatability of 1 arc second. Digital absolute shaft angle encoders (discussed in detail later in this chapter) with electronic angle readouts are available in a wide range and can serve as convenient calibration standards for less accurate devices. The best units have a resolution of about 22 bits and accuracy approaches 0.3 arc seconds.[19] This state-of-the-art accuracy is of course rarely needed, so, to save money, we select an encoder model that is just adequate to our specific calibration needs.

Resistive Potentiometers

Basically, a resistive potentiometer consists of a resistance element provided with a movable contact. (See Fig. 4.4) The contact motion can be translation, rotation, or a combination of the two (helical motion in a multiturn rotational device), thus allowing measurement of rotary and translatory displacements. Translatory devices have strokes from about 0.1 to 20 in and rotational ones range from about 10° to as much as 60 full turns. The cable-extension version[20] allows very long travels (up to 1600 in) and convenient mounting in situations that might be awkward for other configurations. Such devices are also available using digital encoders in place of the potentiometer or with both a potentiometer and a tachometer generator, giving position and velocity data. The resistance element is excited with either dc or ac voltage, and the output voltage is (ideally) a linear function of the input displacement. Resistance elements in common use may be classified as wire-wound, conductive plastic, deposited film, hybrid, or cermet.

If the distribution of resistance with respect to translational or angular travel of the wiper (moving contact) is linear, the output voltage e_o will faithfully duplicate the input motion x_i or θ_i if the terminals at e_o are open-circuit (no current drawn at the output). (For ac excitation, x_i or θ_i amplitude-modulate e_{ex}, and e_o does not look like the input motion.) The usual situation, however, is one in which the

[17]Independent Standards Laboratory, *Instr. Contr. Syst.,* p. 478, March 1961.

[18]Divitron, International Machine and Tool Corp., Warwick, RI, 401-467-6905.

[19]Model 6CMF22AR, Aerospace Controls Corp., Little Rock, AR, 501-490-2048.

[20]Space-Age Control, Inc., Palmdale, CA, 805-273-3000 (www.spaceagecontrol.com); Celesco, Canoga Park, CA, 800-423-5483 (www.celesco.com); E. Newman, "Hooked on Long Sensors," *Machine Design,* April 9, 1993.

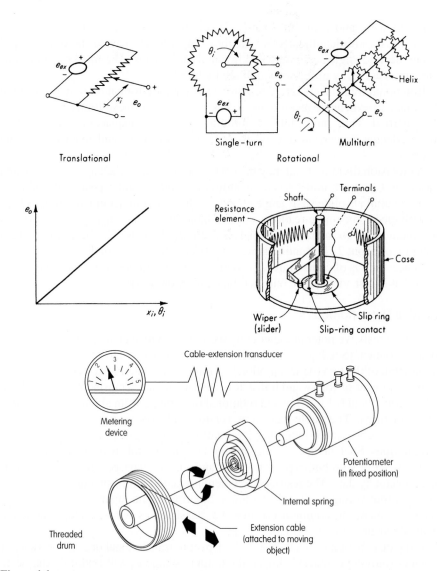

Figure 4.4
Potentiometer displacement transducer.

potentiometer output voltage is the input to a meter or recorder that draws some current from the potentiometer. Thus a more realistic circuit is as shown in Fig. 4.5a. Analysis of this circuit gives

$$\frac{e_o}{e_{ex}} = \frac{1}{1/(x_i/x_t) + R_p/R_m)(1 - x_i/x_t)} \tag{4.1}$$

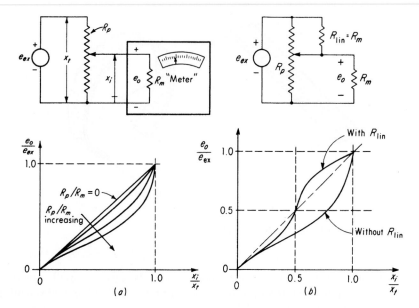

Figure 4.5
Potentiometer loading effect.

which becomes for ideal ($R_p/R_m = 0$ for an open circuit) conditions

$$\frac{e_o}{e_{ex}} = \frac{x_i}{x_t} \qquad (4.2)$$

Thus for no "loading" the input-output curve is a straight line. In actual practice, $R_m \neq \infty$ and Eq. (4.1) shows a nonlinear relation between e_o and x_i. This deviation from linearity is shown in Fig. 4.5a. The maximum error is about 12 percent of full scale if $R_p/R_m = 1.0$ and drops to about 1.5 percent when $R_p/R_m = 0.1$. For values of $R_p/R_m < 0.1$, the position of maximum error occurs in the neighborhood of $x_i/x_t = 0.67$, and the maximum error is approximately $15R_p/R_m$ percent of full scale.

We see that to achieve good linearity, for a "meter" of a given resistance R_m, we should choose a potentiometer of sufficiently *low* resistance relative to R_m. This requirement conflicts with the desire for high sensitivity. Since e_o is directly proportional to e_{ex}, it would seem possible to get any sensitivity desired simply by increasing e_{ex}. This is not actually the case, however, since potentiometers have definite power ratings related to their heat-dissipating capacity. Thus a manufacturer may design a series of potentiometers, say single-turn 2-in-diameter, with a wide range (perhaps 100 to 100,000 Ω) of total resistance R_p, but all these will be essentially the same size and mechanical configuration, giving the same heat-transfer capability and thus the same power rating, say about 5 W at 20°C ambient. If the heat dissipation is limited to P watts, the maximum allowable excitation voltage is given by

$$\max e_{ex} = \sqrt{PR_p} \qquad (4.3)$$

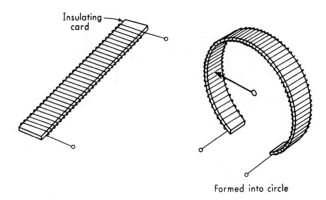

Insulating ——
card

Formed into circle

Figure 4.6
Construction of wirewound resistance elements.

Thus a low value of R_p allows only a small e_{ex} and therefore a small sensitivity. Choice of R_p thus must be influenced by a trade-off between loading and sensitivity considerations. The maximum available sensitivity of potentiometers varies considerably from type to type and also with size in a given type. It can be calculated from the manufacturer's data on maximum allowable voltage, current, or power and the maximum stroke. The shorter-stroke devices generally have higher sensitivity. *Extreme* values are of the order of 15 V/deg for short-stroke rotational types ("sector" potentiometers) and 300 V/in for short-stroke (about $\frac{1}{4}$ in) translational pots. It must be emphasized that these are maximum values and that the usual application involves a much smaller (10 to 100 times smaller) sensitivity. Figure 4.5*b* shows a method for improving linearity without increasing R_m.

The resolution of potentiometers is strongly influenced by the construction of the resistance element. An obvious approach is to use a single slide-wire as the resistance which gives an essentially continuous stepless resistance variation as the wiper travels over it. Such potentiometers are available, but are limited to rather small resistance values since the length of wire is limited by the desired stroke in a translational device and by space restrictions (diameter) in a rotational one. Resistance of a given length of wire can be increased by decreasing the diameter, but this is limited by strength and wear considerations.

To get sufficiently high resistance values in small space, the wirewound resistance element is widely used. The resistance wire is wound on a mandrel or card which is then formed into a circle or helix if a rotational device is desired (see Fig. 4.6). With such a construction the variation of resistance is not a linear continuous change, but actually proceeds in small steps as the wiper moves from one turn of wire to the next (see Fig. 4.7). This phenomenon results in a fundamental limitation on the resolution in terms of resistance-wire size. For instance, if a translational device has 500 turns of resistance wire on a card 1 in long, motion changes smaller than 0.002 in cannot be detected. (This is slightly conservative since the wiper, in going from one turn to the next, goes through an intermediate position in which it is

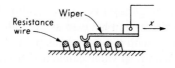

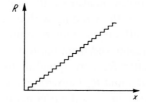

Figure 4.7
Resolution of wirewound potentiometers.

touching *both* turns at once.[21] The resolution thus actually varies from one position to another, with the *worst* value being that given by a simple counting of turns per inch.) The actual limit for wire spacing according to current practice is between 500 and 1,000 turns per inch.[22] For translational devices, resolution is thus limited to 0.001 to 0.002 in, while single-turn rotational devices can trade off increased diameter *D* for increased angular resolution according to the relation

$$\text{Best angular resolution} = \frac{0.12 \text{ to } 0.24}{D} \quad \text{degrees} \quad D \text{ in inches} \quad \textbf{(4.4)}$$

It should be noted that resolution is intimately related to total resistance since the fine wire required to get close wire spacing will naturally have a high resistance. Thus one cannot choose total resistance and resolution independently.

Nonwirewound resistance elements provide improved resolution and life; however, they are more temperature-sensitive, have a high (and variable) wiper contact resistance, and can tolerate only moderate wiper currents. Elements of both cermet (combination of ceramic and metallic materials) and conductive plastic (mixture of plastic resin with proprietary conductive powders) are in the form of flat strips or films and thus present a smooth surface to the wiper. They are conventionally described as having "infinitesimal resolution." However, resolution cannot be actually measured and numerically specified (as it can in wirewound units) because the output-voltage deviations from ideal straight-line behavior are somewhat random rather than the largely reproducible discrete steps of the wirewound pot.

---

[21]H. Gray, "How to Specify Resolution for Potentiometer Servos," *Contr. Eng.,* p. 129, November 1959; C. A. Mounteer, "The Effective Resolution of Wire-Wound Potentiometers," *Giannini Tech. Notes,* G. M. Giannini Co., Pasadena, CA, January–February 1959.

[22]S. A. Davis and B. K. Ledgerwood, "Electromechanical Components for Servomechanisms," p. 53, McGraw-Hill, New York, 1961.

Output smoothness (really *roughness*) is a specification that attempts to quantify this effect in nonwirewound pots and is given as the ratio of the peak amplitude of the random variations to e_{ex}, with typical values being 0.1 percent. While resolution, usually quoted as a theoretical value calculated from wire spacing, is numerically available for wirewound pots, a measurable noise specification (analogous to output smoothness for nonwirewound) called *equivalent noise resistance* R_{en} is often quoted. The test involves measuring the pot's output-voltage fluctuations while the wiper is moved over the winding with a constant wiper current applied, and thus the test includes effects of both resolution and various random-noise effects. The numerical value derived from the test is quoted as a fictitious equivalent contact resistance variation R_{en} corresponding to the largest fluctuations. For a given pot application (e_{ex}, R_p, and R_m known), one can use this R_{en} value to calculate the worst spurious voltage at e_o as the product of wiper current and R_{en}. Typical R_{en} values are 5 to 100 Ω.

A useful feature of conductive plastic elements is that they may be contoured along one edge (see Fig. 4.8) by milling or laser processes to adjust the resistance distribution from the as-molded state. These manufacturing processes are servo-controlled, so that the procedure is cost-effectively applied *individually* to each complete potentiometer to tailor the resistance element to compensate for all other sources of nonlinearity in that particular unit. This manufacturing process also is used to improve the conformity of the resistance element to *desired* nonlinear functions for pots used as computing elements (see Chap. 10).

Application of a narrow track of conductive plastic onto a conventional wirewound element produces the *hybrid potentiometer.* This combines most of the best features of each technology, but at somewhat increased cost.

Figure 4.8
Conductive-plastic potentiometer. *(Courtesy Waters Mfg., Wayland, MA)*

Another approach to increased resolution involves the use of multiturn potentiometers. The resistance element is in the form of a helix, and the wiper travels along a "lead screw." The number of wires per inch of element is still limited, as mentioned above, but an increase in resolution can be obtained by introducing gearing between the shaft whose motion is to be measured and the potentiometer shaft. For example, one rotation of the measured shaft could cause 10 rotations of the potentiometer shaft; thus the resolution of measured-shaft motion is increased by a factor of 10. Multiturn potentiometers are available up to about 60 turns in wirewound, nonwirewound, and hybrid types. For translational devices, various motion-amplifying mechanisms could be used in similar fashion.

Most potentiometers used for motion measurement are intended to give a linear input-output relation and are used as purchased, without calibration. Thus a specification of linearity is essentially equivalent to one of accuracy. Potentiometers are available in a wide range of linearities and corresponding prices. Linearity depends greatly on the uniformity of the resistance winding, but errors in this can be corrected by adding fixed resistances in series and/or parallel at proper locations on the winding. This procedure (called *tapping*) also can correct loading errors so as to give a linear relation for a heavily loaded potentiometer.[23] The best nonlinearities commercially available range from 1 percent of full scale for $\frac{1}{2}$-in-diameter single-turn pots through 0.02 percent for a 2-in-diameter multiturn to 0.002 percent for a 10-in-diameter multiturn. The best nonlinearities of translational pots are about 0.05 to 0.10 percent of full scale. It should be noted that accuracy can be no better (and is generally worse) than one-half the resolution; thus resolution places a limit on accuracy.

Noise in potentiometers refers to spurious output-voltage fluctuations occurring during motion of the slider and includes the effects of resolution. In addition, various mechanical and electrical defects produce noise. In a wirewound pot, motion of the slider over the resistance wires may cause bouncing of the contact at certain speeds, thus resulting in intermittent contact. This phenomenon becomes particularly significant if the speed and wire spacing are such as to produce forces of frequency near the resonant frequency of the spring-loaded contact. By using *two* cantilevered contacts with different beam lengths (and thus different natural frequencies), good contact can be maintained at all operating speeds. Thus if one section is resonating at a certain speed, the other will be off resonance and making continuous contact. Another possibility lies in filling the interior of the potentiometer with a damping fluid or coating the wipers with a layer of elastomeric damping material to limit resonant amplitudes. This also generally increases the shock and vibration tolerance of the unit. Another source of noise is found in dirt and wear products which come between the contact surface and the winding. Even if no dirt or wear products are present, the contact resistance of a moving contact varies during motion, and if any load current is flowing through this contact, a spurious iR voltage appears in the output. This effect also occurs at the slip-ring contact.

---

[23]Ibid., p. 59.

Numerical values of noise voltage quoted in specifications generally include all sources of noise and correspond to a definite speed and current.[24]

The dynamic characteristic of potentiometers (if we consider displacement as input and voltage as output) is essentially that of a zero-order instrument since the impedance of the winding is almost purely resistive at the motion frequencies for which the device is usable. However, the mechanical loading imposed on the measured motion by the inertia and friction of the potentiometer's moving parts should be carefully considered. The friction is usually mostly dry friction, and the manufacturer generally supplies numerical values of the starting and running friction force or torque. These values vary over a wide range, depending on the construction of the potentiometer. Special low-friction rotary pots have starting torques as small as 0.003 oz · in. More conventional instruments may have 0.1 to 0.5 oz · in or more. Translational pots have friction values from less than 1 oz to over 1 lb. Inertia values for both rotary and translatory pots vary widely with size. A typical $\frac{7}{8}$-in-diameter single-turn pot has a moment of inertia of 0.12 g · cm^2, while a 2-in diameter 10-turn pot has about 18 g · cm^2. Moving masses of translatory pots have weights ranging from fractions of an ounce to several ounces.

Finally, selection of potentiometers should take into account various environmental factors such as high or low temperatures, shock and vibration, humidity, and altitude. These may act as modifying and/or interfering inputs that seriously degrade instrument performance. While the life of potentiometers varies greatly with type and environment, representative values might be 2 million rotation cycles for wirewound, 10 million for hybrid, and 50 million for conductive plastic. Recent automotive engine designs often employ many sensing devices, including potentiometric displacement sensors. Design of low-cost sensors for the severe under-the-hood environment has been particularly challenging.[25] Here a term called "dither life" is significant. Dither refers to conditions where case vibration caused by the engine superimposes high-frequency and low-amplitude wiper oscillations on the intended measured motion, a situation conducive to rapid localized wear. Dither life refers to cycles of the low-amplitude vibratory motion, which accumulate much more rapidly than strokes of gross motion. Conductive plastic pots have proved successful in many of these applications, providing dither lives in excess of 100 million cycles.

We conclude this section with a brief discussion of the *magnetoresistive* displacement transducer. This device operates on a principle different from the potentiometer but can replace it in certain applications where its special features apply. Certain materials exhibit the magnetoresistive effect, where the electrical resistance changes in the presence of a magnetic field. Figure 4.9a[26] shows how the principle is applied to a rotary displacement sensor. Two magnetoresistive elements mounted in the stationary housing are connected as shown, with 10 V DC applied between terminals 1 and 3, and output taken between terminals 1 and 2. The input shaft rotates a specially shaped permanent magnet over the resistance elements

[24]PPMA "Conference Report," *Electromech. Des.,* p. 8, April 1964.

[25]W. Wheeler, "Precision Position Sensors in Automotive Applications," *SAE Paper* 780209, 1978.

[26]Midori American Corp., 2555 E. Chapman Ave., Fullerton, CA, 714-449-0997.

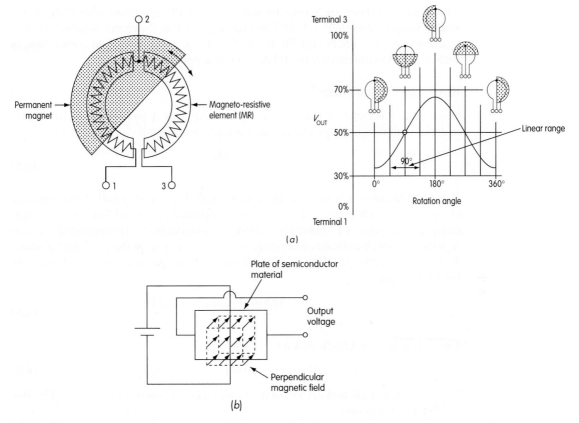

Figure 4.9
(*a*) Magnetoresistive displacement transducer. (*b*) The Hall effect principle.

(without touching them), causing the output voltage to vary with angle as shown. Note that the output voltage is linear with angle only over a restricted range and that there is *no* position that gives zero voltage. The device is also more temperature sensitive than potentiometers, although this can be compensated. These disadvantages may often rule it out; however, the noncontacting feature gives excellent resolution, torque smoothness, long life, and high-speed capability, making it useful in certain applications.

Characteristics similar to the magnetoresistive sensor but using the *Hall effect* are also available.[27] In the Hall effect, Fig. 4.9*b*, a thin plate of semiconductor material is excited with a DC voltage across two terminals. When a magnetic field perpendicular to the plate is applied, an output voltage proportional to the field strength appears at two other terminals. Position sensing is accomplished by suitably moving a permanent magnet relative to the Hall plate. In the referenced motion sensor, which uses two rare-earth magnets, problems of linearity and temperature

---

[27]Hall Effect Rotary Position Sensor, Series HRS100, Clarostat Sensors and Controls, 12055 Rojas Dr., Suite K, El Paso, TX 79936, 800-874-1874 (www.clarostat.com).

sensitivity have been overcome by use of a digitally programmable Hall chip.[28] With a supply voltage of 5-V DC, nonlinearity of 1 to 2 percent is achieved over available ranges of 45, 60, 80, 90, or 180 degrees of rotation. Output voltage ranges such as 5 to 45 percent or 0 to 100 percent of supply are available.

Resistance Strain Gage[29]

Consider a conductor of uniform cross-sectional area A and length L, made of a material with resistivity ρ. The resistance R of such a conductor is given by

$$R = \frac{\rho L}{A} \qquad (4.5)$$

If this conductor is now stretched or compressed, its resistance will change because of dimensional changes (length and cross-sectional area) and because of a fundamental property of materials called *piezoresistance*[30] (pronounced pī-ēzō-resistance), which indicates a dependence of resistivity ρ on the mechanical strain. To find how a change dR in R depends on the basic parameters, we differentiate Eq. (4.5) to get

$$dR = \frac{A(\rho\, dL + L\, d\rho) - \rho L\, dA}{A^2} \qquad (4.6)$$

Since volume $V = AL$, $dV = A\, dL + L\, dA$. Also

$$dV = L(1 + \varepsilon)A(1 - \varepsilon v)^2 - AL \qquad (4.7)$$

where $\varepsilon \triangleq$ unit strain and $v \triangleq$ Poisson's ratio. Since ε is small, $(1 - v\varepsilon)^2 \approx 1 - 2v\varepsilon$ and Eq. (4.7) becomes

$$dV = AL\,\varepsilon\,(1 - 2v) = A\, dL + L\, dA \qquad (4.8)$$

and since $\varepsilon \triangleq dL/L$,

$$A\, dL(1 - 2v) = A\, dL + L\, dA \qquad (4.9)$$

$$-2vA\, dL = L\, dA \qquad (4.10)$$

Substituting in Eq. (4.6) yields

$$dR = \frac{\rho A\, dL + LA\, d\rho + 2v\rho\, A\, dL}{A^2} \qquad (4.11)$$

and thus

$$dR = \frac{\rho\, dL(1 + 2v)}{A} + \frac{L\, d\rho}{A} \qquad (4.12)$$

---

[28]Model MLX90215, Melexis, Inc., 41 Locke Rd., Concord, NH 03301, 603-223-2362 (www.melexis.com).

[29]J. W. Dally and W. F. Riley, "Experimental Stress Analysis," 3d ed., McGraw-Hill, New York, 1991; A. L. Window and G. S. Holister, "Strain Gage Technology," Applied Science Publ., New Jersey, 1982; R. L. Hannah and S. E. Reed, "Strain Gage Users Handbook," Elsevier Appl. Science, New York, 1992.

[30]C. M. Harris and C. E. Crede (eds.), "Shock and Vibration Handbook," vol. 1, pp. 16–35, McGraw-Hill, New York, 1961.

Dividing by Eq. (4.5) gives

$$\frac{dR}{R} = \frac{dL}{L}(1 + 2v) + \frac{d\rho}{\rho} \tag{4.13}$$

and finally

$$\text{Gage factor} \triangleq \frac{dR/R}{dL/L} = \underbrace{1}_{\substack{\text{resistance} \\ \text{change due} \\ \text{to length} \\ \text{change}}} + \underbrace{2v}_{\substack{\text{resistance} \\ \text{change due to} \\ \text{area change}}} + \underbrace{\frac{d\rho/\rho}{dL/L}}_{\substack{\text{resistance change due} \\ \text{to piezoresistance} \\ \text{effect}}} \tag{4.14}$$

Thus if the gage factor is known, measurement of dR/R allows measurement of the strain $dL/L = \varepsilon$. This is the principle of the resistance strain gage. The term $(d\rho/\rho)/(dL/L)$ can also be expressed as $\pi_1 E$, where

$\pi_1 \triangleq$ longitudinal piezoresistance coefficient

$E \triangleq$ modulus of elasticity

The material property π_1 can be either positive or negative. Poisson's ratio is always between 0 and 0.5 for all materials.

The basic principle of the resistance strain gage is implemented in several different ways (with the first two being largely obsolete):

1. Unbonded metal-wire gage
2. Bonded metal-wire gage
3. Bonded metal-foil gage
4. Vacuum-deposited thin-metal-film gage
5. Sputter-deposited thin-metal-film gage
6. Bonded semiconductor gage
7. Diffused semiconductor gage

Strain gages, in general, are applied in two types of tasks: in experimental stress analysis of machines and structures and in construction of force, torque, pressure, flow, and acceleration transducers. The unbonded metal-wire gage employs a set of preloaded resistance wires connected in a Wheatstone bridge, as in Fig. 4.10. At initial preload, the strains and resistances of the four wires are nominally equal, which gives a balanced bridge and $e_o = 0$ (see Chap. 10 for bridge-circuit behavior). Application of a small (full scale $\approx$ 0.04 mm) input motion increases tension in two wires and decreases it in two others (wires *never* go slack), causing corresponding resistance changes, bridge unbalance, and output voltage proportional to input motion. The wires may be made of various copper-nickel, chrome-nickel, or nickel-iron alloys, are about 0.03 mm in diameter, can sustain a maximum force of only about 0.002 N, and have a gage factor of 2 to 4. Electric resistance of each bridge arm is 120 to 1,000 Ω, maximum excitation voltage is 5 to 10 V, and full-scale output typically is 20 to 50 mV.

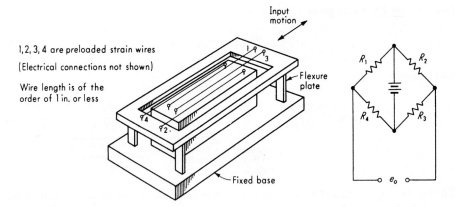

Figure 4.10
Unbonded strain gage.

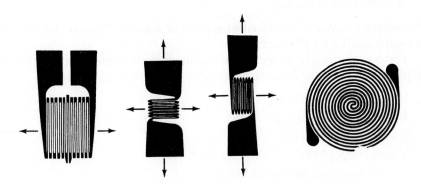

Figure 4.11
Foil strain gages.

The bonded metal-wire gage (today largely superseded by the bonded metal-foil construction) has been applied to both stress analysis and transducers. A grid of fine wire is cemented to the specimen surface, where strain is to be measured. Embedded in a matrix of cement, the wires cannot buckle and thus faithfully follow both the tension and the compression strains of the specimen. Since materials and wire sizes are similar to those of the unbonded gage, gage factors and resistances are comparable.

Bonded metal-foil gages using identical or similar materials to wire gages are used today for most general-purpose stress-analysis tasks and many transducers. The sensing elements are formed from sheets less than 0.0002 in thick by photo etching processes, which allows great flexibility with regard to shape. In Fig. 4.11, for example, the three linear grid gages are designed with "fat" end turns. This local increase in area reduces the transverse sensitivity, a spurious input since the gage is

long. Gages can be applied to curved surfaces; the minimum safe bend radius can be as small as 0.06 in in some gages.

Typical gage resistances are 120, 350, and 1,000 Ω, with the allowable gage current[34] determined by heat-transfer conditions but typically 5 to 40 mA; gage factors are 2 to 4. Resistance of an individual gage is easily measured, but measurement of the gage factor requires cementing the gage to a specimen for which strain can be accurately calculated *from theory.* Since gages cannot be removed and reused, the gage-factor number supplied with a purchased gage has *not* been measured for that individual gage, but rather is an average value obtained by sample-testing the production of that type of gage. Thus we rely on the statistical quality control of the manufacturer in maintaining the accuracy of the gage factor. A $\pm$ 1 percent accuracy is typical, and this is a fundamental limit on accuracy in stress-analysis applications. Note that this does *not* limit accuracy of strain-gage *transducers,* since such transducers are calibrated "end to end" (pressure in to voltage out, say, in a pressure transducer) and one need not even know the gage factor. The maximum measurable strain varies from 0.5 to 4 percent; however, special "postyield" gage devices allow measurement up to 0.1 in/in. Fatigue life of gages varies with conditions; however, 10 million cycles at $\pm$ 1,500 microstrain (45,000 lb/in^2 stress in steel, a common full-scale design value for foil-gage transducers) is typical. Bonded semiconductor gages[35] can work at much lower full-scale strain values (typically 20 $\mu\varepsilon$), which allows design of rugged and fast-responding transducers for low-range inputs where foil gages would require a very compliant elastic element.

Many different adhesives have been developed for fastening strain gages to specimens. Gages and fastening methods are available to cover temperature ranges from $-452°F$ ($-269°C$) to 1500°F (816°C). Extreme high temperatures may require welding or flame-spraying techniques rather than adhesive joining. Some adhesives cure at room temperature; others require baking. Cure times vary from a few minutes to several days. The quality of the adhesive joint obviously is critical to the proper operation of the strain gage, since we rely completely on it to transmit the specimen strain to the gage grid. These questions are particularly critical in extreme temperature and humidity conditions and for long-time installations. Protective and/or waterproofing coatings often are applied to increase reliability.

In addition to single-element gages, gage combinations called *rosettes*[36] (Fig. 4.12) are available in many configurations for specific stress-analysis or transducer applications. While individual gages conceivably could be cemented down in the same patterns, precise relative orientation of the several gages is critical in most of these applications, and this is much more easily obtained in rosette manufacture

---

[34]Optimizing Strain Gage Excitation Levels, TN-502, Measurements Group Inc. (One of a whole series of free technical notes.)

[35]M. Lebo and R. Caris, "Semiconductor Gages or Foil Gages for Transducers," Lebow Assoc., Oak Park, MI, May 1965.

[36]C. C. Perry, "Care and Feeding of Strain Gage Rosettes," *VRE Tech. Ed. News,* Vishay Inc., Malvern, PA, no. 23, September 1978; *Tech Note* 515, Measurements Group, Inc.

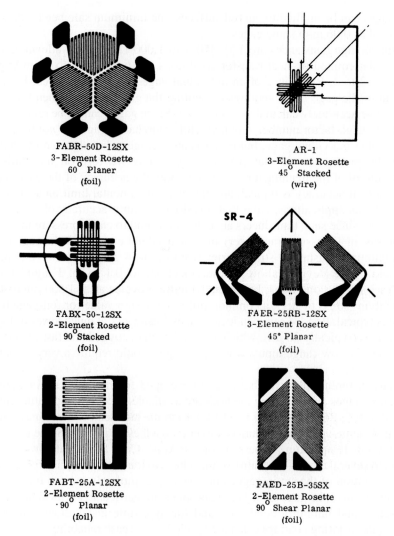

Figure 4.12
Strain-gage rosettes. *(Courtesy BLH Electronics, Waltham, MA.)*

than by the user with single gages. One rosette commonly used in stress analysis solves the problem of a surface stress whose magnitude and direction are totally unknown. Theory shows that measurements with a three-gage rosette allow calculation of all the desired information. Since such measurements are intended to define stress at a *point,* ideally the three gages should be superimposed on that point. This "sandwich" construction (called *stacked rosette*) is feasible; however, it places the top gage farther from the specimen surface and increases its self-heating, since it is better insulated from the underlying metal specimen which acts as a heat sink. If these disadvantages outweigh the advantage of "point" measurement, we can use the planar rosette design, which usually is available (see Fig. 4.12).

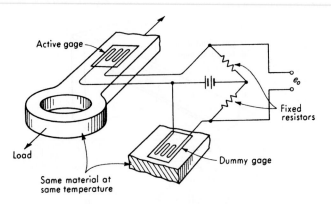

Figure 4.13
Strain-gage temperature compensation.

Temperature is an important interfering input for strain gages since resistance changes with *both* strain and temperature.[37] Since strain-induced resistance changes are quite small, the temperature effect can assume major proportions. Another aspect of temperature sensitivity is found in the possible differential thermal expansion of the gage and the underlying material. This can cause a strain and resistance change in the gage even though the material is not subjected to an external load. These temperature effects can be compensated in various ways. In Fig. 4.13, a "dummy" gage (identical to the active gage) is cemented to a piece of the same material as is the active gage and placed so as to assume the same temperature. The dummy and active gages are placed in adjacent legs of a Wheatstone bridge; thus resistance changes due to the temperature coefficient of resistance and differential thermal expansion will have no effect on the bridge output voltage, whereas resistance changes due to an applied load will unbalance the bridge in the usual way (see Chap. 10 on bridge circuits). Another approach to this problem involves special, inherently temperature-compensated gages. These gages are designed to be used on a specific material and have expansion and resistance properties such that the two effects very nearly cancel and no dummy gage is required.

Temperature also can act as a modifying input in that it may change the gage factor. With metallic gages this effect usually is quite small, except at extremely low or high temperatures. Semiconductor gages are more seriously affected in this way; however, compensation is possible. Although the above temperature problems must be considered carefully in each application, strain gages have been successfully employed from liquid-helium temperature (7°R) to the order of 2000°F. However, these extreme (especially the high temperature) applications require special techniques and yield results of lower accuracy than are obtained in routine room-temperature situations.

Used directly, the bonded strain gage is useful for measuring only very small displacements (strains). However, larger displacements may be measured by

---

[37]"Strain Gage Temperature Effects," TN-128-2, Measurement Group Inc.

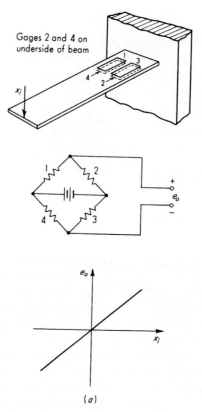

Gages 2 and 4 on
underside of beam

Figure 4.14
Strain-gage displacement transducers and extensometers.

bonding the gage to a flexible element, such as a thin cantilever beam, and applying the unknown displacement to the end of the beam, as in Fig. 4.14a. For such an application, the gage factor need not be accurately known since the overall system can be calibrated by applying known displacements to the end of the beam and measuring the resulting bridge output voltage. The configuration shown is temperature-compensated without the need for dummy gages and has four times the sensitivity of a single gage because of judicious application of bridge-circuit properties. Such transducers may be accurate to 0.1 percent of full scale.

An important application in this general class is found in *extensometers* used in materials-testing machines. Figure 4.14b shows the patent drawing for a type of strain-gage extensometer commercially available[38] in many different models. In the simplified analysis sketch of Fig. 4.15, we see that the extensometer is firmly attached to the test specimen with spring-loaded knife edges which define the gage

---

[38]MTS Systems Corp., Extensometer and Clip Gage Catalog, 14000 Technology Drive, Eden Prairie, MN 55344-2290, 800-944-1687.

Patented Feb. 5, 1974   3,789,508

2 Sheets-Sheet 1

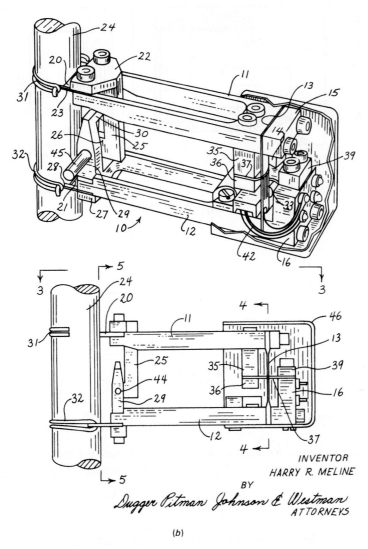

Figure 4.14
(Concluded)

length (typically 1 in) over which the specimen deflection will be measured. When
the specimen is loaded, we want to measure the deflection between the knife edges
and to compute the specimen strain as the ratio of deflection to gage length. Design
criteria for extensometers of this type require that the force exerted on the specimen
by the extensometer be small (compared with the specimen test force) and that the
strain in the gages be large enough (typically 1500 $\mu\varepsilon$) at full-scale deflection to

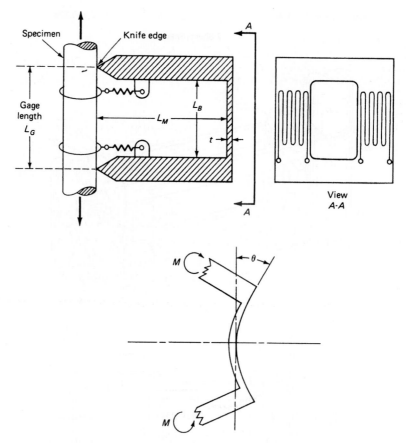

Figure 14.5
Extensometer analysis.

give accurate readings. An analysis model based on standard strength-of-materials formulas for cantilever beams with pure end-moment loading gives

$$t = \frac{2L_B L_M \varepsilon_{des}}{X_{fs}} \qquad F = \frac{2EIX_{fs}}{L_B L_M^2} \qquad (4.15)$$

where
$\varepsilon_{des} \triangleq$ design value of maximum strain in strain gages
$X_{fs} \triangleq$ full-scale displacement of extensometer
$E \triangleq$ elastic modulus of beam material
$I \triangleq$ bending (area) moment of inertia of beam cross section
$F \triangleq$ axial force exerted on specimen by extensometer

The horizontal ("cross") flexural member (37 in Fig. 4.14b) plays no essential role in the transduction of specimen deflection to gage strain described above; however, it contributes other important benefits and is, in fact, the basis of most of the claims for novelty in the patent. In brief, this member suppresses a low-frequency mode of vibration which would otherwise be present and prevent the use of such extensometers in fatigue testing at high (up to 200 Hz) cyclic loading rates. The cross-member is compliant for the intended specimen deflections but very stiff for horizontal relative motions of the two extensometer arms, motions which would otherwise constitute a low-frequency natural-vibration mode of the extensometer.

The dynamic response of bonded strain gages with respect to faithfully reproducing as a resistance variation the strain variation of the underlying surface is very good.[39] The dynamic effects of wave propagation in the cement and strain wires seem to be negligible for frequencies up to at least 50,000 Hz, and so a zero-order dynamic model is generally adequate. The loading effect of the cement and strain wires on the underlying structure is generally negligible except for very thin members, in which the stiffening effect of the strain gage may reduce the measured strain to a value considerably lower than that present without the gage. Compliance techniques can be applied to study this effect.

The voltage output from metallic strain-gage circuits is quite small (a few microvolts to a few millivolts), and so amplification is generally needed. As an example, consider the measurement of a stress level of 1,000 lb/in^2 in steel with a single active gage of 120-Ω resistance and a gage factor of 2.0. If a bridge circuit of all equal arms is used, the maximum allowable bridge voltage for 30-mA gage current is

$$e_{ex} = (240)(0.030) = 7.2 \text{ V} \tag{4.16}$$

The strain ε is $1,000/30 \times 10^6 = 3.33 \times 10^{-5}$ in/in, so that

$$\Delta R = (\text{gage factor})(\varepsilon)(R) = (2)(3.33 \times 10^{-5})(120) = 8.00 \times 10^{-3} \, \Omega \tag{4.17}$$

For the given bridge arrangement,

$$e_o = e_{ex} \frac{1}{4R} \Delta R = \frac{(7.2)(8.00 \times 10^{-3})}{480} = 0.12 \text{ mV} \tag{4.18}$$

Based on limitations of the gage alone, the smallest detectable strain depends on the thermal or Johnson-noise[40] voltage generated in every resistance because of the random motion of its electrons. This random voltage is essentially a white noise of spectral density $4kTR$ volts squared per hertz, where

$$k \triangleq \text{Boltzmann's constant} = 1.38 \times 10^{-23} \text{ J/K} \tag{4.19}$$

$$T \triangleq \text{absolute temperature of resistor, K} \tag{4.20}$$

$$R \triangleq \text{resistance, } \Omega \tag{4.21}$$

Thus if this voltage were measured by a hypothetical noise-free oscilloscope with a bandwidth of Δf Hz, the measured rms voltage would be

$$E_{\text{noise, rms}} = \sqrt{4kT R \, \Delta f} \text{ volts} \tag{4.22}$$

---

[39]P. K. Stein, "Measurement Engineering," vol. 2, Stein Engineering Services, Inc., Phoenix, 1962; M. G. Pottinger, "Effect of Gage Length in Dynamic Strain Measurement," ARL 69-0014, Wright-Patterson AFB, OH, January, 1969.

[40]E. B. Wilson, Jr., "An Introduction to Scientific Research," p. 116, McGraw-Hill, New York, 1952.

As an example, a strain gage of $R = 120 \ \Omega$ at 300 K over a bandwidth of 100,000 Hz would put out an rms noise voltage of 0.45 μV. Comparing this with our earlier calculation of the signal due to 1,000 lb/in² stress, we see that the signal to noise ratio would be 120:0.45 = 267:1. Suppose, however, that we wish to measure 1 lb/in² stress rather than 1,000. The signal is then 0.12 μV, which is less than the noise; therefore the signal would be lost in the noise. Amplification under these conditions is of no use since the signal and noise are both amplified. This simple example does not cover other methods (such as filtering) that have been developed to deal with this limitation, but it should be understood that the limitation is a fundamental one. Similar random fluctuations limit the measurable threshold of all physical variables (see near Eq. (3.54)). In practical strain-gage measurement systems, Johnson noise of resistances other than the strain gage and other sources of noise in transistors usually limit the system resolution. Strain gages located in fluctuating magnetic fields may experience induced noise voltages. If these are not acceptably small, *noninductive* gages using two stacked grids, with the current running in opposite directions to cancel the induced voltages, are available.[41]

To meet the needs of severe environments, difficult installation conditions, and/or high temperatures as typified by pipelines, offshore drilling platforms, tunnels, dams, nuclear containment vessels, steam lines, gas turbines, etc., strain gages enclosed permanently in protective metal housings which are attached by spot welding to the specimen have been developed.[42] These gages incorporate a number of unique features (see Fig. 4.16). The strain filament and heavier (0.007-in diameter) leadout wire are of unitized construction with a carefully controlled taper; this minimizes joint fatigue problems and erratic electric connections. Active and dummy filaments are in close proximity; the dummy gage experiences no strain since the helix angle of its winding matches Poisson's ratio. Both filaments are encased in a strain tube made by welding a tubular shell to a flat flange. The filament is mechanically coupled to the strain tube, but electrically isolated from it by highly compacted magnesium oxide powder, by using a high-speed centrifuge and swaging operation. When the flange is spot-welded to the specimen, the specimen strain is faithfully transmitted to the strain tube and in turn by the powder to the gage filament for both tensile and compressive stresses.

Differential Transformers

Figure 4.17 shows schematic and circuit diagrams for translational and rotational linear variable-differential-transformer (LVDT) displacement pickups. The excitation of such devices is normally a sinusoidal voltage of 3 to 15 V rms amplitude and frequency of 60 to 20,000 Hz. The two identical secondary coils have induced in them sinusoidal voltages of the same frequency as the excitation; however, the amplitude varies with the position of the iron core. When the secondaries are

---

[41]Measurements Group, Inc.

[42]Model AWH, Texas Measurements, Inc., PO Box 2618, College Station, TX 77841, 409-764-0442 (www.straingage.com).

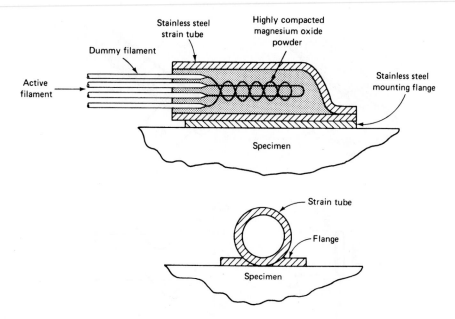

Figure 4.16
Hermetically sealed, weldable strain gage.

connected in series opposition, a null position exists ($x_i \triangleq 0$) at which the net output e_o is essentially zero. Motion of the core from null then causes a larger mutual inductance (coupling) for one coil and a smaller mutual inductance for the other, and the amplitude of e_o becomes a nearly linear function of core position for a considerable range either side of null. The voltage e_o undergoes a 180° phase shift in going through null. The output e_o is generally out of phase with the excitation e_{ex}; however, this varies with the frequency of e_{ex}, and for each differential transformer there exists a particular frequency (numerical value supplied by the manufacturer) at which this phase shift is zero. If the differential transformer is used with some readout system that requires a small phase shift between e_o and e_{ex}, excitation at the correct frequency can solve this problem. If the output voltage is applied directly to an ac meter or an oscilloscope, this phase shift is not a problem.

The origin of this phase shift can be seen from analysis of Fig. 4.18. Applying Kirchhoff's voltage-loop law, if the output is an open circuit (no voltage-measuring device attached), we get

$$i_p R_p + L_p \frac{di_p}{dt} - e_{ex} = 0 \qquad \textbf{(4.23)}$$

Now the voltage induced in the secondary coils is given by

$$\begin{aligned} e_{s1} &= M_1 \frac{di_p}{dt} \\[4pt] e_{s2} &= M_2 \frac{di_p}{dt} \end{aligned} \qquad \textbf{(4.24)}$$

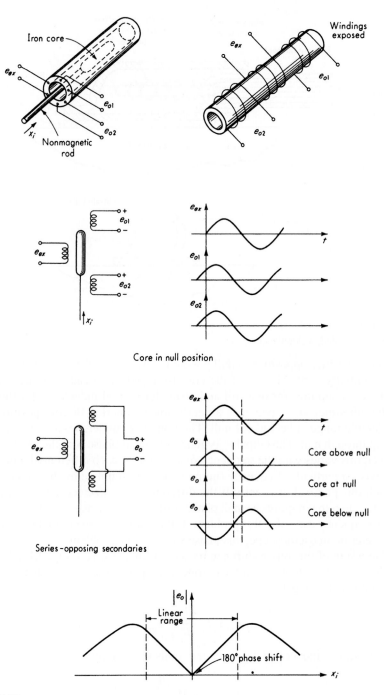

Figure 4.17
Differential transformer.

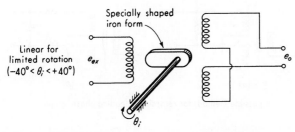

Linear for
limited rotation
$(-40° < \theta_i < +40°)$

Rotational differential transformer

Figure 4.17
(Concluded)

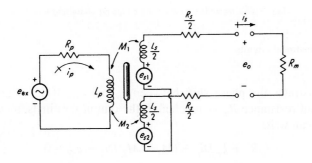

Figure 4.18
Circuit analysis.

where M_1 and M_2 are the respective mutual inductances. The net secondary voltage e_s is then given by

$$e_s = e_{s1} - e_{s2} = (M_1 - M_2) \frac{di_p}{dt} \qquad (4.25)$$

The net mutual inductance $M_1 - M_2$ is the quantity that varies linearly with core motion. We have for a fixed core position

$$e_o = e_s = (M_1 - M_2) \frac{D}{L_p D + R_p} e_{ex} \qquad (4.26)$$

and thus
$$\frac{e_o}{e_{ex}} (D) = \frac{[(M_1 - M_2)/R_p]D}{\tau_p D + 1} \qquad \tau_p \triangleq \frac{L_p}{R_p} \qquad (4.27)$$

In terms of frequency response,

$$\frac{e_o}{e_{ex}} (i\omega) = \frac{\omega(M_1 - M_2)/R_p}{\sqrt{(\omega \tau_p)^2 + 1}} \angle \phi \qquad \phi = 90° - \tan^{-1} \omega \tau_p \qquad (4.28)$$

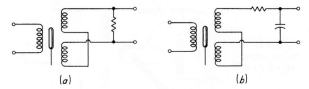

(a)   (b)

Two possible methods for retarding a leading phase angle

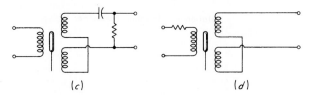

(c)   (d)

Two possible methods for advancing a lagging phase angle

Figure 4.19
Phase-angle-adjustment circuits.

which demonstrates the phase shift between e_o and e_{ex}. If a voltage-measuring device of input resistance R_m is attached to the output terminals, a current i_s will flow, and we can write

$$i_p R_p + L_p D i_p - (M_1 - M_2)D i_s - e_{ex} = 0 \tag{4.29}$$

$$(M_1 - M_2)D i_p + (R_s + R_m)i_s + L_s D i_s = 0 \tag{4.30}$$

which lead to

$$\frac{e_o}{e_{ex}}(D) = \frac{R_m(M_2 - M_1)D}{[(M_1 - M_2)^2 + L_p L_s]D^2 + [L_p(R_s + R_m) + L_s R_p]D + (R_s + R_m)R_p} \tag{4.31}$$

Since the frequency response of $(e_o/e_{ex})(i\omega)$ has a phase angle of $+90°$ at low frequencies and $-90°$ at high, somewhere in between it will be zero, as mentioned earlier. If, for some reason, the excitation frequency cannot be adjusted to this value, the same effect may be achieved for a given frequency by one of the methods shown in Fig. 4.19.

While the output voltage at the null position is ideally zero, harmonics in the excitation voltage and stray capacitance coupling between the primary and secondary usually result in a small but nonzero null voltage. Under usual conditions this is less than 1 percent of the full-scale output voltage and may be quite acceptable. Methods of reducing this null when it is objectionable are available. First, the preferred connection shown in Fig. 4.20a should be used if a balanced (center-tapped) excitation-voltage source is available. The grounding shown tends to reduce capacitance-coupling effects. If a center-tapped voltage source is not available, the arrangement of Fig. 4.20b can be used. With the core at the null position and the output-measuring device connected, the potentiometer is adjusted until the minimum null reading is obtained. The values of R and R_p are not critical, but should be

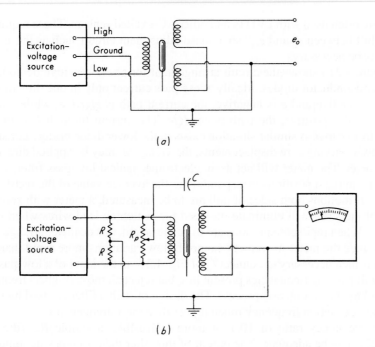

Figure 4.20
Methods for null reduction.

as low as possible without loading (drawing excessive current from) the excitation source.

The output of a differential transformer is a sine wave whose amplitude is proportional to the core motion. If this output is applied to an ac voltmeter, the meter reading can be directly calibrated in motion units. This arrangement is perfectly satisfactory for measurement of static or very slowly varying displacements, except that the meter will give exactly the same reading for displacements of equal amount on *either* side of the null since the meter is not sensitive to the 180° phase change at null. Thus we cannot tell to which side of null the reading applies without some independent check. Furthermore, if rapid core motions are to be measured, the meter cannot follow or record them. Fast-response instruments such as oscilloscopes record the actual waveform of the output as an amplitude-modulated sine wave, which is usually undesirable. What is desired is an output-voltage record that looks like the mechanical motion being measured. To achieve the desired results, demodulation and filtering must be performed; if it is necessary to detect unambiguously the motions on both sides of null, the demodulation must be phase-sensitive. Many different circuits are available for performing these operations. We show here only one arrangement, which is quite simple. To use this approach, all four output leads of the LVDT must be accessible (some have the series-opposition connection *internal* to the case and would thus not be applicable to the following discussion). Section 10.10 discusses *synchronous demodulators,* which are often used with LVDTs and other AC-excited sensors. These require zero phase shift between e_o and e_{ex}, so a phase-adjustment circuit such as those in Fig. 4.19 may be necessary.

Figure 4.21a shows the circuit arrangement for phase-sensitive demodulation using semiconductor diodes. Ideally, these pass current only in one direction; thus when f is positive and e is negative, the current path is *efgcdhe*, while when f is negative and e positive, the path is *ehcdgfe*. The current through R is therefore always from c to d. A similar situation exists in the lower diode bridge. For static or very slowly varying core displacements, the voltage e_o may be applied directly to a dc voltmeter. The meter will act as an electromechanical low-pass filter, with the needle assuming a position corresponding to the average value of the rectified sine wave e_o. If motions both sides of null are to be measured, a meter with zero in the center of the scale will eliminate the need for switching lead wires when e_o goes negative. When rapid core motions are to be measured, this dc meter arrangement is useless since the meter movement cannot follow variations more rapid than about 1 Hz. It is then necessary to connect e_o of Fig. 4.21c to the input of a low-pass filter which will pass the frequencies present in x_i but reject all those (higher) frequencies produced by the modulation process. The design of such a filter is eased by making the LVDT excitation frequency much higher than the x_i frequencies.

If a frequency ratio of 10:1 or more is feasible, a simple RC filter as in Fig. 4.22a may be adequate. The output of this filter then becomes the input to an oscillograph or oscilloscope. For example, suppose we wish to measure a transient x_i whose Fourier transform has dropped to insignificant magnitude for all frequencies higher than 1,000 Hz. Suppose also an LVDT system with an excitation frequency of 10,000 Hz is available. The frequencies produced by the modulation process will thus lie in the band 19,000 to 21,000 Hz. Suppose that we desire the "ripple" due to frequencies at 19,000 Hz and higher to be no more than 5 percent of its unfiltered value. The filter time constant $\tau_f = R_f C_f$ can be calculated then as

$$0.05 = \frac{1}{\sqrt{[(19{,}000)(6.28)\tau_f]^2 + 1}} \tag{4.32}$$

$$\tau_f = 0.00017 \text{ s} \tag{4.33}$$

At the highest motion frequency (1,000 Hz) this filter has an amplitude ratio of 0.68 and a phase shift of $-47°$; thus it will distort the high-frequency portion of the x_i transient considerably. A more selective (sharper cutoff) filter would help this situation. Consider the double RC filter of Fig. 4.22b. The value of τ_f for a 5 percent ripple is now obtained from

$$0.05 = \frac{1}{[(19{,}000)(6.28)\tau f]^2 + 1} \tag{4.32}$$

$$\tau_f = 0.000037 \text{ s} \tag{4.35}$$

Now, at 1,000 Hz the amplitude ratio is 0.94 and the phase angle is $-26°$. Since the phase angle of this filter from $\omega = 0$ to $\omega = 6{,}280$ rad/s is nearly linear and the amplitude ratio is nearly flat (1.0 to 0.94), the waveform of the transient will be

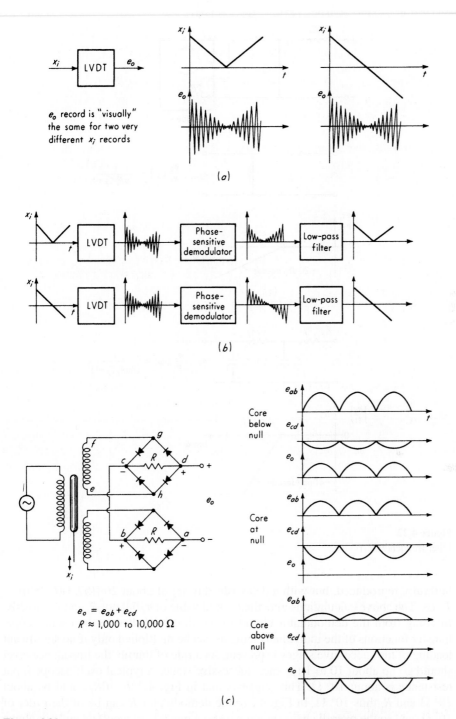

Figure 4.21
Demodulation and filtering.

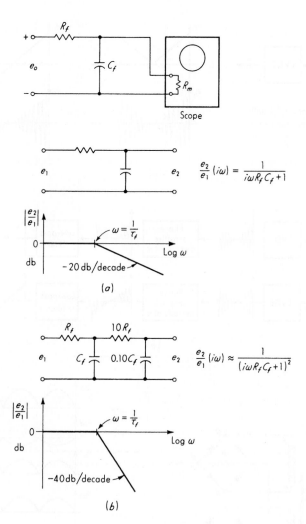

Figure 4.22
Filter frequency response.

faithfully reproduced, but with a delay (dead time) of about $26/[(57.3)(6,280)] = 72\ \mu s$. The above calculations give the desired value of τ_f but not R_f and C_f directly. In going from the demodulator circuit to the filter and then to the oscilloscope, transfer functions of the individual elements can be multiplied only if no significant loading of the successive stages is present. As a rule of thumb, the impedance level should go up about 10 to 1 for each successive stage. A typical oscilloscope input resistance is about $10^6\ \Omega$. This suggests that in Fig. 4.22b, $10R_f$ could be about $10^5\ \Omega$ and R_f thus $10^4\ \Omega$. In Fig. 4.21c the demodulator R can be of the order of $10^3\ \Omega$, and so the overall chain should not show much loading effect and our above calculations should be fairly accurate. In any case, experimental checks of the system should be performed to verify the final design. If R_f of 10^4 Ω is satisfactory, then C_f will be $(37 \times 10^{-6})/10^4 = 0.0037$ μF.

The full-range stroke of commercially available translational LVDTs ranges from about $\pm$ 0.005 to about $\pm$ 25 in, with other sizes available as specials. The nonlinearity of standard units is of the order of 0.5 percent of full scale, with 0.1 percent possible by selection. Sensitivity with normal excitation voltage of 3 to 6 V is of the order of 0.6 to 30 mV per 0.001 in, depending on frequency of excitation (higher frequency gives more sensitivity) and stroke (smaller strokes usually have higher sensitivity). Some special units have sensitivity as high as 1 to 1.5 V per 0.001 in. Since the coupling variation due to core motion is a continuous phenomenon, the resolution of LVDTs is infinitesimal. Amplification of the output voltage allows detection of motions down to a few microinches. There is no physical contact between the core and the coil form; thus there is no friction or wear. There are, however, small radial and longitudinal magnetic forces on the core if it is not centered radially and at the null position. These are in the nature of magnetic "spring" forces in that they increase with motion from the equilibrium point. They are rarely more than $0.1g$ to $0.3g$ and thus are often negligible. Rotary LVDTs have a nonlinearity of about $\pm$ 1 percent of full scale for travel of $\pm$ 40° and $\pm$ 3 percent for $\pm$ 60°. The sensitivity is of the order of 10 to 20 mV/deg. The moving mass (core) of LVDTs is quite small, ranging from less than $0.1g$ in small units to $5g$ or more in larger ones. There is a small radial clearance (air gap) between the core and the hole in which it moves. Motion in the radial direction produces a small output signal, but this undesirable transverse sensitivity is usually less than 1 percent of the longitudinal sensitivity.

The dynamic response of LVDTs is limited mainly by the excitation frequency, since it must be much higher than the core-motion frequencies so as to be able to distinguish between them in the amplitude-modulated output signal. For adequate demodulation and filtering, a frequency ratio much less than 10:1 presents problems. Since few differential transformers are designed to be excited by more than 20,000 Hz, the useful range of motion frequencies is limited to about 2,000 Hz. This is adequate for many applications. Although the transformer principle of the LVDT clearly requires ac excitation, so-called DCDTs are available from manufacturers. The "DC" part of the terminology refers to the fact that the user applies only a dc excitation to the device. The apparent discrepancy is resolved when we discover that microelectronic technology has allowed us to build into the transducer case a complete electronic system including oscillator (produces ac excitation from dc power), demodulator, amplifier, and low-pass filter. LVDTs also have been combined with pneumatic servomechanisms to achieve a noncontacting displacement probe (see Fig. 4.60). LVDT-to-digital converters (similar to the synchro/digital converter of Fig. 4.27) are also available.[43] Figure 4.23[44] shows an LVDT system

---

[43]Model 2S50, Analog Devices, 831 Woburn St., Wilmington, MA 01887, 800-262-5643 (www.analog.com).

[44]Lucus Sensors, 1000 Lucus Way, Hampton, VA 23666, 800-745-8008 (www.schaevitz.com).

Block diagram

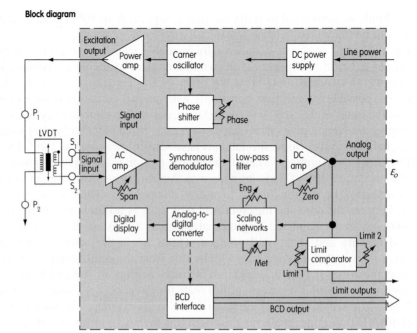

Figure 4.23
LVDT signal conditioning.

with a full complement of available options. In addition to the usual analog output, digital outputs in the form of a visual (meter) display and also a binary-code-decimal electrical signal, which can be sent to other digital systems, are provided. A comparator with two adjustable limits provides "go/no go" control for dimensional gaging systems of manufactured parts.

Synchros and Resolvers

The term "synchro" is applied to a family of ac electromechanical devices which, in various forms, perform the functions of angle measurement, voltage and/or angle addition and subtraction, remote angle transmission, and computation of rectangular components of vectors. In this section we are concerned with only the angle-measuring function; equipment for performing the other functions is covered in later appropriate chapters.

Synchros for angle measurement are most utilized as components of servomechanisms (automatic motion-control feedback systems) where they are used to measure and compare the actual rotational position of a load with its commanded position, as in Fig. 4.24. To perform this function, two different types of synchros, the control transmitter and the control transformer, are used. The error voltage signal e_e is an ac voltage of the same frequency as the excitation and of amplitude proportional (for small error angles) to the error angle $\theta_R - \theta_B$. Its phase changes by $180°$ at the null point; thus the direction of the error is detected. When $\theta_R = \theta_B$,

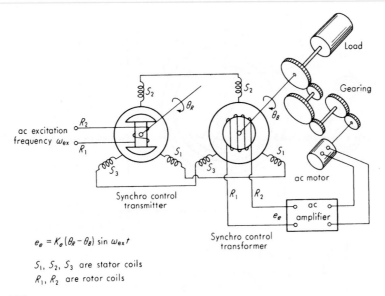

Load

Gearing

S_2

S_2

R_2

ac excitation
frequency ω_{ex}

R_1

θ_R

θ_B

S_1

S_1

S_3

S_3

ac motor

Synchro control
transmitter

R_1 R_2

e_e

ac
amplifier

Synchro control
transformer

$e_e = K_e (\theta_R - \theta_B) \sin \omega_{ex} t$

S_1, S_2, S_3 are stator coils

R_1, R_2 are rotor coils

Figure 4.24
AC servomechanism.

the error voltage (and thus the amplifier output and motor input) is zero and the system stays at rest. If a command rotation θ_R is now put in, $e_e \neq 0$ and the motor will rotate so as to return θ_B to correspondence with θ_R.

The physical constructions of the control transmitter and control transformer are identical except that the transmitter has a salient-pole ("dumbbell") rotor while the transformer has a cylindrical rotor. The construction is similar to that of a wound-rotor induction motor. Figure 4.25a shows the coil arrangement of the transmitter alone. Basically, rotation of the rotor changes the mutual inductance (coupling) between the rotor coil and the stator coils. For a given stator coil, the open-circuit output voltage is sinusoidal in time and varies in amplitude with rotor position, also sinusoidally, as shown in Fig. 4.25b. The three voltage signals from the stator coils uniquely define the angular position of the rotor. When these three voltages are applied to the stator coils of a control transformer, they produce a resultant magnetomotive force aligned in the same direction as that of the transmitter rotor. The rotor of the transformer acts as a "search coil" in detecting the direction of its stator field. If the axis of this coil is aligned with the field, the maximum voltage is induced into the transformer rotor coil. If the axis is perpendicular to the field, zero voltage is induced, giving the null position mentioned above. The output-voltage amplitude actually varies sinusoidally with the misalignment angle, but for small angles the sine and the angle are nearly equal, giving a linear output.

Resolvers basically "resolve" the rotary motion of a rotor into the sine and cosine of the rotor angular position, providing two AC output voltages whose amplitudes are proportional, respectively, to the sine and cosine. This function is directly useful in motion control systems where we need to resolve vectors into their

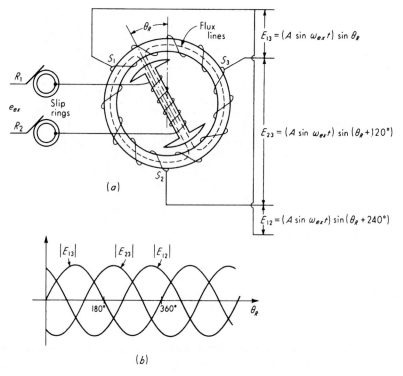

Figure 4.25
Synchro.

rectangular components. In this section, we are instead interested in the use of resolvers to measure the angle itself, rather than the sine and cosine. "Conventional" resolvers use slip rings to transmit signals between the rotating and stationary parts of the device, as in the synchro of Fig. 4.25. More modern "brushless" technology uses a rotary transformer to transmit the signals and thereby eliminate the slip rings, which add friction and electrical noise. Figure 4.26[45] shows a schematic with the sine and cosine outputs connected to a resolver-to-digital converter, a common configuration that provides both digital angle information and an analog (DC voltage) angular velocity signal useful for motion-control-system stabilization. Such arrangements are in competition with shaft-angle encoders, the choice[46] in each application depending on the relative importance of cost, ruggedness, temperature effects, etc. Typical resolver accuracies are ± 7 minutes of arc; sensitivities up to about 0.4 V/degree are available.

[45]Harowe Servo Controls, 110 Westtown Rd., West Chester, PA 19382, 610-692-2700

[46]T. Bullock, "Encoders or Resolvers," *Motion Control,* Sept/Oct 1997, p. 46; M. W. Johnson, "All about Synchros, Resolvers, and Data Acquisition," *Evaluation Engineering,* March 1999, pp. 38–45.

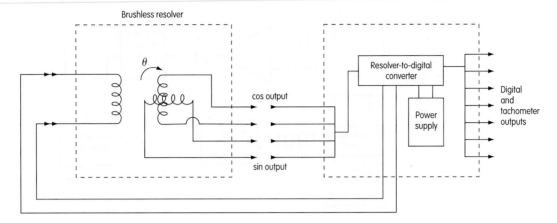

Figure 4.26
Brushless resolver with digital converter.

In motion-control systems that use the popular brushless DC motors, resolvers are generally preferred to synchros for use with the digital converters mentioned above. The resolver is used not only for basic angle sensing but also provides the shaft-position information needed by the motor-control electronics to commutate the motor. The resolver-to-digital converter provides digital angle information to the digital control algorithm and analog velocity information for servo stability augmentation.

While synchros and resolvers could be designed to work at a variety of excitation frequencies, standard commercial units are available for 60,400, and up to 7500 Hz. The physical size ranges from about $\frac{1}{2}$- to 3-in diameter. Sensitivities of synchros are of the order of 1 V/deg rotation while the residual voltage at null is of the order of 10 to 100 mV. For a standard synchro transmitter-transformer pair, the misalignment of the two shafts when rotated from an originally established electrical null to any other null position within a complete rotation is of the order of 10 angular minutes. This type of error puts a basic limit on the positioning accuracy of servosystems using synchros.

Just as in LVDTs, synchros and resolvers require some sort of phase-sensitive demodulation to obtain a signal of the same form as the mechanical-motion input. When used in ac servomechanisms, the conventional two-phase ac servomotor itself accomplishes this function without any additional equipment. In a strictly measurement (as opposed to control) application, some sort of phase-sensitive demodulator is needed if an electric output signal of the same form as the mechanical input is required. The dynamic response is limited by the excitation frequency and demodulator filtering requirements, just as in LVDTs. The mechanical loading of these rotary components on the measured system is mainly the inertia of the rotor.

While synchros were originally developed for use with analog systems, many applications (machine tools, antennas, etc.) today use digital approaches. Even though a digital shaft-angle encoder might seem the logical sensor for an otherwise

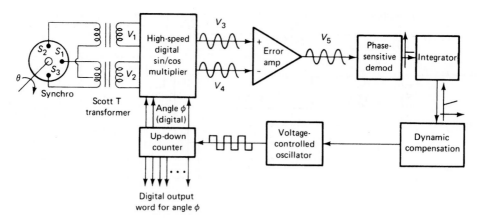

Figure 4.27
Synchro-to-digital converter.

digital system, synchros (and the closely related resolver) have a number of advantages[47] which make them desirable. To interface them with the digital part of the system, we require a synchro-to-digital converter,[48] a piece of solid-state electronic equipment (no moving parts) which we show in block-diagram form in Fig. 4.27. The ac signals from the synchro are first applied to a special transformer, a Scott T, which produces ac voltages with amplitudes proportional to the sine and cosine of the synchro shaft angle θ:

$$V_1 = KE_o \sin \omega t \sin \theta \qquad V_2 = KE_o \sin \omega t \cos \theta \qquad (4.36)$$

[Note that if a resolver (rather than a synchro) were used, the Scott T transformer would be unnecessary since the resolver produces voltages of form (4.36) directly.] The synchro-to-digital converter is an all-electronic feedback system which compares the synchro angle θ with its digital version ϕ. If ϕ is not equal to θ, the converter provides a correction signal to the counter containing ϕ so as to drive its contents toward the correct value. For those familiar with feedback systems, the control system is a Type II (two integrations), so there will be zero steady-state error between command θ and response ϕ for both step and ramp inputs of θ.

The voltages V_1 and V_2 are digitally multiplied by $\sin \phi$ and $\cos \phi$ to give

$$V_3 = KE_o \sin \omega t \sin \theta \cos \phi \qquad (4.37)$$
$$V_4 = KE_o \sin \omega t \cos \theta \sin \phi \qquad (4.38)$$

and then these two voltages are subtracted in the error amplifier to give

$$V_5 = KE_o \sin \omega t (\sin \theta \cos \phi - \cos \theta \sin \phi)$$
$$= KE_o \sin \omega t \sin (\theta - \phi) \qquad (4.39)$$

[47]W. M. Cullum, "Angle-Sensing Transducers for Shaft-to-Digital Conversion," *Tech. Bull.* 121, North Atlantic, 170 Wilbur Place, Bohemia, NY 11716, 516-218-1233.

[48]Ibid.; Data Acquisition Components and Subsystems Catalog, pp. 12–27, Analog Devices, Wilmington, MA (www.analog.com), 1980; G. Boyes, Synchro and Resolver Conversion, Analog Devices, 1980.

The amplitude of the ac voltage V_5 is proportional to $\sin(\theta - \phi) \approx \theta - \phi$, the error between θ and ϕ. This ac error signal is phase-sensitive demodulated to obtain a dc error signal, which is then applied to an op-amp type of integrator. (The next block, dynamic compensation, is not fundamental to the operating principle, but is necessary in most feedback systems to optimize performance.) Since a constant voltage applied to a voltage-controlled oscillator produces a proportional and constant output frequency which will ramp the counter at a fixed rate, this provides the second integration in the control loop. For the Analog Devices SDC1704 operated at 400 Hz and referenced above, the manufacturer gives this transfer function:

$$\frac{\phi}{\theta}(D) = \frac{0.0082D + 1}{3.39 \times 10^{-8} D^3 + 2.73 \times 10^{-5} D^2 + 0.00661D + 1} \tag{4.40}$$

This clearly has zero steady-state error for a step input, while solution of the differential equation (particular solution only) for $\theta = At$ shows zero error for a ramp also. Actually, zero ramp error is maintained only for $A < 12$ r/s. The resolution of this 14-bit device is 1.3 minutes of arc, and the accuracy is ± 2.9 minutes of arc ± 1 LSB (LSB = least significant bit = 1.3 minutes of arc). This device also provides a dc voltage proportional to $\dot{\theta}$, useful in stabilizing a servosystem which uses the synchro for angle information.

Variable-Inductance and Variable-Reluctance Pickups

Closely related to LVDTs and synchros, but in practice distinguished from them by name, is a family of motion pickups variously called variable-inductance, variable-reluctance, or variable-permeance (permeance is the reciprocal of reluctance) pickups or transducers. The terminology used for these pickups is not uniform or necessarily descriptive of their basic principles of operation. We are concerned here mainly with describing some common examples rather than trying to develop a systematic nomenclature.

Figure 4.28a shows the arrangement of a typical translational variable-inductance transducer. Outwardly the physical size and shape are very similar to those of an LVDT. Again, there is a movable iron core which provides the mechanical input. However, only two inductance coils are present; they generally form two legs of a bridge which is excited with 5 to 30 V ac at 60 to 5,000 Hz. With the core at the null position, the inductance of the two coils is equal, the bridge is balanced, and e_o is zero. A core motion from null causes a change in the reluctance of the magnetic paths for each of the coils, increasing one and decreasing the other. This reluctance change causes a proportional change in inductance for each coil, a bridge unbalance, and thus an output voltage e_o. By careful construction, e_o can be made a nearly linear function of x_i over the rated displacement range.

Two alternative methods of forming the bridge circuit are shown in Fig. 4.28b. The total transducer impedance (Z_1 plus Z_2) at the excitation frequency is of the order of 100 to 1,000 Ω. The resistors R are usually about the same value as Z_1 and Z_2, and the input impedance of the voltage-measuring device at e_o should be at least $10R$. If the bridge output must be worked into a low-impedance load, R must be quite small. To get high sensitivity, high excitation voltage is needed; this causes a

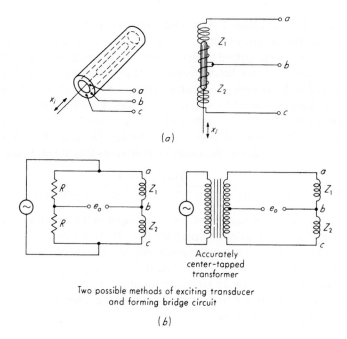

(a)

Two possible methods of exciting transducer
and forming bridge circuit

Accurately
center-tapped
transformer

(b)

Figure 4.28
Variable-inductance pickup.

high power loss (heating) in the resistors R. To solve this problem, a center-tapped transformer circuit may be used. Here the bridge is mainly inductive, and less power is consumed with corresponding less heating.

Such variable-inductance transducers are available in strokes of about 0.1 to as much as 200 in. The resolution is infinitesimal, and the nonlinearity ranges from about 1 percent of full scale for standard units to 0.02 percent for special units of rather long stroke. Sensitivity is of the order of 5 to 40 V/in. Rotary versions using specially shaped rotating cores have a nonlinearity of the order of 0.5 to 1 percent of full scale over a $\pm 45°$ range. The sensitivity is about 0.1 V/deg. Some inductance sensors[49] use a very high (112 kHz) excitation frequency that allows measurement of motion frequencies up to 15 kHz (LVDTs are limited to about 2 kHz).

Figure 4.29 shows another common version of the variable-reluctance principle. This particular application is an accelerometer for measurement of accelerations in the range $\pm 4g$. Since the force required to accelerate a mass is proportional to the acceleration, the springs supporting the mass in Fig. 4.29 deflect in proportion to the acceleration; thus a displacement measurement allows an acceleration measurement. The mass is of iron and thus serves as both an inertial element for transducing acceleration to force and a magnetic circuit element for transducing motion to reluctance.

We consider the complete instrument here since it has several features of general interest with regard to displacement measurement. Ordinarily, such an

[49]Fastar Transducer, Data Instruments, Inc., 100 Discovery Way, Acton, MA 01720, 508-264-9550.

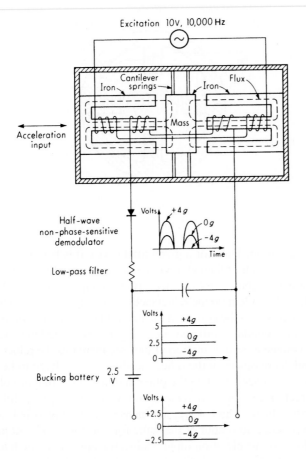

Figure 4.29
Variable-reluctance accelerometer.

instrument would be constructed so that the iron core would be halfway between the two E frames when the acceleration was zero, thus giving zero output voltage for zero acceleration. However, to detect motion on both sizes of zero (corresponding to plus or minus accelerations), a fairly involved phase-sensitive demodulator would be required. It was desired to save the cost, weight, and space of this demodulator, and so another solution (which can also be used with LVDTs and similar devices) was proposed. With zero-acceleration input, the iron core and springs were adjusted so that the core was offset to one side by an amount equal to the spring deflection corresponding to $4g$ acceleration. Thus, with no acceleration applied, the output voltage was not zero but some specific value (2.5 V in this particular case). Then, when $+4g$ of acceleration was applied, the output went to 5.0 V; and when $-4g$ was applied, the output went to zero. In this way a relatively simple demodulator and filter circuit can be used to provide direction-sensitive motion measurement. The main drawback of this scheme (which argues against its use except when necessary) is the loss of linearity. This is because the greatest linearity is found

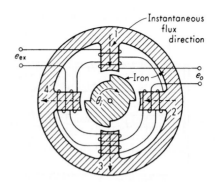

Figure 4.30
Microsyn.

around the null position. Thus for a given total stroke it is better to put one-half of it on each side of null rather than all on one side, as in the above scheme.

Let us return to the basic motion-measuring principle of Fig. 4.29. The primary coils set up a flux dependent on the reluctance of the magnetic path. The main reluctance is the air gap. When the core is in the neutral position, the flux is the same for both halves of the secondary coil; and since they are connected in series opposition, the net output voltage is zero. A motion of the core increases the reluctance (air gap) on one side and decreases it on the other, causing more voltage to be induced into one half of the secondary coil than the other and thus a net output voltage. Motion in the other direction causes the reverse action, with a 180° phase shift occurring at null. The output voltage is half-wave, non-phase-sensitive rectified (demodulated) and filtered to produce an output of the same form as the acceleration input. If the 2.5-V output for zero-acceleration input is objectionable, it can be bucked out with a 2.5-V battery of opposite polarity connected externally to the accelerometer. The actual full-scale motion of the mass in this particular instrument is just a few thousandths of an inch, which gives a displacement sensitivity for the variable-reluctance element of almost 1,000 V/in.

The final variable-reluctance element we consider is the Microsyn,[50] a rotary component shown in Fig. 4.30 and widely used in sensitive gyroscopic instruments. The sketch shows the instrument in the null position where the voltages induced in coils 1 and 3 (which aid each other) are just balanced by those of coils 2 and 4 (which also aid each other but oppose 1 and 3). Motion of the input shaft from the null (say clockwise) increases the reluctance (decreases the induced voltage) of coils 1 and 3 and decreases the reluctance (increases the voltage) of coils 2 and 4, thus giving a net output voltage e_o. Motion in the opposite direction causes a similar effect, except the output voltage has a 180° phase shift. If a direction-sensitive dc output is required, a phase-sensitive demodulator is necessary.

---

[50]P. H. Savet (ed.), "Gyroscopes: Theory and Design," p. 332, McGraw-Hill, New York, 1961.

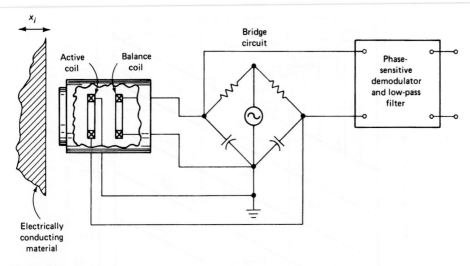

Figure 4.31
Eddy-current noncontacting transducers.

The excitation voltage is 5 to 50 V at 60 to 5,000 Hz. Sensitivity is of the order of 0.2 to 5 V/degree rotation. Nonlinearity is about 0.5 percent of full scale for $\pm 7°$ rotation and 1.0 percent for $\pm 10°$. The null voltage is extremely small, less than the output signal generated by 0.01° of rotation; thus very small motions can be detected. The magnetic-reaction torque is also extremely small. Since there are no coils on the rotor, no slip rings (with their attendant friction) are needed.

Eddy-Current Noncontacting Transducers[51]

In this type of transducer (Fig. 4.31), the probe usually contains two coils, one (active) which is influenced by the presence of a conducting target and a second (balance) which serves to complete a bridge circuit and provide temperature compensation. Bridge excitation is high-frequency (about 1 MHz) alternating current. Magnetic flux lines from the active coil pass into the conductive target surface, producing in the target eddy currents whose density is greatest at the surface and which become negligibly small about three "skin depths" below the surface. Figure 4.32[52] gives formulas for computing skin depth δ and graphs of these formulas for the common excitation frequency of 1 MHz. While thinner targets can be successfully employed, a minimum of three skin depths is recommended to reduce temperature effects. As the target comes closer to the probe, the eddy currents

[51]S. D. Welsby and T. Hitz, "True Position Measurement with Eddy Current Technology," *Sensors,* Nov. 1997, pp. 30–40; S. D. Roach, "Designing and Building an Eddy Current Position Sensor," *Sensors,* Sept. 1998, pp. 56–74.

[52]Appl. Note 108, Kaman Instrumentation Corp., 1500 Garden of the Gods Rd., Colorado Springs, CO 80917-3010, 800-557-6267 (www.kamaninstrumentation.com).

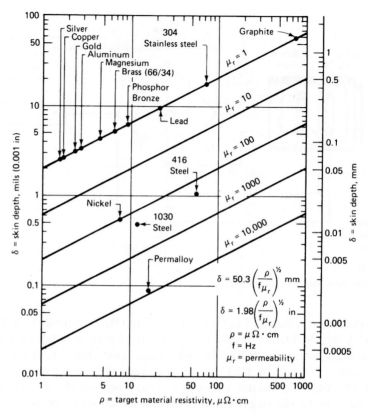

Figure 4.32
Target-material effect on eddy-current transducer.

become stronger, which changes the impedance of the active coil and causes a bridge unbalance related to target position. This unbalance voltage is demodulated, low-pass filtered (and sometimes linearized) to produce a dc output proportional to target displacement. The high excitation frequency not only allows the use of thin targets, but also provides good system frequency response (up to 100 kHz).

Probes are commercially available[53] with full-scale ranges from about 0.25 (probe about 2-mm diameter, 20 mm long) to 30 mm (76 mm-diameter, 40 mm long) nonlinearity of 0.5 percent, and maximum resolution 0.0001 mm. Targets are not supplied with the probes since the majority of applications involve noncontact measurement of existing machine parts (thus the part itself serves as target). Since target material, shape, etc., influence output, generally it is necessary to statically calibrate the system with the specific target to be used. For nonconductive targets, you must fasten a piece of conductive material of sufficient thickness to the surface. Commercially available adhesive-backed aluminum-foil tape[54] is convenient for

---

[53]Kaman Instrumentation Corp.; Bentley Nevada, 1617 Walter St., Minden, NV 89423, 775-782-3611 (www.bentley.com).

[54]Mystik Tape #7453.

this purpose. The recommended measuring range of a given probe begins at a "standoff" distance equal to about 20 percent of the probe's stated range. That is, a probe rated at 0 to 1 mm range should be used at target-probe distances of 0.2 to 1.2 mm.

Flat targets should be about the same diameter as the probe or larger, if possible. Targets larger than the probe have little effect on the output; however, output drops to about 50 percent for target diameter one-half of probe diameter.[55] Curved-surface targets such as the periphery of a circular shaft behave similarly to flat surfaces if the shaft diameter exceeds four transducer diameters.[56] Special four-probe systems[57] for measuring orbital motions of rotating shafts and various centering and alignment operations are available. While rotating-shaft measurements are routinely accomplished, special care may be necessary to deal with "electrical runout"[58] in shafts made of ferromagnetic material (such as steel). This refers to a variation in magnetic permeability around the periphery of the shaft (resulting from inhomogeneities of heat treatment, hardness, etc.) which causes an electrical output even for a perfectly true-running shaft, thus giving false motion readings. If electrical runout is excessive, filtering, differential measurements using two diametrically opposed probes, nickel plating of the shaft, or shaft-surface grinding to remove hardness variations may help.

Capacitance Pickups

A translational or rotational motion may be used in many ways to change the capacitance of a variable capacitor.[59] Two of the most common[60] are shown in Fig. 4.33 and use the parallel-plate configuration with air in the gap. The capacitance is given by:

$$C = \frac{0.225A}{x} \qquad (4.41)$$

where $C \triangleq$ capacitance, pF

                $A \triangleq$ plate area, in^2

                $x \triangleq$ plate separation, in

When the motion is relatively large, the plate separation is usually kept fixed (typically about 0.1 mm) and the area A is changed by moving one plate parallel to the other. For smaller motions, the area is kept fixed and the separation x is varied. An example given in Ref. 60 uses a nominal capacitance of 10 pF and square plates of size 10.63 mm (with a gap of 0.1 mm), and results in Table 4.1.

[55]*Appl. Note* 104, Kaman Instrumentation Corp.

[56]Ibid.

[57]KD-2700 Alignment System, Kaman.

[58]*Appl. Note* 109, Kaman Instrumentation Corp.

[59]L. K. Baxter, "Capacitive Sensors: Design and Applications," IEEE Press, 1997.

[60]T. R. Hicks and P. D. Atherton, "The Nanopositioning Book," Queensgate Instruments Ltd., 1997, chap. 5, (www.queensgate.com).

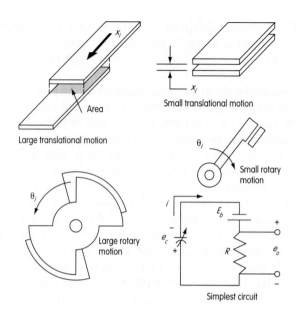

Figure 4.33
Principles of capacitive displacement sensing.

Table 4.1

Configuration	Range	Nominal capacitance	Sensitivity
long-range linear	10.0 mm	10 pF	0.94 fF/μm
long-range rotation	1.5 rad	10 pF	6.9 fF/mrad
short-range linear	100 μm	10 pF	100 fF/μm
short-range rotation	17 mrad	10 pF	580 fF/mrad

The sensitivities of the short-range configurations are about 100 times greater than for the long-range, but the useful range is correspondingly less. In general, the useful range is about equal to the nominal value of the gap (short range) or the plate overlap (long range).

The capacitance variations are themselves not usually directly useful; we need to convert them to signals acceptable as inputs to oscilloscopes, data acquisition boards, etc. A number of different approaches are in use. For the long-range configurations, the capacitance is linear with the displacement (area), and a scheme using AC excitation can be used as follows: We apply a constant-amplitude AC voltage V of frequency f (in Hz) across the capacitor, resulting in an AC current $I = i2\pi fVC$, giving a current signal that is directly proportional to C and thus to plate displacement. For the short-range configurations, the capacitance variation has displacement x in the denominator. To get a linear output signal, we now apply a constant amplitude *current* signal I to the capacitor, making $V = I/(2\pi fC)$; now V is proportional to displacement x. As in most sensor systems that use AC excitation, a phase-sensitive demodulator and low-pass filter are used to get a "DC" output signal. The total system frequency response is then determined by the excitation frequency and the low-pass filter cutoff frequency, typical values ranging from a few hundred Hz to 200 kHz.

Capacitance bridges[61] of various degrees of complexity are also used with capacitive motion sensors. Again, the use of AC excitation means that system dynamic response is set by the excitation frequency and low-pass filter cutoff. Complete bridge/sensor systems are available, some of which are designed to measure extremely small motions. One such product line[62] includes models with ranges from 20 to 1250 μm, frequency response up to 5 kHz, and nonlinearity better than 0.1 percent. As usual with any instrument, due to intrinsic noise levels, resolution can be traded for bandwidth, for example, a unit with 100-μm range, using a 100-Hz bandwidth, has a resolution of 0.2 nm. Figure 4.34a[63] relates resolution to some basic physical constants, while Fig. 4.34b compares noise level with that of a typical laser interferometer, and Fig. 4.34c shows a typical application to feedback control of tiny motions using piezoelectric actuators. Note here the use of low-thermal-expansion materials Zerodur (glass/ceramic) and Invar (metal alloy) to reduce the effect of temperature change, which is vital when dealing with such tiny motions.

A simple circuit used with capacitance microphones and suitable for some other applications is shown in Fig. 4.33. We first note that we *cannot* here measure a *steady* displacement since any steady C gives *no* current and thus no output voltage. This defect is of no consequence in a microphone because only *fluctuating* sound pressures (usually above about 10 Hz) are of interest. If our application does not require response to steady displacement and if dynamic, rather than static, calibration is satisfactory, then this simple circuit may be considered. It can be used with either the "large motion" or the "small motion" type of sensor, though the analysis is somewhat different. We will now analyze the small-motion case, leaving the other for the end-of-chapter problems.

For the small-motion case, the area A is fixed and the displacement measured is the change in the gap x. As usual, we must choose a nominal "stand-off" distance and then assign dimensions so that the capacitance is of the order of 10 pF at this position. The range of the sensor will then be about $\pm$ 50 percent of the stand-off distance, on either side. Equation 4.41 shows that large stand-offs cause lower C values, so A will have to be increased to maintain a nominal design value, such as our 10 pF. Thus, sensors with longer ranges will generally be physically *larger;* this is true of capacitance sensors in general. Let us call the stand-off distance x_0 and the

---

[61]X. Zhao, G. Wilkening, H. Marth and K. Horn, "High-Accuracy Capacitance Sensor for Positioning of Piezoelectric Displacement Actuator," *4th Int. Conf. of New Actuators,* June 15–17, 1994, Bremen, Germany; S. Harb, S. T. Smith and D. G. Chetwynd, "Subnanometer Behavior of a Capacitive Feedback, Piezoelectric Displacement Actuator," *Rev. Sci. Instrum.* 63(92), Feb. 1992, pp. 1680–1689; R. V. Jones and J. C. S. Richards, "The Design and some Applications of Sensitive Capacitance Micrometers," *Jour. of Physics E: Scientific Instruments,* 1973, vol. 6.

[62]Queensgate Instruments Inc., Suite 600, 90 Merrick Ave., East Meadow, NY 11554, 516-357-3900 (www.queensgate.com).

[63]Queensgate Instruments Inc.

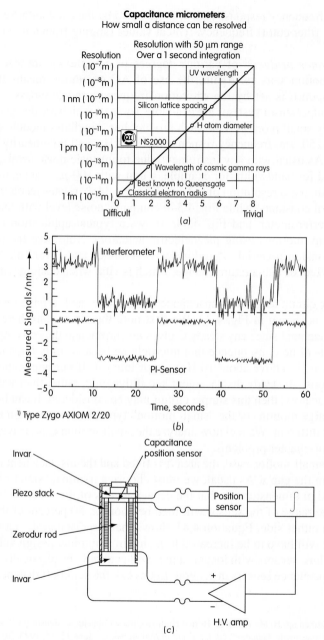

Figure 4.34
Capacitance sensor with nanometer resolution.

displacement away from that point x_i, with x_i called positive if the gap gets smaller. We then have

$$\text{Charge } q = Ce_c = \frac{0.225A}{x_0 - x_i}e_c \tag{4.42}$$

This relation is clearly nonlinear in x_i; if we want to use transfer functions to describe the instrument dynamics, we must linearize the equation, assuming small changes away from the stand-off position. Whether our 50 percent changes mentioned above can be considered "small" depends on how strong the nonlinearity is for a given set of numbers. We could check this by using simulation to model the exact nonlinear relations. Equation 4.42 has two independent variables so we use a multivariable Taylor series for the linearization.

$$q \approx q_0 + \frac{\partial q}{\partial x_i} \cdot x_{i,p} + \frac{\partial q}{\partial e_c} e_{c,p} = \frac{0.225A}{x_0} E_b + 0.225A \left(\frac{e_{c,p}}{x_0} + \frac{E_b}{x_0^2} x_{i,p} \right) \quad \textbf{(4.43)}$$

Since the current is zero at the operating point, the perturbation in the current is the same as the total current. We can then write for the circuit:

$$i = \frac{dq}{dt} \approx \frac{0.225A}{x_0} \left(-R \cdot \frac{di}{dt} \right) + \frac{0.225A}{x_0} \left(\frac{E_b}{x_0} \cdot \frac{dx_{i,p}}{dt} \right) \quad \textbf{(4.44)}$$

$$(RC_0 D + 1)i = \frac{E_b C_0}{x_0} Dx_{i,p} \quad \textbf{(4.45)}$$

Output voltage e_o is given directly by iR, so we finally get

$$\frac{e_o}{x_i}(D) = \frac{K\tau D}{\tau D + 1} \quad \textbf{(4.46)}$$

where $K \triangleq E_b/x_0$ V/in
          $\tau \triangleq RC_0$ s   (C_0 is now in farads, not pF)

As mentioned earlier, we now see analytically from the D in the numerator that this sensor cannot measure static displacements. For sinusoidal inputs with a frequency that is high relative to $1/\tau$, the displacement is accurately measured. We should also note that for "low" frequencies, the output voltage is an accurate measure of the *velocity*. For $C_0 = 10$ pF, $R = 10^{10}$ Ω, $E_b = 10.0$ V, and $x_0 = 0.01125$ in, we have a time constant of 0.1 s and a sensitivity of 888.9 V/in. A full-scale motion of 0.005625 in then gives a 5-V output. Figure 4.35 shows, for the numbers given, accurate measurement ranges for both velocity and displacement.

While the exact behavior of this sensor is governed by a nonlinear differential equation, it turns out that, so long as the motion frequency content is high enough (as indicated by the RC time constant), *the output voltage is perfectly linear with input displacement;* thus the transfer function derived above is not really an approximation when we use it in the accurate frequency range. When computing this range, you should use the C value corresponding to the *largest* air gap (this gives the smallest C and thus the smallest time constant). Problem 4.39 explores this with a simulation study, but the following intuitive physical reasoning makes it plausible. Before any input motion occurs, the capacitor is already charged to the battery voltage, and q is thus E_b/C_0. If we suddenly change x_i, we suddenly change the C value, but the charge must still be the same since a discharging current i requires a *finite* time to drain charge. Thus q is initially fixed, and a change in C causes an *instantaneous* change in capacitor voltage according to $e_c = q/C = C_0 E_b /(0.225A/(x_0 + x_i))$.

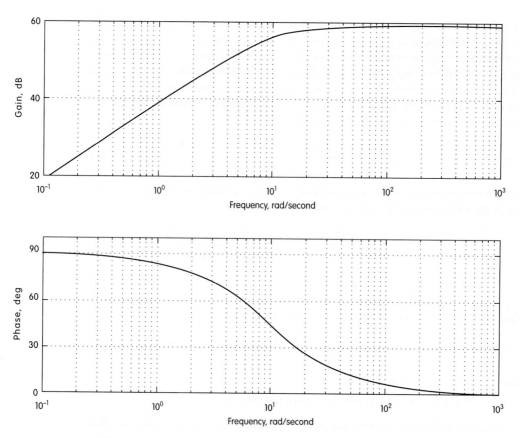

Figure 4.35
Frequency response of capacitive sensor.

This puts x_i in the *numerator*, a *linear* relation. Requiring τ to be "large enough" simply slows the discharge of C enough that the charge (for all practical purposes) *stays* at C_0E_b during any "rapid" motion.

The use of a variable differential (three-terminal) capacitor with a bridge circuit is shown in the differential pressure transducer of Fig. 4.36. Spherical depressions of a depth of about 0.001 in are ground into the glass disks; then these depressions are gold-coated to form the fixed plates of a differential capacitor. A thin, stainless-steel diaphragm is clamped between the disks and serves as the movable plate. With equal pressures applied to both ports, the diaphragm is in a neutral position, the bridge is balanced, and e_o is zero. If one pressure is greater than the other, the diaphragm deflects in proportion, giving an output at e_o in proportion to the differential pressure. For the opposite pressure difference, e_o exhibits a 180° phase change. A direction-sensitive dc output can be obtained by conventional phase-sensitive demodulation and filtering. Balance resistors necessary for initially nulling

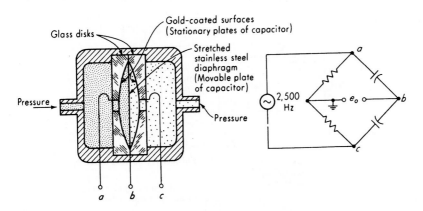

Figure 4.36
Differential-capacitor pressure pickup.

the bridge are not shown in Fig. 4.36. This method (as opposed to that of Fig. 4.33) allows measurement of static deflections. Such differential-capacitor arrangements also exhibit considerably greater linearity than do single-capacitor types.[64]

An ingenious method of circumventing the nonlinear relationship between x and C is shown in Fig. 4.37a. This technique uses an operational amplifier, perhaps the most common building block in linear electronics. The assumptions necessary to an analysis of this circuit depend on the well-known characteristics of op-amps:

1. The input impedance is so high that the amplifier input current may be taken as zero relative to other currents.
2. The gain is so high that if the output voltage of the amplifier is not saturated, the input voltage is extremely small and may be taken as zero relative to other voltages. For example, a typical amplifier has linear output for the range ± 10 V and a gain of 10^7 V/V. Thus the maximum input for linear operation is 10^{-6} V.

Using these assumptions in Fig. 4.37, we can write

$$\frac{1}{C_f} \int i_f \, dt = e_{ex} - e_{ai} = e_{ex} \tag{4.47}$$

$$\frac{1}{C_x} \int i_x \, dt = e_o - e_{ai} = e_o \tag{4.48}$$

$$i_f + i_x - i_{ai} = 0 = i_f + i_x \tag{4.49}$$

[64]N. H. Cook and E. Rabinowicz, "Physical Measurement and Analysis," p. 142, Addison-Wesley, Reading, MA, 1963.

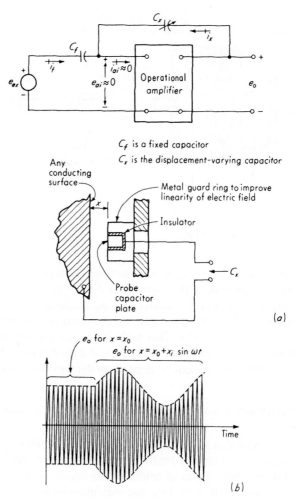

Figure 4.37
(*a*) Feedback-type capacitive pickup. (*b*) Waveform for sinusoidal displacement.
(*Courtesy I. Wayne Kerr Corp.*)

Manipulation then gives

$$e_o = \frac{1}{C_x} \int i_x \, dt = -\frac{1}{C_x} \int i_f \, dt = -\frac{C_f}{C_x} e_{ex} \qquad (4.50)$$

$$e_o = -\frac{C_f \, x e_{ex}}{0.225A} = Kx \qquad (4.51)$$

Equation (4.51) shows that the output voltage is now *directly* proportional to the plate separation $x;$ thus linearity is achieved for both large and small motions. In a commercial instrument, e_{ex} is a 50-kHz sine wave of fixed amplitude. The output e_o is also a 50-kHz sine wave which is rectified and applied to a dc voltmeter calibrated directly in distance units.

For vibratory displacements, e_o will be an amplitude-modulated wave as in Fig. 4.37b. The average value of this wave after rectification is still the mean separation of the plates and still can be read by the same meter as used for static displacements. The vibration amplitude around this mean position is extracted by applying the e_o signal also to a demodulator and a low-pass filter with cutoff at 10 kHz. The output of this filter is applied to a peak-to-peak voltmeter directly calibrated in vibration amplitude and also is available at a jack for connection to an oscilloscope for viewing of the vibration waveform. The instrument is provided with six different probes ranging from 0.0447 to 1.0 inch in diameter of the capacitance plate and covering the full-scale displacement ranges of 0.001, 0.005, 0.01, and 0.5 in. The overall system accuracy is of the order of 2 percent of full scale. Any flat conductive surface may serve as the second plate of the variable capacitor. Thus in vibrating machine parts, the parts themselves often perform this function. The resolution of 0.5 percent of full scale indicates that (with the 0.01-in full-scale probe) it is possible to detect motion as small as 5 μin.

Another linearization technique[65] which is the basis of a line[66] of capacitive pressure transducers and accelerometers is the pulse-width-modulation scheme shown in Fig. 4.38. The sensing capacitors are arranged in the transducer in a differential manner as shown. Voltages e_1 and e_2 switch back and forth at about 400 kHz between the excitation voltage (say 6 V dc) and ground; the output voltage is the average (low-pass filtered) difference between e_1 and e_2. At the transducer null position, $e_1 - e_2$ is a symmetrical square wave of zero average value, which produces no output. Motion changes the relative width (time duration) of the +6- and −6-V portions of $e_1 - e_2$, which gives a positive or negative average value depending on whether x_i was positive or negative. Circuit behavior is as follows. The four solid-state switches are switched simultaneously by the output of two op-amp comparators whenever e_3 or e_4 passes through the fixed reference voltage of 3 V. Note in the diagram that we are observing the circuit at an instant when e_3 is connected to ground ($e_3 = 0$) and capacitor C_1 is discharged, while C_2 is being charged by the 6-V supply through R, with a time constant RC_2. As C_2 charges, e_4 rises exponentially (nearly linearly for early times); and when e_4 reaches 3 V (the reference voltage level), the two inputs of comparator 4 are now equal, causing its output to throw all four switches to their opposite positions. This "instantaneously" discharges C_2 and begins a similar charging process for C_1. Thus an oscillation is established whose period $t_1 + t_2$ (about 2.5 μs in the referenced device) is determined by R, $C_1 + C_2$,

---

[65]S. Y. Lee, "Variable Capacitance Signal Transduction and the Comparison with Other Transduction Schemes," Setra Systems Inc.

[66]Setra Systems Inc., www.setra.com.

and e_{ref}, but whose duty cycle varies as C_1 and C_2 change. In steady state, the output voltage e_o is the average of $e_1 - e_2$. Thus

$$e_o = \frac{e_{ex} t_1}{t_1 + t_2} - \frac{e_{ex} t_2}{t_1 + t_2} = e_{ex} \frac{C_1 - C_2}{C_1 + C_2} \qquad (4.52)$$

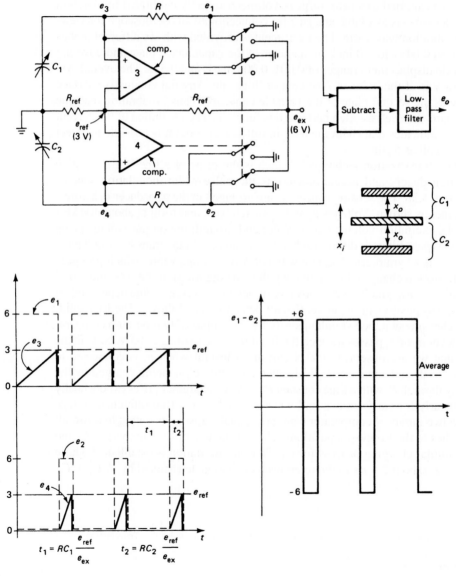

Figure 4.38
Capacitive transducer signal processing.

In the differential capacitor,

$$C_1 = \frac{C_o x_o}{x_o - x_i} \qquad C_2 = \frac{C_o x_o}{x_o + x_i} \tag{4.53}$$

which finally gives the desired linear relation between e_o and x_i:

$$e_o = e_{ex} \frac{x_i}{x_o} \tag{4.54}$$

Linearization of capacitance transducers also can be accomplished by using a feedback system[67] which adjusts capacitor-current amplitude so that it stays constant at a reference value for all displacements x_i. This is accomplished by obtaining a dc signal proportional to capacitor current (from a demodulator), comparing this current with the reference current, and adjusting the voltage amplitude of the system excitation oscillator (typically 100 kHz fixed frequency) until the two currents agree. If the capacitor-current amplitude is kept constant irrespective of capacitor motion, then the capacitor-voltage amplitude is linear with x_i:

$$\frac{i_c}{e_c}(i\omega) = i\omega C \tag{4.55}$$

$$|e_c| = \frac{|i_c|}{\omega C} = \frac{|i_c| x_i}{\omega C_o x_o} = K x_i \tag{4.56}$$

Figure 4.39 shows a capacitive displacement transducer marketed as a high-temperature strain gage. A differential capacitor in the form of three thin rings is used, with the two inner rings supported by a temperature-compensating rod made of the same material as the test specimen. Since the measuring axis is displaced rather far above the surface where strain is to be measured, the outer ring is carried in specially designed flexures, so that the effect of any bending in the specimen is compensated. Maximum strain is $\pm 20,000$ $\mu\varepsilon$, corresponding to a total displacement of ± 0.02 in over the 1-in gage length (a model with $\frac{1}{4}$-in gage length and $\pm 80,000$ $\mu\varepsilon$ is also available). Temperature range is cryogenic to $1500°F$; spurious apparent strain due to temperature is typically ± 300 $\mu\varepsilon$ at $1500°F$. Electronics using ac excitation at 3.39 kHz provides nonlinearity of ± 0.5 percent at 3,000 $\mu\varepsilon$ and $70°F$; linearization details are not given. The gage exerts a load of 20 g on the specimen at 5,000 $\mu\varepsilon$.

A complete line[68] of capacitive displacement sensors includes also special-purpose units for thin gaps and disk-brake wear analysis. The thin-gap sensor replaces mechanical feeler gages and shims with a noncontacting measurement method. A thin wand carries two capacitive sensors, one on each side, for position-compensated gap measurement. For fixed-position applications, a single-sided model is available. Disk-brake sensors[69] are used both on operating vehicles and

---

[67]L. Michelson, "Greater Precision for Noncontact Sensors," *Mach. Des.,* p. 117, Dec. 6, 1979; Lion Precision Corp., www.lionprecision.com.

[68]Capacitec, Inc., 87 Fitchburg Rd., Ayer, MA 01432, 978-772-6033 (www.capacitec.com).

[69]S. Muldoon and R. Sandberg, "Using Capacitive Probes in Automotive Brake Component Testing," *Test Engineering and Management,* Aug.–Sept. 1997, pp. 10–12.

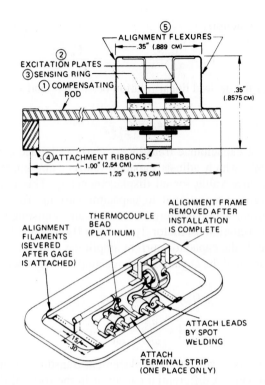

Figure 4.39
Capacitive high-temperature strain gage. Top, capacitive strain gage–cross-section view.
Bottom, capacitive strain gage in alignment frame. *(Courtesy Hi Tec Corp., Westford, Mass.)*

laboratory test setups for dynamic measurements. Single or dual sensors allow
analysis of rotor runout, rotor thickness variation, rotor coning, plate-to-plate orientation, wobble, and thermal expansion at temperatures as high as 1600°F.

Piezoelectric Transducers

When certain solid materials are deformed, they generate within them an electric
charge. This effect is reversible in that if a charge is applied, the material will
mechanically deform in response. These actions are given the name *piezoelectric*
(pī-ēzō-electric) *effect*.[70] This electromechanical energy-conversion principle is
applied usefully in both directions. The mechanical-input/electrical-output direction
is the basis of many instruments used for measuring acceleration, force, and pressure. It also can be utilized as a means of generating high-voltage low-current electric power, such as is used in spark-ignition engines and electrostatic dust filters.
The electrical-input/mechanical-output direction is applied in small vibration shakers, sonar systems for acoustic ranging and direction detection, industrial ultrasonic

---

[70]Harris and Crede, op. cit., chap. 16, pt. 2.

nondestructive test equipment, pumps for ink-jet printers, ultrasonic flowmeters, and micromotion actuators.[71]

The materials that exhibit a significant and useful piezoelectric effect fall into three main groups: natural (quartz, rochelle salt) and synthetic (lithium sulfate, ammonium dihydrogen phosphate) crystals, polarized ferroelectric ceramics (barium titanate, etc.), and certain polymer films.[72] Because of their natural asymmetric structure, the crystal materials exhibit the effect without further processing. The ferroelectric ceramics must be artificially polarized by applying a strong electric field to the material (while it is heated to a temperature above the Curie point of that material) and then slowly cooling with the field still applied. (The *Curie temperature* is the temperature above which a material loses its ferroelectric properties; thus it limits the highest temperature at which such materials may be used.) When the external field is removed from the cooled material, a remanent polarization is retained and the material exhibits the piezoelectric effect.

The piezoelectric effect can be made to respond to (or cause) mechanical deformations of the material in many different modes, such as thickness expansion, transverse expansion, thickness shear, and face shear. The mode of motion effected depends on the shape and orientation of the body relative to the crystal axes and the location of the electrodes. Metal electrodes are plated onto selected faces of the piezoelectric material so that lead wires can be attached for bringing in or leading out the electric charge. Since the piezoelectric materials are insulators, the electrodes also become the plates of a capacitor. A piezoelectric element used for converting mechanical motion to electric signals thus may be thought of as a charge generator and a capacitor. Mechanical deformation generates a charge; this charge then results in a definite voltage appearing between the electrodes according to the usual law for capacitors, $E = Q/C$. The piezoelectric effect is direction-sensitive in that tension produces a definite voltage polarity while compression produces the opposite.

We illustrate the main characteristics of piezoelectric motion-to-voltage transducers by considering only one common mode of deformation, thickness expansion. For this mode the physical arrangement is as in Fig. 4.40*b*. Various double-subscripted physical constants numerically describe the phenomena occurring. The convention is that the first subscript refers to the direction of the electrical effect and the second to that of the mechanical effect, by using the axis-numbering system of Fig. 4.40*a*.

Two main families of constants, the *g* constants and the *d* constants, are considered. For a barium titanate thickness-expansion device, the pertinent *g* constant is g_{33}, which is defined as

$$g_{33} \triangleq \frac{\text{field produced in direction 3}}{\text{stress applied in direction 3}} = \frac{e_o/t}{f_i/(wl)} \tag{4.57}$$

[71]Inchworm Translator, Burleigh Instruments, Fisher, NY, 716-924-9355 (www.burleigh.com).

[72]R. Brown, "Piezo Film for Vibration Monitoring," *Sensor,* pp. 20–26, January 1988; J. V. Chatigny and L. E. Robb, "Piezo Film Sensors," *Sensors,* pp. 6–18, May 1986; Kynar Piezo Film Technical Manual, Pennwalt Corp., King of Prussia, PA.

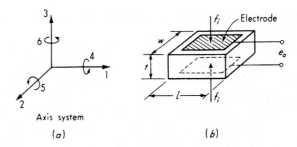

Figure 4.40(a) and (b)
Piezoelectric transducer.

Thus if we know g for a given material and the dimension t, we can calculate the output voltage per unit applied stress. Typical g values are 12×10^{-3} (V/m)/(N/m^2) for barium titanate and 50×10^{-3} for quartz. Thus, for example, a quartz crystal 0.1 in thick would have a sensitivity of 0.88 V/(lb/in^2), illustrating the large voltage output for small stress typical of piezoelectric devices.

To relate applied force to generated charge, the d constants can be defined as

$$d_{33} \triangleq \frac{\text{charge generated in direction 3}}{\text{force applied in direction 3}} = \frac{Q}{f_i} \tag{4.58}$$

Actually, d_{33} can be calculated from g_{33} if the dielectric constant ε of the material is known, since

$$C = \frac{\varepsilon wl}{t} \tag{4.59}$$

$$g_{33} \triangleq \frac{\text{field}}{\text{stress}} = \frac{e_o wl}{t f_i} = \frac{e_o C}{\varepsilon f_i} = \frac{Q}{\varepsilon f_i} = \frac{d_{33}}{\varepsilon} \tag{4.60}$$

$$d_{33} = \varepsilon g_{33} \tag{4.61}$$

The dielectric constant of quartz is about 4.06×10^{-11} F/m while for barium titanate it is $1{,}250 \times 10^{-11}$. For quartz, then,

$$d_{11} = \varepsilon g_{11} = (4.06 \times 10^{-11})(50 \times 10^{-3}) = 2.03 \text{ pC/N} \tag{4.62}$$

(The subscripts 11 are used because in quartz the thickness-expansion mode is along the crystallographic axis conventionally called axis 1.) Sometimes it is desired to express the output charge or voltage in terms of deflection (rather than stress or force) of the crystal, since it is really the *deformation* that causes the charge generation. To do this, we must know the modulus of elasticity, which is 8.6×10^{10} N/m^2 for quartz and 12×10^{10} for barium titanate.

With the above brief introduction as background, we proceed to consider piezoelectric elements as displacement transducers. The ultimate purpose is generally force, pressure, or acceleration measurement, but here we consider only the conversion from displacement to voltage. For analysis purposes it is necessary to consider the transducer, connecting cable, and associated amplifier as a unit. The transducer impedance is generally very high; thus the amplifier is usually a high-impedance type used for buffering purposes rather than voltage gain. Charge amplifiers (see Chap. 10) are commonly used also. The cable capacitance can be significant, especially for long cables. For the transducer alone, if a static deflection x_i is applied and maintained, a transducer terminal voltage will be developed but the charge will slowly leak off through the leakage resistance of the transducer. Since R_{leak} is generally very large (the order of 10^{11} Ω), this decay would be very slow, perhaps allowing at least a quasi-static response. However, when an external voltage-measuring device of low input impedance is connected to the transducer, the charge leaks off very rapidly, preventing the measurement of static displacements. Even relatively high-impedance amplifiers generally do not allow static measurements. Some commercially available[73] systems using quartz transducers (very high leakage resistance) and electrometer input amplifiers (very high input impedance) achieve an effective total resistance of 10^{14} Ω, which gives a sufficiently slow leakage to allow static measurements in some situations.

To put the above discussion on a quantitative basis, we consider Fig. 4.40c. The charge generated by the crystal can be expressed as

$$q = K_q x_i \qquad (4.63)$$

where $\qquad K_q \triangleq$ C/cm $\qquad\qquad (4.64)$

$\qquad\qquad x_i \triangleq$ deflection, cm $\qquad\qquad (4.65)$

The resistances and capacitances of Fig. 4.40d can be combined as in 4.40e. We also convert the charge generator to a more familiar current generator according to

$$i_{cr} = \frac{dq}{dt} = K_q \left(\frac{dx_i}{dt} \right) \qquad (4.66)$$

We may then write

$$i_{cr} = i_C + i_R \qquad (4.67)$$

$$e_o = e_C = \frac{\int i_C \, dt}{C} = \frac{\int (i_{cr} - i_R) \, dt}{C} \qquad (4.68)$$

$$C\left(\frac{de_o}{dt} \right) = i_{cr} - i_R = K_q \left(\frac{dx_i}{dt} \right) - \frac{e_o}{R} \qquad (4.69)$$

$$\frac{e_o}{x_i}(D) = \frac{K \tau D}{\tau D + 1} \qquad (4.70)$$

where $\qquad K \triangleq$ sensitivity $\triangleq \dfrac{K_q}{C}$, V/cm $\qquad\qquad (4.71)$

$\qquad\qquad \tau \triangleq$ time constant $\triangleq RC$, s $\qquad\qquad (4.72)$

---

[73]Kistler Instrument Corp., Amherst, NY 14228, 888-547-8537, www.kistler.com.

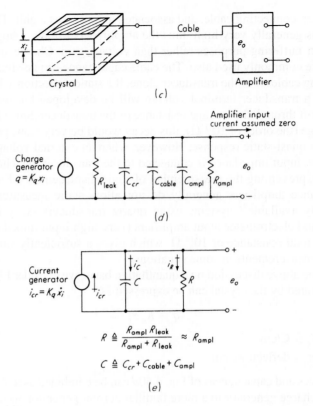

Figure 4.40(c)–(e)
Equivalent circuit for piezoelectric transducer.

We see that, just as in the capacitance pickup of Fig. 4.33, the steady-state response to a constant x_i is zero; thus we cannot measure static displacements. For a flat amplitude response within, say, 5 percent, the frequency must exceed ω_1, where

$$0.95^2 = \frac{(\omega_1 \tau)^2}{(\omega_1 \tau)^2 + 1} \tag{4.73}$$

$$\omega_1 = \frac{3.04}{\tau} \tag{4.74}$$

Thus a large τ gives an accurate response at lower frequencies.

The response of these transducers is further illuminated by considering the displacement input of Fig. 4.41. The differential equation is

$$(\tau D + 1)e_o = (K\tau D)x_i \tag{4.75}$$

Since $x_i = A$ for $0 < t < T$, this becomes

$$(\tau D + 1)e_o = 0 \tag{4.76}$$

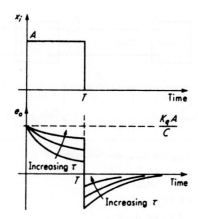

Figure 4.41
Pulse response of piezoelectric transducer.

Now at $t = 0^+$ the displacement x_i is A, and so the charge *suddenly* increases to $K_q A$. Thus our initial condition is

$$e_o = \frac{K_q A}{C} \qquad \text{at } t = 0^+ \tag{4.77}$$

Solving Eq. (4.76) with initial condition (4.77) gives

$$e_o = \frac{K_q A}{C} e^{-t/\tau} \qquad 0 < t < T \tag{4.78}$$

Equation (4.78) holds until $t = T$. At this instant we must stop using it because of the change in x_i. For $T < t < \infty$ the differential equation is

$$(\tau D + 1)e_0 = 0 \tag{4.79}$$

At $t = T^-$, Eq. (4.78) is still valid and

$$e_o = \frac{K_q A}{C} e^{-T/\tau} \tag{4.80}$$

Now, at $t = T$, suddenly x_i drops an amount A, causing a sudden decrease in charge of $K_q A$ and a sudden decrease in e_o of $K_q A/C$ from its value at $t = T^-$. Thus at $t = T^+$, e_o is given by

$$e_o = \frac{K_q A}{C} (e^{-T/\tau} - 1) \tag{4.81}$$

which becomes the initial condition for Eq. (4.79). The solution then becomes

$$e_o = \frac{K_q A}{C} (e^{-T/\tau} - 1)e^{-(t - T)/\tau} \qquad T < t < \infty \tag{4.82}$$

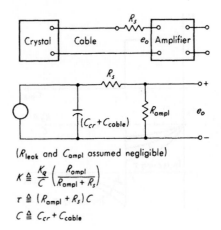

$$K \triangleq \frac{K_q}{C}\left(\frac{R_{ampl}}{R_{ampl} + R_s}\right)$$

$$\tau \triangleq (R_{ampl} + R_s)C$$

$$C \triangleq C_{cr} + C_{cable}$$

Figure 4.42
Use of series resistor to increase time constant.

Figure 4.41 shows the complete process for three different values of τ. It is clear that a large τ is desirable for faithful reproduction of x_i. If the decay and "undershoot" at $t = T$ are to be kept within, say, 5 percent of the true value, τ must be at least $20T$. If an increase of τ is required in a specific application, it may be achieved by increasing either or both R and C. An increase in C is easily obtained by connecting an external shunt capacitor across the transducer terminals, since shunt capacitors add directly. The price paid for this increase in τ is a loss of sensitivity according to $K = K_q/C$. Often this may be tolerated because of the initial high sensitivity of piezoelectric devices. An increase in R generally requires an amplifier of greater input resistance. If sensitivity can be sacrificed, a series resistor connected external to the amplifier, as shown in Fig. 4.42, will increase τ without the need of obtaining a different amplifier.

Although perhaps not quite as versatile with respect to computing and compensating operations as the strain-gage/bridge-circuit combination, piezoelectric elements do provide some similar functions, as exemplified by the three-component force transducer of Fig. 4.43.[74] Two types of thin quartz sensing disks are used here. One type is cut to be sensitive to tension/compression, and the other to shear. One pair of tension/compression disks is used to measure the z component of force, while two pairs of shear disks, each properly oriented, measure the x and y components. Use of pairs of disks rather than single disks simplifies electrode attachment, eliminates the need for insulating layers which reduce rigidity, and increases sensitivity. Note that the disk pairs are electrically in parallel so that the charges produced by each element are summed. This signal-summing action can be extended to more than two elements and is not limited to elements within a single transducer. Thus, say, the x component of force from one transducer could be electrically

[74]G. H. Gautschi, Piezoelectric Multicomponent Force Transducers and Measuring Systems, Kistler Instrument Corp., Amherst, NY, 1978, 888-547-8537 (www.kistler.com).

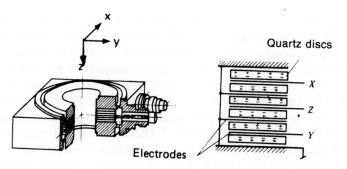

Figure 4.43
Three-component piezoelectric force transducer.

summed with the y component from another. While piezoelectric elements respond to both positive and negative strains (with charge of opposite polarity), it is mechanically difficult to couple tension stresses into the elements, since adhesive connections usually are not considered sufficiently reliable. The usual procedure is thus to install the elements (such as the discs above) under heavy compressive preload, sufficient to preclude relaxation of the preload under any expected external tension and large enough to provide reliable frictional transfer of the x and y shear components without slippage. Here the leakage of charge for steady loads is an advantage, since the preload persists mechanically but produces zero output voltage. Applied positive and negative z loads then produce corresponding plus and minus voltages since the output voltage corresponds to the *change* in strain away from the preload condition. Detailed data on static and dynamic performance characteristics of piezoelectric transducers are deferred to the respective sections on force, pressure, and acceleration measurement, where these data will be more meaningful.

We earlier mentioned briefly the use of piezoelectric technology for micromotion *actuation,* rather than measurement. Since such actuation is often a vital part of a *measurement* system, we now want to expand on such applications. The GyroChip sensor discussed later in this chapter uses a miniature double piezoelectric quartz tuning fork to measure absolute angular velocity. One tuning fork is made to vibrate by applying a sinusoidal drive voltage. Rotation of the sensor causes the second tuning fork to vibrate with a motion related to the angular velocity; this motion causes a piezoelectric output voltage. Tiny translational and rotational motions required to position mirrors, optical fibers, etc. in electro-optical measurement systems are often implemented with piezoelectric actuators. Single- and multiaxis actuators are available in many different configurations and are usually used in a feedback system to overcome the hysteresis present in the piezoelectric materials. Displacement sensors used in these systems may be capacitive, LVDT, eddy-current, or strain-gage types. A typical xy positioning stage[75] using capacitance sensors has a range of 50 μm, resolution of 0.5 nm, and resonant frequency of 500 Hz. Modeling of the dynamic behavior of such systems is discussed in the manufacturers' catalog/handbooks[76] and in my System Dynamics text.[77] Piezoelectric actuation is also a basic component of the various scanning-probe microscopes discussed later in this text. The measurement and control of tiny motions is a vital part of MEMSs (microelectromechanical systems) technology.

Precise tiny motions can also be achieved with "ordinary" electric motors or micrometers by using *elastic actuators*.[78] These are flexural metal devices that accept a "large" motion as input and produce a much smaller motion as output. For example, the input motion could be from an ordinary vernier micrometer of 1-in range and 0.0001-in resolution while the output motion would have a 1-μm range with 1 nm resolution. Such actuators have also been combined with electric motors.[79]

Electro-Optical Devices

We describe several displacement-measuring devices that combine optical and electronic principles. Each is designed for a specific class of applications, but all share the general advantages of optical measurement,[80] noncontacting operation with negligible force exerted on the moving object since the radiation pressure of light is miniscule.

Our first example, the Fotonic sensor,[81] uses fiber optics to measure displacements in the range of microinches to tenths of inches with a relatively simple optical/electronic arrangement. The probe (Fig. 4.44), 0.02 to 0.3 inch in diameter and 3 in long, consists of a bundle of several hundred optical fibers, each a few thousandths of an inch in diameter. The fiber bundle extends for 1 to 3 ft to the electronics chassis, where it is divided into two equal groups of fibers. One group (transmitting fibers) is exposed to a light source and thus carries light to the probe tip, where light is emitted and reflected by the target surface. While targets of high reflectivity give greater output, even rather dull surfaces can be measured. The reflected light is picked up by the other (receiving) group of fibers, transmitted to the electronics package, and focused on a suitable photodetector whose electronics then produces a dc output related to probe-target gap. To minimize cross-talk between adjacent fibers, the fibers are of a core/cladding construction; two glasses of different refractive index are used to obtain total internal reflection.

Probes of this general type display two useful measuring ranges: front slope and back slope. At zero gap, no light can escape from the transmitting fibers and output

---

[76]Polytec PI, Products for Micropositioning. Queensgate Instruments Inc., The Nanopositioning Book.

[77]E. O. Doebelin, "System Dynamics," Marcel Dekker, 1998, pp. 277–282, 306–310, and 686–688.

[78]A. E. Hatheway Inc., Pasadena, CA, 818-795-0514.

[79]Rubicon Actuator, Schaeffer Magnetics Div., Moog, Inc., Chatsworth, CA, 818-341-5156.

[80]"New Dimensions in Optical Gaging," *Inst. Tech.,* p. 9, June 1980.

[81]C. Menadier, C. Kissinger, and H. Adkins, "The Fotonic Sensor," *Inst. & Cont. Syst.,* p. 114, June 1967; Mechanical Technology Inc., Latham, N.Y., 518-785-2464 (www.mechtech.com).

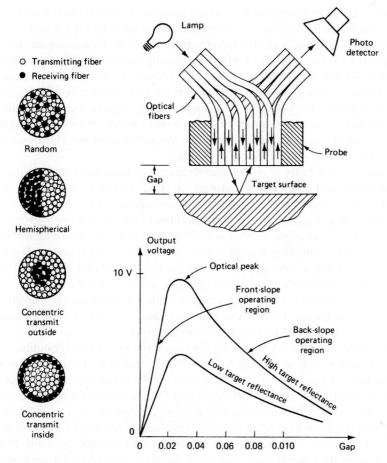

Figure 4.44
Fiber-optic displacement transducer.

is zero. As the gap opens up, more of the target surface is illuminated and reflection increases, giving a very sensitive and nearly linear range of measurement (front slope). As gap increases, finally the entire target is illuminated, giving peak output. Motion beyond this point causes a *reduction* in response since both target illumination and the fraction of reflected light gathered by the sensor decrease roughly according to an inverse square law. This "back slope" region is also useful for measurement, but is less linear and sensitive. Theory shows that maximum sensitivity in the front-slope region is achieved by arranging the fibers in a precise geometric pattern in which each receiving fiber is surrounded by four transmitting fibers. This is not practical from a manufacturing viewpoint, so a random distribution, which has been found to give nearly identical results, is usually used. Three other arrangements (Fig. 4.44) which allow trade-offs among various features are also available, however. A range of probes with stand-off distances from 0.07 to 8.9 mm is available[82] from one manufacturer, with resolution as small as 0.004 μin and frequency response up to 150 kHz. Since the target surface is usually an existing machine part, calibration with that specific surface is usually required; however, automatic compensation for slowly changing surface reflectivity is sometimes possible.[83] Another firm[84] offers sensors that are compensated for changes in reflectance of the measured surface. They also have a *contacting* type probe that places the fiber inside a protective housing, for applications where the optical path might be fouled by hostile environments such as grease or sludge. A theoretical analysis of this general type of fiber optic sensor is available.[85]

Lasers are the basis of many measurement systems used both in industrial manufacturing and research laboratories. Several schemes[86] have been devised using lasers for motion and dimension measurement. One such[87] is shown in Fig. 4.45. Here a single narrow helium-neon laser beam is scanned[88] over a workspace by a five-sided prism rotating at 1800 r/min, giving one measurement scan in $\frac{1}{150}$ s. A special collimating lens produces parallel rays which sweep through the workspace at a linear rate proportional to the prism's rotational speed. Thus the position within the workspace of the cylindrical target (whose diameter is to be measured) is not critical. (If the target moves while it is being scanned, an error will result; however, the rapid measurement cycle reduces this effect. If target motion is oscillatory, averaging can be used to minimize error.)

As a result of the shadow cast by the target, the photodetector output voltage exhibits a "notch" whose width in time is proportional to the target width in space. Measurement of this time interval (by gating an electronic counter) is made more precise by electrically double-differentiating the photodetector signal to produce two narrow spikes. Differentiation also allows ac signal processing and makes the system less sensitive to gradual illumination changes resulting from drift in laser power, smoke in the workspace, etc. A single system oscillator (typically 18 MHz) is used both as the counter clock and the ac frequency reference for the hysteresis-synchronous motor used to rotate the prism. Thus any drifts in oscillator frequency cause motor-speed changes and counter time-base changes which nullify each other. Additional circuitry is provided to reject certain types of "bad" measurements, such as no target present, signal too low for accuracy because of excess smoke in workspace, etc. The system holds the last "good" measurement until such situations

---

[82]Mechanical Technology Inc.

[83]G. J. Philips and F. Hirschfeld, "Rotating Machinery Bearing Analysis," *Mech. Eng.,* July 1980, p. 28.

[84]Philtec, Inc., 1021 St. Margarets Dr., Annapolis, MD 21401, 800-453-6242 (www.philtec.com).

[85]R. O. Cook and C. W. Hamm, "Fiber Optic Lever Displacement Transducer," *Applied Optics,* vol. 18, Oct. 1979, pp. 3230–3241.

[86]G. B. Foster, "Lasers as Dimension Transducers," *Instrum. Tech.,* p. 47, September 1975; T. R. Pryor and J. C. Cruz, "Laser-Based Gauging/Inspection," *Electro-optical Syst. Des.,* p. 26. May 1975.

[87]R. Moore, "Laser-Based Noncontact Gauging System," *Electro-optical Syst. Des.,* p. 46, March 1978; Zygo Corp., Middlefield, CT (www.zygo.com).

[88]G. F. Marshall, "Laser Beam Scanning," Marcel Dekker, New York, 1985.

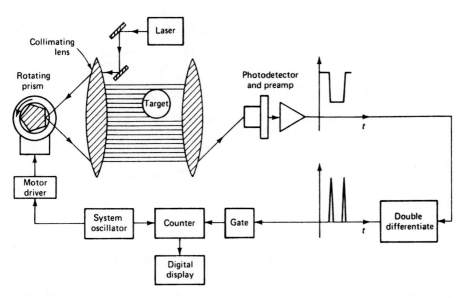

Figure 4.45
Laser dimensional gage.

clear up. While each application exhibits its own characteristics, accuracy on the order of 0.0001 in has been achieved in plastics extrusion, bar-mill diameter control, centerless grinding machine control, and remote diameter measurements of nuclear fuel rods.[89]

We now briefly treat the *laser interferometer,* a very precise motion measuring system. The use of light-interference principles as a measurement tool certainly can be considered a part of classical physics. Michelson, in the 1890s, used the scheme of Fig. 4.46, which is the basis of all subsequent developments. Using the wave model of light, we would expect the observer to see cycles of light and darkness as the motion of the movable mirror shifted the phase of beam 2 with respect to fixed beam 1, causing alternate reinforcement and interference of the two beams. If we know the light wavelength to be, say, 0.5×10^{-6} m, then each 0.25×10^{-6} m of mirror movement corresponds to one complete cycle (light to dark to light) of illumination. By counting the number of illumination cycles, we can calculate the distance between any two positions of the movable mirror.

This seemingly simple measurement principle is fraught with difficulties that have kept it a tedious standards-laboratory procedure of limited application (rather than a practical "machine shop" tool) until the appearance of several laser-based developments. The laser itself provides a much higher-quality monochromatic (single frequency) light source than did earlier "lamps," since its light is "coherent" (stays in phase with itself) over much greater distances and its frequency is very

---

[89]R. Moore, op. cit.

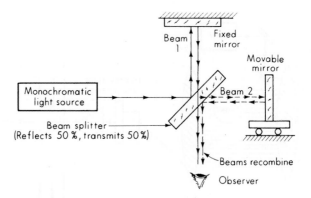

Figure 4.46
Original Michelson interferometer.

stable and precisely known (to about one part in 10^7). When light travels through air, its frequency is unaffected, but its wavelength changes whenever air pressure, temperature, or humidity causes alterations in the air's refractive index and thus in the speed of light. The pressure effect is about 4 parts in 10^7 per millimeter of mercury, while the temperature effect is nearly 11 parts in 10^7 per degree Celsius. Many applications do not warrant corrections for these effects; but if they are deemed necessary, some commercial systems[90] provide temperature, pressure, and humidity sensors that automatically scale the instrument readout to provide continuous correction. Since the lengths of objects being measured are *themselves* temperature-dependent, automatic sensing of this temperature and correction of readings to some standard (often 20°C) are of practical interest and are provided on some systems.

Clever optical, mechanical, and electronic designs have built on the above simple principles to provide measuring systems of much improved portability, precision, range of applicability, and ease of use. Figure 4.47 shows one such system which we briefly discuss (articles[91, 92] below give more details). Our discussion serves not only to explain this specific instrument but to familiarize the reader with several optical and electronic devices that are used in many other measurement systems. This discussion also helps to refresh some optical concepts taught in basic physics courses but perhaps no longer clearly in mind. First, recall that light can be thought of as a rapidly fluctuating sinusoidal *electric field,* described by a vector transverse to the direction of beam propagation. If this vector always lies in a certain plane, then the light is said to be *polarized* in that plane. When two light beams have the *same* plane of polarization, then they can exhibit *interference,* either constructive or destructive. The frequencies of the light waves themselves are generally too high (typically around 10^{14} Hz) to be transduced and manipulated with the usual

[90]Hewlett-Packard Model 5510A.

[91]"A Two-Hundred-Foot Yardstick with Graduations Every Microinch," *Hewlett-Packard J.,* August 1970.

[92]"Remote Laser Interferometry," *Hewlett-Packard J.,* December 1971.

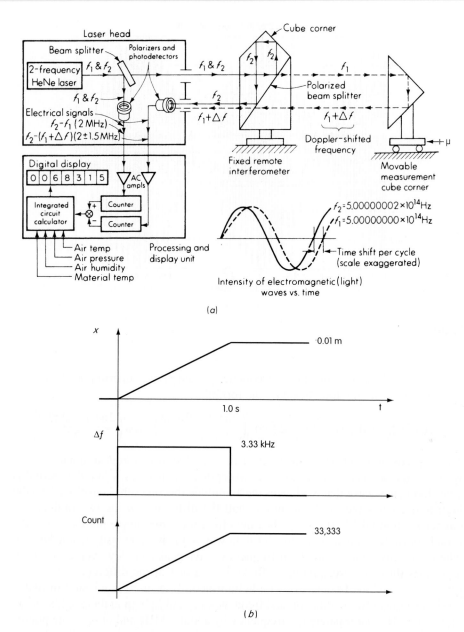

Figure 4.47
Modern laser interferometer.

optoelectronic devices, so various techniques are used to translate the information to lower frequencies. One common method, called *optical heterodyning,* uses a "beat frequency" phenomenon, where two beams of slightly different optical wavelength (and thus frequency) are combined (added) to produce cycles of light/dark illumination at a frequency equal to the *difference* of the optical frequencies.

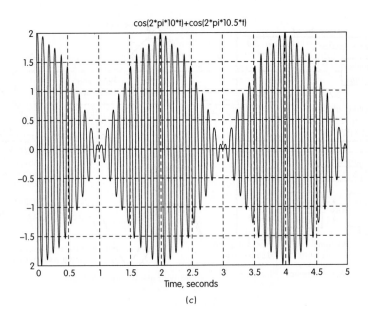

Figure 4.47
(Concluded)

If two light beams of optical frequency f_1 and f_2 are superimposed, we can represent the resulting total beam by the formula

$$A \cos 2\pi f_1 t + A \cos 2\pi f_2 t = 2A \cos \left(\frac{2\pi f_1 t + 2\pi f_2 t}{2} \right) \cdot \cos \left(\frac{2\pi f_2 t - 2\pi f_1 t}{2} \right)$$

The first cosine function to the right of the equal sign is a *high* frequency wave whose frequency is the average of the two optical frequencies, while the second is a *low* frequency wave that modulates the amplitude (becomes the "envelope" of) the high-frequency wave; its frequency is half the difference of the two optical frequencies. Figure 4.47c shows the beating effect for some simple numbers (*not* the actual optical frequencies). In terms of light waves, the light intensity would be high where the "envelope" of the high-frequency wave peaks and low where the envelope goes through zero, giving a "flickering" effect at the difference frequency (twice the envelope frequency). The helium-neon laser used is unusual in that it produces light at two distinct optical frequencies f_1 and f_2, both in the neighborhood of 5×10^{14} Hz, but separated in frequency by about 2 MHz and of opposite polarization. As the beam leaves the laser, it is split in half. One half goes directly to a polarizer and photodetector to create an electric reference signal; the other half proceeds out to the external optics. The reference-beam polarizer gives the two frequencies *equal* polarizations so that they may exhibit ordinary constructive and destructive interference. This interference can be thought of in terms of the intensity-versus-time graph of Fig. 4.47a. There, two waves initially "in phase" get slightly "out of phase" in one cycle, since the frequencies are slightly different. (Note that the time shift per cycle is a very small fraction of one cycle; thus, the waves shown are nearly identical and interfere *constructively,* giving a high illu-mination to the reference-beam photodetector and a large electric output.) *Destruc-tive* interference (giving low illumination and small electric output) will occur 1.25×10^8 cycles $(0.25 \times 10^{-6}$ s) later when the time shift per cycle $(8.000000 \times 10^{-24}$ s) has accumulated a total phase shift of one-half cycle. The "light" on the reference photodetector thus "flickers" at a 2-MHz rate, producing a 2-MHz electric signal.

Turning our attention to the measuring beam, we see that it proceeds out from the laser and encounters first the fixed remote interferometer's polarized beam split-ter, which efficiently reflects the f_2 component around a cube corner (prism) and transmits the f_1 component to the measurement cube corner. If this cube corner is stationary, no frequency change occurs in f_1 between the incident and reflected beam. However, if the cube corner moves, the reflected beam exhibits a Doppler shift of frequency proportional to the velocity v of motion. Recall from physics that the frequency f_v of a wave propagating at speed c between a transmitter and receiver that have a relative approach velocity v is given at the receiver by

$$f_v = f_0 \left(\frac{c + v}{c} \right) = f_0 \left(1 + \frac{v}{c} \right) \qquad \text{frequency change} \triangleq \Delta f = f_0 \frac{v}{c}$$

where $\quad f_0 \triangleq$ frequency at the transmitter

For our case, the speed of light c is 3.00×10^8 m/s and f_0 is 5.00×10^{14}, and since the Doppler shift occurs *twice* (the light enters and then leaves the moving cube corner), the frequency change will be 3.33 MHz for each meter/second of cube corner velocity. When this $f_1 + \Delta f$ beam is recombined with the f_2 beam in the remote interferometer and impinged on the photodetector, again we get interference effects that produce an electric signal of frequency 0.5 to 3.5 MHz, depending on the velocity. The reference and measurement signals are ac-amplified and applied to a subtracting counter that reads zero if there is no motion, but which accumulates counts in proportion to the distance traveled from a chosen reference position. For example (see Fig. 4.47*b*), a motion at 1 cm/s for 1 s (total motion 1 cm) gives 33,000 counts and thus a resolution of about 3×10^{-7} m. This, of course, is exactly half the wavelength λ of the laser light since the cube corner motion of $\frac{1}{2}\lambda$ changes the path length by λ, corresponding to one complete light-to-dark-to-light cycle. Note that interferometers of this general type do not measure absolute position, but rather the *displacement away* from an arbitrarily chosen starting point, that is, the "zero point" could be *any* position where we choose to reset the net count to zero and then allow motion away from that point. In practice, the measured motion is usually driven up against a "hard stop" provided in the machine; we zero the system at that point and then can measure displacement away from the known location of the hard stop.

The "ac" interferometer described above exhibits a number of advantages rela-tive to older "fringe-counting" instruments. The Doppler technique allows use of simpler and more reliable ac (rather than dc) amplifiers, gives unambiguous indi-cation of the *direction* of motion, and allows *velocity* measurement in addition to displacement. Locating the interferometer remote from the laser head reduces

Laser System Measurement Accuracy* Comparison

Environment: Pressure: 760 mm Hg ± 25 mm Hg Relative humidity: 50% ± 10%			
Temperature control	± 0.1°C	± 1.0°C	± 5.0°C
No compensation** (@ 20°C)	± 9.0 ppm	± 9.9 ppm	± 14.0 ppm
HP 10751A/B air sensor (@ 20°C)	± 1.4 ppm	± 1.5 ppm (typical)	± 1.6 ppm
Wavelength tracking compensation***	± 0.14 ppm	± 0.20 ppm	± 0.44 ppm
Measurement in vacuum	± 0.1 ppm	± 0.1 ppm	± 0.1 ppm

  *These accuracy specifications include the laser head term, but exclude
    electronics accuracy and interferometer nonlinearity terms.
 **No compensation means that no correction in compensation value occurs
    during environmental changes.
***System accuracy equals these values (measurement repeatability) plus
    accuracy of initial compensation value.

thermal deformation problems due to the laser's heat. By varying the external optical hardware, the system[93] can be adapted to measure length, flatness, straightness, and squareness to high precision (a few parts per million) over long distances (up to 200 m). In multiaxis machine tools, a single laser beam may be suitably reflected to serve the measurement needs of up to eight axes of motion.[93, 94]

Some recent improvements[95] in the above system include a *wavelength tracker* for better compensation of environmental error sources. The temperature, pressure, and humidity sensors of Fig. 4.47a are all intended to correct for the effect of these variables on wavelength. The tracker eliminates the need for all these sensors by actually measuring the wavelength itself; thus it corrects for *all* error sources which affect wavelength. It also provides index-of-refraction data on the air (or other ambient gas), useful in certain optical systems. The table above compares the achievable accuracies of various compensation schemes. A differential interferometer is combined with an etalon to achieve the wavelength tracking function. In a

[93]A. F. Rudé and M. J. Ward, "Laser Transducer Systems for High-Accuracy Machine Positioning," *Hewlett-Packard J.,* p. 2, February 1976.

[94]W. E. Olson and R. B. Smith, "Electronics for the Laser Transducer," *Hewlett-Packard J.,* p. 7, February 1976.

[95]G. J. Siddall and R. R. Baldwin, "Some Recent Developments in Laser Interferometry," Hewlett-Packard Corp., Palo Alto, CA, 1987.

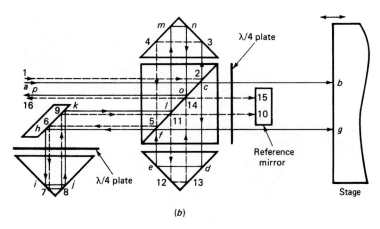

Figure 4.48
Accuracy comparison and differential interferometer.

differential interferometer, the *relative* ("differential") motion of two plane mirrors
is measured as in Fig. 4.48. We again have two laser beams with different frequency
and polarization (as in Fig. 4.47) originating at *a*, 1. The reference beam (dashed
line) follows the path 1, 2, 3, . . . , 16, while the measuring beam follows *a, b,
c, . . . , p*. Both beams first encounter the polarizing beam splitter at 2 where the
reference beam is reflected while the measuring beam passes through, due to their
different polarizations. The measuring beam next passes through the quarter-wave
plate, reflects from the measuring mirror, and again passes through the quarter-wave
plate. The quarter-wave plate is oriented so as to change the polarization of the
measuring beam so that it is now like that of the original reference beam, causing
the measuring beam to be *reflected* at *c,* rather than passing through. Both of the
quarter-wave plates in Fig. 4.48 are used to perform this polarization-adjusting
function, and this should explain the paths actually taken by the two beams on their
way to the exit station 16, *p* where processing to achieve a displacement signal (as
in Fig. 4.47) would take place. Because the difference between the two path lengths
is now the *relative* displacement of reference and measuring mirrors, we have the
differential measurement we want. While this type of interferometer has a number
of applications to basic motion measurement, we now concentrate on its use in the
wavelength tracker.

In the tracker the reference and measuring mirrors are both rigidly fastened to
the ends of a 10-cm-long tubular member, as in Fig. 4.49, so that the measuring path
includes about 40 cm of air *not* in common with the reference path. The length of
the tubular member plus mirrors (called an *etalon*) must be *extremely* stable, but this
is possible with materials such as the glass ceramic Zerodur,[96] giving long-term
stability better than 0.1 ppm/yr. The concept of the wavelength tracker assumes the

---

[96]J. W. Berthold, S. F. Jacobs and M. A. Norton, "Dimensional Stability of Fused Silica, Invar, and Several
Ultralow Thermal Expansion Materials," *Appl. Optics,* vol. 15, no. 8, pp. 1898–1899, August 1976.

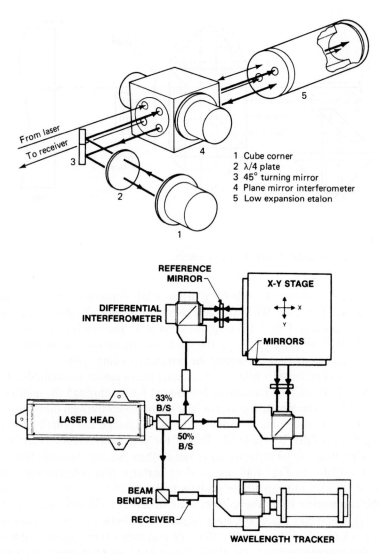

1 Cube corner
2 λ/4 plate
3 45° turning mirror
4 Plane mirror interferometer
5 Low expansion etalon

Figure 4.49
Wavelength tracker and two-axis application.

two mirrors have *zero relative motion;* so if the tracker measuring system *does* indicate a motion, this *must* be due to changes in the refractive index of the air in the tracker's tube, allowing correction for this effect in the motion-measuring axes of the complete system. Figure 4.49 shows how the wavelength tracker can be incorporated into a two-axis motion-measuring scheme by using two differential interferometers.[97]

[97]HP 10717A Wavelength Tracker, 02-5952-7884, Hewlett-Packard Corp., 1987.

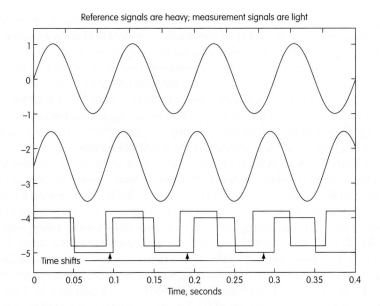

Figure 4.50
Interferometer "phase" (time) shift measurement.

The optical heterodyning concept can be implemented in ways other than that used in the Hewlett-Packard system. One such method uses an *acousto-optical frequency shifter* (Bragg cell).[98] The Bragg cell uses a piezoelectric crystal cemented to a glass block. By exciting the crystal with a (typically) 20-MHz electrical signal, a single laser beam entering the glass will exit as two beams, one of which will have its optical frequency shifted by 20 MHz relative to the other, thus creating the two beams needed for the heterodyne effect. The method for getting the displacement data can also be different from the up/down counter technique; we now discuss a *phase-angle* detection scheme used in a more recent[99] Zygo instrument. The two laser beams (one reflected from the moving target) are combined in the usual way to obtain an electrical beat-frequency signal, which would be at 20 MHz if the target were stationary, but would change its frequency in the usual way when the target moves. When the two optical beams are allowed to interfere, if the target has a fixed velocity, the "phase angle" between the two light waves will change at a fixed rate, proportional to the target velocity. (Actually, the term *phase angle* should strictly be applied only to two waves of the *same* frequency; a more correct terminology when the two waves have slightly *different* frequencies would be *time shift*.) That is, since one wave has a slightly longer period than the other,

[98]A. H. Slocum, "Precision Machine Design," Prentice-Hall, 1992, pp. 176–206; Zygo Corp., Middlefield, CT, 860-347-8506 (www.zygo.com).

[99]F. C. Demarest, "High-Resolution, High-Speed, Low Data Age Uncertainty, Heterodyne Displacement Measuring Interferometer Electronics," *Meas. Sci. Technol.,* vol. 9, 1998, pp. 1024–1030. (available from Zygo Corp.)

each cycle there is a fixed time shift between the zero-crossing points of the two waves. The sine waves are easily electronically converted to square waves, whose zero-crossing points are more easily detected (see Fig. 4.50). Since this time shift is proportional to target velocity, the *accumulated* time shift between two specific times is proportional to the *displacement* of the target over that time interval, which is what we want to measure. Thus the change in "phase angle" between two selected times is a measure of target displacement. This phase angle effect occurs both in the optical signals and in the heterodyned electrical signals, but we can only measure it where the frequencies are low enough, in the electrical signals.

The electrical measurement signal would be at some *fixed* frequency if the target velocity were constant. Suppose the velocity were in the direction that causes an *increase* in the frequency, from 20 Hz to, say, 21 MHz. Then, for each cycle, the rising edge of the measurement signal would occur a little *earlier* relative to a fixed reference wave. By measuring the time of two *consecutive* rising edges and subtracting the second time from the first, we get a measure of the target displacement during that measurement cycle. If we then *accumulate* these displacement increments over any arbitrary time interval, we get the total displacement over that interval. In the referenced instrument, a 1280-MHz clock and an 8-to-1 interpolation scheme give a time resolution of 97.7 ps ($0.7°$ of phase and $\lambda/2048 = 0.31$ nm of position). This displacement measuring scheme works even if the target velocity varies (the usual case) since the incremental displacement measurements are taken at a rate of several million per second.

This instrument was specifically developed for use in machines for step-and-scan photolithography of integrated circuits and other microsystem devices, where as many as eight axes of machine motion must be measured with high speed and accuracy. Since some measurements, such as pitch and yaw angles, require *simultaneous* data, whereas the raw data must be acquired sequentially, it includes means for correcting this time-skew. The use of fiber-optic signal transmission allows the electronics to be remotely located, reducing noise pickup and electrostatic-discharge problems.

The development of self-scanning photodiode arrays[100] has made possible solid-state cameras[101] for applications in pattern recognition, size and position measurement, etc. Figure 4.51a shows a linear array of specific dimensions. However, the manufacturing process allows considerable flexibility in size and geometry; area arrays containing 10,000 elements in a 100×100 matrix with 60-μm center-to-center spacing, for example, allow imaging of two-dimensional objects. On a single silicon chip, such arrays include a row (or rows) of individual light-sensitive photodiodes, each with its own charge-storage capacitor and solid-state multiplex switch. Also contained on the chip is a shift register for serial read-out of the individual element signals. The diodes transduce incident light into

---

[100]EG&G Reticon, CA. www.optimumvision.co.uk/reticon

[101]"The Selection of a Line Scan Camera for Industrial Measurement Applications," Reticon, CA, 1974; "Line-Scan Camera Subsystem Model CCD1300," Brochure 253-11-0007-076/12M, Fairchild Camera and Inst. Corp., Mountain View, CA, 1976; E. Spidell, "Selecting the Optimum Image Sensor: Camera Tubes vs. Solid-State Arrays," *Laser Focus/Electro-Optics,* pp. 158–163, January 1987.

charge, which is stored on the capacitor until readout. The charge is proportional to the product of light intensity and exposure time; however, a saturation level does exist. For an array as in Fig. 4.51a, saturation typically occurs at about $6 \ \mu W \cdot s/cm^2$. Scan frequency (the frequency at which the individual elements are interrogated and read out) can be varied over a wide range (up to about 40 MHz in some models), which allows a trade-off between sensitivity and speed, since slower scan rates allow the same illumination to produce more charge each scan cycle.

Figure 4.51b shows how a linear array can be used to construct a line-scan camera for dimensional measurement. The object to be measured is back-lighted (front lighting is also feasible) so as to produce a light-dark pattern with transitions at the object's edges. Conventional optics focus an image on the photodiode array. Dimensional resolution at the array is, of course, limited by the diode spacing, say 0.001 in. However, if the optics is such that a 5-in object produces a 1-in image, then resolution of the object's dimension is only 0.005 in. If the object is moving (as is often the case in industrial inspection), then additional errors are possible; however, higher scan frequency may alleviate such problems. Figure 4.51b shows

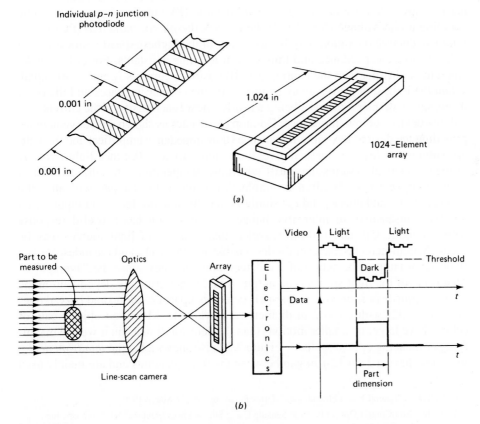

Figure 4.51
Self-scanning diode arrays and cameras. (a) 1024 element array; (b) line-scan camera.

two output signals, "video" and "data." The video signal is a "boxcar" (stepwise-changing) function that shows the time-integrated illumination of each individual picture element (called *pixel*) over one scan cycle. Note that the light (and dark) regions do not give an identical response for each pixel. No illumination field is perfectly uniform; and even if it were, the individual pixel sensitivities cannot be made perfectly equal. To get an unambiguous dimension measurement, the video signal is compared with a judiciously chosen threshold level to produce distinct switching in the "data" signal.

Many useful variations of the above apparatus are possible. Area scanning can be done with an area array or by using an oscillating mirror to deflect the image past a linear array. Wide objects can be measured by using two cameras separated by a known distance. Color applications are possible by using three cameras with proper-color filters. Cameras can be interfaced with microprocessors which have digital representations of objects stored in ROM (read-only memory). The measured objects' "signature" can be compared with a "standard" stored in ROM to check whether they match, thereby identifying the object or perhaps detecting defects in it.

The *lateral-effect photodiode*[102, 103] silicon detector is the basis of a number of useful motion-measuring schemes. In the single-axis device of Fig. 4.52, if terminal G is grounded, we are using the *photovoltaic* (PV) mode of operation, while applying a bias voltage (5 to 15 V) there gives the *photoconductive* (PC) mode. Photoconductive operation may be necessary in the highest-speed applications; it reduces device capacitance and thus system RC time constants. For either mode, when the light spot is at the central ($x = 0$) position, currents i_A and i_C are equal. When the light spot moves away from $x = 0$, one current increases and the other decreases, with their *difference* being a nearly linear function of x if the currents are fed into devices of low impedance. Op-amp electronics as shown are convenient for providing the desired low input impedance (transimpedance amplifier) and the voltage subtraction ($e_A - e_C$) which gives zero output for $x = 0$. Since light *intensity* in many practical systems may drift with time, temperature, etc., the division of $e_A - e_C$ by $e_A + e_C$ is often desirable. Since intensity changes will alter the photocurrents (and thus e_A and e_C) similarly, the division produces an output signal relatively insensitive to intensity changes. Since silicon has a useful response between about 350- and 110-nm wavelengths, a variety of light sources may be used—helium-neon lasers, laser diodes, visible and *IR* LEDs, and incandescent and fluorescent lamps. The action of the detectors is that they sense the "average" or "centroid" of the light spot, so the output signal is not critically dependent on the intensity profile or spot size. Single-axis detectors up to 12 in long and biaxial ones (see Fig. 4.52) with up to 2-in diameter are available. Measured objects can, of course, have larger or smaller motions than the detector dimension if we use magnifying or reducing optical systems. Figure 4.52 also shows a biaxial *quadrant detector.* These have a 2- to 12-μm gap between the four quadrants and are mainly used

[102]B. O. Kelley, "Lateral-Effect Photodiodes," *Laser Focus,* pp. 38–40, March 1976.

[103]W. Light, "Non-Contact Optical Position Sensing Using Silicon Photodetectors," UDT Sensors, Inc., Hawthorne, CA, 310-978-0516 (www.udt.com).

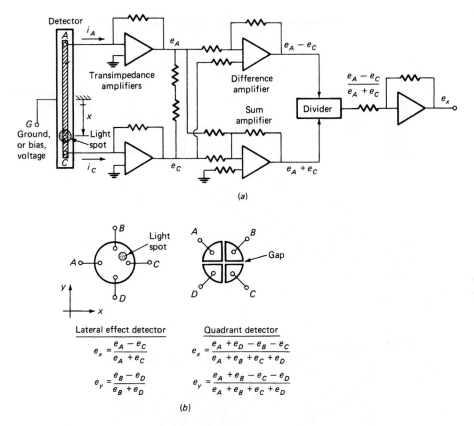

Figure 4.52
Photodiode displacement measurement.

as supersensitive null detectors for small motions near the gap, since they give *no* change in output once the light spot is totally within one quadrant. An auto-collimator (used to measure very small changes in angular position) based on a lateral-effect detector covers a range of 20,600 seconds of arc with 0.2 second of arc resolution, while a similar instrument using a quadrant detector has a range of 300 seconds of arc and 0.01 resolution.[104]

A commercial, ready-to-use motion-measuring system[105] based on a lateral-effect detector uses the "triangulation" principle of Fig. 4.53. Here we are measuring *range* ("z axis") motion rather than *x, y* motion in the target plane. An *IR* laser diode at 850 nm, modulated at 16 kHz, is the light source, projecting a spot onto the surface to be measured, which need not be highly reflective. As the surface moves within the measuring range on either side of the standoff distance, the image of the

[104]Models 1010 and 1020, UDT Sensors, Inc.

[105]Optocator, Selcom (LMI Inc.), Southfield, MI, 248-355-5900 (www.lmint.com).

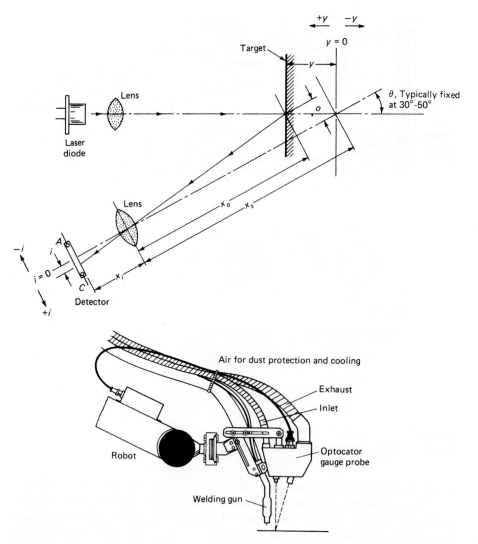

Figure 4.53
Photodiode system analysis.

spot (formed by a suitable lens system) moves laterally over a single-axis detector, producing an output voltage linear with target displacement. The modulated, rather than steady, light source extends life and allows use of high-gain ac-coupled electronics, reducing drift due to background lighting and a temperature-sensitive detector offset voltage. System dynamics are thus limited by the final low-pass filter needed to produce a dc output signal in all such ac systems; however, the high modulation frequency allows quite fast (about 2 kHz) frequency response. Standoff distances of 3.7 to 40 in with corresponding measuring ranges of 0.3 to 20 in are available. Resolution of the 16-bit digitized output is 0.025 percent of the measuring range, while accuracy and linearity are typically ±0.1 percent. Some typical applications in manufacturing include measurement and feedback control of the level of molten iron in the sprue cup of casting molds as well as seam finding in robotic arc welding (Fig. 4.53). By mounting the unit on a motorized slide, it can scan a measured object to develop a profile or contour measurement. By using this scheme, extruded rubber tire treads are continuously monitored for profile at a scan speed of 7 in/s with $\pm$ 0.004-in accuracy.

An approximate geometric analysis of the widely used triangulation method of optical range finding is shown in Fig. 4.53. When the target surface is at the nominal standoff distance ($y = 0$), the detector system's optical axis intersects the axis of the beam projected from the light source, placing the image of the light spot at $i = 0$, giving no output signal. The detector optics are set to be in sharp focus for this position; X_s is the object distance from the lens, and X_i is the image distance in the thin-lens equation[106] of classical optics

$$\frac{1}{f} = \frac{1}{X_i} + \frac{1}{X_s} \tag{4.83}$$

where $f \triangleq$ lens focal length. As the target moves $\pm y$ about the standoff position, the lens is *not* refocused (X_i is fixed). The target light spot is now at X_o and has moved a distance o off the lens axis, causing its image to move a distance i from the detector null position. (Refocusing is not practical for high-speed motions and fortunately is not necessary since the range of y is small compared with that of X_s. This maintains a sufficiently sharp focus as y changes, since any optical system always has some depth of field. Also, as mentioned earlier, lateral-effect photodiodes are especially tolerant of light-spot defocusing.) By assuming X_i fixed at the value given by Eq. (4.83) and $i/o \approx X_i/X_o$ (system nearly in focus), the geometry of Fig. 4.53 leads to

$$i = \frac{fX_s}{X_s - f}\left(\frac{y}{X_s/(\sin\theta - y/\tan\theta)}\right) \tag{4.84}$$

which is a nearly linear relation between i and y since $y/\tan\theta$ is usually small relative to $X_s/\sin\theta$. Of course, any nonlinearity can be taken into account by an overall system static calibration and linearized (if necessary) via common hardware or software schemes.

Various versions of the basic triangulation sensor are available.[107] Different sensor configurations may be needed depending on the target surface's reflectivity. Specular ("mirror like") surfaces reflect the incident laser beam at a specific angle, requiring the detector to be positioned at that angle. Diffuse ("rough") surfaces scatter the light over a wide range of angles, dictating a different sensor design. One

---

[106]D. C. O'Shea, "Elements of Modern Optical Design," p. 23, Wiley, New York, 1985; "Optics Guide 2," Melles Griot Inc., Irvine, CA, p. 8 (www.mellesgriot.com).

[107]W. P. Kennedy, "The Basics of Triangulation Sensors," *Sensors,* May 1998, pp. 76–83; H. Kaplan, "Laser Gauging Enters a Submicron World," *Photonics Spectra,* June 1997, pp. 67–68; D. Morrow and D. Adams, "Laser-Based Angle Measurement for Automobile Power Train Testing," *Sensors,* Nov. 1997, pp. 56–58; W. Chapelle, "Sensing Automobile Occupant Position with Optical Triangulation," *Sensors,* Dec. 1995, pp. 18–22.

maker[108] offers two diffuse models (2.5- and 6.0-in standoff, 5- and 50-μin resolution) and one specular model (1-in standoff and 0.5-μin resolution) with 20 kHz frequency response.

Instead of the lateral-effect photodiode, some[109] sensors use a *pixelized array detector.* Typically, the length of the detector is divided into, say, 256 pixels, each of which produces an output proportional to the light falling on it. The array is electronically scanned to produce a "bar graph" type of output signal, showing the distribution of illumination along the detector's length. This feature has both advantages and disadvantages. The drawbacks include more complex and slower processing and larger physical size. The benefits derive mainly from the ability to perform various useful diagnostic and processing operations on the array data. Direct observation of the bar-graph display allows interpretation of target surface features. Transparent targets give two peaks, one for the front side and one for the back. These would be "smeared" into a single "average" peak position by a lateral-effect photodiode but would be resolved separately by the array detector, allowing several types of useful processing. Since they are charge-coupled devices (CCDs), the array detectors can measure targets with low illumination by "dwelling" longer on each pixel during the scan, allowing more charge to accumulate and thus giving a larger output, though this of course slows the response.

Sensors[110] using "circular triangulation" with a unique detector gather the entire cone of reflected light, giving an averaging effect and insensitivity to surface tilt and "shadowing" of grooves. Another firm[111] that offers conventional triangulation sensors also makes a "confocal displacement meter" operating on a different principle. Here the laser beam is projected onto the target through an objective lens driven by a tuning fork that vibrates along the projection axis. Half-mirrors (beam splitters) direct the reflected beam, which is converged on a pinhole aperture in front of a detector. As the tuning fork vibrates, the beam passes through the aperture only when it is converged precisely at the point of exact focus, where the detector receives peak illumination. A detection trigger, along with a precise tuning fork position sensor, then computes the exact distance to the target surface. Models with standoff distances of 5 to 28 mm, resolution of 0.1 to 0.2 μm, and response time of 2.2 ms are available.

Lateral-effect photodiodes can be used in several ways to measure x, y motions in a plane more or less perpendicular to the instrument's line of sight; Fig. 4.54 shows one commercial version.[112] Here, one or more (up to 120) *IR* LEDs or laser diodes are fastened (wherever direct motion data are desired) to the sensed object and wired to the electronic unit which provides pulsed (up to 10-kHz) power and thus pulsed light output. When several LEDs are used, their flashing sequence can be programmed by the user and is thus known to the processing electronics, so that the correct portion of the detector signal is associated with the proper LED motion.

---

[108]MTI, Latham, NY, 518-785-2464 (www.mechtech.com).

[109]CyberOptics Corp., Minneapolis, MN, 800-746-6315 (www.cyberoptics.com).

[110]Surface Measurements, Inc., (Wolf & Beck, Adelberg, Germany), Burnsville, MN, 612-898-3241.

[111]LT Series, Keyence Corp., Woodcliff Lake, NJ, 201-930-0088 (www.keyence.com).

[112]Selspot II, Selcom (LMI Inc.).

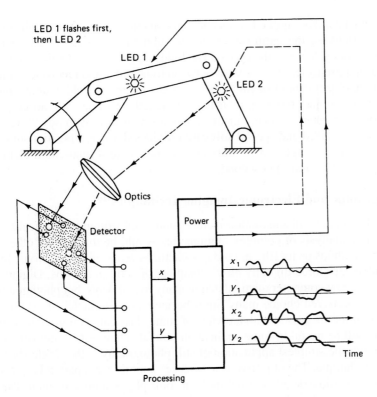

Figure 4.54
Multipoint plane motion measurement.

Thus the *x, y* motion of many individual sensing points can be monitored in a time-sampled fashion. Since the sampling interval is accurately known, velocity and acceleration data are readily calculated from the basic displacement data. For three-dimensional motion, two cameras viewing the object from different positions and angles send data to the digital processing software, which reconstructs the actual motion from the signals and known geometry. Use of infrared light and narrowband optical filters allows the system to function in normal background lighting. Maximum standoff distances ranging from 5 to 200 m are accommodated by selecting LEDs with different output power. Accuracy is about 0.5 percent, and resolution is 0.025 percent of the standoff distance, with the minimum standoff being about 1 ft. The field of view is 22/(lens focal length, mm) times the standoff distance, so the standard 50-mm lens, at a 1-m standoff, could measure motions within a circle of 0.44-m diameter.

Motion and orientation tracking in three dimensions can be accomplished using magnetic field, acoustic, inertial, and optical technologies.[113] A switched-DC magnetic field approach uses a fixed transmitter with *x, y,* and *z* antennas and a sensor (attached to the moving body) with a similar array of antennas (a three-axis fluxgate

---

[113]Ascension Technology Corp., Burlington, VT, 800-321-6596 (www.ascension-tech.com).

magnetometer). The applied magnetic field magnitude is comparable to earth's magnetic field, so the earth-field must be subtracted, which is done during the measurement cycle when the applied DC field is switched *off*. Optical and acoustic methods require a clear line of sight between transmitter and receiver, so the "back side" of an object cannot be measured. The DC magnetic field is unobstructed by the measured object (so long as it is nonmagnetic) so does not suffer from this "occlusion" problem, allowing medical uses inside the human body.[114] The analog signals are digitized and special software provides the x, y, z position and angular orientation (Euler angles, 3×3 orientation matrix, or 4 quaternion elements) at rates up to 120 readings per second.

Photographic and Electronic-Imaging Techniques[115]

The application of still and motion-picture photography often allows qualitative and quantitative analysis of complex motions that would be difficult by other means. Classical methods using film are today sometimes replaced by electronic imaging, with the image stored digitally on videotape or directly in DRAM. Digital images can of course be computer manipulated in many useful ways, but photographic film is still the preferred choice many times because of its unmatched resolution. We begin our brief discussion with a review of some classical photographic techniques that may still be useful in some applications and also have electronic versions.

Perhaps the simplest application of still photography is the single-flash "stop-action" technique. The objective is to "freeze" a motion at a particular phase of its occurrence to allow detailed visual study of some physical phenomenon. The equipment usually employed consists of a still camera, a stroboscopic light source, and some means of triggering a single flash of the strobe light at the desired instant. If the experiment can be performed in a darkened room, the procedure consists of manually opening the camera shutter, allowing the phenomenon to occur, triggering the light at the desired instant, and then manually closing the shutter. If triggering, focus, and exposure are correct, a photo of the phenomenon, "frozen" at the instant of the light flash, is obtained. Such photos can be most helpful in understanding complex physical processes in fluid motion or moving machine parts. Of course, the effective freezing of the motion depends on the flash duration's being sufficiently short compared with the velocity of the motion. Flash durations of the order of 1 to 3 μs are readily available. Thus, for example, a velocity of 300 m/s will cause a "blurring" of 0.03 to 0.09 cm at the object. The actual blurring on the film will be this value times the image/object ratio of the camera setup. If a 1-cm object shows up on the film as 0.1 cm, for example, the above blurring would amount to 0.003 to

[114]J. Harms et al. "Three-dimensional navigated laparoscopic ultrasonography," *Surgical Endoscopy* (2001) 15:1459–1462.

[115]W. G. Hyzer, "Engineering and Scientific High-Speed Photography," Macmillan, New York, 1962; A. A. Blaker, "Handbook for Scientific Photography," 2d ed., Focal Press, 1989; S. F. Ray, "Scientific Photography and Applied Imaging," Focal Press, 1999; S. F. Ray (ed.), "High Speed Photography and Photonics," Focal Press, 1997; S. A. Weiss, "High-Speed Cameras Improve Vehicle Performance and Safety," *Photonics Spectra,* July 1998, pp. 102–109; J. Baran, "Digital Cameras Have an Inspection Edge," *Laser Focus World,* April 1999, pp. 139–142; P. Boroero and R. Rochon, "Match Camera Triggering to Your Application," *Test & Measurement World,* Aug. 1998, pp. 53–58.

0.009 cm on the film, which generally would be considered acceptable. If the experiment cannot be carried out in a darkened room, the opening and closing of the shutter must be synchronized with the flash and the open time of the shutter must be short enough so as not to overexpose the film from the room light.

If a displacement-time record is desired, a multiple-flash still-camera technique may be employed. The setup is essentially the same as above except the strobe light flashes repetitively at a known rate. The result is a multiple-exposure photo showing the moving object in successive positions which are separated by known increments of time. By including a calibrated length scale in the photo (preferably in the plane of the motion) numerical values of displacement at specific time intervals may be measured. Both the single- and multiple-flash methods can be implemented with electronic imaging, using "ordinary" strobe lights or more exotic light sources such as copper vapor lasers.[116] Conventional strobe lights may produce about 15,000 flashes per second with a flash duration of 1 to 20 μs. A high-speed film camera might use 10,000 frames per second with a continuous light source, and an exposure time of 40 μs. In 40 μs, a bullet traveling at 1.6 km/s moves 6.4 cm and a 50-μm diameter spray droplet moving at 20 m/s covers 0.8 mm, causing unacceptable blurring. A copper vapor laser gives a pulsed light source (up to about 32,000 flashes/s) with a duration of about 30 ns, reducing the distance the bullet and droplet have moved to 0.005 cm and 0.6 μm, respectively, giving sharp images. These lasers also have high light output; a 60-W laser can front light a circular area of 1 m diameter.

High-speed motion-picture photography is used to study motions that occur too rapidly for the eye to analyze properly. This is accomplished by taking the pictures at a high camera picture frequency (frames per second) and then projecting the film at a low projector picture frequency. The lowest usable projector frequency is about 16 frames per second since lower frequencies result in flickering because the human eye's persistence of vision is about 0.06 s. The highest usable camera picture frequency depends on the construction of the camera. A "conventional" motion picture camera uses an intermittent action film transport mechanism where each film frame is held *stationary* during exposure. This produces the best resolution and picture steadiness, but high-speed versions are limited to 500 pictures/second (pps) in 16-mm film and 360 pps in 35-mm film. The popular *rotating prism*[117] cameras synchronize a moving image with a continuously moving film and can reach 40,000 pps but with some sacrifice in image quality. Such cameras, using 1000 pps, are widely used for automotive crash testing and are rugged enough to mount on the crashing vehicle. Faster cameras (5000 pps) are used for studies of airbag dynamics. An excellent free CD-ROM showing many application examples (stills and movies) is available.[118]

---

[116]Appl. Note #2, Oxford Lasers, Inc., Acton, MA, 800-222-3632.

[117]S. F. Ray, "High Speed Photography and Photonics," chap. 4, High Speed Cine Systems. This book has an excellent discussion of all types of high-speed imaging. A variety of high-speed cameras, including rotating prism types is available from Photo-Sonics Inc., Burbank, CA, 818-842-2141 (www.photosonics.com). Their publication "Features and Benefits of the Photo-Sonics 16mm-1B Camera System" has much information of general utility.

[118]Oxford Lasers, Inc., Littleton, MA, 800-222-3632 (www.oxfordlasers.com).

The time magnification is defined as the ratio of the camera frequency to the projector frequency; thus if 16 is the projector frequency, magnifications up to about 1200:1 are possible. Aside from picture frequency, the shutter speed of the camera must be sufficiently high to prevent blurring of the individual frames, just as in still photography. For 16-mm film a blur of 0.005 cm is considered acceptable. The shutter-speed requirement can be greatly relaxed by the use of a synchronized short-duration electronic flash as the light source, since the flash duration then controls blur no matter what the shutter speed. Selection of a proper camera picture rate may be judged roughly by the rule that the projection of the complete motion to be visually analyzed should take about 2 to 10 s. Thus a motion occurring in 0.001 s requires a time magnification of 2,000 to 10,000. For vibratory motions the camera picture rate must be several (preferably 5 to 10) times the highest vibration frequency.

For frame rates above about 20,000, camera designs that are simply "speeded-up" versions of normal movie cameras no longer work. Instead, *rotating mirror* or *rotating drum*[119] techniques are used. Rotating mirror cameras, capable of up to 25 million frames per second, relay the light from the target through a spinning mirror and onto a stationary arc of film, producing a sequence of "still" photographs, perhaps 10 or 20 frames. These are not run through a movie projector but rather are simply viewed as a sequence of "frozen" images. This technique is not able to realize the ultimate resolution of the film. If the highest resolution is needed, the rotating drum camera, useful to about 200,000 frames per second, is available. Here the spinning mirror is still used, but the film is held on a rotating drum, again producing a sequence of still photos, as in the rotating mirror camera. As a matter of historical interest, the original ultra-high-speed rotating mirror camera was developed to solve a critical timing problem in the conventional explosive device that triggers the atomic bomb. When this secret information was declassified in 1953, it led to the formation of the Cordin Corp. in 1956.

While electronic imaging has not yet achieved the best resolution obtainable with film, its resolution is adequate for many applications, and it has many other advantages that often make it the method of choice. The basic sensor in most of these systems is an opto-electronic transducer called a *charge-coupled device* (CCD).[120] The CCD has a planar array of pixels that respond individually to incoming light imaged on the sensor area by a lens system. The Kodak Megaplus chip is 9.07×9.16 mm and has 1008×1018 pixels; the smallest resolvable detail is 2 pixels.[121] If the lens magnification is known, the resolution at the target can be computed, assuming the lens does not degrade the resolution. The optical image is transformed into an electronic (voltage) image in a three-step process:

[119]Cordin Corp., Salt Lake City, UT, 801-972-5272 (www.cordin.com).

[120]E. Meisenzahl, "Charge-Coupled Device Image Sensors," *Sensors,* Jan. 1998, pp. 16–27.

[121]"Electronic Imaging Resource Guide," p. 18, Edmund Industrial Optics, Barrington, NJ, 800-363-1992.

1. *Exposure.* Light is converted into electrical charge at each pixel.
2. *Charge transfer.* The packets of charge are moved within the silicon substrate.
3. *Charge-to-voltage conversion and amplification.*

A complete CCD camera adds to the basic sensor the electronics necessary to produce useable images. Several different camera designs are available, depending on the needs of the application.[122] One can assemble a system from components[123] or purchase complete motion analyzer systems.[124] The referenced motion analyzers go up to about 8,000 to 18,000 frames per second. The higher frame rates require a reduction in resolution since the DRAM used to store the images has a limited size. For example, with 1536 Mb of DRAM, 256×256 resolution, and 4500 frames per second, about 23,000 frames can be captured in about 5 s. Since synchronization of the motion analyzer and the dynamic event is never perfect, only a fraction of the total frames will contain the desired images. Once stored, the images can be played back at a variety of speeds, from real-time replay to freeze frame; a few milliseconds can be stretched out to minutes.

CCD cameras are also available for extreme speeds; down to 10-ns intervals (100 million frames per second); image conversion cameras capture information in the picosecond range.[125] The highest speed systems use an image splitter and an array of CCD cameras, suitably multiplexed, as in Fig. 4.55.

Recently, CMOS image sensors have become available to compete with the CCD type, offering higher speed and other advantages.[126] Cameras with 1024×1024 pixels can provide rates up to 1000 pictures/second.

Software is available[127] which will accept inputs from both discrete sensors such as accelerometers and images from cameras, and provide a synchronized display of time-history graphs and photos. For example, we can watch on our computer screen a high-speed movie of an inflating air-bag and simultaneously see a graph of the bag pressure.

Optical holography refers to a wide variety[128] of laser-based optical techniques applicable to vibration measurement, flow visualization, particle diagnostics, etc.

---

[122]G. Fisk, "How to Select a CCD Camera for Your Application," *Laser Focus World,* March 2000, pp. 131–132; S. A. Weiss, "Digital Imaging," *Photonics Spectra,* June 1996, pp. 100–113; D. Zankowsky, "Frame Grabbers Transfer Image Data to Computers," *Laser Focus World,* Aug. 1996, pp. 35–140; C. Williams, "Air-Bag Testing Requires High-Speed Image Capture," *Test and Measurement World,* Feb. 15, 1999, pp. 15–16; R. Chinnock, "Automotive Suppliers Use High-Speed Imaging to Test Restraint Designs," *Vision Systems Design,* Aug. 1997, pp. 24–28; "Hitting the Deck with High-Speed Imaging," *NASA Photonics Tech Briefs,* March 2000, pp. 8a–12a.

[123]"Electronic Imaging Resource Guide," Edmund Industrial Optics.

[124]Roper Scientific MASD (formerly Kodak Motion Analysis Div.), San Diego, CA, 858-535-2909 (www.masdroperscientific.com); Olympus Industrial, Melville, NY, 800-446-5260 (www.olympusipg.com).

[125]Cordin Corp., The Cooke Corp., Tonawanda, NY, 716-833-8274 (www.cookecorp.com).

[126]G. W. Kent, "Video recording captures fast-moving targets," *Laser Focus World,* Oct. 2001, pp. 79–86. Vision Research, Wayne, NJ, 800-737-6588 (www.visiblesolutions.com).

[127]MIDAS System, Xcitex Inc., Cambridge, MA, 617-225-0080 (www.xcitex.com).

[128]R. K. Erf (ed.), "Holographic Nondestructive Testing," Academic, New York, 1974; "Holographic Instrumentation Applications," *NASA SP*-248, 1970.

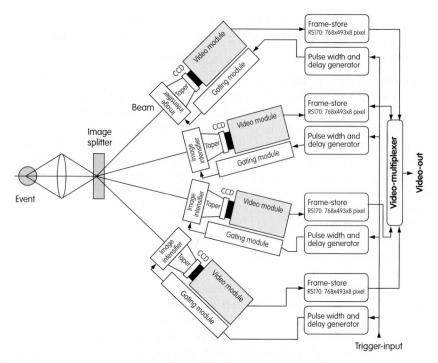

Figure 4.55
Multiple-camera array. *(Courtesy Cooke Corp.)*

Here we briefly describe *holographic interferometry*[129] for vibration measurement, one of the most widely used techniques. Classical (prelaser) interferometry was limited to measurement of small motions of specularly reflecting (polished) flat surfaces, whereas current laser techniques can deal with three-dimensional objects with arbitrarily curved and diffusely reflecting surfaces, in other words, almost any real object.

Figure 4.56 shows the essential features of a simple holographic apparatus. A single laser beam is divided by a beam splitter into two separate beams, which retain a fixed phase relationship with each other over some distance related to the *coherence length* of that particular laser. (Coherence length is the distance over which a laser stays in phase with itself. A long coherence length allows us to measure larger objects. Typically a 20-mW helium-neon laser might have a coherence length of 0.25 m and be capable of measuring objects up to about 0.5 m in all dimensions.) After the laser beam is split, the two beams are each diverged in beam spreaders so that an area (rather than a "point") can be illuminated. (Note that the undiverged beams present *safety hazards* and must be treated with respect.) One

[129]D. J. Monnier, "Practical Applications for Holography in Industry," SESA Fall Meeting, 1973; W. A. Penn, "Holographic Interferometry," Plus Publ. Code No. 119, General Electric Electronics Lab., Syracuse, NY, 1977; D. A. Cain, C. D. Johnson, and G. M. Moyer, "Applications of Optical Holographic Interferometry," Naval Underwater Syst. Cent., Newport, R.I., TM # TD12-134-73, 1973; H. J. Caulfield (ed.), "Handbook of Optical Holography," sec. 10.4, Academic, New York, 1979; R. Jones and C. Wykes, "Holographic and Speckle Interferometry," 2nd ed., Cambridge Univ. Press, 1989.

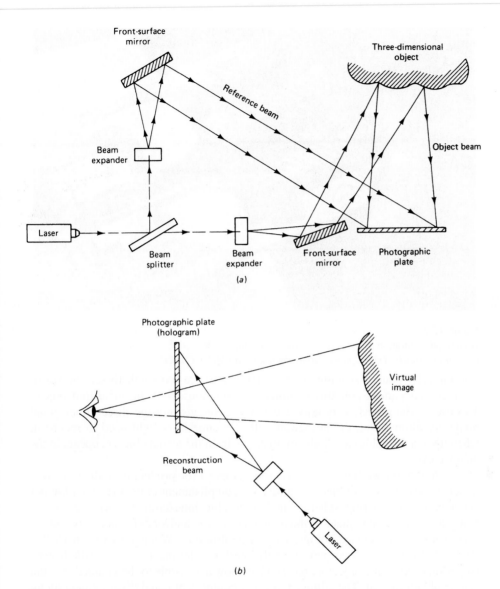

Figure 4.56
Basic holographic apparatus. (*a*) Hologram construction; (*b*) hologram reconstruction.

diverged beam, the reference beam, is directed to illuminate a high-resolution photographic plate, while the other (object) beam illuminates the measured object, which reflects the light to the same photographic plate. If the measured object is stationary, when the object and reference beams recombine at the photographic plate, because of phase shifts resulting from the different path lengths of rays illuminating the object, an intensity pattern called a *hologram* is recorded on the plate, which is then chemically developed as in ordinary photography. The developed plate can be observed with the unaided eye, but bears no resemblance to the object.

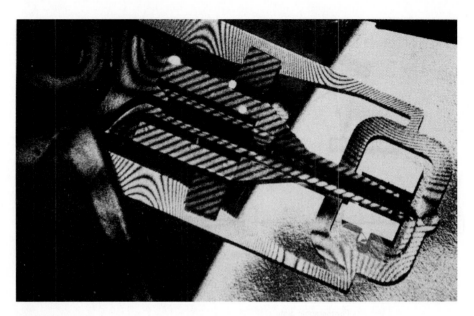

Figure 4.57
Holographic vibration analysis of disk-memory component.
(Courtesy Spectron Development Laboratories, Costa Mesa, Calif.)

The intensity pattern spacings ($\approx 5 \times 10^{-4}$ mm) are too small to see; however, the recorded pattern contains complete information about the measured object. To observe the object, it is now necessary to *reconstruct* the three-dimensional image by illuminating the hologram with the same laser light used to record it, whereupon a virtual image (indistinguishable from the actual object) appears to the unaided eye.

At this point we have a method (holography) for producing realistic (three-dimensional) images of objects—quite an accomplishment in its own right, but not yet a motion-measuring scheme. In holographic interferometry we record *two* holograms on the same plate. If the object has moved and/or deformed between the two exposures, these two holograms will be different. When we reconstruct this double hologram, the two patterns will interfere, causing light and dark fringes (superimposed on the object's image) which now *are* visible to the unaided eye and can be photographed. The fringes represent contours of equal displacement along the viewing axis, with each successive fringe corresponding to $\frac{1}{4}$ to $\frac{1}{2}$ (depending on the geometry of the optical layout) wavelength of light, 6 to 12 μin for a helium-neon laser. To study static and dynamic motions and deformations, various specific techniques[130] (static double exposure, dynamic time average, real time, pulse[131]) have been developed. Figure 4.57 shows typical results obtained in vibration analysis of a machine component.[132]

---

[130]R. K. Erf, op. cit.

[131]R. Levin, "Industrial Holographic Applications," *Electro-Optical Syst. Des.,* p. 31, April 1976.

[132]J. D. Trolinger and R. S. Reynolds, "Stresses during Small Motions," *Ind. Res./Dev.,* pp. 133–136, May 1979.

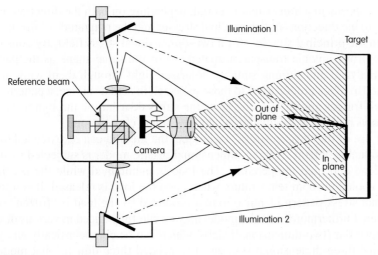

Figure 4.58
3-D motion measurement with speckle interferometry.

As in high-speed photography, electronic imaging can replace the film version of holographic techniques. *Electronic speckle pattern interferometry,*[133] one such method, can produce digital images containing information on surface strain or vibration. Figure 4.58[134] shows the arrangement for getting full-field, three-dimensional measurement of deformations and strains on complex surfaces. Laser light from two different directions illuminates the surface to be measured, and a CCD camera records the speckle images. Since the two beams of laser light change their phase relation linearly with surface deformation, image comparison techniques allow the registration of the three-dimensional motion (two in-plane components and one out-of-plane component) of any point on the measuring area with submicrometer accuracy. The unit measures the surface profile as well as the deformations.

Photoelastic, Brittle-Coating, and Moiré Fringe Stress-Analysis Techniques

Since these methods are really strain- rather than stress-sensitive and since strain is a small displacement, we include a brief treatment in this section on displacement measurement.

Photoelastic methods[135] depend on the property of birefringence under load exhibited by certain natural or synthetic transparent materials. Birefringence (double refraction) under load refers to the phenomenon in which light travels at

[133]J. Tyson II, "Full-Field Vibration and Strain Measurement," *Sensors,* June 1999, pp. 16–22; "Noncontact Full-Field Strain Measurement with 3D ESPI," *Sensors,* May 2000, pp. 62–70.

[134]Trillion Quality Systems, Southeastern, PA, 610-688-0887.

[135]M. M. Frocht, "Photoelasticity," vols. 1 and 2, Wiley, New York, 1941, 1948; J. W. Dally and W. F. Riley, "Experimental Stress Analysis," 3d ed., McGraw-Hill, 1991, Tech. Notes TN-701 to TN-708, Reflections Newsletter, Photolastic Div., Measurements Group, Inc., Raleigh, NC, 919-365-3800 (www.measurementsgroup.com).

different speeds in a transparent material, depending on both the direction of travel relative to the directions of the principal stresses and the magnitude of the difference between the principal stresses, for a two-dimensional stress field. By constructing models (from suitable transparent materials) of the same shape as the part to be stress-analyzed and shining suitably polarized light through them while they are subjected to loads proportional to those expected in actual service, a pattern of light and dark fringes appears which shows the stress distribution throughout the piece and allows numerical calculation of stresses at any chosen point.

By use of the "frozen stress" technique, the method can be extended to three-dimensional problems. A three-dimensional plastic model is subjected to simultaneous load and high temperature. The load is maintained while the specimen is slowly cooled to room temperature whereupon the load is released. It is found that a residual stress pattern identical to that produced by the load is "frozen" into the specimen. Furthermore, now the model may be carefully sliced in various directions to produce flat (two-dimensional) slabs, which can be photoelastically analyzed to determine three-dimensional stresses. The needed three-dimensional models can sometimes be more quickly and economically produced by the *rapid prototyping* method known as stereolithography, rather than conventional casting or machining[136] using new birefringent epoxy resins suitable for this rapid prototyping method. By combining photoelastic and high-speed photographic techniques, the method can be extended to dynamic studies such as the propagation of shock waves through solid bodies.

In *reflective photoelasticity* there is no need to construct a plastic model; the part itself is given a thin "coating" of photoelastic material. Unfortunately, sprayed or brushed coatings, the most convenient application technique, exhibit too much thickness variation and are used only in rough qualitative studies. For accurate work, a thin (0.01 to 0.125 in) sheet of photoelastic plastic is bonded to the part with a reflective adhesive and thus participates in the surface strain when the part is loaded. Polarized light is directed onto the part, reflects from the shiny under-surface, and produces fringe patterns that are analyzed by a reflection polariscope. Flat surfaces are easily fitted with available sheet material. Complex surfaces, however, require casting of a thin flat sheet, careful "form fitting" to the part while the sheet is still "limp," allowing the formed sheet to harden, and then cementing to the part with reflective adhesive.

Photoelastic methods, as compared with bonded resistance strain gages, give an overall picture of the stress distribution in a part. This is very helpful in locating and numerically evaluating stress concentrations and in redesigning the part for optimum material use. Also the method does not disturb the local stress field as a strain gage might.

In the brittle-coating stress-analysis technique,[137] a special lacquerlike material is sprayed on the actual part to be analyzed and the coating allowed to dry.

[136]T. W. Corby, Jr., "Frozen-Stress Photoelastic Analysis of Rapid Prototype Stereolithography Models," Measurements Group Inc.

[137]C. C. Perry and H. R. Lissner, "The Strain Gage Primer," chap. 13, McGraw-Hill, New York, 1955; J. W. Dally and W. F. Riley, op cit.

Application of load causes visible cracking of the brittle coating. The direction of the cracking shows the direction of maximum stress, and the spacing of the cracks indicates magnitude. Under favorable conditions, numerical values of stress can be calculated from measured crack spacing to about 10 percent accuracy. The main features of the method are its simplicity, low cost, and speed in giving an overall picture of stress distribution. Often it is used in conjunction with electric-resistance strain gages, with the brittle coating locating the points of maximum stress and its direction so that strain gages can be applied at the proper places and in the proper orientation for accurate strain measurement.

Moiré fringe methods[138] utilize a fine (200 to 2000 lines/in) square grid of equidistant lines bonded to the surface to be analyzed. When a master (undeformed) grid is superimposed (directly or by projection) on the grid as deformed by loading the part, a fringe pattern appears which can be analyzed to obtain surface displacements or strains.

Displacement-to-Pressure (Nozzle-Flapper) Transducer

The nozzle-flapper-transducer principle is widely used in precision gaging equipment and as a basic component of pneumatic and hydraulic measurement and control apparatus. Figure 4.59 shows the general arrangement. Fluid at a regulated pressure is supplied to a fixed-flow restriction and a variable-flow restriction connected in series. The variable-flow restriction is varied by moving the "flapper" to change the distance x_i. This causes a change in output pressure p_o which, for a limited range of motion, is nearly proportional to x_i and extremely sensitive to it. Thus a pressure-measuring device connected to p_o can be calibrated to read x_i. Ideally (pressure-containing chambers rigid; fluid incompressible) a sudden change in x_i would cause an instantaneous change in p_o. Actually, the dynamics are approximately those of a linear, first-order system for small changes in x_i. The time constant is determined for gases by the compressibility of the gas; for liquids the elastic deformation of the pressure-sensing device often controls.

We analyze the system of Fig. 4.59a for the case of a gaseous medium since a majority of practical applications utilize low-pressure ($p_s \approx 20$ to 30 lb/in^2 gage) air as the working fluid. The principle of conservation of mass is applied to the volume V by stating that during a time interval dt the difference between entering mass and leaving mass must show up as an additional mass storage in V.

It is necessary to obtain expressions for the mass flow rates G_s and G_n. We assume that supply pressure p_s and temperature T_s are constant. Then G_s depends on p_o only; however, the dependence is nonlinear, and so we employ a linearized (perturbation) analysis which will be valid for small changes from an operating point. We can write

$$G_s = G_s(p_o) \approx G_{s,\,0} + \left.\frac{dG_s}{dp_o}\right|_{p_o = p_{o,\,0}} (p_o - p_{o,\,0}) = G_{s,\,0} + K_{sf} p_{o,\,p} \qquad \textbf{(4.85)}$$

---

[138]TN-703, Photolastic Div., Measurements Group Inc.

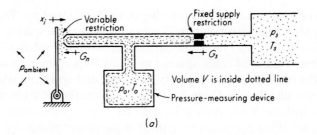

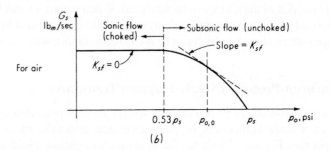

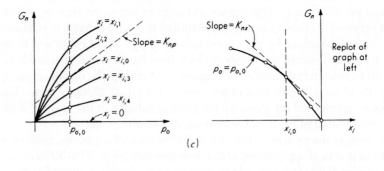

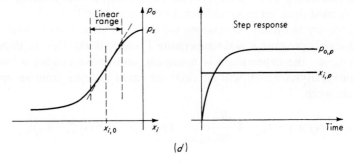

Figure 4.59
Nozzle-flapper transducer.

where $\quad G_{s, 0} \triangleq$ value of G_s at equilibrium operating point $\qquad$ **(4.86)**

$\qquad\quad p_{o, 0} \triangleq$ value of p_o at equilibrium operating point

$\qquad\quad p_{o, p} \triangleq$ small change (perturbation) in p_o from $p_{o, 0}$

$\qquad\quad K_{sf} \triangleq$ value of dG_s/dp_o at $p_{o, 0}$ (a constant) $\qquad$ **(4.87)**

The function $G_s(p_o)$ can be found theoretically from fluid mechanics and thermodynamics (with the help of an experimental orifice-discharge coefficient) or entirely by experiment for a given orifice. Its general shape is given in Fig. 4.59b.

In finding the nozzle mass flow rate G_n, we assume that the process from p_s, T_s to p_o, T_o is a perfect-gas, work-free, adiabatic process. Also, the velocity of the gas at pressure p_o in volume V is assumed zero. Thus the gas at p_o is essentially in a stagnation state, and since the stagnation enthalpy for a perfect-gas, work-free, adiabatic process is constant, the temperature T_o is nearly the same as T_s and remains nearly constant. If T_o may be assumed constant, the nozzle mass flow rate depends on only p_o and x_i. The relationship among G_n, p_o, and x_i can be found from theory (with experimental corrections) or from experiment alone for a specific device. The relationship is again nonlinear, and so a perturbation analysis is in order:

$$G_n(p_o, x_i) \approx G_{n, 0} + \left.\frac{\partial G_n}{\partial p_o}\right|_{\substack{x_{i, 0} \\ p_{o, 0}}} (p_o - p_{o, 0}) + \left.\frac{\partial G_n}{\partial x_i}\right|_{\substack{x_{i, 0} \\ p_{o, 0}}} (x_i - x_{i, 0}) \qquad \textbf{(4.88)}$$

$$G_n \approx G_{n, 0} + K_{np} \, p_{o, p} + K_{nx} \, x_{i, p} \qquad \textbf{(4.89)}$$

Figure 4.59c shows how K_{np} and K_{nx} could be found from experimental data.

The mass storage in volume V can be treated by using the perfect-gas law $p_o V = MRT_o$. We assume V, R, and T_o to be constant. Then

$$p_o = \frac{RT_o}{V} M \qquad \textbf{(4.90)}$$

$$p_{o, 0} + p_{o, p} = \frac{RT_o}{V} (M_0 + M_p) \qquad \textbf{(4.91)}$$

$$\frac{dp_{o, p}}{dt} = \frac{RT_o}{V} \frac{dM_p}{dt} \qquad \textbf{(4.92)}$$

By conservation of mass during a time interval dt,

$$\text{Mass in} - \text{mass out} = \text{additional mass stored}$$

$$(G_{s, 0} + K_{sf} p_{o, p}) \, dt - (G_{n, 0} + K_{np} \, p_{o, p} + K_{nx} \, x_{i, p}) \, dt = dM_p = \frac{V}{RT_o} dp_{o, p} \qquad \textbf{(4.93)}$$

If the operating point $p_{o, 0}$, $x_{i, 0}$ is an equilibrium condition, then $G_{s, 0} = G_{n, 0}$. So we have

$$\frac{V}{RT_o} \frac{dp_{o, p}}{dt} + (K_{np} - K_{sf}) p_{o, p} = (-K_{nx}) x_{i, p} \qquad \textbf{(4.94)}$$

This is clearly a first-order system, and so we define

$$K \triangleq \frac{-K_{nx}}{K_{np} - K_{sf}} \qquad \text{(lbf/in}^2\text{)/in} \tag{4.95}$$

$$\tau \triangleq \frac{V}{RT_o(K_{np} - K_{sf})} \qquad \text{s} \tag{4.96}$$

to give

$$(\tau D + 1)p_{o,p} = Kx_{i,p} \tag{4.97}$$

and

$$\frac{p_{o,p}}{x_{i,p}}(D) = \frac{K}{\tau D + 1} \tag{4.98}$$

To improve speed of response (decrease τ), the volume V should be minimized. Since T_o is usually the ambient temperature, R and T_o are not available for adjustment. An increase in $K_{np} - K_{sf}$ will decrease τ but at the expense of sensitivity, as shown by Eq. (4.95).

A relatively crude device of this type made up for student laboratory use had a nozzle diameter of $\frac{1}{32}$ in and a volume V of the order of 1 in^3. For a supply-orifice diameter of $\frac{1}{32}$ in and a supply pressure of 25 lbf/in^2 gage, this device had a K (at the most sensitive part of its range) of about 2,000 (lbf/in^2)/in and a τ of about 0.12 s. It is quite linear over a range of about ± 0.002 in around $x_{i,0} = 0.004$ in. By changing only the supply-orifice diameter to $\frac{1}{64}$ in, the sensitivity was raised to about 8,000 (lbf/in^2)/in, while τ increased to 0.24 s. The linear range was now about ± 0.0005 in around $x_{i,0} = 0.0015$ in. Since 1 lbf/in^2 = 27.7 in of water, a water manometer used to read p_o gives an easily readable 0.1-in change for an x_i change of only 0.45 $\times$ 10^{-6} in when the sensitivity is 8,000 (lbf/in^2)/in. This illustrates the great sensitivity of this transducer.

A useful approximate expression for the static sensitivity may be easily obtained by assuming incompressible flow. The results are accurate for liquids and a good estimate for gases if pressure changes are not large. The mass flow through the supply orifice is now

$$G_s = \frac{C_d \pi d_s^2}{4} \sqrt{2\rho(p_s - p_o)} \tag{4.99}$$

where
$C_d \triangleq$ discharge coefficient
$d_s \triangleq$ supply-orifice diameter
$\rho \triangleq$ fluid mass density

and we neglect the velocity of approach since the orifice area is very small compared with the upstream passage. The flow area for the nozzle is taken as the peripheral area of a cylinder of height x_i and diameter d_n, the nozzle diameter. This is true only for small values of x_i. The discharge coefficient for this configuration may be different from that for the supply orifice and may vary somewhat with x_i, but here we take it to be the same as C_d in Eq. (4.99). We have then

$$G_n = C_d \pi d_n x_i \sqrt{2\rho(p_o - p_{\text{ambient}})} \tag{4.100}$$

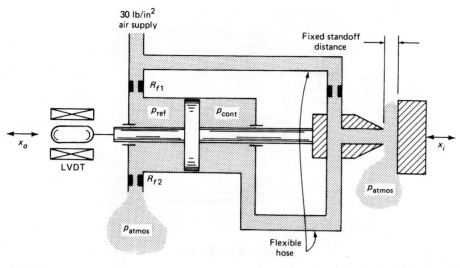

Figure 4.60
Nozzle-flapper/LVDT noncontact gage.

For steady state, $G_n = G_s$ so that (taking $p_{ambient} = 0$ lbf/in.^{2} gage) we get

$$p_o = \frac{p_s}{1 + 16(d_n^2 x_i^2/d_s^4)} \tag{4.101}$$

The sensitivity dp_o/dx_i varies with x_i and is found to have its maximum value at $x_i = 0.14 d_s^2/d_n$. This maximum value is

$$K_{max} = \frac{2.6 d_n p_s}{(d_s)^2} \quad \text{(lbf/in}^2\text{)/in} \tag{4.102}$$

Thus we see that large d_n, large p_s, and small d_s lead to high sensitivity.

An interesting application of the nozzle flapper in noncontact displacement measurement is commercially available[139] and shown in Fig. 4.60. Here the nozzle flapper, whose linear range is severely limited, is used not as the final readout device, but rather as a highly sensitive null detector in a pneumatic servosystem which maintains the measuring probe at a preselected standoff distance from the moving object. Fixed-flow restrictions R_{f1} and R_{f2} are selected so that the constant reference pressure p_{ref} is about 15 lbf/in^2 gage. If the opposing pressure p_{cont} is not equal to this, the piston and attached nozzle will move until the nozzle-flapper output pressure p_{cont} does balance p_{ref}. This will always occur at the same standoff distance, so whenever x_i moves, the servosystem moves x_o an equal amount, maintaining the preselected standoff distance. Any displacement transducer attached to x_o (such as the LVDT shown) thus measures x_i in a noncontacting manner. The

[139]Lucas Control Systems, Hampton, VA, 800-745-8008 (www.schaevitz.com).

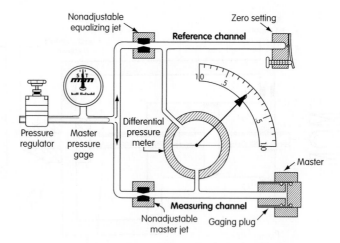

Balanced Air System

With the balanced air system, the air from the supply line passes through a regulator and is divided into two channels. The air in one leg (the reference channel) escapes to atmosphere through the adjustable zero restrictor; the air in the other leg (the measuring channel) escapes to atmosphere through the clearance between the tooling and the workpiece. The two channels are bridged by a precise calibrating meter which responds to any differential in air pressure between them. This bridge system is extremely stable, and is immune to normal air pressure fluctuations. Zero setting, using a nominal size master, equalizes the pressure in both measuring legs. This is independent of the system magnification. Any deviation between workpiece size and setting master size changes the pressure in the measuring leg and produces a corresponding change in the meter reading.

(a)

Figure 4.61
Air-gaging techniques.

standoff distance is typically 0.004 to 0.015, the air force exerted on the measured object is 1g to 10g, and overall repeatability is about 0.0001 in. For time-varying motions, a dynamic analysis[140] of the servosystem allows design for the usual accuracy/stability trade-off typical of feedback-control systems.

Air gaging is widely used in industrial production. Figure 4.61a[141] shows the principle of one system that offers resolution as fine as 5 μin. The Mahr[142] Millipneu system has similar capabilities. An autofocus system for a mask and reticle writing machine used in microdevice manufacture uses an unusual air gage technique as in Fig. 4.61b.[143] It is used in a control system for positioning a lens over delicate wafers with an air gap of 20 to 100 μm, with accuracy of 0.1 μm. Capacitance, laser, and acoustic sensors were considered but discarded in favor of

[140]E. O. Doebelin, "Control System Principles and Design," Wiley, New York, 1985.

[141]Federal Products Co., Providence, RI, 800-343-2050.

[142]Mahr Corp., Cincinnati, OH, 513-489-6116.

[143]Etec Systems Inc., Hayward, CA, 800-388-3832 (www.etec.com).

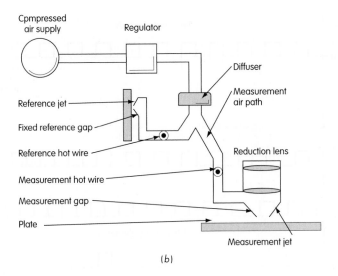

Reference jet

Fixed reference gap

Reference hot wire

Measurement hot wire

Measurement gap

Plate

Cpmpressed air supply

Regulator

Diffuser

Measurement air path

Reduction lens

Measurement jet

(b)

Figure 4.61
(Concluded)

the air gage method.[144] Whereas "classical" air gaging uses pressure sensing, this application uses hot wire anemometers to measure air flow. Detailed analysis, simulation, and experimental testing showed the system to meet all its design criteria, including a bandwidth of about 100 Hz.

Digital Displacement Transducers (Translational and Rotary Encoders)

Since transducers often communicate with digital computers, transducers with digital output would be particularly convenient. However, very few such devices exist, so usually we use an analog transducer to produce a voltage signal and an electronic analog-to-digital converter to realize the desired digital data. Digital transducers called *encoders*[145] do exist, however, for translational and rotary displacement and are in wide use. Three major classes—tachometer, incremental, and absolute—are available in various detail forms; Fig. 4.62 shows their characteristic output signals. A tachometer encoder has only a single output signal which consists of a pulse for each increment of displacement. If motion were always in one direction, a digital counter could accumulate these pulses to determine displacement from a known starting point. However, any reversed motion would produce identical pulses, causing errors. Thus this type of encoder usually is used for speed, rather than displacement, measurement in situations where rotation never reverses.

[144]T. Thomas et al., "Nanometer-Level Autofocus Air Gage," *Precision Engineering,* vol. 22, 1998, pp. 233–242.

[145]J. G. Webster, ed., "The Measurement, Instrumentation, and Sensors Handbook," CRC Press, Boca Raton, 1999, pp. 6-98 to 6-118, 6-126 to 6-127.

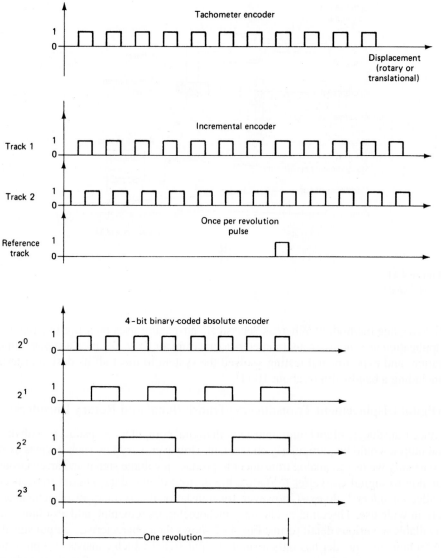

Figure 4.62
Three major classes of encoders.

The incremental encoder[146] solves the reverse-motion problem by employing at least two (and sometimes three) signal-generating elements. By mechanically displacing the two tracks, one of the electric signals is shifted $\frac{1}{4}$ cycle relative to the

---

[146]Techniques for Digitizing Rotary and Linear Motion, Dynamics Research Corp., Wilmington, MA, 800-323-4143 (www.drc.com).

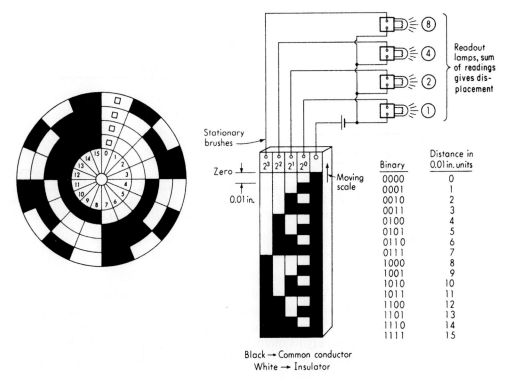

Figure 4.63
Translational and rotary encoders.

other, allowing detection of motion direction by noting which signal rises first. Thus the up-down pulse counter can be signaled to *subtract* pulses whenever motion reverses. A third output, which produces a single pulse per revolution at a distinct point, is sometimes provided as a zero reference. An incremental encoder has the advantage of being able to rotate through as many revolutions as the application requires. However, any false pulses resulting from electric noise will cause errors that persist even when the noise disappears, and loss of system power also causes total loss of position data with no recovery when power is reapplied.

Absolute encoders generally are limited to a single revolution and utilize multiple tracks and outputs, which are read out in parallel to produce a binary representation of the angular position of the shaft. Since there is a one-to-one correspondence between shaft position and binary output, position data are recovered when power is restored after an outage, and transient electric noise causes only transient measurement error.

Encoders of all three types can be constructed as contact devices (stationary brushes rub on rotating code disk, Fig. 4.63) or as noncontacting devices using either magnetic or optical principles. In Fig. 4.63 the readout lamps are shown for explanatory purposes only; the voltages on the four lamp lines could be sent directly

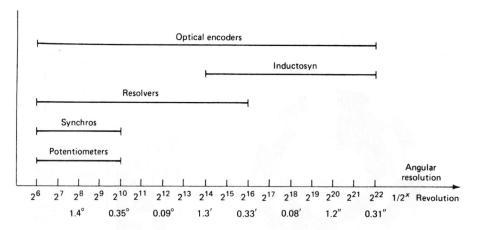

Figure 4.64
Resolution of displacement sensors.

to a computer. If a visual readout *were* desired, these four voltages would be applied to a binary-to-decimal conversion module and then read out decimally on a display, to avoid the need to "mentally sum" lamp readings. For the finest resolution, optical encoders are generally required. Commercial rotary instruments that resolve a fraction of a second of arc are available (see Fig. 4.64); translational encoders may achieve 1-μm resolution. Any transducer with resolution above about 18 bits ($1/2^{18}$) will require extreme care[147] in installation and use. Tachometer and incremental encoders often employ a grating principle in which two glass disks (one fixed, the other rotating), with identical opaque/clear patterns photographically deposited, are mounted side by side with about 0.01-in clearance between. Parallel light is projected through the two disks toward photosensors on the far side. When opaque segments are aligned, a minimum (logical 0) signal is produced while alignment of clear segments gives a maximum (logical 1) signal. Absolute encoders may use the demagnification method in which the light is sharply focused, rather than parallel, and only one disk is employed, with the narrow light beam and photosensor acting in the same fashion as the brushes in a contacting encoder. By use of a semiconductor laser light source, encoders[148] of 81,000 pulses per revolution can be achieved with a 36-mm diameter. External interpolating electronics can increase the resolution to about 10 million pulses per revolution.

Incremental encoders often use some type of *interpolation* scheme to increase resolution beyond the grating spacing. Figure 4.65a[149] shows an example where resolution is increased 20-fold. This company also has an encoder that measures *both* the x and y coordinates of a two-axis stage. Its accuracy over its 2.8-in range is

[147]D. H. Breslow, "Installation and Maintenance of High Resolution Optical Shaft Encoders," Itek Corp., Newton, MA, 1978.

[148]"A Laser Rotary Encoder," *Sensors,* pp. 33–35, June 1988; Laser Rotary Encoder, 7.5K-LRE-ZKG 1087, Canon, Inc., www.usa.canon.com.

[149]Heidenhain Corp., Schaumburg, IL, 847-490-1191.

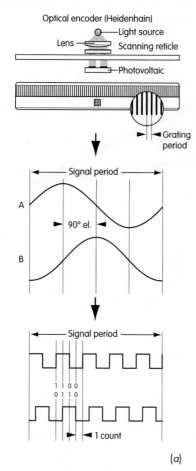

Optical encoder (Heidenhain)
Light source
Lens
Scanning reticle
Photovoltaic
Grating period
Signal period
A
90° el.
B
Signal period
1 1 0 0
0 1 1 0
1 count

(a)

Optical Encoder Operating Principle:

In operation, light is shone through both a graduated reticle and the graduation. Relative motion between the reticle and the graduated scale is detected photoelectrically as a variation in the intensity of light. Thus, sinusoidal output signals are generated by the encoder read head. The diagram shows a transparent grating, however, some of the gratings are reflective, but operate on a similar principle. The output from the encoder is in the form of two sine waves which are electrically 90° out of phase. These signals can then be further reduced through the processes of interpolation and quadrature. The diagram shows a 5x interpolation of the sine waves, which yields, after quadrature (a further subdivision into 4 discrete digital logic states), a final encoder resolution which is $\frac{1}{120}$th of the original grating period.

Figure 4.65
Interpolation increases encoder resolution.

± 2 μm, and it can be repeatably interpolated to 0.01 μm. Figure 4.65*b* shows another company's[150] *xy* encoder that uses interpolation to get 1-nm resolution, 5-nm repeatability, and 200-nm accuracy over its 213 $\times$ 213 mm range. An absolute rotary encoder[151] originally developed for the Smithsonian Astrophysical Observatory's radio telescope has 22-bit resolution and 21-bit accuracy. Each unit is calibrated against a standard with 0.035 arc second resolution, and an accuracy chart provided. Another company[152] uses diode lasers, diffraction gratings, and massive interpolation to get linear resolutions to 0.12 nm and rotary resolution to 2.5 nanoradian, with accuracies up to $\pm$ 0.1 μm and $\pm$ 0.2 arc second.

---

[150]Optra Inc., Topsfield, MA, 978-887-6600 (www.optra.com).

[151]Aerospace Controls Corp., Little Rock, AR, 501-490-2048.

[152]MicroE, Natick, MA, 508-903-5000 (www.micro-e.com).

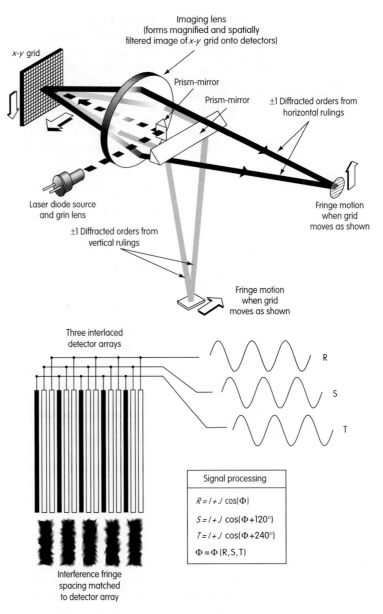

Figure 4.65
(Concluded)

While the code pattern of Fig. 4.63 is most convenient for explaining how motion is represented in the familiar natural binary system, many commercial encoders use different code patterns (such as the Gray code of Fig. 4.66) to avoid errors resulting from small misalignments possible in any real device. For example,

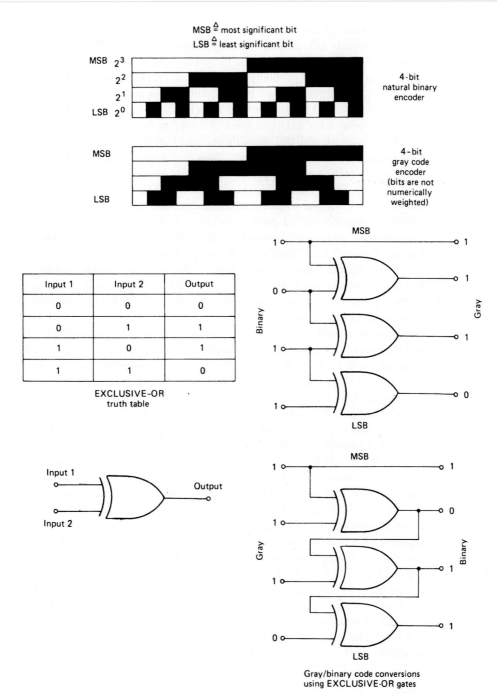

Figure 4.66
Natural and Gray binary codes.

at the midpoint of the natural binary scale, if the shaded area of the 2^2 bit were displaced slightly left, instead of going from 0111 (7) to 1000 (8), the count would go from 0111 (7) to 0011 (3). The Gray code shown in the same figure does not suffer from this type of problem since only one bit changes at each transition. Since the Gray-code output may not be compatible with the readout device, conversion from Gray to natural binary (or vice versa) may be necessary and is easily accomplished by using standard logic gates as shown in Fig. 4.66.

While incremental and absolute encoders are today the major types available for position measurement, a third class has recently appeared[153] that may be preferred for some types of applications. These are called *pseudorandom* encoders, marketed by Gurley as "VIRTUAL ABSOLUTE." Their size and mechanical construction is similar to an incremental encoder, but they have a third track that carries a uniquely coded pattern called a pseudorandom sequence. Whereas a conventional incremental encoder can establish an absolute position only by running the shaft to a "hard stop" whose position is accurately known, or moving the shaft until the "once per rev" pulse is located, the pseudorandom device can locate its absolute position, wherever it might currently be, by making *any* small motion, about 1 or 2 degrees for rotary devices, or 1 mm for translational. Note that we find our position *after* we make the small move, not before. While this mode of operation would not be acceptable for some applications, it is acceptable for others, and the lower price plus other advantages may make the device attractive relative to a true absolute encoder. Note that after a power interruption we *lose* absolute position until we make the required small move.

Figure 4.64 includes data on the Inductosyn, a high-resolution incremental encoder based on the electromagnetic coupling between a fixed scale provided with an ac-excited serpentine conductor (produced by printed-circuit techniques) and a similar but smaller sensing winding which travels over the scale (Fig. 4.67). When alignment is as in Fig. 4.67b, output is at a positive maximum. A displacement of $s/2$ results in minimum output, s gives negative maximum, $3s/2$ gives minimum again, and $2s$ returns the output to positive maximum. The output variation over the $2s$ cycle length is essentially cosinusoidal. A "coarse" digital output is obtained by counting the cycles of spacing $2s$, while fine resolution is obtained by electronically digitizing the analog voltage variation within each cycle. As in other incremental encoders, to detect *direction* of motion, the sensor element includes a second winding displaced $s/2$ from the first, providing a sinusoidal signal. With both a sine and cosine output available, the device behaves essentially as a resolver and can use resolver-to-digital type of electronics[154] similar to those discussed earlier under synchros. Inductosyns are available in both translational (best accuracy ± 0.001 mm) and rotary (best accuracy ± 0.5 second of arc) forms. The standard spacings s available are 0.1 in, 0.05 in, and 2 mm.

[153]J. Gordon, "Encoder Innovations," *Motion Control,* Sept. 2000, pp. 45–46; Gurley Precision Instruments, Troy, NY, 800-759-1844 (www.virtualabsolute.com).

[154]Inductosyn to Digital Converter, Analog Devices, Norwood, MA.

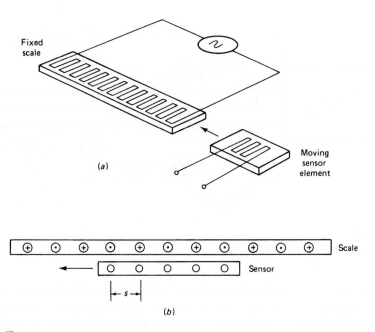

Figure 4.67
Inductosyn transducer.
(Courtesy Farrand Controls, Valhalla, NY, 914-761-2600, www.ruhle.com)

Ultrasonic Transducers

We show some examples of displacement measurement utilizing wave-propagation principles and producing digital output. Such sensors often are called *ultrasonic* since the signals employed are generally outside the frequency range of human hearing. Our first example,[155] intended mainly for full-scale ranges of 1 to 10 ft or more, utilizes a permanent magnet which moves relative to a magnetostrictive wire enclosed in a nonferrous protective tube (see Fig. 4.68). Electronic circuitry drives a current pulse through the wire. At the magnet location, magnetostrictive action generates in the wire a stress pulse which propagates to the receiver location at a fixed speed of about 110,000 in/s. At the receiver location, a pickup coil senses the arrival of the pulse. The time interval between the initiating current pulse and the arrival of the sensed stress pulse is clearly proportional to the displacement x_i, with a sensitivity of about 9 μs/in. By using the two pulses to gate on and off a counter, a digital output reading with resolution of 16 bits can be obtained. To get a continuous analog signal, the pulse is applied repetitively and the transit time interval is used to modulate the width of a rectangular pulse train. Low-pass filtering of this pulse train produces an analog voltage whose amplitude is proportional to the pulse

---

[155]MTS Sensors Div., Cary, NC, 800-633-7609 (www.mts.com); J. Russell, "New Developments in Magnetostrictive Position Sensors," *Sensors,* June 1997, pp. 44–46.

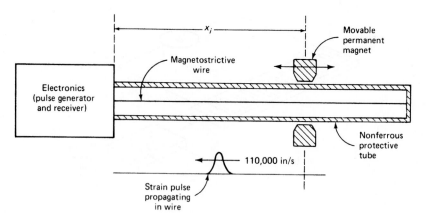

Figure 4.68
Ultrasonic displacement transducer.

width and thus to x_i. The repetition rate is sufficiently high that the low-pass filter can be designed to be flat (-3 dB) to about 200 Hz for a 2-ft-stroke unit. Nonlinearity better than 0.02 percent of full scale is obtained. This transducer is becoming quite popular as the feedback sensor in hydraulically actuated servomechanisms for position control.[156, 157] Here its basic position signal is augmented by deriving digital velocity ($V \approx \Delta X / \Delta t$) and acceleration ($A \approx \Delta V / \Delta t$) signals. These signals allow implementation of derivative control modes on the system controlled variable (load position) which lead to significant improvements in control system performance.

Our next example is based on the automatic focusing system of a popular camera. The motor drive which focuses the lens gets its information from an ultrasonic rangefinder (distance measuring) system. The manufacturer has made available an experimenter's kit and manual[158] which allow study of ultrasonic distance measurement; our description is based on this manual. The heart of the system is an electrostatic transducer (similar in principle to the capacitor microphone of Chap. 6) which is, however, alternately used as a loudspeaker and a microphone. The principle of all such systems, of course, is that of classical sonar where an acoustic signal is sent out and its reflected return timed to allow calculation of the target's distance by use of the known propagation velocity (1100 ft/s for standard air). Whereas simpler ultrasonic ranging systems (such as those used for tank liquid-level measurement) might use a single-frequency signal, the camera rangefinder must deal with targets of variable and unpredictable shape and size. This led to use of a

[156]G. J. Blickley, "Servo Valve Becomes Digital Actuator," *Control Eng.,* June 1986.

[157]B. Benson and D. Deeny, "A Method of Tuning Temposonic Feedback Servo Cylinders in a Digital Control Loop," MTS Systems Corp., Eden Prairie, MN, 1987.

[158]Polaroid Ultrasonic Ranging Experimenter's Kit, Polaroid Corp., Wayland, MA, 781-386-3962 (www.polaroid.com).

multifrequency signal to ensure that a reflected echo would occur reliably. This signal consists of 8 cycles of 60 kHz, 8 cycles of 57 kHz, 16 cycles of 53 kHz, and 24 cycles of 50 kHz for a total of 56 cycles and about 1 ms. By using the transducer as a loudspeaker, this electric signal is applied, causing a similar acoustic signal to propagate toward the target. Before the echo returns, the transducer connections are electronically switched, so that it acts as a microphone to transduce the reflected pressure wave into an electric signal whose time of arrival can be detected and the target distance thereby calculated. The system is designed to measure distances over the range 0.9 to 35 ft with a resolution of 0.1 ft. Since the variation in echo signal strength for targets from 3 to 35 ft is almost a million to one, the microphone's tuned amplifier self-adjusts its gain (16 settings) and bandwidth (8 settings) to maintain signal levels more nearly constant in subsequent circuits. In the kit mentioned above, distance measurements are made and read out digitally on an LED display 5 times per second; however, circuit adjustments can speed this up to about 20 measurements per second.

Sensors using the same basic principle as the Polaroid device just discussed but designed for shorter distances and higher resolution are available.[159] Typical sensors have a range from 2.5 to 40 in, and measure distance with an accuracy of about 0.1 percent of the range (about 0.001 in at 2.5 in range).

4.4 RELATIVE VELOCITY: TRANSLATIONAL AND ROTATIONAL

We consider here devices for measuring the velocity of translation, along a line, of one point relative to another and the plane rotational velocity about a single axis of one line relative to another.

Calibration

The measurement of rotational (angular) velocity is probably more common than that of translational velocity. Since translation generally can be obtained from rotation by suitable gearing or mechanisms, we consider mainly the calibration of rotational devices. For angular and linear velocities, perhaps the most convenient calibration scheme uses a combination of a toothed wheel, a simple magnetic proximity pickup and an electronic EPUT (events per unit time) meter (see Fig. 4.69). The angular rotation is provided by some adjustable-speed drive of adequate stability. The toothed iron wheel passing under the proximity pickup produces an electric pulse each time one tooth passes. These pulses are fed to the EPUT meter, which

---

[159]Cleveland Motion Controls, Cleveland, OH, 800-831-8072 (www.cmccontrols.com); Ultrasonic Arrays, Woodinville, WA, 425-481-6611 (www.ultrasonicarrays.com); Massa Products Corp., Hingham, MA, 781-749-4800 (www.massa.com); D. P. Massa, "Choosing an Ultrasonic Sensor for Proximity or Distance Measurement," *Sensors,* Feb. 1999, pp. 35–37; March 1999, pp. 28–42; J. Billings, "Ultrasonics on the Automotive Assembly Line," *Sensors,* March 2000, pp. 52–54; A. Cittadine, "MEMS Reshapes Ultrasonic Sensing," *Sensors,* Feb. 2000, pp. 17–27; D. Campbell, "Ultrasonic Noncontact Dimensional Measurement," *Sensors,* July 1986, pp. 37–41.

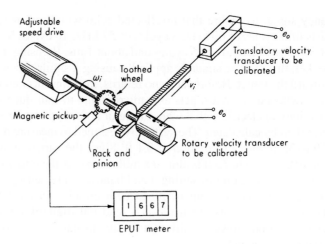

Figure 4.69
Velocity-calibration setup.

counts them over an accurate period (say 1.00000 s), displays the result visually for a few seconds to enable reading, and then repeats the process. The stability of the rotational drive is easily checked by observing the variation of the EPUT meter readings from one sample to another. The inaccuracy of pulse counting is ± 1 pulse plus the error in the counter time base, which is of the order of 1 ppm. The overall accuracy achieved depends on the stability of the motion source, the angular velocity being measured, and the number of teeth on the wheel. If the motion source were *absolutely* stable (no change in velocity whatever), very accurate measurement could be achieved simply by counting pulses over a long time, since then the average velocity and the instantaneous velocity would be identical. If the motion source has some drift, however, the time sample must be fairly short. For example, a shaft rotating at 1,000 r/min with a 100-tooth wheel produces 1,667 pulses in a 1-s sample period. The inaccuracy here would be 1 part in 1,667 (the 1-ppm time-base error is totally negligible), or 0.06 percent. If the shaft rotated at 10 r/min, the error would be 6 percent. Slow rotations can be measured accurately by such means if the toothed wheel is placed on a shaft which is sufficiently geared up from the shaft driving the transducer being calibrated. The toothed wheel and magnetic pickup can of course be replaced by a tachometer encoder, and this may be necessary if very slow rotation is to be calibrated because we now need a large number of pulses per revolution in order to get accuracy. Tachometer encoders with several thousand pulses per revolution (much more if interpolation is used) allow accurate measurement when we cannot get enough resolution from a toothed wheel (gear).

The above procedure uses relatively simple equipment and generally provides entirely adequate accuracy. Other simpler and less accurate procedures can be employed if they are adequate for their intended purpose. These usually consist of simply comparing the reading of a velocity transducer known to be accurate with the reading of the transducer to be calibrated when both are experiencing the same velocity input.

Velocity by Electrical Differentiation of Displacement Voltage Signals

The output of any displacement transducer may be applied to the input of a suitable differentiating circuit to obtain a voltage proportional to velocity (see Chap. 10 for differentiating circuits). The main problem is that differentiation accentuates any low-amplitude, high-frequency noise present in the displacement signal. Thus a deposited-film potentiometer would be preferable to the wirewound type, and de-modulated and filtered signals from ac transducers may cause trouble because of the remaining ripple at carrier frequency. Workable systems using electrical differenti-ation are possible, however, with adequate attention to details.

Average Velocity from Measured Δx and Δt

Often a value of average velocity over a short distance or time interval is adequate, and a continuous velocity/time record is not required. A useful basic method is somehow (optically, magnetically, etc.) to generate a pulse when the moving object passes two locations whose spacing is accurately known. If the velocity were constant, any spacing could be used, with large spacing, of course, leading to greater accuracy. If the velocity is varying, the spacing Δx should be small enough that the average velocity over Δx is not very different from the velocity at either end of Δx. The same technique is applicable to rotational motion.

Figure 4.70a shows the application of a variable-reluctance proximity pickup such as was used in Fig. 4.69. When magnetic material passes close in front of the face of the pickup, the reluctance of the magnetic path changes with time, generat-ing a voltage in the coil. These pickups are simple and cheap and give a large output voltage (often several volts) under typical operating conditions. The output voltage increases with velocity and closeness of the external moving iron to the pickup. Display of the two pulses on a single sweep of an oscilloscope with a calibrated time base allows measurement of the average velocity. Greater accuracy may be achieved by applying the voltage pulses to an electronic time-interval meter. In some velocity-measurement or proximity-detecting applications, pickups of this type are unsatisfactory because signal size decreases with velocity and is zero for a stationary object. We might then use proximity pickups of the eddy-current or Hall-effect type. Eddy-current proximity pickups are similar in principle to the eddy-current displacement transducers discussed earlier, except that a switching (rather than a linearly proportional) output is desired. When a conductive metal object approaches, the pickup's oscillator experiences a reduction in amplitude sufficient to trigger a switching signal.[160] Hall-effect devices utilize the Hall principle of Fig. 4.71a, where a suitable material (originally gold, but today a semiconductor) supplied with constant current produces an output voltage e_o when a transverse magnetic field is applied. Proximity pickups based on this principle may utilize a permanent magnet[161] attached to the moving object or else require a ferrous target

---

[160]Mini-Prox II, Electro Corp., Sarasota, FL, 941-355-8411 (www.electrocorp.com).

[161]"Handbook for Applying Solid State Hall Effect Sensors," Honeywell Micro Switch Division, Freeport, IL, 1976.

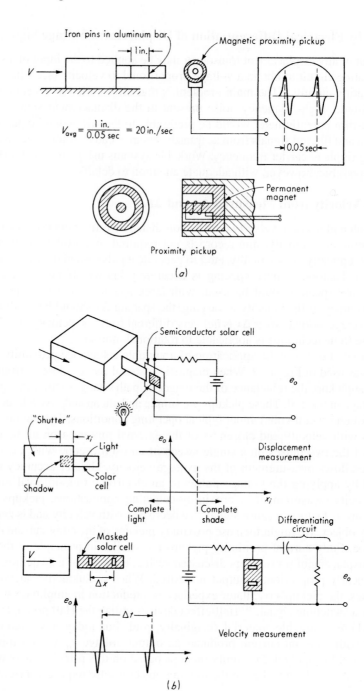

Figure 4.70
Velocity measurement as $\Delta x/\Delta t$.

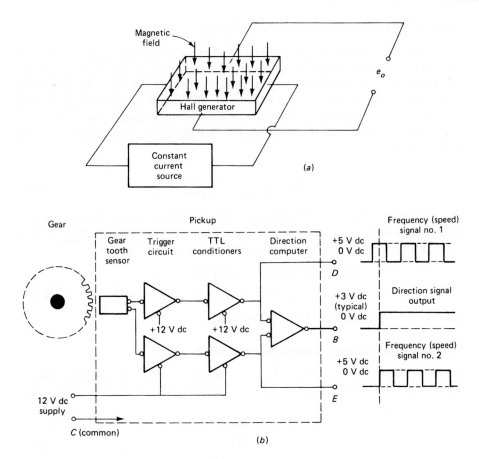

Figure 4.71
Hall-effect proximity pickup.

whose approach changes the reluctance of an internal magnetic circuit whose flux the Hall sensor feels.[162] In either case, the Hall element output voltage triggers a standard TTL logic signal as output. Very low velocities can be measured by electronically timing the interval between passage of each pair of adjacent gear teeth, holding the previous velocity value until updated by the next tooth passage. Figure 4.71b shows a version using two mechanically offset Hall sensors which provide also a direction-of-rotation signal.

Electro-optical techniques also can be employed to generate the pulses necessary in such measurement schemes; Fig. 4.70b shows one version that uses solar-cell transducers. Photography of the motion, by using a stroboscopic lamp flashing at a known rate, also provides velocity data of this type.

---

[162]Allegro Microsystems, Worcester, MA, 508-835-5000 (www.allegromicro.com).

Mechanical Flyball Angular-Velocity Sensor

A classical rotary speed-measuring device still in use today, especially as a measuring element of industrial speed-control systems for engines, turbines, etc., is the flyball. Figure 4.72 shows the general arrangement schematically. Since the centrifugal force varies as the square of input velocity ω_i, the output x_o will not vary linearly with speed if an ordinary linear spring is used. For *small* changes in ω_i, a linearized model may be used to show that the transfer function between ω_i and x_o is essentially of the form

$$\frac{x_o}{\omega_i}(D) = \frac{K}{D^2/\omega_n^2 + 2\zeta D/\omega_n + 1} \tag{4.103}$$

The nonlinear static relation between ω_i and x_o for large speed changes may be acceptable in some systems. Where it is not, a nonlinear spring with $F_{\text{spring}} = K_s x_o^2$ can be used to get a linear overall characteristic since, at balance,

$$\text{Centrifugal force} = \text{spring force}$$

$$F_c = K_c \omega_i^2 = F_{\text{spring}} = K_s X_o^2 \tag{4.104}$$

and thus $x_o = \sqrt{K_c/K_s}\,\omega_i$, a linear relationship.

Mechanical Revolution Counters and Timers

When continuous reading and an electric output signal are not required, a variety of mechanical revolution counters (with or without built-in timers) are available. They

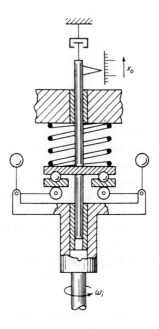

Figure 4.72
Flyball velocity pickup.

are generally supplied with a variety of rubber-tipped wheels which transmit by friction the motion to be measured to the counter input shaft.

Tachometer Encoder Methods

The wide availability, good performance, and modest cost of tachometer encoders makes them popular choices for many speed measurements. They may be used in a number of ways:

1. Send the pulse train to a frequency counter
2. Send the pulse train to data acquisition software in a computer, set up as a "virtual" counter
3. Send the pulse train to data acquisition software in a computer, set up for FFT analysis
4. Send the pulse train to an analog *frequency-to-voltage* (f/v) converter
5. Send the pulse train to a "Δt" type of f/v converter

Method 1 has already been discussed under Calibration at the beginning of this section. Method 2 simply implements method 1 using software. The DASYLAB software discussed in Chap. 13 is easily set up for this task by connecting a few icons on a worksheet screen.

Method 3 can be used even if the input signal is not the usual clean pulse train, but rather a complex vibration or sound signal coming from a rotating machine. The basic idea is that the frequency spectrum of the signal will have a noticeable *peak* at a frequency corresponding to the rotational speed. An FFT spectrum analysis can detect this peak and read out its frequency. This technique is available in stand-alone instruments[163] and can be set up in software, such as DASYLAB. For pulse trains, such as from a tachometer encoder, we can discuss here some of the details. Suppose we want to measure a time-varying speed near 1.0 rev/s with an accuracy of about 1 percent and need to get 20 speed readings over a 20-s time span. For any FFT analysis, the frequency resolution is given by the reciprocal of the sample length. If our peak frequency were 1.0 Hz, for the 1 percent accuracy, we would need a resolution of 0.01 Hz, which requires a sample length of 100 s, much too long. Since we need a reading every second, we need to use an encoder with enough pulses/revolution to raise the frequency. With, say, 1000 pulses/revolution, we get a frequency of 1000 Hz at 1 rev/s. Our 1 percent resolution now requires a sample length of 0.1 s, so we could get as many as 200 speed readings in 20 s. To conservatively avoid aliasing, we might sample at 10,000 samples/s, giving 1000 (we would actually use 1024, a common sample size for FFTs) readings in 0.1 s.

Method 4 uses available f/v converters (see Chap. 10) to produce an output DC voltage proportional to input frequency. Such devices use a final low-pass filter that limits the speed of response to changing frequency. To allow this filter to be fast enough, the input frequency must be high enough, so we again need an encoder with sufficiently high pulses/revolution.

---

[163]Model FT-500, Ono Sokki, Inc., Addison, IL, 630-627-9700 (www.onosokki.com).

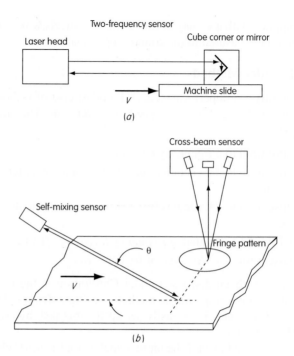

Figure 4.73
Laser velocity sensors.

Method 5[164] is basically different from all the others and overcomes the need to accumulate "long" samples in order to get reliable speed information. In this instrument, instead of accumulating a large number of pulses and dividing this by the sample time to get frequency, we digitally measure (with a high-frequency crystal clock) the *time duration* of each and every pulse and derive the "instantaneous" frequency by taking the reciprocal of this time. Thus, for a 1000 pulse/revolution encoder rotating at 1 rev/s, we would get a frequency reading about every 0.001 s. The referenced unit covers the range from 1 Hz to 20 kHz in 4 subranges, 1 Hz to 1 kHz, 5 Hz to 5 kHz, 10 Hz to 10 kHz, and 20 Hz to 20 kHz, having a resolution of 1/1023 in each range. Due to needed processing time, the response is a little slower than one input cycle. On the 1 Hz to 1 kHz range, for example, the extra delay is 0.375 ms, so a 1000 Hz signal is read out every 0.001375 s (1.375 ms).

Laser-Based Methods

The laser interferometers discussed in Fig. 4.47 are basically velocity-sensing devices, so this is one possibility (Fig. 4.73a). However, this requires attaching a mirror or cube corner to the moving object, so that configuration is practical mainly for machine or instrument slides or stages. A second approach (Fig. 4.73b) uses the same method as the laser-doppler anemometers for measuring the velocity of

---

[164]Model FV-801, Ono Sokki, Inc., www.onosokki.net.

flowing fluids discussed in Chap. 7 ("particles" moving through light and dark fringes formed by two intersecting laser beams). There the velocity of the particles entrained in the fluid is measured, but the technique works well also with solid objects and is used, for example, to measure the speed of material being processed as moving webs or "wires" (textiles, metal rolling processes, paper, plastic films, glass, etc.). The surface roughness of most such materials causes doppler-shifted light to be "scattered" in all directions, and a sufficient amount returns in the direction of the source ("back-scattering") to allow a measurement. A typical unit[165] has a standoff distance of 12 in with a 2-in depth of field and measures velocities from 0.01 to 120 m/s with an accuracy of about 0.2 percent of reading. The update rate is as fast as 32 ms. Angular alignment of the laser beams with the measured surface is not critical but the standoff distance must be maintained within the stated depth of field to ensure that the fringe pattern lies on the measured surface. Integration of the velocity signal can be used to measure the *length* of material that has passed the measuring station between two selected time instants, which is useful, for example, in controlling a cutoff machine.

A third type of laser sensor is the *self-mixing laser interferometer*[166] (Fig. 4.73b). Here, the angle θ is critical and must be maintained close to a set value, usually about 20°. Then, an error of 1° causes a speed error of only 0.6 percent. The standoff distance (slant range) is less critical than in the crossed-beam sensors. There is only one laser beam (a diode laser is used), and we again depend on back-scattered, doppler-shifted light from the target surface to return to the laser head. Here the laser *power* (light intensity) is modulated at the doppler frequency, and the photodiode that is part of most diode laser packages is used to produce an output signal whose frequency f_d is proportional to target speed v, according to $v = (\lambda f_d)/(2 \cos \theta)$, where λ is the laser wavelength. A unit with a range of 0.15 to 15 m/s has a speed resolution of better than 1 percent and accuracy of 0.1 to 1 percent.

Radar (Microwave) Speed Sensors

The doppler techniques used with lasers at light-wave frequencies (about 5×10^{14} Hz) can also be employed at the lower frequencies called microwave or radar[167] (typically 3.55×10^{10} Hz). The signal reflected from a target (moving with a velocity component V along the sensor beam direction) is mixed with the transmitted signal in a diode, producing an output signal with frequency F_d given by $F_d = 2VF_t/(3 \times 10^8)$, where F_t is the transmitted microwave frequency. For $F_t = 35.5$ GHz, this gives a calibration factor of 2377 Hz/(m/s) or 105.8 Hz/MPH. Vehicle ground-speed measurement (classically done with a "fifth-wheel" sensor) is

---

[165]TSI Inc., St. Paul, MN, 800-436-3877 (www.tsi.com).

[166]P. J. M. Suni and C. J. Grund, "Noncontact Laser Speed Measurements," *Measurements and Control,* Sept. 1999, pp. 112–120, LightWorks, LLC, Berthoud, CO, 970-532-5780 (www.lightworksllc.com).

[167]H. S. Williams, "Microwave Motion Sensors for Off-Road Vehicle Velocity Data and Collision Avoidance," *Sensors,* Dec. 1999, pp. 34–41; R. Badmann, "Measuring Vehicle Ground Speed with a Radar Sensor," *Sensors,* Dec. 1996, pp. 30–31; F. Ervin III, "Improvements in Vehicle-Testing Technology," *Sensors,* Dec. 1999, pp. 50–54; S. Soloman, "Sensors Handbook," McGraw-Hill, NY, 1999, pp. 1.53–1.63.

a common application; here the microwave sensor is attached to the car, pointing forward with the beam at about 30° with respect to the ground. Modern diesel-electric locomotives use wheel-slip control systems to optimize traction. This requires measurement of both wheel rotational speed and vehicle ground speed. Mobile railcar movers,[168] farm tractors, and construction machinery also use such systems to optimize traction and braking. For vehicles that can have significant pitch angular motion, two sensors (one pointing forward and one backward) are used to remove the error caused by pitch angle. Since the wavelength of microwaves is much larger than light wavelengths (about 10 mm versus 0.1 μm), microwave sensors require larger targets than optical sensors. A typical transmitter/receiver unit[169] (a 2-in diameter cylinder about 4 in long) can track a car at 1000 ft, a person at 100 ft, and a 1-in diameter aluminum sphere at 10 ft. Accuracy depends on alignment, vibration, clutter, etc., but it can approach 0.5 percent.

Stroboscopic Methods

Rotational velocity may be conveniently measured by using electronic stroboscopic lamps which flash at a known and adjustable rate. The light is directed onto the rotating member which itself usually has spokes, gear teeth, or some other feature enabling "lock on." If not, a simple black-and-white paper target can be attached. The frequency of lamp flashing is adjusted until the "target" appears motionless. At this setting, the lamp frequency and motion frequency are identical, and the numerical value can be read from the lamp's calibrated dial to an inaccuracy of about ± 1 percent of the reading (0.01 percent in some units with crystal-controlled time base). The range of lamp frequency of a typical unit[170] is 110 to 25,000 flashes per minute. Speeds greater than 25,000 r/min can be measured by the following technique. Synchronism can be achieved at any flashing rate r that is an integral submultiple of the speed to be measured, n. The flashing rate is adjusted until synchronism is achieved at the largest possible flashing rate, say r_1. Then the flashing rate is slowly decreased until synchronism is again achieved at a rate r_2. The unknown speed n is then given by

$$n = \frac{r_1 r_2}{r_1 - r_2}$$ **(4.105)**

For very high speeds, r_1 and r_2 are close together, giving poor accuracy. Accuracy can be improved by reducing the flashing rate below r_2 until synchronism is again achieved. This procedure can be continued until synchronism is obtained N times $(r_1, r_2, r_3, \ldots, r_N)$. The speed n is given by

$$n = \frac{r_1 r_N (N - 1)}{r_1 - r_N}$$ **(4.106)**

This procedure can extend the upper range to about 250,000 r/min.

---

[168]Trackmobile Inc., LaGrange, GA, 706-884-6651 (www.trackmobile.com).

[169]Model DRS1000, application notes 1000 and 1001, GMH Engineering, Orem, UT, 801-225-8970 (www.gmheng.com).

[170]Monarch Instrument, Amherst, NH, 800-999-3390 (www.monarchinstrument.com).

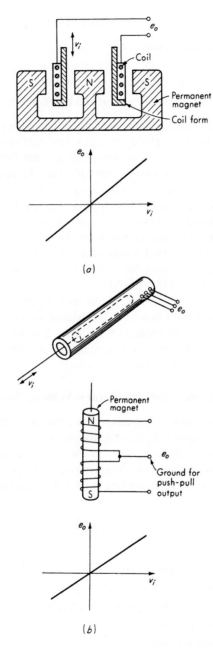

Figure 4.74
Moving-coil velocity pickup.

Translational-Velocity Transducers
(Moving-Coil and Moving-Magnet Pickups)

The moving-coil pickup of Fig. 4.74a is based on the law of induced voltage

$$e_o = (Blv_i)10^{-8} \tag{4.107}$$

where e_o = terminal voltage, V
          B = flux density, G
          l = length of coil, cm
          v_i = relative velocity of coil and magnet, cm/s

Since B and l are constant, the output voltage follows the input velocity linearly and reverses polarity when the velocity changes sign. Such pickups are widely used for the measurement of vibratory velocities. Since the flux density available from permanent magnets is limited to the order of 10,000 G, an increase in sensitivity can be achieved only by an increase in the length of wire in the coil. To keep the coil small, this requires fine wire and thus high resistance. High-resistance coils require a high-resistance voltage-measuring device at e_o to prevent loading. A typical pickup of about 500-Ω resistance has a sensitivity of 0.15 V/(in/s) and a full-scale displacement of 0.15 in with a nonlinearity of $\pm$ 1 percent. A more sensitive coil used in a seismometer (instrument to measure earth shocks) has 500,000-Ω resistance and a sensitivity of 115 V/(in/s).

The transducer shown in Fig. 4.74b uses a permanent-magnet core moving inside a form wound with two coils connected as shown. Units are available[171] in full-range strokes from about 0.5 to 24 in. Sensitivity varies from about 0.5 to 0.05 V/(in/s), nonlinearity is less than 2.5 percent, and frequency response (99 percent flat) from 90 to 1500 Hz.

DC Tachometer Generators for Rotary-Velocity Measurement

An ordinary dc generator (using either a permanent magnet or separately excited field) produces an output voltage roughly proportional to speed. By emphasizing certain aspects of design, such a device can be made an accurate instrument for measuring speed rather than a machine for producing power. The basic principle is again Eq. (4.107), which when applied to the rotational configuration of a dc generator becomes

$$e_o = \frac{n_p n_c \theta N}{60 n_{pp}} 10^{-8} \tag{4.108}$$

where $e_o \triangleq$ average output voltage, V
          $n_p \triangleq$ number of poles
          $n_c \triangleq$ number of conductors in armature
          $\phi \triangleq$ flux per pole, lines
          $N \triangleq$ speed, r/min
          $n_{pp} \triangleq$ number of parallel paths between positive and negative brushes

The voltage e_o is a dc voltage proportional to speed which reverses polarity when the angular-velocity reverses. A small superimposed ripple voltage is present

---

[171]Trans-Tek Inc., Ellington, CT, 800-828-3964 (www.transtekinc.com).

because of the finite number of conductors. While low-pass filtering is effective in reducing ripple at high speeds, this is not usually practical at low speeds and approaches such as gearing should be tried. (Special designs[172] for low speeds claim useful output for speeds as low as 0.5 r/day.) A typical high-accuracy unit[173] (permanent magnet) has a sensitivity of 7 V per 1,000 r/min, a rated speed of 5,000 r/min, nonlinearity of 0.07 percent over a range 0 to 3,600 r/min, ripple voltage 2 percent of average voltage for speeds above 100 r/min, friction torque of 0.2 in · oz, rotor inertia of 7 g · cm², output impedance of 2,800 Ω, and a total weight of 3 oz.

A special dc tachometer[174] of unique design for use where a limited ($\pm$ 15°) angular travel is acceptable exhibits a very high sensitivity. A 1-in-diameter model gives 500 V per 1,000 r/min while a 3-in-diameter gives 30,000 V per 1,000 r/min. The nonlinearity is $\pm$ 9 percent for $\pm$ 15° travel, and the operating torque is 500 g · cm. In this generator the permanent magnet rotates while the coil is stationary, and no commutator is needed because of the limited travel.

AC Tachometer Generators for Rotary-Velocity Measurement

An ac two-phase squirrel-cage induction motor can be used as a tachometer by exciting one phase with its usual ac voltage and taking the voltage appearing at the second phase as output. With the rotor stationary, the output voltage is essentially zero. Rotation in one direction causes at the output an ac voltage of the same frequency as the excitation and of an amplitude proportional to the instantaneous speed. This output voltage is in phase with the excitation. Reversal of rotation causes the same action, except the phase of the output shifts 180° (see Fig. 4.75). While squirrel-cage rotors sometimes are used, the most accurate units employ a drag-cup rotor. This does not change the basic operating characteristics.

A typical high-accuracy unit[175] is excited by 115-V 400-Hz voltage, has a sensitivity of 2.8 V per 1,000 r/min, nonlinearity of 0.05 percent from 0 to 3,600 r/min, negligible rotor friction, rotor inertia of 7 g · cm², and a total weight of 6.7 oz. Most commercial ac tachometers are designed to be used in ac servomechanisms which conventionally operate on either 60 or 400 Hz, and so they are generally designed for operation at these frequencies. For general-purpose motion measurement, the frequency response of such units is limited (as are all ac or "carrier"-type devices) by the carrier frequency to about one-tenth to one-fifth of the carrier frequency. It is usually possible, however, to excite a tachometer designed for 400 Hz at considerably higher frequencies if necessary, although some of or all the performance characteristics may change value.

Eddy-Current Drag-Cup Tachometer

Figure 4.76 shows schematically an eddy-current tachometer. Rotation of the magnet induces voltages into the cup which thereby produce circulating eddy

---

[172]Model T8453-01, Sierracin/Magnedyne, Carlsbad, CA.

[173]General Precision Inc., Kearfott Div., Little Falls, NJ.

[174]Armstrong Whitworth Equipment, Hucclecote, Gloucester, England.

[175]General Precision Inc., Kearfott Div., Little Falls, NJ.

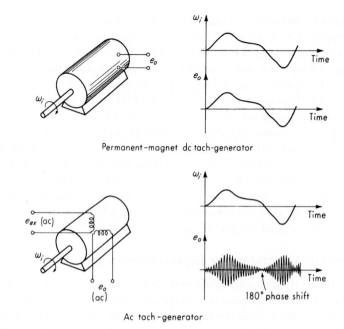

Figure 4.75
Tachometer generators.

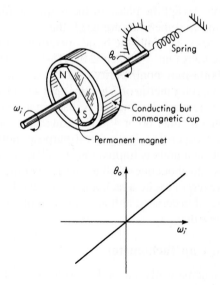

Figure 4.76
Drag-cup velocity pickup.

currents in the cup material. These eddy currents interact with the magnet field to produce a torque on the cup in proportion to the relative velocity of magnet and cup.

This causes the cup to turn through an angle θ_o until the linear spring torque just balances the magnetic torque. Thus in steady-state the angle θ_o is directly proportional to ω_i, the input velocity. If an electric signal is desired, any low-torque displacement transducer can be used to measure θ_o. Dynamic operation is governed by the rotary inertia of parts moving with θ_o, spring stiffness, and the viscous damping effect of the eddy-current coupling between magnet and cup, leading to a second-order response of the form

$$\frac{\theta_o}{\omega_i}(D) = \frac{K}{D^2/\omega_n^2 + 2\zeta D/\omega_n + 1} \tag{4.109}$$

Nonlinearity of the order of 0.3 percent can be achieved in such units.

4.5 RELATIVE-ACCELERATION MEASUREMENTS

The fourth edition of this book showed one commercial transducer for relative acceleration, but, as of this writing, this device was no longer marketed, as far as I know. Since many well-proven transducers for *absolute* acceleration are available, the lack of this device is probably not a great practical problem. Because the moving-coil or moving-magnet sensors of Fig. 4.74 provide rather "clean" velocity signals, these can often be electrically differentiated to get useable acceleration signals.

4.6 SEISMIC- (ABSOLUTE-) DISPLACEMENT PICKUPS

While the most general motion of a rigid body involves three-dimensional translation and rotation, the total *vector* quantities usually are not amenable to direct measurement. So we employ an *array of uniaxial transducers* oriented along selected axes to measure orthogonal components of the vectors, which are then combined by calculation to define the total vector magnitudes and directions. So-called "seismic" sensors or pickups (based on a spring/mass system) are widely used in all types of shock and vibration measurements. Important examples include structural dynamics studies that yield dynamic models (natural frequencies, mode shapes, and damping) for vehicle and machine-tool frames, bridges, and buildings, and also for monitoring the "health" of machinery,[176] alerting factory personnel to incipient machine faults such as bearing failures. Vibration variables of interest are displacement, velocity, and acceleration of selected points on the structure or machine. In a particular application, one of these variables may be more significant than the others, so seismic pickups are available for each of the three. The frequency range of the vibration also influences the choice of instrument since high-frequency vibration has small (perhaps immeasurable) displacement and large acceleration, while

---

[176]J. S. Mitchell, "The History of Condition Monitoring and Condition Based Maintenance," *Sound and Vibration*, Nov. 1999, pp. 21–28.

low-frequency vibration has large displacement but very small acceleration. These statements follow directly from the equation for basic vibratory motion

$$x = X_0 \sin \omega t \quad v = \omega X_0 \cos \omega t \quad a = -\omega^2 X_0 \sin \omega t$$

For example, a large (say 20 g) acceleration at 5000 Hz would have a displacement of only 0.2 μm, while a large (say 1 mm) displacement at 1 Hz would have only $0.004g$ of acceleration. While we might at this point guess that *velocity* pickups would strike a happy medium in this dilemma, other factors not yet discussed make the situation less simple,[177] and in fact, arguments among expert practitioners and instrument manufacturers continue. Finally, vibrations of shafts supported in fluid-film bearings are often measured by *nonseismic,* noncontact *relative* displacement sensors,[178] such as the eddy-current probes discussed earlier (Fig. 4.31). Seismic pickups attached to the bearing housing would receive very little motion because the "soft" fluid allows shaft motion without much housing motion.

Figure 4.77a, b show the general construction of a seismic-displacement pickup for uniaxial translatory or rotary motions. These devices are used almost exclusively for measurement of vibratory displacement in those (many) cases where a fixed reference for relative-displacement measurement is not available. That is, the vibration of a body can be measured with any of the relative-motion transducers discussed earlier in this chapter, but only if one end of the transducer can be attached to a stationary reference. For measurements on moving vehicles, such references are not generally available, and in many other situations, measurement of absolute motion is easier and more desirable. The basic principle of seismic- (absolute-) displacement pickups is simply to measure (with any convenient relative-motion transducer) the relative displacement of a mass connected by a soft spring to the vibrating body. For frequencies above the natural frequency, this relative displacement is also very nearly the absolute displacement since the mass tends to stand still.

To obtain a quantitative measure of performance for such systems, we analyze the configuration of Fig. 4.77a. The rotational configuration is completely analogous. Newton's law may be applied to the mass M as follows:

$$K_s x_o + B\dot{x}_o = M\ddot{x}_M = M(\ddot{x}_i - \ddot{x}_o) \tag{4.110}$$

where x_i and x_M are the absolute displacements and we have chosen our reference for x_o such that x_o is zero when the gravity force (weight of M) is acting along the x axis statically. Manipulation gives

$$\frac{x_o}{x_i}(D) = \frac{D^2/\omega_n^2}{D^2/\omega_n^2 + 2\zeta D/\omega_n + 1} \tag{4.111}$$

[177]M. Gilstrap, "Acceleration vs. Velocity as Absolute Vibration Measurement Transducers," *Sensors,* Aug. 1984, pp. 16–20; R. M. Barrett, Jr., "Low Frequency Machinery Monitoring: Measurement Considerations," Wilcoxon Research, Inc., Gaithersburg, MD, 1993; P. Kapteyn, "Low Frequency Vibration Sensing," *Sound and Vibration,* Feb. 1984, pp. 6–12.

[178]Bently Nevada Corp., 775-782-3611 (www.bently.com).

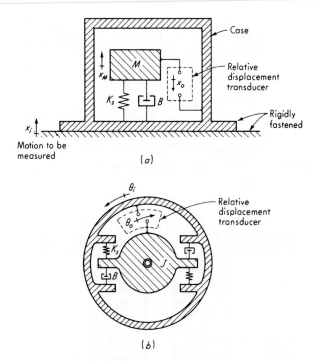

Figure 4.77
Translational and rotational seismic pickups.

where
$$\omega_n \triangleq \sqrt{\frac{K_s}{M}}$$

$$\zeta \triangleq \frac{B}{2\sqrt{K_s M}}$$

Since the pickup is intended mainly as a vibration sensor, the frequency response is of prime interest:

$$\frac{x_o}{x_i}(i\omega) = \frac{(i\omega)^2/\omega_n^2}{(i\omega/\omega_n)^2 + 2\zeta i\omega/\omega_n + 1} \qquad (4.112)$$

This is graphed in Fig. 4.78. We note that there is no response to static displacement inputs and that ω_n should be much less than the lowest vibration frequency ω for accurate displacement measurement. For those frequencies much above ω_n, $(x_o/x_i)(i\omega) \to 1 \angle 0°$, indicating perfect measurement. The characteristics of the relative-displacement transducer in converting x_o to a voltage e_o also must be considered. Since the force in spring K_s is directly proportional to x_o, if strain gages are used, they can be applied directly to this spring, which may be in the form of a cantilever beam. Since a low ω_n is desired, either a large mass or a soft spring (or both) is necessary. To keep size (and thereby loading on the measured system) to a

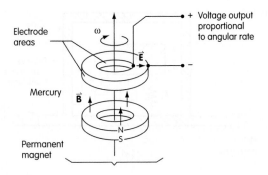

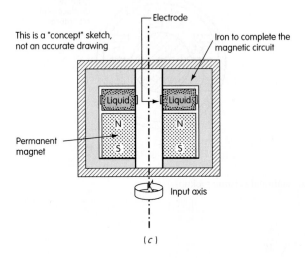

(c)

Figure 4.77
(*Continued*)

minimum, soft springs are preferred to large masses. Intentional damping in the range $\zeta = 0.6$ to 0.7 is often employed to minimize resonant response to slow transients.

A commercial unit[179] uses a flexure/mass system with natural frequency about 7 Hz. Eddy-current damping (equivalent to viscous damping but less temperature sensitive) is provided by a metal plate moving through a permanent-magnet air gap. An LVDT relative displacement sensor gives useable signals down to the μinch range. The mechanical and electrical behavior is sufficiently good to allow electrical *dynamic compensation* below the resonant frequency, extending the flat amplitude ratio ($\pm$ 3 percent) to 3 Hz.

---

[179]Indikon Co., Inc., Somerville, MA, 617-625-3604 (www.indikon.com); P. Kapteyn, "Low Frequency Vibration Sensing," *Sound and Vibration*, Feb. 1984, pp. 6–12.

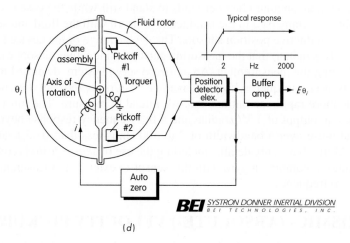

Model 8301 Inertial angular displacement sensor block diagram

(d)

Figure 4.77
(Concluded)

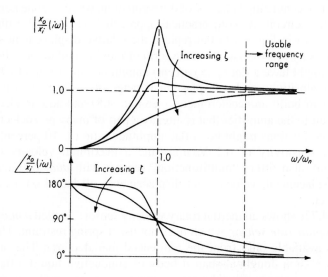

Figure 4.78
Seismic-displacement-pickup frequency response.

Figure 4.77*d* shows an inertial angular displacement sensor[180] with extreme sensitivity and resolution, which is used in the most critical applications, such as

---

[180]Model 8301, Systron Donner, Concord, CA, 1-800-227-1625 (www.systron.com); H. D. Morris, "The Inertia Angular Displacement Sensor," *Theory and Application*, 1987.

super-precise pointing of satellite cameras making images of the earth. A liquid inertial mass in an annular chamber tends to stand still while the case oscillates. A lightweight vane immersed in the fluid tends to move with the fluid and serves as a target for two inductive position sensors. The design range of the device is so small (as small as 10 μradian) that the measuring gap of the inductive sensors would often close, defeating the measurement. An autozero servo system, designed to respond very slowly, uses a torquer to keep the average position at zero but does not "fight" against the more rapid oscillations that we intend to measure. The most sensitive version has an output of 1 V/μradian and a flat frequency response beyond 2 Hz. The output noise over a bandwidth of 2 to 500 Hz is less than 0.03 μradian. The article by Morris has many details, including the use of *sensor fusion* (complementary filtering) to combine a gyro with this sensor to extend the measurement capability to zero frequency.

4.7 SEISMIC- (ABSOLUTE-) VELOCITY PICKUPS

Again the application here is limited to vibratory velocities, and the basic configuration is exactly the same as in Fig. 4.77a. First, a voltage signal from the displacement pickup could be electrically differentiated, but this approach suffers from the general noise-accentuating behavior of differentiation, whether done numerically or with an analog circuit. A more practical approach[181] is to build a displacement pickup as shown in Fig. 4.77a but replace the relative displacement sensor with a relative velocity sensor, usually some form of moving-coil device (Fig. 4.74). A typical unit might have a flat ($\pm$ 5 percent) amplitude ratio from 8 to 700 Hz with 0.1 V/(in/s) sensitivity and a displacement range of $\pm$ 0.25 in. Another viable scheme[182] is to build an accelerometer (see Sec. 4.8) and add an electrical *integrating* circuit to the amplifier that is already a part of many piezoelectric accelerometers. Specifications might be: a flat amplitude ratio ($\pm$ 10 percent) from 2 to 3500 Hz, a sensitivity of 0.1 V/(in/s) and a maximum velocity of 50 in/s. This company also offers "multifunction" sensors with *dual* outputs (acceleration and displacement, velocity and displacement), all based on a piezoelectric accelerometer.

Figure 4.77c shows an inertial rotary velocity sensor[183] called *a magnetohydrodynamic angular rate sensor,* which does not use a spring restraint. This patented principle is available only from the referenced manufacturer. The only internal moving "part" is a doughnut-shaped body of conductive liquid, often mercury, which serves as a rotary inertial mass, coupled to the housing by viscous drag. If the housing is rotated about the sensitive axis very slowly, the fluid tends to rotate with the housing, without any relative motion. If the housing rotates rapidly in an

[181]CEC Vibration Products, Pomona, CA, 800-468-1345 (www.cecvp.com).

[182]Wilcoxon Research, Gaithersburg, MD, 800-945-2696 (www.wilcoxon.com).

[183]ATA Sensors, Albuquerque, NM, 505-823-1320 (www.atasensors.com); M. Hawes, "Testing Crash Dummies with an Angular Motion Sensor," *Sensors,* Sept. 1989, pp. 32–40; D. R. Laughlin, et al., "Inertial Angular Rate Sensors: Theory and Application," *Sensors,* Oct. 1992.

oscillatory fashion, the liquid inertia tends to stand still, giving a relative motion between housing and liquid that is the same as the absolute motion of the housing. Then the conductive fluid acts as a "wire" cutting through the magnetic field, generating a voltage proportional to angular velocity that can be picked off at electrodes touching the inner and outer surfaces of the fluid doughnut. While the "low" frequencies are lost, a wide range of flat amplitude ratio and zero phase angle exists, typically from about 1 to over 1000 Hz. The sensor output voltage is sometimes connected to a transformer that steps up the voltage to get even higher sensitivities and maintain low noise. An alternative design uses op-amp circuitry to boost the sensitivity and/or process the velocity signal to get displacement or acceleration signals. A basic sensor (ARS-04) gives 1 μV/(rad/s) in a case size about 0.5 in³, while an op-amp unit (ARS-03) has 0.5 V/(rad/s) and a $\pm$ 20 rad/s range in about 1.5 in³. While the basic sensor stays flat beyond 1000 Hz, if such high frequencies are not needed, the op-amp circuitry is used to give a low-pass filtering effect beyond 1000 Hz, thus reducing high-frequency noise.

4.8 SEISMIC- (ABSOLUTE-) ACCELERATION PICKUPS (ACCELEROMETERS)

The most important pickup for vibration,[184] shock, and general-purpose absolute-motion measurement is the accelerometer. This instrument is commercially available in a wide variety of types and ranges to meet correspondingly diverse application requirements. The basis for this popularity lies in the following features:

1. Frequency response is from zero to some high limiting value. Steady accelerations can be measured (except in piezoelectric types).

2. Displacement and velocity can be easily obtained by electrical integration, which is much preferred to differentiation.

3. Measurement of transient (shock) motions is more readily achieved than with displacement or velocity pickups.

4. Destructive forces in machinery, etc., often are related more closely to acceleration than to velocity or displacement.

The preference for accelerometers, compared with seismic displacement and velocity pickups, is testified to by the large number of manufacturers and models available. Having stated this preference, we hasten to admit that velocity pickups are favored by some vibration analysts, particularly for "machinery health" monitoring applications in the low-frequency ranges. In such applications, high-frequency vibration may also be present but not contribute useful information. In fact, high-frequency acceleration can saturate and overload the amplifier built into many

[184]G. Buzdugan, E. Mihailescu, and M. Rades, "Vibration Measurement," M. Nijhoff, Boston, 1986; V. Wowk, "Machinery Vibration: Measurement and Analysis," McGraw-Hill, New York, 1991; K. G. McConnell, "Vibration Testing: Theory and Practice," Wiley, New York, 1995; P. L. Walter, "The History of the Accelerometer," *Sound and Vibration*, March 1997, pp. 16–22.

accelerometers. The amplifier's response to overload is a *low-frequency* phenomenon, so a spectrum analyzer may show a large (and *false*) vibration component (sometimes called "ski-slope"[185]) at low frequency, which may be accepted as valid data if the overloading is not recognized.

The basic accelerometer configuration is again shown in Fig. 4.77a. The operating principle is as follows: Suppose the acceleration $\ddot{x}_i$ to be measured is constant. Then, in steady state, the mass M will be at rest relative to the case, and thus its absolute acceleration will also be $\ddot{x}_i$. If mass M is accelerating at $\ddot{x}_i$, there must be some force to cause this acceleration, and if M is not moving relative to the case, this force can come only from the spring. Since spring deflection x_o is proportional to force, which in turn is proportional to acceleration, x_o is a measure of acceleration $\ddot{x}_i$. Thus absolute-acceleration measurement is reduced to the measurement of the force required to accelerate a known mass (sometimes called the "proof" mass). This dependence on mass leads to problems (mainly in inertial guidance systems, not in vibration measurement) since a mass also experiences forces due to gravitational fields. That is, an accelerometer cannot distinguish between a force due to acceleration and a force due to gravity.

The majority of accelerometers may be classified as either deflection type or null-balance type. Those used for vibration and shock measurement are usually the deflection type whereas those used for measurement of gross motions of vehicles (submarines, aircraft, spacecraft, etc.) may be either type, with the null-balance being used when extreme accuracy is needed.

Deflection-Type Accelerometers

A large number of practical accelerometers have the configuration of Fig. 4.77a and differ only in details, such as the spring element used, relative-motion transducer employed, and type of damping provided. Since the desired input is now $\ddot{x}_i$, we can rewrite Eq. (4.111) as

$$\frac{x_o}{D^2 x_i}(D) = \frac{x_o}{\ddot{x}_i}(D) = \frac{K}{D^2/\omega_n^2 + 2\zeta D/\omega_n + 1} \tag{4.113}$$

where

$$K \triangleq \frac{1}{\omega_n^2} \quad \text{cm/(cm/s}^2) \tag{4.114}$$

Since output voltage $e_o = K_e x_o$ for many motion transducers, Eq. (4.113) has the correct form for the acceleration-to-voltage transfer function also. We see that the accelerometer is an ordinary second-order instrument; thus all our previous work on this type is immediately applicable. The frequency response extends from zero to some fraction of ω_n, depending on the accuracy required and the damping. Because sensitivity $K = 1/\omega_n^2$, high-frequency response must be traded for sensitivity. Since the dynamic characteristics of second-order instruments have been thoroughly

[185]R. M. Barrett, Jr., "Low Frequency Machinery Monitoring: Measurement Considerations," Catalog W-14, p. 10, Wilcoxon Research, Inc., Gaithersburg, MD, 800-945-2696.

discussed, here we concentrate mainly on the specific characteristics of commercially available instruments.

Deflection-type accelerometers can be classified as follows:

1. Classical technology
   (*a*) DC response
         Potentiometer
         LVDT
         Bonded strain gage
         Unbonded strain gage
         Capacitance
   (*b*) No DC response
         Piezoelectric
2. MEMS (microelectromechanical system) technology
   (*a*) DC response
         Piezoresistance
         Capacitance

Classical (non-MEMS) technology assembles the accelerometer "manually" from discrete parts. MEMS (see Chap. 14) uses integrated circuit technology to produce tiny complete instruments by the hundreds, at low cost, on a silicon wafer. So far, MEMS accelerometers are used mainly in consumer products such as automobiles, where the large market and modest performance matches MEMS capabilities. For the more demanding engineering-test and machinery-monitoring applications, MEMS performance may or may not be adequate and should be carefully evaluated. By "DC response," we mean the capability of measuring from zero frequency (steady acceleration) up to some higher limiting frequency. DC response is important in some applications (for example, vehicle performance measurements) but is not needed in most shock and vibration studies.

Non-MEMS accelerometers with DC response differ mainly in the relative displacement transducer used. Potentiometer types are used mainly for low frequencies, a typical unit[186] has a natural frequency of about 100 Hz, 1.5 percent full scale (FS) accuracy, and a threshold of 0.8 percent FS. A family[187] of unbonded strain gage accelerometers has ranges of $\pm 5g$ to $\pm 500g$, with natural frequencies from 300 to 2900 Hz, sensitivity about 20 mV FS, cross-axis sensitivity of 1 percent, and infinitesimal resolution. Bonded semiconductor strain gage units are available[188] in similar ranges but with somewhat higher sensitivities (about 100 mV FS). While LVDT accelerometers were referenced in the 1990 edition of this text, manufacturers of LVDT accelerometers could not be found for this edition.

Piezoelectric accelerometers are in wide use for shock and vibration measurements. In general, they do not give an output for constant acceleration because of

---

[186]Humphrey Inc., San Diego, CA, 619-565-6631 (www.humphreyinc.com).

[187]CEC Vibration Products, Pomona, CA, 800-468-1345 (www.cecvp.com).

[188]Sensotec, Inc., Columbus, OH, 800-848-6564 (www.sensotec.com).

the basic characteristics of piezoelectric displacement sensing. But they do have large output-voltage signals, small size, and can have very high natural frequencies, which are necessary for accurate shock measurements. No intentional damping is provided, with material hysteresis being the only source of energy loss. This results in a very low (about 0.01) damping ratio, but this is acceptable because of the very high natural frequency. The transfer function is a combination of Eqs. (4.70) and (4.113):

$$\frac{e_o}{\ddot{x}_i}(D) = \frac{[K_q/(C\omega_n^2)]\tau D}{(\tau D + 1)(D^2/\omega_n^2 + 2\zeta D/(\omega_n + 1)} \tag{4.115}$$

The low-frequency response is limited by the piezoelectric characteristic $\tau D/(\tau D + 1)$ while the high-frequency response is limited by mechanical resonance. The damping ratio ζ of piezoelectric accelerometers is not usually quoted by the manufacturer, but can be taken as zero for most practical purposes. The accurate (5 percent high at the high-frequency end and 5 percent low at the low-frequency end) frequency range of such an accelerometer is $3/\tau < \omega < 0.2\omega_n$. Accurate low-frequency response requires large τ, which is usually achieved by use of high-impedance voltage amplifiers or charge amplifiers. Systems designed for response below about 1.0 Hz and subject to temperature transients may exhibit errors because of the pyroelectric effect present in most piezoelectric materials. Here a charge output is produced in response to a temperature input. For systems with negligible response at low frequencies, these temperature-induced signals (because they are "slow" transients) produce little output. But serious errors may occur if τ has been made large to measure low-frequency accelerations and if the accelerometer has not been designed to minimize thermal effects.

The design details of piezoelectric accelerometers can be varied to emphasize selected features of performance desired for particular applications. No single configuration is ideal for all situations since trade-offs exist here just as in all engineering design. Figure 4.79 shows several designs from a single manufacturer. The basic compression design (Fig. 4.79a) is the simplest and most rugged and has the best mass/sensitivity ratio; but because the housing acts as an integral part of the spring/mass system, this type is more sensitive to spurious inputs. [For piezoelectric accelerometers these include temperature,[189] acoustic noise, "base bending"[190] (surface strains caused by bending at the mounting surface), cross-axis motion, and magnetic fields.] The spring is generally preloaded to work the piezoelectric material at a more linear portion of its charge-strain curve. This preload also allows measurement of both positive and negative accelerations without putting the piezo material in tension. That is, the initial preload causes an output voltage of a certain polarity. However, this immediately leaks off, and the polarity of subsequent

---

[189]C. F. Vezzeti and P. S. Lederer, An Experimental Technique for the Evaluation of Thermal Transient Effects on Piezoelectric Accelerometers, *NBS Tech. Note* 855, January 1975.

[190]Tech. Data A510, 508A, 509A, Endevco Corp., San Juan Capistrano, CA, 877-362-3826 (www.endevco.com).

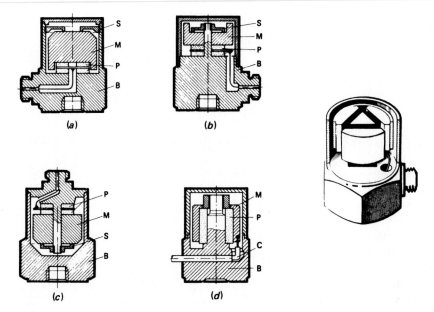

Figure 4.79
Piezoelectric accelerometer designs. (*a*) Peripheral-mounted compression design. (*b*)
Center-mounted compression design. (*c*) Inverted center-mounted compression design.
(*d*) Shear design. (*Courtesy Bruel and Kjaer Instruments, Marlboro, MA*)

acceleration-induced voltages will follow the direction of the motion since the
charge polarity depends on the *change* in strain, not its total value. The preload is
made large enough that it is never relaxed by even the largest input acceleration.

Reduced response to spurious inputs is achieved in the "single-ended compression" or center-mounted designs of Fig. 4.79*b* and *c*. The inverted design of
Fig. 4.79*c* is especially tolerant of base-bending strains, as is the shear design of
Fig. 4.79*d*. Shear designs using bolted stacks of flat plate elements have been introduced recently to gain further improvements in performance. The Delta Shear unit
of Fig. 4.79 is a variation which uses a shrink-fitted collar, rather than bolts, to hold
the elements fast.

Microcircuit electronics has allowed the design of piezoelectric accelerometers
with charge amplifiers (Integrated Circuit Piezoelectric (ICP); see Chap. 10) built
into the instrument housing. A single two-conductor cable which carries both amplifier power and the measurement signal connects the instrument to a simple
constant-current power supply, which also provides a high-level (a few volts) output
signal directly to oscilloscopes or signal analyzers. This arrangement allows greater
sensitivity with a smaller, higher-frequency accelerometer, reduces cable-noise
effects and length limitations, and lowers costs. These advantages are traded for a
reduced temperature range (the microcircuit electronics is more temperature-limited
than the accelerometer itself) and less versatile signal conditioning (the built-in

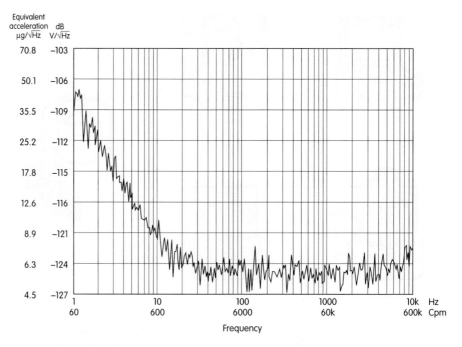

Figure 4.80
Accelerometer noise floor.

amplifiers allow little or no adjustment). The *noise floor* of an accelerometer/ amplifier combination may need to be considered, especially at low frequencies, where the magnitude of acceleration may be quite small and noise might swamp the signal. Figure 4.80[191] shows a typical root-mean-square (rms) frequency spectrum, $(\mu g)/(\sqrt{Hz})$, for the sensor random noise. Using such data, if a spectrum analysis of some measured acceleration used a filter bandwidth of, say, 4 Hz centered on 10 Hz, the expected noise signal would have a magnitude of about $(9 \ \mu g)(\sqrt{4}) = 18 \ \mu g$ rms. If the true acceleration were close to this value, it could not be reliably measured. Catalog W-14 suggests that this signal-to-noise ratio should be at least about 10. If the signal processing does not use narrow-band filtering but instead records the "total" (wide-band) signal, then a corresponding *total* rms noise value would be of interest; typically about 700 μg rms for the unit just described. Such considerations of noise floor may be pertinent in any sensor application; however, data such as that found in Fig. 4.80 is not often available.

Piezoelectric accelerometers are available in a very wide range of specifications from a large number of manufacturers. The sensitivity-frequency response trade-off

[191]Machinery Monitoring Catalog W-14, p. 9, Wilcoxon Research, Inc., Gaithersburg, MD, 800-945-2696 (www.wilcoxon.com); F. Schloss, "Accelerometer Noise," *Sound and Vibration,* March 1993, pp. 22–23; D. Formenti and T. Norsworthy, "Accelerometer Dynamic Range," *Sound and Vibration,* June 1999, pp. 6–10.

is apparent in typical specifications; a shock accelerometer may have 0.004 pC/g and natural frequency of 250,000 Hz, while a unit designed for low-level seismic measurements has 1,000 pC/g and 7,000 Hz. The smallest units (needed to reduce mass-loading errors in measurements on light structures) are about 3×3 mm with 0.5-g mass including cable. Response to spurious thermal and base-bending inputs for the best isolated-shear designs may be 200 times less than for designs not optimized in this respect. Triaxial units as small as a 7-mm cube with 1-g mass are available. Uncooled instruments usable over a temperature range of -450 to $+1500°F$ typically exhibit a ± 10 percent sensitivity change from -100 to $+1500°F$ and may be designed to survive radiation environments typical of nuclear reactors. A system of accelerometers, cabling, and signal conditioning designed for economical implementation of multichannel vibration tests uses inexpensive piezo film transducers as plug-in modules to create uniaxial, biaxial, or triaxial sensing points for benign laboratory environments.[192]

Piezoelectric accelerometers tend to have somewhat larger cross-axis sensitivities than other types; however, this is typically held to about 2 to 4 percent and usually is not a critical factor. Some manufacturers indicate the location of the cross axis of *least* transverse sensitivity, allowing the user to orient the instrument so as to minimize this effect in each installation. Accelerometers may be mounted with threaded studs (the preferred method), with cement or wax adhesives, or with magnetic holders. The main effect of the various mounting methods is a reduction of the natural frequency below that of the unmounted accelerometer itself, as a result of elastic and inertial characteristics of the mounting. The unmounted resonant frequency is measured by suspending the accelerometer freely from its cable, exciting the piezo element electrically (sinusoidally), and finding the frequency at which current and voltage are in phase. Mounted natural frequency is the frequency of peak voltage output when the accelerometer is mechanically vibrated sinusoidally. For a solid steel stud (the optimum mounting), the mounted natural frequency is typically about 60 percent of the unmounted. Experience has shown that for measurements above 5 kHz, the steel-stud mounting is significantly improved when a drop of light oil is applied between the mating surfaces before torquing the connection tight. Apparently a stiffer elastic coupling is achieved by the oil filling microscopic voids at the interface. When electrically insulated studs are used to reduce electric noise problems arising from ground loops, the lower stiffness of the insulating material causes reduced frequency response. Data on this and other mounting effects are available in the literature.[193]

Figure 4.81 shows an accelerometer design using the capacitance displacement transducer of Fig. 4.38. A thin diaphragm with spiral flexures provides the spring, proof mass, and moving plate of the differential capacitor. Plate motion between the electrodes "pumps" air parallel to the plate surface and through holes in the plate to

---

[192]Structcel System, PCB Piezotronics, Depew, NY, 716-684-0001 (www.pcb.com).

[193]G. K. Rasamen and B. M. Wigle, "Accelerometer Mounting and Data Integrity," *Sound & Vib.,* pp. 8–15, November 1967; S. V. Bowers et al., "Real-World Mounting of Accelerometers for Machinery Monitoring," *Sound and Vibration,* Feb. 1991, pp. 14–23.

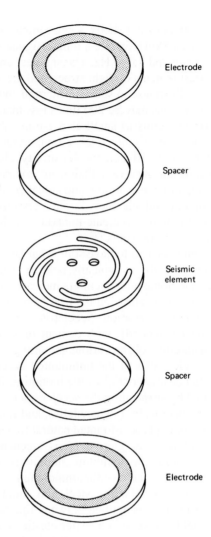

Figure 4.81
Capacitive accelerometer design.
(Courtesy Setra, Boxborough, MA, 800-257-3872, www.setra.com)

provide squeeze-film damping. Since air viscosity is less temperature-sensitive than oil, the desired 0.7 damping ratio changes only about 15 percent per 100°F. A family of such instruments with full-scale ranges from $\pm 2g$ (200 Hz flat response) to $\pm 600g$ (3000 Hz), cross-axis sensitivity less than 1.0 percent, and full-scale output of ± 1.5 V are available in a case size about a 1 in^3 weighing 1.7 oz.

MEMS-type accelerometers had their "discovery period" from 1974 to 1985 and reached full commercialization by 1997,[194] whereas the discovery period for

---

[194]R. H. Grace, "The New MEMS and Their Killer Apps," *Sensors,* July, 2000, pp. 4–8.

MEMS pressure sensors was 1954 to 1985, with full commercialization by 1990. MEMS accelerometers (usually 50g range) are widely used in automotive air bag actuation systems, replacing earlier sensors that only gave an on/off type of switching signal when a crash of critical severity occurred. Other automotive applications include antilock braking, stability augmentation, active suspension, and car theft alarms. The theft alarm (theft by towing the car away) uses the "gravity sensitivity" of all accelerometers to measure car "tilt." If the angle changes by more than 0.5°/min, the alarm sounds.[195] MEMS accelerometers use the usual mass/spring system, but in miniaturized form and produced by micromachining processes that result in low cost. Capacitive deflection sensing is used in one manufacturer's units[196] to provide single- or dual-axis sensors in low (2 to 10g) and high (25 to 50g) ranges. Air-bag accelerometers can tolerate relatively high noise floors that would be unacceptable in machinery-monitoring applications. To meet these more stringent needs, revised designs and manufacturing processes have been implemented.[197] A ± 5g unit has 10-kHz bandwidth and a noise floor of (225 μg)/ ($\sqrt{Hz}$) rms (earlier 50g units had 6500). This noise density is also *constant* with frequency, compared to piezoelectric units, where noise *increases* at low frequencies and causes difficulty when we integrate to get velocity or displacement signals. The sensor itself (mass/spring) is 0.5×0.4 mm, 2 μm thick, and has a mass of only 0.5 μg (the size increases significantly when the unit is packaged for practical use). The deflection due to 1g is 1 nm, and the minimum resolvable deflection is 210 fm. An on-chip temperature sensor allows compensation for temperature effects.

Piezoresistive technology is used in a 2000g shock accelerometer[198] with 0.2 mV/g sensitivity. The mounted resonant frequency is 28 kHz, giving flat (± 5 percent) frequency response to 5000 Hz. Packaged for practical use, the unit is $10 \times 12 \times 5$ mm and weighs 0.8 g. A Wheatstone bridge with two 500-Ω resistors and two boron-doped strain gages uses 10-V DC excitation. Some shock accelerometers use mechanical filtering (see Fig. 10.29).

Although angular accelerometers based on the configuration of Fig. 4.77b are available, the selection of manufacturers and models is much less extensive than for the translational units. Angular accelerometers of the servo type are available from several manufacturers and are discussed in the next section. It is also possible to use an array of translational accelerometers to measure rotational motions. Figure 4.82 shows the most general situation, where nine translational accelerometers are attached to a fixture which is itself then fastened to the rigid body whose rotational

---

[195]H. Weinberg, "Accelerometers—Fantasy & Reality," *Analog Dialogue,* vol. 33-8, 1999, pp. 25–26; Analog Devices, Inc., Cambridge, MA, 800-262-5643 (www.analog.com/imems). Discusses the practicality of many consumer product applications.

[196]Analog Devices, Inc.

[197]J. Doscher, "ADXL105: A Lower Noise, Wider Bandwidth Accelerometer Rivals Performance of More Expensive Sensors," *Analog Dialogue,* vol. 33-6, 1999, pp. 27–29.

[198]J. Suminto, "Measuring Acceleration in Automotive Safety Systems," *Sensors,* Aug. 1996, pp. 16–22; "A Rugged, High Performance Piezoresistive Accelerometer," Endevco, San Juan Capistrano, CA, 949-493-8181 (www.endevco.com).

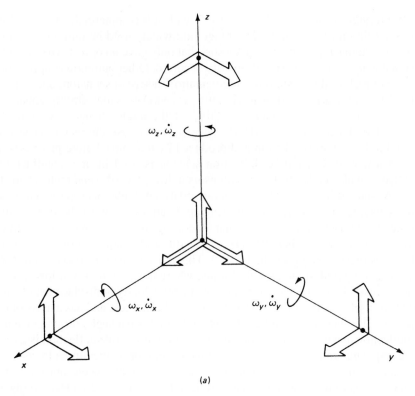

(a)

Figure 4.82
Nine-accelerometer array. *(Courtesy Endevco Corp., San Juan Capistrano, CA)*

motion is to be measured. By suitable processing of the signals from the nine accelerometers, it has been shown[199] that one can compute the three perpendicular components of both angular velocity and angular acceleration for the body-fixed axes shown. The PiezoBEAM[200] accelerometer uses piezoelectric bending beams as shown in Fig. 4.83 to produce two output signals: one proportional to translational acceleration and another to angular acceleration, each about a single defined axis. Charge converters internal to the accelerometer convert charge to voltage. Then external summing amplifiers add (translational signal) and subtract (angular signal) to give outputs of $\pm$ 10g, 1.0 V/g, $\pm$ 18,000 rad/s^2, 0.5 to 50 mV/(rad/s^2), flat response 1.0 to 2000 Hz, and thresholds of 300 μg rms and 0.6 rad/s^2 rms. The instrument's performance in practical testing has been documented.[201]

[199]A. I. King, A. J. Padgaonkar, and K. W. Krieger, "Measurement of Angular Acceleration of a Rigid Body Using Linear Accelerometers," *Biomechanics Res. Center Rep.,* Wayne State University, Detroit, MI, December 6, 1974.

[200]Kistler, Amherst, NY, 888-547-8537 (www.kistler.com); M. D. Insalaco, "The Art of Fabricating a Rotational Accelerometer," *Sensors,* Sept. 2000, pp. 114–119.

[201]A. L. Wilks et al., "Acceleration Measurements on a Free-Free Beam," *Sensors,* Sept. 1989, pp. 59–66.

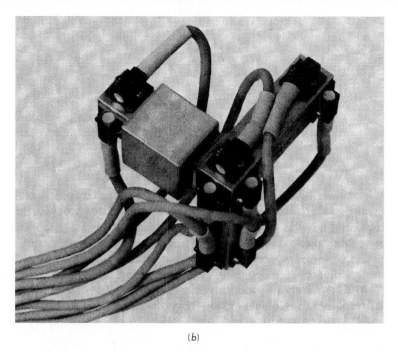

(b)

Figure 4.82
(*Concluded*)

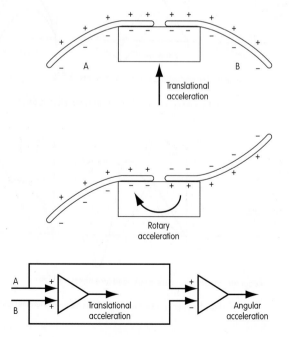

Figure 4.83
PIEZOBEAM accelerometer measures both translational and angular accelerations.

We conclude this section with Fig. 4.84, which provides guidance for accelerometer selection to meet dynamic-response requirements.

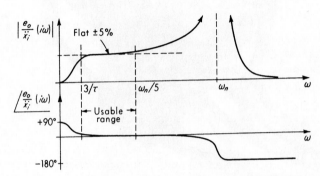

Piezoelectric accelerometer frequency response

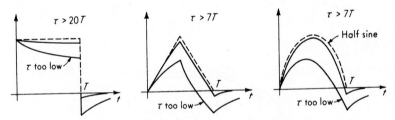

Requirements for accurate peak measurements ±5%

Low-frequency response problems (piezoelectric only)

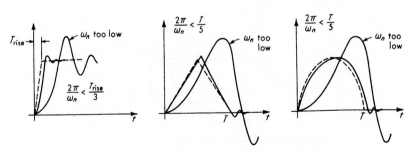

Requirements for accurate peak measurements ±10%

High-frequency response problems (all accelerometers)

Figure 4.84
Accelerometer selection criteria.

Null-Balance- (Servo-) Type Accelerometers[202]

So-called servo-accelerometers using the principle of feedback have been developed for applications requiring greater accuracy than is generally achieved with instruments using mechanical springs as the force-to-displacement transducer. In these null-balance instruments, the acceleration-sensitive mass is kept very close to the zero-displacement position by sensing this displacement and generating a magnetic force which is proportional to this displacement and which always opposes motion of the mass from neutral. This restoring force plays the same role as the mechanical spring force in a conventional accelerometer. Thus we may consider the mechanical spring to have been replaced by an electrical "spring." The advantages derived from this approach are the greater linearity and lack of hysteresis of the electrical spring as compared with the mechanical one. Also, in some cases, electrical damping (which can often be made less temperature-sensitive than mechanical damping) may be employed. There is also the possibility of testing the static and dynamic performance of the device just prior to a test run by introducing electrically excited test forces into the system. This convenient and rapid remote self-checking feature can be quite important in complex and expensive tests where it is extremely important that all systems operate correctly before the test is commenced. These servo-accelerometers usually are used for general-purpose motion measurement and low-frequency vibration. Also they are particularly useful in acceleration-control systems since the desired value of acceleration can be put into the system by introducing a proportional current i_a from an external source.

Figure 4.85 illustrates in simplified fashion the operation of a typical instrument designed to measure a translational acceleration $\ddot{x}_i$. (Angular acceleration also can be measured by these techniques by using an obvious mechanical modification. Perhaps not quite obvious is the *liquid rotor* angular accelerometer, similar in construction to the liquid rotor displacement sensor of Fig. 4.77d. The essential difference is that the servo system now nulls the vane deflection for *all* frequencies, and the acceleration output signal is a voltage developed across a readout resistor by the torquer current. The magnetic torque equals the inertia torque, so the current is proportional to the angular acceleration. Full-scale ranges from $\pm$ 0.1 to 200 rad/s^2 are available; natural frequencies are 10 to 200 Hz.[203]) In Fig. 4.85 the acceleration $\ddot{x}_i$ of the instrument case causes an inertia force f_i on the sensitive mass M, tending to make it pivot in its bearings or flexure mount. The rotation θ from neutral is sensed by an inductive pickup and is amplified, demodulated, and filtered to produce a current i_a directly proportional to the motion from null. This current is passed through a precision stable resistor R to produce the output-voltage signal and is applied to a coil suspended in a magnetic field. The current through the coil

---

[202]E. J. Jacobs, "New Developments in Servo Accelerometers," *Proc. Inst. of Environ. Sci., 14th Ann. Meeting,* April 29, 1968, St. Louis (Endevco Corp., Reprint); C. E. Bosson and D. W. Busse, "Autonetics Microminiature Digital Accelerometer," *Proc. Inst. of Navig., Natl. Space Nav. Mtg.,* Boston, April 21–22, 1966 (Autonetics Corp., Reprint); H. K. P. Neubert, "Instrument Transducers," chap. 5, Oxford University Press, London, 1963.

[203]Systron Donner, models 4590, 4591, and 4597F.

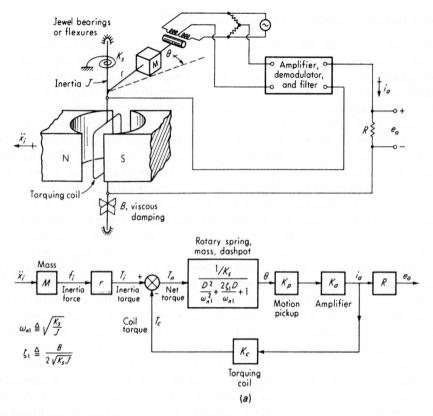

Figure 4.85
Servo-type accelerometer.

produces a magnetic torque on the coil (and the attached mass M) which acts to return the mass to neutral. The current required to produce a coil magnetic torque that just balances the inertia torque due to $\ddot{x}_i$ is directly proportional to $\ddot{x}_i$, thus e_o is a measure of $\ddot{x}_i$. Since a nonzero displacement θ is necessary to produce a current i_a, the mass is not returned exactly to null, but comes very close because a high-gain amplifier is used. Analysis of the block diagram reveals the details of performance as follows:

$$\left(Mr\ddot{x}_i - e_o\frac{K_c}{R}\right)\frac{K_pK_a/K_s}{D^2/\omega_{n1}^2 + 2\zeta_1 D/\omega_{n1} + 1} = \frac{e_o}{R} \tag{4.116}$$

$$\left(\frac{D^2}{\omega_{n1}^2} + \frac{2\zeta_1 D}{\omega_{n1}} + 1 + \frac{K_cK_pK_a}{K_s}\right)e_o = \frac{MrRK_pK_a}{K_s}\ddot{x}_i \tag{4.117}$$

Now, by design, the amplifier gain K_a is made large enough so that $K_cK_pK_a/K_s \gg 1.0$; then

$$\frac{e_o}{\ddot{x}_i}(D) = \frac{K}{D^2/\omega_n^2 + 2\zeta D/\omega_n + 1} \tag{4.118}$$

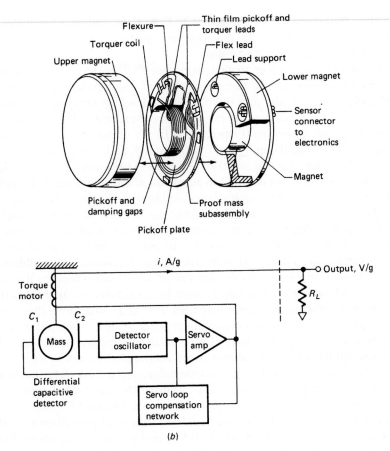

Figure 4.85
(Concluded)

where
$$K \triangleq \frac{MrR}{K_c} \qquad \text{V/(m/s}^2) \tag{4.119}$$

$$\omega_n \triangleq \omega_{n1} \sqrt{\frac{K_p K_a K_c}{K_s}} \qquad \text{rad/s} \tag{4.120}$$

$$\zeta \triangleq \frac{\zeta_1}{\sqrt{K_p K_a K_c / K_s}} \tag{4.121}$$

Equation (4.119) shows that the sensitivity now depends on only M, r, R, and K_c, all of which can be made constant to a high degree. This demonstrates again the usefulness of high-gain feedback in shifting the requirements for accuracy and stability from many components to a chosen few where the requirements can be met. As in all feedback systems, the gain cannot be made arbitrarily high because of dynamic instability; however, a sufficiently high gain can be achieved to give excellent performance. Turning to Eq. (4.120), we see that ω_n is increased from the basic spring-mass frequency ω_{n1} by the factor $\sqrt{K_p K_a K_c / K_s}$, another benefit of high-gain feedback. However, ζ is decreased by the same factor, and so ζ_1 must be made sufficiently high to compensate for this.

Figure 4.85*b* shows a commercial instrument[204, 205] based on this principle that uses a micromachined quartz flexure suspension, differential-capacitance angle pickoff, and air squeeze film plus servo lead compensation for system damping. Of the various available models, a 30*g*-range unit has threshold and resolution of 1 μg, frequency response flat within 0.1 percent to 10. Hz (2 percent at 100. Hz, natural frequency of 500. Hz), damping ratio of 0.3 to 0.8, and transverse sensitivity of 0.001 g/g. The output current is 1.3 mA/g; a 250-Ω readout resistor would give about $\pm$ 10 V full scale. In addition to acceleration sensing in many aircraft- and missile-control systems, this unit has been used to measure tilt angles (with respect to gravity) in oil well drilling and axle angular bending in aircraft weight and balance systems, using the gravity sensitivity inherent in all accelerometers. Its micro-*g* sensitivity and low noise allow it to be used for condition assessment[206] of large structures such as dams, offshore drilling rigs, freeways, large buildings, and spacecraft launch vehicles, *using the ambient vibration environment* as the exciting input signal. Most vibration testing is done by artificially exciting the structure using vibration shakers or impact hammers. For large structures, this can be a complex, expensive, and time-consuming process, so it is often not done, even though the results are valuable for validating analytical models or detecting incipient failures. All the structures mentioned are continually excited by their environment (traffic going over bridges, microearthquakes, ocean waves, etc.) but at such a low level that the response motions are very small. The servo-accelerometer listed allows accurate measurement of these small signals. Digital FFT analysis used a sampling rate of 200 samples/second for 256 s, with band-pass filtering (0.033 to 25 Hz), an accelerometer sensitivity of 10 V/*g*, and an additional gain of 200 V/V. Other versions of this instrument (QA-3000) include a built-in temperature sensor that is used with experimental modeling of errors caused by temperature, allowing implementation of correction formulas for such errors. For example, a scale factor error of 150 ppm/°C can be reduced to 100 ppm over the *entire* temperature range (-55 to $+95$°C). The temperature testing results are fitted by fourth-degree polynomials, using multiple-regression methods.[207]

Accelerometers for Inertial Navigation

Inertial navigation is accomplished in principle by measuring the absolute acceleration (usually in terms of three mutually perpendicular components of the total-acceleration vector) of the vehicle and then integrating these acceleration signals

[204]W. Cady, Q-Flex, "Accelerometer Construction and Principle of Operation," TN-103, Honeywell, Redmond, WA, 425-885-8936 (www.inertialsensor.com).

[205]Model QA 700, *Tech. Data,* Honeywell.

[206]Z. H. Duron, S. Rubin and H. Ozisik, "Condition Assessment of Large Structures," *Sound and Vibration,* Aug. 1998, pp. 16–20; J. O. Lassiter, "Microgravity Acceleration Measurements," *Sound and Vibration,* Feb. 1996, pp. 16–19.

[207]E. O. Doebelin, "Engineering Experimentation," McGraw-Hill, New York, 1995, chap. 4.

twice to obtain displacement from an initial known starting location. Thus instantaneous position is always known without the need for any communication with the world outside the vehicle. To keep the accelerometers' sensitive axes always oriented parallel to their original starting positions, elaborate stable platforms using gyroscopic references and feedback systems are necessary. Because the accelerometers are sensitive also to gravitational force, this force must be computed and corrections applied continuously. Because the inertial navigation system measures absolute motion, systems for navigation over the earth's surface (such as for submarines) must include means for compensating for the earth's own motions. An alternative design concept ("strap-down system") does not require the stable platform but rather mounts the gyros and accelerometers directly to the vehicle frame. This approach simplifies the hardware but requires elaborate computation to transform the body-fixed measured data to the required space-fixed coordinate system. Early systems could not use this approach because of the excessive size, limited speed, and limited memory of available computers. Once again, advancing computer technology allows a "software" solution to a practical design problem.

While accelerometers for such navigation systems must operate on essentially the same basic principles that we considered above, their extreme performance requirements and the desire to obtain integrals of the acceleration rather than acceleration itself lead to special techniques and configurations. The desire for compatibility with the required data-processing computers (often digital) also influences the designer's choice of alternatives. The details of these applications are beyond the scope of this book, but may be found in other references.[208]

Mechanical Loading of Accelerometers on the Test Object[209]

The attachment of an accelerometer to a vibrating system results in a change in the motion measured as compared with the undisturbed case. We can apply general impedance principles to this problem to calculate the significance of this effect in any particular instance. In doing so, a useful simplification, which is adequate in most cases, is to regard the entire accelerometer as one rigid mass equal to the total mass of the instrument. This approximation generally holds since accelerometers are used below their natural frequency, and thus there is little relative motion of the proof mass and the instrument case. For simple structures, such as beams, the shift in resonant frequency due to added mass can be estimated from theory.[210]

Laser Doppler Vibrometers

While laser doppler vibrometers are *not* seismic or absolute motion sensors, they are used to make vibration measurements in many of the same applications as seismic

[208]C. F. Savant et al., "Principles of Inertial Navigation," McGraw-Hill, New York, 1961; P. H. Savet (ed.), "Gyroscopes: Theory and Design," McGraw-Hill, New York, 1961; S. Merhav, "Aerospace Sensor Systems and Applications," Springer, New York, 1996.

[209]"Weight Limitations of Accelerometers for High Frequency Vibration Measurements," Wilcoxon Research, Bethesda, MD, *Tech. Bull.,* no. 1, August 1963.

[210]W. T. Thomson, "Theory of Vibrations with Application," 3d ed., p. 24, Prentice-Hall, Englewood Cliffs, NJ, 1985.

devices, so they deserve some mention in this section. Compared to accelerometers, they have the usual advantages of *noncontacting* measurement (no mass loading of the tested device, usable for very small objects, applicable to very hot objects, etc.). We have already covered the laser-doppler principle (Fig. 4.47), but there the emphasis was on measurements of "large and slow" motions such as machine tool and coordinate measuring machine slides. Here, we concentrate on "small and fast" motions characteristic of shock and vibration. Recall that the laser-doppler principle gives basically a *velocity*-sensitive device; displacement or acceleration measurement requires proper data processing. The brief discussion below is based on one manufacturer's publications.[211]

We first need to note that, compared with Fig. 4.47, there is usually no need to attach mirrors or cube corners to the vibrating object; most surfaces can be measured "as is." Also, a *single-frequency* laser whose beam is split into measuring and reference beams is used, with a Bragg cell (acousto-optical) modulator used to frequency shift the reference beam. This produces the needed frequency difference in the two beams and also allows detection of the *direction* of motion, vital for vibration sensing. To cover a wide range of frequencies and amplitudes in optimum fashion, an instrument is provided with a set of velocity and displacement "decoders" of different designs. In the referenced unit, three velocity and three displacement encoders are available. Full-scale ranges of 20 to 20,000 mm/s, resolution of 0.3 to 5 μm/s, and frequency response ranges from 25 to 1500 kHz, with maximum accelerations from 150 to 9,600,000g, can be accommodated by the velocity decoders. For the displacement decoders, full-scale ranges of 0.008 to 82 mm, resolutions from 0.002 to 20 μm, frequency response ranges from 25 to 250 kHz, with maximum velocity from 0.06 to 10 m/s, are available. A special "ultrasonic" decoder for the frequency range of 50 kHz to 20 MHz and full-scale range of 150 nm is useful for applications such as nondestructive testing and calibration of piezoceramic sensors.

A *fiber-optic* version is useful for hard-to-reach targets and short stand-off distances; it also offers the best spatial resolution for microscopic test objects. An *in-plane* version measures motion perpendicular to the axis of the two converging laser beams, while a *rotational* version measures torsional vibration. A *scanning* version is particularly useful for structural dynamics studies that require measurement not only of natural frequencies but also the *mode shapes* of the tested structure. Here a vibration exciting device such as an electrodynamic shaker is attached at a fixed point on the structure. An *xy* galvanometer scanning system that can deflect the laser beam to any point on the measured structure is provided in the vibrometer. The force-input point and the motion-output point thus defined allow measurement of a transfer function between these two locations. The input force might be a "chirp" signal covering the frequency range of interest. Under computer control, the system rapidly moves the motion-measurement point over a selected grid of locations on the test object, all the while computing the motion/force transfer function at each point. Sophisticated FFT software then computes the mode

---

[211]Polytec PI, Inc., Auburn, MA, 508-832-3456 (www.polytecpi.com).

shapes at the various natural frequencies and displays a color-coded picture. A video camera superimposes a frozen video image of the structure to facilitate orientation of the displayed mode shape to the actual measured surface.

4.9 CALIBRATION OF VIBRATION PICKUPS

While the response of vibration pickups to interfering and modifying inputs such as temperature, acoustic noise, and magnetic fields is often of interest, here we are concerned with the response to the desired input of displacement, velocity, or acceleration. An excellent reference giving a more complete treatment of this subject is available.[212] Here we briefly touch on some of the main points.

The calibration methods in wide use may be classified into three broad types: constant acceleration, sinusoidal motion, and transient motion. Constant-acceleration methods (which are suitable only for calibrating accelerometers) include the tilting-support method and the centrifuge. The tilting-support method utilizes the accelerometer's inherent sensitivity to gravity. Static "accelerations" over the range $\pm 1g$ may be accurately applied by fastening the accelerometer to a tilting support whose tilt angle from vertical is accurately measured. This method requires that the accelerometer respond to static accelerations; therefore most piezo-electric devices cannot be calibrated in this way. The accuracy of the method depends on the accuracy of angle measurement and the knowledge of local gravity. The accuracy is of the order of $\pm 0.0003g$. In the centrifuge method, the sensitive axis of the accelerometer is radially disposed on a rotating horizontal disk so that it experiences the normal acceleration of uniform circular motion. Static accelerations in the range 0 to $60,000g$ are achievable with an accuracy of ± 1 percent. The allowable weight of the pickup varies from 100 lb at $100g$ to 1 lb at $60,000g$.

The sinusoidal-motion method is exemplified by the calibration facility of the National Inst. of Standards and Technology.[213] This consists of a modified electro-dynamic vibration shaker which has been carefully designed to provide uniaxial pure sinusoidal motion and which is equipped with an accurately calibrated moving-coil velocity pickup to measure its table motion. If a motion is known to be purely sinusoidal, knowledge of its velocity and frequency enables accurate calculation of the displacement and acceleration. (The motion frequency is easily obtained with high accuracy by electronic counters.) This technique is thus useful for displacement, velocity, or acceleration pickups. The particular equipment referred to above can calibrate pickups (obtaining both amplitude ratio and phase angle) of a weight up to 2 lb over the frequency range 8 to 2,000 Hz. The acceleration range available is 0 to $25g$, velocity range is 0 to 50 in/s, and displacement range is 0 to 0.5 in. Accuracy is ± 1 percent from 8 to 900 Hz and ± 2 percent from 900 to 2,000 Hz.

---

[212]R. R. Bouche, "Calibration of Shock and Vibration Measuring Transducers," Shock and Vibration Info. Ctr., Naval Res. Lab., Code 8404, Washington, 1979.

[213]R. R. Bouche, "Improved Standard for the Calibration of Vibration Pickups," *Exp. Mech.,* April 1961; W. Tustin, "Improved Vibration Calibration at NIST," *Test Engineering & Management,* Dec.-Jan. 1998–1999, pp. 22–23.

Calibration of primary standards such as the NIST moving-coil velocity pickup above is accomplished by the so-called absolute method, with the reciprocity[214] and laser interferometer[215] techniques being the most common. These are both rather involved procedures reserved for standards-laboratory use. Figure 4.86 shows a typical apparatus for the laser method as used to calibrate a high-precision piezo-electric accelerometer which is then utilized as a "transfer standard" to conveniently calibrate other accelerometers by a comparison method. The laser method actually measures displacement amplitude x_o and relies on the purity of the shaker's sinusoidal motion, plus accurate frequency measurement, to infer acceleration amplitude as $\omega^2 x_o$. The particular accelerometer shown is made with a mirror-finish surface to reflect the laser light; thus there is no need to attach a separate mirror when calibration is performed. When the accelerometer has a displacement amplitude x_o, the number of fringes counted by the phototransistor observing the interferometer output is $8x_o/\lambda$ for each complete vibration cycle, which gives 12.64 cycles/μm when a laser with $\lambda = 0.6328$ μm is used. For a typical calibration frequency of 160 Hz, a 10 m/s^2 peak acceleration corresponds to a displacement of 9.9 μm, which gives 125.1 fringes per cycle and 20.016-kHz fringe frequency. To "automate" the needed division of the 20.016-kHz fringe signal by the shaker frequency, a ratio-type counter is used as shown, and it gives a direct readout of the 125.1 motion-amplitude signal.

Once a precision standard accelerometer is calibrated by one of the absolute methods, it can serve as a comparison standard for calibrating "working" transducers. By providing a threaded mounting hole in each end of the standard accelerometer, we can attach it to the calibration shaker and then attach the transducer to be calibrated directly to the standard, ensuring that both will feel the same motion when they are shaken sinusoidally over the desired frequency range. This is the most common method of calibration for working transducers.

Ready-made calibration systems are available from several manufacturers. One such system[216] uses an air-bearing guided table driven by a linear DC motor, giving a 6-in pp stroke for the low frequency range (0 to 200 Hz). A special high-frequency electrodynamic shaker with 0.30-in pp stroke is used for up to 20,000 Hz. Shock calibrations from 10 to 10,000g use a pneumatically operated projectile ("POP") system, while shock calibrations from 5000 to 100,000g use a Hopkinson bar method.[217] Further calibration information is found in other references.[218]

---

[214]Bouche, "Calibration of Shock and Vibration Measuring Transducers," pp. 95–109.

[215]"Accelerometer Calibration, BR 0173," Bruel and Kjaer Instruments, Marlboro, MA.

[216]"Automatic Accelerometer Calibration System," Endevco, San Juan Capistrano, CA (www.endevco.com).

[217]J. Dosch and L. Jing, "Hopkinson Bar Acceptance Testing for Shock Accelerometers," *Sound and Vibration,* Feb. 1999, pp. 16–21.

[218]E. L. Petersen and M. McElroy, "Accurate Calibration of Vibration Transducers from 0.25 Hz at 11,500 Hz," *Cal Lab,* Nov./Dec. 1995, pp. 30–37; F. Schloss, "Improved Reciprocity Calibration of Accelerometers to Very High Frequencies," *Sound and Vibration,* Aug. 1993, pp. 6–8; W. Tustin and F. Lin, "Accelerometer Calibration Using a Laser Doppler Displacement Meter," *Sound and Vibration,* April 1995, pp. 12–14; R. Lally et al., "Six Axis Dynamic Calibration of Accelerometers," PCB Piezotronics, Depew, NY, 716-684-0001.

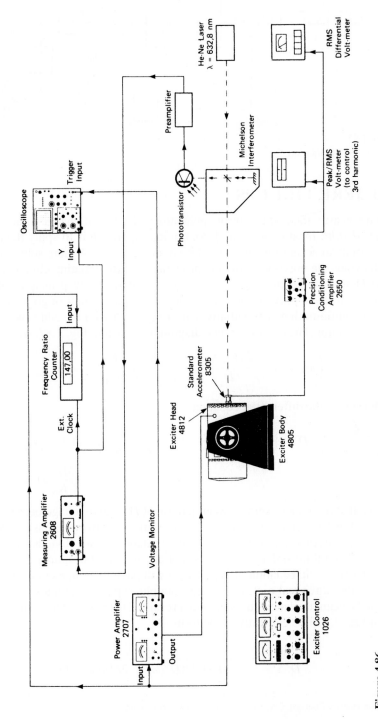

Figure 4.86

Accelerometer calibration by laser interferometer. (*Courtesy Bruel and Kjaer Instruments, Marlboro, MA*)

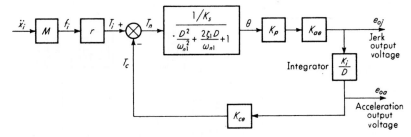

Figure 4.87
Jerkmeter block diagram.

4.10 JERK PICKUPS

In some measurement and control applications, the rate of change of acceleration, or *jerk, d^3x/dt^3*, must be measured. An obvious approach is to apply the electrical output from an accelerometer to a differentiating circuit. A more subtle technique which avoids the noise-accentuating problems of differentiating circuits is applied in the Donner Jerkmeter.[219] By ingenious use of feedback principles, this null-balance instrument provides both acceleration and jerk signals of good quality. The physical configuration is essentially that of Fig. 4.85 with the addition of an electronic integrator. The resulting block diagram is shown in Fig. 4.87. Analysis gives

$$\left(MrD^2 x_i - \frac{K_i K_{ce} e_{oj}}{D} \right) \frac{K_p K_{ae}/K_s}{D^2/\omega_{n1}^2 + 2\zeta_1 D/\omega_{n1} + 1} = e_{oj} \qquad \textbf{(4.122)}$$

which leads to the differential equation

$$\left(\frac{K_s}{K_i K_p K_{ae} K_{ce} \omega_{n1}^2} D^3 + \frac{2\zeta_i K_s}{K_i K_p K_{ae} K_{ce} \omega_{n1}} D^2 + \frac{K_s}{K_i K_p K_{ae} K_{ce}} D + 1 \right) e_{oj}$$
$$= \frac{Mr}{K_i K_{ce}} D^3 x_i \qquad \textbf{(4.123)}$$

We note that the relationship between jerk input $D^3 x_i$ and voltage output e_{oj} is that of a third-order differential equation. The static sensitivity is easily seen to be $Mr/(K_i K_{ce})$ V/(cm/s³); however, the dynamic behavior is not obvious since we have not considered third-order instruments. For any given set of numerical values, the cubic characteristic equation of Eq. (4.123) will have three numerical roots. These will be either three real roots or one real root plus a pair of complex conjugates. In an actual design, the latter situation often prevails. This means that the transfer function of Eq. (4.123) is generally of the form

$$\frac{e_{oj}}{D^3 x_i}(D) = \frac{K}{(\tau D + 1)(D^2/\omega_n^2 + 2\zeta D/\omega_n + 1)} \qquad \textbf{(4.124)}$$

[219]Systron Donner, Concord, CA, 800-227-1625 (www.systron.com).

where $K \triangleq \dfrac{Mr}{K_i K_{ce}}$

and τ, ζ, and ω_n can be found if numerical values for all the constants are given. The frequency response for Eq. (4.124) is easily plotted by using logarithmic methods. Actually, the frequency response for Eq. (4.123) is quite revealing. It is clear that, for sufficiently low frequencies, $e_{oj} \approx [Mr/(K_i K_{ce})]\ddot{x}_i$ for *any* values of the system constants since the first three terms involve ω^3, ω^2, and ω as factors and thus go to zero as $\omega \to 0$. To increase the usable frequency range, the *coefficients* of the ω^3, ω^2, and ω terms must be made small. This can be done by making $K_s/(K_i K_p K_{ae} K_{ce})$ small, which corresponds to making the gain of the feedback loop large. A limit is placed on the gain, however, since excessive gain will cause dynamic instability and resultant destruction of the instrument.

The Routh stability criterion[220] shows that for a cubic characteristic equation of the form $a_3 D^3 + a_2 D^2 + a_1 D + a_0 = 0$, stability is ensured if all the a's are positive and $a_0 a3 < a_1 a_2$. In our case this becomes

$$\frac{K}{K_i K_p K_{ae} K_{ce} \omega_{n1}^2} < \frac{2\zeta_1 K_s^2}{(K_i K_p K_{ae} K_{ce})^2 \omega_{n1}} \tag{4.125}$$

or

$$\zeta_1 > \frac{K_i K_p K_{ae} K_{ce}}{2 K_s \omega_{n1}} \tag{4.126}$$

This shows that if $K_s/(K_i K_p K_{ae} K_{ce})$ is made small to increase the usable frequency range, then the damping ζ_1 must be correspondingly increased to retain adequate stability. This required damping effect also can be obtained by adding proper electrical compensating networks to the circuit. This approach is often used. No matter how the damping is achieved, however, a trade-off will always be necessary between frequency range and stability. Systron Donner also has a version for *angular* jerk.

4.11 PENDULOUS (GRAVITY-REFERENCED) ANGULAR-DISPLACEMENT SENSORS

In a number of applications, the measurement of angular displacement relative to the local vertical (gravity vector) is a useful technique. Examples include sensing elements for control systems of road-paving and scraping machines; drainage-tile-laying machines; alignment of construction forms, piles and bridges; and attitude control of vehicles (such as submarines) and torpedoes or missiles. These relatively simple instruments (basically plumb bobs with electrical output) can sometimes replace more complex and expensive gyroscopic instruments which perform similar functions. Their main disadvantages relative to gyros are their sensitivity to interfering translatory acceleration inputs and their dependence on a gravity field. (They do not work in essentially gravity-free space.)

---

[220]E. O. Doebelin, "Control System Principles and Design," pp. 187–191.

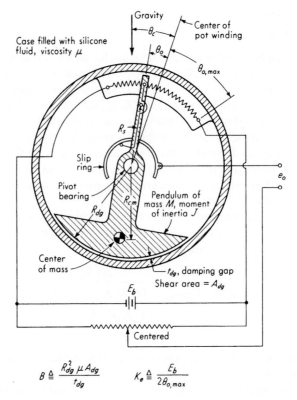

Figure 4.88
Pendulum displacement sensor.

Figure 4.88 shows a typical configuration of a single-axis pendulum-type sensor. The desired input to be measured is the case rotation angle θ_c. The damping effect is not essential to the theoretical operation of the device, but is included in most practical instruments to reduce oscillations at pendulum frequency caused by transient interfering inputs. A variety of electrical displacement transducers may be employed, depending on the required characteristics; a potentiometer is shown for simplicity. The following assumptions are justifiable for most purposes in simplifying the analysis:

1. Angles are small enough that the sine and the angle are nearly equal and the cosine is nearly 1.
2. The inertia effect of the fluid on the pendulum motion is negligible.
3. The damping effect of the fluid is limited to the damping gap.
4. All dry-friction effects in pot wipers, bearings, and slip rings may be neglected for dynamic analysis.
5. The buoyant force on the pendulum is negligible.

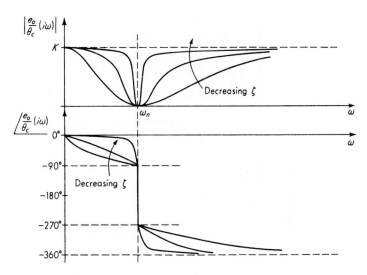

Figure 4.89
Frequency response of pendulum sensor.

The analysis is left for the problems at the end of this chapter. However, the results are as follows:

$$K \triangleq K_e \tag{4.127}$$

$$\omega_n \triangleq \sqrt{\frac{MgR_{cm}}{J}} \tag{4.128}$$

$$\zeta \triangleq \frac{B}{2\sqrt{J(MgR_{cm})}} \tag{4.129}$$

$$\frac{e_o}{\theta_c}(D) = \frac{K(D^2/\omega_n^2 + 1)}{D^2/\omega_n^2 + 2\zeta D/\omega_n + 1} \tag{4.130}$$

The frequency response of this system is shown in Fig. 4.89. Note the "notch-filter" effect at ω_n.

The pendulum sensor is unfortunately sensitive to horizontal accelerations, and so its application is ruled out where such accelerations are large enough to cause a significant output signal. A simple analysis shows that a steady horizontal acceleration A_x will cause an output voltage $K_e A_x/g$. Pendulums with potentiometer sensors[221] have accuracy and resolution about 0.3°, natural frequency about 2 Hz, and damping ratio of 0.1 to 0.7. Any low-range translational *accelerometer* that has a DC response can also be used as a "tiltmeter." The output will be nonlinear with the tilt angle, but this can be linearized in various ways.[222]

---

[221]Humphrey, San Diego, CA, 619-565-6631 (www.humphreyinc.com).

[222]Crossbow Technology, Inc., San Jose, CA, 408-965-3300 (www.xbow.com).

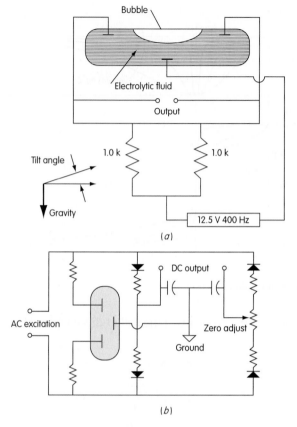

Figure 4.90
Electrolytic tilt sensor.

Electrolytic tilt sensors[223] are an electronic version of the familiar carpenter's bubble level and were first used in the guidance system of Germany's V2 rockets toward the end of World War II (1944). Figure 4.90*a* shows the principle of operation. A glass vial with three attached electrodes is partially filled with a conductive liquid, forming two liquid "resistors" whose resistances are equal when the vial is level. When the vial is tipped relative to gravity, the liquid stays level, and consequently, one resistance increases and one decreases. The two liquid resistors are connected in a Wheatstone bridge with two ordinary fixed resistors and an AC excitation voltage source. (If DC excitation were used, positive and negative liquid ions would migrate to the electrodes, eventually rendering the liquid nonconductive. In any signal-processing circuitry, a DC voltage must *never* be allowed to appear

---

[223]The Fredericks Co., Huntington Valley, PA, 215-947-2500 (www.frederickscom.com); Spectron Glass & Electronics, Inc., Hauppage, NY, 516-582-5600 (www.spectronsensors.com); Lucas Sensing Systems Inc., 800-745-8008 (www.schaevitz.com); W. B. Powell and D. Pheifer, "The Electrolytic Tilt Sensor," *Sensors,* May 2000, pp. 120–125.

across the vial electrodes.) Any tilt angle will unbalance the bridge, producing an AC bridge output voltage related to the tilt angle. Figure 4.90*b* (Fredericks Co.) shows one scheme for obtaining a DC output which changes polarity with the direction of tilt. Full-scale ranges from 1 to 360° are available. Low-scale ranges can have sensitivities of 250 mV/(arc min). Both linear and nonlinear responses are possible, as are dual-axis units. The dynamic response is similar to a second-order system because of the liquid "sloshing" effect. Damping can be controlled by adjusting liquid viscosity and settling times are the order of 1 s. An alternative technology[224] uses a *dielectric* (insulating) liquid and a capacitance sensing scheme. Tilt angle causes a differential change in capacitance of two capacitors, which in turn changes the frequency of a digital oscillator chip. A typical instrument has a range of ± 60° with a resolution of 0.001° and flat response (± 3 dB) to 0.5 Hz.

Since accelerometers are sensitive to gravity forces, they can be used to measure tilt angles. By combining a three-axis accelerometer (measures roll and pitch angles) with a three-axis magnetometer (measures yaw angle), it is possible to measure all three position angles in a compact sensor package.[225] Note that an accelerometer *cannot* detect the yaw (sometimes called azimuth) angle since the gravity vector has no component in this direction. For most locations on earth, earth's magnetic field is essentially in the horizontal plane and in a "northerly" direction, so azimuth rotation is easily detected. The *magnetometer* measures the three vector components of the earth's magnetic field that exist at the measurement site, and this allows calculation of the azimuth angle. For the three-dimensional rotations involved, the accelerometer and magnetometer output signals must be processed through the appropriate trigonometric conversions to get angles that are directly related to angular orientation. Software to perform these calculations is provided by the sensor manufacturer.

4.12 GYROSCOPIC (ABSOLUTE) ANGULAR-DISPLACEMENT AND VELOCITY SENSORS

While gyroscopic instruments have been used in limited numbers and applications (gyrocompasses in ships and aircraft, turn-and-bank indicators in aircraft, etc.) since around World War I, developments during and after World War II brought them to an extreme degree of refinement, and their use is now common in military and commercial applications.[226] In most of these applications, the gyro is used to measure the motion of some vehicle, either as a permanent part of a vehicle measurement-control system or as a test instrument temporarily attached to gather data. Less common, but still important, applications include oil well drilling, where a gyro package is included in the drill string, to measure and control angular attitude and thus drilling direction, miles below the earth's surface. The "classical"

---

[224]Accustar Electronic Clinometer, Schaevitz Sensors, Hampton, VA, 800-745-8008 (www.schaevitz.com).

[225]Applied Physics Systems, Mountain View, CA, 650-965-0500 (www.appliedphysics.com).

[226]Sidney Lees (ed.), "Air, Space, and Instruments," p. 32, McGraw-Hill, New York, 1963.

gyroscope always contains a spinning wheel, but the term is today also used for devices that measure absolute angular motion using *other* principles that have nothing to do with spinning wheels (ring laser, vibrating tuning fork, fiber optic, fluidic vortex, hot-wire jet-deflection). Spinning-wheel instruments still dominate the overall market, but certain niche markets have been taken over by some of the other technologies (for example, ring laser gyros for inertial navigation of commercial jet aircraft and vibrating tuning fork for automotive stability augmentation). All of these technologies will be discussed, but analytical treatment is concentrated on the spinning-wheel instruments.

Perhaps the simplest gyro configuration is the free gyro shown in Fig. 4.91. These instruments are used to measure the absolute angular displacement of the vehicle to which the instrument frame is attached. A single free gyro can measure rotation about two perpendicular axes, such as the angles θ and ϕ. This can be accomplished because the axis of the spinning gyro wheel remains fixed in space (if the gimbal bearings are frictionless) and thus provides a reference for the relative-motion transducers. If the angles to be measured do not exceed about $10°$, the readings of the relative-displacement transducers give directly the absolute rotations with good accuracy. For larger rotations of both axes, however, there is an interaction effect between the two angular motions, and the transducer readings do *not* accurately represent the absolute motions of the vehicle. The free gyro is also limited to relatively short-time applications (less than about 5 min) since gimbal-bearing friction causes gradual drift (loss of initial reference) of the gyro

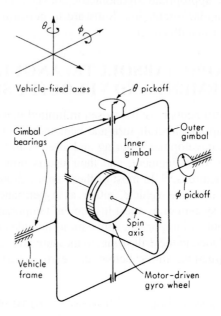

Figure 4.91
Free gyroscope (two-axis position gyro).

spin axis. A constant friction torque T_f causes a drift (precession) of angular velocity ω_d given by

$$\omega_d = \frac{T_f}{H_s} \tag{4.131}$$

where H_s is the angular momentum of the spinning wheel. It is clear that a high angular momentum is desirable in reducing drift. A typical application would be found in the guidance system of a short-range missile, where the short lifetime (less than a minute) can tolerate the drift. A representative example[227] uses compressed gas to spin up the rotor in 0.25 s, whereafter it "coasts," giving a useful life of about a minute during which the drift is about 1° in the first 15 s. Potentiometer pickoffs have ranges of $\pm 60°$ for the outer gimbal (optional to 355°) and $\pm 40°$ for the inner gimbal. The gyro is cylindrical: 2-in diameter and 4 in long, including the gas chamber. The angular momentum is 2.5×10^5 g · cm²/s.

Rather than using free gyros to measure two angles in one gyro (thus requiring two gyros to define completely the required three axes of motion), high-performance systems utilize the so-called single-axis or constrained gyros. Here a single gyro measures a single angle (or angular rate); therefore three gyros are required to define the three axes. This approach avoids the coupling or interaction problems of free gyros, and the constrained (rate-integrating) gyros can be constructed with exceedingly small drift. We consider here two common types of constrained gyros: the rate gyro and the rate-integrating gyro. The rate gyro measures absolute angular velocity and is widely used to generate stabilizing signals in vehicle control systems. The rate-integrating gyro measures absolute angular displacement and thus is utilized as a fixed reference in navigation and attitude-control systems. The configuration of a rate gyro is shown in Fig. 4.92; the rate-integrating gyro is functionally identical except that it has no spring restraint.

While a general analysis of gyroscopes is exceedingly complex, useful results for many purposes may be obtained relatively simply by considering small angles only. This assumption is satisfied in many practical systems. Figure 4.93 shows a gyro whose gimbals (and thus the angular-momentum vector of the spinning wheel) have been displaced through small angles θ and ϕ. We apply Newton's law

$$\sum \text{torques} = \frac{d}{dt}(\text{angular momentum}) \tag{4.132}$$

to the x and y axis components of angular momentum. This angular momentum is made up of two parts, one part due to the spinning wheel and another part (due to the motion of the wheel, case, gimbals, etc.) which would exist even if the wheel were not spinning. The latter part depends on (for the x axis) the angular velocity $d\phi/dt$ and the moment of inertia I_x of everything that rotates when the outer gimbal

---

[227]"Two Axis Gyroscope, Gas Activated, Model TAG 000," Allied Signal, Cheshire, CT, 203-250-3704 (www.alliedsignal.com). This company supplies a CD-ROM that catalogs many gyro instruments and also includes a 60-page "textbook" on gyro theory and application.

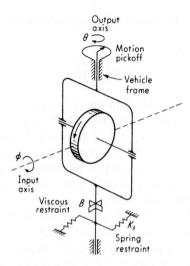

Figure 4.92
Single-axis restrained gyro.

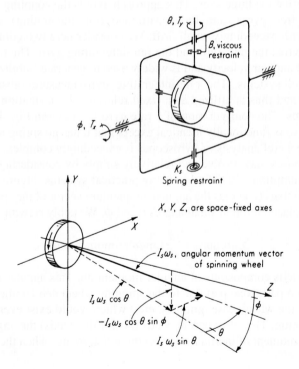

Figure 4.93
Gyro analysis.

turns in its bearing. For the y axis, the angular momentum depends on $d\theta/dt$ and the moment of inertia I_y of everything that rotates when the inner gimbal turns in its bearing. The external applied torques T_x and T_y are included to allow for the possibility of bearing friction and for intentionally applied torques from small electromagnetic "torquers" which are used in some systems to cause desired precessions for control or correction purposes. The inertias I_x and I_y (which are about *space-fixed* axes) actually change when θ and ϕ change, but this effect is negligible for small angles. Also, the exact equations would contain terms in the *products* of inertia as well as the moments of inertia, but these are again negligible because of the small angles and the inherent symmetry of gyro structures. With the above qualifications, we may write for the x axis

$$T_x = \frac{d}{dt}\left(H_s \sin\theta + I_x \frac{d\phi}{dt}\right) \tag{4.133}$$

and for the y axis

$$T_y - B\frac{d\theta}{dt} - K_s\theta = \frac{d}{dt}\left(-H_s \cos\theta \sin\phi + I_y \frac{d\theta}{dt}\right) \tag{4.134}$$

We now assume H_s is a constant (the gyro wheel is driven by a constant-speed motor) and $\cos\theta = 1$, $\sin\theta = \theta$, $\sin\phi = \phi$ to get

$$T_x = H_s\frac{d\theta}{dt} + I_x\frac{d^2\phi}{dt^2} \tag{4.135}$$

$$T_y - B\frac{d\theta}{dt} - K_s\theta = -H_s\frac{d\phi}{dt} + I_y\frac{d^2\theta}{dt^2} \tag{4.136}$$

These are two simultaneous linear differential equations with constant coefficients relating the two inputs T_x and T_y to the two outputs θ and ϕ. Writing these equations in operator form, we can treat them as algebraic equations to solve for ϕ and θ as desired. For ϕ we get

$$\phi = \frac{(I_y D^2 + BD + K_s)\,T_x - (H_s D)\,T_y}{D^2\,(I_x I_y D^2 + BI_x D + H_s^2 + I_x K_s)} \tag{4.137}$$

Since ϕ depends on both T_x and T_y, transfer functions can be obtained by considering each input separately and then using superposition. Letting $T_y = 0$, we get

$$\frac{\phi}{T_x}(D) = \frac{I_y D^2 + BD + K_s}{D^2\,(I_x I_y D^2 + BI_x D + H_s^2 + I_x K_s)} \triangleq G_1(D) \tag{4.138}$$

and letting $T_x = 0$ gives

$$\frac{\phi}{T_y}(D) = -\frac{H_s}{D(I_x I_y D^2 + BI_x D + H_s^2 + I_x K_s)} \triangleq G_2(D) \tag{4.139}$$

This leads to the block diagram of Fig. 4.94*a*. Similar analysis for θ gives

$$\frac{\theta}{T_x}(D) = \frac{H_s}{D(I_x I_y D^2 + BI_x D + H_s^2 + I_x K_s)} \triangleq G_3(D) = -G_2(D) \tag{4.140}$$

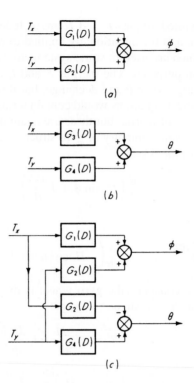

Figure 4.94
Gyro block diagrams.

and

$$\frac{\theta}{T_y}(D) = \frac{I_x}{I_x I_y D^2 + BI_x D + H_s^2 + I_x K_s} \triangleq G_4(D) \tag{4.141}$$

which leads to the block diagram of Fig. 4.94b. An overall block diagram may then be constructed as in Fig. 4.94c.

The above results are of quite general applicability to gyro systems of various configurations as long as the small-angle requirement is met. For single-axis rate and rate-integrating gyros, considerable simplification of the above results is possible. In these applications (see Fig. 4.92) we consider the input to be the *motion* ϕ. A torque T_x also exists and would be felt by the vehicle; however, it does not enter our calculations since we assume the motion ϕ is given. The angle θ is an indication of the angle ϕ (rate-integrating gyro) or angular velocity ϕ (rate gyro); thus we should like to have transfer functions relating θ to ϕ. The torque T_y (neglecting bearing friction) is zero for this application unless a torquer is used for some special purpose. Then the desired θ-ϕ relation may easily be obtained by solving Eqs. (4.138) and (4.140) for T_x and setting them equal. The result is

$$\frac{\theta}{\phi}(D) = \frac{H_s D}{I_y D^2 + BD + K_s} \tag{4.142}$$

For a rate gyro, then, we have the second-order response

$$\frac{\theta}{D\phi}(D) = \frac{\theta}{\dot{\phi}}(D) = \frac{K}{D^2/\omega_n^2 + 2\zeta D/\omega_n + 1} \tag{4.143}$$

where $\quad K \triangleq \dfrac{H_s}{K_s} \quad$ rad/(rad/s) $\tag{4.144}$

$$\omega_n \triangleq \sqrt{\frac{K_s}{I_y}} \quad \text{rad/s} \tag{4.145}$$

$$\zeta \triangleq \frac{B}{2\sqrt{I_y K_s}} \tag{4.146}$$

A high sensitivity is achieved by large angular momentum H_s and soft spring K_s, although low K_s gives a low ω_n. Natural frequencies of commercially available rate gyros are of the order of 10 to 100 Hz. The damping ratio is usually set at 0.3 to 0.7. Large angular momentum is obtained in small size by using high-speed (often 24,000 r/min) motors to spin the gyro wheel. Full-scale ranges of about $\pm$ 10 to $\pm$ 1000 degree/s are readily available. Resolution of a $\pm$ 10 degree/s range instrument is of the order of 0.005 degree/s. In some high-performance rate gyros, the mechanical spring is replaced by an "electrical spring" arrangement similar to that used in the servo-accelerometer of Fig. 4.85. A two-axis unit[228] that uses this servo principle employs several clever design features to produce high performance at low cost. The two-axis feature is a result of a spherical permanent-magnet rotor supported in a hydrodynamic gas bearing. Laser-diode optical pickoffs measure the rotor deflection in each of the two axes and supply error signals to servo electronics that torque the two axes to maintain the desired null position. Torquing current then is a measure of angular rate. Clever magnetic design allows the rotor/stator combination to provide both spin axis drive torque and also two-axis rebalance torques. Whereas "conventional" rate gyros have over 130 parts, a $1500 cost, and complex assembly procedures, the Minitact has only 14 parts, simplifying assembly and lowering the cost.

The *dynamically tuned* gyro[229] is a unique spinning-wheel instrument developed as a replacement for the classical floated rate-integrating gyro. This two-axis instrument does not use any viscous damping effects, but achieves its performance with a clever *negative spring constant* concept.

To measure all three components (roll, pitch, and yaw) of angular velocity in a vehicle, an arrangement of three rate gyros, such as in Fig. 4.95, may be employed. It should be pointed out that in a rate gyro only the output angle θ must be kept small. This requires use of a very sensitive motion pickoff, but such are available. The input angle ϕ may be indefinitely large since no matter how large it gets the spin angular-momentum vector is always perpendicular (except for a small error resulting from nonzero θ) to the input angular-velocity vector. Thus a roll-rate gyro in an aircraft will function correctly even if the aircraft rolls completely over. Of

---

[228]"Minitact," Allied Signal; H. T. Califano, "Minitact Gyroscope-The Low Cost Alternative."

[229]S. Merhav, pp. 237–256; Litton, Salt Lake City, UT, 801-323-6555 (www.littongcs.com).

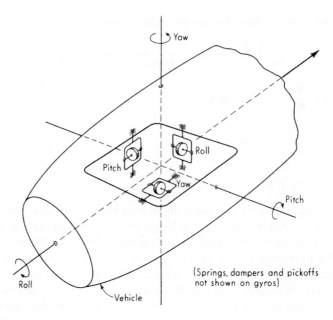

Figure 4.95
Three-axis rate-gyro package.

course, the angular-velocity components measured are those about the *vehicle* axes rather than about space-fixed axes, but these are usually the velocities desired for stabilization signals in control systems.

To obtain a rate-integrating gyro, we merely remove the spring restraint from the configuration of Fig. 4.92. Equation (4.142) then becomes

$$\frac{\theta}{\phi}(D) = \frac{K}{\tau D + 1} \tag{4.147}$$

where

$$K \triangleq \frac{H_s}{B} \quad \text{rad/rad} \tag{4.148}$$

$$\tau \triangleq \frac{I_y}{B} \quad \text{s} \tag{4.149}$$

We see that the output angle θ is a direct indication of the input angle ϕ according to a standard first-order response form. High sensitivity again requires high H_s, a typical value being 5×10^5 g · cm^2/s. Low damping B also increases sensitivity, but at the expense of speed of response, as shown by Eq. (4.149). For a K of 15, a τ of 0.006 s is typical of high-performance units. Increase of K to 440 raises τ to 0.17 s. The rate-integrating gyro is the basis of highly accurate inertial navigation systems, where it is used as a reference to maintain so-called stable platforms in a fixed attitude within a vehicle while the vehicle moves arbitrarily. This is done by using the motion signals from the gyros to drive servomechanisms which maintain the platform in a fixed angular orientation. The system accelerometers are mounted on this

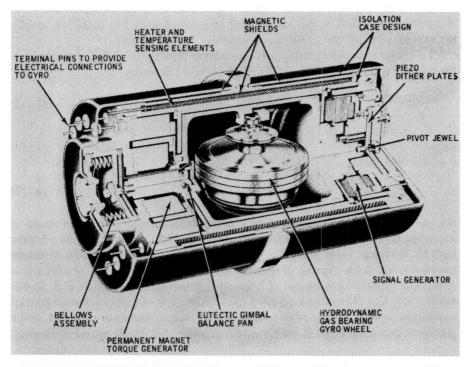

Figure 4.96
Floated, rate-integrating gyro. (*Courtesy Honeywell, Inc., Minneapolis, MN*)

platform, and their double-integrated output is an accurate measure of vehicle motion along the three orthogonal axes since the platform always moves parallel to its initial orientation.

The rate-integrating gyro has been subjected to extremely intensive and extensive engineering development to bring its performance characteristics to remarkable levels while maintaining small size and weight and great resistance to rugged environments. This has been accomplished by painstaking attention to minute mechanical, thermal, and electrical details which would be completely negligible in a less sophisticated device. Figure 4.96 shows a floated rate-integrating gyro which incorporates a number of clever design features necessary to high performance. The inner gimbal is in the form of a hollow cylinder containing the gyro wheel and is mounted in precision pivot jewel bearings. A small radial gap between this cylinder and its housing is filled with damping fluid to form the damper *B*. Temperature feedback-control systems regulate electric heaters to maintain damping-fluid viscosity constant. This heating must be carefully applied so as to minimize convection currents in the damping fluid which would exert spurious torques on the inner gimbal. The density of the damping fluid (which completely surrounds the hollow cylinder) is adjusted so as to obtain neutral buoyancy (this explains the "floated gyro" terminology), thereby removing the radial load from the jewel bearings and reducing their already-low friction even more. A further 10:1 bearing-friction reduction is achieved by vibrating ("dithering") the pivot bearings at low amplitude and high frequency by means of electrically driven piezo crystals. (Magnetic noncontact bearings[230] are also sometimes used to reduce friction.) A further advantage of the floated construction is an increase in shock and vibration tolerance since case shocks are transmitted to the inner gimbal through the damping fluid rather than the rather fragile jewel bearings.

At original assembly, the inner gimbal is very carefully statically balanced since any unbalance creates a spurious gravity torque. When such a gyro is built into, say, an ICBM, it may be inactive for long periods, and various environmental and aging effects can cause a loss of balance. To correct for such unbalance without removing the gyro from the missile or even disturbing its hermetic seals, the eutectic gimbal balance pan is provided. If the gyro is activated by running its wheel up to speed, the gyro becomes a very sensitive detector of unbalance since the inner gimbal will slowly precess under the action of the unbalanced gravity torque and this motion will be detected by the θ displacement transducer (signal generator). To rebalance the gimbal, an electric current is applied to melt a eutectic alloy in the balance pan so that the molten metal migrates to move its mass center and thereby regain balance. When gimbal precession stops, rebalance has been obtained and the metal is allowed to solidify. Gyro wheel bearings of the hydrodynamic gas type contribute to long life and low noise since they ride on a film of gas with no metal-to-metal contact, except momentarily when starting or stopping. The permanent-magnet torque generator allows precise electrical application of inner gimbal torques for self-testing and correction procedures. An "isoelastic" design for the inner gimbal bearing suspension ensures that elastic deflections caused by case linear accelerations will be in the same direction as the acceleration. This means that the inertia torque will not have a moment about the center of mass, which would cause a spurious gyro output.

Developments in absolute angular-motion sensors have included both exotic versions of the classical "spinning wheel" concept and totally new devices devoid of gyroscopic effects but often still called "gyros" because they perform the same measurement function as the traditional mechanical instrument. Electrostatic gyros[231] utilize feedback-type electrostatic systems to levitate a 1-cm-diameter beryllium sphere spinning at 146,000 r/min without physical contact inside a chamber evacuated to 10^{-7} torr. The gyro is a free-rotor type with two-axis angle-measuring capability for large angles, which makes it especially suitable for strapdown navigation systems. A deliberate mass unbalance in the spinning sphere causes motions related to sphere angular attitude which are picked up by a capacitance-displacement sensor that uses eight electrodes arranged octahedrally over the spherical housing's inner surface. These electrodes are timeshared between

[230]R. H. Frasier, P. J. Gilinson, and G. A. Oberbeck, "Magnetic and Electric Suspensions," MIT Press, Cambridge, MA, 1974.

[231]R. R. Duncan, "Micron A Strapdown Inertial Navigator Using Miniature Electrostatic Gyros," Autonetics Div., Rockwell Intl., Anaheim, CA, 1973; S. Merhav, "Aerospace Sensor Systems and Applications," Springer, New York, 1996, pp. 258–261.

the suspension servosystem and the angle-measuring system, with 20 percent of the duty cycle being devoted to angle measurement. Extreme accuracy is obtained by calibrating each gyro to obtain numerical values for a sophisticated error model containing 56 sources of error. This model is stored by the navigation computer and used to process the gyro angle data. Here again we see the use of computing power (made feasible by microcomputer technology) to overcome basic electromechanical hardware limitations in instrument design.

Optical (ring-laser) "gyros"[232] do not use gyroscopic principles at all. Rather, they measure phase shifts between two laser beams directed around a loop by mirrors fastened to the object whose rotation is to be measured. One beam travels with the rotation while the other travels against it, causing a phase shift proportional to rotation when the beams are recombined and compared. The practical realization of this basic concept uses two active optical oscillators internal to the loop, rather than an external light source. These oscillators (lasers) are arranged so that rotation produces a frequency difference between them. In a typical instrument with a triangular optical path of 8.4-in perimeter and laser wavelength of 0.6328 μm, a 1 rad/s rotation rate produces a 65-kHz output frequency. Since the time integral of the output-pulse rate is proportional to input angle, a simple summing of pulses gives a rate-integrating-gyro type of response with a resolution of 1 count for 3.1 seconds of arc of rotation. The development of the no-moving-parts ring-laser gyro appears to be synergistic with that of the microcomputer since the two together make the use of so-called strapdown guidance systems very attractive. These systems, rather than using complex servo systems to maintain the gyro mounting platform in a fixed attitude, strap the gyro directly to the vehicle frame and compute, from the gyro signals, the axis-transformation data required to convert from the vehicle reference frame to the necessary inertial coordinates. The low-cost, compact microcomputers thus allow us to substitute computing power for electromechanical hardware.

Almost all commercial and military jet aircraft now use inertial navigation systems based on ring-laser gyros. The gyro manufacturers[233] generally do not sell these gyros as separate units but rather incorporate them into complete navigation systems that they design for specific projects. For these and government security reasons, one cannot usually obtain detailed specification sheets for these gyros. A huge (1.2 × 1.2 m, 600 kg) version[234] built of Zerodur in Germany and installed in a subterranean cave in New Zealand is used to measure the earth's rotation with 0.1 ppm resolution.

The *interferometric-fiber-optic gyro*[235] (IFOG) uses essentially the same principle as the ring-laser gyro, but the carefully sealed, gas-filled laser cavities and precision mirrors are replaced by a long coil of optical fiber. Merhav discusses both

[232]S. Merhav, op. cit., chap. 8.

[233]Honeywell Sensor Guidance Products, Minneapolis, MN, 612-957-4051 (www.honeywell.com); Litton Guidance & Control Systems, Salt Lake City, UT, 801-323-6555 (www.littongcs.com).

[234]A. Strass, "Ring-Laser Gyro Provides New Geological Information," *Laser Focus World,* March 1997, pp. 48–52.

[235]S. Merhav, op. cit., chap. 7; H. Lefevre, "The Fiber-Optic Gyroscope," Artech House, Boston, 1993; K. Hotate, "Fiber Optic Gyros Put a New Spin on Navigation," *Photonics Spectra,* Apr. 1997, pp. 108–112.

"open-loop" and "closed-loop" versions. Closed-loop versions[236] are expensive but have high enough performance for many inertial navigation applications. The simpler construction and lower cost of the open-loop version[237] are accompanied by lower performance; however, it is adequate for many applications. The phase shift ΔS[238] between the counter-propagating light beams is the basic output of this gyro:

$$\Delta S = \frac{2\pi L D}{c\lambda} \, \Omega \qquad (4.150)$$

where

$L \triangleq$ length of fiber coil (typically about 100 meters)

$D \triangleq$ effective coil diameter

$c \triangleq$ velocity of light

$\lambda \triangleq$ mean optical wave length

$\Omega \triangleq$ angular velocity about sensitive axis

This phase shift is extremely small and requires careful processing (included in commercial units) to get a useable instrument. The output of the ring-laser gyro is the frequency difference Δf[239]:

$$\Delta f = \frac{4A}{\lambda L} \, \Omega \qquad (4.151)$$

where L is now the perimeter of the (often triangular) light path and A is the area enclosed by this perimeter. Measurement of this frequency again requires careful opto-electronic processing.

The MEMS technology that provides low-cost pressure transducers and accelerometers has also been applied to providing miniature replacements for spinning-wheel-type gyroscopes. Most of these use some version of the Coriolis principle, also employed in some mass-flow meters (see Chap. 7). Figure 4.97*b* shows the GyroChip,[240] which uses micromachined piezoelectric quartz tuning forks to implement a Coriolis scheme. The drive tines are driven in resonance at about 10 kHz. Although they are really *distributed* vibrating systems, think of them as lumped, with a concentrated mass m at the tip of each tine. Their steady vibration provides the V (velocity) term in the Coriolis acceleration given by $2\omega \times V$, which results in an inertia force $m \cdot 2\omega \times V$ which acts *perpendicular* to V. When an input rate ω_i exists, the inertia forces of the two drive-tine masses cause an exciting torque that

[236]Fibersense Technology Corp., Canton, MA, 781-830-9690 (www.fibersense.com).

[237]S. M. Bennett et al., "Fiber Optic Rate Gyros as Replacements for Mechanical Gyros," KVH Industries, Middletown, RI, 401-847-3327 (www.kvh.com).

[238]S. M. Bennett, op. cit.

[239]S. Merhav, p. 379.

[240]Systron Donner, Concord, CA; A. M. Madni et al., "A Miniature Yaw Rate Sensor for Intelligent Chassis Control," 925-671-6400 (www.systron.com); A. M. Madni and R. C. Geddes, "A Micromachined Quartz Angular Rate Sensor for Automotive and Advanced Inertial Application," *Sensors,* Aug. 1999, pp. 26–34; S. D. Orlosky and H. D. Morris," A Quartz Rotational Rate Sensor," *Sensors,* Feb. 1995.

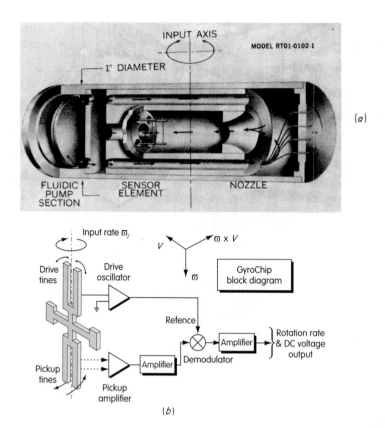

Figure 4.97
Hot-wire jet-deflection sensor and Gyrochip.

is transmitted through the support to the pickup tines, causing them to vibrate at the same frequency as the drive tines. The pickup tine *resonance,* however, is chosen about 300 Hz *above* this frequency, so the pickup tine motion exists in the "low" frequency range where the spring effect dominates and displacement is *proportional* to the driving torque. This displacement is thus proportional to input angular velocity and is measured piezoelectrically. Standard ranges include $\pm$ 50 to $\pm$ 1000°/s, bandwidth is about 60 Hz, and full-range output is $\pm$ 2.5 V. Sensor noise has a flat PSD from 1 to 100 Hz and is quoted as 0.01 (degree/second)/$\sqrt{\text{Hz}}$. Since we sometimes want to include noise in dynamic models of sensors, let us now use this device as an example of the general method.

The noise value quoted on the specifications sheet is not really a PSD, so let us first convert to that standard quantity, which would be in (degree/second)^2/Hz and thus would numerically be 0.0001. The specification sheet also gives the sensitivity of this unit as 0.0352 V/(degree/second), so the PSD in these units is now (3.52×10^{-4})V^2/Hz. If we had a sensor transfer function and wanted to simulate this with SIMULINK, including the noise voltage, we could proceed as follows. SIMULINK's random-signal generator does not allow us to directly enter a number

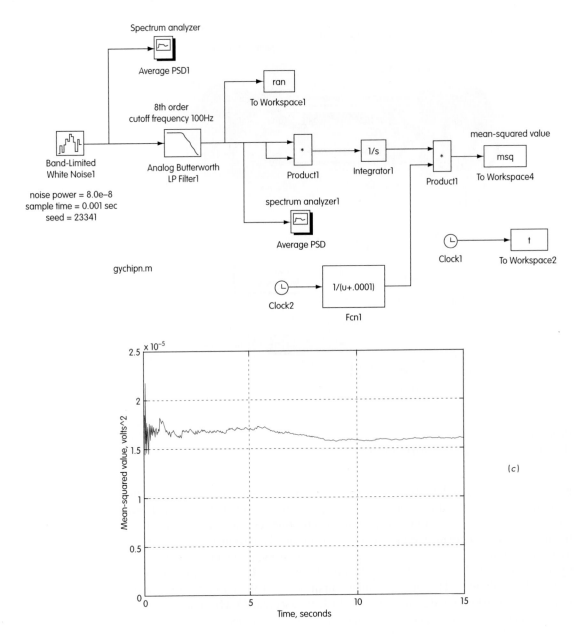

Figure 4.97
(Continued)

for a PSD value. Rather, we must enter a value for something called "noise power" and another value for "sample time." The choice of sample time sets the frequency range of the noise, that is, smaller times give higher frequency range. Our approach is to choose a value that gives a frequency range a little beyond the 100 Hz we really want and then low-pass filter to give a smoothly varying signal with a bandwidth of

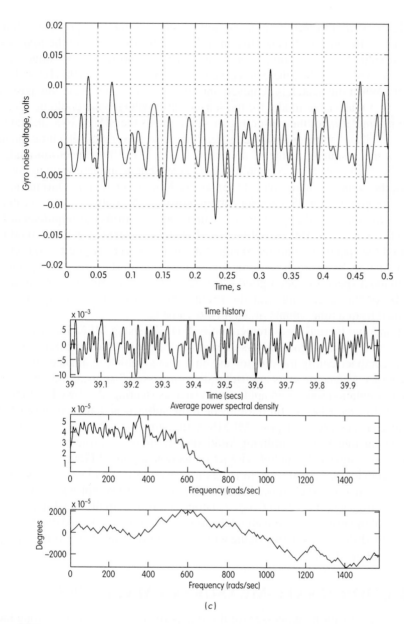

Figure 4.97
(Concluded)

about 100 Hz. In SIMULINK, to get white noise with a bandwidth of about f Hz, you need to set the sample time to about $1/(10f)$, in our case, $\frac{1}{1000} = 0.001$ sec. The SIMULINK help notes for "band-limited white noise" gives some relations between "noise power," sample time, and covariance (mean squared value) and also states that "noise power" is "the height of the PSD of the white noise." Unfortunately,

PSD has *several* accepted definitions, and the manual does not give the definition being used. Because of this confusion, I decided to just compute the *total* mean-squared value by "brute force" and then adjust the "noise power" value by trial and error until I matched the desired value of $(3.52 \times 10^{-4})^2$ (100 Hz) = 1.24×10^{-5} V^2. When this match is achieved, the signal coming out of the low-pass filter should be a reasonable model for the sensor noise voltage, and could be used in a dynamic simulation of the sensor. Figure 4.97c shows the SIMULINK diagram and some graphical results for a trial run where the noise power was set a little too high, giving a mean-squared value of about 1.6×10^{-5}. The two spectrum analyzers shown are "advertised" as computing PSD values, but again they seem to not quite do this; they don't seem to divide by Δf, as the definition requires. Thus the numerical values graphed for PSD are not correct; however this only affects the scaling of the plot, the *shape* of the spectrum is accurate, and confirms our requirement for a flat PSD out to about 100 Hz. (If these PSD numbers were correct, we could have just used the spectrum analyzer to adjust the noise power.) If you try to duplicate this simulation, be aware that, for random signals, one must *average* a *large* number of individual spectra to get a good estimate.[241] The spectrum analyzer used *does* provide for this averaging, and the results shown used about 65 spectra, each computed from a time sample of length 0.6 seconds.

Strapdown inertial measurement units[242] (IMU) using MEMS accelerometers and rate gyros, plus magnetometers (measuring the earth's magnetic field) are available in compact configurations for use in various vehicle guidance and control applications. Magnetometers are useful in "compassing" applications and also for angular orientation sensing in applications such as steering of oil-well drilling. The referenced website has useful tutorials on these subjects and also on sensor fusion techniques necessary in adapting MEMS instruments for critical applications.

Angular-rate sensors utilizing fluid (nongyroscopic) principles include the fluidic vortex device[243] and hot-wire jet-deflection sensor.[244] Figure 4.97[245] shows the latter device, which utilizes vibrating piezo crystals to circulate a steady laminar flow of gas in a sealed chamber. The gas jet passes over two hot-wire flow sensors (see hot-wire anemometers in Chap. 7) connected differentially in a bridge circuit. Rotation of the instrument case about the input axis causes a lateral deflection of the gas stream, differential cooling of the wires, and a bridge-unbalance voltage proportional to angular rate.

4.13 COORDINATE-MEASURING MACHINES

Modern manufacturing processes for discrete parts require fast and accurate measuring devices to check critical dimensions against their specified values. In a numerically controlled machining center with an automatic tool changer, one of the tool

---

[241]E. O. Doebelin, "System Modeling and Response," Wiley, 1980, pp. 268–270.

[242]Crossbow Technology, Inc., San Jose, CA, 408-965-3300 (www.xbow.com).

[243]G. P. Wachtell, "Fluidic Vortex Angular Rate Sensor," *USAAV LABS Rep.* 70-25, Fort Eustis, VA, 1970.

[244]*Eng. Bull.* 1075, Humphrey, San Diego, CA, 858-565-6631 (www.humphreyinc.com).

[245]Ibid.

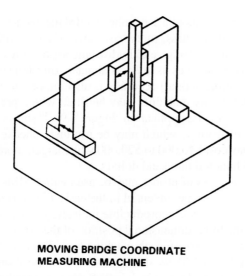

**MOVING BRIDGE COORDINATE
MEASURING MACHINE**

Figure 4.98
Axis arrangements in coordinate-measuring machines.

positions may be occupied by a gaging probe, so that each part can be measured, and corrections made, as that part is being machined. This is called *in-process gaging*. When this approach is not possible or desirable, *off-line gaging* may be used. Here the parts are not measured until they leave the machining process and reach a separate gaging station. Each part (or perhaps a statistically selected partial sample of the total production) is then measured, and statistical data on average values and deviations are accumulated. These data are periodically compared with part specifications, and machine-setting corrections are made when necessary. Many gaging techniques ranging from simple, manually operated ones to complex, computer-controlled systems have been developed for in-process[248] and off-line gaging.[247] In this section we briefly describe a versatile off-line technique based on locating critical part features in a Cartesian coordinate system.

A typical coordinate-measuring machine[246] (CMM) has a probe mounted on a set of three mutually perpendicular slides such that the probe can be positioned at any desired *x, y, z* location within the machine's working space. Different arrangements of the slides offer specific advantages, so manufacturers offer CMMs in a variety of configurations; Fig. 4.98 shows one. Each slide has an accurate motion transducer with a digital readout. Machine cost and performance are related to the accuracy and resolution of the slide motion transducers, with optical encoders, Inductosyns, and laser interferometers in common use and providing a range of resolutions from about 1 to 0.01 μm (40 to 0.4 μin). Since the machine working

[246]S. D. Murphy, "In-Process Measurement and Control," Marcel Dekker, New York, 1995.

[247]C. W. Kennedy, E. G. Hoffman, and S. D. Bond, "Inspection and Gaging," 6th ed., Industrial Press, New York, 1988.

[248]J. A. Bosch, "Coordinate Measuring Machines and Systems," Marcel Dekker, New York, 1995.

volume may be as large as about a 2-m cube, careful mechanical design of the slides is essential to ensure straightness, squareness, etc., of the coordinate motions. Both precision rolling-contact bearings and hydrostatic air bearings have been successfully used to achieve accuracies as good as about 1 μm for a slide with 1-m travel. "Volumetric" (three-dimensional) accuracy is also important but more difficult to specify and measure. Many machines may be fitted with a precision *rotary table,* allowing even more versatile application. Figure 4.99[249] shows a sample measured part and the kinds of features which may be routinely checked. Small CMMs of modest accuracy may cost $10,000 to $20,000 while larger, more accurate machines range up to several hundred thousand dollars.

Although various types of probes may be used with CMMs first we discuss the widely used *touch-trigger* type, invented in the early 1970s. Here the probe is used only to provide a sensitive and reproducible indication of when the probe touches the measured part; the three-dimensional position of the probe is actually read from the slide-position transducers on each machine axis. Thus to measure the width w of the part shown in Fig. 4.100, we drive the slide from position 1 in the negative x direction until contact is made, at which instant the x, y, z readings will be "frozen" so that we (or a machine memory) can record them. Then we move the slide to position 2 (same y and z readings as position 1) and drive it in the positive x direction until a "touch" signal again freezes the readings. Knowing the diameter of the probe's spherical trip, we can easily calculate w from the difference between the two x readings. In an actual CMM, the probe may be positioned anywhere in the working space *manually* (air bearings and z-axis counterweights allow you to grasp the probe body in one hand and easily move all three slides where you wish), by using a *joystick control* to manually command electric motor drives at preselected speeds or by using computer-commanded electric motor servodrives to accomplish the desired moves automatically under program control. This last *computer numerical control* capability provides a powerful measurement tool when it is combined with software to automate the various geometric calculations needed to extract part features from slide-position readings. A common and important use of CMMs is to check, for conformance to specifications, the first machined part in a production run from some numerically controlled machine tool. This verification of the part-programming process and all other aspects of machining is necessary before you can confidently proceed with the production run.

As in so many other measurement systems, the addition of powerful computing capability to basic CMM functions has greatly increased the versatility, ease of use, speed, and accuracy of these machines. For example, parts of basically rectangular shape need not be carefully aligned with the fixed x, y, z coordinate system of the CMM before measurements are taken. Rather, the part is fixed in *any* convenient position, and three mutually perpendicular part faces are chosen to define a part coordinate system with axes x_p, y_p, z_p. By touching the probe to three widely spaced points on each part surface, the equations of the three planes which define the part axes can be numerically calculated and the intersection point of these planes

---

[249]L K Tool Division, Cincinnatti Milacron, Tempe, AZ.

Holes	19
Shoulders	6
Center distances	15
Angles	9
Radii	10
Total checks	59
Total number of checked points	96
Inspection cycle	4 min 55 s

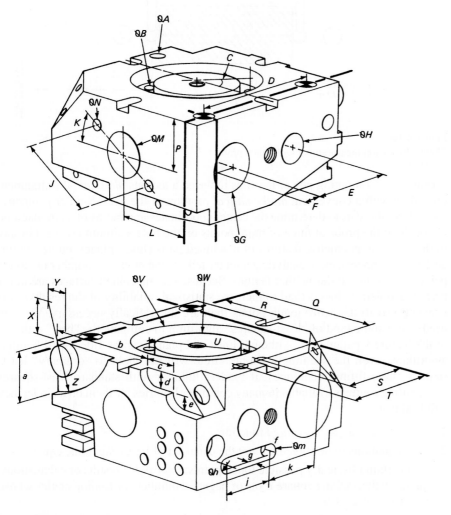

Figure 4.99
Measurements and timing on a typical manufactured part.

(origin of x_p, y_p, z_p) can be solved for. Once the part coordinate system has thus been located in the machine's axis system, all further part measurements can be automatically converted from machine coordinates to the more convenient part

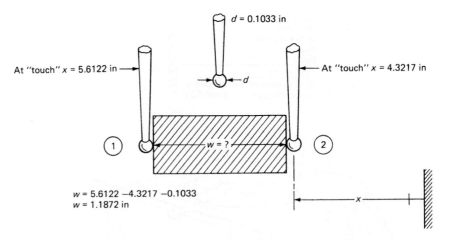

Figure 4.100
Simple dimension measurement with touch probe.

coordinates. Computing power has thus replaced a tedious *physical* part-alignment procedure with a convenient *mathematical* rotation of the axes. Similar geometric algorithms allow us to determine the center coordinates and radius of a circular hole by touching the probe at three or more points on the hole's circumference. For any of the common geometric features of machined parts (lines, planes, circles, cylinders, cones, spheres, etc.), analytic geometry tells us the *minimum* number of "touch points" necessary to define that feature. Because real machined surfaces are never perfect geometric objects (and to increase statistical reliability of the results), it is conventional to use *more* touch points than are theoretically necessary. We then need to use some kind of "least-squares" algorithm to compute the "best" estimate of the feature's parameters. Although lines and planes can be fitted with familiar algorithms, even the seemingly simple circle will *not* yield to "ordinary" least-squares curve-fitting techniques. If readers familiar with standard least-squares curve-fitting routines available in many computer libraries try them on a circle, they will find that

1. The circle is a *double-valued* function.
2. The equations to be solved are *not* the usual set of linear algebraic equations.
3. For certain physically possible touch points, *impossible* results or calculations arise, such as taking square roots of negative numbers or finding angles whose cosines exceed 1.0.

These problems *defeat* the conventional linear or nonlinear regression techniques, and thus *special* algorithms must be developed.[250] Algorithms actually used on CMMs are considered proprietary by the CMM manufacturers, and I was unable to obtain any details.

---

[250]I. Kasa, "A Circle Fitting Procedure and Its Error Analysis," *IEEE Trans. Inst. Meas.*, vol. IM-25, pp. 8–14, March 1976.

We now want to give a little more detail on *probe types* [passive ("hard"), switching ("touch trigger"), analog-proportional, analog-nulling] and *machine operating modes* (free-floating manual, driven manual, direct computer-controlled). Perhaps the simplest operation is free-floating manual mode with a passive probe. For free-floating operation, the machine slides must be nearly friction-free (as with hydrostatic air bearings), since the human operator simply grasps the probe holder and moves all three axes, to position the probe at the feature to be measured. The "probe" is nothing but a rigid piece of metal whose shape is appropriate to the feature being gaged. For simple width measurement of a rectangular part, the probe could just be a cylinder of known diameter which is manually pressed against the flat part surface (once for each side). Clearly such a method will suffer from variability in the force applied by the operator; however, it is entirely satisfactory for many applications. The centers of holes are quite nicely located with a conical probe, which is self-centering as it is gently lowered into the hole.

Perhaps the most widely used probe is the touch-trigger[251] type, usable in all three machine operating modes. As explained briefly earlier, this probe is an on/off switching type which "freezes" the readings of the three slide motion sensors as the probe tip touches and is deflected by the part surface. Its most common form is shown in Fig. 4.101. The probe stylus is kinematically located in a single unique position by the six contacts of the three cylindrical rods with the six balls, with a light spring preload maintaining this position when no external forces are applied to the stylus. The six contacts are electrically wired in series, as shown, and a constant-current source of about 0.5 mA is connected. The total resistance of the six contacts in the neutral position is on the order of a few ohms, making the voltage e_o a few millivolts. When the probe's spherical tip is deflected against the spring preload by contact with a measured part, one or more of the contact resistances increase *very* greatly with tiny deflections. When the total resistance exceeds about 3000 Ω, voltage e_o passing through 1.5 V trips a circuit which freezes all three slide-position readouts, recording the position of the probe at the instant of touch. A uniquely favorable feature of the probe is its three-dimensional nature; tip deflections in $\pm x$, $\pm y$, $+ z$ directions will *all* cause triggering, thus the probe may approach the measured part from various directions. Note that $- z$ forces are opposed, not by the light spring preload, but by the very stiff rod-and-ball contacts. Thus the $- z$ direction *cannot* be used for gaging; however, this is rarely a problem.

An important detail of probe operation which was ignored in Fig. 4.100 is probe bending and "pretravel." The probe does not actually trigger at the instant of touch since it does require a small, but finite, force and deflection to increase the electric resistance to the 3000-Ω trigger point. Also bending deflection of the probe (minimized by using short, stiff probes whenever possible) causes a small unmeasured deflection between touching and triggering. Fortunately, these effects are largely repeatable and may be corrected by calibration. For example, in Fig. 4.100 before

---

[251]Renishaw Inc., Schaumberg, IL (www.renishaw.com).

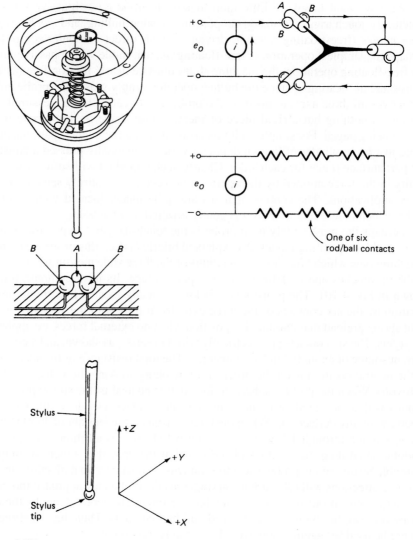

Figure 4.101
Details of touch-trigger probe.

measuring an *unknown w,* one would measure a precisely known *w* (such as a gage block), to find the "effective working diameter" d_e of the probe from the equation

$$d_e = \text{measured size} - \text{actual size}$$

Then this one d_e value can be used to correct for all three effects (ball diameter, pretravel, and bending) by using the formula

$$\text{Actual size} = \text{measured size} - d_e$$

In practice, d_e is usually found by touching a calibration *sphere* at about 10 points on the sphere's surface and using a special algorithm to compute d_e. This more complicated scheme is better since it "exercises" the probe's characteristics in *many* directions, making the d_e value more correct for a *general* measurement.

Although the construction and operation of the probe seem quite simple, continuous high accuracy and reliability depend on subtle design details and use of appropriate materials. Lubricants with special electrical, lubricating, and corrosion-resistant properties, for example, are used on the ball-rod contacts. Probes used for in-process gaging with, say, a lathe (rather than in a CMM) use higher spring preloads to prevent false triggering due to vibration, etc. This degrades some other probe characteristics, such as accuracy, but these trade-offs are necessary to optimize the probe's *overall* performance for a specific application. Special materials are also used for the rods, balls, and stylus tips. Rod and ball materials are considered proprietary, but stylus tips are often synthetic ruby, an aluminum oxide ceramic noted for hardness, smoothness, and dimensional stability. Touch-trigger probes also have favorable *overtravel* characteristics. That is, in touching the probe to the part, some motion *past* the touch point is unavoidable, especially if one is moving the probe manually. For example, a probe with 0.1-μm repeatability, 0.15-g trigger force, and 10-μm tip deflection to trigger may have a safe overtravel of $\pm$ 2.5 mm and a maximum probe approach velocity of 7 mm/s. High probe-approach velocity speeds machine operation and increases throughput but degrades accuracy. *Driven manual* operation (operator controls probe motion with a joystick) and *direct computer control* (probe motion is commanded by computer) obviously give better control of probe velocity than does free-floating manual operation.

Analog probes[252] give a voltage output proportional to probe deflection from the null position. They are available in both contact and noncontact (optical) versions. Analog probes are capable of higher resolution and accuracy than the on/off touch-trigger probes, but tend to be more fragile and expensive with poorer overtravel. Duplicating the three-dimensional capabilities of touch-trigger types leads to rather complicated mechanical configurations. Note that to obtain a number for, say, the *x* location of the probe tip, we must now *add* the reading of the probe to that of the slide-motion sensor. An alternative to this is the *nulling* operation of the probe/CMM system. Here the probe's signal (plus or minus from null) is used as an *error* signal in a feedback loop which controls the CMM slide position, causing the machine slide to automatically drive to a position where the probe signal is zero (or some selected bias position). For "three-dimensional" operation, this can lead to quite complicated servosystems.

Although most CMMs are used to measure discrete part features, as in Fig. 4.99 ("point-to-point" mode), form documentation for "sculptured surfaces" such as turbine blades ("contouring" mode) is also possible, though it presents special problems.[253] Here analog nulling type probe systems have some advantages in smoothly following the part surface under servo control as data are taken. In overall design

---

[252]EMD, Inc., East Budd Lake, NJ, 973-691-4755 (www.emdsceptre.com). This website includes a very interesting history of CMM probes.

[253]B. Van den Berg, "Closed Loop Inspection of Sculptured Surfaces in a Computer Integrated Environment," *Proc. 8th International Conf. on Automated Inspection and Product Control,* pp. 145–156, 1987.

philosophy, CMM manufacturers have adopted different viewpoints. One approach strives for perfection in manufacture and assembly of each mechanical component, to achieve overall machine accuracy. Other designers have opted for less stringent requirements on components, followed by an individual calibration of each machine to obtain numerical values for a software error-correction scheme built into the CMM's computer. This latter approach is often feasible since the major machine errors[254] tend to be systematic (reproducible) rather than random. Recently CMM manufacturers and users collaborated in producing a national standard[255] devoted to defining methods for evaluating CMM performance. I found this document extremely useful and recommend it highly to anyone interested in CMM design and operation. Up to this point we have concentrated on the CMM itself; however, the *ambient thermal environment* in which the machine operates is critical, as emphasized in B89.1.12M and explored in detail in another standard.[256] Large, high-accuracy machines present the greatest thermal problems, as related in a recent article.[257] Theoretical finite-element studies of thermally induced deflections showed that when the temperature of the machine's bridge was 0.6°C higher than its base, the bridge bowed upward by 10.4 μm, or 8 times the desired measuring tolerance. This led to the design of a sophisticated air-flow and temperature-control system which maintains large volumes of the machine room air at 20 $\pm$ 0.06°C, a considerable feat of engineering in itself. While thermal problems are a major factor, barometric pressure, humidity, gaseous and particulate concentrations, lighting, and floor vibration may also require careful attention. Finally, the integration of CMMs into the total manufacturing process is receiving increasing study and promises great benefits as computer-aided design (CAD), computer-aided manufacture (CAM), and computer-aided inspection (CAI) are all coordinated.[258]

NIST is developing a "super CMM" to advance the state of the art in this important technology. They also study the evaluation of software for CMMs.[259] Three-D measurement of free-form surfaces can be accomplished optically, using triangulation of 2-D images, with accuracy of about 20 μm.[260]

4.14 SURFACE-FINISH MEASUREMENT

The "gross" part dimensions determined by the CMMs of Sec. 4.13 are not the only significant measurements required on manufactured parts. Surface characteristics,

---

[254]R. B. Zipin, "Measuring Machine Accuracy," Sheffield Measurement Division, Dayton, OH.

[255]"Methods for Performance Evaluation of Coordinate Measuring Machines," ANSI/ASME B89.1.12M-1990, ASME, New York, 1985.

[256]*ASME* B89.6.2.

[257]W. Hobson and M. L. Majlak, "Special Room for Special CMM," *Quality,* pp. 22–27, November 1987.

[258]W. Tandler, "High Performance Coordinate Measuring Systems," Society of Manufacturing Engineers, *Paper* M584–713, 1984.

[259]E. D. Teague, "The National Institute of Standards and Technology Molecular Measuring Machine Project: Metrology and Precision Engineering Design," *J. Vac. Sci. Technol.* B7(6), Nov./Dec. 1989, pp. 1898–1902. Cathleen Diaz, "Algorithm Testing and Evaluation Program for Coordinate Measuring Systems: Testing Methods," *NISTIR,* p. 5686, July 1995.

[260]CogniTens Ltd., Ramat Hasharon, Israel, www.cognitens.com.

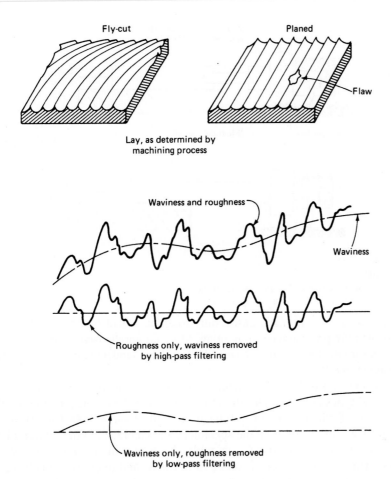

Figure 4.102
Definitions of surface-finish features.

on a more microscopic scale, may be vitally important with regard to friction, wear, fatigue resistance, cleanliness in food processing, leakage in seals, paintability, scattering losses on optical surfaces, substrate surface flaws in microcircuits, and visual appearance.[261] The study of surface properties is related to the larger area called *tribology,* which includes all aspects of friction, wear, and lubrication. The deviation of an actual surface from the theoretical ideal is described in terms of *lay, flaws, waviness,* and *roughness* (Fig. 4.102). Lay is the directional aspect of the dominant surface pattern, usually determined by the production process. Flaws are isolated defects which occur at infrequent intervals. Waviness plus roughness is called *profile* and is the total vertical excursion, at a cross section, from a fixed datum, with waviness being the long-period component and roughness the short-period

[261]R. D. Young, "The National Measurement System for Surface Finish," *NBS* IR75-927, 1976.

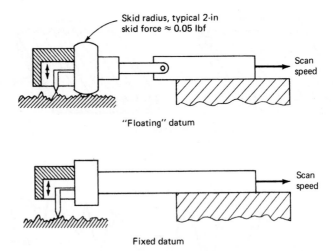

Figure 4.103
Stylus instruments with fixed or floating references.

(closely spaced) one. In many cases, only roughness is of interest, so the simpler instruments may provide data on it alone. More comprehensive instruments allow one to measure the total excursion (profile) roughness and/or waviness by selection of instrument modes.

Most surface-finish measurements[262] are made with stylus instruments; however, optical types are available (though expensive) when noncontacting methods are necessary. Typical diamond styli have tip radii of about 0.0001 in, exert about 0.0001 to 0.001 lbf of force on the specimen, and can reach to the bottom of most scratch marks more than 0.0002 in wide. High-resolution units for very fine surfaces such as gage blocks use 0.00005-in radius and reduced stylus force. The stylus size cannot be reduced much below these values because the tip pressure begins to exceed the yield point of most materials, causing plastic deformation of the measured surface. This places a fundamental limitation on the horizontal resolution of these instruments. To measure roughness along a chosen line on a surface, the stylus is dragged along this line at a fixed slow speed (0.0001 to 0.1 in/s), using sufficient stylus force and slow enough scan speed that the stylus faithfully follows the surface. The stylus may be vertically guided by an attached shoe or skid, whereby the datum for the measurement changes as the shoe follows the local waviness and the stylus measures mainly roughness. If the *total* profile is to be measured, a shoeless probe is used and an absolute external datum is fixed by the traversing mechanism (see Fig. 4.103). When roughness along a *curved* line must

[262]"Surface Texture," *ASME Standard,* B46.1-1985; H. Dagnall, "Exploring Surface Texture," Rank Taylor Hobson Ltd., Rolling Meadows, IL, 708-290-8090 (www.taylor-hobson.com); D. J. Whitehouse, "Handbook of Surface Metrology," IOP Publishing, 1994; D. J. Whitehouse and W. L. Wang, "Dynamics and Trackability of Stylus Systems," *Proc. Instn. Mech. Engrs.,* vol. 210, pp. 159–165, 1996; X. Liu et al., "Improvement of the Fidelity of Surface Measurement by Active Damping Control," *Meas. Sci. Techno.,* vol. 4, 1993, pp. 1330–1340.

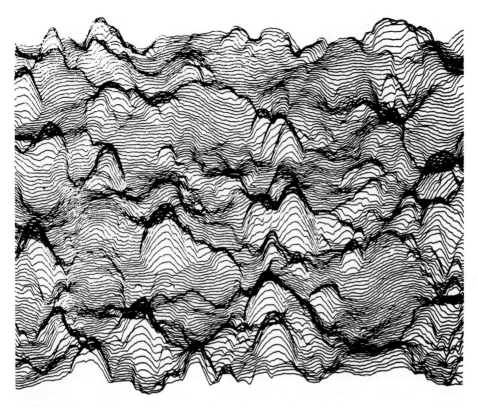

Figure 4.104
Three-dimensional microtopography of machined surface.

be measured (bearing balls, gear teeth, etc.), suitable curved external datum mechanisms are used.

Although a few instruments use the moving-coil velocity transducer to obtain a stylus-motion signal, such devices give *no* signal unless the surface is being scanned and they cannot be statically calibrated. A preferred approach is to use an LVDT or similar displacement transducer, which can be made very sensitive for the small full-scale deflections needed ($<$ 0.050 in). Since most instruments have a built-in graphic recorder, it is conventional to quote sensitivities in terms of magnification between inches of stylus motion and inches of recorder pen motion. For the vertical (roughness) axis, maximum magnification approaches 1 million, while the limited horizontal resolution (caused by the stylus-tip size limitation mentioned earlier) leads to maximum magnification of about 500 on the horizontal (scan) axis. Since the vertical and horizontal magnifications are often quite different, direct visual interpretation of the graphs must keep this in mind. Some instruments allow scanning over a horizontal plane by "indexing" successive x scans a precise amount in the y direction and "stacking" these traces vertically on an XY plotter. Figure 4.104 shows such a measurement on a "flat" surface produced by the electric-discharge machining process, using a z magnification of 500 and x, y magnification of 200.

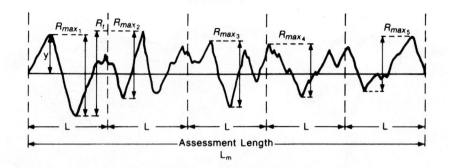

$$R_a = \frac{1}{L} \int_0^L |y(x)| \, dx$$

$$R_q = \sqrt{\frac{1}{L} \int_0^L y^2(x) \, dx}$$

$$R_{max} = max \, | R_{max, n} |$$

R_t = Maximum peak – minimum valley within L_m

$$R_{tm} = \left(R_{max_1} + R_{max_2} + R_{max_3} + R_{max_4} + R_{max_5} \right) / 5$$

$$R_{pm} = \left(R_{p_1} + R_{p_2} + R_{p_3} + R_{p_4} + R_{p_5} \right) / 5$$

Figure 4.105
Definitions of roughness parameters.

When the stylus is moved horizontally at a fixed and known speed, the surface profile "becomes a function of time," and we can use standard time-function filtering methods to perform the *spatial* filtering mentioned earlier as a means of separating roughness and waviness. For example, a sinusoidal profile of wavelength 0.005 in when scanned at 0.05 in/s becomes a 10-Hz time signal. Usually the filters are adjustable, allowing a selection of cutoff wavelengths, a typical selection being 0.003, 0.01, 0.03, 0.1, and 0.3 in. For roughness measurement, wavelengths *longer* than the quoted value are filtered out; for waviness, those shorter.

For many practical measurements, the roughness profile is processed to obtain values for certain roughness *parameters,* which can then be compared numerically with goals or specifications to decide whether the surface meets some functional requirement. Relating machine functional requirements (such as wear life, friction torque, heat generation, scoring and galling, etc.) to numerical values of roughness parameters is at present more art and experience than science and so perhaps is a fruitful area for research. The most common parameters are R_a (the average absolute value of the profile excursions from the mean line) and R_q (the rms value), as defined in Fig. 4.105.[263] Instruments offer a selection of sampling lengths L, with a typical set being 0.021, 0.07, 0.21, 0.70, and 2.1 in. Usually R_a or R_q is measured 5 times (over five consecutive values of L), and an average is computed. Some other parameters used less frequently (R_{max}, R_t, R_{tm}, R_{pm}) are also defined in the figure. In addition to functional criteria, roughness parameters, such as R_a, are related to production time/cost (Fig. 4.106).[264]

---

[263]M. Brock, "Fourier Analysis of Surface Roughness," *B&K Tech. Rev.,* pp. 3–45, November 3, 1983; Brull & Kjaer, Marlborough, MA.

[264]Ibid.

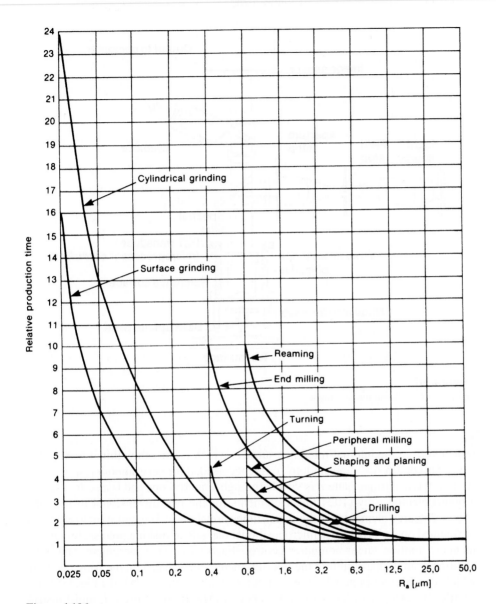

Figure 4.106

Roughness related to machining process and cost.

We conclude this section with a brief description of an optical surface-finish instrument (see Fig. 4.107).[265] The instrument uses enhanced phase-measurement interferometry. Light reflected from the object under test interferes with light

---

[265]Quantitative Micro-Surface Measurement Systems," 1/86-5M, Wyco Corp., Tucson, AZ, 520-741-1044 (www.wyko.com).

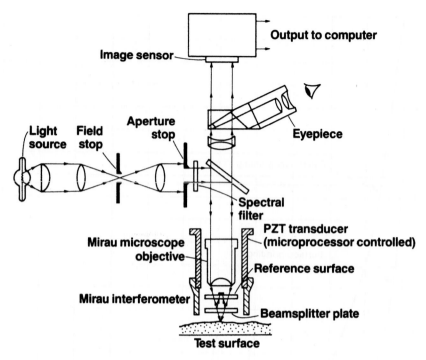

Figure 4.107
Optical surface-finish measurement.

reflected from the internal-reference surface, which is smooth within 4 to 8 Å (0.4 to 0.8 nm) or 1.5 Å ("supersmooth" reference needed for very smooth test objects). An interference fringe pattern is visible through a microscope and is also focused on a photodetector array (1024 linear elements for the two-dimensional instrument, 256 × 256 area elements for the three-dimensional one). The entire interferometer is mounted on a piezoelectric micropositioner capable of making precise, tiny increments of motion under computer control. Each successive increment causes an additional small phase shift in the interferometer and a change in the fringe pattern, which can be interpreted by the computer in terms of measured object-surface profile. By using a variety of interferometers (Michelson, Mirau, Linnik) instruments with a range of specifications are available. The profile length ranges from about 9 to 0.0667 mm; however, a computer-controlled *XY* specimen table can extend this. The spatial sampling interval is from 27 mm to 0.2 μm; the working distance is 12 mm to 0.2 mm. The maximum surface heights are 8 to 0.54 μm, vertical resolution is 1 to 3 Å, maximum surface slopes are 0.26° to 54°. Extensive software and color graphics are available to compute and display a wide variety of surface characteristics.

4.15 MACHINE VISION[266]

In the late 1970s, machine (or computer) vision systems for laboratory automation began to be practical, and this field has since shown accelerated growth. Figure 4.108[267] classifies typical applications of this technology. The dividing line between optical gaging and measuring systems, such as those discussed earlier in this chapter, and "true" machine vision is not always clear-cut, since available definitions allow some room for interpretation. Machine vision attempts to duplicate some portion of the capabilities of the human eye and brain system for forming and interpreting images, and studies of human perception have been a significant factor in the development of machine vision.[268] A useful definition of machine vision requires that at least three basic functions be present: image formation, image analysis, and image interpretation.

Image formation involves a proper combination of illumination type and camera (sensor). Back-lighting provides a silhouette image of maximum contrast, which is useful for dimensional measurements and detection of the presence or absence of features such as holes. Front lighting may be needed to deal with features such as labels or bar codes. Detection of three-dimensional features may require side lighting. Other possibilities include polarized light and "structured" light. Structured light projects a geometrically defined light/dark pattern onto the measured object so as to enhance certain features of the image. Perhaps the most used type of structured light is a simple flat beam or "sheet" of laser light, as provided by a rotating or oscillating mirror[269] or (without moving parts) cylindrical lenses.[270] Projecting such a sheet of light onto a target object produces a line of reflected light, which follows the contours of the object and whose image can be processed to obtain various useful measurements. For example, several such sensors can be stationed around the periphery of an automobile door to check for proper fit during assembly by measuring the (one-dimensional) gap (Fig. 4.109*a*). "Contouring" of three-dimensional objects is achieved in other applications, as shown in Fig. 4.109*b*. The light sheet optically "slices through" the measured body, creating a cross-sectional image which can be observed by a suitable camera and analyzed to obtain desired part dimensions. *Motion* of the target object through the light sheet

---

[266]B. Jahne (ed.), "Handbook of Computer Vision and Applications," Academic Press, San Diego, 1999; P. West, "A Roadmap for Building a Machine Vision System," Imagenation Corp., Portland, OR, 800-366-9131 (www.imagenation.com); D. Zankowsky, "Machine Vision Systems Move to Center Stage," *Laser Focus World,* Apr. 1996, pp. 127–137; R. Grosklaus, "Taking the Mystery out of Machine Vision," *Sensors,* Apr. 1997, pp. 22–27; R. L. Cromwell, "Sensors and Processors Enable Robots to See and Understand," *Laser Focus World,* Mar. 1993, pp. 67–78; "Electronic Imaging Resource Guide," Edmund Industrial Optics, Barrington, NJ, 800-363-1992 (www.edsci.com).

[267]TechTran Consultants, "Machine Vision Systems," 2nd ed., McGraw-Hill, New York, 1985.

[268]Ibid.

[269]D. C. O'Shea, "Elements of Modern Optical Design," pp. 81–84, Wiley, New York, 1985.

[270]D. C. O'Shea, op. cit., pp. 282–311.

Applications		Feature measurement capabilities					Image model		Performance		
		Image shapes	Distance (range)	Orientation	Motion	Surface shading	Two dimensions	Three dimensions	High resolution	High speed	High discrimination
Inspection	Dimensional accuracy	●					●			●	
	Hole location and number	●					●			●	
	Component verification	●		●			●			●	
	Component defects	●					●			●	
	Surface flaws					●	●				●
	Surface contour accuracy		●					●	●		●
Part identification	Part sorting	●					●			●	
	Palletizing	●		●			●			●	
	Character recognition	●					●			●	
	Inventory monitoring	●					●			●	
	Conveyer picking—no overlap	●		●			●			●	
	Conveyer picking—overlap	●		●		●	●			●	●
	Bin picking	●	●	●		●		●		●	●
Guidance and control	Seam weld tracking		●	●		●		●		●	●
	Part positioning	●	●	●				●		●	
	Processing/machining	●	●	●				●		●	
	Fastening/assembly	●	●	●				●		●	
	Collision avoidance	●	●	●	●			●		●	

Figure 4.108
Machine vision applications and features.

can be used to produce successive slices which reveal the shape and size of the entire object. This motion must be provided and accurately measured, via precision positioners available commercially. In some cases, on-line gaging can use a motion that is already part of the manufacturing process, such as a conveyer or assembly-line motion.

Light sources include incandescent and fluorescent lamps, fiber optics, lasers, arc lamps, and strobe lights. Proper lighting also requires judicious positioning of the illuminated object. Early vision systems used vidicon cameras as image sensors to convert light to electric signals and then to a two-dimensional image of the three-dimensional object. More recently, solid-state cameras using charge-coupled device (CCD) or charge-injected device (CID) photosensors have become popular. When

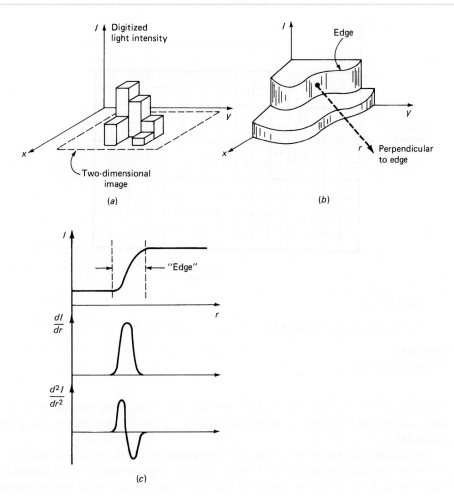

Figure 4.110
Definition of edges.

crossing of the second derivative d^2I/dr^2 may be better for pinning down its location. Computer programs can use combinations of these (or other) criteria to decide whether a pixel is an edge pixel and then to convert the gray-scale image to a binary one which sharply defines part features for measurement. The widely used Sobel[274, 275] edge detector computes for each pixel a numerical value G useful for deciding whether the pixel P is an edge pixel, via the scheme of Fig. 4.111 on a 3×3 "window" of pixels. The formula for G shows that it will be zero for uniform light intensity and will get larger as the rate of change of intensity in either or both the x and y directions increases. It will thus "detect" edges irrespective of the

[274]W. K. Pratt, "Digital Image Processing," pp. 487, 498, Wiley, New York, 1978.

[275]R. O. Duda and P. E. Hart, "Pattern Classification and Scene Analysis," p. 271, Wiley, New York, 1973.

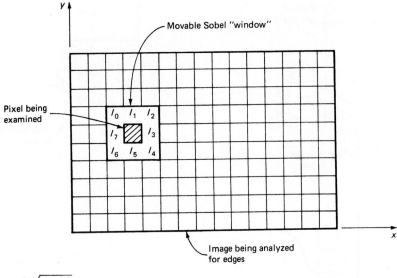

$$G \triangleq \sqrt{G_x^2 + G_y^2}$$

$$G_x \triangleq (I_2 + 2I_3 + I_4) - (I_0 + 2I_7 + I_6) \qquad G_y \triangleq (I_0 + 2I_1 + I_2) - (I_6 + 2I_5 + I_4)$$

Figure 4.111
Sobel edge detection.

"direction" r (Fig. 4.110) of the edge. One thus scans the Sobel window over the entire image, and wherever G exceeds a selected threshold value, that is called an *edge pixel*.

In searching an image of an assembled mechanism or in a bin of assorted parts for the presence of a particular part (Fig. 4.112), the template-matching technique is widely used. A replica of the object of interest is compared with all the unknown objects in the field of view, and if the match is sufficiently close, we conclude that we have found the sought object. Since a template match can never be perfect because of image noise, spatial and intensity quantization, and uncertainties in the exact shapes of manufactured parts, we need to define some measure $D(m, n)$ of the difference between the template and the image intensities when the template is displaced by $\Delta x = m$, $\Delta y = n$ from a selected image location $x = j$, $y = k$. As we scan the template over the image region of interest, $D(m, n)$ takes on various values, one of which will be the smallest. If this smallest value of D is less than some preselected threshold value, then we decide that we have a match and also know the location (m, n values) of the sought object. A mean-square type of difference criterion[276] is often used:

$$D(m, n) = \sum_{j} \sum_{k} [F(j, k) - T(j - m, k - n)]^2 \tag{4.152}$$

[276]Pratt, op, cit., p. 552.

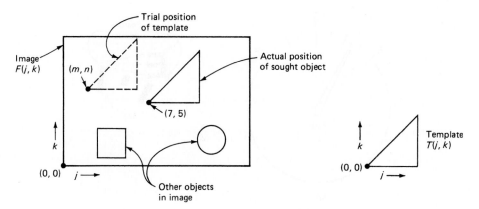

Figure 4.112
Template matching.

where $F(j, k)$ is the intensity (at location j, k) of the image being searched, $T(j, k)$ is the intensity of the template, and m and n define a trial location of the template (see Fig. 4.112). When we try $m = 7$ and $n = 5$ in this example, a perfect match occurs and $D = 0$, whereas all other m and n combinations give D larger values. (In a practical case with quantized images, the "sought object" and the template would not be simple line drawings; the sides of the triangle would have some "width," and the template width would be chosen a little larger than the object width, to provide some tolerance.) While Eq. (4.152) and Fig. 4.112 reveal the basic concept of template matching, experience shows the D calculation is not reliable in the face of real-world complications such as uncontrolled overall scene illumination, etc. Some kind of normalization procedure which produces a *relative,* rather than *absolute,* value is often of help in such situations. Here the normalized cross correlation[277] has been found useful:

$$\tilde{R}_{FT}(m, n) \triangleq \frac{\sum_j \sum_k F(j, k) \, T(j - m, k - n)}{\sum_j \sum_k [F(j, k)]^2} \qquad \textbf{(4.153)}$$

It can be shown that when m and n are chosen to give a perfect template match, $\tilde{R}_{FT}$ assumes its maximum value, which is always 1.0. A practical template-matching criterion is that $\tilde{R}_{FT}$ be larger than some selected value.

Before completing this section with a brief discussion of an actual manufacturing application of machine vision, we want to again caution the reader about the need to carefully consider all aspects of any proposed application, since many subtle traps await the unwary.[278] Our example uses machine vision to help automate the

---

[277]Pratt, op. cit., p. 553.

[278]R. C. Stafford, "Induced Metrology Distortions Using Machine Vision Systems," pp. 3–38 to 3–48; S. H. Lapidus and A. C. Englander, Understanding How Images Are Digitized, "Vision '85," pp. 4–23 to 4–33.

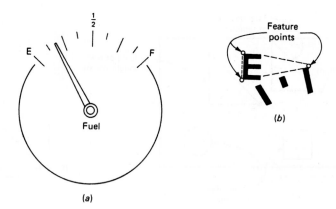

Figure 4.113
Fuel gage and feature points.

assembly, calibration, and quality control of panel instruments for automobiles, specifically a fuel-gage display.[279] The final step in the manufacture of such devices involves fastening the pointer to the rotating shaft so that it aligns with a preprinted "empty" mark when the electrical input is set at the empty value, selecting the proper range resistor for each gage so that the pointer aligns with the "full" mark when the electric full-tank signal is applied, and finally inspecting each gage for proper position at the empty and full conditions and recording any deviations from normal for statistical process control purposes. Automation of these operations without vision systems had been difficult because the many different gage styles in use required many different hardware fixtures, with attendant long changeover times. Standardization on a single gage configuration was rejected as too restrictive on styling design. The vision system accommodates nearly unlimited gage variations by assigning needed modifications to computer software (easily changed), rather than hardware fixtures.

With the preprinted gage dial fastened securely in a fixture, first the vision system determines the location of three "feature points" near the empty mark, as in Fig. 4.113*b*. Actually, two points should be sufficient; use of three gives some desired redundancy. Next the triangle formed by the three feature points on the dial is compared with a triangle template stored in computer memory and characteristic of that dial pattern. The system now knows where the empty mark on this particular dial is located, and the system signals some stepping-motor drives to position the pointer correctly and then stake it permanently to the gage shaft, the shaft having been properly positioned by applying the "empty" electric signal to the gage. The vision system also needs to measure the pointer position to accomplish this. Then the "full" electric signal is applied, causing the shaft and pointer to rotate toward the F mark on the dial. Economic manufacture results in gage electrical movements (really voltmeters) which exhibit rather large variance in sensitivity, so most gages

---

[279]D. E. Lemke and D. A. Langdon, "Gagesite-Computerized Vision for Automotive Instrument Calibration," *SAE Paper* 830327, 1983.

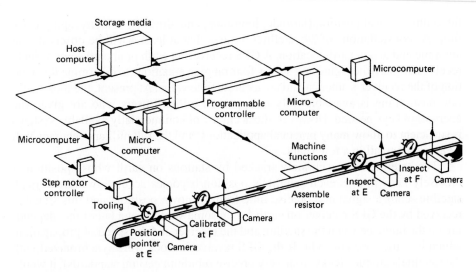

Figure 4.114
Integration of machine vision into manufacturing system.

will *not* go accurately to the F mark when the proper signal is applied. This intentionally allowed variation is corrected (individually for each gage) at the final assembly and calibration stage which we are describing, by noting the position error (the vision system does this also) and then selecting a compensating resistor to correct the sensitivity. This resistor is automatically assembled into the gage. Now mechanically and electrically complete, the gage is passed to an inspection station, where similar vision systems check for deviations from normal and record data for statistical process control. For example, if the selected resistors are *not* bringing the pointer within tolerance for the "full" condition, the manufacturing process for the gage electrical movements may have drifted off specification, requiring compensating changes in the program for selecting calibration resistors.

This system uses a solid-state camera with 100×100 pixels and a Canon zoom lens with a frontal collector lens for magnification. The location accuracy is $\pm\frac{1}{2}$ pixel, which corresponds to about ± 0.0025 in at the gage face. Gage location takes about 40 to 80 ms while pointer location requires about 20 ms. Figure 4.114 gives an overview of the entire operation.

4.16 THE GLOBAL POSITIONING SYSTEM (GPS)

One of the most significant developments of the space program is the artificial satellite system, and within this, the global positioning system[280] has important applications to motion measurement. Developed by the U.S. Department of Defense, GPS is a space-based radio positioning utility that provides accurate,

[280]Hewlett Packard Application Note 1272, 1995; "Guide to GPS Positioning," Canadian GPS Associates, Canadian Institute of Surveying and Mapping, Box 5378, Postal Station F, Ottawa, Ontario, Canada, K2C311, 613-224-9851 (www.agilent.com).

three-dimensional position (latitude, longitude, and altitude), velocity, and precise time. A "constellation" of 24 satellites assures that at least 4 will be observable at any time and at any earth location. A GPS receiver on earth or aboard some vehicle receives the satellite signal and solves four ranging equations to determine the location of the receiver's antenna and an accurate time value. At present, the most accurate positioning is available only to approved military users who are given the encryption keys needed. However, the accuracy of commercially available systems is sufficient to allow many practical applications, and the full military accuracy will be available to all in a few years.

The satellites are accurately tracked by stations on earth whose position is precisely known, thus each satellite's position becomes accurately known. Each satellite sends a signal at a precise time from a known position, and this signal is received by the GPS receiver on earth (or on some vehicle) at a later time, depending on the range between the satellite and the receiver and the speed of propagation, which is c, the speed of light. If the GPS receiver's clock were synchronized with the satellite's atomic clocks (four very precise rubidium/cesium standards), it would require only three satellites, giving three propagation times, to compute the unknown location of the GPS receiver. It is not practical to synchronize the GPS clock, so this timing error (called the clock bias CB) is allowed to exist but is corrected by using one more satellite to provide a fourth equation. This also transfers the accuracy of the satellite clocks to the GPS clock, so we now have both accurate position data and time data. The form of the range equation for a single satellite is given below. The other three equations would have the same form.

$$(x_{s1} - x_r)^2 + (y_{s1} - y_r)^2 + (z_{s1} - z_r)^2 = (\Delta t_1 c - CB \cdot c)^2$$
$$= (x \text{ component of range})^2 \qquad \textbf{(4.154)}$$

In this equation, the satellite coordinates s_{s1}, y_{s1}, z_{s1}, the propagation time Δt_1, and the speed of light c are all known, while the four unknowns are the 3 GPS receiver coordinates x_r, y_r, z_r and the clock bias CB. Similar equations for the other 3 satellites complete the set of 4 equations needed to solve for the four unknowns. *Differential GPS* operation uses *two* GPS receivers, one of whose locations is precisely known. When this mode is feasible, position accuracies of 3 to 5 meters are possible. GPS can be used by itself in many applications, but is also combined with other systems to realize the best features and negate the bad features of each using the concept of *sensor fusion* (*complementary filtering*). An example is the combination of GPS with an inertial measurement unit using 3 accelerometers and 3 rate gyros.[281] The inertial system has fast response, but becomes inaccurate with time, whereas the GPS maintains long-term accuracy but is slow.

Since the GPS signals are "free" for anyone to use, many commercial applications are available. For example, the measurement of vehicle speed on a test track (usually accomplished with a "fifth-wheel" device) can now be done more accurately using a GPS-based system.[282]

---

[281]DYMO-3, A-DAT Corp., Livonia, MI, 734-458-1701 (www.a-dat.com).

[282]VBOX, Kistler Michigan, www.racelogic.com.

PROBLEMS

4.1 Derive an equation analogous to (4.1) for the system of Fig. 4.5*b*.

4.2 Derive Eq. (4.1).

4.3 The output of a potentiometer is to be read by a recorder of 10,000-Ω input resistance. Nonlinearity must be held to 1 percent. A family of potentiometers having a thermal rating of 5 W and resistances ranging from 100 to 10,000 Ω in 100-Ω steps are available. Choose from this family the potentiometer that has the greatest possible sensitivity and meets the other requirements. What is this sensitivity if the potentiometers are single-turn (360°) units?

4.4 If a potentiometer changes resistance because of temperature changes, what effect does this have on motion measurements?

4.5 A 10-in-stroke wirewound translational potentiometer is excited with 100 V. The output is read on an oscilloscope with a "sensitivity" of 0.5 mV/cm. It would appear that measurements to the nearest 0.0001 in are easily possible. Explain why this is not so.

4.6 In Fig. P4.1 a potentiometer whose moving part weighs 0.01 lbf measures the displacement of a spring-mass system subjected to a step input. The measured natural frequency is 30 Hz. If the spring constant and mass M of the system are unknown, can the true natural frequency be deduced from the above data? Suppose an additional 0.01-lbf weight is attached to the potentiometer and the test repeated, giving a 25-Hz frequency. Calculate the true natural frequency of the system, that is, the frequency before the potentiometer was attached.

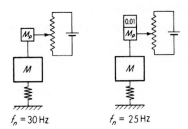

Figure P4.1

4.7 Explain why increasing the cross-sectional area of the end loops in foil-type strain gages reduces transverse sensitivity.

4.8 If, in the discussion following Eq. (4.22), dynamic strains only in the range 0 to 10,000 Hz need to be measured, explain how and to what extent the noise voltage may be reduced.

4.9 In a Wheatstone bridge, leg 1 is an active strain gage of Advance alloy and 120-Ω resistance, leg 4 is a similar dummy gage for temperature compensation, and legs 2 and 3 are fixed 120-Ω resistors. The maximum gage current is to be 0.030 A.

(a) What is the maximum permissible dc bridge excitation voltage? (Use this value in the remaining parts of this problem.)

(b) If the active gage is on a steel member, what is the bridge-output voltage per 1,000 lb/in^2 of stress?

(c) If temperature compensation were *not* used, what bridge output would be caused by the active gage's increasing temperature by 100°F if the gage were bonded to steel? What stress value would be represented by this voltage? Thermal-expansion coefficients of steel and Advance alloy are 6.5 × 10^{-6} and 14.9 × 10^{-6} in/(in · F°), respectively. The temperature coefficient of resistance of Advance is 6 × 10^{-6} Ω/(Ω · F°).

(d) Compute the value of a shunt calibrating resistor that would give the same bridge output as 10,000 lb/in^2 stress in a steel member.

4.10 From Eq. (4.31), find an expression for the frequency at which zero phase shift occurs.

4.11 Perform an analysis similar to that leading to Eq. (4.31), assuming output loaded with R_m, for (a) the circuit of Fig. 4.19a, (b) the circuit of Fig. 4.19b, (c) the circuit of Fig. 4.19c, (d) the circuit of Fig. 4.19d.

4.12 In Fig. 4.21c, let x_i be a periodic motion with a significant frequency content up to 500 Hz, and let the excitation frequency be 10,000 Hz. The output voltage e_o is connected to an oscillograph galvanometer, which is a second-order system with $\zeta = 0.65$ and a natural frequency of 1,000 Hz. Will this combination result in a satisfactory measurement system? Justify your answer with numerical results.

4.13 In Eq. (4.46), suppose a flat amplitude ratio within 5 percent down to 20 Hz is required. What is the minimum allowable τ? If $A = 0.5$ in^2 and x_o is 0.005 in, what value of R is needed?

4.14 A piezoelectric transducer has a capacitance of 1,000 pF and K_q of 10^{-5} C/in. The connecting cable has a capacitance of 300 pF while the oscilloscope used for readout has an input impedance of 1 MΩ paralleled with 50 pF.

(a) What is the sensitivity (V/in) of the transducer alone?

(b) What is the high-frequency sensitivity (V/in) of the entire measuring system?

(c) What is the lowest frequency that can be measured with 5 percent amplitude error by the entire system?

(d) What value of C must be connected in parallel to extend the range of 5 percent error down to 10 Hz?

(e) If the C value of part d is used, what will the system high-frequency sensitivity be?

4.15 A piezoelectric transducer has an input

$$x_i = \begin{cases} At & 0 \le t < T \\ 0 & T < t < \infty \end{cases}$$

Solve the differential equation to find e_o. For $t = T^-$, find the error [(ideal value of e_o) − (actual value of e_o)]. Approximate this error by using the truncated series

$$e^{-T/\tau} \approx 1 - \frac{T}{\tau} + \frac{1}{2}\left(\frac{T}{\tau}\right)^2$$

Express this approximate error as a percentage of the ideal value of e_o. What must T/τ be if the error is to be 5 percent? For this value of T/τ, evaluate the error caused by truncating the series. (Use the theorem on the remainder of an alternating series.)

4.16 Analyze the nozzle-flapper displacement pickup of Fig. P4.2, using the simple incompressible relations. Explain the advantages of this configuration.

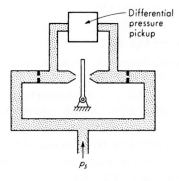

Figure P4.2

4.17 Prove Eqs. (4.105) and (4.106).

4.18 In the variable-capacitance velocity pickup shown in Fig. P4.3, prove that the current i is directly proportional to the angular velocity $d\theta/dt$. Since

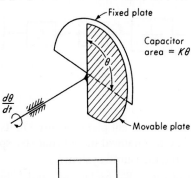

Figure P4.3

voltage signals are more readily manipulated, how might the current signal be transduced to a proportional voltage? Does your method of doing this affect the basic operation? What must be required if the basic operation is to be only slightly affected?

4.19 Construct logarithmic frequency-response curves for a seismic-displacement pickup with $\zeta = 0.3$ and $\omega_n/(2\pi) = 10$ Hz. For what range of frequencies is the amplitude ratio flat within 1 dB?

4.20 Make a comprehensive list and explain the action of modifying and/or interfering inputs for the systems of the following:

(a) Fig. 4.4	(g) Fig. 4.60
(b) Fig. 4.10	(h) Fig. 4.72
(c) Fig. 4.17	(i) Fig. 4.74
(d) Fig. 4.33	(j) Fig. 4.76
(e) Fig. 4.40c	(k) Fig. 4.85
(f) Fig. 4.59	(l) Fig. 4.92

4.21 Derive $(\theta_o/\theta_i)\,(D)$ for Fig. 4.77b.

4.22 Construct logarithmic frequency-response curves for a piezoelectric accelerometer with $K_q/(C\omega_n^2) = 0.001$ V/(in/s^2), $\tau = 0.10$ s, $\omega_n/(2\pi) = 10{,}000$ Hz, and $\zeta = 0$. Will this accelerometer be satisfactory for shock measurements of half-sine pulses with a duration of 0.05 s? If not, suggest needed changes.

4.23 Explain, giving a sketch, how the principle of the system of Fig. 4.85 can be adapted to the measurement of angular acceleration. Your device must *not* be sensitive to translational acceleration.

4.24 In the system of Fig. 4.85, if K_s and B are made zero, $(\theta/T_n)\,(D) = 1/(JD^2)$. Obtain $(e_o/\ddot{x}_i)\,(D)$ for this situation. What is the defect in this system? To remedy this, electric "damping" may be introduced by adding a circuit with a transfer function as shown in Fig. P4.4. Obtain $(e_o/\ddot{x}_i)\,(D)$ for this arrangement. Why must $\tau_1 > \tau_2$?

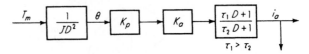

Figure P4.4

4.25 When a seismic-displacement pickup is used in its proper frequency range, what is an adequate mechanical model (masses, springs, dashpots) for its loading effect on the measured system?

4.26 In the Jerkmeter of Fig. 4.87, replace ω_{n1} and ζ_1 by their values in terms of J, B, and K_s, thus rewriting Eq. (4.123). Suppose a sensitivity $Mr/(K_i K_{ce}) = 0.05$ V/(ft/s^3) is required. Assume temporarily that $B = K_s = 0$, and neglect the fact that this system would be unstable. Assume also that J is due mainly to M; thus take $J = Mr^2$.

(a) Find the numerical value of $K_p K_{ae}/r$ needed to give a flat amplitude ratio for $(e_o/D^3 x_i)(i\omega)$ within 5 percent over the range 0 to 5 Hz.

(b) If $r = 0.1$ ft and $K_p = 57.3$ V/rad, find K_{ae}.

(c) Suppose now that $M = 0.01$ lbm. Find $K_i K_{ce}$.

(d) If B and K_s are not zero, find the value of BK_s needed to put the system just on the margin of instability. (Use the Routh criterion.)

(e) Let the design value of BK_s be 10 times the value of part d. If $K_s = 0.275$ ft · lbf/rad, what is B?

(f) With the above values of B and K_s, recheck the amplitude ratio at 5 Hz. Does it meet the 5 percent requirement?

(g) To correct the situation found in part f, make $K_{ae} = 2,490$. Recheck the amplitude ratio at 5 Hz. Recheck the stability.

(h) To regain the stability lost in part g, reduce M to 0.001 lbm. Recheck the amplitude ratio at 5 Hz.

(i) Recheck the overall system static sensitivity. How much external amplification is now needed to return to the required 0.05 V/(ft/s³)?

4.27 Derive Eqs. (4.127) to (4.130). Give a physical explanation of the "notch-filter" effect. Explain the apparent discontinuity in phase angle.

4.28 Find the steady-state response of the system of Fig. 4.88 to a constant horizontal acceleration.

4.29 A commercial version of the system of Fig. 4.88 has a pendulum made as shown in Fig. P4.5. Where is the center of buoyancy of this pendulum? Where is the center of mass? Will the buoyant force tend to cause an output? Why? Derive a relation showing the requirements for completely unloading the pivot bearing. (This "floating" reduces bearing friction and thus system threshold.)

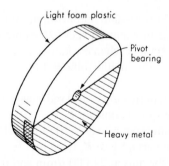

Light foam plastic

Pivot bearing

Heavy metal

Completely immersed in liquid

Figure P4.5

4.30 Equation (4.139) can be written as

$$(I_x I_y D^2 + BI_x D + H_s^2 + I_x K_s)\phi = -\frac{H_s}{D}T_y$$

For a gyro with no spring or damping, suppose T_y is a unit impulse. Solve Eqs. (4.139) and (4.141) for ϕ and θ. The combined sinusoidal motion of ϕ and θ causes the spin axis to rotate in space. This motion is called *nutation*. Describe it qualitatively and define its frequency. What is the effect on the above results if damping is present?

4.31 A rate gyro is mounted on a long, thin missile, which is quite flexible in bending, as in Fig. P4.6. Bending vibrations cause the slope at the gyro location to go through sinusoidal oscillations of 0.1° amplitude at 50 Hz. What maximum angular velocity will the gyro feel because of vibration? If the gross (rigid-body) rotation of the missile (which is what the gyro is *intended* to measure) is 10 rad/s, what percentage of the total gyro signal is due to the vibration? Where could one relocate the gyro to minimize this problem? If the fastest rigid-body motions expected are 1 Hz, what other solution is possible?

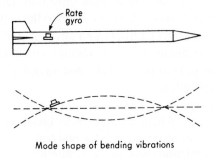

Mode shape of bending vibrations

Figure P4.6

4.32 In a rate gyro, the steady-state output/input ratio θ/ϕ is not strictly linear because the angular-momentum vector does not remain perpendicular to the angular-velocity vector ϕ when θ rotates away from zero.
   (a) What is the maximum allowable θ if this nonlinearity is not to exceed 1 percent?
   (b) If a rate gyro requires $\omega_n = 100$ rad/s and if $I_y = 0.0013$ in $\cdot$ lbf $\cdot$ s^2, what must the spring constant K_s be?
   (c) If the maximum input rate ϕ is 10 rad/s, what must H_s be if θ_{max} is the value found in part *a*? Use K_s from part *b*.
   (d) If the spin motor runs at 24,000 r/min and if the wheel is a solid cylinder of a length equal to its diameter and is made of a material with a specific weight of 0.3 lbf/in^3, what are the required dimensions of the wheel? Use all necessary numbers from the previous parts of the problem.

4.33 The float-type wave height gage of Fig. P4.7 operates on the buoyant-force principle. The float has cross-sectional area A and mass M, and B represents the only significant frictional effect in the system. Derive (x_o/h_i) (D). Suppose waves pass the gage at the rate of 0.2 wave per second and that the wave profile contains significant Fourier harmonics up to the fifth. We wish to have 95 percent dynamic accuracy, based on the frequency-response/amplitude ratio. Get an expression for the maximum allowable float mass.

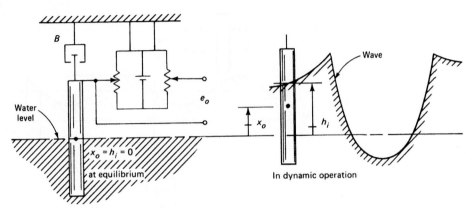

Figure P4.7

4.34 Modify the gyro analysis of Fig. 4.93 to account for a spurious input in the form of a vehicle rotation θ_f about the θ axis. Call the absolute gimbal rotation θ_g, and note that the θ pickoff output would now be proportional to $\theta_g - \theta_f$. Use these results to obtain disturbance-input transfer functions (θ/θ_f) (D) for rate and rate-integrating gyros.

4.35 In a surface-roughness instrument (Fig. P4.8), the stylus of a displacement transducer is free to move vertically as the carriage is moved over the surface at constant velocity V. If V is too large, the stylus will bounce off the surface; too small V wastes time. To study this problem, consider the sinusoidal surface of wavelength L and amplitude A (Fig. P4.8b). Assume the weight of the stylus to be the only effect tending to maintain contact; neglect all friction. For a given A and L find the maximum allowable V to prevent bouncing. Find this V if $A = 2.5 \times 10^{-4}$ cm and $L = 2.5 \times 10^{-2}$ cm.

The configuration of Fig. P4.8c is suggested as a method of increasing the allowable V by a factor of about 6. Investigate this conjecture and list any drawbacks to this scheme.

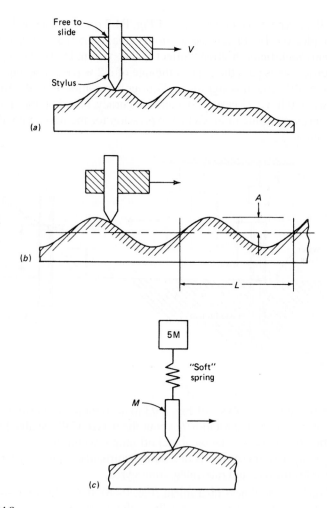

Figure P4.8

4.36 Perform a linear dynamic analysis on the pneumatic noncontact displacement gage of Fig. 4.60, using a sufficiently complete model to allow the possibility of instability. Then perform an accuracy-stability trade-off study.

4.37 A servo accelerometer (Fig. 4.85a) is intended to respond to motion $\ddot{x}_i$ only; however, it may have some response to other motions.

   (a) If $\ddot{x}_i \equiv 0$ but a cross-axis acceleration $\ddot{y}_i$ (perpendicular to $\ddot{x}_i$ and perpendicular to the paper in Fig. 4.85) occurs, will the instrument produce a spurious output? Explain.

   (b) If a constant $\ddot{x}_i$ is simultaneously present with a constant $\ddot{y}_i$, will the instrument output have an error due to $\ddot{y}_i$? Calculate this error.

   (c) If $\ddot{x}_i = \ddot{y}_i \equiv 0$ but the instrument frame experiences a steady angular acceleration $\ddot{\theta}_f$, will a spurious output occur? Derive $(e_o/\ddot{\theta}_f)$ (D).

4.38 Derive Eq. (4.15) and then design an extensometer for a deflection range of ± 0.020 in and maximum gage strain of 1500 microstrain. Check your design for maximum F and modify it if F is too large.

4.39 Derive the *exact* nonlinear differential equation for the capacitive displacement sensor analyzed on a linearized basis just after Eq. (4.42). Perform a SIMULINK (or other available) simulation of this model and show that, within the frequency range where accurate displacement measurements are expected, the sensor behaves in a perfectly linear fashion.

BIBLIOGRAPHY

1. J. W. Dally and W. F. Riley, "Experimental Stress Analysis," 2d ed., McGraw-Hill, New York, 1978.

2. D. N. Keast, "Measurements in Mechanical Dynamics," McGraw-Hill, New York, 1970.

3. J. F. Blackburn (ed.), "Potentiometers, Components Handbook," chap. 8, McGraw-Hill, New York, 1948.

4. D. H. Parkes, "The Application of Microwave Techniques to Noncontact Precision Measurement," *ASME Paper* 63-WA-346, 1963.

5. J. G. Collier and G. F. Hewitt, "Film-Thickness Measurements," *ASME Paper* 64-WA/HT-41, 1964.

6. J. T. Broch, "Mechanical Vibration and Shock Measurements," Brüel and Kjaer Instruments, Marlboro, MA, 1972.

7. C. M. Harris and C. E. Crede (eds.), "Shock and Vibration Handbook," McGraw-Hill, New York, 1988.

8. D. F, Wilkes and C. E. Kreitler, "The Long Period Horizontal Air Bearing Seismometer," Sandia Corp., Albuquerque, NM, SCTM74-62(13), 1962.

9. "Intermediate Accuracy Gyroscope Design Criteria," *Monograph,* A. D. Little, Inc., Cambridge, MA, June 1970.

10. P. K. Stein, "Strain Gages, Measurement Engineering," Stein Engineering Services, Inc., Phoenix, AZ, 1964.

11. H. Dagnall, "Let's Talk Roundness," Rank Taylor Hobson Ltd., Leicester, UK, 1984.

12. R. F. Hill and M. Skunda, "Measuring Tool Wear Radiometrically," *Mech. Eng.,* February 1972.

13. J. B. Bryan et al., "Thermal Effects in Dimensional Metrology," *ASME Paper* 65-Prod-13, 1965.

14. D. G. Fleming et al. (eds.), "Handbook of Engineering in Medicine and Biology," sec. B, Instruments and Measurements, CRC Press, Boca Raton, FL, 1978.

15. D. M. Anthony, "Engineering Metrology," Pergamon Press, New York, 1987.

16. W. R. Moore, "Foundations of Mechanical Accuracy," Moore Special Tool Co., Bridgeport, CT, 1970 (distributed by MIT Press).

17. J. G. Webster (ed.), "The Measurement, Instrumentation and Sensors Handbook," CRC Press, Boca Raton, FL, 1999.

18. S. Soloman, "Sensors Handbook," McGraw-Hill, New York, 1998.

5

CHAPTER

Force, Torque, and Shaft Power Measurement

5.1 STANDARDS AND CALIBRATION

Force is defined by the equation $F = MA$; thus a standard for force depends on standards for mass and acceleration. Mass is considered a fundamental quantity, and its standard is a cylinder of platinum-iridium, called the International Kilogram, kept in a vault at Sèvres, France. Other masses (such as national standards) may be compared with this standard by means of an equal-arm balance, with a precision of a few parts in 10^9 for masses of about 1 kg. Tolerances on various classes of standard masses available from NIST may be found in its publications.[1]

Acceleration is not a fundamental quantity, but rather is derived from length and time, two fundamental quantities whose standards are discussed in Chap. 4. The acceleration of gravity, g, is a convenient standard which can be determined with an accuracy of about 1 part in 10^6 by measuring the period and effective length of a pendulum or by determining the change with time of the speed of a freely falling body.[2] The actual value of g varies with location and also slightly with time (in a periodic predictable fashion) at a given location. It also may change (slightly) unpredictably because of local geological activity. The so-called standard value of g refers to the value at sea level and 45° latitude and is numerically 980.665 cm/s^2. The value at any latitude ϕ degrees may be computed from

$$g = 978.049(1 + 0.0052884 \sin^2 \phi - 0.0000059 \sin^2 2\phi) \qquad \text{cm/s}^2 \qquad \textbf{(5.1)}$$

---

[1]T. W. Lashof and L. B. Macurdy, "Precision Laboratory Standards of Mass and Laboratory Weights," *Natl. Bur. Std. (U.S.), Circ.* 547, sec. 1, 1954; P. E. Pontius, "Mass and Mass Values," *Natl. Bur. Std. (U.S.), Monograph* 133, 1974.

[2]A. Bray, G. Barbato, and R. Levi, "Theory and Practice of Force Measurement," Academic Press, New York, 1990, chap. 3; W. Torge, "Gravimetry," de Gruyter, Berlin, 1989.

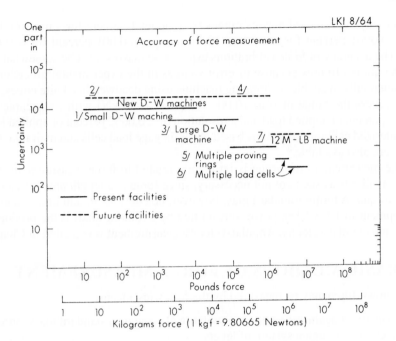

Figure 5.1
Force standards. ["Future facilities" are now available.]

while the correction for altitude h in meters above sea level is

$$\text{Correction} = -(0.00030855 + 0.00000022 \cos 2\phi)h$$
$$+ 0.000072 \left(\frac{h}{1,000}\right)^2 \quad \text{cm/s}^2 \quad \quad \textbf{(5.2)}$$

Local values of g also may be obtained from the National Ocean Survey, National Oceanic and Atmospheric Administration.

When the numerical value of g has been determined at a particular locality, the gravitational force (weight) on accurately known standard masses may be computed to establish a standard of force. This is the basis of the "deadweight" calibration of force-measuring systems. The National Bureau of Standards (now NIST) capability (Fig. 5.1)[3] for such calibrations is an inaccuracy of about 1 part in 5,000 for the range of 10 to 1 million lbf. Above this range, direct deadweight calibration is not presently available. Rather, proving rings[4] or load cells of a capacity of 1 million lbf or less are calibrated against deadweights, and then the unknown force is applied to a multiple array of these in parallel. The range 1 to 10 million lbf is covered by such arrangements with somewhat reduced accuracy. At the low-force end of the scale,

---

[3]"Accuracy in Measurements and Calibrations," *Natl. Bur. Std. (U.S.), Tech. Note* 262, 1965.

[4]"Proving Rings for Calibrating Testing Machines," *Natl. Bur. Std. (U.S.), Circ.* C454, 1946.

the accuracy[5] of standard masses ranges from about 1 percent for a mass of 10^{-5} lbm to 0.0001 percent for the 0.1 to 10 lbm range to 0.001 percent for a 100-lb mass. The accuracy of *force* calibrations using these masses must be somewhat less than the quoted figures because of error sources in the experimental procedures.[6] A commercially available[7] calibrating machine using deadweights, knife edges, and levers covers the range of 0 to 10,000 lbf (or 0 to 50 kN) with an accuracy of ±0.005 percent of applied load and a resolution of ±0.0062 percent of applied load. Computerized calibration systems based on strain gage load cells and hydraulic load frames are also available.[8]

The measurement of torque is intimately related to force measurement; thus torque standards as such are not necessary, since force and length are sufficient to define torque. A torque standard may, however, be *convenient,* and one was under development in 1998.[9] The power transmitted by a rotating shaft is the product of torque and angular velocity. Angular-velocity measurement was treated in Chap. 4.

5.2 BASIC METHODS OF FORCE MEASUREMENT

An unknown force may be measured by the following means:

1. Balancing it against the known gravitational force on a standard mass, either directly or through a system of levers
2. Measuring the acceleration of a body of known mass to which the unknown force is applied
3. Balancing it against a magnetic force developed by interaction of a current-carrying coil and a magnet
4. Transducing the force to a fluid pressure and then measuring the pressure
5. Applying the force to some elastic member and measuring the resulting deflection
6. Measuring the change in precession of a gyroscope caused by an applied torque related to the measured force
7. Measuring the change in natural frequency of a wire tensioned by the force

---

[5]R. M. Schoonover and F. E. Jones, "Examination of Parameters That Can Cause Errors in Mass Determination," *CAL LAB,* July/Aug. 1998, pp. 26–31.

[6]"Calibration of Force-Measuring Instruments for Verifying the Load Indication of Testing Machines," *ASTM Std.* E-74, 1974.

[7]W. C. Dillon Co. (www.dillonnews.com).

[8]Gold Standard System, Interface, Inc., Scottsdale, AZ, 800-947-5598 (www.interfaceforce.com); C. Ferrero et al., "Main Metrological Characteristics of IMGC Six-Component Dynamometer," *RAM,* vol. 2, 1986, pp. 21–28; R. Hellwig, "Precision Force Transducer for International Comparison Measurements on Force Standard Machines," *RAM,* vol. 3, 1987, pp. 17–22; HBM, Norcross, GA, 888-816-9006 (www.hbm-home.com).

[9]F. A. Davis, "Design of the 1st UK National Standard Static Torque Calibration Machine," National Physical Laboratory, Queens Road, Teddington, Middlesex, United Kingdom, TW 11 OLW, 0181-943-6194.

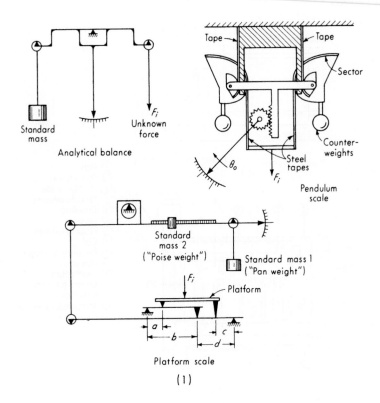

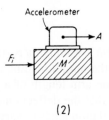

Figure 5.2
Basic force-measurement methods.

In Fig. 5.2, method 1 is illustrated by the analytical balance, the pendulum scale, and the platform scale. The analytical balance, while simple in principle, requires careful design and operation to realize its maximum performance.[10] The beam is designed so that the center of mass is only slightly (a few thousandths of an inch) below the knife-edge pivot and thus barely in stable equilibrium. This makes the beam deflection (which in sensitive instruments is read with an optical

[10]L. B. Macurdy, "Performance Tests for Balances," *Inst. & Cont. Syst.,* pp. 127–133, September 1965.

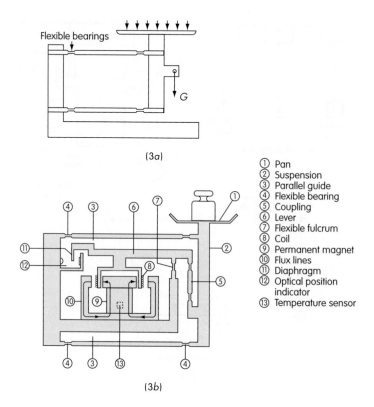

Flexible bearings

(3a)

① Pan
② Suspension
③ Parallel guide
④ Flexible bearing
⑤ Coupling
⑥ Lever
⑦ Flexible fulcrum
⑧ Coil
⑨ Permanent magnet
⑩ Flux lines
⑪ Diaphragm
⑫ Optical position indicator
⑬ Temperature sensor

(3b)

Figure 5.2
(Continued)

micrometer) a very sensitive indicator of unbalance. For the low end of a particular instrument's range, often the beam deflection is used as the output reading rather than attempting to null by adding masses or adjusting the arm length of a poise weight. This approach is faster than nulling but requires that the deflection-angle unbalance relation be accurately known and stable. This relation tends to vary with the load on the balance, because of deformation of knife edges, etc., but careful design can keep this to a minimum. For highly accurate measurements, the buoyant force due to the immersion of the standard mass in air must be taken into account. Also, the most sensitive balances must be installed in temperature-controlled chambers and manipulated by remote control to reduce the effects of the operator's body heat and convection currents. Typically, a temperature difference of $1/20°C$ between the two arms of a balance can cause an arm-length ratio change of 1 ppm, significant in some applications. Commercially available analytical balances may be classified as follows:[11]

[11]F. Baur, "The Analytical Balance," *Ind. Res.,* p. 64, July–August 1964.

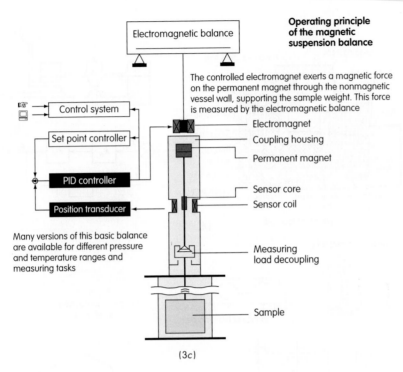

Electromagnetic balance

Operating principle of the magnetic suspension balance

The controlled electromagnet exerts a magnetic force on the permanent magnet through the nonmagnetic vessel wall, supporting the sample weight. This force is measured by the electromagnetic balance

Control system

Set point controller

PID controller

Position transducer

Many versions of this basic balance are available for different pressure and temperature ranges and measuring tasks

Electromagnet

Coupling housing

Permanent magnet

Sensor core
Sensor coil

Measuring load decoupling

Sample

(3c)

Figure 5.2

(*Continued*)

Description	Range, g	Resolution, g
Macro analytical	200–1,000	10^{-4}
Semimicro analytical	50–100	10^{-5}
Micro analytical	10–20	10^{-6}
Micro balance	less than 1	10^{-6}
Ultramicro balance	less than 0.01	10^{-7}

The pendulum scale is a deflection-type instrument in which the unknown force is converted to a torque that is then balanced by the torque of a fixed standard mass arranged as a pendulum. The practical version of this principle utilizes specially shaped sectors and steel tapes to linearize the inherently nonlinear torque-angle relation of a pendulum. The unknown force F_i may be applied directly as in Fig. 5.2 or through a system of levers, such as that shown for the platform scale, to extend the range. An electrical signal proportional to force is easily obtained from any angular-displacement transducer attached to measure the angle θ_o.

The platform scale utilizes a system of levers to allow measurement of large forces in terms of much smaller standard weights. The beam is brought to null by a proper combination of pan weights and adjustment of the poise-weight lever arm

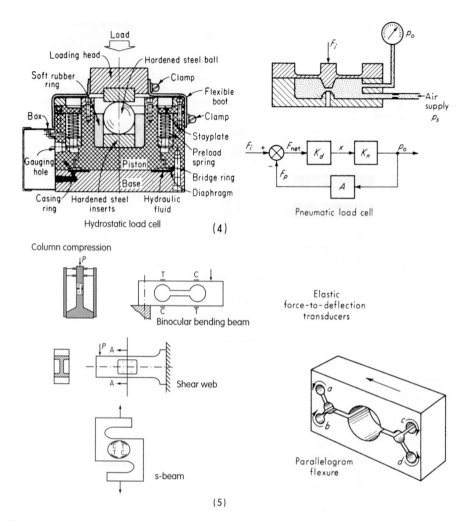

Figure 5.2

(Concluded)

along its calibrated scale. The scale can be made self-balancing by adding an electrical displacement pickup for null detection and an amplifier-motor system to position the poise weight to achieve null. Another interesting feature is that if $a/b = c/d$, the reading of the scale is independent of the location of F_i on the platform. Since this is quite convenient, most commercial scales provide this feature by use of the suspension system shown or others that allow similar results.

While analytical balances are used almost exclusively for "weighing" (really determining the *mass* of) objects or chemical samples, platform and pendulum scales are employed also for force measurements, such as those involved in shaft power determinations with dynamometers. All three instruments are intended mainly for static force measurements.

Method 2, the use of an accelerometer for force measurement, is of somewhat limited application since the force determined is the *resultant* force on the mass. Often *several* unknown forces are acting, and they cannot be separately measured by this method.

The electromagnetic balance[12] (method 3) utilizes a photoelectric (or other displacement sensor) null detector, an amplifier, and a torquing coil in a servo-system to balance the difference between the unknown force F_i and the gravity force on a standard mass. Its advantages relative to mechanical balances are ease of use, less sensitivity to environment, faster response, smaller size, and ease of remote operation. Also, the electric output signal is convenient for continuous recording and/or automatic-control applications. Balances with built-in microprocessors[13] allow even greater convenience, versatility, and speed of use by automating many routine procedures and providing features not formerly feasible. Automatic tare-weight systems subtract container weight from total weight to give net weight when material is placed in the container. Statistical routines allow immediate calculation of mean and standard deviation for a series of weighings. "Counting" of small parts by weighing is speeded by programming the microprocessor to read out the parts count directly, rather than the weight. Accurate weighing of live laboratory animals (difficult on an ordinary balance because of animal motion) is facilitated by averaging scale readings over a preselected time. Interfacing the balance to (external or built-in) printers for permanent recording also is eased by the microprocessor. Figure 5.2, part $3a$,[14] shows a design available in range from 22 to 405 grams, with resolutions from 2 to 100 μg. Part $3a$ shows schematically the parallelogram flexure system that guides the motion produced by an applied force (weight), while part $3b$ shows details of the complete system (except for the servo and readout electronics). A flexure-pivot lever system (up to 15:1) puts large input forces within the range of a relatively small magnetic force coil. The signal from the optical displacement sensor is the error signal in the servo system, which provides a coil current (and thus magnetic force) to balance the unknown input force and restore the deflection to near zero. All motions are constrained with flexure bearings (rather than rolling or sliding bearings) to give the nearly frictionless performance required for resolutions as small as 2 μg. Temperature effects (observed mainly in the magnetic field strength) are compensated in software; the temperature is measured using the signal from the temperature sensor. The seven-digit readout testifies to the extreme resolution of these instruments. Figure 5.2, part $3c$,[15] shows a version that allows the weighed sample to be immersed in an atmosphere of controlled temperature, pressure, and fluid composition, completely sealed off from the weighing balance, for sensitive density, sorbtion, and chemical studies.

---

[12]L. Cahn, "Electromagnetic Weighing," *Instrum. Contr. Syst.,* p. 107, September 1962; Cahn Instrument Div. (www.thermocahn.com).

[13]B. Ludewig, "Microprocessor Balance," *Am. Lab.,* pp. 81–83, May 1979.

[14]Mettler-Toledo, Inc., Hightstown, NJ, 800-638-8537 (www.mico.mt.com).

[15]Rubotherm GMBH, S. Natick, MA, 508-655-3950 (www.rubotherm.com).

Method 4 is illustrated in Fig. 5.2 by hydrostatic[16] and pneumatic load cells. Hydraulic cells are completely filled with oil and usually have a preload pressure of the order of 30 lb/in^2. Application of load increases the oil pressure, which is read on an accurate gage. Electrical pressure transducers can be used to obtain an electrical signal. The cells are very stiff, deflecting only a few thousandths of an inch under full load. Capacities to 100,000 lbf are available as standard while special units up to 10 million lbf are obtainable. Accuracy is of the order of 0.1 percent of full scale; resolution is about 0.02 percent. A hydraulic totalizer[17] is available to produce a single pressure equal to the sum of up to 10 individual pressures in multiple-cell systems used for tank weighing, etc. (see Chap. 10).

The pneumatic load cell shown uses a nozzle-flapper transducer as a high-gain amplifier in a servoloop. Application of force F_i causes a diaphragm deflection x, which in turn causes an increase in pressure p_o since the nozzle is more nearly shut off. This increase in pressure acting on the diaphragm area A produces an effective force F_p that tends to return the diaphragm to its former position. For any constant F_i, the system will come to equilibrium at a specific nozzle opening and corresponding pressure p_o. The static behavior is given by

$$(F_i - p_o A)K_d K_n = p_o \tag{5.3}$$

where $K_d \triangleq$ diaphragm compliance, in/lbf   **(5.4)**

$K_n \triangleq$ nozzle-flapper gain, (lb/in^2)/in   **(5.5)**

Solving for p_o, we get

$$p_o = \frac{F_i}{1/(K_d K_n) + A} \tag{5.6}$$

Now K_n is not strictly constant, but varies somewhat with x, leading to a nonlinearity between x and p_o. However, in practice, the product $K_d K_n$ is very large, so that $1/(K_d K_n)$ is made negligible compared with A, which gives

$$p_o = \frac{F_i}{A} \tag{5.7}$$

which is linear since A is constant. As in any feedback system, dynamic instability limits the amount of gain that actually can be used. A typical supply pressure p_s is 60 lb/in^2, and since the maximum value of p_0 cannot exceed p_s, this limits F_i to somewhat less than 60 A. A line of commercial pneumatic weighing systems[18] using similar principles (combined with lever/knife-edge methods) is available in standard ranges to 110,000 lbf.

While all the previously described force-measuring devices are intended mainly for static or slowly varying loads, the elastic deflection transducers of method 5 are

---

[16]A. H. Emery Co. (www.emerywinslow.com).

[17]Ibid.

[18]"An Introduction to the Darenth Gnu-Weigh Pneumatic Weighing System," Darenth Americas, Bridgeville, DE, 1980. (A google search in 2002 could not find this company.)

widely used for both static and dynamic loads of frequency content up to many thousand hertz. While all are essentially spring-mass systems with (intentional or unintentional) damping, they differ mainly in the geometric form of "spring" employed and in the displacement transducer used to obtain an electrical signal. The displacement sensed may be a gross motion, or strain gages may be judiciously located to sense force in terms of strain. Bonded strain gages have been found particularly useful in force measurements with elastic elements. In addition to serving as force-to-deflection transducers, some elastic elements perform the function of resolving vector forces or moments into rectangular components. As an example, the parallelogram flexure of Fig. 5.2 (part 5) is extremely rigid (insensitive) to all applied forces and moments except in the direction shown by the arrow. A displacement transducer arranged to measure motion in the sensitive direction thus will measure only that component of an applied vector force which lies along the sensitive axis. Perhaps the action of this flexure may be most easily visualized by considering it as a four-bar linkage with flexure hinges at the thin sections *a, b, c,* and *d.*

Because of the importance of elastic force transducers in modern dynamic measurements, we devote a considerable portion of this chapter to their consideration. Although they may differ widely in detail construction, their dynamic-response form is generally the same, and so we treat an idealized model representative of all such transducers in the next section. Discussion of methods 6 and 7 is deferred to the end of the chapter since they are not as common as methods 1 through 5.

5.3 CHARACTERISTICS OF ELASTIC FORCE TRANSDUCERS

Figure 5.3 shows an idealized model of an elastic force transducer. The relationship between input force and output displacement is easily established as a simple second-order form:

$$F_i - K_s x_o - B\dot{x}_o = M\ddot{x}_o \tag{5.8}$$

$$\frac{x_o}{F_i}(D) = \frac{K}{D^2/\omega_n^2 + 2\zeta D/\omega_n + 1} \tag{5.9}$$

where

$$\omega_n \triangleq \sqrt{\frac{K_s}{M}} \tag{5.10}$$

$$\zeta \triangleq \frac{B}{2\sqrt{K_s M}} \tag{5.11}$$

$$K \triangleq \frac{1}{K_s} \tag{5.12}$$

Note that devices of this type are also (unintentional) accelerometers and produce a spurious output in response to base vibration inputs (see Prob. 5.1).

For transducers that do not measure a gross displacement but rather use strain gages bonded to the "spring," the output strain ε may be substituted for x_o if K_s

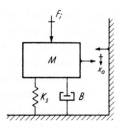

Figure 5.3
Elastic force transducer.

is reinterpreted as force per unit strain rather than force per unit deflection. Many elastic elements have been analyzed, and the results "cataloged," for stress (strain),[19] deflection,[20] and natural frequency.[21] These results are in the form of formulas involving material properties and dimensions, and so are useful for preliminary design of transducers. For unusual shapes, or to check details such as stress concentrations, standard finite-element software can be used, but recall here that this approach does *not* provide any formulas, only specific numerical results for specific dimensions. Most force transducers (with few exceptions)[22] have *no* intentional damping; ζ is due entirely to parasitic effects and is *not* possible to predict theoretically. While theoretical and finite-element results are often adequate for *machine* design, since transducers must be accurate to 1 percent or better, they must *always* be calibrated after construction and before use; this calibration will easily get the actual damping ratio.

Since the dynamic response of second-order instruments has been fully discussed, we concentrate mainly on details peculiar to specific force transducers. For transducers based on strain gages, gage manufacturers provide helpful handbooks.[23] Measurements Group Inc. also sells software called "TRANSCALC" that facilitates design for 14 different load-cell geometries and diaphragm pressure transducers. This software can be downloaded from the Internet for a free 10-day trial.

While simple geometries such as the cantilever beam can be easily designed using available formulas from texts, such formulas are *not* available for some of the more popular forms, such as the "binocular." The TRANSCALC software includes this form, but the formulas are said to be proprietary. An analysis of this form in the

---

[19]W. C. Young and R. G. Budynas, "Roark's Formulas for Stress and Strain," McGraw-Hill, New York, 2001. This is also available as a software product.

[20]Ibid.

[21]R. D. Blevins, "Formulas for Natural Frequency and Mode Shape," Krieger, Malabar, FL, 1995.

[22]Tedea Huntleigh, Canoga Park, CA, models 240, 1410, and 9010, 818-716-0593 (www.tedeahuntleigh.com).

[23]"The Route to Measurement Transducers, HBM Inc., Marlboro, MA, 888-235-4243 (www.hbminc.com/gageland); "Strain Gage Based Transducers," Measurement Group Inc., Raleigh, NC, 919-365-3800 (www.measurementsgroup.com).

open literature does not appear to be available. Of course, *any* complex geometry can be subjected to finite-element software for analysis, but such studies do *not* provide design formulas, only numerical results for a *given* set of dimensions and material properties. I decided to try a "strength of materials" type of analysis myself and, after some study, came up with a simple analysis based on certain assumptions. To check this analysis, I then also did a finite-element analysis, and finally actually built one unit for laboratory testing. Since these various steps might be part of *any* load-cell design process, the following example, which briefly summarizes the procedure, is provided.

Figure 5.4*a* shows the meshed finite-element model but will also serve as an "analysis sketch" for the strength of materials study. The vertical support at the left is not an inherent part of such load cells; it was included to allow space under the load cell for a deflection-measuring dial indicator. The first assumption made was to think of the device as a four-bar linkage, where the usual pin joints are replaced by elastic hinges at the four thin sections. If this assumption is essentially correct, then the "link" at the right (where the vertical load *F* is applied) would have a simple (mainly vertical) translation *x*. The symmetry of the device suggests that each of the four hinges contributes an *equal* elastic resisting moment, to balance the applied loading. (Figure 5.4*b* shows the finite-element deflection results that were later obtained and which seem to validate these assumptions.)

I next tried a "direct" approach from strength of materials beam theory, but was unable to evaluate the needed bending moments or convince myself of the insensitivity to load application point that is claimed for this design. Finally I tried an energy-based method which led to useful results. The assumption here was that the applied force would move vertically through a certain deflection *y,* thereby doing a definite amount of work $Fy/2$. This work would show up as an equal amount of stored elastic energy in the four hinges, which would each feel a torque *T* and rotate through the same angle θ. Since the angle is sure to be small, $\tan\theta \; y/L \approx \theta$, where *L* is the distance between hole center-lines (2.0 inches in the example). Equating the energies gives the result $Fy/2 = 4T\theta/2$, and thus the bending moment at each hinge is $T = FL/4$. This result not only allows us to compute stresses and strains, but also verifies the claim that the load cell gives essentially the same output irrespective of the horizontal location of the applied force. (This claim really applies only to locations somewhat to the right of the right-most gage.)

Using standard beam bending formulas now gives

$$\varepsilon = \frac{\sigma}{E} = \frac{6T}{bt^2 E} = \frac{3FL}{2bt^2 E} \tag{5.13}$$

where *t* is the thickness at the thinnest point and *b* is the depth, (0.04 in and 0.50 in in the example). This formula is useful for choosing dimensions when the full scale load *F* and design strain (often about 1500 $\mu\varepsilon$) are given (our example has $F = 5.0$ lb). The TRANSCALC software does *not* compute the load cell stiffness; the company told me that their customers rarely request this bit of information. It *would* of course be of interest for any dynamic application since it, together with the

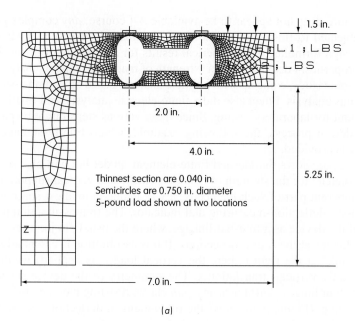

1.5 in.

L 1 ; L B S

2 ; L B S

2.0 in.

4.0 in.

5.25 in.

Thinnest section are 0.040 in.
Semicircles are 0.750 in. diameter
5-pound load shown at two locations

7.0 in.

(a)

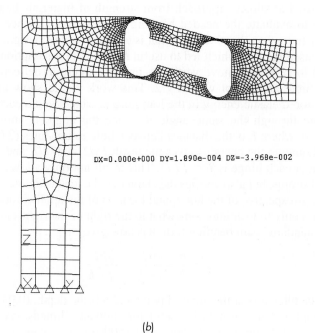

DX=0.000e+000 DY=1.890e-004 DZ=-3.968e-002

(b)

Figure 5.4
Finite-element model of binocular force transducer.

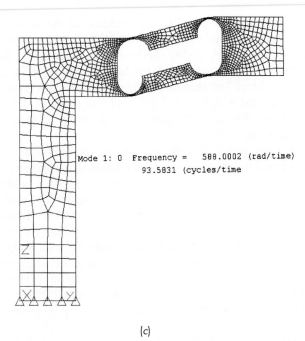

Mode 1: 0 Frequency = 588.0002 (rad/time)
93.5831 (cycles/time

(c)

Figure 5.4
(Concluded)

attached mass, determines the natural frequency. With a few more assumptions and some additional analysis, we can estimate the stiffness by computing the angle θ. The formula for deflection angle of a beam is

$$\Delta\theta = \int \frac{M}{EI}\, dx \tag{5.14}$$

where the integral is taken over the length of the beam and M is the bending moment. We will take the bending moment as the constant T but express the area moment of inertia in terms of x since the beam thickness t varies with x. Note that in computing stress and strain at the gage locations, we treated t as a constant equal to the thickness at the *thinnest* point. Strain gages have a finite gage length, so the measured strain will be somewhat less than the *peak* value that we have estimated since the beam gets thicker away from the midpoint. We cannot use this approach for the deflection. Expressing t (and then I) as a function of x gives

$$\Delta\theta = \frac{12T}{bE} \int_{-R}^{R} \frac{dx}{\left(R + t\sqrt{R^2 - x^2}\right)^3} \tag{5.15}$$

where t is now the constant equal to the thickness at the thinnest point. I could not find an analytical formula for this integral so it was computed numerically for the assumed dimensions, giving the numerical value 3101; then the angle is 0.01879 rad, the deflection under the 5-pound load is 0.03759 in. and the stiffness is 133 lbf/in. The "block" of aluminum under the load is the major moving mass; it is about 1.22 in^3 and weighs 0.122 lbf. Using $\omega_n = \sqrt{K_s/M}$, we get a natural frequency of 100 Hz. The actual frequency will probably be somewhat lower since the two links that rotate will add some mass.

Using the Algor finite-element software (www.algor.com), the first natural frequency was found to be 93.6 Hz, with the mode shape shown in Fig. 5.4c. This mode shape looks much like the *static* deflection curve; first modes often do. The second and third modes were 626 and 2588 Hz, respectively, with more complex mode shapes. For transducer design, only the first mode is generally of interest because we can get good dynamic accuracy only for frequencies below the first resonant peak. Algor also gave the deflection under the load as 0.0397 in, close to our estimate of 0.0376 in. Figure 5.5 shows strains in the neighborhood of one of the gage locations; all four were very similar, justifying our symmetry assumption. The peak values of 0.00193 are close to our estimates of 0.00189.

For the laboratory-tested transducer the natural frequency observed was 97 Hz and the stiffness was 121 lbf/in. Gage strains at the 5-lb load were 0.002073, 0.001838, 0.001827, and 0.001971, which average to 0.00193. It appears that the design formulas developed are useful for this type of transducer, whether it is implemented with strain gages or some gross-deflection sensor such as an LVDT.

Bonded-Strain-Gage Transducers

A typical construction for a strain-gage load cell for measuring compressive forces is shown in Fig. 5.6. (Cells to measure both tension and compression require merely the addition of suitable mechanical fittings at the ends.) The load-sensing member is short enough to prevent column buckling under the rated load and is proportioned to develop about 1,500 $\mu\varepsilon$ at full-scale load (typical design value for all forms of foil gage transducers). Materials used include SAE 4340 steel, 17-4 PH stainless steel, and 2024–T4 aluminum alloy, with the last being quite popular for "home-made" transducers. Foil-type metal gages are bonded on all four sides; gages

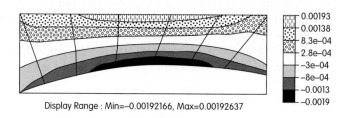

Display Range : Min=−0.00192166, Max=0.00192637

Figure 5.5
Finite-element analysis results for strain near gage locations.

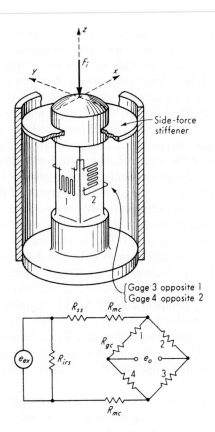

Figure 5.6
Strain-gage load cell.

1 and 3 sense the direct stress due to F_i, and gages 2 and 4 the transverse stress due to Poisson's ratio μ. This arrangement gives a sensitivity $2(1 + \mu)$ times that achieved with a single active gage in the bridge. [See Eq. (10.8) for bridge-circuit behavior.] It also provides primary temperature compensation since all four gages are (at least for steady temperatures) at the same temperature. Furthermore, the arrangement is insensitive to bending stresses due to F_i being applied off center or at an angle. This can be seen by replacing an off-center force by an equivalent on-center force and a couple. The couple can be resolved into x and y components which cause bending stresses in the gages. If the gages are carefully placed so as to be symmetric, the bending stresses in gages 1 and 3 will be of opposite sign, and by the rules of bridge circuits the net output e_o due to bending will be zero. Similar arguments hold for gages 2 and 4 and for bending stresses due to F_i being at an angle. The side-force stiffener plate also reduces the effects of angular forces, since it is very stiff in the radial (x, y) direction but very soft in the z direction.

   The deflection under full load of such load cells is of the order of 0.001 to 0.015 in, indicating their high stiffness. Often the natural frequency is not quoted since it is determined almost entirely by the mass of force-carrying elements external to the transducer. This is especially true in the many applications where the load cell is used for weighing purposes. The high stiffness also implies a low sensitivity. To increase sensitivity (in low-force cells where it is needed) without sacrificing column stability and surface area for mounting gages, a hollow (square on the outside, round on the inside) load-carrying member may be employed.

To achieve the high accuracy (0.3 to 0. 1 percent of full scale), required in many applications, additional temperature compensation is needed.[24] This is accomplished by means of the temperature-sensitive resistors R_{gc} and R_{mc} shown in Fig. 5.6. These resistors are permanently attached internal to the load cell so as to assume the same temperature as the gages. The purpose of R_{gc} is to compensate for the slightly different temperature coefficients of resistance of the four gages. The purpose of R_{mc} is to compensate for the temperature dependence of the modulus of elasticity of the load-sensing member. That is, although we wish to measure force, the gages sense strain; thus any change in the modulus of elasticity will give a different strain (and thus a different e_o) even though the force is the same. Since all metals change modulus somewhat with temperature, this effect causes a sensitivity drift. The resistance R_{mc} compensates for this by changing the excitation voltage actually applied to the bridge by just the right amount to counteract the modulus effect. Two additional (non-temperature-sensitive) resistors are often found in commercial load cells. They are R_{ss}, which is adjusted to standardize the sensitivity for a nominal e_{ex} to a desired value, and R_{irs}, which is used to adjust the input resistance to a desired value.

When adequate sensitivity cannot be achieved by use of tension/compression members, configurations employing bending stresses may be helpful. These generally provide more strain per unit applied force, but at the expense of reduced stiffness and thus natural frequency. Of the many possibilities, two are shown in Fig. 5.7a. The cantilever-beam gage arrangement provides four times the sensitivity of a single gage, temperature compensation, and insensitivity to x and y components of force if identical gages and perfect symmetry are assumed. (While simple to analyze and construct, the cantilever beam is actually not much used in commercial load cells, forms such as the binocular and s-beam being preferred when a bending configuration is wanted.) Figure 5.7b shows some recent techniques[25] which reduce costs, simplify manufacture, and improve performance of weighing scales. The "folded cantilever" elastic element produces equal tensile and compressive stresses at adjacent locations on the *same* surface, allowing use of prefabricated planar gage

[24]J. Dorsey, "Homegrown Strain-Gage Transducers," *Exp. Mech.,* pp. 255–260, July 1977; *VRE Tech. Ed. News,* no. 21, Vishay Research and Education, Malvern, PA, February 1978; J. Dorsey, "Linearization of Transducer Compensation," Vishay Intertechnology, Romulus, MI, 1977; "SR-4 Strain Gage Handbook," BLH Electronics, Waltham, MA, 1980; L. Clegg, "Trends in Bonded Foil Strain Gage Force Transducers," Interface, Inc., 1996 (www.interfaceforce.com).

[25]W. J. Ort, "A Fabricated Platform Load Cell, VD 84 004a"; New Developments in Foil Strain Gages for Transducers, VD 5.85-2.0 dr, Hottinger Baldwin Measurements, Framingham, MA, 1984 (www.hbiminc.com).

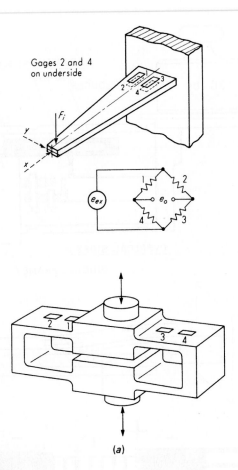

Figure 5.7
Strain-gage beam transducers.

assemblies which incorporate a full bridge of gages, intergage wiring, and zero-balance networks. The geometric nonlinearity of this element is *zero* for a perfectly centered loading point and changes algebraic sign as this point is placed to either side. This controllable nonlinearity can be used to advantage, to compensate for nonlinearities elsewhere in a complete system. The gage's "unit construction" offers many manufacturing and performance advantages. Manual gage placement and lead soldering are largely replaced by automated methods. Only one "assembly" (rather than four individual gages) must be located, and inherent temperature compensation is better since the four gages are fabricated as a unit. The temperature coefficient of the gage factor for the foil material has been tailored to compensate the elastic modulus temperature coefficient of the elastic member, eliminating the need for modulus-compensating resistors. Creep of the elastic element is also compensated by including intentionally "creepy" elements in the gage-foil pattern. These elements have a creep effect opposite to that of the elastic element and may be

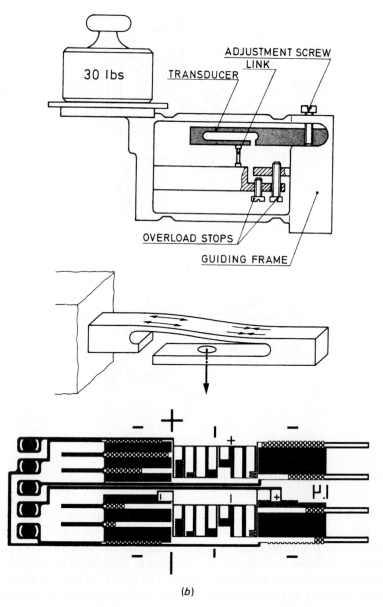

Figure 5.7
(Concluded)

selectively put into action by cutting shorting loops in the foil. Each gaged element is tested for creep to determine the amount of correction needed; then sufficient shorting loops are cut to achieve this. The planar construction makes feasible moisture protection with thin *metallic* films that are far superior to the usual organic

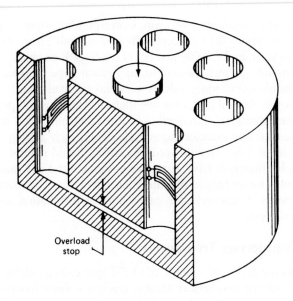

Figure 5.8
Shear-web force transducer.

materials. Finally, the new construction permits a more automated quality-test procedure.

Transducers using shear loading[26] can be designed to be very compact (in the load application direction) and have little sensitivity to off-axis forces and moments, good symmetry for tension/compression, long fatigue life, simple overload protection, and high stiffness. Figure 5.8 shows one possible design, where four gages (only two show) oriented at 45° with the load axis are cemented to the shear webs between the holes to pick up the tension and compression strains produced by the applied shear stress. With two gages in compression and two in tension, a full-bridge circuit can be used. The nature of the overload stop allows "stacking" of a high-range (say 100,000 lbf) and a low-range (say 10,000 lbf) cell to provide a dual-range transducer whose accuracy is maintained at loads below 10 percent of high range. That is, at loads below 10,000 lbf we use the output of the low-range cell, but as load goes above 10,000 lbf, this cell "bottoms out" (safely) and we take our readings from the high-range cell.

Commercial strain-gage force transducers of the various types are available for full-scale loads of a few pounds to hundreds of thousands of pounds and in several accuracy (and price) grades. The lowest-accuracy grade typically has an overall (combined nonlinearity, hysteresis, nonrepeatability, etc.) error of about 1 percent of full scale, temperature effects of ±0.005 percent of full scale/°F on zero shift

---

[26]A. Umit Kutsay, "Flat Load Cells," *Inst. & Cont. Syst.*, pp. 123–125, February 1966; "Interface Load Cell—A New Dimension in Force Measurement," Interface Inc., Scottsdale, AZ, 1980 (www.interfaceforce.com).

and ± 0.01 percent of reading/°F on span. Corresponding figures for the highest-accuracy grade are about 0.15 percent full-scale overall error, ± 0.0015 percent full-scale/°F zero shift, and ± 0.008 percent of reading/°F on span.

When maximum output is desired for any strain-gage transducer, one should consider the possible use of low-modulus materials (such as aluminum) to increase strain per unit force, several gages (or one high-resistance gage) per bridge leg (if space allows), the intentional introduction of stress concentrations at the gage locations, and/or use of semiconductor gages.[27] However, such techniques also present associated problems. Low modulus reduces stiffness and natural frequency, and some low-modulus materials have excessive hysteresis and low fatigue life. Stress concentrations also lower fatigue life, and their effect may be difficult to calculate for design purposes. Semiconductor gages present installation and temperature compensation problems.

Differential-Transformer Transducers

Figure 5.9 shows the design of a family of LVDT load cells available in ranges from 10 g to 10 kg which use two helical flexures (each machined from one solid piece) as the elastic element. An external threaded ring allows convenient zero adjustment by changing the axial position of the spring-loaded LVDT coil form. Full-scale deflections are about 0.020 to 0.030 in, nonlinearity is better than 0.2 percent full scale, and operating temperature range is -65 to $+200$°F.

Piezoelectric Transducers

These force transducers have the same form of transfer function as piezoelectric accelerometers. They are intended for dynamic force measurement only, although some types (quartz pickup with electrometer charge amplifier) have sufficiently large τ to allow short-term measurement of static forces and static calibration. Just as in the piezoelectric accelerometers of Chap. 4, force transducers are available with or without built-in microelectronics. Typical units have about 1 percent nonlinearity and very high stiffness (10^6 to 10^8 lbf/in) and natural frequency (10,000 to 300,000 Hz), though, as with any force transducer, the *system* natural frequency will be lower because of the unavoidable mass of necessary attached components. A single transducer often is useful over a very wide range of forces and preloads because the 1 percent nonlinearity applies to any calibrated range and the natural "leak-off" of output voltage for steady loads gives a convenient zero reading. Useful temperature ranges and sensitivity to temperature inputs depend on piezoelectric material used, design details, and whether built-in electronics are employed. Piezoelectric force pickups tend to be sensitive to side loading, and most manufacturers recommend special precautions to minimize this, but do not always quote numerical values of cross-axis sensitivity. One manufacturer[28] offering specially designed

---

[27]M. Lebow and R. Caris, "Semiconductor Gages or Foil Gages for Transducers," Lebow Associates Inc., Troy, MI, 1965 (www.lebow.com).

[28]Wilcoxon Research, Gaithersburg, MD (www.wilcoxon.com).

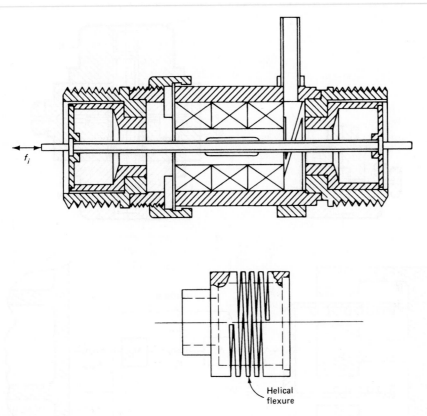

Helical
flexure

Figure 5.9
LVDT force transducer. *(Courtesy Schaevitz Sensors, Hampton, VA, www.schaevitz.com)*

pickups resistant to side loading quotes transverse sensitivity as less than 7 percent
of axial sensitivity.

Figure 5.10*a* shows construction details of two piezoelectric load cells. The
smaller one is permanently preloaded to measure in the range 1,000-N tension to
5,000-N compression without relaxing the built-in compressive preload. The larger
unit comes with external preloading nuts which may be adjusted to cover the range
4,000-N tension to 16,000-N compression. An analysis[29] by the manufacturer
suggests that such transducers may be modeled as a spring (piezoelectric elements)
sandwiched between two end masses. Figure 5.10*b* shows such a transducer being
used to measure the force applied by a vibration shaker to some structure being
vibration-tested (K_{m1} and K_{m2} represent the stiffness of the mounting screws). If the
impedance of the structure (including K_{m2}) is called Z_s, then since $Z_s \triangleq f_s/v_s$, the
force actually applied to the structure is $Z_s v_s$. The force F_m *measured,* however, is
that in K_t, which is proportional to the relative displacement of M_{t1} and M_{t2} since
this is the deflection of the piezoelectric element. Under dynamic conditions, F_m is

[29]W. Braender, "High-Frequency Response of Force Transducers," *B & K Tech. Rev.,* no. 3, Bruel and Kjaer
Instruments, Marlboro, MA, 1972 (www.bkhome.com).

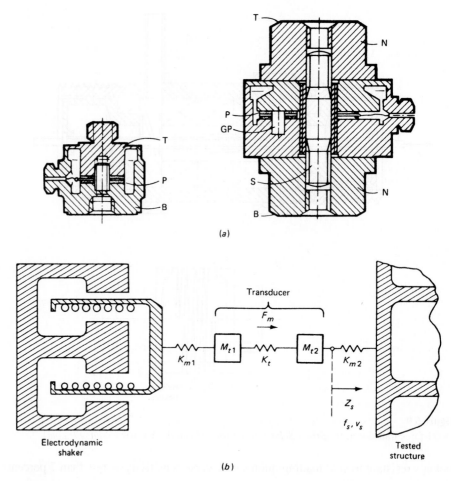

Figure 5.10
Piezoelectric load cells and dynamic error analysis. (a) Sectional drawing of the force
transducer. T = top; P = piezoelectric disks; GP = guide pin; S = preloading screw;
N = preloading nut; B = base. *(Courtesy Bruel and Kjaer Instruments, Marlboro, MA)*

not necessarily equal to F_s, and a dynamic analysis as shown below is useful in
developing criteria for accurate measurement:

$$F_m - v_s Z_s = M_{t2} Dv_s \qquad \frac{F_m}{F_s}(D) = \frac{M_{t2} D}{Z_s(D)} + 1 \qquad \textbf{(5.16)}$$

If $M_{t2} \equiv 0$, there would be no error no matter what Z_s might be; thus transducers
with small M_{t2} are clearly preferable (the small unit of Fig. 5.10a has $M_{t2} \approx 3g$).
For a "springlike" structure, $Z_s(D) = K_s/D$, and

$$\frac{F_m}{F_s}(i\omega) = 1 - \frac{M_{t2}}{K_s}\omega^2 \qquad \textbf{(5.17)}$$

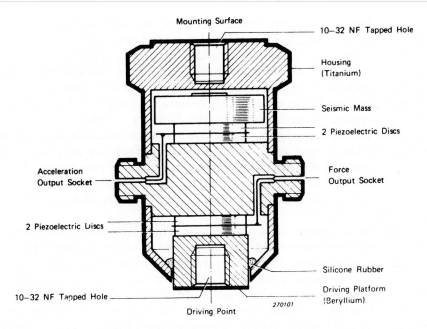

Figure 5.11
Piezoelectric impedance head. *(Courtesy Bruel and Kjaer Instruments, Marlboro, MA)*

which shows how accuracy varies with frequency ω. Other assumed forms of Z_s are easily investigated for accuracy criteria.

Studies of structural impedance actually are quite useful and common in dynamic analysis of mechanical systems.[30] Thus manufacturers have developed the *impedance head,* a dual sensor which combines a separate load cell and an accelerometer into a single, compact package (see Fig. 5.11). While impedance basically involves force and *velocity,* generally accelerometers are used since they are more convenient and the integration necessary to obtain velocity is easily accomplished electronically or numerically. Again, low mass between the force-sensing crystals and the driving point is desired, but we now also strive for high *stiffness* between the accelerometer base and the driving point, so that the measured acceleration accurately reflects that of the driving point. When this acceleration measurement is accurate (by using either an impedance head or *separate* force and acceleration sensors), an accuracy-improving trick called *mass cancellation* is possible. Here the "true" value of $F_s = v_s Z_s$ is calculated from Eq. (5.16) by electrically subtracting from the directly measured F_m a value $M_{t2} \dot{v}_s$, using the known value of M_{t2} and the measured acceleration $\dot{v}_s$.

Variable-Reluctance/FM-Oscillator Digital Systems

While the electrical signal from any force transducer can be converted to digital form by suitable equipment (see Chap. 10), some types of pickups are specifically

---

[30]E. O. Doebelin, "System Modeling and Response," Wiley, New York, 1980.

designed with this in mind. We here discuss briefly one such type which has been used in rocket-engine testing where digital data provide advantages of accuracy and ease of performing computations automatically. The elastic member is a proving ring with a two-arm variable-reluctance bridge displacement transducer. The signal from this transducer is used to change the frequency of a frequency-modulated (FM) oscillator. The frequency change (from some base value) of this oscillator is directly proportional to displacement (and thus to force). A digital reading of force is accomplished by applying the output signal of the oscillator to an electronic counter over a known time interval. The total number of pulses accumulated is thus a digital measure of force. In a typical unit, a change of force from minus full scale to plus full scale causes the frequency to change from 10,000 to 12,500 Hz. For a 1-s counting period, a full-scale force will thus cause a counter reading 1,250 above the base value of 11,250.

Computing advantages of such a system arise from the desire to know the total impulse of the rocket engine and to be able to add and subtract various forces in multicomponent test stands. The total impulse is the integral of force with respect to time; this is simply the total number of counts over the desired integrating time. When several forces must be added or subtracted, each has its own force pickup and 10,000 to 12,500 Hz oscillator. The oscillator output signals are combined in an electronic adder unit, which produces a single output whose frequency is the sum (or difference) of the input frequencies. The output of the adder unit is a signal with a frequency of 30,000 to 37,500 Hz; counting this over a timed interval gives the algebraic sum of the measured forces. Integration of this sum signal can be easily accomplished by the same method used for a single signal. Figure 5.12 shows block diagrams of these systems. By applying the dc output voltage to a voltage-to-frequency converter (see Chap. 10), these digital methods can be extended to any load cell with dc output.

Loading Effects

Since force is an effort variable, the pertinent loading parameter is either stiffness or impedance, and the associated flow variable is either displacement or velocity. Stiffness is perhaps more convenient for elastic force sensors since impedance would be infinite for the static case. The generalized input stiffness S_{gi} of the system of Fig. 5.3 is given by

$$S_{gi}(D) = \frac{F_i}{x}(D) = K_s\left(\frac{D^2}{\omega_n^2} + \frac{2\zeta D}{\omega_n} + 1\right) \tag{5.18}$$

Recall that for small error due to loading, S_{gi} must be sufficiently large compared with S_{go}, the generalized output stiffness of the system being measured. The frequency characteristic $S_{gi}(i\omega)$ is shown in Fig. 5.13 for a particular (small) value of ζ. Note that, near ω_n, $S_{gi}(i\omega)$ becomes very small. However, force pickups are generally used only for $\omega \ll \omega_n$; thus in most cases $S_{gi}(i\omega)$ is adequately approximated as simply K_s.

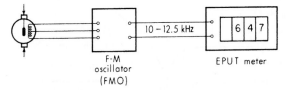

F-M
oscillator
(FMO)

EPUT meter

Force measurement

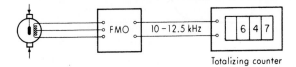

Totalizing counter

Total impulse measurement

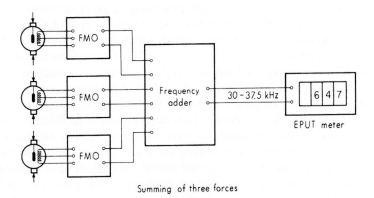

Summing of three forces

Figure 5.12
Digital force measurement, integration, and summing.

5.4 RESOLUTION OF VECTOR FORCES AND MOMENTS INTO RECTANGULAR COMPONENTS

In a number of important practical applications, the force or moment to be measured is not only unknown in magnitude but also of unknown and/or variable direction. Outstanding examples of such situations are "balances" for measuring forces on wind-tunnel models,[31] dynamometers (force gages) for measuring cutting forces in

---

[31]P. K. Stein, "Measurement Engineering," vol. 1, p. 431, Stein Engineering Services, Inc., Phoenix, AZ, 1964; C. C. Perry and H. R. Lissner, "The Strain Gage Primer," p. 212, McGraw-Hill, New York, 1955; A. Pope and K. L. Goin, "High Speed Wind Tunnel Testing," chap. 7, R. E. Krieger Publ. Co., Huntington, NY, 1978; L. Bernstein, "Force Measurements in Short-Duration Hypersonic Facilities," *AGARD*-AG-214 (AGARD ograph no. 214); A. Stromberg and E. Schmitz, "Wind-Tunnel Scales for Measuring Aerodynamic Forces on the Hypersonic Configuration ELAC I," *RAM,* vol. 10, 1996, no. 1, pp. 9–12, HBM, Inc. (www.hbminc.com).

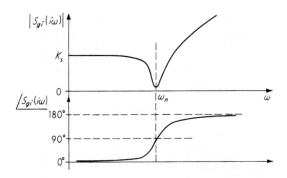

Figure 5.13
Stiffness of force transducers.

machine tools,[32] and thrust stands[33] for determining forces of rocket engines. Elastic force transducers of either the bonded-strain-gage or gross-deflection variety are employed in these applications. Ingenious use of various types of flexures for isolating and measuring different force components characterizes the design of these devices. Depending on the degree to which the force or moment direction is unknown, force-resolving systems of varying degrees of complexity may be devised. The most general situation (measurement of three mutually perpendicular force components and three mutually perpendicular moment components) is regularly accomplished with high accuracy. Two major classes of application lead to two main design philosophies in multiaxis force measurement. For large test objects (such as huge rocket engines), the approach is to use a multiplicity (usually six) of *single-axis* load cells, suitably configured with flexure pivots so as to isolate each load cell from all forces except one. For small objects (since space is at a premium), we usually try to design a *single* load cell so arranged as to measure *all* the desired force components, perhaps by a clever selection of elastic element shape and judicious placement of strain gages.[34]

Figure 5.14 shows a six-component thrust stand used in testing rocket engines. Load cells 1, 2, and 3 are mounted at the corners of an equilateral triangle, and load cells 4, 5, and 6 are in the sides of a concentric, smaller equilateral triangle. The engine to be tested is rigidly fastened at the common center of both triangles and produces a force of unknown magnitude and direction (which can be expressed in terms of components F_x, F_y, and F_z) and a moment of unknown magnitude and direction (which can be expressed in terms of components M_x, M_y, and M_z). The rocket forces are transmitted from the mounting plate to the rigid foundation

[32]E. G. Loewen, E. R. Marshall, and M. C. Shaw, Electric Strain Gage Tool Dynamometers, *Proc. Soc. Exp. Stress Anal.,* vol. 8, no. 2, 1951.

[33]Integrated Aerosystems (www.pacpress.com).

[34]A. Rupp and V. Grubisic, "Reliable and Efficient Measurement of Multi-Axial In-Service Loads on Cars and Commercial Vehicles," *RAM,* vol. 10, 1996, no. 1, HBM Inc. (www.hbminc.com)

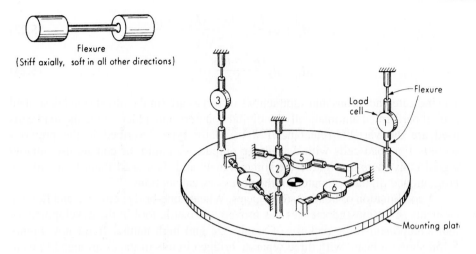

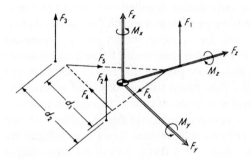

Figure 5.14
Six-component load frame.

through the six load cells and their associated flexures. The action of the suspension system is most clearly seen if we consider each flexure as a ball-and-socket joint. Actually ball-and-socket joints are not used because of their lost motion and friction. A static analysis of the force system gives the following results, which allow calculation of the rocket forces and moments from the measured load-cell forces and stand dimensions:

$$F_x = F_1 + F_2 + F_3 \tag{5.19}$$

$$F_y = \frac{F_5 - F_4 - (F_4 - F_6)}{2} \tag{5.20}$$

$$F_z = \frac{\sqrt{3}\,(F_5 - F_6)}{2} \tag{5.21}$$

$$M_x = \frac{-d_1(F_4 + F_5 + F_6)}{2\sqrt{3}} \tag{5.22}$$

$$M_y = d_2 \frac{F_1 - F_2 - (F_3 - F_1)}{2\sqrt{3}} \tag{5.23}$$

$$M_z = d_2 \frac{(F_3 - F_2)}{2} \tag{5.24}$$

The indicated additions and subtractions of forces are (in the thrust stand described above) performed automatically by digital counters and adders since the load cells used are the variable-reluctance/FM oscillator type described in the previous section. For load cells with dc voltage output, we could, of course, use analog/digital converters to digitize the force signals and then send them to a suitably programmed digital computer for the necessary calculations.

A combination of bonded strain gages, Wheatstone-bridge circuits, and flexible elements of various geometries has proved a versatile tool in the development of multicomponent-force pickups of small size and high natural frequency. Figure 5.15a shows a beam with three separate bridge circuits of gages arranged to measure the three rectangular components of an applied force. All bridge circuits are temperature-compensated and respond only to the intended component of force; however, the point of application of the force must be at the center of the beam cross section. An eccentricity in the y direction, for example, would give F_z a moment arm, causing bending stresses in the y direction that would be indistinguishable from those due to F_y. If side loads F_x and F_y are present, the end of the beam will deflect, causing just such eccentricities; thus beam stiffness must be adequate to keep deflection sufficiently low. This stiffness tends, of course, to reduce sensitivity. If "cross-talk" among the various axes becomes unacceptable, a calibration and correction scheme may be necessary. For example, if three strain gage bridges are used to measure the x, y, and z components of a force, producing three output voltages e_{ox}, e_{oy}, and e_{oz}, we might model the interactions with the equations

$$e_{ox} = k_{xx} f_x + k_{yx} f_y + k_{zx} f_z \tag{5.25}$$
$$e_{oy} = k_{xy} f_x + k_{yy} f_y + k_{zy} f_z \tag{5.26}$$
$$e_{oz} = k_{xz} f_x + k_{yz} f_y + k_{zz} f_z \tag{5.27}$$

The k coefficients would be found by a calibration where we would apply several sets of *known x, y,* and *z* forces and measure the resulting output voltages, generating a set of equations in the unknown k's. Since there are nine coefficients, a minimum of nine equations are needed. However, since we would also like to "cover" the full-scale ranges of each force, in conventional practice, we use perhaps 10 points in each range and also exercise many *combinations* of forces. This could lead to very expensive and time-consuming calibration and suggests the use of standard "design of experiments" techniques and multiple regression[35] which can give accurate k values with modest effort. While the above equations properly show the voltages as dependent variables and the forces as independent variables, when we

---

[35]E. O. Doebelin, "Engineering Experimentation," McGraw-Hill, New York, 1995, chap. 4; A. Bray, G. Barbato and R. Levi, "Theory and Practice of Force Measurement," Academic Press, New York, 1990, pp. 289–302.

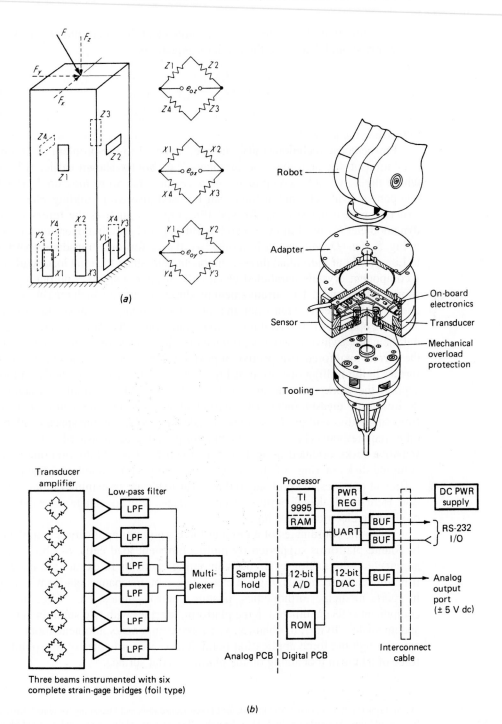

Figure 5.15
Resolution of vector forces.

use the calibration, the voltages will be known and the forces unknown. We therefore really should determine the c's in the equations

$$f_x = c_{xx}e_{ox} + c_{xy}e_{oy} + c_{xz}e_{oz} \tag{5.28}$$

$$f_y = c_{yx}e_{ox} + c_{yy}e_{oy} + c_{yz}e_{oz} \tag{5.29}$$

$$f_z = c_{zx}e_{ox} + c_{zy}e_{oy} + c_{zz}e_{oz} \tag{5.30}$$

rather than the above k's.

Piezoelectric techniques also are well adapted to the problem of component resolution of vector forces and moments,[36] as briefly shown earlier (Fig. 4.43), where the x, y, and z components of a vector force were measured. The three-component load cell shown there can be used also as a building block in constructing measuring systems for specific purposes, such as the three-component dynamometer for machining research in milling and grinding operations illustrated in Fig. 5.16a. (*Dynamometer* is another word used for force transducer, particularly in Europe.) Four individual three-component cells share the total load, with all four x-component crystals paralleled electrically to give a single x output signal, and similarly for y and z. This arrangement produces an output which measures the force components correctly even when the force application point moves (as it *would,* in a milling operation). The reading of *each* of the four cells varies with force position, but their *sum* (which is what is produced by the electric parallel connection) gives the correct total force, irrespective of position. This scheme is carried even one step further in the milling operation of Fig. 5.16b, where two complete dynamometers support the workpiece and are electrically paralleled (externally) to achieve the desired force measurement. The individual sensing elements can be constructed from single disks of quartz or rings of disks, depending on the space available and other requirements (Fig. 5.17). When a torque must be measured, a ring of shear-sensitive disks oriented as in Fig. 5.17b can be used. Multicomponent devices combine disks or rings of the desired sensitivity in a "sandwich" fashion under sufficient compressive preload so that shear loads are transmitted reliably by friction. Here the "leakage" of piezoelectric voltages is an advantage since the preloads produce no electrical zero shift.

When the coordinates of a moving force's point of application in a plane must be found, platforms supported by multicomponent load cells, as above, must be augmented with analog or digital computing elements to extract the needed data from the load-cell signals. A good example is a biomechanics platform[37] employed in sports medicine, orthopedics, posture control studies, etc. Here, four three-component cells are utilized in the platform, together with two summing amplifiers and an analog divider, to obtain x, y, z force components, moment about a vertical axis through the force-application point, and x, y coordinates of this point for the force of a human foot acting on the platform's flat surface.

---

[36]G. H. Gautschi, "Piezoelectric Multicomponent Force Transducers and Measuring Systems," Kistler Instrument Corp., Amherst, NY, 1978; G. A. Spescha, "Piezoelectric Multicomponent Force and Moment Measurement," Kistler Instrument Corp., 1970 (www.kistler.com).

[37]Model 9261A, Kistler Instrument Corp., Amherst, NY.

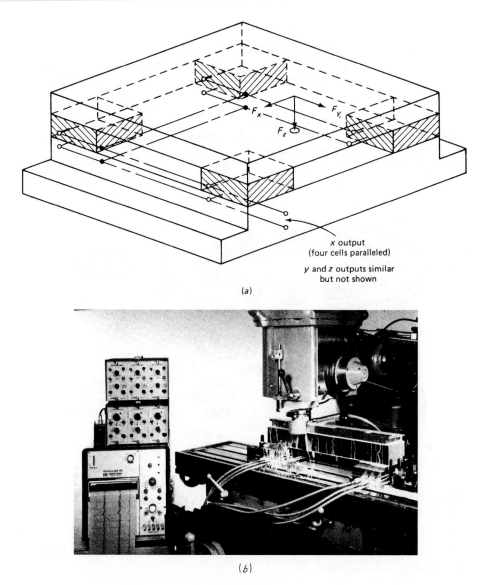

F_x

F_y

F_z

x output
(four cells paralleled)

y and z outputs similar
but not shown

(a)

(b)

Figure 5.16
Multicomponent force transducer for machining studies.
(Courtesy Kistler Instruments, Amherst, NY)

Robotic manufacturing and assembly operations may be improved by adding various sensing functions, such as force sensing at the end of the robot's arm. A six-axis (x, y, z force; x, y, z torque) sensor, complete with signal-processing hardware and software to interface with the robot controller, is available[38] for connection between the robot arm and end-of-arm tooling (Fig. 5.15*b*). Simpler force sensors

---

[38]ATI Industrial Automation, Garner, NC, 919-772-0115 (www.ati-ia.com).

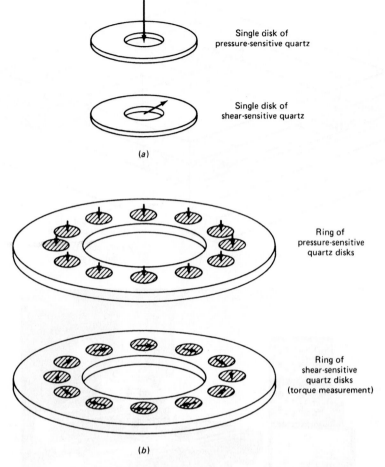

Single disk of
pressure-sensitive quartz

Single disk of
shear-sensitive quartz

(a)

Ring of
pressure-sensitive
quartz disks

Ring of
shear-sensitive
quartz disks
(torque measurement)

(b)

Figure 5.17
Component separation with quartz wafers. *(Courtesy Kistler Instruments, Amherst, NY)*

can also perform useful functions in robotic systems, sometimes obviating the need for expensive vision systems.[39]

5.5 TORQUE MEASUREMENT ON ROTATING SHAFTS

Measurement of the torque carried by a rotating shaft is of considerable interest for its own sake and as a necessary part of shaft power measurements. Torque transmission through a rotating shaft generally involves both a source of power and a sink (power absorber or dissipator), as in Fig. 5.18. Torque measurement may be

---

[39]R. Butters, "Making Sense of Force Sensing," *Sensors,* pp. 9–14, June 1988.

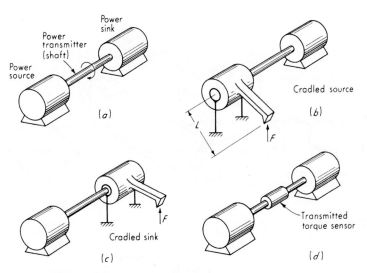

Figure 5.18
Torque measurement of rotating machines.

accomplished by mounting either the source or the sink in bearings ("cradling") and measuring the reaction force F and arm length L; or else the torque in the shaft itself is measured in terms of the angular twist or strain of the shaft (or a torque sensor coupled into the shaft).

The cradling concept is the basis of most shaft power dynamometers. These are utilized mainly for measurements of steady power and torque, by using scales or load cells to measure F. A free-body analysis of the cradled member reveals error sources resulting from friction in the cradle bearings, static unbalance of the cradled member, windage torque (if the shaft is rotating), and forces due to bending and/or stretching of power lines (electric, hydraulic, etc.) attached to the cradled member. To reduce frictional effects and to make possible dynamic torque measurements, the cradle-bearing arrangement may be replaced by a flexure pivot with strain gages to sense torque,[40] as in Fig. 5.19. The crossing point of the flexure plates defines the effective axis of rotation of the flexure pivot. Angular deflection under full load is typically less than 0.5°. This type of cross-spring flexure pivot is relatively very stiff in all directions other than the rotational one desired, just as in an ordinary bearing. The strain-gage bridge arrangement also is such as to reduce the effect of all forces other than those related to the torque being measured. Speed-torque curves for motors may be obtained quickly and automatically with such a torque sensor by letting the motor under test accelerate an inertia from zero speed up to maximum while measuring speed with a dc tachometer.[41] The torque and speed signals are applied to an *XY* recorder to give automatically the desired curves.

[40]Lebow Products, Troy, MI, 800-803-1164 (www.lebow-siebe.com).

[41]B. Hall, "Motor Tests Using *X-Y* Recorders," *Electro-Technol.,* p. 116, May 1964.

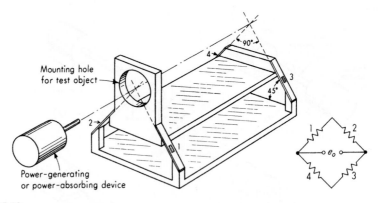

Figure 5.19
Strain-gage torque table.

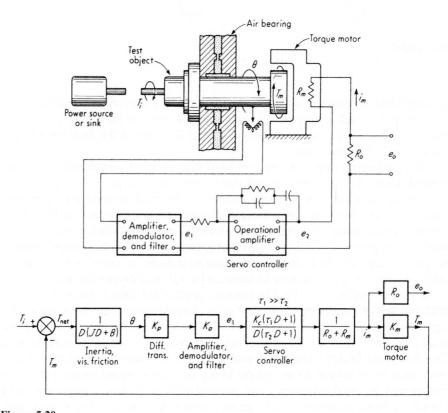

Figure 5.20
Feedback torque sensor. *(Courtesy McFadden Electronics, South Gate, CA)*

Another variation (see Fig. 5.20) on the cradle principle is found in a null-balance torquemeter using feedback principles to measure small torques in the range 0 to 10 oz · in. In this device the test object is mounted on a hydrostatic air-

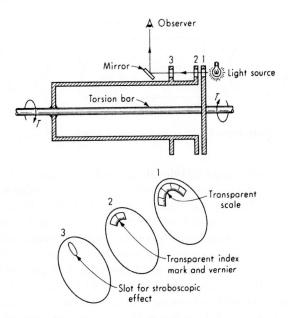

Figure 5.21
Torsion-bar dynamometer.

bearing table to reduce bearing friction to exceedingly small values. Any torque on the test object tends to cause rotation of the air-bearing table, but this rotation is immediately sensed by a differential-transformer displacement pickup. The output from this pickup is converted to direct current and amplified to provide the coil current of a torque motor, which applies opposing torque to keep displacement at zero. The amount of current required to maintain zero displacement is a measure of torque and is read on a meter. The servoloop uses integral control[42] to give zero displacement for any constant torque. Approximate derivative control is used also to give stability. The threshold of this air-bearing system is less than 0.0005 oz · in while the torque/current nonlinearity is 0.001 oz · in. The overall system behaves approximately as a second-order system with a natural frequency of about 10 Hz and damping ratio of 0.7 when no test object is present on the table.

The use of elastic deflection of the transmitting member for torque measurement may be accomplished by measuring either a gross motion or a unit strain. In either case, a main difficulty is the necessity of being able to read the deflection while the shaft is rotating. Figure 5.21 illustrates a torsion-bar torquemeter, using optical methods of deflection measurement. The relative angular displacement of the two sections of the torsion bar can be read from the calibrated scales because of the stroboscopic effect of intermittent viewing and the persistence of vision.

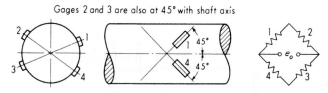

Figure 5.22
Strain-gage torque measurement.

Even though they require additional equipment to transmit power and signal between rotating shaft and stationary readout, strain-gage torque sensors are very widely used. Figure 5.22 shows the basic principle. (Chap. 11 gives data on slip rings, rotary transformers, and radio-telemetry systems to deal with the data transmission problem.) This arrangement (given accurate gage placement and matched gage characteristics) is temperature-compensated and insensitive to bending or axial stresses. The gages must be precisely at 45° with the shaft axis, and gages 1 and 3 must be diametrically opposite, as must gages 2 and 4. Accurate gage placement is facilitated by the availability of special rosettes in which two gages are precisely oriented on one sheet of backing material. In some cases the shaft already present in the machine to be tested may be fitted with strain gages. In other cases a different shaft or a commercial torquemeter must be used to get the desired sensitivity or other properties. Of the various configurations shown in Fig. 5.23, one manufacturer[43] uses the hollow cruciform for low-range units and the solid, square shaft for high-range ones. Placement of the gages on a square, rather than round, cross section of the shaft has some advantages. The gages are more easily and accurately located and more firmly bonded on a flat surface. Also, the corners of a square section in torsion are stress-free and thus provide a good location for solder joints between lead wires and gages. These joints are often a source of fatigue failure if located in a high-stress region. Also, for equivalent strain/torque sensitivity, a square shaft is much stiffer in bending than a round one, thus reducing effects of bending forces and raising shaft natural frequencies.

The torque of many machines, such as reciprocating engines, is not smooth even when the machine is running under "steady-state" conditions. If we wish to measure the average torque so as to calculate power, the higher frequency response of strain-gage torque pickups may be somewhat of a liability since the output voltage will follow the cyclic pulsations and some sort of averaging process must be performed to obtain average torque. If exceptional accuracy is not needed, the low-pass filtering effect of a dc meter used to read e_o may be sufficient for this purpose. In the cradled arrangements of Fig. 5.18 (employed in many commercial dynamometers for engine testing, etc.), the inertia of the cradled member and the low-frequency response of the platform or pendulum scales used to measure F perform the same averaging function.

[43]Lebow Products, Troy, MI.

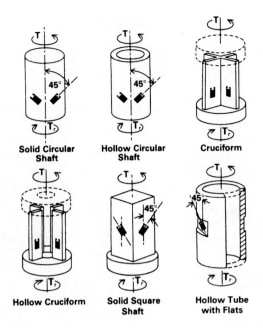

Figure 5.23
Various torque sensor designs.
(Courtesy Lebow Products, Troy, MI, Torque Sensor and Dynamometer Catalog, no. 250, 1979.)

Commercial strain-gage torque sensors are available with built-in slip rings and speed sensors. A family[44] of such devices covers the range 10 oz · in to 3×10^6 in · lbf with full-scale output of about 40 mV. The smaller units may be used at speeds up to 24,000 r/min, the largest to 350 r/min. Torsional stiffness of the 10 oz · in unit is 112 in · lbf/rad while a 600,000 in · lbf unit has 4.0×10^6 in · lbf/rad. Nonlinearity is 0.1 percent of full scale while temperature effect on zero is 0.002 percent of full scale/F° and temperature effect on sensitivity is 0.002 percent/F° over the range 70 to 170°F. Figure 5.24[45] shows some construction details for a line of German transducers that use strain gage, LVDT, or a unique "inductive meander" technology to generate the electrical output. Either slip-ring or rotating-transformer methods are used to deal with the rotating-shaft problem. Free-floating, foot-mounted, reaction, and pulley versions are available. For any rotating-torque sensor, the mounting is critical, and various forms of misalignment couplings must be properly used. Figure 5.25[46] from another manufacturer shows how two of their units are built into a chassis dynamometer, useful in a wide variety of automotive tests.

The dynamic response of elastic deflection torque transducers is essentially slightly damped second-order, with the natural frequency usually determined by the stiffness of the transducer and the inertia of the parts connected at either end.

---

[44]Lebow Products, Troy, MI.

[45]Staiger Mohilo, Schlenker Enterprises Ltd., Hillside, IL, 800-992-2777.

[46]S. Himmelstein and Co., Hoffman Estates, IL, 708-843-3300.

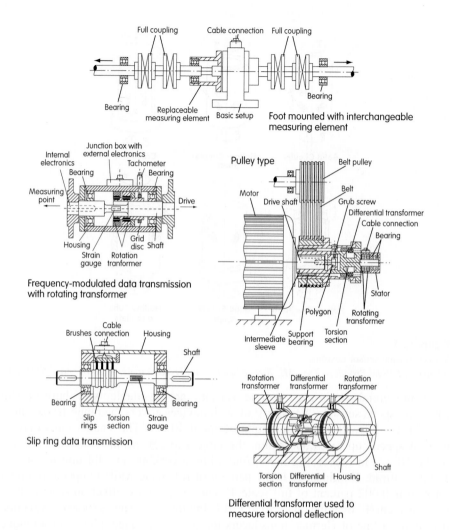

Figure 5.24
Shaft torque sensor details.

Damping of the transducers themselves usually is not attempted, and any damping present is due to bearing friction, windage, etc., of the complete test setup.

5.6 SHAFT POWER MEASUREMENT (DYNAMOMETERS)

The accurate measurement of the shaft power input or output of power-generating, -transmitting, and -absorbing machinery is of considerable interest. While the basic measurements, torque and speed, have already been discussed, their practical

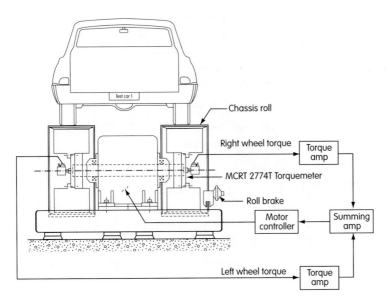

Figure 5.25
Shaft torque sensors used in chassis dynamometer.

application to power measurement is considered briefly here. The term *dynamome-ter* [47] generally is used to describe such power-measuring systems, although it is also used as a name for elastic force sensors. The type of dynamometer employed depends somewhat on the nature of the machine to be tested. If the machine is a power generator, the dynamometer must be capable of absorbing its power. If the machine is a power absorber, the dynamometer must be capable of driving it. If the machine is a power transmitter or transformer, the dynamometer must provide both the power source and the load, unless a four-square [48] method is feasible.

Perhaps the most versatile and accurate dynamometer is the dc electric type. Here a dc machine is mounted in low-friction trunnion bearings (see Fig. 5.18*b*) and provided with field and armature control circuits. This machine can be coupled to either power-absorbing or power-generating devices since it may be connected as either a motor or a generator. When it is employed as a generator, the generated power is dissipated in resistance grids or recovered for use. The dc dynamometer can be adjusted to provide any torque from zero to the maximum design value for speeds from zero to the so-called base speed of the machine. At this speed the maximum torque develops the maximum design horsepower. At speeds above base speed, torque must be progressively reduced so as to maintain horsepower less than the design maximum. The controllability of the dc dynamometer lends it particularly to modern automatic load and speed programming applications.

---

[47]M. Plint and A. Martyr, Engine Testing: Theory and Practice, Butterworth, 1995, chaps. 7, 8, 9, 15; Mustang Dynamometers, www.mustangdynl.com; Dynamometer Directory, www.dynamometer.org.

[48]E. O. Doebelin, "Measurement Systems," 4th ed., p. 421.

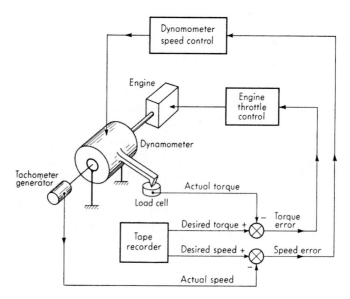

Figure 5.26
Servocontrolled dynamometer.

Figure 5.26 illustrates such a situation. Tape recordings of engine torque and speed measured under actual driving conditions for an automobile are utilized to reproduce these conditions in the laboratory engine test. Two feedback systems control engine speed and torque. A tachometer generator speed signal from the dynamometer is compared with the desired speed signal from the tape recorder; if the two are different, the dynamometer control is automatically adjusted to change speed until agreement is reached. Actual engine torque is obtained from a load cell on the dynamometer and compared with the desired torque from the tape recorder. If these do not agree, the error signal actuates the engine throttle control in the proper direction. Both systems operate simultaneously and continuously to force engine speed and torque to follow the tape-recorder commands.

By adapting recent developments in variable-frequency ac motor control, a trunnion-mounted induction motor becomes a versatile dynamometer [49] for both driving and absorbing applications. Features include fast response, flexible control, simplified maintenance, and energy conservation, since up to 85 percent of the energy absorbed is returned to the ac power line.

Dynamometers capable only of absorbing power (see Fig. 5.27) include the eddy-current brake [50] (inductor dynamometer) and various mechanical brakes employing dry friction (Prony brake) and fluid friction (air and water brakes). The eddy-current brake is easily controllable by varying a dc input, but it cannot produce

[49]C. S. Lassen, "Conserving Energy in Engine Testing with Adjustable Frequency Regenerative Dynamometers," Eaton Corp., Kenosha, WI, 1981.

[50]J. B. Winther, "Dynamometer Handbook of Basic Theory and Applications," Eaton Corp., 1974.

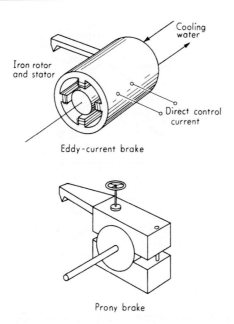

Figure 5.27
Absorption dynamometers.

any torque at zero speed and only small torque at low speeds. However, it is capable of higher power and speed than a dc dynamometer. The power absorbed is carried away by cooling water circulated through the air gap between rotor and stator. The Prony brake is a simple mechanical brake in which friction torque is manually adjusted by varying the normal force with a handwheel. Torque is available at zero speed, but operation may be jerky because of the basic nature of dry friction. Water and air brakes utilize the churning action of paddle wheels or vanes rotating inside a fluid-filled casing to absorb power. A flow of air or water through the device is maintained for cooling purposes. No torque is available at zero speed, and only small torques are available at low speeds. High speed and high power can be handled well, however, with some water brakes being rated at 10,000 hp at 30,000 r/min. A recently developed air dynamometer[51] is especially suited for testing aircraft and industrial gas-turbine engines. It uses a variable-geometry radial outflow compressor, with an adjustable shroud for the impeller blades to control the resisting torque. The available brochure explains its many advantages over eddy-current and hydraulic dynamometers, including small size and weight, ambient air (rather than water) cooling, portability, lack of a high-speed gearbox, and lower cost. Two models cover a speed range from 0 to 24,000 RPM and horsepower to 18,000.

---

[51]VAROC Air Dynamometer, Concepts NREC, White River Junction, VT, 802-296-2321 (www.conceptseti.com).

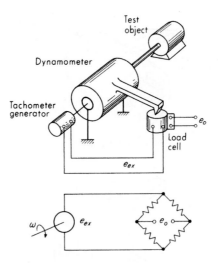

Figure 5.28
Instantaneous power measurement.

In all the above power measurement applications, torque and speed are measured separately and then power is calculated manually. This calculation can be performed automatically in a number of ways since the basic operation (multiplication) can be accomplished physically in various ways. An interesting scheme using the properties of bridge circuits is shown in Fig. 5.28. Speed is measured with a dc tachometer generator, and this voltage is applied as the excitation of a strain-gage load cell used to measure torque. Since bridge output is directly proportional to excitation voltage and directly proportional to torque, the voltage e_o is actually an instantaneous power signal.

5.7 GYROSCOPIC FORCE AND TORQUE MEASUREMENT

The use of gyroscopic principles (method 6) in the measurement of force and torque is today not widespread, and we provide here only a reference[52] for those with an interest in the idea.

5.8 VIBRATING-WIRE FORCE TRANSDUCERS

Method 7 in the list of basic force-sensing schemes is based on the well-known formula from classical physics for the first natural frequency ω of a string of length L and mass per unit length m_1, which is tensioned by the force F to be measured:

$$\omega = \frac{1}{2L}\sqrt{\frac{F}{m_1}} \tag{5.31}$$

---

[52]E. O. Doebelin, "Measurement Systems," 4th ed., pp. 429–431.

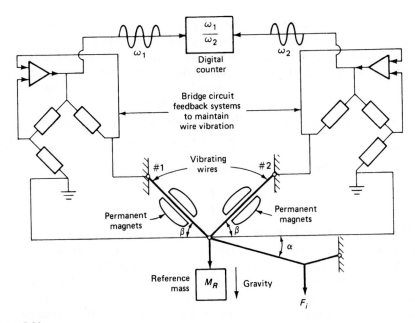

Figure 5.29
Vibrating-wire force transducer.

Since ω varies smoothly with F, the measuring principle is basically analog; however, the frequency is easily measured with conventional digital counters, so the transducer is sometimes described as a digital device. Temperature sensitivity (variation in L) and the nonlinear ω/F relation cause some inconvenience in practical use.[53]

The temperature and linearity problems above have been overcome by clever system design in a sensor used as part of industrial material feeding controllers[54] (Fig. 5.29). While the wire-tensioning force *determines* the natural frequency of vibration, some vibration *exciting system* must be provided in any vibrating-wire sensor to make up friction losses and to maintain wire vibration at a fixed amplitude. The "wires" (actually thin tapes of electrically conducting but nonmagnetic metal) are placed in the fields of permanent magnets. In this way, they serve as both velocity-to-voltage transducers (used in one leg of a bridge circuit to provide an oscillatory input voltage to an amplifier) and current-to-force transducers, which accept an oscillatory output current from the amplifier and produce vibration-exciting transverse forces on the wires.

By proper design, these feedback systems maintain constant-amplitude wire vibrations at a frequency determined by the force carried by each wire. When the measured force is zero, a fixed reference mass M_R applies a preload tension F_o which is equal in each wire, causing both wires to vibrate at the same frequency and

[53]K. S. Lion, "Instrumentation in Scientific Research," pp. 83–85, McGraw-Hill, New York, 1959.

[54]M. Gallo and G. Bussian, "The Benefits of Mass Measurement for Gravimetric Feeding of Bulk Solids in Industrial Environments," K-Tron International, Inc., Pitman, NJ, 856-589-0500 (ww.ktron.com).

making the frequency ratio $\omega_1/\omega_2 = 1.0$. If temperature changes, both wires experience the *same* length and frequency changes, but the frequency *ratio* is unaffected, which provides temperature compensation. To make the output signal ω_1/ω_2 linear with F_i, a force proportional to F_i is applied through the (nonvibrating) wire at an angle α such that an increase in F_i increases F_1 (the tension in wire 1) but decreases F_2. Choice of angles β and α allows us to adjust the relation between ΔF_1 and ΔF_2 over a wide range. Analysis shows that the formulas for F_1 and F_2 are of the form

$$F_1 = F_o + k_1 F_i \qquad F_2 = F_o - k_2 F_i \qquad (5.32)$$

where k_1 and k_2 can be adjusted by using α and/or β. Then we get

$$\frac{\omega_1}{\omega_2} = \sqrt{\frac{F_o + k_1 F_i}{F_o - k_2 F_i}} \qquad (5.33)$$

The function $(F_o + k_1 F_i)/(F_o - k_2 F_i)$ of F_i is nonlinear with an increasing slope, while the square root function has a decreasing slope; so the two nonlinearities, while not canceling, do *oppose* each other. By proper choice of k_1 and k_2 it is found that ω_1/ω_2 becomes linear with F_i to a close approximation. [Measurement of the frequency ratio ω_1/ω_2 is a standard digital counter technique in which the ω_2 signal is substituted for the usual internal clock oscillator of the counter and the number of cycles of ω_1 occurring during a preselected (and fixed) number of cycles of ω_2 is measured.] The referenced instrument is quoted as having ± 0.003 percent repeatability, ± 0.03 percent linearity, and stability of ± 0.03 percent over 18 months.

The vibrating wire principle, in somewhat different form, is also the basis of a line[55] of geotechnical instruments for measuring strain, load, pressure, temperature, and tilt.

PROBLEMS

5.1   For the force transducer of Fig. 5.3, let the "foundation" have an interfering input displacement $X_i(t)$. Obtain the necessary transfer functions for a block diagram, showing how instrument output is produced by the superimposed effects of desired input f_i and interfering input X_i. If an accelerometer sensitive to $\ddot{X}_i$ (either built into the force transducer or as a separate instrument) is present, show how a summing amplifier could be used to obtain an output signal compensated for the spurious vibration input.

5.2   The torque wrench in Fig. P5.1 is claimed to produce an output voltage e_o proportional to the torque applied by force f to the nut, *irrespective* of the point of force application L_f, as long as $L_f > 3L$. Investigate the validity of this claim.

---

[55]geokon, Inc., Lebanon, NH, 603-448-1562 (www.geokon.com).

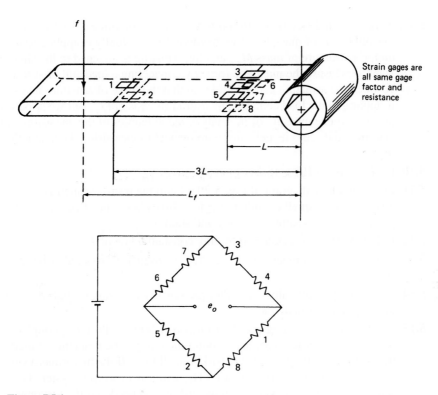

Figure P5.1

5.3 An object with a volume of 10 in³ is weighed on an equal-arm balance. The standard mass required for balance is 1 lbm and has a volume of 3 in³. What is the value of the correction necessary for air buoyancy?

5.4 A brass balance beam has a length of exactly 1 m at 60°F, and the pivot is perfectly centered to give an equal-arm balance. If one end of the beam comes to 80°F and there is a uniform temperature gradient to the other end at 60°F, what inequality in arm length results?

5.5 What general form of dynamic response would you expect from the systems of Fig. 5.2, part 1? Why?

5.6 Prove that the reading is independent of location of F_i if $a/b = c/d$ in the platform scale of Fig. 5.2, part 1.

5.7 If, in Fig. 5.2, part 2, $M = 1$ lbm and $A = 20g$, what is the net force on M? If a friction force which is unknown but less than 1 lbf may be present, what error may be expected in F_i?

5.8 Carry out a linear dynamic analysis and an accuracy/stability trade-off study of the pneumatic load cell of Fig. 5.2, part 4. Use $(x/F_{net})(D) = K_d/(D^2/\omega_n^2 + 2\zeta D/\omega_n + 1)$ and $(p_o/x)(D) = K_n/(\tau D + 1)$.

5.9   A translational electric motor provides a magnetic force f_m to drive a machine slide of mass M_s. The magnetic force actually is applied to the motor's armature of mass M_m, and the armature is attached to the slide. We want to measure the magnetic force by connecting a load cell of stiffness K_s between the armature and the slide, so that the force of armature on the slide is applied to the load cell. This measured force is *not* f_m, but we hope it will be an adequate measure of it. Analyze the dynamics of this measurement system and discuss its behavior with respect to measuring the magnetic force.

5.10 Derive Eqs. (5.19) to (5.24).

5.11 From the block diagram of Fig. 5.20, obtain the transfer function $(e_o/T_i)(D)$, assuming τ_2 is small enough to neglect. Investigate the effect of system parameters on dynamic accuracy and stability.

5.12 In Fig. 5.20 prove that $\theta = 0$ for any constant value of T_i.

5.13 Prove that the arrangement of Fig. 5.22 is insensitive to axial or bending stresses.

5.14 Prove that for equivalent strain/torque sensitivity, a square shaft is stiffer in bending than a round one.

5.15 A torque sensor with torsional stiffness of 1,000 in $\cdot$ lbf/rad is coupled between an electric motor and a hydraulic pump. The moments of inertia of the motor and pump are each 0.01398 in $\cdot$ lbf $\cdot$ s^2. If the motor has a small oscillatory torque component at 60 Hz, will this measuring system be satisfactory? Explain what torsional stiffness is needed if the response at 60 Hz is to be no more than 105 percent of the static response. The amount of damping is unknown.

5.16 Suppose the tachometer generator in the system of Fig. 5.28 puts out 6 V/1,000 r/min and the load cell produces 0.05 mV/(lbf $\cdot$ V excitation). What will be the power calibration factor for e_o in horsepower per millivolt if the arm length is 1 ft?

5.17 In the system of Fig. P5.2:
1. For $F = 0$ and heat off, $R_1 = R_2 = R_3 = R_4$ and $e_o = 0$.
2. The gage factor of the gages is $+ 2.0$, and the temperature coefficient of resistance of the gages is positive.
3. The modulus of elasticity of the beam decreases with increased temperature.
4. The thermal-expansion coefficient of the gage is greater than that of the beam.
5. Assume the gage temperature is the same as that of the beam *immediately* beneath it.

At time $t = 0$, an upward force F is applied and maintained constant thereafter. After oscillations have died out, at a later time t_1 the radiant-heat source is turned on and left on thereafter. Sketch the general form of e_o

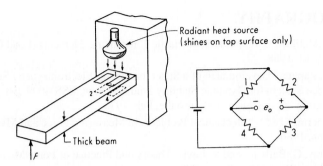

Figure P5.2

versus t, justifying clearly by detailed reasoning the shape you give the curve.

5.18 It is necessary to design a strain-gage thrust transducer for small experimental rocket engines which are roughly in the shape of a cylinder 6 in in diameter by 12 in long. The following information is given:
1. Weight of motor and mounting bracket, 20 lbf.
2. Maximum steady thrust, 50 lbf.
3. Oscillating component of thrust, $\pm$ 10 lbf maximum.
4. Oscillating components of thrust up to 100 Hz must be measured with a flat amplitude ratio within $\pm$ 5 percent.
5. A recorder with a sensitivity of 0.1 V/in, frequency response flat to 120 Hz, and input resistance of 10,000 Ω is available.
6. Thrust changes of 0.5 lbf must be clearly detected.
7. Gages with a resistance of 120 Ω and a gage factor of 2.1 are available. They are 0.5 $\times$ 1.0 inch in size.
8. An amplifier (to be placed between transducer and recorder) is available with a gain up to 1,000.

Design the transducer so as to require a minimum of amplifier gain. If damping is employed, calculate the required damping coefficient B, but do not design the damper. Use the cantilever-beam arrangement of Fig. P5.3.

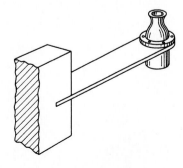

Figure P5.3

BIBLIOGRAPHY

1. G. F. Malikov, "Computation of Elastic Tensometric Elements (Load Cells)," *NASA* TT F-513, 1968.

2. R. N. Zapata, "Development of a Superconducting Electromagnetic Suspension and Balance System for Dynamic Stability Studies," *Rept.* ESS-4009-101-73U, University of Virginia, Charlottesville, Feb. 1973.

3. H. Wieringa (ed.), "Mechanical Problems in Measuring Force and Mass," M. Nijhoff, Boston, 1986.

4. A. Bray, G. Barbato, and R. Levi, "Theory and Practice of Force Measurement," Academic Press, New York, 1990.

5. V. B. Braginsky and A. B. Manukin, "Measurement of Weak Forces in Physics Experiments," U. of Chicago Press, Chicago, 1977. A fascinating study of how simple "mass/spring" systems are used to measure unbelievably small forces in exotic physics experiments, such as detection of gravity waves.

6. S. P. Wolsky and E. J. Zdanuk, "Ultra Micro Weight Determination in Controlled Environments," Wiley Interscience, New York, 1969.

7. W. Torge, "Gravimetry," de Gruyter, Berlin, 1989. Includes design details of gravity-measuring instruments which resolve *tiny* changes in local gravity by sensing correspondingly tiny force changes.

8. F. E. Jones and R. M. Schoonover, "Handbook of Mass Measurement," CRC Press, New York 2002.

9. P. A. Parker and R. DeLoach, "Structural Optimization of a Force Balance Using a Computational Experiment Design," AIAA-2002-0504.

CHAPTER 6

Pressure and Sound Measurement

6.1 STANDARDS AND CALIBRATION

Pressure is not a fundamental quantity, but rather is derived from force and area, which in turn are derived from mass, length, and time, the latter three being fundamental quantities whose standards have been discussed earlier. Pressure "standards" in the form of very accurate instruments are available, though, for calibration of less accurate instruments. However, these "standards" depend ultimately on the fundamental standards for their accuracy. The basic standards[1] for pressures ranging from medium vacuum (about 10^{-1} mmHg) up to several hundred thousand pounds per square inch are in the form of precision mercury columns (manometers) and dead-weight piston gages. For pressures in the range 10^{-1} to 10^{-3} mmHg, the McLeod vacuum gage is considered the standard. For pressures below 10^{-3} mmHg, a pressure-dividing technique allows flow through a succession of accurate orifices to relate the low downstream pressure to a higher upstream pressure (which is accurately measured with a McLeod gage).[2]

This technique can be further improved by substituting a Schulz hot-cathode or radioactive ionization vacuum gage for the McLeod gage. Each of these must be calibrated against a McLeod gage at one point (about 9×10^{-2} mmHg), but their known linearity is then used to extend their accurate range to much lower pressures.[3]

---

[1] D. P. Johnson and D. H. Newhall, "The Piston Gage as a Precise Pressure-Measuring Instrument," *Instrum. Contr. Syst.,* p. 120, April 1962; "Errors in Mercury Barometers and Manometers," *Instrum. Contr. Syst.,* p. 121, March 1962; "2″ Range Hg Manometer," *Instrum. Contr. Syst.,* p. 152, September 1962.

[2] J. R. Roehrig and J. C. Simons, "Calibrating Vacuum Gages to 10^{-9} Torr," *Instrum. Contr. Syst.,* p. 107, April 1963.

[3] J. C. Simons, "On Uncertainties in Calibration of Vacuum Gages and the Problem of Traceability," *Transactions of 10th National Vacuum Symposium,* p. 246, Macmillan, New York, 1963.

Gage and Pressure Measurement Uncertainties

Type of Instrument	Range	Uncertainty
Gas-operated PG	1.4 kPa to 17 MPa	±57 ppm
Oil-operated PG	700 kPa to 100 MPa	±63 ppm
	100 MPa to 280 MPa	±60 to ±150 ppm
Oil-operated PG	40 to 400 MPa	±186 ppm

The inaccuracies of the above-mentioned pressure standards are summarized graphically in Fig. 6.1a,[4] with Fig. 6.1b[5] giving more recent data. Since the above-mentioned pressure standards are also pressure-measuring instruments (of the highest quality and used under carefully controlled conditions), their operating principles and characteristics are not discussed here since they are adequately covered later.

6.2 BASIC METHODS OF PRESSURE MEASUREMENT

Since pressure usually can be easily transduced to force by allowing it to act on a known area, the basic methods of measuring force and pressure are essentially the same, except for the high-vacuum region where a variety of special methods not directly related to force measurement are necessary. These special methods are described in the section on vacuum measurement. Other than the special vacuum techniques, most pressure measurement is based on comparison with known dead-weights acting on known areas or on the deflection of elastic elements subjected to the unknown pressure. The deadweight methods are exemplified by manometers and piston gages while the elastic deflection devices take many different forms.

6.3 DEADWEIGHT GAGES AND MANOMETERS

Figure 6.2 shows the basic elements of a deadweight or piston gage. Such devices are employed mainly as standards for the calibration of less accurate gages or transducers. The gage to be calibrated is connected to a chamber filled with fluid whose pressure can be adjusted by some type of pump and bleed valve. The chamber also connects with a vertical piston-cylinder to which various standard weights may be applied. The pressure is slowly built up until the piston and weights are seen to "float," at which point the fluid "gage" pressure (pressure above atmosphere) must equal the deadweight supported by the piston, divided by the piston area.

For highly accurate results, a number of refinements and corrections are necessary. The frictional force between the cylinder and piston must be reduced to a minimum and/or corrected for. This is generally accomplished by rotating either the piston or the cylinder. If there is no axial relative motion, this rotation should reduce the axial effects of dry friction to zero. There must, however, be a small clearance between the piston and the cylinder and thus an axial flow of fluid from

---

[4]"Accuracy in Measurements and Calibrations," *NBS Tech. Note* 262, 1965.

[5]*NBS Spec. Publ.* 250, 1987.

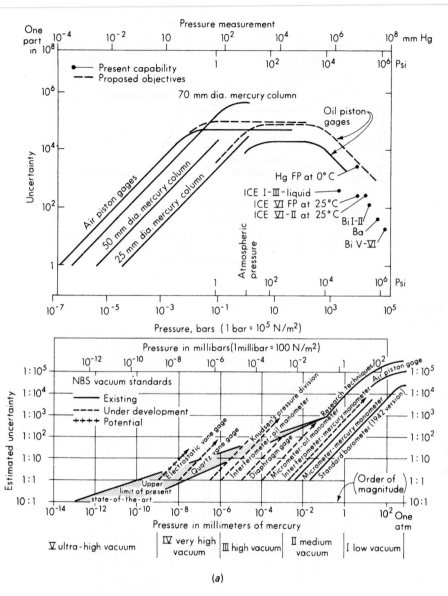

Figure 6.1
Pressure/vacuum standards.

the high-pressure end to the low-pressure end. This flow produces a viscous shear force tending to support part of the deadweight. This effect can be estimated from theoretical calculations.[6] However, it varies somewhat with pressure since the

[6] R. J. Sweeney, "Measurement Techniques in Mechanical Engineering," p. 104, Wiley, New York, 1953.

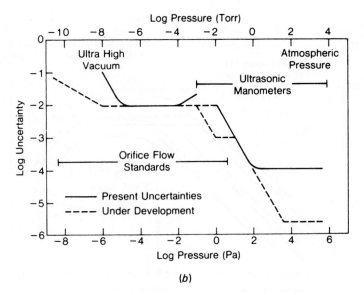

Figure 6.1
(Concluded)

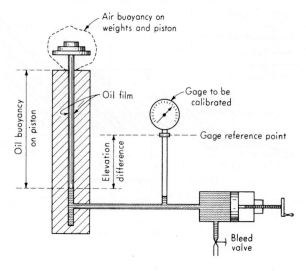

Figure 6.2
Deadweight gage calibrator.

piston and cylinder deform under pressure, thereby changing the clearance. The clearance between the piston and cylinder also raises the question of which area is to be used in computing pressure. The effective area generally is taken as the average of the piston and cylinder areas. Further corrections are needed for temperature effects on areas of piston and cylinder, air and pressure-medium buoyancy effects, local gravity conditions, and height differences between the lower end of the piston and the reference point for the gage being calibrated. Special designs and techniques allow use of deadweight gages for pressures up to several hundred thousand pounds per square inch. An improved design, the controlled-clearance piston gage,[7] employs a separately pressurized cylinder jacket to maintain the effective area constant (and thus achieve greater accuracy) at high pressures, which cause expansion and error in uncompensated gages.

Since the piston assembly itself has weight, conventional deadweight gages are not capable of measuring pressures lower than the piston weight/area ratio ("tare" pressure). This difficulty is overcome by the tilting-piston gage[8] in which the cylinder and piston can be tilted from vertical through an accurately measured angle, thus giving a continuously adjustable pressure from 0 lb/in² gage up to the tare pressure. The described gage uses nitrogen or other inert gas as the pressure medium and covers the range 0 to 600 lb/in² gage, having two interchangeable piston-cylinders and 14 weights. The accuracy is 0.01 percent of reading in the range 0.3 to 15 lb/in² gage and 0.015 percent of reading in the range 2 to 600 lb/in² gage. The tilting feature is used for the ranges 0 to 0.3 and 0 to 2.0 lb/in² gage; higher pressures are obtained in increments by the addition of discrete weights.

Some piston gages have been highly instrumented and automated to allow more convenient and rapid use. One such line[9] includes sensors for relative humidity, barometric pressure, ambient temperature, piston/cylinder temperature, piston rotation speed and acceleration, and piston drop rate. These readings are manipulated in the software to provide a readout of the calibration pressure. A typical formula[10] showing the relations is

$$\text{Gauge pressure} = \frac{Mg_1\left(1 - \dfrac{\rho_{\text{air}}}{\rho_{\text{mass}}}\right) + \pi DT}{A_{(20,\,0)} \cdot [1 + (\alpha_p + \alpha_c)\cdot(\theta - 20)]\cdot(1 + \lambda P)} - (\rho_{\text{fluid}} - \rho_{\text{air}})\cdot g_1\, h \quad \textbf{(6.1)}$$

where
M = the total mass load
g_1 = the local acceleration of gravity
ρ = density
D = piston diameter (computed from $A_{(20,\,0)}$, the piston/cylinder effective area at 20°C and 0 gage pressure)

---

[7]D. H. Newhall and L. H. Abbot, "Controlled-Clearance Piston Gage," *Meas. & Data*, January–February 1970.

[8]Ruska Instrument Corp., Houston, TX, www.ruska.com.

[9]DH Instruments, Inc., Phoenix, AZ, 602-431-9100 (www.dhinstruments.com).

[10]DH Instruments, "Precision Pressure Measurement Handbook," Ametek, Largo, FL, 727-536-7831 (www.ametek.com/tci). The handbook contains the following relevant entries: "Uncertainty Analysis for Pressure Defined by a PG7601, 7102 or 7302 Piston Gage"; "PG7000 Differential Mode for Defining Low and Negative Differential Pressure at Various Static Pressures"; "Increasing the Accuracy of Pressure Measurement Through Improved Piston Gage Effective Area Determination"; "Fundamental Differential Pressure Calibrations."

T = gage-fluid surface tension

α_p, α_c = thermal expansion coefficients of piston and cylinder, respectively

θ = the piston/cylinder temperature

λ = piston/cylinder elastic deformation coefficient

h = the height difference between piston gage reference level and the reference level of the unit under calibration

This equation makes clear how various error sources must be corrected to achieve the highest possible accuracy.

A very convenient pressure standard[11] (although not really a deadweight gage) combines a precision piston gage with a magnetic null-balance laboratory scale[12] (Fig. 6.3). The gas or liquid pressure to be measured (generated and regulated by a system external to the pressure standard) is applied to one end of a rotating piston; the other end of the piston is supported by the "weighing platform" of the laboratory scale, which measures the pressure force and gives a digital readout of 40,000 counts full scale. Tungsten carbide piston/cylinders allow clearances less than 1 μm; the high hardness and elastic modulus maintain precision in the face of potential wear and pressure expansion. Piston-cylinder pairs are easily interchanged to give five full-scale ranges from 80 to 1,200 lb/in^2. Since deadweights are *not* utilized to measure the pressure force, periodic recalibration against a set of four precision masses is required. However, this is made quick and easy by using the scale's auto-tare feature and a simple screwdriver span adjustment. Instrument uncertainty on the 80 lb/in^2 range (other ranges are proportional) is $\pm(0.004 + 10^{-4}p)$ lb/in^2, where p is the actual pressure in pounds per square inch and repeatability is ± 1 count.

Deadweight gages may be employed for absolute- rather than gage-pressure measurement by placing them inside an evacuated enclosure at (ideally) 0 lb/in^2 absolute pressure. Since the degree of vacuum (absolute pressure) inside the enclosure must be known, this really requires an additional independent measurement of absolute pressure.

The manometer in its various forms is closely related to the piston gage, since both are based on the comparison of the unknown pressure force with the gravity force on a known mass. The manometer differs, however, in that it is self-balancing, is a deflection rather than a null instrument, and has continuous rather than stepwise output. The accuracies of deadweight gages and manometers of similar ranges are quite comparable; however, manometers become unwieldy at high pressures because of the long liquid columns involved. The U-tube manometer of Fig. 6.4 usually is considered the basic form and has the following relation between input and output for static conditions:

$$h = \frac{p_1 - p_2}{\rho g} \tag{6.2}$$

[11]Model 20400, DH Instruments, Inc.; P. Delajoud and M. Girard, "The Development of a Digital Read-Out Primary Pressure Standard," DH Instruments, Pittsburgh, PA, 1981, www.dhinstruments.com.

[12]Mettler Instrument Corp., Hightstown, NJ, www.mt.com/pro.

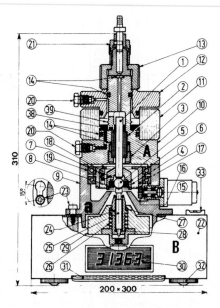

Figure 6.3
Pressure standard using electromagnetic balance. The digital standard is made up of
a piston-cylinder measuring element (A + a) and an electronic dynamometer
(B) manufactured Mettler Instrument.

Measuring element (A + a):
1. Piston in tungsten carbide
2. Cylinder in tungsten carbide
3. Cylinder retaining nut
4. Piston head
5. Ball in tungsten carbide
6. Ball bearing to center the ball (5)
7. Drive bearing
8. Retaining ring for ball (5)
9. Rotation mechanism housing
10. Piston-cylinder housing
11. Acrylic sight glass
12. Cover
13. Retaining nut
14. O-ring seals
15. Electric drive motor
16. Drive pinion
17. Toothed drive wheel
18. Drive bearing pin
19. Toothed wheel bearings
20. Purge screws
21. Quick-connect system (standard threads available)

Electronic dynamometer B:
22. Housing
23. 3 pins giving a quick release facility for the measuring assembly (A + a). (*After 15° rotation on the pins a locking mechanism secures the measuring element to the dynamometer.*)
24. Force-limiting guide
25. Coupling rod
26. Force-limiting spring
27–28. 2 vibration dampers
29. Force receiving plate
30. 40,000 points, 5-digit display
31. Auto-zero bar
32. 2 leveling screws
33. Bubble level

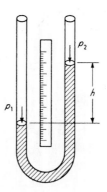

Figure 6.4
U-tube manometer.

where $g \triangleq$ local gravity and $\rho \triangleq$ mass density of manometer fluid. If p_2 is atmospheric pressure, then h is a direct measure of p_1 as a gage pressure. Note that the cross-sectional area of the tubing (even if not uniform) has no effect. At a given location (given value of g) the sensitivity depends on only the density of the manometer fluid. Water and mercury are the most commonly used fluids. To realize the high accuracy possible with manometers, often a number of corrections must be applied. When visual reading of the height h is employed, the engraved scale's temperature expansion must be considered. The variation of ρ with temperature for the manometer fluid used must be corrected and the local value of g determined. Additional sources of error are found in the nonverticality of the tubes and the difficulty in reading h because of the meniscus formed by capillarity. Considerable care must be exercised in order to keep inaccuracies as small as 0.01 mmHg for the overall measurement.[13]

A number of practically useful variations on the basic manometer principle are shown in Fig. 6.5. The *cistern* or *well-type manometer* is widely utilized because of its convenience in requiring reading of only a single leg. The well area is made very large compared with the tube; thus the zero level moves very little when pressure is applied. Even this small error is compensated by suitably distorting the length scale. However, such an arrangement, unlike a U tube, is sensitive to nonuniformity of the tube cross-sectional area and thus is considered somewhat less accurate.

Given that manometers inherently measure the pressure *difference* between the two ends of the liquid column, if one end is at zero absolute pressure, then h is an indication of absolute pressure. This is the principle of the *barometer* of Fig. 6.5. Although it is a "single-leg" instrument, high accuracy is achieved by setting the zero level of the well at the zero level of the scale before each reading is taken. The pressure in the evacuated portion of the barometer is not really absolute zero, but rather the vapor pressure of the filling fluid, mercury, at ambient temperature. This is about 10^{-4} lb/in^2 absolute at 70°F and usually is negligible as a correction.

---

[13]A. J. Eberlein, "Laboratory Pressure Measurement Requirements for Evaluating the Air Data Computer," *Aeronaut. Eng. Rev.,* p. 53, April 1958.

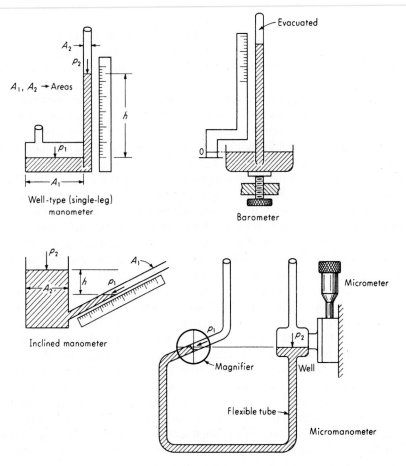

Figure 6.5
Various forms of manometers.

To increase sensitivity, the manometer may be tilted with respect to gravity, thus giving a greater motion of liquid along the tube for a given vertical-height change. The *inclined manometer* (draft gage) of Fig. 6.5 exemplifies this principle. Since this is a single-leg device, the calibrated scale is corrected for the slight changes in well level so that rezeroing of the scale for each reading is not required.

The accurate measurement of extremely small pressure differences is accomplished with the *micromanometer,* a variation on the inclined-manometer principle. In Fig. 6.5 the instrument is initially adjusted so that when $p_1 = p_2$, the meniscus in the inclined tube is located at a reference point given by a fixed hairline viewed through a magnifier. The reading of the micrometer used to adjust well height is now noted. Application of the unknown pressure difference causes the meniscus to move off the hairline, but it can be restored to its initial position by raising or lowering the well with the micrometer. The difference in initial and final micrometer readings gives the height change h and thus the pressure. Instruments using water as the working fluid and having a range of either 10 or 20 in of water can be read to about 0.001 in of water.[14] In another instrument in which the inclined tube (rather than the well) is moved and which uses butyl alcohol as the working fluid, the range is 2 in of alcohol, and readability is 0.0002 in. This corresponds to a resolution of 6×10^{-6} lb/in^2.

While manometers usually are read visually by a human operator, various schemes for rapid and accurate automatic readout are available, mainly for calibration and standards work using gaseous media. The sonar manometer[15] employs a piezoelectric transducer at the bottom of each 1.5-in-diameter glass tube to launch ultrasonic pulses, which travel up through the mercury columns, are reflected at the meniscus, and return to the bottom to be received by the transducers. The pulse from the shorter column turns on a digital counter, while that from the longer one turns it off. Thus a digital reading is obtained that is proportional to the difference in column height and thus to pressure. Resolution is 0.0003 inHg, and accuracy is 0.001 inHg or 0.003 percent of reading, whichever is greater. Since temperature effects on sonic velocity and column length cause additive errors, a feedback control system keeps instrument temperature at $95 \pm 0.05°$F.

Another instrument[16] employs two large mercury cisterns (one fixed, one vertically movable by an electromechanical servosystem) connected by flexible tubing to create a U-tube manometer. Each cistern has a capacitor formed by a metal plate, the mercury surface, and a small air gap between them. The two capacitors are connected in an electric circuit which exhibits a null reading when the air gaps in both cisterns are equal and produces an error voltage when they are not. This error voltage causes the servosystem to drive the movable cistern to an elevation where balance is again achieved. A digital counter on the servosystem motor shaft reads out position to the nearest 0.0001 inHg. System accuracy is $\pm$0.0003 inHg $\pm$0.003 percent of reading. For manometers such as the two above, accessory automatic systems for generating and regulating the pressures of the gaseous (usually air or nitrogen) calibration media usually are available.

Manometer Dynamics

While manometers are utilized mainly for static measurements, their dynamic response is sometimes of interest. The general problem of the oscillations of liquid columns is an interesting (and rather difficult) question in fluid mechanics and has received considerable attention in the literature.[17] Here we take a somewhat simplified view of the problem which nonetheless gives results of practical interest.

---

[14]Meriam Instrument Co., Cleveland OH, www.meriam.com.

[15]D. E. Van Dyck, "Sonic Measurement of Manometer Column Height," Wallace & Tiernan Div., Belleville, NJ.

[16]Schwein Engineering, Pomona, CA.

[17]J. F. Ury, "Viscous Damping in Oscillating Liquid Columns, Its Magnitude and Limits," *Int. J. Mech. Sci.,* vol. 4, p. 349, 1962; P. D. Richardson, "Comments on Viscous Damping in Oscillating Liquid Columns," *Int. J. Mech. Sci.,* vol. 5, p. 415, 1963.

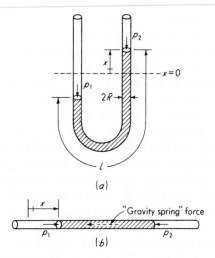

Figure 6.6
Manometer model.

In the U-tube configuration of Fig. 6.6a, the unknown pressures p_1 and p_2 are exerted by a gas whose inertia and viscosity may be considered negligible compared with those of the manometer liquid. If the pressures vary with time, the reading of the manometer varies with time; we are interested in the fidelity with which the manometer reading follows the pressure variation. The motion of the manometer liquid in the tube is caused by the action of various forces. If we consider the manometer liquid in its entirety as a free body and search for forces acting on it, the following forces come to mind:

1. The gravity force (weight) distributed uniformly over the whole body of fluid
2. A drag force due to motion of the fluid within the tube and related to the wall shearing stress
3. The forces on the two ends of the free body due to pressures p_1 and p_2
4. Distributed normal pressure of the tube on the fluid
5. Surface-tension effects at the two ends of the body of fluid

A detailed analysis of all these effects would lead to rather complex and unwieldy mathematics. Fortunately, useful results may be obtained by a simplified analysis. The initial step is the assumption that the system shown in Fig. 6.6b is dynamically equivalent to that of Fig. 6.6a. The "gravity spring" force of Fig. 6.6b is explained as follows: In Fig. 6.6a, whenever $x \neq 0$, there is an unbalanced gravity force acting on the liquid column tending to restore the level to $x = 0$. The magnitude of this force is $-2\pi R^2 x \gamma$, where $\gamma \triangleq$ manometer-fluid specific weight, in pounds force per cubic inch. We see that this force is proportional to the displacement x and always opposes it; thus it has all the characteristics of a spring force. When the liquid column is "straightened out" in Fig. 6.6b, we must include this "gravity spring" force in our equivalent system if we are to preserve the analogy of the two configurations. In comparing Fig. 6.6b with Fig. 6.6a, we note that any effects of flow curvature in the 180° bend are lost. But these will probably be small if the diameter of the bend is large compared with the inside diameter of the manometer tube and if the total length L is large compared with the bend length. We also neglect any surface-tension effects at the ends of the column. This is usually a good assumption if the column is long relative to its diameter.

In addition to the gravity-spring force and the pressure forces due to p_1 and p_2, the liquid column is subjected to a drag force at the interface between the liquid and the wall of the tube. This drag force is equal to the wall shearing stress times the surface area of the liquid column. The motion of the liquid in the tube may be thought of as an unsteady pipe flow. We assume that at any instant of time the wall shearing stress can be computed from the instantaneous velocity of the liquid by *using the formulas commonly used for steady pipe flows.*

The flow of liquid in the tube may occur in the laminar, transition, or turbulent regimes. First let us assume laminar flow prevails. The pressure drop Δp due to pipe friction for both laminar and turbulent flow is given by

$$\Delta p = f\,\frac{\gamma L V_{\text{av}}^2}{2gd} = f\,\frac{L}{d}\frac{\rho V_{\text{av}}^2}{2g_0} \tag{6.3}$$

where
$g_0 \triangleq$ mass unit conversion factor, lbm/slug
$\rho \triangleq$ fluid density, lbm/ft^3
$f \triangleq$ friction factor
$\gamma \triangleq$ fluid specific weight, local, lbf/ft^3
$V_{\text{av}} \triangleq$ average velocity
$g \triangleq$ local gravity acceleration, ft/s^2
$d \triangleq$ diameter of pipe
$L \triangleq$ pipe length

The wall shearing stress τ_0 is given by

$$\tau_0 = \Delta p\,\frac{d}{4L} \tag{6.4}$$

Thus

$$\tau_0 = f\,\frac{\gamma V_{\text{av}}^2}{8g} = f\,\frac{\rho V_{\text{av}}^2}{8g_0} \tag{6.4a}$$

For laminar flow the friction factor is given by

$$f = \frac{64\mu g}{d\gamma V_{\text{av}}} = \frac{64}{dV_{\text{av}}\,\rho/(\mu g_0)} \tag{6.5}$$

so that

$$\tau_0 = \frac{4\mu V_{\text{av}}}{R} \tag{6.6}$$

where $R \triangleq d/2$.

This result also can be obtained directly from the laminar velocity distribution

$$V = V_c \left[1 - \left(\frac{r}{R} \right)^2 \right] \tag{6.7}$$

where $V \triangleq$ velocity at radius r and $V_c \triangleq$ center-line velocity.

The velocity gradient is

$$\frac{dV}{dr} = -\frac{V_c}{R^2}(2r) \tag{6.8}$$

which becomes at the wall

$$\left. \frac{dV}{dr} \right|_{r=R} = -\frac{2V_c}{R} \tag{6.9}$$

Shearing stress is given by

$$\tau = \mu \frac{dV}{dr} \tag{6.10}$$

and so the magnitude of the wall shearing stress τ_0 is

$$\tau_0 = \frac{4\mu V_{av}}{R}$$

since $V_c = 2V_{av}$ for laminar flow in circular pipes.

We are now in a position to apply Newton's law to the system of Fig. 6.6b. The average flow velocity V_{av} corresponds to $\dot{x}$, the first derivative of x with respect to time. Considering the entire body of liquid as a free body and taking the effective mass of the moving liquid as four-thirds of its actual mass, based on the kinetic energy of steady laminar flow, we can write for motion in the x direction

$$\pi R^2 (p_1 - p_2) - 2\pi R^2 \gamma x - 2\pi RL \frac{4\mu \dot{x}}{R} = \frac{4}{3} \frac{\pi R^2 L \gamma}{g} \ddot{x} \tag{6.11}$$

This reduces to

$$\frac{2\ddot{x}}{3g/L} + \frac{4\mu L}{R^2 \gamma} \dot{x} + x = \frac{1}{2\gamma} p \tag{6.12}$$

where we have defined $p \triangleq p_1 - p_2$. In operator form, this becomes

$$\left(\frac{2D^2}{3g/L} + \frac{4\mu L}{R^2 \gamma} D + 1 \right) x = \frac{1}{2\gamma} p \tag{6.13}$$

The operational transfer function relating output x to input p is

$$\frac{x}{p}(D) = \frac{1/(2\gamma)}{2D^2/(3g/L) + [4\mu L/(R^2 \gamma)] D + 1} \tag{6.14}$$

which is of the form
$$\frac{x}{p}(D) = \frac{K}{D^2/\omega_n^2 + 2\zeta D/\omega_n + 1} \tag{6.15}$$

where
$$K \triangleq \frac{1}{2\gamma} \quad \text{in/(lb/in}^2) \tag{6.16}$$

$$\omega_n \triangleq \sqrt{\frac{3g}{2L}} \quad \text{rad/s} \tag{6.17}$$

$$\zeta \triangleq 2.45\,\mu\, \frac{\sqrt{gL}}{R^2\gamma} \tag{6.18}$$

We note from the above that the manometer is a second-order instrument. The numerical values of the parameters are usually such that $\zeta < 1.0$; that is, the instrument is underdamped.

Since laminar flow was assumed in carrying out the above analysis, we should try to estimate a typical Reynolds number to see under what conditions laminar flow occurs. As a numerical example, we take a mercury manometer with

$L = 26.5$ in
$R = 0.13$ in
$\mu = 2.18 \times 10^{-7}$ lbf $\cdot$ s/in^2
$\gamma = 0.491$ lbf/in^3

Suppose that we wish to check our theoretical results by measuring ζ and ω_n experimentally for a step-function input. Computing ζ and ω_n from Eqs. (6.17) and (6 18) gives $\zeta = 0.007$ and $\omega_n = 4.7$ rad/s. Since $\zeta = 0.007$ represents a *very lightly* damped system, we can estimate the maximum flow velocity by assuming no damping at all. A second-order system with no damping executes pure sinusoidal oscillations when subjected to a step-function input. Thus its motion would be given by

$$x = X \sin \omega_n t \tag{6.19}$$

where X is the size of the step function. The velocity $\dot{x}$, which is the same as the average flow velocity, would be

$$\dot{x} = \omega_n X \cos \omega_n t \tag{6.20}$$

and its maximum value would thus be $\omega_n X$. The Reynolds number for steady pipe flow is given by

$$N_R = \frac{2\gamma R V_{av}}{g\mu} \tag{6.21}$$

and the critical value for transition from laminar to turbulent flow is 2,100. Since $V_{av} = \omega_n X$, it should be clear that there is a maximum-size step function that can be used without exceeding $N_R = 2,100$. This limiting value X_m is given by

$$2,100 = \frac{2\gamma R \omega_n X_m}{g\mu} \tag{6.22}$$

which in this example gives

$$X_m = \frac{(2,100)(386)(2.18 \times 10^{-7})}{(2)(0.491)(0.13)(4.7)} = 0.30 \text{ in} \tag{6.23}$$

Thus, to ensure laminar flow at all times during the oscillation, the step input can be no larger than 0.30 in. Suppose we wish to measure ζ and ω_n by simple visual methods, ζ from the size of the first overshoot and ω_n by counting and timing cycles. This requires much larger step inputs for reasonable accuracy. Therefore we must investigate the effect of the presence of turbulent flow on our analysis.

Suppose that we decide that a step input X_m of 5 inHg will be sufficiently large to allow accurate measurements of ζ and ω_n. The maximum Reynolds number then would be

$$N_R = \frac{5}{0.30}(2,100) = 35,000 \tag{6.24}$$

For steady turbulent flow in smooth pipes with $3,000 < N_R < 100,000$, the Blasius equation for friction factor is

$$f = \frac{0.316}{(N_R)^{0.25}} \tag{6.25}$$

The turbulent wall shearing stress is then given by Eq. (6.4) as

$$\tau_0 = \frac{0.0378\gamma^{0.75}\,\mu^{0.25}\,V_{av}^{1.75}}{g^{0.75}\,R^{0.25}} \tag{6.26}$$

Using the numerical values of this particular example, we get

$$\tau_0 = 9.18 \times 10^{-6}\,V_{av}^{1.75} \qquad \text{lbf/in}^2 \tag{6.27}$$

For laminar flow the comparable expression is

$$\tau_0 = 6.71 \times 10^{-6}V_{av} \qquad \text{lbf/in}^2 \tag{6.28}$$

The most significant difference between Eqs. (6.27) and (6.28) is that (6.27) represents a nonlinear relation between shear stress and velocity whereas (6.28) is linear. This means that when the force due to wall shearing stress is substituted into Newton's law, the result is a nonlinear differential equation because of the term $(\dot{x})^{1.75}$. This nonlinear equation cannot be solved analytically, so the presence of turbulent flow leads to mathematical difficulties.

In working with oscillations of systems with nonlinear damping terms similar to the $(\dot{x})^{1.75}$ of this example, engineers have developed an approximate method of analysis which is quite useful. This approach is based on the observation that while the linear damping term $\dot{x}$ and the nonlinear term, such as $(\dot{x})^{1.75}$, are quite different mathematically, the general *form* of the oscillation in the two systems is not radically different in experimental tests. If the linear system is excited by a sinusoidal exciting force, it will respond with a sinusoidal motion, whereas the nonlinear system's motion will not be purely sinusoidal. However, observation shows that the deviation from pure sinusoidal motion is usually quite small. Using these facts as a basis, then, we might reason as follows: If a system with nonlinear damping is executing steady oscillations of fixed amplitude, during each cycle the damping force will dissipate a certain amount of energy. If we know from experience that the waveform of the nonlinear oscillation is nearly sinusoidal, we can compute approximately the energy dissipation per cycle. This is done as follows: Suppose there exists a steady oscillation of amplitude Y and frequency ω. If we assume the waveform to be sinusoidal, we can write

$$y = Y \sin \omega t \tag{6.29}$$

and

$$\dot{y} = Y\omega \cos \omega t \tag{6.30}$$

Now, in general, the instantaneous power is the product of instantaneous force and instantaneous velocity. Thus the power dissipation due to damping is the product of velocity and damping force. If the damping force is a known function of velocity $f(\dot{y})$, we can write

$$\text{Instantaneous power dissipation} = \dot{y}f(\dot{y}) \tag{6.31}$$

and the energy dissipated per cycle will be given by

$$\int_{\text{one cycle}} \dot{y}f(\dot{y}) \, dt \tag{6.32}$$

For a linear damping the function $f(\dot{y})$ is just $B\dot{y}$, where B is a constant. So the energy dissipation per cycle is

$$\int_0^{2\pi/\omega} (Y\omega \cos \omega t)(BY\omega \cos \omega t) \, dt = \pi B\omega Y^2 \tag{6.33}$$

For the nonlinear damping due to turbulent flow, the function $f(\dot{y})$ is $C(\dot{y})^{1.75}$, where C is a constant. The energy dissipation per cycle for this nonlinear damping is

$$\int_0^{2\pi/\omega} (Y\omega \cos \omega t)C(Y\omega \cos \omega t)^{1.75} \, dt \tag{6.34}$$

This is equal to

$$C(Y\omega)^{2.75} \int_0^{2\pi/\omega} (\cos \omega t)^{2.75} \, dt \tag{6.35}$$

In evaluating the integral in (6.35), we must use physical reasoning, because when cos ωt becomes *negative,* the quantity $(\cos \omega t)^{2.75}$ is not defined in terms of real numbers. Physical reasoning, however, tells us that the physical processes occurring during the first quarter-cycle $[0 \leq t \leq \pi/(2\omega)]$ give exactly the same energy dissipation as those occurring during the other three quarters of the cycle. Thus, we can integrate over only the first quarter and multiply by 4 to get the total energy dissipation. During the first quarter-cycle, cos ωt is always positive, and so no mathematical difficulties arise. This amounts to saying that, to agree with the known physical facts, integral (6.34) should really be written as

$$\int_0^{2\pi/\omega} |Y\omega \cos \omega t| [C(Y\omega)^{1.75}|(\cos \omega t)|^{1.75}] \, dt \qquad \textbf{(6.36)}$$

with the absolute-value signs as shown.

By defining $\theta \triangleq \omega t$, integral (6.36) can be written as

$$CY^{2.75}\omega^{1.75} \int_0^{2\pi} (\cos \theta)^{2.75} \, d\theta \qquad \textbf{(6.37)}$$

Since this integral is not available analytically, we use SIMULINK to obtain it numerically, which gives

$$2.77C\omega^{1.75}Y^{2.75} \qquad \textbf{(6.38)}$$

Having obtained the above results, we now define the *equivalent linear damping* as that linear damping which would dissipate exactly the same energy per cycle as the nonlinear damping at a given frequency and amplitude. Thus we set (6.33) equal to (6.38) and get

$$\pi B_e \omega Y^2 = 2.77C\omega^{1.75}Y^{2.75} \qquad \textbf{(6.39)}$$

$$B_e \triangleq \text{equivalent linear damping}$$

$$B_e = 0.882C(\omega Y)^{0.75} \qquad \textbf{(6.40)}$$

Since C is the constant that multiplies $(\dot{y})^{1.75}$, in the manometer example we have

$$\text{Damping force} = 2\pi RL\tau_0 = \frac{0.237R^{0.75}\gamma^{0.75}\mu^{0.25}L\dot{x}^{1.75}}{g^{0.75}} \qquad \textbf{(6.41)}$$

and thus

$$C = \frac{0.237\gamma^{0.75}R^{0.75}L\mu^{0.25}}{g^{0.75}} \qquad \textbf{(6.42)}$$

Now Eq. (6.18) can be written as

$$\zeta = \frac{2.45\mu L\sqrt{g}}{R^2\gamma\sqrt{L}} = \frac{0.0974B\sqrt{g}}{R^2\gamma\sqrt{L}} \qquad \textbf{(6.43)}$$

since $B = 8\pi\mu L$ for the linear system. We can now define the equivalent linear damping ratio ζ_e by substituting B_e from Eqs. (6.40) and (6.42) in (6.43):

$$\zeta_e \triangleq \frac{0.0203\sqrt{L}[\mu/(\gamma g)]^{0.25}}{R^{1.25}}(\omega Y)^{0.75} \qquad \textbf{(6.44)}$$

This shows clearly the dependence of ζ_e on the frequency and amplitude of the oscillation. For turbulent flow the value of ω_n is also somewhat different since the velocity profile tends to be more nearly square than parabolic. If the velocity is assumed uniform over the cross section, the effective mass becomes equal to the actual mass since all particles have the same velocity. If the turbulent damping, though larger than the laminar, is assumed to be still quite small, it is reasonable to expect that it will have little effect on the frequency. We therefore compute ω_n for turbulent flow by neglecting the nonlinear damping completely where it would appear in Eq. (6.11). We then get

$$\omega_n = \sqrt{\frac{2g}{L}}$$

(6.45)

for turbulent flow.

Many assumptions were made in the above analysis. The formulas for steady pipe flow were employed for an unsteady situation. In an oscillating flow, velocity actually goes to zero twice each cycle, no matter how great the amplitude or frequency; thus one wonders whether part of such a cycle is turbulent and part laminar. In the analysis above, turbulent equations were used for the whole cycle. The nonlinear differential equation containing $(\dot{x})^{1.75}$ actually has no closed-form analytical solution. Thus what is the meaning of a ζ_e and an ω_n attached to such a process? Such questions and others may be at least partially resolved by more complex analyses or experimental studies. To provide some idea of the degree of validity of our simplified analysis, some experimental results are given. They were obtained at The Ohio State University by undergraduate students who study manometer dynamics in a simple experiment in a measurement course.

The experiment consists in part of suddenly releasing an air pressure applied to a mercury manometer and observing the resulting oscillations. The process is slow enough that ζ_e can be estimated from the size of the first overshoot and ω_n by counting and timing cycles with a stopwatch. The manometers used have the numerical values quoted earlier in this section. A step pressure input of 10 inHg ($x = 5$ in) is utilized; thus turbulent flow may be expected. If laminar flow were assumed, the theoretical values would be $\zeta = 0.007$ and $\omega_n = 4.7$ rad/s. For turbulent flow, ω_n becomes 5.4 rad/s. To calculate ζ_e from Eq. (6.44), the frequency and amplitude of the oscillation must be known. For a step input the frequency is the damped natural frequency rather than ω_n, however, these are practically identical for the small damping present, and so we use ω_n. We experimentally measure ζ_e from the first overshoot, and so the proper amplitude to employ in the theoretical calculation might be an average of the initial amplitude and that at the first overshoot. Only the initial amplitude (5 in) is known, though, and so this is used. Equation (6.44) then gives $\zeta_e = 0.0905$. By timing and counting cycles (about 8 cycles of the decaying oscillation can be measured easily) the experimental value of ω_n is 5.2 rad/s. This lies between the values calculated for turbulent and laminar flow and thus is not unreasonable, since several of the 8 cycles used were of quite low amplitude because of the decay of the oscillation. The first overshoot is about 4.05 in, giving an experimental value of ζ of 0.067. Therefore the theoretical estimate of 0.0905 is fairly good. If we now use the experimental value of ω (5.2) and the average amplitude of the first half-cycle (4.53) in Eq. (6.44), then the predicted ζ_e is 0.082, which compares even more favorably. Based on even these limited results, a certain amount of confidence in the theoretical predictions is established.

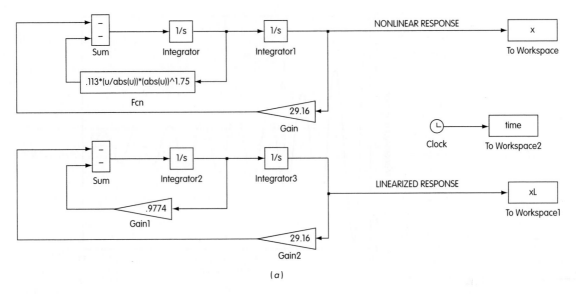

Figure 6.7
Manometer dynamic simulation and step response results.

Using our SIMULINK digital simulation, we can easily "solve" the nonlinear manometer differential equation. For the numerical values given, by substituting Eq. (6.27) in (6.11) and using the actual (rather than the $\frac{4}{3}$ times the actual) mass appropriate for turbulent flow, we get

$$\frac{d^2x}{dt^2} + 0.1113 \cdot \frac{\dfrac{dx}{dt}}{abs\left(\dfrac{dx}{dt}\right)}\left(\frac{dx}{dt}\right)^{1.75} + 29.16x = 59.39(p_1 - p_2) \quad \textbf{(6.46)}$$

Using the form $(dx/dt)/(abs(dx/dt))$ to get the correct algebraic sign on the nonlinear damping term as velocity changes direction, this equation is simulated in Fig. 6.7a. We start x off at 5 in, as in the experimental test. For comparison, we also simulate a *linear* manometer with $\zeta = 0.0905$ and $\omega_n = 5.4$. The responses of Fig. 6.7b show that the linearized approximate model (dashed curve) agrees almost perfectly with the nonlinear for the first cycle, but damps down too quickly later, as we would expect, since the nonlinear damping *decreases* with amplitude while the linear stays fixed. Since the first overshoot is modeled almost perfectly by the linearized system, the discrepancies noted earlier between experimental and linearized theoretical results are clearly *not* due to inadequacies in the linearization technique. Rather, they must be charged to deficiencies in the original nonlinear model, inaccuracy in numerical values of parameters, and/or errors in the experimental results.

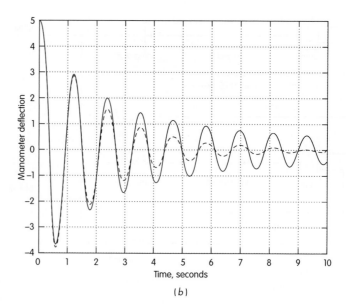

Time, seconds

(b)

Figure 6.7
(Concluded)

6.4 ELASTIC TRANSDUCERS

While a wide variety of flexible metallic elements conceivably might be used for pressure transducers, the vast majority of practical devices utilize one or another form of Bourdon tube,[18] diaphragm,[19] or bellows[20] as their sensitive element, as shown in Fig. 6.8. The gross deflection of these elements may directly actuate a pointer/scale readout through suitable linkages or gears, or the motion may be transduced to an electrical signal by one means or another. Strain gages bonded directly to diaphragms or to diaphragm-actuated beams are widely used to measure local strains that are directly related to pressure.

The Bourdon tube is the basis of many mechanical pressure gages and is also used in electrical transducers by measuring the output displacement with potentiometers, differential transformers, etc. The basic element in all the various forms is a tube of noncircular cross section. A pressure difference between the inside and outside of the tube (higher pressure inside) causes the tube to attempt to attain a circular cross section. This results in distortions which lead to a curvilinear translation of the free end in the C type and spiral and helical types and an angular rotation

---

[18]D. M. Considine (ed.), "Process Instruments and Controls Handbook," sec. 3, McGraw-Hill, New York, 1957; R. W. Bradspies, "Bourdon Tubes," *Giannini Tech. Notes,* Giannini Corp, Duarte, CA, January–February 1961.

[19]Considine, ibid.; "Pressure Capsule Design," *Giannini Tech. Notes,* November 1960; C. K. Stedman, "The Characteristics of Flat Annular Diaphragms," *Statham Instrum. Notes,* Statham Instruments Inc., Los Angeles; ref. 25 (p. 509).

[20]Considine, op. cit.; R. Carey, "Welded Diaphragm Metal Bellows," *Electromech. Des.,* p. 22, August 1963.

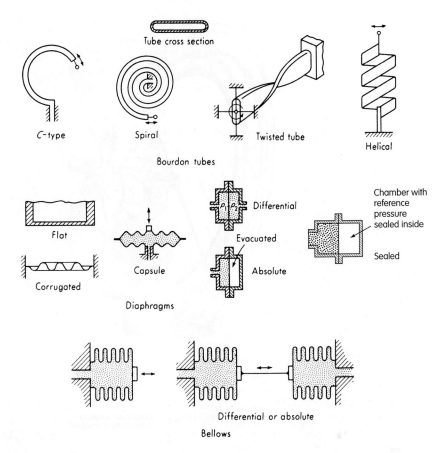

Figure 6.8
Elastic pressure transducers.

in the twisted type, which motions are the output. The theoretical analysis of these effects is difficult, and practical design at present still makes use of considerable empirical data. The C-type Bourdon tube has been utilized up to about 100,000 lb/in². The spiral and helical configurations are attempts to obtain more output motion for a given pressure and have been used mainly below about 1,000 lb/in². The twisted tube shown has a crossed-wire stabilizing device which is stiff in all radial directions but soft in rotation. This reduces spurious output motions from shock and vibration. Figure 6.9 shows construction of a higher accuracy (0.1 percent) C-type Bourdon test gage. Also shown is an optional bimetal temperature compensator which maintains accuracy over the range − 25 to + 125°F. This device corrects for both thermal zero shift and span shift.

Electrical output pressure transducers are available in forms combining various elastic elements with most of the displacement transducers described in Chap. 4, i.e., potentiometers, strain gages, LVDTs, capacitance pickups, eddy-current probes,

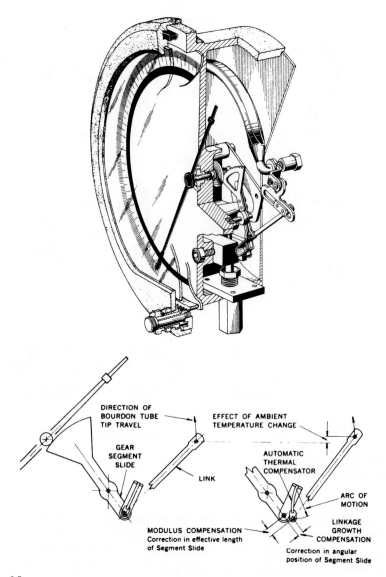

Figure 6.9
Bourdon-tube gage construction. *(Courtesy Heise Gage, Dresser Industries, Newton, CT.)*

reluctance and inductance pickups, piezoelectric elements, etc. Space does not permit detailed discussion of all types, so we choose specific examples that illustrate features we feel to be particularly significant and interesting.

Figure 6.10 shows a sensor utilizing an infrared LED (light-emitting diode) and two photodiodes to optically measure displacement (0.020 in full scale) of the pressure-sensitive elastic element. The reference and measurement photodiodes are on the same chip and thus are equally affected by temperature changes. Changes in

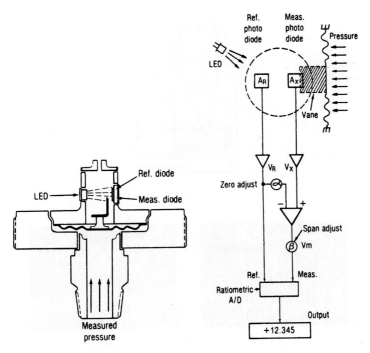

Figure 6.10
Electro-optic pressure transducer. *(Courtesy Heise Gage, Dresser Industries.)*

LED output due to temperature or age also cancel, since both diodes share the same illumination and a ratiometric integrating analog/digital converter is employed to obtain a digital output sensitive to only diode-illuminated areas A_R and A_X and pot settings α and β. While diode signals exhibit nonlinearity which varies from unit to unit, this nonlinearity and all others are linearized in the analog/digital converter using a look-up table resident in a pair of PROMs (programmable read-only memories). Thus each sensor has its individual nonlinearities linearized by programming its PROM at calibration time. This design is characteristic of recent trends which, rather than making a great effort to obtain linearity in basic sensors, accept their nonlinearity (as long as it is repeatable) and then use programmable electronic compensation tailored to each unit to achieve overall linearity. The referenced unit is available in ranges of 0 to 50 in H_2O to 60,000 lb/in^2 gage and 5 lb/in^2 absolute to 60,000 lb/in^2 absolute with accuracy 0.1 percent of span and temperature effect on span of ±0.004 percent/°F from 0 to 160°F. An automatic-zero feature in the analog/digital converter minimizes thermal zero shifts. Interestingly, the manufacturer also uses the same optical sensor as a null detector in a force-balance type of pressure transducer. Here the diode nonlinearity need not be linearized since the servosystem always returns the sensor to the same balance point.

Piezoelectric pressure transducers share many common characteristics with piezoelectric accelerometers and force transducers, discussed earlier. Generally they

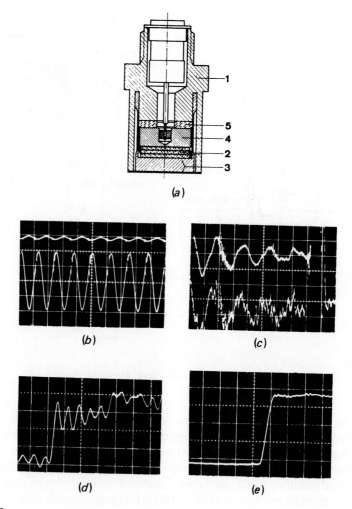

Figure 6.11
Piezoelectric transducer with acceleration compensation. (*a*) 1 = housing;
2 = pressure-sensitive quartz disks; 3 = diaphragm assembly; 4 = compensating mass;
5 = compensating quartz disk.

have high natural frequencies and little response to spurious acceleration inputs [typically 0.02 (lb/in^2)/g]. However, transducers designed for low pressures and mounted on objects with severe vibration may require an acceleration-compensated design. Figure 6.11[21] shows how a compensating accelerometer is built right into the pressure transducer and its signal combined with that of the pressure-sensing

[21]K. H. Martini, "Practical Application of Quartz Transducers," Kistler Instruments Corp., Amherst, NY, 1971, www.kistler.com.

crystals to cancel the vibration effect. About a 5:1 or 10:1 improvement is available in this way. A higher degree of compensation is technically possible but is not attempted in practice since a *transverse* vibration sensitivity (which *cannot* be compensated) limits performance at this level. Figure 6.11*b* compares the output of a compensated (upper trace) and an uncompensated transducer when both are mounted on the same sinusoidal vibration shaker; Fig. 6.11*c* shows a similar comparison when the pressure transducers are each fastened to a strongly vibrating diesel-engine exhaust pipe to measure exhaust pressure. The compensation is clearly quite effective in these examples. However, it may *not* be effective in suppressing the pressure transducer's natural resonance, if the accelerometer resonance does not adequately match the pressure transducer's, as can be seen in Fig. 6.11*d,* where a compensated transducer is mounted in the wall of an aero-dynamic shock tube. The oscillatory signal at the far left actually occurs even *before* the air shock wave strikes the transducer diaphragm and is due to mechanical exci-tation of the transducer resonance caused by the *mechanical* shock wave in the tube wall, which travels *faster* than the airborne wave. To obtain the improved response of Fig. 6.11*e,* a transducer with about four times higher natural frequency was mounted with a plastic adapter between the wall and the transducer. The plastic adapter serves as a mechanical shock filter, while the higher-natural-frequency transducer is inherently less acceleration sensitive.

Another solution to the transducer resonance phenomenon involves careful matching of the compensating accelerometer's resonance with that of the pressure elements. When this is successful, both acceleration effects and transducer "ring-ing" response to sharp pressure changes are reduced. Transducers[22] based on this approach may have 500-kHz resonant frequency, 1-μs rise time and less than 15 percent overshoot for step inputs, and acceleration sensitivity of 0.002 (lb/in^2)/g for a 500 lb/in^2 full-scale unit with integral microelectronic amplifier (10-V full-scale output).

Pressure transducers based on foil-type metal strain gages may either apply the strain gages directly to a flat metal diaphragm or use a convoluted diaphragm to apply force to a strain-gage beam. Flat diaphragms exhibit nonlinearity at large deflections, where a stretching action adds to the basic bending, causing a stiffening effect. This nonlinearity in the stresses follows closely the nonlinearity in the center deflection y_c for which the following theoretical result is available:

$$p = \frac{16Et^4}{3R^4(1 - v^2)}\left[\frac{y_c}{t} + 0.488\left(\frac{y_c}{t}\right)^3\right] \tag{6.47}$$

where $p \triangleq$ pressure difference across diaphragm

               $E \triangleq$ modulus of elasticity

               $t \triangleq$ diaphragm thickness

               $v \triangleq$ Poisson's ratio

               $R \triangleq$ diaphragm radius to clamped edge

---

[22]PCB Piezotronics, Buffalo, NY, www.pcb.com.

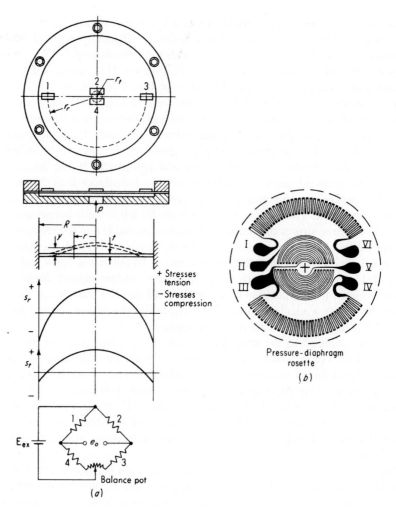

Figure 6.12
Diaphragm-type strain-gage pressure pickup.

By designing for a sufficiently small value of y_c/t, a desired nonlinearity may be achieved. However, note that small y_c/t also gives small strains and output voltage. Sputtered-film strain-gage transducers have shown particularly good stability characteristics.[23]

Such a diaphragm, clamped at the edges and subjected to a uniform pressure difference p, has at any point on the low-pressure surface a radial stress s_r and a tangential stress s_t given by the following formulas (see Fig. 6.12):

---

[23]"Strain Gage Pressure Transducer Stability," CEC Instruments Div., Pasadena, CA, 1987, www.cecvp.com.

$$s_r = \frac{3pR^2\nu}{8t^2}\left[\left(\frac{1}{\nu}+1\right)-\left(\frac{3}{\nu}+1\right)\left(\frac{r}{R}\right)^2\right]$$ (6.48a)

$$s_t = \frac{3pR^2\nu}{8t^2}\left[\left(\frac{1}{\nu}+1\right)-\left(\frac{1}{\nu}+3\right)\left(\frac{r}{R}\right)^2\right]$$ (6.48b)

where $\nu \triangleq$ Poisson's ratio. The deflection at any point is given by

$$y = \frac{3p(1-\nu^2)(R^2-r^2)^2}{16Et^3}$$ (6.49)

Equations (6.48) to (6.49) all give linear relations between stress or deflection and pressure and are accurate only for sufficiently small pressures. Equation (6.47) may be used to estimate the degree of nonlinearity. The stress situation on the diaphragm surface is fortunate because both tension and compression stresses exist simultaneously. This allows use of a four-active-arm bridge in which all effects are additive (giving large output) and gives temperature compensation also. Gages 2 and 4 are placed as close to the center as possible and oriented to read tangential strain, since it is maximum (positive) at this point. Gages 1 and 3 are oriented to read radial strain and placed as close to the edge as possible, since radial strain has its maximum negative value at that point. The laws of bridge circuits show that the pressure effects on all four gages are additive. In computing the overall sensitivity, Eq. (6.48) cannot be employed directly to determine the strains "seen" by the gages since the diaphragm surface is in a state of *biaxial* stress and *both* the radial and tangential stress contribute to the radial or tangential strain at any point. The general biaxial stress-strain relation gives

$$\varepsilon_r = \frac{s_r - \nu s_t}{E}$$ (6.50)

$$\varepsilon_t = \frac{s_t - \nu s_r}{E}$$ (6.51)

Once the gage strains are calculated, the individual gage ΔR's are obtained from the gage factors, and then e_o can be determined from the bridge-circuit sensitivity equations.

To facilitate construction and miniaturization of such pressure transducers, the discrete individual gages may be replaced by a rosette available in various sizes from several strain-gage manufacturers. Figure 6.12b[24] shows how such rosettes are configured to take advantage of radial strains at the diaphragm edge and tangential strains near the center, while the solder tabs are located in a low-strain region to increase reliability of the solder joints. For the referenced rosette, the manufacturer provides this design formula:

$$\frac{e_o}{E_{ex}} = 820\frac{pR^2(1-\nu^2)}{Et^2}\qquad \frac{mV}{V}$$

---

[24]"Design Considerations for Diaphragm Pressure Transducers," TN–129–3, Micro-Measurements, Romulus, MI, 1974, www.vishay.com.

When such rosettes are utilized, the radius R can be made quite small and usually the diaphragm is "machined from solid," rather than using the "assembled from pieces" construction of Fig. 6.12a, thereby improving accuracy and hysteresis.

If the transducer is to be utilized for dynamic measurements, its natural frequency is of interest. A diaphragm has an infinite number of natural frequencies, however, the lowest is the only one of interest here. For a clamped-edge diaphragm vibrating in a fluid of density ρ_f, the lowest natural frequency is given by[25]

$$\omega_n = \frac{10.21}{CR^2} \sqrt{\frac{Et^2}{12\rho_d(1 - v^2)}} \qquad \text{rad/s} \qquad \textbf{(6.52)}$$

$$C \triangleq \sqrt{1 + 0.669 \frac{\rho_f}{\rho_d} \frac{R}{t}}$$

where $\rho_d \triangleq$ mass density of diaphragm material. A number of factors may make the actual operating value of ω_n different from the prediction of Eq. (6.52). The edge clamping is never perfectly rigid; any softness tends to lower ω_n. If the diaphragm is not perfectly flat, tightening the clamping bolts may cause a slight (perhaps imperceptible) "wrinkling," tending to stiffen the diaphragm and raise ω_n. If the diaphragm is used to measure liquid pressures, the inertia of the liquid tends to lower ω_n, especially if a small-diameter tube connects the pressure source to the diaphragm. When it is employed with gases, the volume of gas "trapped" behind the diaphragm may act as a stiffening spring, raising ω_n.

A pressure transducer of the discrete gage type constructed for use in a transducer research project at The Ohio State University serves as a numerical example for the above discussion. This transducer used a phosphor-bronze diaphragm with

$$E = 16 \times 10^6 \text{ lbf/in}^2 \qquad R = 1.830 \pm 0.002 \text{ in} \qquad \textbf{(6.53)}$$
$$v = \tfrac{1}{3} \qquad\qquad\qquad r_t = 0.15 \pm 0.01 \text{ in}$$
$$\rho = 0.00083 \text{ lbf} \cdot \text{s}^2/\text{in}^4 \qquad r_r = 1.52 \pm 0.01 \text{ in}$$
$$t = 0.0454 \pm 0.0003 \text{ in}$$

The strain gages had a gage factor of 1.97 ± 2 percent and a resistance of $119.5 \pm 0.3 \ \Omega$. The bridge was excited with 7.5 V dc, and the transducer was designed for a full-scale range of 10 lb/in². The theoretically calculated sensitivity was 0.516 mV/(lb/in²). Static calibration gave 0.513 mV/(lb/in²) and indicated a maximum nonlinearity of 2 percent of full scale. The theoretically calculated natural frequency was 924 Hz, and the experimental value (with atmospheric air on both sides of the diaphragm) was 897 Hz.

Figure 6.13 shows construction details for a transducer in which a thin diaphragm (welded to the housing) serves as a seal and pressure-gathering member; however, the majority of the pressure force is taken by strain-gage bending beams in the form of spokes on a wheel. Nonlinearity as small as 0.05 to 0.1 percent can be achieved with designs of this type. Just as in the strain-gage force transducers of

---

[25]M. Di Giovanni, "Flat and Corrugated Diaphragm Design Handbook," p. 196, Marcel Dekker, New York, 1982.

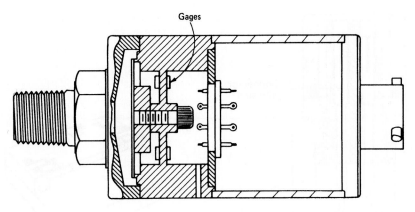

Gages

Figure 6.13
Bonded foil strain-gage pressure transducer. *(Courtesy Sensotec Inc., Columbus, OH.)*

Chap. 5, various levels of temperature compensation can be provided, depending on application needs.

Strain-gage pressure transducers using MEMS technology (see Sect. 14.1 for more details) are now quite common, particularly in high-volume, low-cost applications. Several examples of these miniature sensors will now be discussed. Figure 6.14 shows construction of a piezoresistive (semiconductor "strain gage") pressure transducer[26] of flush-diaphragm design with full-scale ranges from 0 to 2 to 0 to 200 bars. The measuring cell is a welded assembly of two silicon chips forming a sealed chamber containing the reference pressure (or vacuum for absolute pressure sensing). One chip is bored out on one side to form a diaphragm of thickness suited to the desired range, while the reverse side has a complete bridge of strain-sensitive regions diffused into the silicon surface. Use of constant-current bridge excitation (see figure) enhances primary temperature compensation, which together with external discrete compensating resistors brings the overall thermal zero shift to less than 0.01 percent of full-scale/°C and the sensitivity shift to less than 0.02 percent/°C. Pressure is transmitted from the thin metal diaphragm through an oil fill (which also provides damping) to the silicon capsule. Natural frequency is greater than 70 kHz, transducer compliance is 0.001 mm^3/bar, and acceleration sensitivity is 10^{-4} bar/g.

A line of piezoresistive pressure transducers[27] available in ranges from 2 to 50,000 lb/in^2 applies the fluid pressure *directly* to a silicon diaphragm with a complete Wheatstone bridge of strain elements diffused into the diaphragm. Chemical etching processes are employed to produce a "sculptured" diaphragm (rather than the usual flat configuration) with thick and thin sections carefully designed to

[26]H. R. Winteler and G. H. Gautschi, "Piezoresistive Pressure Transducers," Kistler Instruments, Amherst, NY, 1979.

[27]R. M. Whittier, "Basic Advantages of the Anisotropic Etched, Transverse Gage Pressure Transducer," *Prod. Dev. News,* vol. 16, no. 3, Endevco Corp., San Juan Capistrano, CA, 1980.

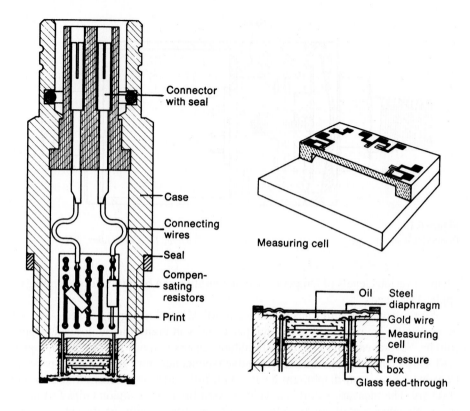

Measuring cell

Current excitation

The pressure transducers are excited by constant current. The voltage rise due to the increase in resistance with temperature compensates for the decrease of the gage factor with temperature. The graph shows the typical relative changes of the resistance R, the gage factor G and the output voltage U_{out} in function of temperature.

a) $\dfrac{\Delta R}{R}$

b) $\dfrac{\Delta G}{G} = \dfrac{\Delta U_{out}}{U_{out}}$ with constant voltage excitation

c) $\dfrac{\Delta U_{out}}{U_{out}}$ with constant current excitation

Figure 6.14

Diffused semiconductor strain-gage transducer.

(Courtesy Kistler Instruments, Amherst, NY, model 4041.)

concentrate strain at the locations where the gages will be diffused into the silicon. A further innovation employs the piezoresistive material in a transverse, rather than longitudinal, configuration, which gives improved linearity since the transverse gages exhibit compensating nonlinearities for the tension and compression legs of the bridge. While the transducers are specified for use with dry nonconductive gases, short-term applications in water[28] have been successful. A $\pm$ 2 lb/in^2 gage unit is only 0.092 inch in diameter; has 1 percent nonlinearity (3 percent at 3 times full-scale pressure), burst pressure of $\pm$ 40 lb/in^2 gage, 157 mV/(lb/in^2) sensitivity, resonant frequency of 45,000 Hz, acceleration sensitivity of 0.0005 (lb/in^2)/g longitudinal (0.00003 lateral), compensated temperature range 0 to 200°F (3 percent zero shift, 4 percent sensitivity shift); and uses a 10-V constant-voltage excitation.

Diffused strain-gage pressure transducers intended for high-volume, low-cost applications such as appliances, automotive parts, etc., are available also. Figure 6.15a[29] shows the typical construction of the basic unit and an oil-filled version with a flexible silicone rubber barrier ("sock"), required for protection from certain working fluids. Often such transducers are designed to employ auto-referencing[30] techniques in order to achieve good accuracy in the face of zero shifts resulting from time and/or temperature. This requires that the transducer periodically be exposed (momentarily) to some *known* reference pressure (perhaps the most common example would be a gage pressure unit, which could be vented to atmosphere, giving a known 0 lb/in^2 gage input). The automatic-reference circuit measures and "remembers" the voltage output for zero pressure input, and when the transducer is reconnected to its usual pressure signal, the circuit adds or subtracts this value, thereby correcting for bias errors from all sources. (If two different reference pressures can be employed, the technique can correct for slope errors too.) Figure 6.15b[31] shows block diagrams for "hardware" and "software" (microprocessor) versions of automatic-referencing (consult the reference for details).

Fiber optic sensors have been proposed for measuring various physical phenomena, but many principles have been restricted to laboratory use, rather than robust commercial applications. An exception is a pressure-sensing technique[32] related to that of the FOTONIC displacement sensor of Fig. 4.44, but with many refinements[33] needed to achieve a practical pressure sensor of reasonable cost for difficult environments such as long-term natural-gas engine-cylinder pressure sensing. A near-infrared LED sends light down a single fiber to be reflected from the inner surface of a specially shaped diaphragm and into a single return fiber to a PIN Si photodiode and the remote electronic signal conditioner. The intensity of the reflected light varies with the diaphragm deflection (typically 0.0008 in max), using

---

[28]K. Souter and H. E. Krachman, "Measurement of Local Pressure Resulting from Hydrodynamic Impact," TP269, Endevco Corp., San Juan Capistrano, CA, 1978.

[29]"Pressure Transducer Handbook," National Semiconductor Corp., Santa Clara, CA, 1977.

[30]Ibid.; "Auto Zeroing Circuits," *Bull.* 107-057, Data Instruments Inc., Lexington, MA.

[31]"Pressure Transducer Handbook," op. cit.

[32]M. Wlodarczyk et al., "Long-Life Fiber Optic Pressure Sensors for Harsh Environment Applications," 1999, Optrand Inc., Ann Arbor, MI, 734-451-3480 (www.optrand.com).

[33]US patent #5,600,125, 1997. (Gives details on some of the techniques used.)

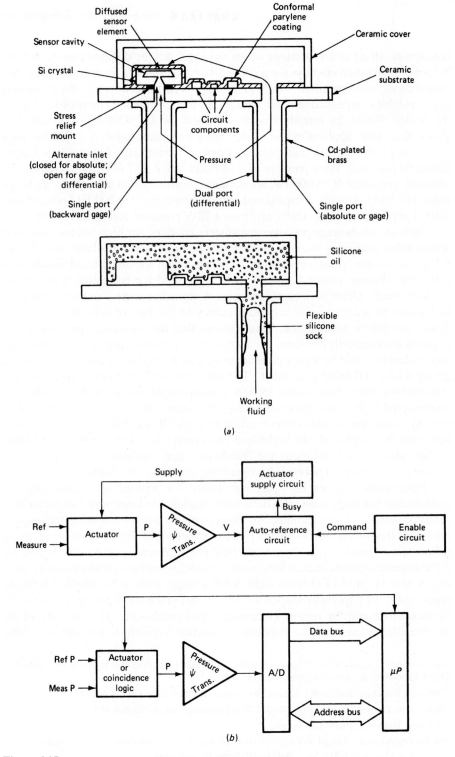

Figure 6.15
Diffused sensor transducers and autoreference techniques.

the "front-slope" of a curve similar to that of Fig. 4.44. Diaphragm diameters of 1.7 to 8 mm cover the pressure range from 100 to 30,000 psi, and a frequency response to 30 kHz is available in some models.

Most designs of pressure transducers are available in gage (psig), absolute (psia), differential (psid), and sealed (psis) reference versions. A sealed reference transducer is similar to the absolute pressure version, except that the sealed reference chamber contains a nonzero pressure rather than 0 lb/in² absolute. Since this reference pressure will change with temperature, applications must take this fact into account. Some transducer designs expose sensitive electrical elements to the measured fluid and thus may be limited to dry noncorrosive gases.

Differential-pressure transmitters are utilized widely in process instrumentation and control systems for flow, liquid level, density, viscosity, etc.[34] They present particularly difficult problems of transducer design since they often need to sense small differences (0 to 5 in H_2O) in large pressures, withstand full-line pressure (perhaps 2,000 lb/in²) overload, and interface with difficult process liquids. Mechanical, pneumatic, and electrical output devices are available.[35] Here we briefly describe some electrical output devices that provide the standard 4 to 20 mA current signal popular in process systems.[36]

Figure 6.16 shows a unit whose sensing diaphragm forms the moving plate of a differential capacitor; the motion is transduced to a proportional direct current by using an ac-excited bridge circuit, demodulation, and a feedback-type current regulator. Stainless-steel (316 SS) isolating diaphragms protect the sensor from process fluids while transmitting the measured pressures through silicone-fluid fills to the sensing diaphragm. Overpressure causes the sensing diaphragm to (safely) bottom out on the glass backup plates, causing less than 0.25 percent of span error when normal operation resumes after a 2,000 lb/in² overload.

Figure 6.17 shows a unit using diffused silicon "strain gage" technology. A single silicon crystal is cut to $\frac{1}{4}$-in² by $\frac{1}{10}$ in thick, and the center of one side is ground out and etched to form an integral diaphragm of thickness suitable for the desired pressure range (3 inH_2O to 10,000 lb/in²). Then boron is diffused into the other side of the silicon chip to form four complete Wheatstone-bridge circuits, each having four strain-sensitive elements properly located on the diaphragm. A computerized test system checks each of the four bridges and selects for further processing only the best one. (This approach is typical of integrated-circuit manufacturing technology in which the most cost/effective approach often involves high-rate production with a certain percentage of defectives, which are later weeded out by automated test equipment.) Isolation diaphragms and silicone-fluid fills are employed to protect the sensor from process fluids. Overload protection is provided by valving (actuated by an overload bellows) which equalizes pressure on both sides of the silicon diaphragm when the differential pressure approaches unsafe values.

---

[34]V. N. Lawford, "Differential-Pressure Instruments: The Universal Measurement Tools," *Instrum. Tech.,* pp. 30–40, December 1974; M. Slomiana, "Using Differential Pressure Sensors for Level, Density, Interface, and Viscosity Measurements," *In Tech,* pp. 63–68, September 1979.

[35]M. Slomiana, "Selecting Differential Pressure Instrumentation," *In Tech,* pp. 32–40, August 1979.

[36]R. F. Wolny, "Applying Electronic Δ*P* Transmitters," *Instrum. Tech.,* pp. 47–53, July 1978.

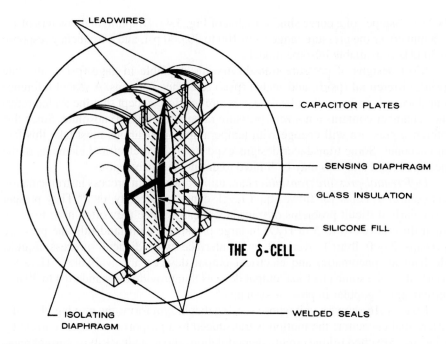

LEADWIRES

CAPACITOR PLATES

SENSING DIAPHRAGM

GLASS INSULATION

SILICONE FILL

THE δ-CELL

WELDED SEALS

ISOLATING DIAPHRAGM

Figure 6.16
Capacitive differential-pressure transmitter. *(Courtesy Rosemount Inc., Minneapolis, MN.)*

Multivariable "smart" transmitters[37] may include three sensors, A/D conversion, and microprocessor computing power in a single unit. For example,[38] a capacitance differential pressure sensor, a piezoresistive microsensor for pressure, and an RTD for measuring temperature are used to compute mass flow rate and to provide correction terms for the sensors, using a microprocessor once the analog sensor signals have been digitized. These methods allow accuracies of about 0.075%, which extend the rangeability to about 100-to-1, important in flowmeters where the square-root relation between flow rate and differential pressure makes flow rangeability equal to the square root of delta-p rangeability. Output includes an analog 4-20 mA signal and a HART (Highway Addressable Remote Transducer) communication channel. The HART channel can use the same two wires as the 4-20 mA signal since it uses frequency-shift keying (FSK) to transmit the information. In FSK, information is sent in binary form using bursts of two frequencies, say, 1200 Hz means a 0 and 2400 Hz means a 1. These frequencies are high enough relative to the analog signal that the two do not interfere. The HART system also allows commands to be sent to the transmitter, allowing for the programming of things such as zero and span setpoints, and electronic damping, without opening the transmitter case. Either a HART hand-held communicator device plugged in at the

[37]P. Cleaveland, "Pressure Transmitters; A Unit for Every Application," *I & CS,* Nov. 1999, pp. 41–46.

[38]Model 3051, Rosemount Inc., Chanhassen, MN, 800-903-3728 (www.rosemount.com).

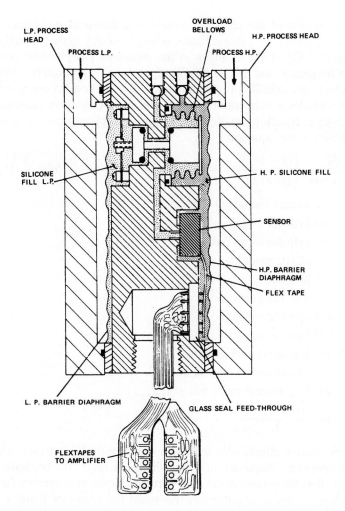

Figure 6.17
Diffused strain-gage differential-pressure transmitter.
(Courtesy Honeywell Inc., Ft. Washington, PA.)

transmitter or a remote supervisory computer can "listen or talk" to the transmitter. Transmitter-power requirements are low enough that it can be powered from solar cells, making remote installations such as natural gas fields convenient.

6.5 VIBRATING-CYLINDER AND OTHER RESONANT TRANSDUCERS

Gas-pressure transducers of very high accuracy and long-term stability, suitable even as calibration standards, have been produced by using the *vibrating-cylinder* principle. Here one of the vibration modes of a thin-walled cylinder is kept in continuous oscillation by using a limit-cycling feedback system exactly as in the densitometer of Fig. 7.47. Changes in the measured gas pressure cause related changes in frequency, and this "digital" output signal is accurately measured with standard electronic counting/timing methods. For highest accuracy the pressure transducer includes a temperature sensor and microprocessor calibration model using a 15-term equation developed through multiple-regression methods. The design is based on the equation[39]

$$\omega^2 = \frac{Eg}{(1 - \nu^2)\gamma r^2}\left[\frac{(1 - \nu^2)\lambda^4}{(n^2 + \lambda^2)^2} + \frac{t^2}{12r^2}(n^2 + \lambda^2)^2 + \frac{(1 - \nu^2)r}{Et}\left(n^2 + \frac{\lambda^2}{2}\right)\right]p \quad \textbf{(6.54)}$$

where
$\omega \triangleq$ natural frequency
$L \triangleq$ cylinder length
$r \triangleq$ cylinder mean radius
$E \triangleq$ elastic modulus
$p \triangleq$ pressure
$t \triangleq$ wall thickness
$\gamma \triangleq$ specific weight
$\nu \triangleq$ Poisson's ratio
$g \triangleq$ acceleration of gravity
$n \triangleq$ circumferential mode no. $= 2, 3, \ldots$
$m \triangleq$ longitudinal mode no. $= 1, 2, \ldots$
$\lambda \triangleq \dfrac{\pi r(m + 0.3)}{L}$

Note that pressure p effects only the rightmost term in this equation. One usually desires an operating frequency of a few thousand hertz (for accurate frequency counting in short times) and about a 20 percent change in frequency for full-scale pressure changes. Space requirements constrain the choice of L and r. Computer numerical experimentation with Eq. (6.54) quickly shows that t must be quite thin to get good sensitivity of ω to p. If one chooses NI-SPAN C as the material (for its temperature stability, corrosion resistance, strength), $L = 2.0$ in, $r = 0.40$ in, and one desires a frequency change from about 4000 to 5000 Hz for a 0 to 20 lb/in^2 absolute pressure range, then computer trial and error leads to $m = 1$, $n = 4$, and $t = 0.003$ in. With these values plus others for the magnetic forcing coil, motion transducer, and electronics, a simulation can be run to check out total system operation in a fashion very similar, again, to the densitometer example mentioned above. Commercial[40] transducers (Fig. 6.18c) have 0.01 percent (of reading) accuracy over 20 to 100 percent of their range and 12-month stability of 0.01 percent of full scale.

---

[39]J. S. Mixon and R. W. Herr, *NASA* TR R-145, 1962.

[40]SOL/AERO/84/P187, Solartron 3087 Series Digital Air Pressure Transducers, Farnborough, Hampshire, England, 1989.

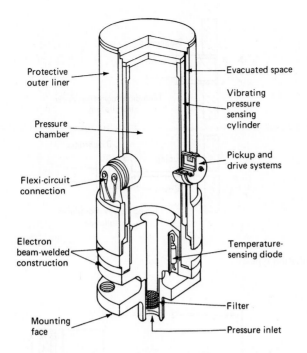

Figure 6.18

Related high-accuracy, frequency-modulated transducers use quartz[41] or silicon[42] elastic elements maintained in steady vibration by servosystems similar to that of the vibrating cylinder just described. Again, the natural frequency of the elastic element is changed when pressure applied to a diaphragm changes the load. Extensive temperature and pressure calibration establishes a matrix of correction coefficients, and an embedded temperature sensor allows microprocessor correction calculations.

6.6 DYNAMIC EFFECTS OF VOLUMES AND CONNECTING TUBING

We have mentioned the possible strong effect of fluid properties and "plumbing" configurations on the dynamic behavior of pressure-measuring systems. In this

[41]D. B. Juanarena et al., "A Monolithic Quartz Resonator Pressure Transducer for High Accuracy and Stability Pressure Measurement," 1997, Pressure Systems, Hampton, VA, 757-865-1243 (www.psih.com).

[42]T. Nishikawa et al., "Take Pressure Sensing to New Levels," *Chemical Engineering,* Dec. 1994; Johnson Yokogawa Corp., Newnan, GA, 770-253-7000 (www.yca.com); Druck Inc., New Fairfield, CT, 203-746-0400 (www.druckinc.com).

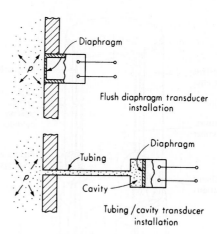

Figure 6.19
Transducer installation types.

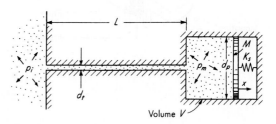

Figure 6.20
Transducer tubing model.

section some of these problems are investigated. First, it should be pointed out that if maximum dynamic performance is to be attained, then a flush-diaphragm transducer mounted directly at the point where a pressure measurement is wanted should be used if at all possible (see Fig. 6.19). Any connecting tubing or volume chambers will degrade performance to some extent. The fact that this degradation is studied in this section indicates that in many practical circumstances a flush-diaphragm transducer is not applicable.

Liquid Systems, Heavily Damped, Slow-Acting

In the system of Fig. 6.20, the spring-loaded piston represents the flexible element of the pressure pickup. For the present analysis, the only pertinent characteristic of the pressure pickup is its volume change per unit pressure change, the compliance C_{vp} [in^3/(lb/in^2)]. This can be calculated or measured experimentally. (The transducer of Fig. 6.14, for example, has $C_{vp} = 0.001$ mm^3/bar.) For systems that are heavily damped (a criterion for judging this is given shortly) or subjected to relatively slow pressure changes, the inertia effects of both the fluid and the moving parts of the pickup are negligible compared with viscous and spring forces. We show that under these conditions the measured pressure p_m follows the desired pressure p_i with a first-order lag. For steady laminar flow in the tube we have

$$p_i - p_m = \frac{32\mu L V_{t,\,av}}{d_t^2}$$

where $\mu \triangleq$ fluid viscosity and $V_{t,\,av} \triangleq$ average flow velocity in tube. While this equation is exact only for steady flow, it holds quite closely for slowly varying velocities. During a time interval dt, a quantity of liquid enters the chamber. This is given by

$$dV = \text{volume entering} = \frac{\pi d_t^2}{4} V_{t,\,av}\, dt = \frac{\pi d_t^4 (p_i - p_m)\, dt}{128 \mu L} \tag{6.54}$$

Any volume added or taken away results in a pressure change dp_m given by

$$dp_m = \frac{dV}{C_{vp}} = \frac{p_i - p_m}{\tau}\, dt \tag{6.55}$$

and thus

$$\tau \frac{dp_m}{dt} + p_m = p_i \tag{6.56}$$

where

$$\tau \triangleq \frac{128 \mu L C_{vp}}{\pi d_t^4} \tag{6.57}$$

Thus the response is seen to follow a standard first-order form. To keep τ small, the tubing length should be short, the diameter large, and C_{vp} small.

The viewpoint of this analysis is that any sudden changes in p_i will cause much more gradual changes in p_m, and thus the pickup spring-mass system will not be able to manifest its natural oscillatory tendencies. Under such conditions an overall first-order response from the tubing/transducer system may be expected. To obtain some numerical estimate of the conditions required for such behavior, we later carry out an analysis which includes inertia effects. This will lead to a second-order type of response, and when ζ of this model is greater than about 1.5, the simpler first-order model may be employed with fair accuracy, at least (in terms of frequency response) for frequencies less than ω_n.

The model used above predicts that, for a change in p_i, p_m starts to change immediately. This cannot be true since a pressure wave propagates through a fluid at finite speed (the velocity of sound) for small disturbances. There is thus a dead time τ_{dt} equal to the distance traversed divided by the speed of sound. For liquids and reasonable tube lengths usually this delay is small enough to ignore completely. The speed of sound v_s in a fluid contained in a nonrigid tube is given by

$$v_s = \sqrt{\frac{E_L}{\rho_L}} \sqrt{\frac{1}{1 + 2R_t \cdot E_L/(tE_t)}} \tag{6.58}$$

where $E_L \triangleq$ bulk modulus of fluid
$\rho_L \triangleq$ mass density of fluid
$R_t \triangleq$ tube inside radius
$t \triangleq$ tube-wall thickness
$E_t \triangleq$ tube-material modulus of elasticity

If this dead time is significant in a practical problem, it may be included in the transfer function as

$$\frac{p_m}{p_i}(D) = \frac{e^{-\tau_{dt} D}}{\tau D + 1} \tag{6.59}$$

with fair accuracy.

Finally, it should also be pointed out that the result [Eq. (6.56)] of this analysis also may be applied to systems using gases rather than liquids if the elastic pressure-sensing element is sufficiently soft that its volume change per unit pressure change is much larger than that due to gas compressibility.

Liquid Systems, Moderately Damped, Fast-Acting

When the motions of the liquid and the pickup elastic element are rapid, their inertia is no longer negligible. An analysis[43] of this situation using energy methods is available. In the system of Fig. 6.20, any change in pressure p_m must be accompanied by a volume change; this in turn requires an inflow or outflow of liquid through the tube. If the tube is of small diameter compared with the equivalent piston diameter of the pickup, the tube flow will be at a much higher velocity than the piston velocity, and the kinetic energy of the liquid in the tube may be a large (sometimes major) part of the total system kinetic energy. This increase in kinetic energy is equivalent to adding mass to the piston and ignoring the fluid inertia, and the analysis below calculates just how much mass should be added to give the same effect as the fluid inertia. This added mass lowers the system natural frequency and thereby degrades dynamic response.

To find the equivalent piston/spring configuration for a given transducer, the volume change per unit pressure change must be equal for both systems. This gives

Transducer volume change = equivalent piston volume change

$$pC_{vp} = \frac{\pi^2 d_p^4 p}{16 K_s} \tag{6.60}$$

$$\frac{d_p^4}{K_s} = \frac{16 C_{vp}}{\pi^2} \tag{6.61}$$

---

[43]G. White, "Liquid Filled Pressure Gage Systems," *Statham Instrum. Notes,* Statham Instruments Inc., Los Angeles, January–February 1949.

Thus d_p and K_s for the equivalent system can have any values that satisfy Eq. (6.61). Also, the natural frequencies of both systems with no fluid present must be equal. This gives

$$\omega_{n,\,t} = \sqrt{\frac{K_s}{M}} \qquad (6.62)$$

$$\frac{K_s}{M} = \omega_{n,\,t}^2 \qquad (6.63)$$

where $\omega_{n,\,t} \triangleq$ transducer natural frequency. Again, any values of K_s and M that satisfy Eq. (6.63) may be used. Thus, to define the equivalent system, only C_{vp} and $\omega_{n,\,t}$ need be known; they can be found from experiment if theoretical formulas are unavailable.

Since we have just shown how the equivalent system is related to the real system, now we can proceed with an analysis of the equivalent system. The volume change dV is related to the piston motion dx by

$$dV = \frac{\pi d_p^2 \, dx}{4} \qquad (6.64)$$

Thus
$$\frac{dV}{dt} = \frac{\pi d_p^2}{4} \frac{dx}{dt} \qquad (6.65)$$

and
$$\frac{\pi}{4} d_t^2 V_{t,\,av} = \frac{\pi d_p^2}{4} \frac{dx}{dt} \qquad (6.66)$$

where $V_{t,\,av} =$ average flow velocity in tube. We then get

$$V_{t,\,av} = \left(\frac{d_p}{d_t}\right)^2 \frac{dx}{dt} \qquad (6.67)$$

Next we assume laminar flow in the tube, with the parabolic velocity profile characteristic of steady flow:

$$V_t = V_{t,\,c}\left[1 - \left(\frac{r}{R}\right)^2\right] \qquad (6.68)$$

where $V_t \triangleq$ velocity at radius r; $V_{t,\,c} \triangleq$ centerline velocity and $R \triangleq$ tube inside radius. The kinetic energy (KE) of an annular element of fluid (density ρ) of thickness dr at radius r is

$$d(\text{KE}) = \frac{(2\pi r \, dr)L\rho V_t^2}{2} \qquad (6.69)$$

Substitution of Eq (6.68) and integration give the fluid kinetic energy as

$$\text{KE} = \frac{\pi\rho L V_{t,\,av}^2 \, d_t^2}{6} \qquad (6.70)$$

For a square velocity profile the kinetic energy would be

$$\text{KE}_s = \frac{\pi \rho L V_{t,\text{av}}^2 \, d_t^2}{8} \tag{6.71}$$

The actual velocity profile will be somewhere between parabolic and square. Even if laminar flow exists, the velocity profile is nonparabolic except for steady flow.[44] Turbulent flow gives a rather square profile. We assume Eq. (6.70) to hold here; however, Eq. (6.71) can be carried through with equal ease to "bracket" the correct value. The rigid mass M_e, attached to M, which would have the same kinetic energy as the fluid, is given by

$$\frac{M_e}{2}\left(\frac{dx}{dt}\right)^2 = \frac{\pi \rho L d_p^4}{6 d_t^2}\left(\frac{dx}{dt}\right)^2 \tag{6.72}$$

$$M_e = \frac{\pi \rho L d_p^4}{3 d_t^2} \tag{6.73}$$

The natural frequency of the transducer/tubing system is then

$$\omega_n = \sqrt{\frac{K_s}{M + M_e}} = \sqrt{\frac{1}{M/K_s + M_e/K_s}} = \sqrt{\frac{1}{1/\omega_{n,t}^2 + 16\rho L C_{vp}/(3\pi d_t^2)}} \tag{6.74}$$

To keep ω_n as high as possible, L and C_{vp} must be as small as possible and d_t as large as possible. In many practical cases, $M_e \gg M$, which allows simplification of Eq. (6.74) to

$$\omega_n = \sqrt{\frac{3\pi d_t^2}{16\rho L C_{vp}}} \tag{6.75}$$

Next we calculate the damping ratio of the transducer/tubing system. The transducer itself is assumed to have negligible damping; thus the only damping is due to the fluid friction in the tube. Again we assume the validity of the steady laminar-flow relations to calculate the pressure drop due to fluid viscosity as $32\mu L V_{t,\text{av}}/d_t^2$. The force on the piston due to this pressure drop is the damping force $B(dx/dt)$; thus

$$\frac{\pi d_p^2}{4}\frac{32\mu L V_{t,\text{av}}}{d_t^2} = B\frac{dx}{dt} \tag{6.76}$$

and, since $V_{t,\text{av}} = (dx/dt)(d_p/d_t)^2$,

$$B = 8\pi\mu L\left(\frac{d_p}{d_t}\right)^4 \tag{6.77}$$

---

[44]C. K. Stedman, "Alternating Flow of Fluid in Tubes," *Statham Instrum. Notes,* Statham Instruments Inc., January 1956.

Then, using the general formula for the damping ratio of a spring-mass-dashpot system, we get

$$\zeta = \frac{B}{2\sqrt{K_s(M + M_e)}} = \frac{64\mu L C_{vp}}{\pi d_t^4 \sqrt{1/\omega_{n,t}^2 + 16\rho L C_{vp}/(3\pi d_t^2)}} \quad (6.78)$$

If $M_e \gg M$, this simplifies to

$$\zeta = \frac{16\sqrt{3/\pi}\,\mu\,\sqrt{LC_{vp}/\rho}}{d_t^3} \quad (6.79)$$

The above theory has been partially checked experimentally in an M.Sc. thesis by Fowler.[45] (A later, more detailed, investigation by NASA[46] confirmed the essential features of the theory and made useful extensions for those cases where the compliance of the liquid itself and/or that of entrained gas bubbles is not negligible compared with transducer compliance.) The transducer used was the diaphragm strain-gage instrument whose numerical parameters are presented in Eq. (6.53). The liquid used was water, and ω_n and ζ were found from step-function tests using a bursting cellophane diaphragm to obtain a sudden 10 lb/in^2 release of pressure. Tubing of 0.042- to 1.022-in inside diameter and 4 to 32 inch in length was studied. Conditions were such that turbulent flow probably existed much of the time; thus use of Eq. (6.71) was indicated. The value of C_{vp} was calculated from theory as 0.00441 in^3/(lb/in^2) while an experimental calibration gave 0.00500. For all cases in which oscillation occurred (and thus ω_n could be accurately measured), it was found that ω_n was accurately predicted within about 5 to 10 percent. To show the severity of the performance degradation, the natural frequency (which was 897 Hz with no tubing and in air) became about 60 Hz with a 1.022-in diameter tube 10 inch in length and about 3 Hz with a 0.092-in tube 32 in long. However, the prediction of ζ was much less satisfactory, being invariably too low. (Fortunately, ζ is of little importance in those many practical applications which exhibit light damping.) The small-diameter long tubes (which encourage laminar flow) gave the best correlation with theory, but even there errors of 100 percent occurred. For example, the 0.042-in tube 10 in long had a ζ of about 1.0 while theory predicted 0.66. The poor agreement undoubtedly can be charged to turbulent flow and energy losses from expansion and contraction (end effects) at the tubing ends.

Fowler developed corrections for these effects that yielded much better agreement. However, the turbulent flow and end effects are nonlinear damping mechanisms; thus the meaning of ζ is confused for both theoretical predictions and experimental measurements. The use of smaller step functions should give lower velocities and thus more nearly laminar flow. This effect was checked for a

[45]R. L. Fowler, "An Experimental Study of the Effects of Liquid Inertia and Viscosity on the Dynamic Response of Pressure Transducer-Tubing Systems," M.Sc. Thesis, The Ohio State University, Mechanical Engineering Department, 1963.

[46]R. C. Anderson and D. R. Englund, Jr., "Liquid-Filled Transient Pressure Measuring Systems: A Method for Determining Frequency Response," *NASA* TN D-6603, December 1971.

0.092-in-diameter, 10-in-long tube. A 10 lb/in² step function gave $\zeta = 0.34$; a 5 lb/in² step function gave 0.22. The theoretical value was 0.065, suggesting that the theory might be quite accurate for sufficiently small inputs. Detailed discussion of these problems may be found in the reference.

When ζ calculated from Eq. (6.78) or (6.79) is 1.5 or greater, the simpler first-order model of Eq. (6.56) may be adequate. To show the relationships of these two analyses, recall that in the second-order form $K/[D^2/\omega_n^2 + 2\zeta D/\omega_n + 1]$, the inertia effect resides in the D^2 term. Neglecting this gives $K/(2\zeta D/\omega_n + 1)$. If we calculate $2\zeta/\omega_n$ from Eqs. (6.74) and (6.78), it is numerically equal to τ of Eq. (6.57).

Gas Systems with Tube Volume a Small Fraction of Chamber Volume

When, in the configuration of Fig. 6.20, a gas is the fluid medium, the compressibility of the gas in volume V becomes the major spring effect when the pressure pickup is at all stiff. It is reasonable, then, to assume that the volume V is enclosed by rigid walls ($C_{vp} = 0$). The majority of practical problems involve rather low frequencies. This allows treatment as a lumped-parameter system with fluid properties considered constant along the length of the tube for small disturbances. The validity of this viewpoint is shown by noting that pressure-wave propagation in the gas follows the general law of wave motion:

$$\lambda = \text{wavelength} = \frac{\text{velocity of propagation}}{\text{frequency}} = \frac{c}{f} \tag{6.80}$$

Our lumped-parameter assumption says that the gas in the tube moves as a unit, as opposed to wave motion *within* the tube. For any given tube length L, for sufficiently low frequencies, the lumped-parameter approach becomes valid. For example, the velocity of pressure waves in standard air is about 1,120 ft/s. If an oscillation of 10-Hz frequency exists, its wavelength must be 112 ft/cycle. This means that the spacewise variation of fluid pressure due to wave motion has a wavelength of 112 ft (see Fig. 6.21). For a tube of, say, 1-ft length, the variation in pressure (and thus in density, etc.) from one end of the tube to the other resulting from

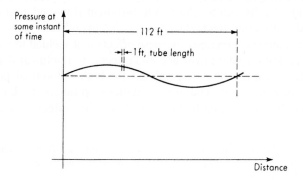

Figure 6.21
Justification of lumped-parameter analysis.

wave motion is very small. That is, there is negligible *relative* motion of particles in the tube; they all move together, as a unit. While the above aspect of wave motion is neglected, the dead time due to the finite speed of propagation can be relatively easily taken into account (approximately) by multiplying the second-order transfer function (which we obtain shortly) by $e^{-\tau_{dt}D}$, where

$$\tau_{dt} = \frac{L}{v_s} \tag{6.81}$$

$$\text{Sound velocity} = v_s = \sqrt{\frac{kp}{\rho}} \tag{6.82}$$

where $k \triangleq$ ratio of specific heats, $\rho \triangleq$ mass density, and $p \triangleq$ average pressure.

The analysis[47] we carry out below is valid only for small pressure changes; the system becomes nonlinear for large disturbances. Steady-laminar-flow formulas are used for calculating fluid friction. However, the effective mass is taken equal to the actual mass (rather than the four-thirds actual mass given by the parabolic velocity profile). We follow the reference in this respect; use of the $\frac{4}{3}$ factor is easily incorporated if we wish to bracket a more correct value. Numerically the effect is rather small in any case.

The analysis consists merely of applying Newton's law to the "slug" of gas in the tube. We assume that initially $p_i = p_m = p_0$ when p_i changes slightly in some way. From here on, the symbols p_i and p_m are taken to mean the *excess* pressures over and above p_0. The force due to the pressure p_i is $\pi p_i d_t^2/4$. The viscous force due to the wall shearing stress is $8\pi\mu L\dot{x}_t$, where x_t is the displacement of the slug of gas in the tube. If the slug of gas moves into the volume V an amount x_t, the pressure p_m will increase. We assume this compression occurs under adiabatic conditions. The adiabatic bulk modulus E_a of a gas is given by

$$E_a \triangleq -\frac{dp}{dV/V} = kp \tag{6.83}$$

The displacement x_t causes a volume change $dV = \pi d_t^2 x_t/4$. This, in turn, causes a pressure excess $p_m = \pi E_a d_t^2 x_t/(4V)$. The force due to this pressure excess is $\pi^2 E_a d_t^4 x_t/(16V)$. Newton's law then gives

$$\frac{\pi p_i d_t^2}{4} - 8\pi\mu L\dot{x}_t - \frac{\pi^2 E_a d_t^4 x_t}{16V} = \frac{\pi d_t^2 L\rho}{4}\ddot{x}_t \tag{6.84}$$

and since $p_m = \pi E_a d_t^2 x_t/(4V)$,

$$\frac{4L\rho V}{\pi E_a d_t^2}\ddot{p}_m + \frac{128\mu LV}{\pi E_a d_t^4}\dot{p}_m + p_m = p_i \tag{6.85}$$

---

[47]G. J. Delio, G. V. Schwent, and R. S. Cesaro, "Transient Behavior of Lumped-Constant Systems for Sensing Gas Pressures," *NACA, Tech. Note* 1988, 1949.

This is clearly the standard second-order form, and so we define

$$\omega_n = \frac{d_t}{2}\sqrt{\frac{\pi E_a}{LV\rho}} = \frac{d_t c}{2}\sqrt{\frac{\pi}{LV}} \qquad (6.86)$$

$$c \triangleq \text{speed of sound} = \sqrt{kgRT} = \sqrt{1.4 \cdot 32.2 \cdot 53.4 \cdot 530} = 1128 \text{ ft/s}$$

$$\zeta = \frac{32\mu}{d_t^3}\sqrt{\frac{LV}{\pi\rho E_a}} = \frac{32\mu}{cd_t^3}\sqrt{\frac{LV}{\pi\rho}} \qquad (6.87)$$

Because k and R are essentially fixed for a given gas (except at exotic temperatures and pressures) and g is a constant, the natural frequency depends mainly on tube length L and diameter d_t, chamber volume V, and temperature T. In Eq. (6.86), I have calculated c for air at 70°F as an example, to show at least one set of numerical values using one choice of unit systems. We also note that if we had, say, a step pressure change from 50 psia and 70°F to 100 psia, assuming reversible adiabatic compression, the new temperature would be 646°R and the c value would go from 1128 to 1246 ft/s. That is, the natural frequency is not really a constant, meaning the system is nonlinear. However, the departure from linearity is small enough that the linear model has little error even for the rather large pressure change used in the example. The damping ratio is also not constant since c, μ, and ρ change during a pressure change, but again the effect is slight for modest pressure changes. The accuracy can be improved slightly by using *average* values of p and T in the calculations for c, μ, and ρ.

When L becomes very short, Eq. (6.86) predicts a very large ω_n. In practice, this will not occur since, even when $L = 0$, there is some air (close to the opening in the volume) which has appreciable velocity and therefore kinetic energy. Theory shows that this end effect may be taken into account by using for L in Eq. (6.86) an effective length L_e given by

$$L_e = L\left(1 + \frac{8}{3\pi}\frac{d_t}{L}\right) \qquad (6.88)$$

In most cases, the term $[8/(3\pi)](d_t/L)$ is completely negligible compared with 1.0. However, if $L = 0$ (tube degenerates into simply a hole in the side of volume V), we can still compute an ω_n since then $L_e = [8/(3\pi)]d_t$. The computation of damping for this case is not straightforward and is not discussed here.

Gas Systems with Tube Volume Comparable to Chamber Volume

When the volume of the tube becomes a significant part of the total volume of a system, compressibility effects are no longer restricted to the volume chamber alone and the above formulas become inaccurate. More refined analyses[48] give the following formulas for ζ and ω_n:

$$\omega_n = \frac{c}{L\sqrt{\dfrac{V}{V_t} + \dfrac{1}{2}}} \tag{6.89}$$

$$\zeta = \frac{16\mu L}{d_t^2 \sqrt{kp\rho}} \sqrt{\frac{1}{2} + \frac{V}{V_t}} \tag{6.90}$$

where $V_t \triangleq$ tubing volume. In these formulas, if $V/V_t \gg \frac{1}{2}$ (tubing volume negligible compared with chamber volume), the term $\frac{1}{2}$ may be neglected and the formulas become identical to (6.86) and (6.87).

When absolute gas pressures get very low, as in vacuum processing and high-altitude flight, sensor dynamic models must take into account so-called *slip flow,* where the fluid no longer sticks to the tube wall. A study[49] motivated by the needs of proposed hypersonic flight vehicles provides a model for such conditions.

While in general a flush-diaphragm installation is preferred, in some rocket-engine testing a short length of tubing purged with a steady helium flow has been found desirable to reduce fouling and temperature damage. A piezoelectric transducer[50] (resonant frequency 250,000 Hz) of this type using a tube of about 0.15-in diameter and 1-in length has flat response $\pm$ 10 percent to about 10,000 Hz. Helium is used since it has a value of k/ρ about nine times that of air, which gives faster response.

The Infinite-Line Pressure Probe[51]

The configuration of Fig. 6.22, called an *infinite-line probe,* in some cases can be used to achieve a gas pressure-measuring system of very high frequency response. The basic idea behind this design is that, for a sufficiently long ("infinite") line, pressure fluctuations at the measuring station will have attenuated to small enough values at the far end that their reflections back to the measuring station will be very small, giving negligible measurement errors. The far-end "termination" of the line may be an open end, a closed end, or a terminating volume, depending on the needs of the particular experiment. The reference gives calculation methods, typical theoretical results, and experience with a specific probe using a silicon diaphragm pressure transducer of 1.2-mm diameter and 500-kHz natural frequency which gave useful measurements to about 100 kHz.

---

[48]S. A. Whitmore and C. T. Leondes, "Pneumatic Distortion Compensation for Aircraft Surface Pressure Sensing Devices," *J. Aircraft,* vol. 28, no. 12, Dec. 1991, pp. 828–836.

[49]S. A. Whitmore et al., "A Dynamic Response Model for Pressure Sensors in Continuum and High Knudsen Number Flows with Large Temperature Gradients," *NASA* TM 4728, 1996.

[50]Helium-Bleed Rocket Probe, Kistler Instruments, Amherst, NY.

[51]D. R. Englund and W. B. Richards, "The Infinite Line Pressure Probe," *ISA Trans.,* vol. 24, no. 2, 1985.

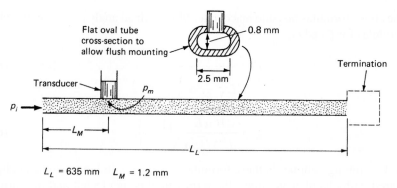

Flat oval tube
cross-section to
allow flush mounting

0.8 mm

2.5 mm

Termination

Transducer

p_m

p_i

L_M

L_L

L_L = 635 mm L_M = 1.2 mm

Figure 6.22
Infinite-line pressure probe.

Conclusion

The results of this section are to be thought of as practical working relations. The general problem treated here is quite complex and has been the subject of many intricate analyses and experimental studies, some of which we have referenced in the footnotes. Most of the difficulties encountered are in the area of very high frequencies, where the lumped-parameter models used in this section are inadequate and give faulty predictions.[52]

6.7 DYNAMIC TESTING OF PRESSURE-MEASURING SYSTEMS

To determine the regions of accuracy of theoretical predictions or to find accurate numerical values of system dynamic characteristics for critical applications, recourse must be made to experimental testing. This commonly takes the form of impulse, step, or frequency-response tests, with step-function tests being perhaps the most common. A comprehensive review[53] of this subject is available. Here we can mention only a few high points.

For step-function tests of systems in which natural frequencies are not greater than about 1,000 Hz, the bursting of a thin diaphragm subjected to gas pressure is often satisfactory. A general rule for step testing is that the rise time of the "step" function must be less than about one-fourth of the natural period of the system tested if it is to excite the natural oscillations. Thus a 1,000-Hz system requires a step with a rise time of 0.25 ms or less. Figure 6.23a shows schematically the principle of such devices. The pressures p_1 and p_2 are each individually adjustable. The volume containing p_2 is much smaller than that containing p_1; thus when the thin

[52]T. W. Nyland et al., "On the Dynamics of Short Pressure Probes," *NASA* TN D-6151, February 1971.

[53]J. L. Schweppe et al., "Methods for the Dynamic Calibration of Pressure Transducers," *Natl. Bur. Std. (U.S.), Monograph* 67, 1963.

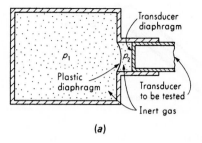

(a)

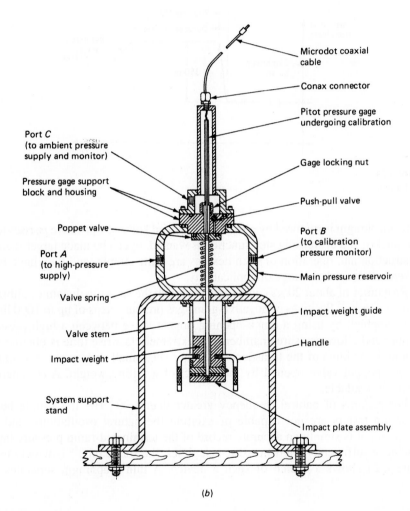

(b)

Figure 6.23
Step-test apparatus.

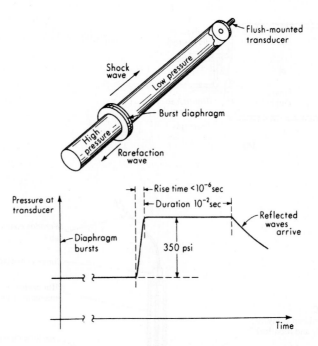

Figure 6.24
Shock tube.

plastic diaphragm is ruptured by a solenoid-actuated knife, the pressure p_2 rises to p_1 very quickly. If a decreasing step function is wanted, p_2 can be made larger than p_1. Construction and operation of such devices are quite simple, and they have been utilized widely in their range of applicability.

Rise times of about 20 μs can be achieved with the Aronson dynamic calibrator of Fig. 6.23b.[54, 55] Positive- or negative-going gas pressure steps of up to 1000 lb/in^2 can be applied, by using a quick-opening poppet valve between a high-pressure chamber and a low-pressure chamber. The extremely fast rise time is obtained by keeping the volume of the transducer chamber very small and by using a simple, flat-face poppet valve opened by the impact of a falling weight. A commercial model[56] is available.

For pickups of natural frequency greater than 1,000 Hz, the simple burst-diaphragm testers are not capable of exciting the natural oscillations, and the pickup output is simply an accurate record of the terminated-ramp pressure input. To achieve sufficiently short pressure-rise times, the shock tube[56] is widely used. Figure 6.24 shows a sketch of such a device. A thin diaphragm separates the

[54]P. M. Aronson and R. H. Waser, Naval Ordnance Laboratory, White Oak, MD, NOLTR 63-143, 1963.

[55]U.S. patent 3273376.

[56]Model 907A, PCB Piezotronics, Depew, NY, shock tube Model 901A10, www.pcb.com.

high-pressure and low-pressure chambers, and the transducer to be tested is mounted flush with the end of the low-pressure chamber. When the diaphragm is caused to burst, a shock wave travels toward the low-pressure end at a speed that may greatly exceed the speed of sound (5,000 ft/s is not unusual). From one side of this shock front to the other, there is a pressure change of the order of 2:1 over a distance which may be of the order of 10^{-4} in. (At the same time, a rarefaction wave travels from the diaphragm toward the high-pressure end.) When the shock front reaches the end of the tube where the transducer is mounted, it is reflected as a shock wave with more than twice the pressure difference of the original shock wave. The transducer is thus exposed to a very sharp ($\sim 10^{-8}$ s) pressure rise which is maintained constant for a short interval before various reflected waves arrive to confuse the picture. The length of this interval may be controlled to a certain extent by proper proportioning and operation of the shock tube. Some numerical characteristics of a typical shock tube[57] are as follows: high-pressure chamber 7 ft long, low-pressure chamber 15 ft long, tubing inside dimensions 1.4 in^2 (wall thickness $\frac{1}{4}$ in), maximum high pressure 600 lb/in^2, operating fluid air, maximum pressure step 350 lb/in^2, burst diaphragm 0.001- to 0.005-in-thick Mylar plastic, and duration of constant pressure 0.01 s. For a pressure pickup of, say, 100,000-Hz natural frequency, a pressure-rise time of less than 0.25×10^{-5} s is required. This is readily met by the tube described above. The 0.01-s step duration would give time for about 1,000 cycles of oscillation of a 100,000-Hz pickup—more than adequate to determine the dynamic characteristics. Very fast transducers often have a response more complex than simple second-order. Then the frequency response may be a useful characterization. This may be derived from the measured step response using the method explained in the text near Fig. 3.116.

A simple impulse-type test method applicable to flat, flush-diaphragm transducers utilizes a small steel ball dropped onto the diaphragm. The impact excites the natural oscillations, which are recorded and analyzed for natural frequency and damping ratio. Although this input is a concentrated force rather than a uniform pressure, the results have been found[58] to correlate quite well with shock-tube tests. Another impulse method employs the shock wave created by discharging 25,000 V across a spark gap. The spark gap and transducer are located about 3 in apart in open air. A pressure impulse of 0.2-μs rise time and 100 lb/in^2 peak can be obtained.

Figure 6.25a shows one method of constructing a frequency-response tester using liquid as the pressure medium. The vibration shaker applies a sinusoidal force of adjustable frequency and amplitude to the piston/diaphragm to create sinusoidal pressure in the liquid-filled chamber. Such vibration shakers are readily available in industry and cover a wide range of force and frequency. The average pressure about which oscillations take place may be adjusted by regulating the bias pressure on the air side of the cylinder. Since usually it is not possible to predict accurately and

---

[57]R. Bowersox, "Calibration of High-Frequency-Response Pressure Transducers," *ISA J.,* p. 98, November 1958.

[58]W. C. Bentley and J. J. Walter, "Transient Pressure Measuring Methods Research," Princeton Univ. Aeron. Eng. Dept., *Rept.* 595g, p. 103, 1963.

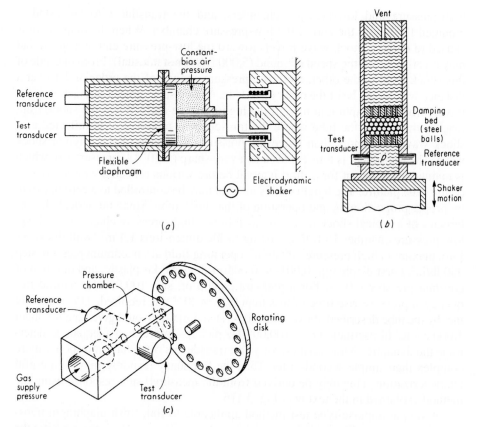

Figure 6.25
Sinusoidal test apparatus for liquid and gas.

reproducibly the pressure actually produced by such an arrangement as frequency and/or amplitude is varied, it is customary to mount a reference transducer at a location where it will experience the same pressure as the transducer under test. The reference transducer must have a known flat frequency response beyond any frequencies to be tested. This can be determined by some independent method, such as a shock tube. In testing another transducer, one merely calculates the amplitude ratio and phase shift between the reference and test transducer to determine the test-transducer frequency response. A different version[59] of the shaker method uses an open, vertical fluid column (Fig. 6.25b) to generate the test pressure.

An interesting special-purpose version of this technique, which employs microprocessor technology, is the blood pressure systems analyzer of Fig. 6.26. At the

[59]C. F. Vezzetti et al. "A New Dynamic Pressure Source for the Calibration of Pressure Transducers," *NBS Tech. Note* 914, June 1976; K. H. Cate and S. D. Young, "A Dynamic Pressure Calibration Standard," *NASA* Langeley LAR-13443, 1986.

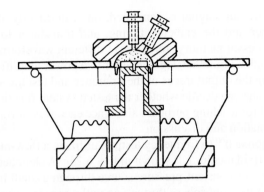

DOME VIEW: **The pressure generator consists of a magnet and voice coil which applies pressure to the fluid through a piston and rubber seal. Current in the coil is accurately converted to pressure in the dome.**

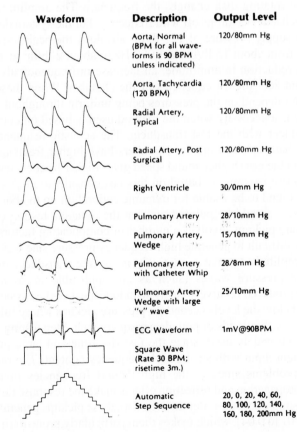

Waveform	Description	Output Level
	Aorta, Normal (BPM for all wave-forms is 90 BPM unless indicated)	120/80mm Hg
	Aorta, Tachycardia (120 BPM)	120/80mm Hg
	Radial Artery, Typical	120/80mm Hg
	Radial Artery, Post Surgical	120/80mm Hg
	Right Ventricle	30/0mm Hg
	Pulmonary Artery	28/10mm Hg
	Pulmonary Artery, Wedge	15/10mm Hg
	Pulmonary Artery with Catheter Whip	28/8mm Hg
	Pulmonary Artery Wedge with large ''v'' wave	25/10mm Hg
	ECG Waveform	1mV@90BPM
	Square Wave (Rate 30 BPM; risetime 3m.)	
	Automatic Step Sequence	20, 0, 20, 40, 60, 80, 100, 120, 140, 160, 180, 200mm Hg

Figure 6.26
Microprocessor-based pressure calibrator.
(Courtesy Bio-Tek Instruments, Inc., Shelburne, VT, www.dninevada.com.)

dome where the pressure signal is generated, one can directly attach a reference pressure transducer and the catheter, tubing, and transducer to be dynamically tested. A microprocessor memory stores the 11 different waveforms and outputs the one selected through a digital/analog converter to a power amplifier and the shaker coil. By comparing the output traces of the reference and test transducers on a two-channel recorder, one can decide whether the tested system functions properly. The apparatus is useful for teaching purposes also since each waveform can be produced quickly for examination and discussion.

A different approach[60] to frequency testing, utilizing a flow-modulating principle and gas as the fluid medium, is shown in Fig. 6.25c. A chamber is supplied with compressed gas from a constant-pressure source through a small inlet passage. The gas is exhausted to the atmosphere through an outlet passage whose area is modulated approximately sinusoidally with time. This is accomplished by rotating a disk containing holes in front of the exhaust port so that outflow periodically is cut on and off. This produces a periodic (nearly sinusoidal) variation in chamber pressure, which is measured by both a reference transducer and the test transducer. Varying the speed of the rotating disk changes the frequency. The amplitude of pressure oscillation of such a device drops off with frequency. For the system described, with helium gas at a supply pressure of 121 lb/in^2 absolute, the peak-to-peak pressure amplitude goes from about 15 lb/in^2 at 1,000 Hz to about 2 lb/in^2 at 11,000 Hz. In addition to this reduction in amplitude, an increase in frequency also brings into play the resonant acoustical frequencies of the chamber. When these resonances occur, one cannot depend on the pressures being uniform throughout the chamber. This uniformity is a necessity when the method used is based on comparison of a reference transducer with the test transducer. The acoustic resonant frequencies depend on the chamber size (smaller chambers have higher frequencies) and the speed of sound in the gas (higher sound speed gives higher frequencies). The use of helium in the above example is based on this last consideration. A system of the above type was found to be usable for dynamic calibration up to about 10,000 Hz.

In summary, it should be pointed out that the dynamic testing of very-high-frequency pressure pickups involves a number of complicating factors. It has been found extremely difficult to generate high-frequency pressure sine waves that are of large enough amplitude to give a relatively noise-free transducer output signal. Small-amplitude pressure waves (such as are applicable to sound-measuring systems) can be produced relatively easily with loudspeaker-type systems, but their amplitude is far below the levels needed for pressure pickups whose full-scale range is tens, hundreds, or thousands of pounds per square inch. Step testing with a shock tube thus has been widely used, since the fast rise time and large pressure steps result in a transient input with strong high-frequency content. The pickups themselves present problems since at the high natural frequencies involved, many complex wave-propagation and reflection effects make the response deviate considerably from the simple second-order model. Also, these pickups generally have little damping ($\zeta \approx 0.01$ to 0.04), which makes them particularly prone to ringing at their natural frequency if any sharp transients occur.

---

[60]W. C. Bentley and J. J. Walter, op. cit., p. 63.

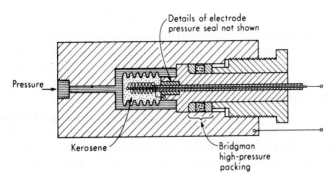

Figure 6.27
Very-high-pressure transducer.

6.8 HIGH-PRESSURE MEASUREMENT[61]

Pressures up to about 100,000 lb/in² can be measured fairly easily with strain-gage pressure cells or Bourdon tubes. Bourdon tubes for such high pressures have nearly circular cross sections and thus give little output motion per turn. To get a measurable output, the helical form with many turns generally is used. Inaccuracy of the order of 1 percent of full scale may be expected with a temperature error of an additional 2 percent/100 F°. Strain-gage pressure cells can be temperature-compensated to give 0.25 percent error over a large temperature range.

For fluid pressures above 100,000 lb/in², electrical gages based on the resistance change of Manganin or gold-chrome wire with hydrostatic pressure are generally utilized.[62] Figure 6.27 shows a typical gage. The sensitive wire is wound in a loose coil, one end of which is grounded to the cell body and the other end brought out through a suitable insulator. The coil is enclosed in a flexible, kerosene-filled bellows, which transmits the measured pressure to the coil. The resistance change, which is linear with pressure, is sensed by conventional Wheatstone-bridge methods. Pertinent characteristics of the common wire materials are as follows:

	Pressure sensitivity, $(\Omega/\Omega)/(lb/in^2)$	Temperature sensitivity, $(\Omega/\Omega)/F°$	Resistivity, $\Omega \cdot cm$
Manganin	1.69×10^{-7}	1.7×10^{-6}	45×10^{-6}
Gold chrome	0.673×10^{-7}	0.8×10^{-6}	2.4×10^{-6}

Although its pressure sensitivity is lower, gold chrome is preferred in many cases because of its much smaller temperature error. This is particularly significant since the kerosene used in the bellows will experience a transient temperature change when sudden pressure changes occur, because of adiabatic compression or

[61]R. K. Kaminski, "Measuring High Pressure above 20,000 PSIG," *Instrum. Tech.*, pp. 59–62, August 1968; W. H. Howe, "The Present Status of High Pressure Measurement and Control," *ISA J.*, p. 77, March 1955; I. L. Spain and J. Paauwe (eds.), "High Pressure Technology," vol. 1. chap. 8, Marcel Dekker, New York, 1977.

[62]W. H. Howe, op. cit.

expansion. The response of the wire resistance to pressure changes is practically instantaneous; however, the accompanying temperature change will cause a transient error if temperature sensitivity is too high. Gages of the above type are commercially available with full scale up to 200,000 lb/in^2 and inaccuracy of 0.1 to 0.5 percent. They also have been utilized successfully for much higher pressures on a special-application basis.

The measurement of local contact pressures between rolling elements in gears, cams, and bearings may be accomplished by depositing a thin strip of Manganin or gold chrome onto the surface as a pressure transducer. Studies[63] of such a technique, using a Manganin element 0.002 in wide and 3×10^{-6} in thick, have been reported.

6.9 LOW-PRESSURE (VACUUM) MEASUREMENT

While the Pascal is the standard SI unit for *all* pressure measurements, the vacuum field continues to use well-established nonstandard units such as the *torr* and the *millibar.* A torr is the pressure equivalent of 1 mmHg at standard conditions; conversions use the relation

$$1 \text{ mbar} = 0.75006 \text{ torr} = 0.10000 \text{ kPa} = 0.014504 \text{ psi}$$

A wide variety of vacuum gages are available.[64] For the higher pressure ranges of vacuum, gages can still use the familiar concept of force per unit area as their operating principle; these are called *absolute* gages, and their readings do *not* depend on the gas being measured. For lower pressures, other principles must be used, and all these *are* sensitive to the specific gas, that is, two different gases at the *same* pressure will give *different* readings, so readings must be corrected for each gas. Figure 6.28 (based on data from a reference[65]) shows some of the different gage types. Figure 6.29[66] shows where various types of vacuum gages might be used in a typical vacuum process.

Diaphragm Gages

While we have already discussed transducers based on the deflection of a diaphragm, those used in the vacuum regime are sufficiently unique and important to warrant a separate discussion. We concentrate on transducers with electrical output. Here, the transduction technique is almost exclusively that of capacitance change.[67] Recent high-accuracy models exceed the performance shown in Fig. 6.28,

[63]J. W. Kannel and T. A. Dow, "The Evolution of Surface Pressure and Temperature Measurement Techniques for Use in the Study of Lubrication in Metal Rolling," *ASME Paper* 74-Lub5-7, 1974.

[64]J. H. Leck, "Total and Partial Pressure Measurement in Vacuum Systems," Blackie, Glasgow, London, 1989.

[65]T. A. Delchar, "Vacuum Physics and Techniques," Chapman & Hall, London, 1993, chap. 5.

[66]S. P. Hansen, "Vacuum Pressure Measurement," *Vacuum and Thinfilm,* May 1999, pp. 24–29.

[67]J. J. Sullivan, "Development of Variable Capacitance Pressure Transducers for Vacuum Applications," *J. Vac. Sci. Technol.* vol. A 3 (3), May/June 1985, pp. 1721–1730; D. Jacobs, "Advances in Capacitance Manometers for Pressure Measurement," *Vacuum and Thinfilm,* Feb. 1999, pp. 30–35; J. J. Sullivan, "Advances in Vacuum Measurement Almost Meet Past Projections," *R&D Magazine,* Aug. 1995, pp. 31–34.

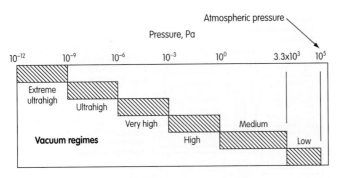

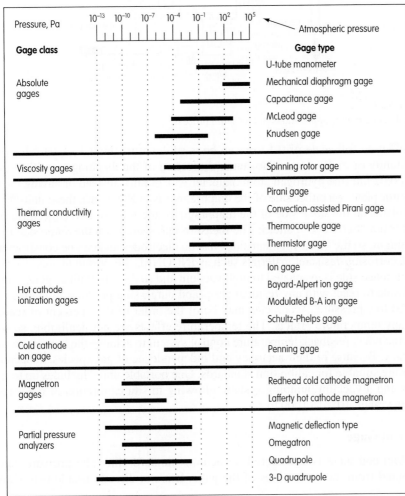

Figure 6.28
Vacuum gage types and ranges.

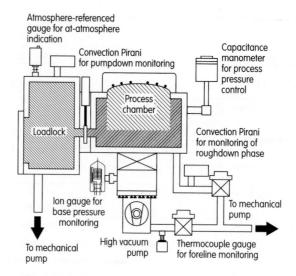

Figure 6.29

Typical vacuum gage applications in a vacuum process.

a unit[68] with full-scale of 0.1 torr can be used down to about 1×10^{-6} torr with uncertainty of $\pm 10^{-7}$ torr. The center deflection of the diaphragm at 10^{-6} torr is about 0.02 nm (the hydrogen atom diameter is 0.25 nm)! Instead of having the two capacitor plates on either side of the diaphragm (as in Fig. 6.16), these units[69] have both plates on the *same* side of the diaphragm in the form of two concentric rings that form a "bulls-eye" pattern. The measured pressure is on the *other* side of the diaphragm, so the "measurement side" has no electrodes and can be constructed of clean, corrosion-resistant materials such as INCONEL. For absolute pressure units, the reference side is evacuated to high vacuum and sealed off, with a "getter" material inside to react with any residual gases and outgassing products to maintain the desired low pressure. Accuracies may be of the order of 0.12 percent of reading, plus any temperature effects. The temperature effects are not negligible, so some units include a feedback temperature control system to achieve the highest possible accuracy. Because of their accuracy and independence of gas species, capacitance gages are suitable for calibration standards. One portable unit[70] includes up to three gages and a vacuum pumping system, allowing in situ calibration of other gages over the range 10^{-5} to 1000 torr.

McLeod Gage

The McLeod gage is considered a vacuum standard since the pressure can be computed from the dimensions of the gage. It is not directly usable below about

[68]MKS Model 624, MKS Instruments, Andover, MA, 800-227-8766 (www.mksinst.com).

[69]D. Jacobs, op cit.

[70]MKS Instruments, model PVS6D.

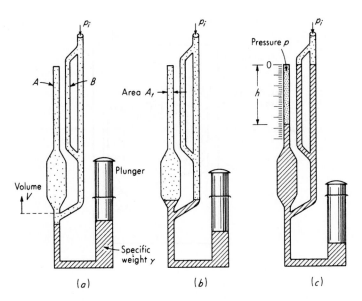

Figure 6.30
McLeod gage.

10^{-4} torr; however, pressure-dividing techniques (see Sec. 6.1) allow its use as a calibration standard for considerably lower ranges. The multiple-compression technique[71] is also being studied to extend its range. The inaccuracy of McLeod gages is rarely less than 1 percent and may be much higher at the lowest pressures.

Of the many variations of McLeod gages, here we consider only the most basic. The principle of all McLeod gages is the compression of a sample of the low-pressure gas to a pressure sufficiently high to read with a simple manometer. Figure 6.30 shows the basic construction. By withdrawing the plunger, the mercury level is lowered to the position of Fig. 6.30a, admitting the gas at unknown pressure p_i. When the plunger is pushed in, the mercury level goes up, sealing off a gas sample of known volume V in the bulb and capillary tube A. Further motion of the plunger causes compression of this sample, and motion is continued until the mercury level in capillary B is at the zero mark. The unknown pressure is then calculated, by using Boyle's law, as follows:

$$p_i V = p A_t h \tag{6.91}$$

$$p = p_i + h\gamma \tag{6.92}$$

$$p_i = \frac{\gamma A_t h^2}{V - A_t h} \approx \frac{\gamma A_t h^2}{V} \qquad \text{if } V \gg A_t h \tag{6.93}$$

---

[71]W. Kreisman, "Extension of the low Pressure Limit of McLeod Gages," *NASA, CR*-52877.

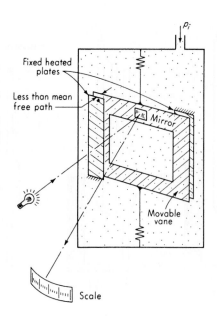

Figure 6.31
Knudsen gage.

In using a McLeod gage, it is important to realize that if the measured gas contains any vapors that are condensed by the compression process, then the pressure will be in error. Except for this effect, the reading of the McLeod gage is not influenced by the composition of the gas. Only the diaphragm and the Knudsen gage share this desirable feature of composition insensitivity. The main drawbacks of the McLeod gage are the lack of a continuous output reading and the limitations on the lowest measurable pressures. When it is employed to calibrate other gages, a liquid-air cold trap should be used between the McLeod gage and the gage to be calibrated to prevent the passage of mercury vapor.

Knudsen Gage

Although the Knudsen gage is little utilized at present, we discuss it briefly since it is relatively insensitive to gas composition and thus gives promise of development into a standard for pressures too low for the McLeod gage. In Fig. 6.31 the unknown pressure p_i is admitted to a chamber containing fixed plates heated to absolute temperature T_f, which temperature must be measured, and a spring-restrained movable vane whose temperature T_v also must be known. The spacing between the fixed and movable plates must be less than the mean free path of the gas whose pressure is being measured. The kinetic theory of gases shows that gas molecules rebound from the heated plates with greater momentum than from the cooler movable vane, thus giving a net force on the movable vane which is measured by the deflection of the spring suspension. Analysis shows that the force is directly proportional to pressure for a given T_f and T_v, following a law of the form

$$p_i = \frac{KF}{\sqrt{T_f/T_v - 1}}$$

where F is force and K is a constant. The Knudsen gage is insensitive to gas composition except for the variation of accommodation coefficient from one gas to another. The accommodation coefficient is a measure of the extent to which a rebounding molecule has attained the temperature of the surface. This effect results, for example, in a 15 percent change in sensitivity between helium and air. Knudsen gages at present cover the range from about 10^{-8} to 10^{-2} torr.

Momentum-Transfer (Viscosity) Gages

For pressures less than about 10^{-2} torr, the kinetic theory of gases predicts that the viscosity of a gas will be directly proportional to the pressure. The viscosity may be measured, for example, in terms of the torque required to rotate, at constant speed, one concentric cylinder within another. (For pressures greater than about 1 torr, the viscosity is independent of pressure.) The variation of viscosity with pressure is different for different gases; thus gages based on this principle must be calibrated for a specific gas. While gages based on viscosity principles can measure to about 10^{-7} torr, such ranges are characteristic of laboratory-type equipment requiring great care in its use.

A typical commercial gage, shown schematically in Fig. 6.32, is calibrated for dry air and covers the range from 0 to 20 torr. The range from 0 to 0.01 torr occupies about 10 percent of the total scale. The scale is nonlinearly calibrated because most of the range is above 10^{-2} torr and viscosity here is not proportional to pressure. To enable readings above 1 torr, where viscosity tends to become pressure-independent, bladed wheels rather than smooth concentric cylinders are used in this gage. These wheels cause a turbulent momentum exchange which is pressure-dependent above 1 torr, extending the useful range to 20 torr. To reach the quoted lower limit (10^{-7} torr) of viscosity gages, the construction of Fig. 6.32 is not employed. One approach[72] measures the deceleration of a levitated spinning ball immersed in the gas. The referenced unit is used mainly as a calibration standard over the range 5×10^{-7} to 0.01 torr.

Thermal-Conductivity Gages

Just as for viscosity, when the pressure of a gas becomes low enough that the mean free path of molecules is large compared with the pertinent dimensions of the apparatus, a linear relation between pressure and thermal conductivity is predicted by the kinetic theory of gases. For a viscosity gage, the pertinent dimension is the spacing between the relatively moving surfaces. For a conductivity gage, it is the spacing

---

[72]MKS Instruments, model SRG-2CE.

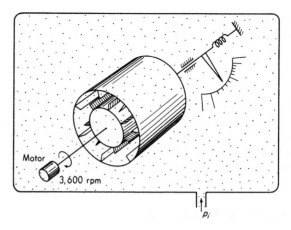

Figure 6.32
Momentum gage. *(Courtesy General Electric Co.)*

between the hot and cold surfaces. Again, when the pressure is increased sufficiently, conductivity becomes independent of gas pressure. The transition region between dependence and nondependence of viscosity and thermal conductivity on pressure is approximately the range 10^{-2} to 1 torr for apparatus of a size convenient to construct.

The application of the thermal-conductivity principle is complicated by the simultaneous presence of another mode of heat transfer between the hot and cold surfaces, namely, radiation. Most gages utilize a heated element supplied with a constant energy input. This element assumes an equilibrium temperature when heat input and losses by conduction and radiation are just balanced. The conduction losses vary with gas composition and gas pressure; thus for a given gas, the equilibrium temperature of the heated element becomes a measure of pressure, and this temperature is what is actually measured. If the radiation losses are a major part of the total, then pressure-induced conductivity changes will cause only a slight temperature change, giving poor sensitivity. Analysis shows that radiation losses may be minimized by using surfaces of low emissivity and by making the cold-surface temperature as low as practical. Since conduction and radiation losses depend on *both* the hot- and cold-surface temperatures, the cold surface may be maintained at a known constant temperature if overall accuracy warrants this measure. A further source of error is in the heat-conduction loss through any solid supports by which the heated element is mounted. The relative importance of the above-mentioned effects varies with the details of construction of the gage. The most common types of conductivity gages are the thermocouple, resistance thermometer (Pirani), and thermistor.[73]

---

[73]V. Comello, "Using Thermal Conductivity Gages," *R&D Magazine,* July 1997, p. 57; "When to Choose a Thermocouple Gage," *R&D Magazine,* May 2000.

Dual-Gage Technique[79]

When large amounts of water vapor (common during early stages of evacuation) or helium (from a leak-detection system) are present, simultaneous readings from an ionization gage and a thermocouple gage (both of which will read incorrectly) can be processed to not only get a correct vacuum reading but also determine the amount of moisture or helium present.

6.10 SOUND MEASUREMENT

The measurement of airborne and waterborne sound is of increasing interest to engineers. Airborne-sound measurements are important in the development of less noisy machinery and equipment, in diagnosis of vibration problems, and in the design and test of sound recording and reproducing equipment. In large rocket and jet engines, the sound pressures produced by the exhaust may be large enough to cause fatigue failure of metal panels because of vibration ("acoustic fatigue"). Waterborne sound has been applied in underwater direction and range-finding equipment (sonar). Since most sound transducers (microphones and hydrophones) are basically pressure-measuring devices, it is appropriate to consider them briefly in this chapter.

The basic definitions of sound are in terms of the magnitude of the fluctuating component of pressure in a fluid medium. The *sound pressure level* (SPL) is defined by

$$\text{SPL} \triangleq \text{sound pressure level} \triangleq 20 \log_{10} \frac{p}{0.0002} \text{ decibels (dB)} \qquad \textbf{(6.95)}$$

where $\qquad\qquad p \triangleq$ root-mean-square (rms) sound pressure, μbar $\qquad\qquad$ **(6.96)**

and $\qquad\qquad 1\ \mu\text{bar} = 1\ \text{dyn/cm}^2 = 1.45 \times 10^{-5}\ \text{lb/in}^2 \qquad\qquad$ **(6.97)**

The rms value of the fluctuating component of pressure is employed because most sounds are random signals rather than pure sine waves. The value $0.0002\ \mu$bar is an accepted standard reference value of pressure against which other pressures are compared by Eq. (6.95). Note that when $p = 0.0002\ \mu$bar, the sound pressure level is 0 dB. This value was selected somewhat arbitrarily, but represents the average threshold of hearing for human beings if a 1,000-Hz tone is used. That is, the 0-dB level was selected as the lowest pressure fluctuation normally discernible by human beings. Since 0 dB is about 3×10^{-9} lb/in^2, the remarkable sensitivity of the human ear should be apparent. The decibel (logarithmic) scale is used as a convenience because of the great ranges of sound pressure level of interest in ordinary work. For example, a noisy office may have an SPL of 74 dB (1 μbar). The average human threshold of pain is 144 dB. Sound pressures close to large rocket engines are on the order of 170 dB (1 lb/in^2). One atmosphere (14.7 lb/in^2) is 194 dB. The span from the lowest to the highest pressures of interest is thus of the order of 10^{-9} to 1, a tremendous range.

---

[79]C. A. Schalla, "Making Accurate Vacuum Readings," *Mach. Des.*, pp. 122–124, February 26, 1981.

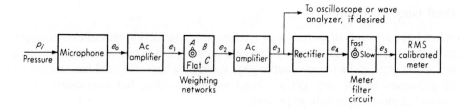

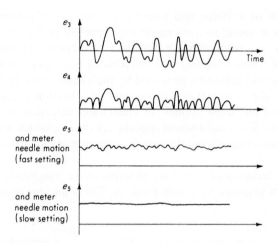

Figure 6.36
Sound-level meter.

Sound-Level Meter

The most commonly utilized instrument for routine sound measurements is the sound-level meter.[80] This is actually a measurement *system* made up of a number of interconnected components. Figure 6.36 shows a typical arrangement. The sound pressure p_i is transduced to a voltage by means of the microphone. Microphones generally employ a thin diaphragm to convert pressure to motion. The motion is then converted to voltage by some suitable transducer, usually a capacitance, piezo-electric, or moving-coil type. Microphones often have a "slow leak" (capillary tube) connecting the two sides of the diaphragm, to equalize the average pressure (atmospheric pressure) and prevent bursting of the diaphragm. This is necessary because the (slow) hour-to-hour and day-to-day changes in atmospheric pressure are much greater than the sound-pressure fluctuations to which the microphone must respond.

[80]C. Thomsen, "Sound Level Meters—Their Use and Abuse," *Sound & Vib.,* pp. 28–31, March 1979;
J. R. Hassall and K. Zaveri, "Acoustic Noise Measurements," 4th ed., Bruel and Kjaer Instruments, Inc., Marlboro, MA, 1979; "Acoustics Handbook," *Appl. Note* 100, pp. 63–81, Hewlett-Packard Co., Palo Alto, CA, 1968.

(Note that the eustachian tube of the human ear serves a similar function.) The presence of this leak dictates that microphones will not respond to constant or slowly varying pressures. This is usually no problem since many measurements involve a human response to the sound, and this is known to extend down to only about 10 to 20 Hz. Thus the microphone frequency response need go only to this range, not to zero frequency.

The output voltage of the microphone generally is quite small and at a high impedance level; thus an amplifier of high input impedance and gain is used at the output of the microphone. This can be a relatively simple ac amplifier, since response to static or slowly varying voltages is not required. Capacitor microphones often use for the first stage a FET-input amplifier built right into the microphone housing. This close coupling reduces stray capacitance effects by eliminating cables at the high-impedance end.

Following the first amplifier are the weighting networks. They are electrical filters whose frequency response is tailored to approximate the frequency response of the average human ear. Figure 6.37a[81] displays "equal-loudness contours" obtained from measurements on human beings, showing that the frequency response of the human ear is both "nonflat" and nonlinear. Each curve is labeled with a loudness unit called a *phon,* with 0 phon corresponding to the threshold of hearing. The ordinate (sound-pressure level in decibels) tells what pressure amplitude must be applied at any given frequency so that the human observer will perceive a sensation of equal loudness. For example, at a 50-phon loudness level, a 58-dB SPL at 100 Hz sounds as loud as a 50-dB SPL at 1,000 Hz, which demonstrates the nonflatness of the ear's frequency response. Its nonlinearity is manifested by the need for a *family* of curves for various loudness levels, rather than just a single curve. Since the main use of a sound-level meter is *not* the accurate measurement of pressure, but rather the determination of the loudness perceived by human beings, a flat instrument frequency response is not really wanted. The weighting networks of Fig. 6.36 are electrical filters designed to approximate the human ear's response at three different loudness levels, so that instrument readings will reflect perceived loudness. Usually three filters: *A* (approximates 40-phon ear response), *B* (70-phon), and *C* (100-phon) are provided, and Fig. 6.37b shows the frequency response of these filters (dashed lines show tolerances allowed on "precision sound-level meters"). Some meters also provide a "flat" setting if true pressure measurements are wanted; if not, the *C* network is a good approximation. Actually, many practical measurements are made by employing the *A* scale since it is a simple approach which has given good results in many cases and has been written into many standards and codes. Figure 6.38 shows a table from OSHA (Occupational Safety and Health Administration) rules on allowable noise exposure for industrial workers. Readings taken with a weighting network are called *sound level* rather than *sound-pressure level.*

The output of the weighting network is further amplified and an output jack provided to lead this signal to an oscilloscope (if observation of the waveform is

---

[81]B Katz, "Primer on Sound Level Meters and Acoustical Calibration," Bruel & Kjaer Instruments, Inc.

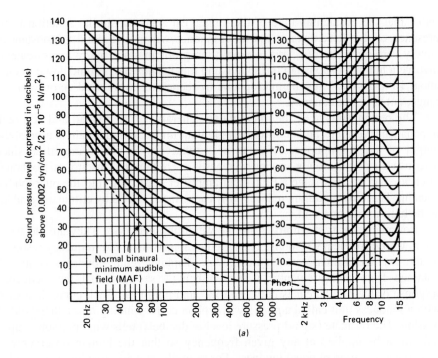

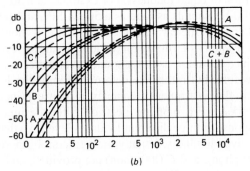

Figure 6.37
Response of human ear and weighting networks.

Hours per day of noise exposure	A-weighted noise level, dB
8	90
4	95
2	100
1	105
0.5	110
0.25	115

Figure 6.38
Allowable noise exposure for industrial workers.

desired) or to a spectrum analyzer (if the frequency content of the sound is to be determined). If only the overall sound magnitude is desired, the rms value of e_3 must be found. While true rms voltmeters are available, their expense is justifiable only in the highest-grade sound-level meters. Rather, the *average* value of e_3 is determined by rectifying and filtering, and then the meter scale is *calibrated* to read rms values. This procedure is exact for pure sine waves since there is a precise relation between the average value and the rms value of a sine wave. For nonsinusoidal waves this is not true, but the error is generally small enough to be acceptable for relatively unsophisticated work. The filtering is accomplished by both a simple low-pass RC filter and the low-pass meter dynamics. Some meters have a slow-fast response switch which changes the filtering. The slow position gives a steady, easy-to-read needle position, but masks any short-term variations in the signal. If these short-term variations are of interest, they may be visually observed on the meter by switching it to fast response. While the meter is actually reading the rms value of e_3 (and thus of p_i), it is calibrated in decibels since Eq. (6.95) establishes a definite relation between sound pressure in microbars and decibels.

Microphones

While the design of microphones is a specialized and complex field with a large technical literature, here we will only point out some of the main considerations. Frequency response is still of major interest; however, the effects on frequency response of sound wavelength and direction of propagation are aspects of dynamic behavior not regularly encountered in other measurements. The *pressure response* of a microphone refers to the frequency response relating a uniform sound pressure applied at the microphone diaphragm to the output voltage of the microphone. The pressure response of a given microphone may be estimated theoretically or measured experimentally by one of a number of accepted methods.[82]

What is usually desired is the *free-field response* of the microphone. That is, what is the relation between the microphone output voltage and the sound pressure that existed at the microphone location *before* the microphone was introduced into the sound field? The microphone disturbs the pressure field because its acoustical impedance is radically different from that of the medium (air) in which it is immersed. In fact, for most purposes, the microphone (including its diaphragm) may be considered as a rigid body. Sound waves impinging on this body give rise to complex reflections that depend on the frequency, the direction of propagation of the sound wave, and the microphone size and shape. When the wavelength of the sound wave is very large compared with the microphone dimensions (low frequencies), the effect of reflections is negligible for any angle of incidence between the diaphragm and the wave-propagation direction, and the free-field response is the same as the pressure response. At very high frequencies, where the wavelength is much smaller than the microphone dimensions, the microphone acts as an infinite

---

[82]P. V. Bruel and G. Rasmussen, "Free Field Response of Condenser Microphones," *B & K Tech. Rev.,* Bruel & Kjaer Instruments Inc., no. 1, January 1959; no. 2, April 1959.

wall and the pressure at the microphone surface [for waves propagating perpendicular to the diaphragm ($0°$ angle of incidence)] is twice what it would be if the microphone were not there. For waves propagating parallel to the diaphragm ($90°$ incidence angle), the average pressure over the diaphragm surface is zero, giving no output voltage. Between the very low and very high frequencies, the effect of reflections is quite complicated and depends on sound wavelength (frequency), microphone size and shape, and angle of incidence.

For simple geometric shapes such as spheres and cylinders, theoretical results are available.[83] Experiments on actual microphones give results such as those shown in Fig. 6.39. Note that for sufficiently low frequencies (below a few thousand hertz) there is little change in pressure because of the presence of the microphone; also the angle of incidence has little effect. This flat frequency range can be extended by reducing the size of the microphone; however, smaller size tends to reduce sensitivity. MEMS microphones are currently in the R&D stage[84] but should be commercially available before long. Their small size makes high-frequency measurements easier, and promises many other advantages. The size effect is directly related to the relative size of the microphone and the wavelength of the sound. The wavelength λ of sound waves in air is roughly $13,000/f$ inches, where f is the frequency in hertz. When λ becomes comparable to the microphone-diaphragm diameter, significant reflection effects can be expected. For example, a 1-in-diameter microphone would not be expected to have good response much above 13,000 Hz. (These limitations can be relaxed to some extent by clever use of acoustical mechanical techniques.[85])

The lower part of Fig. 6.39 shows a curve labeled "random incidence." This refers to the response to a diffuse sound field where the sound is equally likely to come to the microphone from any direction, the waves from all directions are equally strong, and the phase of the waves is random at the microphone position. Such a field may be approximated by constructing a room with highly irregular walls and placing reflecting objects of various sizes and shapes in it. A source of sound placed in such a room gives rise to a diffuse sound field at any point in the room. Microphones calibrated under such conditions are of interest because many sound measurements take place in enclosures which, while not giving perfect random incidence, certainly do not give pure plane waves. Microphone calibrations may give the pressure response and the free-field response for selected incidence angles, usually 0 and $90°$. Figure 6.40 shows typical curves.

Microphones used for engineering measurements are usually piezoelectric, capacitor, or electret types. Figure 6.41a[86] shows construction details of a piezoelectric unit which uses PZT (lead zirconate titanate) as a bending beam coupled to

---

[83]L. Beranek, "Acoustic Measurements," chap. 3, Wiley, New York, 1949.

[84]D. P. Arnold, et al., "A Piezoresistive Microphone for Aeroacoustic Measurements," 2001 ASME International Congress, Nov. 11–16, 2001, New York.

[85]G. Rasmussen, "Miniature Pressure Microphones," *B & K Tech. Rev.,* no 1, Bruel & Kjaer Instruments Inc., 1963.

[86]Bruel & Kjaer Instruments Inc.

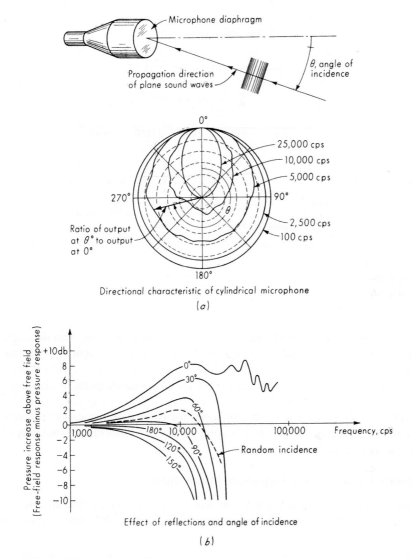

Figure 6.39
Microphone response characteristics.

the center of a conical diaphragm of thin metal foil. An electret type is shown in Fig. 6.41b.[87] These are related to the capacitor types discussed in detail in the next section; however, they require no polarizing voltage since their charge is permanently "built into" the polymer film which forms the diaphragm. Since the unsupported polymer film would sag and creep excessively, a backup plate with

---

[87]*GR Today,* p. 8. General Radio, Concord, MA, Autumn 1972.

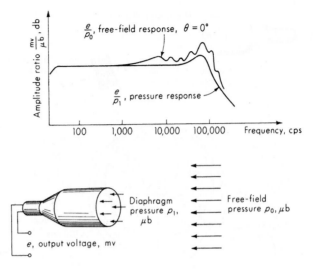

Figure 6.40
Free-field and pressure response.

"raised points" is used. Such microphones[88] are less expensive than the capacitor type, can be used under high-humidity conditions (where the capacitor type may arc over), and result in instruments of smaller size and power consumption. A version[89] which preserves the desirable features of an all-metal diaphragm also has been developed.

The selection and use of microphones for critical applications require some background in acoustics, which is beyond the scope of this text; fortunately useful references[90] are available.

Pressure Response of a Capacitor Microphone

Of the several types of microphones in common use, generally the capacitor type is considered capable of the highest performance. Figure 6.42 shows in simplified

---

[88]R. W. Raymond and S. V. Djuric, "The Latest in Instrumentation Quality Microphones," *Sound & Vib.,* pp. 4–6, May 1974.

[89]E. Frederiksen, N. Eirby, and H. Mathiasen, "Prepolarized Condenser Microphones for Measurement Purposes," *B & K Tech. Rev.,* no. 4, pp. 3–25, Bruel & Kjaer Instruments Inc., 1979.

[90]A. J. Schneider, "Microphone Orientation in the Sound Field," *Sound & Vib.,* pp. 20–25, February 1970; W. R. Kundert, "Everything You've Wanted to Know about Measurement Microphones," *Sound & Vib.,* pp. 10–26, March 1978; A. P. G. Peterson and E. E. Gross, Jr., "Handbook of Noise Measurement," General Radio Corp., 1972; Microphone Calibration, Brochure BR-0092, Bruel & Kjaer Instruments Inc., 1980; "Condenser Microphones and Microphone Preamplifiers, Theory and Application Handbook," Bruel & Kjaer Instruments Inc., 1976; G. Rasmussen, "A New Generation of Precision Measurement Microphones," *Sound & Vib.,* March 1996, pp. 18–22; M. Cooper and D. Formenti, "Differences in Measurement and Studio Microphones," *Sound & Vib.,* Feb. 2000, p. 14; "Do You Also Have a 4-dB Error at 10 kHz?," *Bruel & Kjaer Magazine,* no. 1, 2000, pp. 14–16.

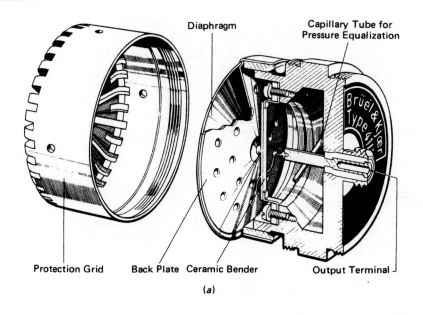

Protection Grid   Back Plate   Ceramic Bender   Output Terminal

(a)

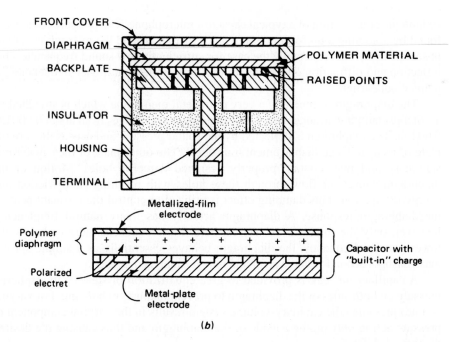

(b)

Figure 6.41
Piezoelectric and electret microphones.

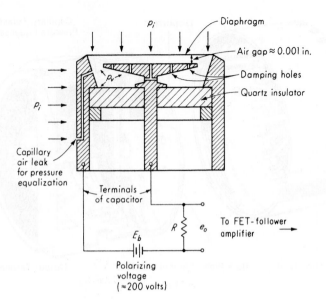

Figure 6.42
Capacitor microphone.

fashion the construction of a typical capacitor microphone. The pressure response is found by assuming a uniform pressure p_i to exist all around the microphone at any instant of time. This is actually the case for sufficiently low sound frequencies, but reflection and diffraction effects distort this uniform field at higher frequencies, as pointed out earlier.

The diaphragm is generally a very thin metal membrane which is stretched by a suitable clamping arrangement. Diaphragm thickness ranges from about 0.0001 to 0.002 in. The diaphragm is deflected by the sound pressure and acts as the moving plate of a capacitance displacement transducer. The other plate of the capacitor is stationary and may contain properly designed damping holes. Motion of the diaphragm causes air flow through these holes with resulting fluid friction and energy dissipation. This damping effect is utilized to control the resonant peak of the diaphragm response. A diaphragm actually has many natural frequencies; however, only the lowest is of interest here. For frequencies near or below the lowest natural frequency, the diaphragm behaves essentially as a simple spring-mass-dashpot second-order system and may be analyzed as such.

A capillary air leak is provided to give equalization of steady (atmospheric) pressure on both sides of the diaphragm to prevent diaphragm bursting. For varying (sound) pressures the capillary-volume system results in the varying component of pressure acting *only* on the outside of the diaphragm and thus causing the desired diaphragm deflection.

The variable capacitor is connected into a simple series circuit with a high resistance R and "polarized" with a dc voltage E_b of about 200 V. This polarizing voltage acts as circuit excitation and determines the neutral (zero-pressure) diaphragm position because of the electrostatic attraction force between the capacitor plates. For a constant diaphragm deflection, no current flows through R and no output voltage e_o exists; thus there is no response to static pressure differences across the diaphragm. For dynamic pressure differences, a current *will* flow through R and an output voltage exists. The voltage e_o usually is applied to the input of a FET-follower amplifier, which always has a gain less than 1; thus the purpose of the amplifier is not to increase the voltage level. Rather, it has a high input impedance (>1 GΩ) to prevent loading of the microphone, which has a high output impedance. Since the output impedance of the amplifier is low ($<100\ \Omega$), its output signal may be coupled into long cables and low-impedance loads without loss of signal magnitude.

The first step in the analysis involves determination of the effective force tending to deflect the diaphragm in terms of the pressure p_i. The relation between p_i and the pressure p_v in the microphone internal volume V may be obtained from Eq. (6.85). We neglect the inertia term since in microphones the viscous effect predominates and the filtering effect of the capillary is significant only at low frequencies. Thus we get

$$\tau_l \dot{p}_v + p_v = p_i \qquad (6.98)$$

where

$$\tau_l \triangleq \text{leak time constant} \triangleq \frac{128\,\mu L V}{\pi E_a d_t^4} \qquad (6.99)$$

Now the deflection of the diaphragm is due to the *difference* between p_i and p_v. Operationally,

$$p_v = \frac{p_i}{\tau_l D + 1}$$

and thus

$$p_i - p_v = p_i\left(1 - \frac{1}{\tau_l D + 1}\right) = \frac{\tau_l D}{\tau_l D + 1}p_i \qquad (6.100)$$

The total force f_d on the diaphragm is $A_d\,(p_i - p_v)$, where A_d is the diaphragm area. Thus

$$\frac{f_d}{p_i}(D) = \frac{A_d \tau_l D}{\tau_l D + 1} \qquad (6.101)$$

The frequency response of this shows clearly that $f_d \to 0$ as frequency $\to 0$; thus slow pressure changes do not result in forces tending to burst the diaphragm. However, the time constant τ_l must be small enough that $(f_d/p_i)(i\omega) \approx A_d$ for all frequencies above about 10 Hz, the lowest-frequency sound pressures usually of interest.

The next step requires study of the electromechanical-energy-conversion process in a moving-plate capacitor. Although the moving "plate" (diaphragm) of the microphone is not flat, we analyze the situation for a flat plate for simplicity. One can always find a flat-plate capacitor that is equivalent to the diaphragm capacitor in the sense that the linearized capacitance variation with plate separation is the

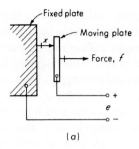

Figure 6.43
(*a*) Moving-plate capacitor.

same (at least for small motions) for both. (Analysis of a "dished"-shape capacitor is available.[91]) Considering Fig. 6.43*a*, we recall that

$$\text{Energy stored by a capacitor} = \frac{q^2}{2C} = \frac{Ce^2}{2} \qquad (6.102)$$

where $q \triangleq$ charge, $e \triangleq$ voltage, and $C \triangleq$ capacitance. We wish to show that the two plates attract each other with a force f. The capacitance of a parallel-plate capacitor whose area A is large compared with the plate separation x is given very closely by

$$C = \frac{\varepsilon A}{x} \qquad (6.103)$$

where $\quad \varepsilon \triangleq$ permittivity of material between plates

$\qquad = 8.86 \times 10^{-12}$ F/m for vacuum or dry air

We now suppose the capacitor is charged and then open-circuited so that q must remain constant. If the plates are separated an additional amount dx, we may write

$$\text{Original energy } (x = x_0) = \frac{q^2}{2C_0} = \frac{q^2 x_0}{2\varepsilon A} \qquad (6.104)$$

$$\text{Final energy } (x = x_0 + dx) = \frac{q^2}{2C_f} = \frac{q^2(x_0 + dx)}{2\varepsilon A} \qquad (6.105)$$

The energy change is thus $q^2 dx/(2\varepsilon A)$. Since energy is conserved in this system, it must have required a force f on the plate to cause the motion dx, since then mechanical work $f\, dx$ would have been done and converted to electrical energy $(q^2/dx)/(2\varepsilon A)$. The force f thus may be calculated from

$$\frac{q^2\, dx}{2\varepsilon A} = f\, dx$$

$$f = \frac{q^2}{2\varepsilon A} = \frac{\varepsilon A e^2}{2x^2} \qquad (6.106)$$

[91]H. K. P. Neubert, "Instrument Transducers," pp. 274–278, Oxford University Press, London, 1963.

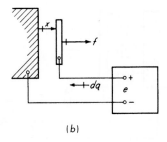

(b)

Figure 6.43
(b) Microphone model.

For air, with e in volts and A and x in any consistent units, this becomes

$$f = 0.99 \times 10^{-12} \frac{Ae^2}{x^2} \quad \text{lbf} \tag{6.107}$$

As an example, if $e = 200$ V, $A = 1$ in^2, and $x = 0.001$ in, the force is 0.04 lbf.

If the capacitor is connected to an external circuit as in Fig. 6.43b, we can show Eq. (6.106) still holds as follows: The work done in moving a charge dq through a potential difference e is $e\,dq$. Then, by conservation of energy,

$$f\,dx + e\,dq = d(\text{stored energy}) = d\left(\frac{Ce^2}{2}\right) \tag{6.108}$$

Then, $\quad f = -e\frac{dq}{dx} + \frac{d}{dx}\left(\frac{Ce^2}{2}\right) = -e\frac{d}{dx}(Ce) + \frac{d}{dx}\left(\frac{Ce^2}{2}\right) \tag{6.109}$

$$f = -e\left(C\frac{de}{dx} + e\frac{dC}{dx}\right) + Ce\frac{de}{dx} + \frac{e^2}{2}\frac{dC}{dx} \tag{6.110}$$

$$f = -\frac{e^2}{2}\frac{dC}{dx} = -\frac{e^2}{2}\left(-\frac{\varepsilon A}{x^2}\right) = \frac{\varepsilon A e^2}{2x^2} \tag{6.111}$$

Next we model the microphone as in Fig. 6.43c. The mass M and spring K_s must be such as to give the same natural frequency as the lowest natural frequency of the diaphragm. The dashpot B must be such as to give the same resonant peak as in the microphone's measured pressure response. The capacitor plate area and air gap (with no external forces acting) must be such as to give the same capacitance as is measured for the microphone under similar conditions. The spring constant K_s and capacitor dimensions also must be such that the force f_d causes a capacitance variation equal (at least for small motions) to that caused in the actual microphone by a pressure difference $p_i - p_v = f_d/A_d$. If all the above conditions are met, the simplified model of Fig. 6.43c will respond essentially in the same way as the microphone itself. While the equivalent system described is defined in terms of experimental measurements on an existing microphone, microphone designers have theoretical formulas for estimating these parameters *before* a new microphone is built.

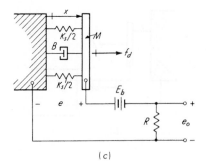

(c)

Figure 6.43
(c) Capacitor with external circuit.

Assuming that the equivalent system is a reasonable model, we can proceed with the analysis. With no force f_d applied and with the capacitor uncharged, the mass will assume an equilibrium position x_{fl}, where x_{fl} is the free length of the springs. If now the polarizing voltage E_b is applied, the moving plate will experience an attractive force and will move to a new position x_0 such that the spring force and electrostatic force just balance (see Fig. 6.44). Now, when pressure force f_d is applied, motion will take place around x_0 as an operating point. To find x_0, we can write

$$K_s(x_{fl} - x_0) = \frac{\varepsilon A E_b^2}{2x_0^2} \tag{6.112}$$

This equation in x_0 has two positive solutions, x_0 and x_0', for a practical case. The solution (equilibrium position) x_0' is unstable in the sense that any slight motion away from x_0' results in *further* motion away from this point. The desired (stable) equilibrium position is x_0, where small disturbances from equilibrium give rise to forces tending to restore equilibrium. Thus the microphone must be designed to operate at x_0 rather than x_0'.

We apply Newton's law to the mass M to get

$$-B\frac{dx}{dt} + K_s(x_{fl} - x) - \frac{\varepsilon A e^2}{2x^2} + f_d = M\frac{d^2x}{dt^2} \tag{6.113}$$

The electrostatic-force term makes this differential equation nonlinear. For small changes in e and x from the equilibrium operating point, this nonlinear term may be linearized approximately with good accuracy. This may be done by employing only the linear terms of a Taylor-series expansion of the nonlinear function. That is, if in general

$$z = z(x, y)$$

then $\qquad z \approx z(x_0, y_0) + \frac{\partial z}{\partial x}\bigg|_{\substack{x=x_0 \\ y=y_0}} (x - x_0) + \frac{\partial z}{\partial y}\bigg|_{\substack{x=x_0 \\ y=y_0}} (y - y_0) \qquad$ **(6.114)**

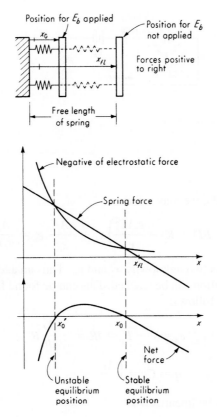

Figure 6.44
Determination of equilibrium points.

In this specific case, the nonlinear function is e^2/x^2; thus

$$\frac{e^2}{x^2} \approx \frac{E_b^2}{x_0^2} + E_b^2\left(-\frac{2}{x_0^3}\right)(x - x_0) + \frac{1}{x_0^2} 2E_b(e - E_b) \qquad (6.115)$$

We now define $x_1 \triangleq x - x_0$ and $e_0 \triangleq e - E_b$ to get

$$\frac{e^2}{x^2} \approx \frac{E_b^2}{x_0^2} - \frac{2E_b^2}{x_0^3} x_1 + \frac{2E_b}{x_0^2} e_o \qquad (6.116)$$

Also $\qquad K_s(x_{fl} - x) = K_s(x_{fl} - x_1 - x_0) = - K_s x_1 + \dfrac{\varepsilon A E_b^2}{2x_0^2} \qquad (6.117)$

Now Eq. (6.113) may be written as

$$-B\frac{dx_1}{dt} - K_s x_1 + \frac{\varepsilon A E_b^2}{2x_0^2} - \frac{\varepsilon A}{2}\left(\frac{E_b^2}{x_0^2} + \frac{2E_b}{x_0^2} e_o - \frac{2E_b^2}{x_0^3} x_1\right) + f_d = M\frac{d^2 x_1}{dt^2} \qquad (6.118)$$

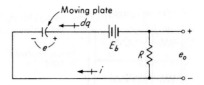

Figure 6.45
Circuit analysis.

Bringing in Eq. (6.101), we may write

$$\left(MD^2 + BD + K_s - \frac{\varepsilon AE_b^2}{x_0^3}\right)x_1 + \frac{\varepsilon AE_b}{x_0^2}e_o = \frac{A_d\tau_l D}{\tau_l D + 1}p_i \qquad (6.119)$$

This equation contains two unknowns, x_1 and e_o. Thus an additional equation must be found before a solution can be reached. This can be found from an analysis of the circuit of Fig. 6.45 as follows:

$$e_o = e - E_b = iR = -\frac{dq}{dt}R \qquad (6.120)$$

$$q = Ce = \frac{\varepsilon Ae}{x} \qquad (6.121)$$

Equation (6.121) may be linearized as

$$q \approx \frac{\varepsilon AE_b}{x_0} - \frac{\varepsilon AE_b}{x_0^2}x_1 + \frac{\varepsilon A}{x_0}e_o \qquad (6.122)$$

Then, approximately,

$$\frac{dq}{dt} = -\frac{\varepsilon AE_b}{x_0^2}\frac{dx_1}{dt} + \frac{\varepsilon A}{x_0}\frac{de_o}{dt} \qquad (6.123)$$

and

$$\frac{e_o}{R} = -\frac{dq}{dt} = \frac{\varepsilon AE_b}{x_0^2}\frac{dx_1}{dt} - \frac{\varepsilon A}{x_0}\frac{de_o}{dt} \qquad (6.124)$$

thus finally giving

$$-\frac{\varepsilon AE_b R}{x_0^2}Dx_1 + \left(1 + \frac{\varepsilon AR}{x_0}D\right)e_o = 0 \qquad (6.125)$$

Since we are primarily interested in e_o rather than x_1, Eq. (6.125) may be combined with (6.119) to eliminate x_1 and get

$$\left[\frac{M\tau_e}{K_e}D^3 + \left(\frac{M}{K_e} + \frac{B\tau_e}{K_e}\right)D^2 + \left(\frac{B}{K_e} + \tau_e + \frac{\tau_e^2 E_b^2}{x_0^2 RK_e}\right)D + 1\right]e_o$$

$$= \frac{A_d E_b \tau_e}{K_e x_0}\frac{\tau_l D^2}{\tau_l D + 1}p_i \qquad (6.126)$$

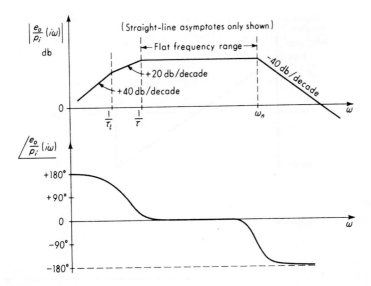

Figure 6.46
Microphone frequency response.

where
$$K_e \triangleq K_s - \frac{\varepsilon A E_b^2}{x_0^3} \qquad (6.127)$$

$$\tau_e \triangleq \frac{\varepsilon A R}{x_0} \qquad (6.128)$$

The cubic left-hand side is not readily factored until numerical values are known. In general, one gets two complex roots and one real root. This leads to a transfer function of the form

$$\frac{e_o}{p_i}(D) = \frac{K\tau_l \tau D^2}{(\tau_l D + 1)(\tau D + 1)(D^2/\omega_n^2 + 2\zeta D/\omega_n + 1)} \qquad (6.129)$$

The frequency response of the microphone is then as shown in Fig. 6.46. The sensitivity in the flat range is typically of the order of 1 to 5 mV/μbar, while the low-frequency cutoff is about 1 to 10 Hz, though lower values are possible. The upper limit of frequency can be extended well beyond the range of human hearing; 100,000 Hz is not unattainable.

When microphones are employed to measure relatively long-duration transient sounds, such as the "sonic boom" (Fig. 6.47) caused by the overflight of supersonic aircraft, adequate response at low frequencies becomes critical.[92] This problem is conveniently studied using digital simulation, and we again use SIMULINK for this purpose. If, in transfer function (6.129), we choose to define the output as e_o/K, then

---

[92]J. J. Van Houten and R. Brown, "Investigation of the Calibration of Microphones for Sonic Boom Measurement," *NASA* CR-1075, June 1968.

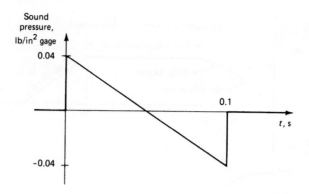

Figure 6.47
Sonic boom pressure signature.

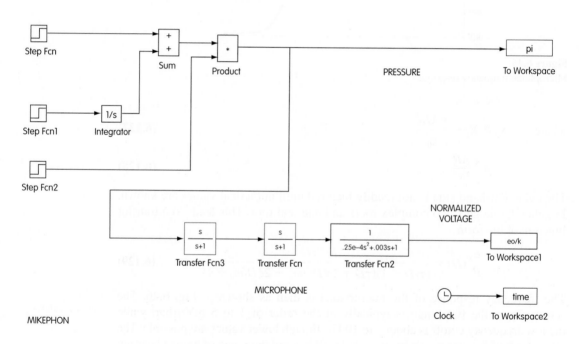

Figure 6.48
SIMULINK simulation of microphone response so sonic boom pressure.

the superimposed graphs of input p_i and output will directly show how well the microphone is measuring pressure, that is, a perfect instrument would show the two graphs lying exactly one upon the other. In the SIMULINK block diagram of Fig. 6.48, we form the pressure wave of Fig. 6.47 using suitably delayed step functions, an integrator, and a multiplier. The transfer function could be done in a single large

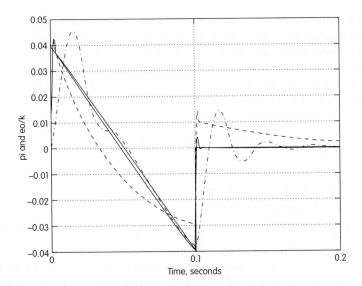

Figure 6.49
Response to sonic boom of one "good" and two "bad" microphone designs.

block, but we prefer to split it up into three parts as shown. We also will make three runs, showing one good design and two poor ones.

In the first run (solid curve of Fig. 6.49), we use $\omega_n = 2000$ rad/sec, $\zeta = 0.6$, $\tau_l = \tau = 1.0$ s, a "good" design for this application. Those familiar with microphones will note that $\omega_n = 2,000$ is very low, many microphones being good to 15,000 Hz. Use of the more realistic large value unfortunately causes simulation problems associated with "stiff" mathematical models.[93] A stiff mathematical model has both some very slow components (τ, $\tau_l \approx 1.0$ s) and some very fast ($\omega_n >$ about 10,000 in this problem) which create numerical problems in the integrating algorithm. Fortunately, if ω_n is "large enough" to accurately follow p_i, then making it even larger has essentially no effect on the response. The value $\omega_n = 2,000$ is just a convenient value; anything in the range of 1,000 to 10,000 gives almost identical results. We see that the output voltage is an almost perfect reproduction of the pressure. The next run (dashed line) uses the same "good" natural frequency but uses time constants that are too short (both 0.1 s) for a transient which lasts 0.1 s. The accuracy here is clearly unacceptable. Finally (dot-dash line), we use adequate time constants (both 1.0 s) but a natural frequency that is too low (200 rad/s) and a poor damping ratio of 0.3. Again, the accuracy is poor.

Acoustic Intensity

Traditionally many acoustical studies have been based on sound *pressure* measurements with microphones as described above. Recent instrumentation developments

---

[93]E. O. Doebelin, "System Modeling and Response," pp. 218–221.

now often make sound *intensity* measurements a preferred technique for sound power determination, noise-source location, sound absorption, and transmission loss. A brief introduction to sound-intensity measurement follows below.

While sound pressure is a scalar quantity, sound intensity or energy flux is a vector which describes the amount and direction of time-averaged flow of acoustic power per unit area at a given position[94]:

$$I_r \triangleq \lim_{T \to \infty} \frac{1}{T} \int_0^T p(t) v_r(t) \, dt \qquad \text{W/m}^2 \qquad \textbf{(6.131)}$$

where
$$I_r \triangleq \text{acoustic intensity}$$
$$p(t) \triangleq \text{pressure at selected position}$$
$$T \triangleq \text{averaging time}$$
$$v_r(t) \triangleq \text{particle velocity in selected direction } r, \text{ at selected position}$$

Measurement of p with conventional microphones is straightforward but direct measurement of v_r (by using hot-wire anemometers, etc.) presents many practical difficulties, so an indirect method using an approximation based on fluid-mechanics theory is employed. From the linearized momentum (Euler) equation with zero mean flow we have

$$\rho \frac{\partial v_r}{\partial t} + \frac{\partial p}{\partial r} = 0 \qquad \rho \triangleq \text{fluid mass density} \qquad \textbf{(6.132)}$$

Approximating the partial derivative as $\Delta p / \Delta r$ leads to

$$v_r \approx \frac{1}{\rho \, \Delta r} \int_0^t (p_2 - p_1) \, dt \qquad \textbf{(6.133)}$$

where p_1 and p_2 are pressures measured by two microphones located along the direction vector r at r_1 and r_2 $(r_2 > r_1)$. Further approximating p as $(p_1 + p_2)/2$ gives

$$I_r \approx \frac{1}{2\rho \, \Delta r} \frac{1}{T} \int_0^T \left(\frac{p_1 + p_2}{2} \right) (p_2 - p_1) \, dt \qquad \textbf{(6.134)}$$

where T is a suitably long averaging time.

In designing an intensity probe, the needed microphone spacing might be implemented in several ways (Fig. 6.50a); however, tests[95] show (Fig. 6.50b) that the face-to-face configuration, with a solid spacer between, is superior. The spacer prevents reflections between the two microphones but still permits the two pressures to be "felt" since slits in the microphone's protective grids allow penetration of the pressure to the diaphragm. For accurate velocity measurements, the spacing Δr must be small compared with the sound wavelength. This, together with other

---

[94]S. Gade, "Sound Intensity," *B & K Tech. Rev.* nos. 3, 4, 1982, Bruel and Kjaer Instruments Inc.
[95]Ibid., no. 4, p. 11.

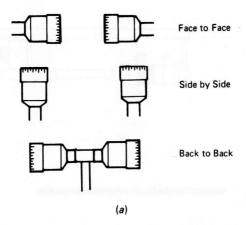

Face to Face

Side by Side

Back to Back

(a)

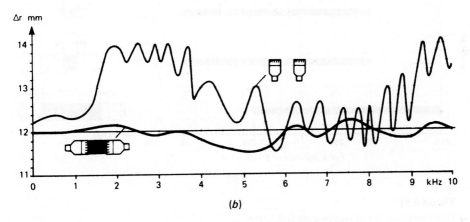

(b)

Figure 6.50
Microphone configurations and respective errors for acoustic intensity probes.

accuracy problems, may require use of different spacers (and/or microphones) to cover a wide frequency range (Fig. 6.51).

Just as with sound-pressure measurements, frequency-spectrum analysis of sound intensity is usually desired, so commercial instruments provide this capability. A few references[96, 97] give more details on specific instruments and applications.

---

[96]Product Data, Type 3360, 2-044 0029-2A; "Application Brief, Sound Intensity Measurement Instrumentation," Bruel & Kjaer Instruments Inc., 1987; *Appl. Notes* B0 0203-11, "Gated Sound Intensity Measurements on a Diesel Engine."

[97]R. P. Kendix, "Using Sound Intensity Analyses for Machinery Diagnostics," *Sound & Vib.,* pp. 26–30, March 1988.

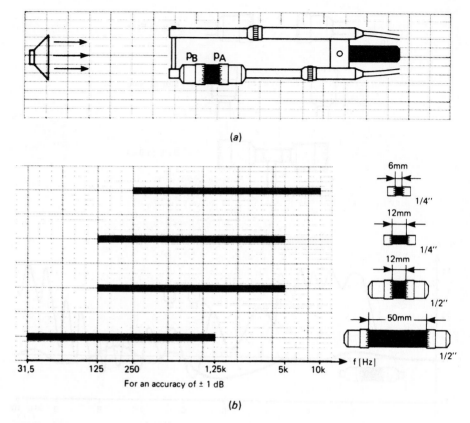

Figure 6.51
Probe orientation and size/frequency relations.

Acoustic Emission

Acoustic emission methods[98] are utilized in materials research to study deformation and fracture processes, such as dislocation pileup and crack initiation, and in non-destructive testing[99] to evaluate structural integrity, monitor pressure vessels for incipient failure, etc. Acoustic emission transducers (usually piezoelectric) are fastened to the specimen surface and detect high-frequency (20 kHz to 2 MHz) stress waves caused by the phenomena under study and propagated "acoustically" through the material. The above references should be sufficient to get the interested reader started on an in-depth study.

---

[98]R. E. Herzog, "Forecasting Failures with Acoustic Emission," *Mach. Des.,* pp. 132–137, July 14, 1973; Acoustic Emission, STP 505, *ASTM,* Philadelphia, PA, 1972; Acoustic Emission Bibliography 1970–72, STP 571, *ASTM,* Philadelphia, PA, 1975; T. Licht, "Acoustic Emission," *B&K Tech. Rev.,* no. 2, Bruel & Kjaer Instruments Inc., 1979.

[99]A. Vary, "Nondestructive Evaluation Technique Guide," *NASA* SP-3079, 1973.

microprocessor computing power to periodically recalibrate not only the transducers but also amplifiers and analog/digital converters against three precision pressure values. This three-point calibration corrects for transducer nonlinearity (which otherwise would be a problem in transducers of this type) and thermal zero and sensitivity shifts. Systems of up to 512 channels with acquisition rates to 14,000 measurements per second and throughput of 500 measurements per second are available.

By using *temperature-compensated* microsensors, individually characterized at 9 pressures and 10 temperatures, calibration intervals can be extended from several hours to several months.[104] A "quick zero" feature that checks the zero point more often helps to extend the calibration interval.

6.12 SPECIAL TOPICS

Pressure Distribution

In varied applications (gasketed joints, clutch and brake pressure plates, bolted joints, tire tread patterns, human foot and other biomechanical pressures, roll pressures in coating machines and printing presses, etc.), it is useful to know the *distribution* of the pressure over the contacting surfaces. In theoretical predictions of such pressures, the assumption of a *uniform* pressure is often made to allow a simple analysis, knowing full well that this is not exactly the case. Functional problems often are related to the *nonuniformity* of the pressure, and we then need to study the pressure distribution. This turns out to be one of the most difficult measurements we encounter. (Note that in contrast with the rest of this chapter, we are now usually dealing with pressure between *solid* (rather than fluid) objects.)

One approach[105] interposes between the contacting surfaces a thin (about 0.006 in) but tough plastic film that contains microcapsules (of varying sizes) of color-forming material, and a color-developing layer. When the pressure is applied, the microcapsules break, the larger ones breaking at low pressure and the smaller ones at high pressure. The result is an image of the pressure surface in which the color density (say, "shades of red") is related to the local pressure. One can "read" the actual pressure by visually comparing with calibration charts or more quickly and precisely by using a special electro-optic densitometer (spatial resolution 2 mm) made by Fuji. The film is available in 5 ranges: from 28 to 85 psi to 7,100 to 18,500 psi. The quoted accuracy is about ±10%. One should of course be careful to consider whether the presence of the film significantly *changes* the pressure that would be there without the film. While solid-to-solid applications are most common, the film has been used to study the pressure distribution created by the impingement of high-velocity water jets, used to remove mill scale from steel ingots.

[104]C. A. Matthews, "Intelligent Pressure Measurement in Multiple Sensor Arrays," *Sensors,* March 1996, pp. 26–30 (Scanivalve Corp.).

[105]Fuji Photo Film Co., Ltd., distributed by Sensor Products, Inc., East Hanover, NJ, 201-884-1755.

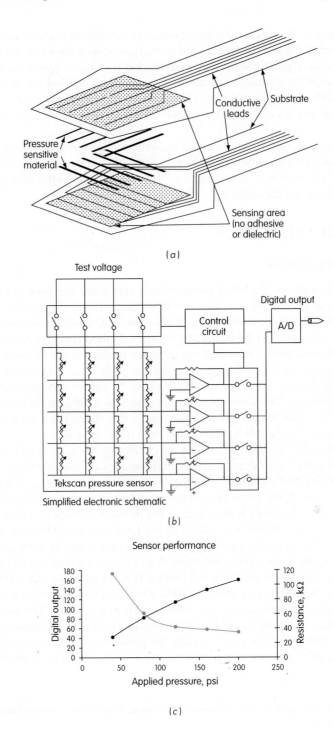

Figure 6.53
Sensor matrix used to measure pressure distribution.

Photoluminescent paints and special lighting have been used to map both pressure and temperature distributions in wind-tunnel models.[106] The procedure is quite complex; consult the references for details. In addition to the *normal* pressure measured by the above paints, there is also interest in the *shear* stress distribution on the aerodynamic surface. Shear-sensitive liquid crystal coatings have been used[107] to make this measurement.

If a continuous electrical output capable of dynamic sensing is desired, a "crossbar" type of multisensor system is available.[108] In Fig. 6.53*a,* strips of piezo-resistive ink are deposited in a "row and column" pattern and sandwiched between two plastic sheets with a similar pattern of conductive leads. The total thickness of the "sandwich" is 0.004 in, the spacing between rows and columns can be as small as 0.5 mm. At each crossing point of the conductive ink lines there is formed a pressure-sensitive resistor that senses the local pressure at that point by changing its electrical resistance. Pressure ranges go from 0 to 15 kPa to 0 to 175 MPa. Figure 6.53*b* shows the readout electronics arrangement. A standard system can read out 250,000 individual cells per second. Software provides a three-dimensional image showing local pressure as a function of *xy* coordinates. The relation between local pressure, sensor resistance, and digital output signal is shown in Fig. 6.53*c.*

Overpressure Protection for Gages and Transducers

In Sect. 6.6, we were concerned that our instrument would not adequately respond to fast-changing pressures. We now need to warn that *damage* or *failure* can be caused by transient pressure "spikes" that we might have no interest in measuring. A prime example is found in fluid-power hydraulic systems where transducers are used as part of a control system. For example, in a die-casting machine whose nominal working pressure was about 1300 psi, pressure spikes of up to 20,000 psi and 0.25 msec duration routinely occurred during normal operation and led to transducer failure.[109] The reference gives guidance on how to reduce such spikes, but protective devices called *snubbers*[110] may still often be needed. These are relatively simple and inexpensive fluid resistance devices of various forms: porous plugs, orifices, pistons, or capillary tubes. When properly sized, these can protect the transducer but still allow adequate speed of response.

---

[106]M. J. Morris and J. F. Donovan, "Application of Pressure- and Temperature-Sensitive Paints to High-Speed Flows," *AIAA,* 94-2231, 1994; L. Vandendorpe, "Aerospace Paints with a Purpose," *R&D Magazine,* May 1997, pp. 69–70.

[107]D. C. Reda and M. C. Wilder, "The Shear-Sensitive Liquid Crystal Coating Method," *Sensors,* Oct. 1998, pp. 38–48.

[108]Tekscan, Inc., South Boston, MA, 800-248-3669 (www.tekscan.com).

[109]B. H. Shapiro, "Coping with Transducer-Crippling Pressure Spikes," *Hydraulics and Pneumatics,* Nov. 1989, pp. 80–81; P. Shipley, "Taking the Pressure off," *Flow Control,* Sept. 1998.

[110]Dynamic Fluid Components, West Union, SC, 800-988-1276 (www.dynamicfc.com); Ray Pressure Snubber, OMS, Charlotte, NC, 888-768-2237 (www.raysnubber.com).

PROBLEMS

6.1   For the system of Fig. 6.2:
   (a) By what factor must the actual weight of steel weights be multiplied to correct for air buoyancy?
   (b) What correction must be applied to the platform weight to account for oil buoyancy if the piston is immersed 5 in and has a diameter of 0.2 in? Take oil specific weight as 50 lbf/ft^3.
   (c) If, in part *b*, air, rather than oil, is the pressure medium, what would the correction be when the gage pressure is 100 lb/in^2 gage and temperature is 70°F? Make an estimate, assuming constant air temperature and pressure varying linearly from the high-pressure end of the piston to atmospheric at the top end.

6.2   A well-type mercury manometer employed to measure water flow rate is shown in Fig. P6.1 for zero flow rate ($p_1 = p_2$). Derive a relation between $p_1 - p_2$ and h for this configuration.

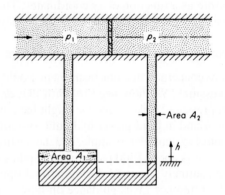

Figure P6.1

6.3   For the inclined manometer of Fig. 6.5, derive a relation between $p_1 - p_2$ and the displacement reading along the calibrated scale.

6.4   Estimate the largest step change that will give linear behavior in a water manometer with $L = 26.5$ in and $R = 0.13$ in. What are ζ and ω_n for this manometer? If a step change five times the value found above is applied, estimate ζ_e and ω_n for this situation.

6.5   In Eq. (6.44), get an expression for $d\zeta_e/\zeta_e$ by taking the log of both sides and then differentiating. If L, μ, γ, g, R, ω, and Y are each in error by 1 percent, what is the percentage error in ζ_e?

6.6   From Eq. (6.47), plot p versus y_c/t for $0 \leq y_c/t \leq 1$ and $E = 25 \times 10^6$ lb/in^2, $t = 0.04$ in, $R = 4.0$ in, and Poisson's ratio = 0.26.

6.7 Design a pressure pickup and bridge circuit such as that in Fig. 6.12 to meet the following requirements:

Maximum pressure = 100 lb/in² gage
Natural frequency in vacuum = 10 Hz minimum
Maximum nonlinearity by Eq. (6.46) = 3 percent
Full-scale output = 10 mV minimum
Diaphragm material, stainless steel.
Strain gages with 350-Ω resistance, gage factor of 2, and size 0.3 by 0.3 in

6.8 A pressure pickup as in Fig. 6.12 has the following characteristics:

R = 3.0 in	$E = 28 \times 10^6$ lb/in²	gage resistance = 120 Ω
r_r = 2.5 in	$\mu = 0.26$	gage factor = 2.0
r_t = 0.5 in	$\gamma = 0.3$ lbf/in³	battery voltage = 5.0 V
t = 0.05 in		

(*a*) Calculate the sensitivity in mV/(lb/in²).
(*b*) What is the natural frequency in vacuum?
(*c*) Based on Eq. (6.46), what is the maximum allowable pressure for 2 percent nonlinearity? What is the voltage output at this point?

6.9 A tubing/transducer pressure measurement installation is of the type "liquid-filled, heavily damped, slow acting." Tubing length is 10 in, diameter 0.3 in, and fluid viscosity is 0.0000100 lbf-sec/in². The installation must meet *all* the following requirements:
a. At least 95% accuracy within 0.05 s after a step input.
b. Steady-state error of no more than 2.0 psi for a ramp input of 100 psi/s.
c. Amplitude accuracy no worse than 90% for a sine wave input of frequency 20 Hz.
Find the maximum allowable transducer compliance.

6.10 For a *flush-diaphragm installation,* compare (with the vacuum value) diaphragm natural frequencies if the measured fluid is air or water, using a *steel* diaphragm.

6.11 For the differential-pressure installation of Fig. P6.2, derive an expression for the time constant if the system is "liquid-filled, heavily damped, slow-acting." The liquid is the same on both sides of the diaphragm.

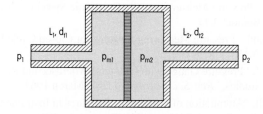

Figure P6.2

6.12 For the system of Fig. P6.2, but now for the liquid-filled fast-acting model, derive an expression for the natural frequency.

6.13 From Eq (6.57) compute τ for a system with $C_{vp} = 0.4$ cm^3/(lb/in^2), $d_t = 0.10$ in, $L = 10$ ft, and $\mu = 0.001$ lbf $\cdot$ s/ft^2. Using Eq. (6.58) find the dead time associated with this system if the tube is of steel with a wall thickness of 0.02 in, $E_L = 100,000$ lb/in^2, and the fluid specific weight is 0.03 lbf/in^3. Is this dead time significant relative to τ?

6.14 A pressure transducer has a natural frequency in vacuum of 5,000 Hz and $C_{vp} = 0.0003$ in^3/(lb/in^2). It is used with a liquid of specific weight 0.04 lbf/in^3 and viscosity 0.0005 lbf $\cdot$ s/ft^2. The tubing inside diameter is 0.2 in, and its length is 5 ft. Find ω_n and ζ of the combined transducer/tubing system.

6.15 The pressure pickup of Prob. 6.14 has an internal volume of 0.004 in^3. If it is used with the same tubing as in Prob. 6.14 but the fluid medium is changed to air at 100 lb/in^2 absolute and 100°F, what are the values of ζ and ω_n ?

6.16 If the transducer of Prob. 6.15 is used with tubing of 1-in length and 0.1-in inside diameter, what will ω_n and ζ be?

6.17 Compute the resistance change of 100-Ω coils of Manganin and gold chrome for 50,000 lb/in^2 pressure and 100°F temperature changes.

6.18 Design a capillary leak for a microphone such that frequencies of 10 Hz and above will be measured with an amplitude-ratio error of no more than 10 percent. Assume standard atmospheric air and a leak length L of 1 in. Find the required leak diameter d_t. The microphone has an internal volume of 0.5 in^3. What will be the amplitude ratio for atmospheric pressure drifts of 2 cycles/h frequency?

6.19 For a liquid-filled transducer/tubing system, modify the text analysis to get a formula for ω_n if the tubing is made of two sections of different lengths and diameters.

6.20 For a liquid-filled transducer/tubing system, modify the text analysis to account for a gas bubble of volume V_g trapped at the transducer diaphragm.

6.21 Derive formulas for the compliance C_{vp} of U-tube, well-type, and inclined manometers.

BIBLIOGRAPHY

1. P. Smelser, "Pressure Measurements in Cryogenic Systems," National Bureau of Standards, Boulder, CO.

2. R. E. Engdahl, "Pressure Measuring Systems for Closed Cycle Liquid Metal Facilities," *NASA*, CR-54140.

3. D. Baganoff, "Pressure Gauge with One-tenth Microsecond Risetime for Shock Reflection Studies," *Rev. Sci. Instrum.*, p. 288, March 1964.

4. A. S. Iberall, "Attenuation of Oscillatory Pressures in Instrument Lines," *Natl. Bur. Std. (U.S.), Res. Paper* RP2115, *ASME Trans.*, July, 1950.

5. D. Alpert, "Theoretical and Experimental Studies of the Underlying Processes and Techniques of Low Pressure Measurement," *NASA,* N-64-17582, 1964.

6. K. Chijiiwa and Y. Hatamura, "Miniature Gages for Soil, Grains and Powders," *Bull. JSME,* vol. 15, no. 82, pp. 455–465, 1972.

7. D. S. Pallett and M. A. Cadoff, "The National Measurement System for Acoustics," *Sound & Vib.,* pp. 20–31, October 1977.

8. D. G. Fleming (ed.), "Indwelling and Implantable Pressure Transducers," CRC Press, Boca Raton, FL, 1977.

9. A. Noordergraaf, "Circulatory System Dynamics," Academic, New York, 1978.

10. R. P. Benedict, "Fundamentals of Temperature, Pressure and Flow Measurement," 2d ed., Wiley, New York, 1977.

11. R. Clayton, "Setting Zero on an Absolute Pressure Transducer," *I&CS,* Nov. 1998, p. 49–51.

12. M. L. Dunbar and K. Sager, "A Novel, Media-Compatible Pressure Sensor for Automotive Applications," *Sensors,* Jan. 2000, pp. 28–34.

13. W. S. Czaranocki and J. P. Schuster, "The Evolution of Automotive Pressure Sensors," *Sensors,* May 1999, pp. 52–65.

14. P. Van Vessem and D. Williams, "Rediscovering the Strain Gauge Pressure Sensor," *Sensors,* Apr. 1999, pp. 36–40.

15. G. Bitko et al., "Improving the MEMS Pressure Sensor," *Sensors,* July 2000, pp. 62–67.

16. J. G. Webster (ed.), "The Measurement, Instrumentation, and Sensors Handbook," CRC Press, Boca Raton, FL, 1999, chap. 26.

17. S. Soloman, "Sensors Handbook," McGraw-Hill, New York, 1999.

7
CHAPTER

Flow Measurement

7.1 LOCAL FLOW VELOCITY, MAGNITUDE AND DIRECTION

In many experimental studies of fluid flow phenomena, it is necessary to determine the magnitude and/or direction of the flow-velocity at a point in the fluid and how this varies from point to point. That is, a description of the flow field is desired. Various methods of *flow visualization* allow us to gain an overall view of flow patterns. Sometimes the qualitative information available from direct visual observation is sufficient; however, most methods also allow quantitative analysis. Once (by flow visualization or other sources of information) localized regions of particular interest have been pinpointed, it may be necessary to insert *velocity probes* to obtain accurate point measurements. Such probes (pitot-static tubes, hot-wire anemometers, and laser-doppler velocimeters are the most common) always involve a sensing volume of finite size. Thus true "point" measurements are impossible. However, sensing volumes can be made sufficiently small to provide data of practical utility.

Flow Visualization[1]

The majority of flow visualization schemes are based on one of two basic principles: the introduction of tracer particles or the detection of flow-related changes in fluid optical properties. In liquids, colored dyes and gas bubbles are common tracers. A line of hydrogen bubbles, for example, can be formed in water by applying a short electric pulse to a straight wire immersed in the flow. Photography with steady illumination shows the bubbles as short streaks whose length can be measured to obtain velocity data, while stroboscopic light gives a series of dots whose spacing gives similar information. For gas flows, smoke, helium-filled "soap"

---

[1]W. Merzkirch, "Making Fluid Flows Visible," *Am Sci.,* vol. 67, pp. 330–336, May–June 1979; W. Merzkirch, "Flow Visualization," Academic, New York, 1987; T. Asanuma (ed.), "Flow Visualization," Hemisphere, New York, 1979; "Flow Visualization," vol. 3, Hemisphere, New York, 1984.

bubbles, or gas molecules made luminous by an ionizing electric spark have served as tracers.

Shadowgraph, schlieren, and interferometer techniques[2] employ, in different ways, the variation in refractive index of the flowing gas with density. For compressible flows (Mach number above about 0.3), density varies with velocity sufficiently to produce measurable effects. In shadowgraph and schlieren methods, light and dark patterns related to flow conditions are produced by the bending of light rays as they pass through a region of varying density. The optical apparatus for these two methods is relatively simple to use, and they are widely utilized, mostly for qualitative studies. For quantitative results, the more difficult interferometer approach may be necessary. Here the light/dark patterns are formed by interference effects resulting from phase shifts between a reference beam and the measuring beam. For no flow, a regular grid of light/dark fringes is present. When flow occurs, this grid is distorted and numerical values of density can be calculated from the fringe displacements.

An important "visualization" technique that also produces *quantitative* results is *partical image velocimetry*[3] (PIV). First developed in the 1980s, it is widely used today, with hardware and software available from several manufacturers. Our brief discussion below is based on a reference[4] that also gives useful information on laser doppler anemometry (LDA) and thermal (hot-wire/film) anemometry. As in LDA, the liquid or gas flow must be seeded with near-neutrally-buoyant particles; the velocity of these particles is what is actually measured. Using cylindrical lenses, a periodically pulsed laser beam is expanded into a two-dimensional "light sheet," defining a plane in the flow where velocity will be measured. The light sheet actually has some thickness, controlled by a spherical lens to be about 1 mm, so that there is not too much out-of-plane motion. (A related technique[5] allows for three-dimensional measurements.) Using a multiple-exposure method, the locations of the particles at two times separated by a known time interval are used to compute the magnitude and direction of the particle's velocity (see Fig. 7.1*a*). Complete systems can use either film or electronic cameras to record the needed images; Fig. 7.1*b* shows a CCD camera, the type most used.

Applications range from low-velocity natural convection flows up through supersonic flows. A complete system consists of a laser light source with optics, an image recording medium, a programmable time delay and sequence generator, camera interface, computer, and image acquisition/analysis software. While Fig. 7.1*a* shows the basic principle, a software *cross-correlation* method is actually

---

[2]A. Seiff, "Shadow, Schlieren, and Interferometer Photographs, Ballistic Range Technology," chap. 8, *AGARDOGRAPH*-138, 1971.

[3]I. Grant, "Partical Image Velocimetry: A Review," *Proc. Instn. Mech. Engrs.,* vol. 211, part C, 1997, pp. 55–76.

[4]"Advances in Fluid Flow Diagnostics," (printed notes for seminar, 1998), TSI Inc., St. Paul, MN, 800-874-2811 (www.tsi.com).

[5]Y. G. Guezennec et al., "Algorithms for Fully Automated Three-Dimensional Particle Tracking Velocimetry," *Experiments in Fluids,* vol. 17, 1994, pp. 209–219. (Also SAE paper 940279, 1994.)

PIV principle

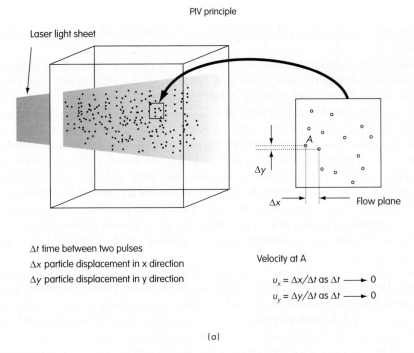

Δt time between two pulses
Δx particle displacement in x direction
Δy particle displacement in y direction

Velocity at A
$$u_x = \Delta x / \Delta t \text{ as } \Delta t \longrightarrow 0$$
$$u_y = \Delta y / \Delta t \text{ as } \Delta t \longrightarrow 0$$

(a)

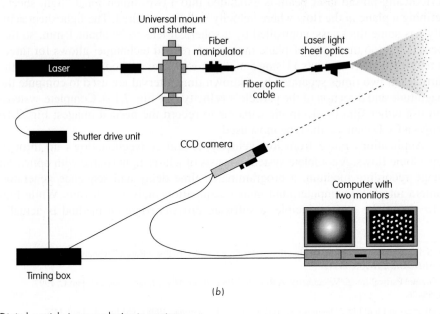

Digital particle image velocimetry system.

(b)

Figure 7.1
Particle-image velocimetry. *(a) (Courtesy TSI Inc.); (b) (Dantec Measurement Technology, Inc., Mahway, NJ, 201-512-0037, www.dantecmt.com.)*

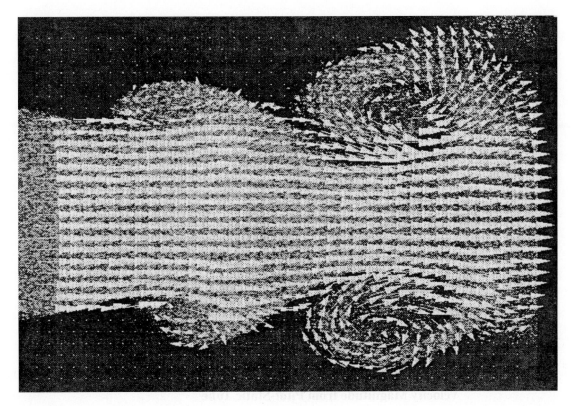

Figure 7.2
PIV image of transient vortex structure in a circular jet. *(Courtesy TSI, Inc.)*

used to get the particle displacements from image to image. Figure 7.2 shows the transient vortex structure in a circular jet, as revealed by PIV measurement. The accuracy of PIV data depends on many factors; details are available in the literature.[6] Under good conditions and with proper equipment and technique, the accuracy is adequate for many applications. When seed particles are unacceptable, *molecular tagging velocimetry,*[7] a technique more complex than PIV, may be applicable. Here fluid molecules themselves are turned into tracers when excited by photons of appropriate wavelength.

Versions of the holographic techniques[8] discussed in Chap. 4 have been developed for flow measurements. Combinations[9] of several visualization methods are sometimes helpful in difficult problems.

[6]"How Accurate Is PIV Data?" *Dantec Newsletter,* vol. 5, no. 2, 1998.

[7]M. M. Koochesfahani and D. G. Nocera, "Molecular Tagging Velocimetry Maps Fluid Flows," *Laser Focus World,* June 2001, pp. 103–108.

[8]W. A. Benser, "Holographic Flow Visualization within a Rotating Compressor Blade Row," *NASA* TMX-71788, August 1975.

[9]R. Sedney, C. W. Kitchens. Jr., and C. C. Bush, "Combined Techniques for Flow Visualization," *AIAA Paper* 76–55, 1976.

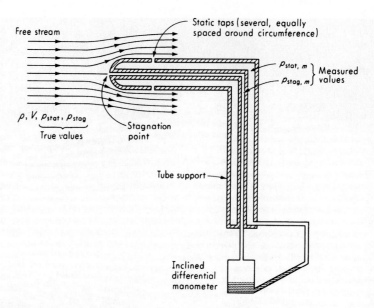

Figure 7.3
Pitot-static tube.

Velocity Magnitude from Pitot-Static Tube

In some situations the direction of the velocity vector is known with sufficient accuracy without taking any measurements. If the direction is not known, it may be found in several ways discussed later. Let us assume the direction is known, so that a pitot-static tube may be properly aligned with this direction, as in Fig. 7.3. Assuming steady one-dimensional flow of an incompressible frictionless fluid, we can derive the well-known result

$$V = \sqrt{\frac{2(p_{\text{stag}} - p_{\text{stat}})}{\rho}} \tag{7.1}$$

where
$V \triangleq$ flow velocity

$\rho \triangleq$ fluid mass density

$p_{\text{stag}} \triangleq$ stagnation or total pressure, free stream

$p_{\text{stat}} \triangleq$ static pressure, free stream

In an actual pitot-static tube, deviations from the ideal theoretical result of Eq. (7.1) arise from a number of sources. If ρ is accurately known, the errors can be traced to inaccurate measurement of p_{stag} and p_{stat}.

The static pressure is usually the more difficult to measure accurately. The difference between true (p_{stat}) and measured ($p_{\text{stat, }m}$) values of static pressure may be due to the following:

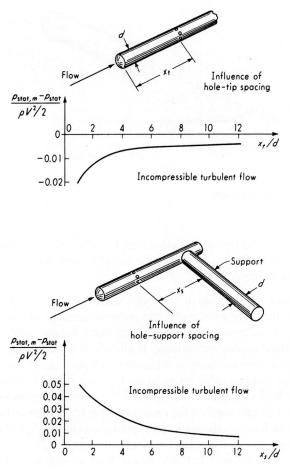

Figure 7.4
Static-pressure errors.

1. Misalignment of the tube axis and velocity vector. This exposes the static taps to some component of velocity.

2. Nonzero tube diameter. Streamlines next to the tube must be longer than those in undisturbed flow, which indicates an increase in velocity. This is accompanied by a decrease in static pressure, which makes the static taps read low. A similar (and possibly more severe) effect occurs if a tube is inserted in a duct whose cross-sectional area is not much larger than that of the tube.

3. Influence of stagnation point on the tube-support leading edge. This higher pressure causes the static pressure upstream of the leading edge also to be high. If the static taps are too close to the support, they will read high because of this effect. Note that this error and that of item 2 above tend to cancel. By

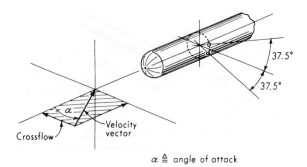

$\alpha \triangleq$ angle of attack

Figure 7.5
Probe insensitive to misalignment.

proper design, effective cancellation may be achieved[10] (see also Prandtl pitot tube, Fig. 2.17b). Figure 7.4 shows the nature of both these errors as revealed by experimental tests.[11]

An important application of the pitot-static tube is found in aircraft and missiles.[12] Here the stagnation- and static-pressure readings of a tube fastened to a vehicle are used to determine the airspeed and Mach number while the static reading alone is utilized to measure altitude. If altitude is to be measured with an error of 100 ft, the static pressure must be accurate to 0.5 percent.[13] To achieve this accuracy, methods for compensating errors of the type mentioned in items 1, 2, and 3 of the above have been developed and reported.[14] An interesting and useful result of these studies is a simple method for reducing error from angular misalignment. It was found that by locating the static-pressure taps as in Fig. 7.5, the measurement was essentially insensitive to angle of attack for the range $-2° < \alpha < 12°$ and Mach numbers in the range 0.4 to 1.2. While this method, as shown, is effective only for misalignment in the particular plane shown, it can be extended to arbitrary directions of misalignment by designing the probe with a single vane to rotate about its longitudinal axis and automatically locate the taps 37.5° from the cross-flow stagnation point.[15] It is also possible to design multiple-vaned probes[16] mounted on a gimbal system with complete rotational freedom. These probes act in a fashion similar to a weather vane (except that they have angular freedom about two axes)

[10]V. S. Ritchie, "Several Methods for Aerodynamic Reduction of Static Pressure Sensing Errors for Aircraft at Subsonic, Near-Sonic, and Low Supersonic Speeds," *NASA, Tech. Rept.* R-18, 1959.

[11]R. G. Folsom, "Review of the Pitot Tube," *Trans. ASME,* p. 1450, October 1956.

[12]W. Gracey, "Measurement of Aircraft Speed and Altitude," Wiley, New York, 1981; E. A. Haering, Jr., "Airdata Measurement and Calibration," *NASA TM* 104316, 1995.

[13]Ritchie, op. cit.

[14]Ibid.

[15]F. J. Capone, "Wind-Tunnel Tests of Seven Static-Pressure Probes at Transonic Speeds," *NASA, Tech. Note* D-947, 1961.

[16]Ibid.

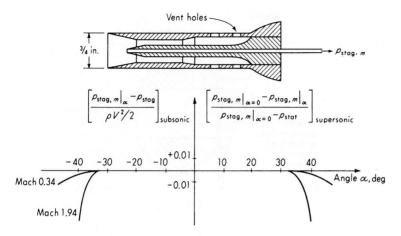

Figure 7.6
Special stagnation probe.

and thus align themselves with the velocity vector. A conventional ring of evenly spaced static taps then gives accurate readings. By measuring the rotations of the gimbals, such probes also provide information on the *direction* of the velocity.

While errors in the stagnation pressure are likely to be smaller than those in the static pressure, several possible sources of error are present, namely:

1. Misalignment. This situation prevents formation of a true stagnation point at the measuring hole since the velocity is not zero. Tubes of special design have been developed which exhibit considerable tolerance to misalignment.[17] Figure 7.6 shows an example of such a tube which has an error less than 1 percent of the velocity pressure $\rho V^2/2$ for misalignments up to $\pm 38°$ for velocities from low subsonic to Mach 2. Conventional tubes not specifically designed for misalignment insensitivity may show 1 percent errors at only 5 or 10°.

2. Two- and three-dimensional velocity fields. When the velocity is not uniform, a probe of finite size intercepts streamlines of different velocities and the stagnation pressure measured corresponds to some sort of average velocity (see Fig. 7.7a). For the two-dimensional situation of Fig. 7.7a, if we knew the displacement δ, we could assign the measured stagnation pressure (and thus velocity) to a specific point in the flow. Some limited data[18] on this problem are available.

3. Effect of viscosity. Equation (7.1) assumes the fluid to be frictionless. At sufficiently low Reynolds number, the viscosity of the fluid exerts a

[17]W. Gracey, D. E. Coletti, and W. R. Russell, "Wind-Tunnel Investigation of a Number of Total-Pressure Tubes at High Angles of Attack, Supersonic Speeds," *NACA, Tech. Note* 2261, January 1951; W. Gracey, W. Letko, and W. R. Russell, "Wind-Tunnel Investigation of a Number of Total-Pressure Tubes at High Angles of Attack, Subsonic Speeds," *NACA, Tech. Note* 2331, April 1951.

[18]Folsom, op. cit., p. 1451.

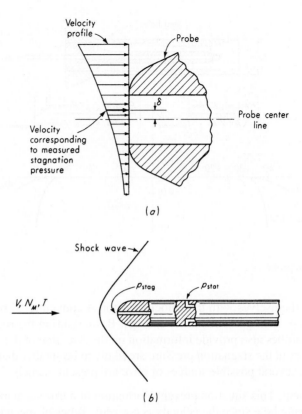

Figure 7.7
Nonuniform-velocity profile and supersonic probe.

noticeable additional force at the stagnation hole, causing the stagnation pressure to be higher than predicted by Eq. (7.1). This effect can be taken into account by introducing a correction factor C as follows:

$$p_{\text{stag}, m} = p_{\text{stat}} + \frac{C\rho V^2}{2} \tag{7.2}$$

For negligible viscosity effects, $C = 1.0$ and Eq. (7.2) is the same as (7.1). For a given probe, the factor C is a function of Reynolds number only and may be found theoretically for simple probe shapes such as spheres and cylinders. A typical result[19] for a cylindrical probe is

$$C = 1 + \frac{4}{N_R} \qquad 10 < N_R < 100 \tag{7.3}$$

where Reynolds number $\triangleq N_R \triangleq V\rho r/\mu$, $r \triangleq$ probe radius, and $\mu \triangleq$ fluid viscosity. Equation (7.3) shows that the effect is about 4 percent of the

---

[19]Ibid., p. 1453.

velocity pressure $\rho V^2/2$ at $N_R = 100$. At $N_R = 10$, however, the effect is 40 percent. Theory and experimental tests[20] show that viscosity corrections are rarely needed for $N_R > 500$, no matter what the shape of the probe.

When a pitot-static tube is employed in a compressible fluid, Eq. (7.1) no longer applies, although it may be sufficiently accurate if the Mach number is low enough. For subsonic flow (Mach number $N_M < 1$) the velocity is given by[21]

$$V = \sqrt{\frac{2k}{k-1}\frac{p_{\text{stat}}}{\rho_{\text{stat}}}\left[\left(\frac{p_{\text{stag}}}{p_{\text{stat}}}\right)^{(k-1)/k} - 1\right]} \tag{7.4}$$

where
$$k \triangleq \frac{\text{specific heat at constant pressure}}{\text{specific heat at constant volume}} = \frac{C_p}{C_v} \tag{7.5}$$

Measurement of free-stream density ρ_{stat} requires knowledge of static temperature, which may itself be a difficult measurement. Equation (7.4) may be rewritten as

$$p_{\text{stag}} = p_{\text{stat}}\left[1 + \frac{k-1}{2}\left(\frac{V}{c}\right)^2\right]^{k/(k-1)} \tag{7.6}$$

where
$$c \triangleq \text{acoustic velocity} = \sqrt{\frac{kp_{\text{stat}}}{\rho_{\text{stat}}}} = \sqrt{kgRT} \tag{7.7}$$

and
$$g \triangleq \text{gravitational acceleration}$$
$$T \triangleq \text{free-stream static temperature}$$
$$R \triangleq \text{gas constant}$$

The right side of Eq. (7.6) may be expanded in a power series to give

$$p_{\text{stag}} = p_{\text{stat}} + \left(\rho_{\text{stat}}\frac{V^2}{2}\right)\left(1 + \frac{N_M^2}{4} + \frac{2-k}{24}N_M^4 + \cdots\right) \tag{7.8}$$

where
$$N_M \triangleq \frac{V}{c} \tag{7.9}$$

Since the Mach number of an incompressible fluid is zero, Eq. (7.8) shows that p_{stag} is higher for compressible than for incompressible flow. Also, if N_M is sufficiently small, Eq. (7.8) is closely approximated by Eq. (7.1).

For supersonic flow ($N_M > 1$), a compression shock wave forms ahead of the pitot tube. Between this shock wave and the tube end, the velocity is subsonic. This subsonic velocity is then reduced to zero at the tube stagnation point (see Fig. 7.7b). Analysis[22] gives the following formula for computing the free-stream Mach number and thus the velocity:

$$\frac{p_{\text{stag}}}{p_{\text{stat}}} = N_M^2\left(\frac{k+1}{2}\right)^{k/(k-1)}\left[\frac{2kN_m - k + 1}{N_M^2(k+1)}\right]^{1-1/(k-1)} \tag{7.10}$$

---

[20]Ibid.

[21]R. C. Binder, "Advanced Fluid Dynamics and Fluid Machinery," p. 51, Prentice-Hall, Englewood Cliffs, NJ, 1951.

[22]Ibid., p. 52.

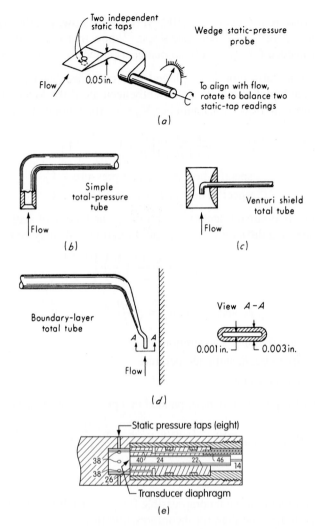

Figure 7.8
Pressure probes.

The measurement of stagnation and static pressures may be combined in a single probe, as in Fig. 7.3, or two separate probes, one for stagnation and the other for static, may be employed. Figure 7.8 shows several examples[23] of commonly used forms. The wedge static-pressure probe of Fig. 7.8a also can be used to measure velocity direction in a single plane. When the two static taps read equal pressures, the wedge is aligned with the flow. This probe is usable for both subsonic and supersonic flow. At Mach 0.9 the sensitivity to misalignment is about 1.5 in of water

---

[23]Aero Research Instrument Co., Chicago.

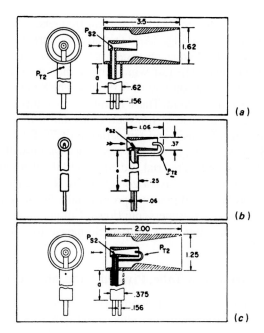

Figure 7.9
Boost-venturi pitot tubes. *(Courtesy United Sensor Control Corp., Watertown, MA.)*

per angular degree. The total (stagnation) probes of Fig. 7.8*b* and *c* also are intended for both sub- and supersonic flow. The simple tube is insensitive to misalignment up to about $\pm 20°$ while the venturi shielded tube is good to $\pm 50°$. The boundary-layer probe is usable up to Mach 1.0 and is insensitive to misalignment up to $\pm 5°$. Boundary-layer thickness can be measured with such a probe with an error of the order of 0.002 in. The probe and associated pressure-measuring equipment have a long time lag because of the small flow passage (0.001 in) at the probe tips.

At low velocities, where the pitot-static tube becomes insensitive, a larger Δp signal may be obtained by using a version called a "boost venturi" (Fig. 7.9). Here p_{stag} is the normal value; the measured p_{stat} is *not* the free-stream value, but instead is artificially lowered by locally accelerating the flow in a venturi section. Figure 7.9*a* and *b* shows two different designs while Fig. 7.9*c* combines them to get the maximum effect, about a 25:1 increase in Δp, compared with the value for a normal pitot-static tube. These instruments require some care in use since they must be individually calibrated and are sensitive to variations in Mach number and/or Reynolds number.

Long, small-diameter tubing used with pitot-static tubes leads to a slow dynamic response. Some modern aircraft use pressure data in their control systems and thus require a fast response; a small time lag (fast response) is also needed in some wind-tunnel testing. These needs can sometimes be met using tiny pressure transducers[24] (0.062-in diameter, 225 kHz natural frequency) located very close to the pressure-sensing hole, as in Fig. 7.8e[25] (a static pressure probe). For a chamber volume of 0.00567 in^3, the frequency response (-3 dB) was estimated at about 800 Hz from a shock tube step test which the author interpreted as first-order dynamic behavior with a time constant of 0.196 ms. The 8 pressure tap holes had a diameter of 0.02 in; length 0.123 in. These 8 "tubes" are equivalent to a single tube of diameter $\sqrt{8}d_t$, allowing use of Eq. 6.86 (a *second-order* model). I used the reference's air pressure of 7 psia and temperature of 70°F to calculate a natural frequency of 4098 Hz. Since the theoretical model does *not* usually give a good estimate of damping, I adjusted this by trial and error until the theoretical step response matched the measured ($\zeta = 2.5$ gave a good match). Reference 25 also describes other fast-response probes: total pressure probes and a 5-hole probe which allows measurement of flow angles in supersonic flows. Commercial versions, including an 18-hole spherical (9.5 mm dia.) flow-angle-measuring probe have recently appeared.[26]

Velocity Direction from Yaw Tube, Pivoted Vane, and Servoed Sphere

In addition to laboratory studies of flow processes in fluid machinery, ducting, etc., flow-velocity direction information is of interest in flight vehicles[27] where angle-of-attack measurements are utilized in attitude measurement and control, stability augmentation, and gust alleviation systems.

So-called yaw tubes[28] of one form or another conventionally are employed to determine the direction of local flow velocity. Perhaps the simplest form, useful for finding the angular inclination in one plane only, is shown in Fig. 7.10a. Taps 1 and 3 are connected to a differential-pressure instrument that reads zero when the tube is aligned with the flow. A central tap 2 is often included to read the stagnation pressure after alignment is attained (valid only if the angle of attack is zero). The claw tube of Fig. 7.10b operates on similar principles, but may be utilized in regions where the flow direction changes greatly, since its sensing holes may be located very close together. The two-axis probes of Fig. 7.10c and d conceivably could be designed to allow rotation about each axis; however, the complexity and size of such a design are generally prohibitive. Thus probe operation consists of rotation about the probe axis to balance taps 1 and 3. Then pressures 2 and 4 are each measured, and calibration charts give the angle of attack. Tap 5 does not read stagnation pressure directly; this can be obtained from calibration charts. Any of these probes may be made automatically self-aligning by using the pressure difference $p_1 - p_3$ as

[24]Kulite Semiconductor Products, Inc., Leonia, NJ, 201-461-0900 (www.kulite.com).

[25]a. A. R. Porro, "Results of Dynamic Testing of Static Pressure Probes," NASA Lewis (now Glenn) Interoffice Memo, Oct. 23, 1996; b. NASA/TM-2001-211096, Pressure Probe Designs for Dynamic Pressure Measurements in a Supersonic Flow Field, July 2001.

[26]Aeroprobe, Dantec Dynamics, www.dantecdynamics.com, 2002.

[27]H. H. Koelle, "Handbook of Astronautical Engineering," pp. 13–33, McGraw-Hill, New York; E. A. Haering, Jr., op. cit., 1961.

[28]Aero Research Instrument Co.

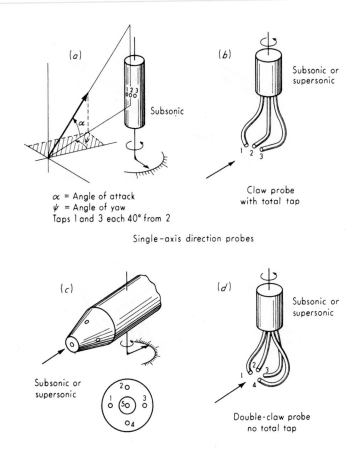

α = Angle of attack
ψ = Angle of yaw
Taps 1 and 3 each 40° from 2

Single-axis direction probes

Two-axis direction probes

Figure 7.10
Flow-direction probes.

the error signal in a servosystem which rotates the probe until a null is achieved. The details of a system of this type are shown in Fig. 7.12.

Determination of angles of attack and yaw aboard flight vehicles is often accomplished with vane-type probes[29] as in Fig. 7.11. These devices are essentially one- or two-axis weather vanes with suitable damping to reduce oscillation and with motion pickups to provide electrical angle signals. A dynamic analysis of these devices is available.[30] Limitations of this type of device for certain high-speed, high-altitude applications have led to the development of the servoed-sphere type of

[29]Space Age Control, Inc., Palmdale, CA, 661-273-3000 (www.spaceagecontrol.com).

[30]G. J. Friedman, "Frequency Response Analysis of the Vane-Type Angle of Attack Transducer," *Aero/Space Eng.,* p. 69, March 1959; P. S. Barna and G. R. Crossman, "Experimental Studies of Flow Direction Sensing Vanes," *NASA CR*-2683, May 1976.

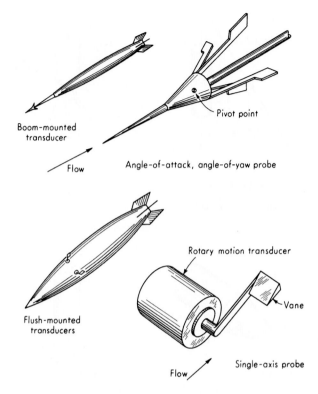

Figure 7.11
Vane-type probes.

sensor shown in Fig. 7.12. The one shown was developed[31] for the X-15 rocket research aircraft. A servo-driven sphere is continuously and automatically aligned with the velocity vector by means of two independent servosystems using the differential-pressure signals $p_1 - p_2$ and $p_3 - p_4$ as error signals. A fifth tap measures the stagnation pressure. Block diagrams and frequency response of a single axis are given in Fig. 7.12. The angle-of-attack axis is designed for the range -10 to $+40°$ while the sideslip axis covers $\pm20°$. Static inaccuracy of angle measurement is 0.25°.

A five-tap sensor similar to that of Fig. 7.10c but requiring no rotation is available for aircraft applications (see Fig. 7.13[32]). The installation shown combines the five-tap angle sensor, integral pitot-static taps (P_{p2}, P_{s2}), and a separate side-mounted pitot-static tube for pilot instruments (P_{p1}, P_{s1}). Angle of attack α is related to pressure difference $P_{\alpha1} - P_{\alpha2}$; however, the relation varies with altitude and airspeed. Most of this variation can be accounted for by dividing $P_{\alpha1} - P_{\alpha2}$ by a normalizing factor $K_1[P_3 - P_{\beta1} + (P_{\beta1} - P_{\beta2})/2]$, this quotient now being closely

[31]Northrop Corp., Nortronics Div., *Rept.* NORT60-46.

[32]*Bull.* 1013, 1014, Rosemount Inc., Minneapolis, MN.

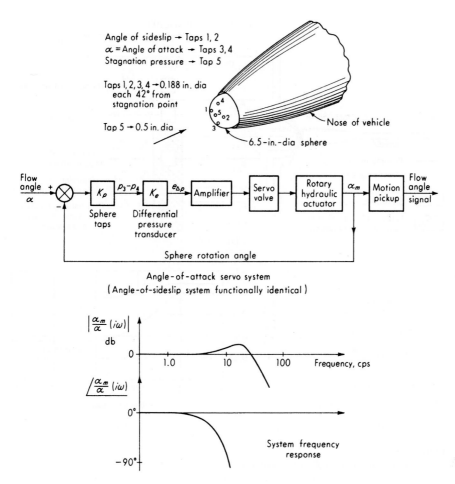

Figure 7.12
Servoed-sphere probe.

proportional to α. The "constant" K_1 still varies somewhat with Mach number (Fig. 7.13c); however, this correction can be included in the system air-data computer. Sideslip angle is measured by an identical scheme. Absolute- and differential-pressure transducers of the capacitance type are employed to perform the desired subtractions. Overall systems of this type, using aerodynamic compensation methods to optimize sensor design for a particular aircraft, can achieve errors as small as 0.25° for flow angles up to ±40° and wide ranges of subsonic and supersonic flight speeds. Additional information on the general class of five-tap sensors is available.[33]

---

[33]M. R. Hale, "The Analysis and Calibration of the Five-Hole Spherical Pitot," *ASME Paper* 67-WA/FE-24, 1967; T. J. Dudzinski and L. N. Krause, "Flow-Direction Measurement with Fixed-Position Probes," *NASA TM X-1904*, October 1969.

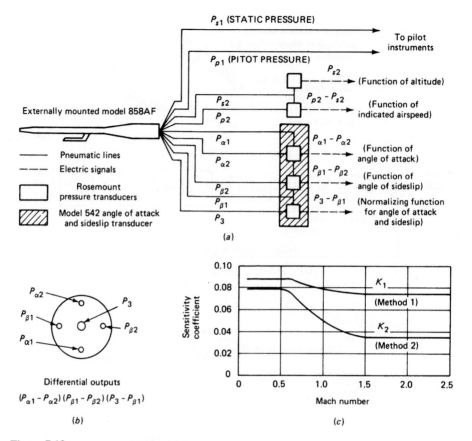

Figure 7.13
Five-tap pitot tube for aircraft instrumentation.

Dynamic Wind-Vector Indicator

Figure 7.14 shows a transducer that measures the magnitude and direction of flow velocity in terms of the drag force exerted on a hollow sphere. The drag force F_d on a body is given by

$$F_d = C_d \frac{A\rho V^2}{2} \tag{7.11}$$

where $C_d \triangleq$ drag coefficient of body
$= 0.567$ for these transducers
$A \triangleq$ projected area of body

Clearly, if C_d, A, and ρ are known, then V may be found by measuring F_d. If all directional components of V are to be equally effective in producing drag force, a body with spherical symmetry must be employed. If this is done, measurement of the x, y, and z components of F_d completely defines the magnitude and direction

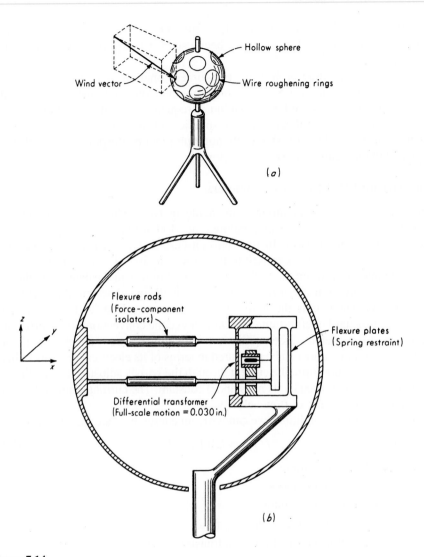

Figure 7.14
Wind-vector indicator.

of V. Since the drag coefficient of a smooth sphere is somewhat dependent on the Reynolds number (and thus V), wire roughening rings are attached to the sphere surface to ensure turbulence. The drag coefficient of the roughened sphere is constant over the entire design range of V for a given transducer. The separation of the total drag force into three rectangular components is accomplished by mounting the sphere on a force-resolving flexure assembly. Figure 7.14b shows the x-axis mechanism of this structure. (The y and z axes are identical except for their orientation.) The force components are applied to flexure-plate springs to produce proportional motions which are then measured by suitable displacement transducers. Sensors of this type have been used for both wind and ocean current measurement.

A miniaturized drag probe[34] for one-dimensional air velocity measurements, using a flat plate target 1.5×1.5 mm mounted on a commercially available semiconductor strain-gage beam, exhibited frequency response to 10 kHz. A two-dimensional version[35] utilized a hollow spherical glass target of 0.153-in diameter mounted on the stylus of a stereo phonograph pickup. Response to 26 kHz was estimated theoretically for 700 ft/s air flow.

Hot-Wire and Hot-Film Anemometers

Hot-wire anemometers commonly are made in two basic forms: the constant-current type and the constant-temperature type. Both utilize the same physical principle but in different ways. In the constant-current type, a fine resistance wire carrying a fixed current is exposed to the flow velocity. The wire attains an equilibrium temperature when the i^2R heat generated in it is just balanced by the convective heat loss from its surface. The circuit is designed so that the i^2R heat is essentially constant; thus the wire temperature must adjust itself to change the convective loss until equilibrium is reached. Since the convection film coefficient is a function of flow velocity, the equilibrium wire temperature is a measure of velocity. The wire temperature can be measured in terms of its electrical resistance. In the constant-temperature form, the current through the wire is adjusted to keep the wire temperature (as measured by its resistance) constant. The current required to do this then becomes a measure of flow velocity.

For equilibrium conditions we can write an energy balance for a hot wire as

$$I^2 R_w = hA(T_w - T_f) \tag{7.12}$$

where

$I \triangleq$ wire current

$R_w \triangleq$ wire resistance

$T_w \triangleq$ wire temperature

$T_f \triangleq$ temperature of flowing fluid

$h \triangleq$ film coefficient of heat transfer

$A \triangleq$ heat-transfer area

Now h is mainly a function of flow velocity for a given fluid density. For a range of velocities, this function (sometimes called *King's law*) has the general form

$$h = C_0 + C_1\sqrt{V} \tag{7.13}$$

For the measurement of average (steady) velocities a "manual balance" constant-temperature mode of operation can be used. Figure 7.15 shows a possible

[34]L. N. Krause and G. C. Fralick, "Miniature Drag-Force Anemometer," *NASA* TM-3507, June 1977.

[35]D. Y. Cheng and P. Wang, "Viscous Force Sensing Fluctuating Probe Technique," *AIAA Paper* 73–1044, 1973.

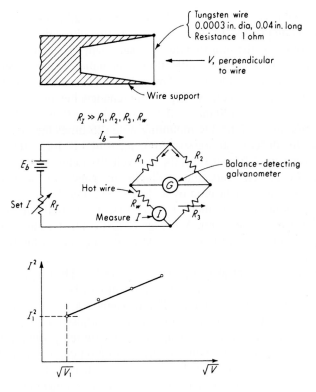

Figure 7.15
Hot-wire anemometer.

circuit arrangement. For accurate work, a given hot-wire probe must be calibrated in the fluid in which it is to be used. That is, it is exposed to *known* velocities (measured accurately by some other means), and its output is recorded over a range of velocities. When velocities are too low to allow direct use of a pitot-static tube as the calibration standard, a number of alternative schemes are possible. By using a flow passage with two sections of widely differing areas connected in series, a pitot tube may be employed in the small-area (high-velocity) section, the hot wire in the large-area (low-velocity) section, and the known area ratio used to infer velocities at the hot wire from those measured at the pitot tube. Another approach[36] utilizes a choked-flow orifice (to establish a mass flow proportional to supply pressure) together with a flow passage with two areas of known ratio, providing a free jet whose velocity is closely proportional to supply pressure. For the lowest velocities, a rig which moves the probe through still fluid (a rotating arm[37] is perhaps most convenient) substitutes the (easy) measurement of solid-body velocity for the more

---

[36]"New Anemometer Calibration Equipment," *DISA Inform.*, no. 13, pp. 37–39, Disa Electronics, Franklin Lakes, NJ, May 1972.

[37]J. Anhalt, "Device for In-Water Calibration of Hot-Wire and Hot-Film Probes," *DISA Inform.*, no. 15, pp. 25–26, Disa Electronics, October 1973.

difficult fluid-velocity measurement. A commercial calibrator[38] uses a stagnation chamber with a compressed air supply to provide flow through a carefully designed nozzle discharging to the atmosphere. Measurement of stagnation pressure and temperature, and nozzle pressure drop allows calculation of the velocity. The reference explains a number of important corrections and features that must be included to achieve the desired accuracy. Three interchangeable nozzles provide three ranges: 0.5 to 60 m/s, 5 to 120 m/s, and 5 m/s to Mach 1.0. A fourth (lowest) range (0.02 to 0.5 m/s) uses the nozzle mounting hole (25 times the area of the largest nozzle). Since the differential pressure to be measured is, for example, only about 0.0002 in of water for 1-ft/s air flow, the low velocity ranges require very sensitive and accurate transducers; capacitance diaphragm gages are often used. On the lowest range, velocity is measured by a "low-speed transducer" rather than using a differential pressure measurement. I phoned Dantec to find out what this meant and was told it was a trade secret. Accuracy of this computer-automated calibrator is quoted as ± 0.5 percent of reading ± 0.02 m/s; a single calibration point takes about 15 s.

In calibration, V is set at some known value V_1. Then R_I is adjusted to set hot-wire current I at a value low enough to prevent wire burnout but high enough to give adequate sensitivity to velocity. The resistance R_w will come to a definite temperature and resistance. Then the resistor R_3 is adjusted to balance the bridge. This adjustment is essentially a measurement of wire temperature, which is held fixed at all velocities. The first point on the calibration curve is thus plotted as I_1^2, $\sqrt{V_1}$. Now V is changed to a new value, causing wire temperature and R_w to change and thus unbalancing the bridge. Then R_w, and thus wire temperature, is restored to its original value by adjusting I (by means of R_I) until the bridge balance is restored (R_3 is *not* changed). The new current I and the corresponding V may be plotted on the calibration curve, and this procedure is repeated for as many velocities as desired.

Once calibrated, the probe can be employed to measure unknown velocities by adjusting R_I until bridge balance is achieved, reading I, and obtaining the corresponding V from the calibration curve. This assumes that the measured fluid is at the same temperature and pressure as for the calibration. Correction methods for varying temperature and pressure are fairly simple, but are not discussed here. For the above constant-temperature mode of operation, Eqs. (7.12) and (7.13) can be combined to give

$$I^2 = \frac{A(T_w - T_f)(C_0 + C_1 \sqrt{V})}{R_w} \triangleq C_2 + C_3 \sqrt{V} \tag{7.14}$$

indicating that the calibration curve of Fig. 7.15 should be essentially a straight line. This is borne out by experimental tests.

While the above described measurement of steady velocities is of some practical interest, perhaps the main application of hot-wire instruments is the

---

[38]T. V. Stannov, "An Accurate, Easy to Use Low-Turbulence Calibrator for Hot-Wire Anemometers," *Dantec Information,* no. 14, Jan. 1995; "Using Streamline Calibrator with Various Gases," *Appl. Note* 04, May 1995, Dantec Measurement Technology Inc.

measurement of rapidly fluctuating velocities, such as the turbulent components superimposed on the average velocity. Both constant-current and constant-temperature techniques are used; first we consider the constant-current operation. Figure 7.16 shows the basic arrangement. The current can be assumed constant at a value I even if R_w changes, since $R_1 \gg R_w$. Let us suppose the velocity is constant at a value V_0. This will cause R_w to assume a constant value, say R_{w0}, and a voltage IR_{w0} will appear across R_w. Now, we let the velocity V fluctuate about the value V_0 so that $V = V_0 + v$, where v is the fluctuating component. This will result in R_w varying so that $R_w = R_{w0} + r_w$, where r_w is the varying component. Now, during a time interval dt, we may write for the wire (neglecting conduction and radiation effects)

Electrical energy generated $-$ energy lost by convection $=$ energy stored in wire

$$(7.15)$$

The energy lost by convection is given by

$$A(T_w - T_f)(C_0 + C_1\sqrt{V})\,dt \qquad (7.16)$$

while the wire temperature T_w may be related to its resistance by

$$T_w = K_{tr}(R_{w0} + r_w) \qquad (7.17)$$

where K_{tr} is the reciprocal of a temperature coefficient of resistance. The term $C_0 + C_1\sqrt{V}$ may be approximately linearized for small changes in V with good accuracy as follows:

$$f(V) = C_0 + C_1\sqrt{V} \approx (C_0 + C_1\sqrt{V_0}) + \left.\frac{\partial f}{\partial V}\right|_{V=V_0}(V - V_0) \qquad (7.18)$$

$$C_0 + C_1\sqrt{V} \approx (C_0 + C_1\sqrt{V_0}) + K_v v \qquad (7.19)$$

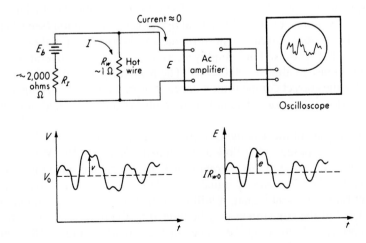

Figure 7.16
Velocity-fluctuation measurement.

Equation (7.15) then becomes

$$I^2(R_{w0} + r_w)\, dt - A(T_w - T_f)(C_0 + C_1\sqrt{V_0} + K_v v)\, dt = MC\, dT_w \quad \textbf{(7.20)}$$

where $M \triangleq$ mass of wire and $C \triangleq$ specific heat of wire.

Now,

$$I^2 R_{w0} + I^2 r_w - A[K_{tr}(R_{w0} + r_w) - T_f](C_0 + C_1\sqrt{V_0} + K_v v) = MCK_{tr}\frac{dr_w}{dt} \quad \textbf{(7.21)}$$

and since

$$I^2 R_{w0} - A(K_{tr}R_{w0} - T_f)(C_0 + C_1\sqrt{V_0}) = 0 \quad \textbf{(7.22)}$$

because this represents the initial equilibrium state, we get

$$I^2 r_w - AK_{tr}r_w(C_0 + C_1\sqrt{V_0}) - AK_{tr}r_w K_v v - A(K_{tr}R_{w0} - T_f)K_v v = MCK_{tr}\frac{dr_w}{dt}$$

$$\textbf{(7.23)}$$

The term $AK_{tr}K_v r_w v$ may be neglected relative to the other terms since it contains the product $r_w v$ of two small quantities. Now the voltage across R_w is $IR_w = I(R_{w0} + r_w)$. The fluctuating component of this is Ir_w, which we call e. Equation (7.23) then leads to

$$\frac{e}{v}(D) = \frac{K}{\tau D + 1} \quad \textbf{(7.24)}$$

where

$$K \triangleq \frac{-K_v AI(K_{tr}R_{w0} - T_f)}{K_{tr}A(C_0 + C_1\sqrt{V_0}) - I^2} \quad \text{V/(m/s)} \quad \textbf{(7.25)}$$

$$\tau \triangleq \frac{MCK_{tr}}{K_{tr}A(C_0 + C_1\sqrt{V_0}) - I^2} \quad \text{s} \quad \textbf{(7.26)}$$

We see that the voltage follows the flow velocity with a first-order lag. The time constant τ cannot be reduced much below 0.001 s in actual practice, which would limit the flat frequency response to less than 160 Hz. This is quite inadequate for turbulence studies since frequencies of 50,000 Hz and more are of interest. This limitation is overcome by use of electrical dynamic compensation (see Chap. 10 for a general discussion). Circuits whose frequency response just makes up the deficiency in the hot wire itself are employed, as in Fig. 7.17. The overall system then has a flat response to almost 100,000 Hz. The main difficulty in applying this technique is that the correct compensation depends on τ, whose value is not known and varies with flow conditions. The next paragraph explains a method of overcoming this difficulty.

The basic idea of the scheme is to force a square-wave current through the hot wire while it is exposed to the flow to be studied (see Fig. 7.18). We show that the output-voltage response to this *current* signal has exactly the same time constant as the response to the *flow-velocity* signal. Thus if the compensation can be adjusted to be correct for the current signal, it will be correct for the velocity input also. The "correctness" of the adjustment may be judged by the degree to which the output voltage corresponds to a square wave. In the circuit of Fig. 7.18, a good approximation to linear behavior may be expected for small input signals (current or

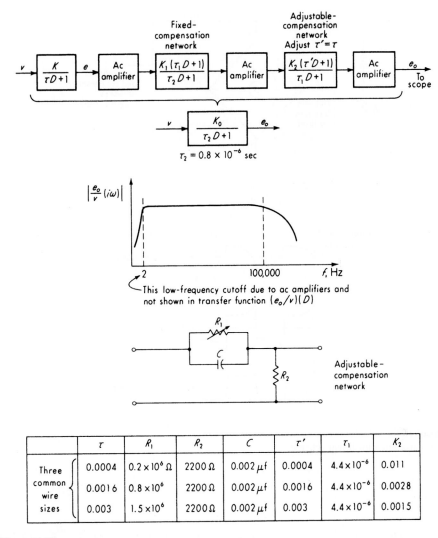

Figure 7.17
Dynamic compensation.

velocity); thus the superposition principle will apply, and the effects of current and velocity inputs may be considered separately. If the square-wave current is turned off, R_l adjusted to give the desired hot-wire current I_0, and R_b adjusted to balance the bridge, then $R_a/R_{w0} = R_r/R_b$ and the voltage $E_{B1,\,B2} = 0$. Now we let the square-wave current i_1 be turned on, causing a current i, which we calculate to be $i_1(R_a + R_r)/(R_a + R_r + R_w + R_b)$, to flow through R_w and a current $i_2 = i_1(R_w + R_b)/(R_a + R_r + R_w + R_b)$ to flow through R_a and R_r. Equation (7.15) may be applied to this situation to give

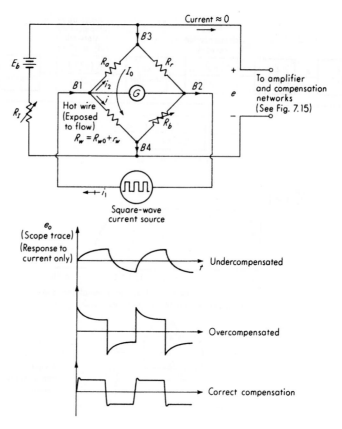

Figure 7.18
Compensation adjustment scheme.

$$(I_0 + i)^2(R_{w0} + r_w) - A[K_{tr}(R_{w0} + r_w) - T_f](C_0 + C_1\sqrt{V_0}) = MCK_{tr}\frac{dr_w}{dt} \quad \textbf{(7.27)}$$

Now, since $I_0^2 R_{w0} = A(K_{tr}R_{w0} - T_f)(C_0 + C_1\sqrt{V_0})$ and because we neglect $2I_0 ir_w$ and $i^2(R_{w0} + r_w)$, since they are products of small quantities, Eq. (7.27) reduces to

$$\frac{r_w}{i}(D) = \frac{K_i}{\tau D + 1} \quad \textbf{(7.28)}$$

where $\tau \triangleq \dfrac{MCK_{tr}}{K_{tr}A(C_0 + C_1\sqrt{V_0}) - I_0^2}$ s $\textbf{(7.29)}$

$K_i \triangleq \dfrac{2I_0 R_{w0}}{K_{tr}A(C_0 + C_1\sqrt{V_0}) - I_0^2}$ Ω/A $\textbf{(7.30)}$

Let us now calculate e, the varying component of the voltage appearing across $B3$ and $B4$, which will be the input to the amplifiers and compensating networks:

$$e = -R_a i_2 + (R_{w0} + r_w)i + I_0 r_w \approx -R_a i_2 + R_{w0} i + I_0 r_w \qquad ir_w \approx 0 \qquad (7.31)$$

$$e = -R_a \frac{R_w + R_b}{R_a + R_r} i + R_{w0} i + I_0 r_w$$

$$= \frac{-R_a R_w - R_a R_b + R_{w0} R_a + R_{w0} R_r}{R_a + R_r} i + I_0 r_w \qquad (7.32)$$

Now $R_{w0} R_a \approx R_w R_a$ and $R_{w0} R_r = R_a R_b$ (balanced-bridge relation); thus

$$e \approx I_0 r_w \qquad (7.33)$$

Thus, finally,

$$\frac{e}{i}(D) = \frac{K_e}{\tau D + 1} \qquad (7.34)$$

where

$$K_e \triangleq I_0 K_i \qquad (7.35)$$

We see now that the response of the voltage e to impressed current signals has the identical time constant τ as the response to flow-velocity signals. Thus the compensating networks may be adjusted to optimize the response to current inputs and ensure optimum response for flow-velocity inputs. Since this adjustment is made while the probe is exposed to flow, the output will contain a superposition of current response and velocity response, resulting in a sometimes confusing picture, rather than the simple waveforms of Fig. 7.18. Usually, however, the compensation adjustment can be made satisfactorily.

The operation of the constant-temperature type of instrument for steady velocities is explained earlier in relation to Fig. 7.15. This mode of operation can be extended to measure both average and fluctuation components of velocity by making the bridge-balancing operation automatic, rather than manual, through the agency of a feedback arrangement. The advantages of this scheme are such that most instruments now employ it, that is, the majority of hot-wire and hot-film anemometers now use the feedback (self-balancing) version of the constant-temperature system. We earlier analyzed the constant-current system mainly to define the wire time constant, which is needed in the analysis of the feedback-type constant-temperature system. Constant-current operation is provided as a selectable option on some constant-temperature systems, but this mode is rarely used. A simplified functional schematic of the feedback-type system is shown in Fig. 7.19. With zero flow velocity and the bridge excitation shut off ($i_w = 0$), the hot wire assumes the fluid temperature. Then the variable resistor R_3 is adjusted manually so that $R_3 > R_w$, thereby unbalancing the bridge. When the excitation current is turned on, the unbalanced bridge produces an unbalance voltage e_e, which is applied to the input of a high-gain current amplifier supplying the bridge excitation current. The current now flowing through R_w increases its temperature and thus its resistance. As R_w increases, it approaches R_3 and the bridge-unbalance voltage e_e decreases. If the current amplifier had an infinite gain, the bridge current required to heat R_w to match R_3 precisely could be produced with an infinitesimally small error voltage e_e,

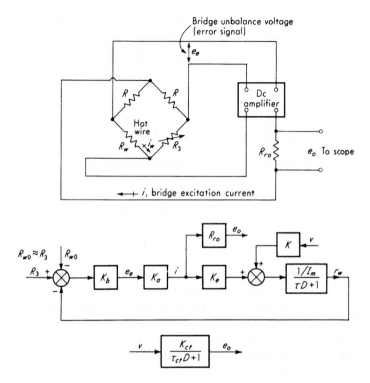

Figure 7.19
Constant-temperature anemometer.

and thus perfect bridge balance would be attained automatically. The current i (which is a measure of flow velocity V) produces an output voltage in passing through the readout resistor R_{ro}. Since an actual amplifier has finite gain, the bridge-unbalance voltage cannot go to zero; but it can be extremely small. This also means that R_w will be very close, but not exactly equal, to R_3.

The above-described self-adjusting equilibrium will come about at *any* steady-flow velocity, not just zero, with the equilibrium current in each case being a measure of velocity. If the velocity is fluctuating rapidly, perfect continuous bridge balance requires an infinite-gain amplifier. Lacking this, practical systems exhibit some lag between velocity and current. System dynamic response for small velocity fluctuations can be obtained by superimposing previously obtained results for response to velocity (wire-current constant) and response to current (flow-velocity constant), since in the constant-temperature mode both current and velocity are changing simultaneously. The effect of velocity on r_w (the varying component of R_w) is obtained from Eq. (7.24) as

$$\frac{r_w}{v}(D) = \frac{K/I_m}{\tau D + 1} \tag{7.36}$$

where I_m is the constant current about which fluctuations take place. From Eq. (7.34), the effect of current on r_w is given by

$$\frac{r_w}{i}(D) = \frac{K_e/I_m}{\tau D + 1} \tag{7.37}$$

The total effect of i and v on r_w is then, by superposition,

$$(\tau D + 1)r_w = \frac{Kv + K_e i}{I_m} \tag{7.38}$$

Now a change in r_w causes a bridge-unbalance voltage change according to

$$e_e = \frac{I_m R}{R + R_{w0}} r_w \triangleq - K_b r_w \tag{7.39}$$

We assume that the amplifier has no lag and produces output current proportional to input voltage as given by

$$i = K_a e_e \tag{7.40}$$

The block diagram of Fig. 7.19 embodies the above relations, which may now be manipulated to give the relation between input v and output e_o:

$$\left(\frac{K_{eo}}{R_{ro}} + Kv\right)\frac{(1/I_m(-K_b K_a R_{ro})}{\tau D + 1} = e_o \tag{7.41}$$

This gives finally

$$\frac{e_o}{v}(D) = \frac{K_{ct}}{\tau_{ct} D + 1} \tag{7.42}$$

where

$$\tau_{ct} \triangleq \frac{\tau}{1 + K_e K_b K_a/I_m} \tag{7.43}$$

and

$$K_{ct} \triangleq \frac{-KK_b K_a R_{ro}/I_m}{1 + K_e K_b K_a/I_m} \tag{7.44}$$

Note that τ_{ct}, the time constant of the constant-temperature anemometer system, is always less than τ (the time constant of the wire itself) and in actual practice is *much* less, since a very high value of amplifier gain K_a is used. A more comprehensive analysis[39] of the system, which models the amplifier as a second-order dynamic system and adds a capacitor (used to "tune" system for optimum response) in parallel with R, leads to a third-order closed-loop differential equation. This allows the usual Routh-criterion analysis for stability and allowable loop gain. As in all feedback systems, too high a loop gain will cause instability; however, careful design allows sufficiently high gain to make τ_{ct} of the order of $\frac{1}{100}$ of τ or less. A typical

---

[39]P. Freymuth, "Feedback Control Theory for Constant-Temperature Hot-Wire Anemometers," *Rev. Sci. Instrum.,* vol. 38, no. 5, pp. 677–681, May 1967; J. A. Borgos, "A Review of Electrical Testing of Hot-Wire and Hot-Film Anemometers," *TSI Quart.,* TSI Inc., St. Paul, MN, August/September 1980.

instrument has flat (within 3 dB) frequency response to 17,000 Hz when the average flow velocity is 30 ft/s, 30,000 Hz for 100 ft/s, and 50,000 Hz at 300 ft/s.

The stated preference for the feedback-type constant-temperature instrument is based on a number of considerations. In the constant-current type, the current must be set high enough to heat the wire considerably above the fluid temperature for a given average velocity. If the flow suddenly drops to a much lower velocity or comes to rest, the hot wire will burn out since the convection loss cannot match the heat generation before the wire temperature reaches the melting point. The constant-temperature type does not have this drawback because the feedback system *automatically* sets wire current to maintain the desired (safe) wire temperature for every velocity.

A further advantage of the constant-temperature method lies in the nature of the dynamic compensation. In the constant-current method, the compensating network must be reset (using the square-wave current) whenever the average velocity changes appreciably. Furthermore, if velocity fluctuations about the average are large (more than, say, 5 percent of the average velocity), the dynamic compensation will not be complete since the value of τ varies with V and thus the compensating network time constant τ' should be continuously and instantaneously varied, which it is not. The feedback arrangement of the constant-temperature system provides more nearly correct compensation for large velocity fluctuations. However, for large changes in average velocity, installation of a different probe, and/or changes in the connecting cable, the system must be "retuned" to optimize the frequency response. A square-wave circuit is provided to guide the adjustment of the tuning knobs.

While large velocity changes result in nonlinear response in both types of instrument, the nonlinearity of the constant-temperature feedback system appears algebraically on the output signal, where it is easily linearized with an external electronic linearizer. Usually these linearizers utilize a polynomial function through the fourth power, with the coefficient of each power individually adjustable to suit the probe and flow conditions. These units are fairly simple to adjust and are in wide use, but no similar apparatus has been devised for the constant-current type. Figure 7.20[40] shows system output voltage, both nonlinear and linearized, for a specific instrument, making clear the variation of sensitivity (high at low velocities, low at high velocities) typical of the basic thermal anemometer principle.

The hot-wire anemometer may be employed to measure the direction of the average flow velocity in several ways.[41] It has been found that a single wire, as in Fig. 7.21a, responds essentially to the component of velocity perpendicular to it, if the angle between wire and velocity vector is between 90 and about 25°. For this range, then, the V in our derivations may be replaced by $V \sin \theta$. (For $\theta < 25°$ the heat loss is greater than predicted by $V \sin \theta$; for $\theta = 0$ it is about 55 percent of that for $\theta = 90°$.) With the arrangement of Fig. 7.15 and a rotatable probe as in Fig. 7.21a, the flow-direction angle (in a single plane) could be found by determining

[40]"Hot Film and Hot Wire Anemometry: Theory and Application," *Bull.* TB5, TSI Inc., St. Paul, MN.

[41]H. H. Lowell, "Design and Applications of Hot-Wire Anemometers for Steady-State Measurements at Transonic and Supersonic Speeds," *NACA, Tech. Note* 2117, 1950.

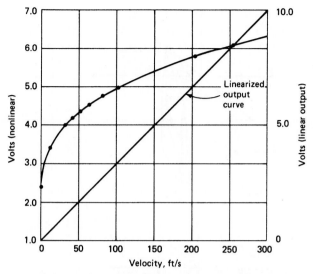

Calibration for a 0.002-in (0.051-mm) diameter hot-film sensor in atmospheric air; 0 – 300 ft/s (0 – 91m/s)

Figure 7.20
Nonlinear and linearized responses of hot-film sensor.

the probe-rotation angle which gives a maximum value of I. This method is quite inaccurate, however, since $\sin \theta$ changes very slowly with θ when θ is near 90°. A better procedure, if the flow angle is roughly known, is as follows: The wire is set at about 50° from the flow direction, and I is measured. Then the probe is rotated in the opposite direction until an angle is found at which the same I is measured as before. The bisector of the angle between the two locations then determines the flow direction. This method is more accurate since the rate of change of I with θ is a maximum near 50°. Even greater convenience and accuracy are achieved by use of a so-called V array of hot wires, as in Fig. 7.21b. The two hot wires R_{w1} and R_{w2} are assumed identical and form a V with an included angle, which is typically 90°. They are connected into a bridge as shown. When the probe is rotated, a bridge null occurs when the flow-velocity vector is aligned with the bisector of the V. In a specific case,[42] the sensitivity of this arrangement was sufficient to determine velocity direction within about 0.5°.

Practical problems in the application of hot-wire anemometers are found in the limited strength of the fine wires and the calibration changes caused by dirt accumulations. Unless the flow is very clean, significant calibration changes can occur in a relatively few minutes of operation. Larger dirt particles striking wires may actually break them. At high speeds, wires may vibrate because of aerodynamic loads and flutter effects. Applications are mainly in gas flows; however, special

---

[42]Ibid.

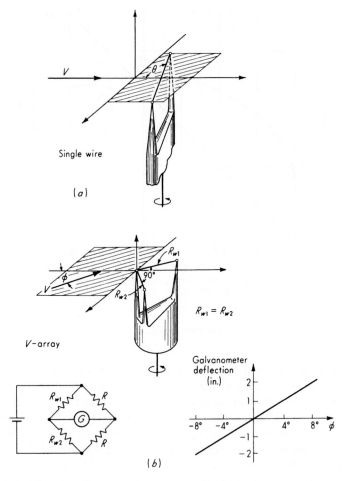

Figure 7.21
Flow-direction measurement.

probes have been developed and successful results obtained in both electrically conducting and nonconducting liquids, such as water and oil. In compressible flows, the film coefficient h is really related to ρV; thus calibrations such as that of Fig. 7.15 must be adjusted when a probe calibrated at one density is employed at another. (The sensitivity to ρV makes the hot-wire useful as a *mass-flow meter,* and it is used as such in automobile engine control systems, where an inlet air mass-flow signal from the hot wire is used to adjust the fuel injection to maintain a desired air to fuel ratio.[43] General-purpose hot-wire gas mass-flow sensors based on MEMS technology are available.[44]) Further complications arise when fluid temperature

---

[43]H. Heisler, "Advanced Engine Technology," SAE, 1995, pp. 430–432.

[44]Honeywell, www.honeywell.com/sensing.

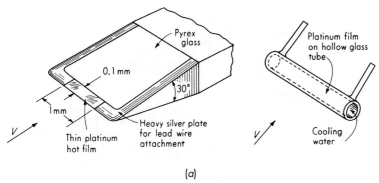

(a)

Figure 7.22 (a)
Hot-film sensors.

varies, either as a steady change from calibration conditions or dynamically during measurement; however, various compensation techniques[45] are available.

A variation of the hot-wire anemometer intended to extend its utility is the hot-film transducer. Here the resistance element is a thin film of platinum deposited on a glass base. The film takes the place of the hot wire; the required circuitry is basically similar to that used in constant-temperature hot-wire approach. The film transducers have great mechanical strength and may be used at very high temperatures by constructing them with internal cooling-water passages. Various configurations of sensors are possible; Fig. 7.22a shows two possibilities.

In addition to measurement of velocity magnitude and direction, hot-wire and hot-film instruments may be adapted to measurements of fluid temperatures, turbulent shear stresses, and concentrations of individual gases in mixtures.[46] By using a probe with two wires or films in an X configuration and two channels of electronics, velocity fluctuations in two perpendicular directions and the directional correlation (Reynolds stress) can be measured. For three-dimensional flows where flow direction is completely unknown, a system[47] employing six film sensors on three mutually perpendicular rods obtains velocity magnitude and direction in all eight octants with no ambiguity.

As with many other instruments, hot-wire anemometers are now available interfaced to personal computers with specialized software to automate, simplify, and improve many of the basic operations such as linearization and dynamic compensation. While the feedback-type constant-temperature instrument had more convenient dynamic compensation than that used with the older constant-current units, it still included a square-wave test circuit that required manual "tuning" to get the best response for given flow conditions. Now, the most recent computerized

[45]"Temperature Compensation of Thermal Sensors," *Tech. Bull.* 16, TSI Inc., St. Paul, MN; R. E. Drubka, J. Tan-atichat. and H. M. Nagib, "Analysis of Temperature Compensating Circuits for Hot-Wires and Hot-Films," *DISA Inform.,* no. 22, pp. 5–14, Disa Electronics, December 1977.

[46]S. Corrsin, "Extended Applications of the Hot-Wire Anemometer," *NACA, Tech. Note* 1864, 1949.

[47]System 1080, TSI Inc., St. Paul, MN.

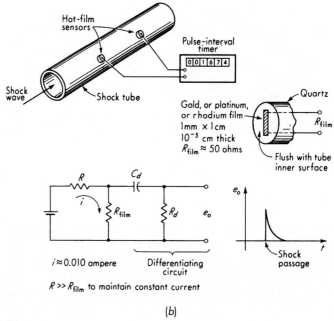

Figure 7.22 (*b*)
Thin-film velocity sensor.

instruments[48] feature totally automatic tuning of the dynamic compensation. The probe, cable, and flow velocity can be changed, and the system automatically tunes itself for optimum dynamic response.

An unusual application[49] used a three-dimensional hot wire mounted on a servo-controlled three-dimensional traversing rig to "fly" the sensor over a terrain model in a wind tunnel. The object was to simulate an actual aircraft crash where an airplane approached an airport next to a fjord (with steep mountains on each side) in the Faroe Islands in 1996. The hot-wire data showed a strong vertical turbulence upon entering the fjord and up to 1.5 mi from the airport. Subsequent flight-simulator tests showed that these wind conditions made a safe landing virtually impossible.

In addition to texts[50] devoted entirely to thermal anemometry, a voluminous literature of papers exists and has been documented in an excellent bibliography[51] organized to facilitate the reader's search for any particular topic.

---

[48]SMARTUNE system, TSI Inc., St. Paul, MN, 800-874-2811.

[49]Dantec Newsletter, vol. 5, no. 1, 1998.

[50]A. E. Perry, "Hot-wire anemometry," Oxford University Press, New York, 1982; C. G. Lomas, "Fundamentals of Hot-wire anemometry," Cambridge University Press, New York, 1986; S. Mol'yakov and T. Kachenko, "The measurement of Turbulent Fluctuations," Springer, New York, 1983; H. H. Bruun, "Hot-wire Anemometry," Oxford University Press, 1995.

[51]P. Freymuth, A Bibliography of Thermal Anemometry, 2nd ed., 1990, TSI Inc., St. Paul, MN.

Hot-Film Shock-Tube Velocity Sensors

In shock-tube experiments, the propagation velocity of the shock wave down the tube often must be measured. Of the various means available for making this measurement, hot-film temperature sensors are in wide use. The passage of the shock wave past a particular section of the tube is accompanied by a step change in gas temperature. By locating thin resistance films flush with the inside of the tube, the instant of wavefront passage may be detected as a temperature (and therefore resistance) change. If two such film sensors are mounted a known distance apart, the average wave velocity may be computed from the time interval between the two sensor responses. The films are operated at constant current by the simple circuit of Fig. 7.22b, and a differentiating circuit is employed to sharpen the pulses for greater timing accuracy. With systems of this type,[52] shock-wave velocities have been measured with an accuracy of 1 percent for shock Mach numbers as high as 7.5 to 10.

Laser Doppler Anemometer (LDA)[53]

This instrument measures local flow velocity and is thus in competition with pitot tubes, hot-wire anemometers, and PIV. Its main features include:

1. Measurement of velocity is direct rather than by inference from pressure (pitot tube) or heat-transfer coefficient (hot-wire).
2. No "physical object" need be inserted into the flow; thus flow is undisturbed by measurement.
3. Sensing volume can be very small (a cube 0.2 mm on a side is not unusual).

Disadvantages involve the need for transparent flow channels, the necessity for tracer particles in the fluid, and the cost and complexity of the apparatus. In brief, the operating principle involves focusing laser beams at the point where velocity is to be measured and then sensing with a photodetector the light scattered by tiny particles carried along with the fluid as it passes through the laser focal point. The velocity of the particles (assumed to be identical to the fluid velocity) causes a doppler shift of the scattered light's frequency and produces a photodetector signal directly related to velocity.

Actually, *artificial* tracer particles are not always necessary; the microscopic particles normally present in liquids may suffice; however, gas flows often need to be seeded. Under extreme conditions, particles may not perfectly follow the flow,

---

[52]"Shock Tubes, Handbook of Supersonic Aerodynamics," sec. 18 *NAVORD Rept.* 1488, vol. 6, pp. 543, 558.

[53]L. E. Drain, "The Laser Doppler Technique," Wiley-Interscience, New York, 1980; T. S. Durrani and C. A. Greated, "Laser Systems in Flow Measurement," Plenum, New York, 1976; H. D. Thompson and W. H. Stevenson, "Laser Velocimetry and Particle Sizing," Hemisphere, New York, 1979; F. Durst, A. Melling, and J. H. Whitelaw, "Principles and Practice of Laser Anemometry," Academic, New York, 1976.

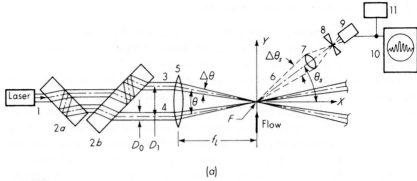

Figure 7.23 (*a*)
Layout of laser doppler velocimeter.

but studies have shown highly accurate following in many practical cases. A simple analysis[54] accurate for many gas flows, which models the particle as a spherical body of radius r and density ρ immersed in a fluid of viscosity μ, uses Stokes' law for the viscous force to obtain the differential equation

$$\frac{4}{3}\pi r^3 \rho \frac{dv_p}{dt} = 6\pi\mu r(v_f - v_p) \tag{7.45}$$

This gives a first-order dynamic response with time constant $\tau \triangleq 2\rho r^2/(9\mu)$ relating particle velocity v_p to fluid velocity v_f. For water droplets of (typical) 1-μm diameter in air, we get $\tau \approx 3 \mu$s, which allows flat (within 1 percent) frequency response to about 7,400 Hz. More correct (and complex) models for particle motion, needed when Eq. (7.45) becomes inaccurate, are available.[55]

Laser doppler velocimeters (LDVs) (another name for laser doppler anemometers (LDAs) have been operated in several different configurations; Figs. 7.23a to 7.24a[56] give some details on the popular dual-beam (differential doppler) or "fringe" mode. While a careful physical explanation of the device rests on a detailed study of the doppler-shift effect, many workers in the field find the interference-fringe explanation of Fig. 7.23b useful. Coated optical flats (*2a, 2b*) cause laser beam 1 to be split into two equal-intensity, parallel beams, 3 and 4. Lens 5 causes these beams to cross and focus at common point *F*. Light 6, scattered from particles moving through the fringe pattern, is selectively collected by the lens/pinhole aperture combination 7 and 8 and then detected by photosensor 9. Light intensity versus time is displayed on scope 10 and simultaneously processed by doppler burst signal

[54]L. E. Drain, op. cit., p. 183.

[55]A. Melling and J. H. Whitelaw, "Seeding of Gas Flows for Laser Anemometry," *DISA Inform.*, Disa Electronics, Franklin Lakes, NJ, no. 15, October 1973; N. S. Berman, "Particle-Fluid Interaction Corrections for Flow Measurements with a Laser Doppler Flowmeter," *Rept.* for contract NAS 8-21397, Eng. Res. Ctr., Arizona State University, Tempe.

[56]D. B. Brayton et al., "Project Squid Conf. on Laser Doppler Velocimeter," Purdue Univ., pp. 52–100, March 9–10, 1972.

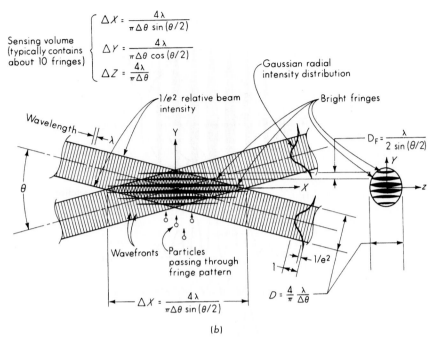

Sensing volume (typically contains about 10 fringes)

$$\Delta X = \frac{4\lambda}{\pi \Delta \theta \, \sin(\theta/2)}$$

$$\Delta Y = \frac{4\lambda}{\pi \Delta \theta \, \cos(\theta/2)}$$

$$\Delta Z = \frac{4\lambda}{\pi \Delta \theta}$$

Gaussian radial intensity distribution

$1/e^2$ relative beam intensity

Bright fringes

Wavelength λ

$$D_F = \frac{\lambda}{2 \sin(\theta/2)}$$

Wavefronts Particles passing through fringe pattern

$$\Delta X = \frac{4\lambda}{\pi \Delta \theta \, \sin(\theta/2)}$$

$$D = \frac{4}{\pi} \frac{\lambda}{\Delta \theta}$$

$1/e^2$

(b)

Figure 7.23 (b)
Interference fringes in sensing volume.

processor 11 to derive flow-velocity data. The frequency f of the electric signals (shown in Fig. 7.24a) produced by a particle moving across the dark and light fringe pattern with a velocity component V normal to the fringes is given by

$$f = \frac{2V \sin(\theta/2)}{\lambda} \tag{7.46}$$

For typical laser-light wavelength $\lambda = 5 \times 10^{-5}$ cm and $\theta = 30°$, we get $f = 10{,}340V$ Hz, where V is in centimeters per second. Note that the method measures the velocity *component* perpendicular to the fringe pattern. We can rotate the fringe pattern 90° to measure the other component of a two-dimensional velocity vector.

In addition to the dual-beam forward-scatter ("fringe") mode of Fig. 7.23a, equipment is commercially available[57] to implement other modes of application suited to particular measurement problems. The *reference beam* mode requires less critical optical alignment, but has difficulty handling low particle concentrations. Various *back-scattering* schemes require optical access from only one side of the flow, but again require high particle concentrations. Simultaneous measurement of several velocity components at a point may be achieved by either *polarization* schemes using a single laser or *two-color* systems employing two lasers of different wavelength. When velocity components actually change sign (reverse), ambiguities

[57]TSI Inc., St. Paul, MN; Dantec Measurement Technology Inc., Mahwah, NJ, www.dantecdynamics.com.

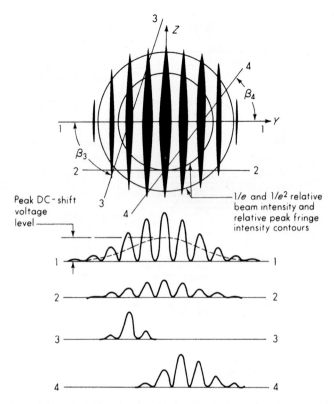

Signal amplitude vs. particle position near $X = 0$ for a number of particle trajectories. The indicated fringe width is proportional to the local peak fringe intensity.

(a)

Figure 7.24
Electronic signals from velocimeter and comparison of processors.

present in "ordinary" systems may be resolved by using *frequency-shifting* schemes employing acousto-optic (Bragg) cells. Several types of signal processors are utilized. *Frequency trackers* work best with high particle concentrations, while *counter-type* instruments may be needed for high-velocity flows, which generally have a few particles in the measurement volume. Two more sophisticated processors (burst spectrum and photon correlation) are compared with the more basic types in Fig. 7.24b.[58] Fiber-optic[59] probes and links permit application in cramped spaces, easier scanning, and more flexible positioning.

[58]L. Lading, "Spectrum Analysis of LDA Signals," *Dantec Inform.,* no. 5, pp. 2–8, September 1987, Dantec Electronics, Allendale, NJ.

[59]S. Kaufman, "Fiber Optics in Laser Doppler Velocimetry," *Lasers and Applications,* pp. 72–73, July 1986, FP4875M, TSI Inc., St. Paul, MN.

	Counter	Tracker	BSA	Photon correl.
Single-burst detection high S/N	● ● ●		● ● ●	
Single burst detection low S/N	●		● ● ●	
Many particles in the measuring volume high S/N	●	● ● ●	● ● ●	● ●
Many particles in the measuring volume low S/N, high background		● ●	● ● ●	● ●
Many particles in the measuring volume low S/N, low background		● ●	● ●	● ● ●

(*b*)

Figure 7.24
(*Concluded*)

The use of opto-electronic methods in the laser-doppler anemometer may lead to the supposition that it inherently has a faster response than the hot-wire, that depends on heat transfer effects which intuitively seem slow. This is in fact not the case; the hot-wire easily measures turbulent fluctuation of 100 kHz while the LDA might typically be limited to 3 kHz.[60] The reference gives the details, but the basic idea is that, to follow a high-frequency velocity fluctuation accurately, we must have high "data density." That is, there must be lots of particles passing through the measurement volume per unit time, so that we measure many points on a single cycle of the oscillation. This requirement turns out to be difficult to meet.

7.2 GROSS VOLUME FLOW RATE

The total flow rate through a duct or pipe often must be measured or controlled. Many instruments (flowmeters)[61] have been developed for this purpose. They may

---

[60]A. Host-Madsen and C. Casperson, "The Limitations in High Frequency Turbulence Spectrum Estimation Using the Laser Doppler Anemometer," 1996, Dantec Measurement Technology Inc.

[61]R. W. Miller, "Flow Measurement Engineering Handbook," McGraw-Hill, New York, 1983. This reference has a wealth of practical information, including many numerical examples, useful charts, and tables.

be categorized in various ways; a useful overall classification divides devices into those which measure volume flow rate [(ft^3 or m^3)/time] and those which measure mass flow rate [(lbm or kg)/time]. The total flow occurring during a given time interval also may be of interest. This measurement requires integration with respect to time of the instantaneous flow rate. The integrating function may be performed in various ways, sometimes being an integral part of the flowmetering concept and other times being performed by a general-purpose integrator more or less remote from the flowmeter.

Calibration and Standards

Flow-rate calibration depends on standards of volume (length) and time or mass and time. Primary calibration,[62] in general, is based on the establishment of steady flow through the flowmeter to be calibrated and subsequent measurement of the volume or mass of flowing fluid that passes through in an accurately timed interval. If steady flow exists, the volume or mass flow rate may be inferred from such a procedure. Any stable and precise flowmeter calibrated by such primary methods then itself becomes a secondary flow-rate standard against which other (less accurate) flowmeters may be calibrated conveniently. As in any other calibration, significant deviations of the conditions of use from those at calibration will invalidate the calibration. Possible sources of error in flowmeters include variations in fluid properties (density, viscosity, and temperature), orientation of the meter, pressure level, and particularly flow disturbances (such as elbows, tees, valves, etc.) upstream (and to a lesser extent downstream) of the meter. A commercial calibrator for precise primary calibration of flowmeters using liquids is shown in Fig. 7.25.[63] These units use a convenient dynamic weighing[64] scheme, are available in models to cover the range 0.5 to 150,000 lbm/h, and have an overall accuracy of ±0.1 percent.

Another type of calibrator, called a *ballistic flow prover,*[65] is particularly useful for fast-response, high-resolution flowmeters such as turbine, positive-displacement, vortex-shedding, etc., types, where steady state can be achieved quickly and the integration of the flow rate to get total flow is accomplished accurately by accumulating the meter pulse-rate output in a counter. (The integration gives accurate total flow even if the flow rate is not perfectly steady.) One such calibrator[66] uses a Teflon-sealed air-driven free piston traveling down a precision-honed tube to dispense a precise volume of calibration fluid through the attached flowmeter to be calibrated. Precise time and displacement measurements on the moving piston are utilized in a microprocessor data-reduction system to achieve a claimed

[62]M. R. Shafer and F. W. Ruegg, "Liquid-Flowmeter Calibration Techniques," *Trans. ASME,* p. 1369, October 1958.

[63]*Bull.* CA6600, Cox Instrument, Detroit, MI, 1978, www.cox-instrument.com.

[64]L. O. Olsen, "Introduction to Liquid Flow Metering and Calibration of Liquid Flowmeters," *NBS Tech. Note* 831, p. 30, 1974.

[65]Ibid., p. 35.

[66]"Ballistic Flow Prover," May 1980; "Ballistic Flow Calibrators," TD-016; *The Flow Factor,* vol. 6, no. 2, Flow Technology Inc., Phoenix, AZ, 800-528-4225, www.ftimeters.com, August 1978.

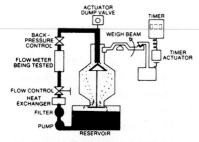

Running operation before test—Fluid contained in the reservoir is pumped through a closed hydraulic circuit. First, it enters the filter and then the heat exchange equipment, which controls temperature within It then passes through the control valves, the meter under test, the back-pres-sure valve, the weigh tank, then back into the reservoir.

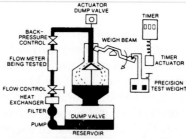

Weighing cycle in operation—The weighing cycle is continued as a precision weight is placed on the weigh pan, again deflecting the beam. The uniquely designed cone-shaped deflector at the inlet of the weigh tank permits the even distribution of the metered fluid.

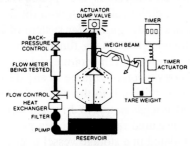

Start of preliminary fill (tare time)—When the control valves have been adjusted for desired flow, a tare weight is placed on the weigh pan. Then the cycle start button is pushed, resetting the timer, closing the dump valve that starts the filling of the weigh tank.

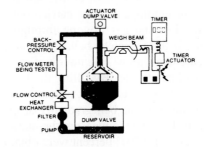

End of weighing cycle—As the tank fills, the weigh pan rises, until it again trips the timer actuator, stopping the timer and indicating the time within a thousandth of a second. By combining the precision test weight with the timed interval, the actual flow rate in pounds/hour is easily and accurately determined. From these basic mass units, other flow units can be accurately calculated.

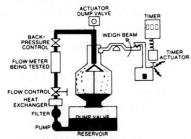

End of prefill, start of weighing cycle—As the weigh tank fills, the weigh pan rises, tripping the timer actuator, and the electronic timer begins counting in milliseconds, starting the actual weighing cycle. The preliminary fill, balanced out by the tare weight before actual weighing begins, permits a net measurement of the new fluid added after the preliminary fill. The preliminary fill method permits measurement of only a portion of the cycle, eliminating the mechanical errors in the start and stop portions and allowing dynamic errors to be self-cancelling.

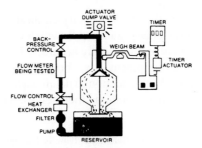

Emptying for recycling—After the beam movement trips the timer, the weigh tank automatically empties in less than 25s, even at maximum flow. The calibrator is now ready for the next flow setting. This fast recycling cuts total calibrating time as much as 50 percent.

Figure 7.25
Flow-rate calibration by dynamic weighing.

flow-rate accuracy of ± 0.02 percent. The small flow volume (about 5 gal for a 700 gal/min full-scale unit) involved in checking a single flow rate allows rapid calibration, 20 repeats of a single point typically being achieved in 8 minutes for a 300 gal/min flow rate. Units are available with maximum flow rates of 100 to 6,000 gal/min, each unit being usable over about a 1,000:1 range. A similar type of calibrator (except that the piston is driven by an electric motor) is available for gas flows (see Fig 7.26b).[67] Detailed uncertainty analyses are available[68] for these calibrators. This company offers nine different types of calibrators[69] for liquid and gas flows, both for sale and as an ISO 9002-certified calibration service.

The calibration of flowmeters to be used with gases often can be carried out with liquids as long as the pertinent similarity relations (Reynolds number) are maintained and theoretical density and expansion corrections are applied. If this procedure is felt to be of insufficient accuracy, a direct calibration with the actual gas to be employed can be carried out by means of the *gasometer* system of Fig. 7.26a. Here the gas flowing through the flowmeter during a timed interval is trapped in the gasometer bell, and its volume is measured. Temperature and pressure measurements allow calculation of mass and conversion of volume to any desired standard conditions. By filling the bell with gas, raising it to the top, and adding appropriate weights, such a system may be used as a gas *supply* to drive gas through a flowmeter as the bell gradually drops at a measured rate. By using a precision analytical balance to measure the mass accumulated in a storage vessel over time, accuracies of ± 0.02 percent were obtained[70] for flow rates up to 9 kg/s (20 lbm/s). A volumetric method (corrected for pressure and temperature) is also explained in this reference. The size of the equipment makes it more economical for large flow rates; however, the accuracy is less (± 0.08 percent).

Mass flowmeters (often the thermal type) for gases are used in large numbers in processes for manufacturing microelectronics and MEMS devices. Accurate calibration of the low flow rates used there is possible with a unique laminar-flow calibrator.[71] The references give a detailed explanation of the design and use of such a standard.

When the above primary calibration methods cannot be justified, comparison with a secondary standard flowmeter connected in series with the meter to be calibrated may be sufficiently accurate. Turbine flowmeters and their associated digital counting equipment have been found particularly suitable for such secondary standards. With attention to detail, such standards can closely approach the accuracy of the primary methods themselves. The Navy Primary Standards Laboratory at

---

[67]"Aerotrak Gas Flow Calibrator," Flow Technology, Inc.

[68]*Repts.* 088-1 and 188-1, Flow Technology, Inc., 1988.

[69]"Calibrator Selection Guide," *Bull.* CSG-951, 1995, Flow Technology, Inc.

[70]B. T. Arnberg and C. L. Britton, "Two Primary Methods of Proving Gas Flow Meters," *NASA CR-72896*, 1971.

[71]P. Delajoud and M. Girard, "A High Accuracy, Portable Calibration Standard for Low Mass Flow," 1994; "The Need For Evolution in Standards and Calibration to Improve Process Measurement and Control of Low Gas Flows," 1996; P. Delajoud and M. Bridge, "MOLBLOC Uncertainty Analysis," 1993; all published by DH Instruments, Inc., Tempe, AZ, 480-967-1555 (www.dhinstruments.com).

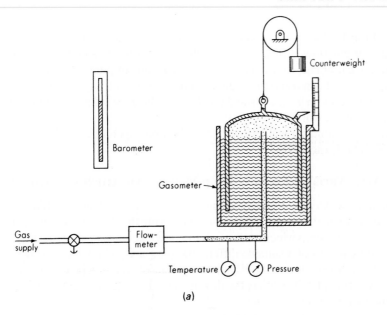

(a)

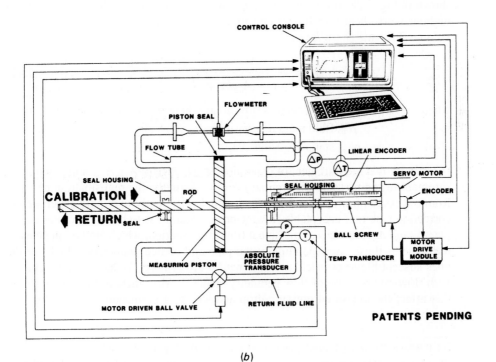

(b)

Figure 7.26
Gas-flow calibration.

Pensacola, Florida, has such a system[72] with an inaccuracy of the order of 0.2 percent. For gas flow, a flow nozzle discharging air to the atmosphere can be very accurately calibrated[73] for mass flow rate by means of pitot-tube traverses at the discharge. Then this nozzle can be connected in series with any flowmeter to be calibrated and used as an accurate standard. This can be done with very small error up to Mach number about 0.9.

Figure 7.27[74] summarizes the flow-rate calibration capabilities of the National Bureau of Standards (now NIST).

Constant-Area, Variable-Pressure-Drop Meters ("Obstruction" Meters)

Perhaps the most widely used flowmetering principle involves placing a fixed-area flow restriction of some type in the pipe or duct carrying the fluid. This flow restriction causes a pressure drop which varies with the flow rate; thus measurement of the pressure drop by means of a suitable differential-pressure pickup allows flow-rate measurement. In this section we briefly discuss the most common practical devices that utilize this principle: the orifice, the flow nozzle, the venturi tube, the Dall flow tube, and the laminar-flow element.

The *square-edge orifice* is undoubtedly the most widely employed flowmetering element, mainly because of its simplicity, its low cost, and the great volume of research data available for predicting its behavior. A typical flowmetering setup is shown in Fig. 7.28. If one-dimensional flow of an incompressible frictionless fluid without work, heat transfer, or elevation change is assumed, theory[75] gives the volume flow rate Q_t (m³/s) as

$$Q_t = \frac{A_{2f}}{\sqrt{1 - (A_{2f}/A_{1f})^2}} \sqrt{\frac{2(p_1 - p_2)}{\rho}} \tag{7.47}$$

where $A_{1f}, A_{2f} \triangleq$ cross-section flow areas where p_1 and p_2 are measured, m²
  $\rho \triangleq$ fluid mass density, kg/m³
  $p_1, p_2 \triangleq$ static pressures, Pa

We see that measurement of Q requires knowledge of A_{1f}, A_{2f}, and ρ and measurement of the pressure differential $p_1 - p_2$. Actually, the real situation deviates from the assumptions of the theoretical model sufficiently to require experimental correction factors if acceptable flowmetering accuracy is to be attained. For example, A_{1f} and A_{2f} are areas of the actual flow cross section, which are *not,* in general, the same as those corresponding to the pipe and orifice diameters, which are the ones susceptible to practical measurement. Furthermore, A_{1f} and A_{2f} may change with flow rate because of flow geometry changes. Also, there are frictional losses that affect the measured pressure drop and lead to a permanent pressure loss. To

[72]R. P. Bowen, "Designing Portability into a Flow Standard," *ISA J.,* p. 40, May 1961.

[73]R. J. Sweeney, "Measurement Techniques in Mechanical Engineering," p. 220, Wiley, New York, 1953.

[74]*NBS Spec. Publ.* 250, NBS Calibration Services, 1987.

[75]R. W. Miller, op. cit., chap. 9.

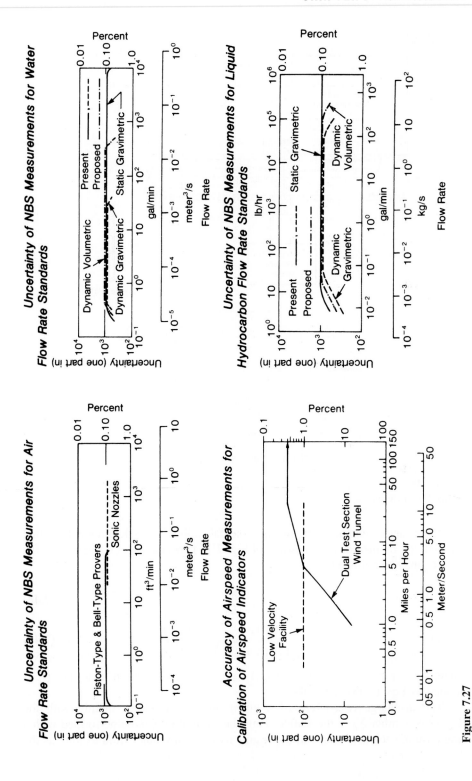

Figure 7.27
Flow-rate calibration capabilities.

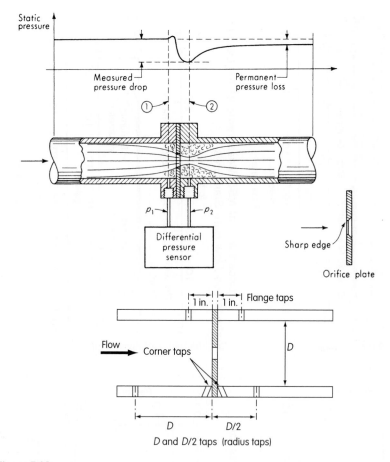

Figure 7.28
Orifice flowmetering.

take these factors into account, an experimental calibration to determine the actual
flow rate Q_a by methods such as that of Fig. 7.25 is necessary. A discharge coefficient C_d may be defined by

$$C_d \triangleq \frac{Q_a}{Q_c} \tag{7.48}$$

and thus

$$Q_a = \frac{C_d A_2}{\sqrt{1 - (A_2/A_1)^2}} \sqrt{\frac{2(p_1 - p_2)}{\rho}} \tag{7.49}$$

where $A_1 \triangleq$ pipe cross-section area, $A_2 \triangleq$ orifice cross-section area, and Q_c is
defined by Eqs. (7.48) and (7.49).

The discharge coefficient of a given installation varies mainly with the
Reynolds number N_R at the orifice. Thus the calibration can be performed with a
single fluid, such as water, and the results used for any other fluid as long as the

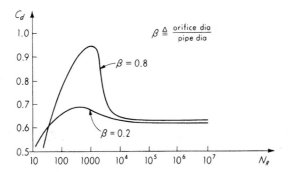

Figure 7.29
Variation in discharge coefficient.

Reynolds numbers are the same. Variation of C_d with N_R follows typically the trend of Fig. 7.29. While the above discussion would seem to indicate that each installation must be individually calibrated, fortunately this is not the case. If one is willing to construct the orifice according to certain standard dimensions[76] and to locate the pressure taps at specific points, then quite accurate (about 0.4 to 0.8 percent error) values of the discharge coefficient may be obtained based on many past experiments. Such data are available for pipe diameters 2 in and greater, β ratios (see Fig. 7.29) of 0.2 to 0.7, and Reynolds numbers above 10,000. Installations exceeding these limits should be calibrated individually if high accuracy is required. It also should be noted that the standard calibration data assume no significant flow disturbances such as elbows, bends, tees, valves, etc., for certain minimum distance upstream of the orifice. The presence of such disturbances close to the orifice can invalidate the standard data, causing errors of as much as 15 percent. Information on the minimum distances is available in Miller's book; Fig. 7.30 shows a typical example.[77] If the minimum distances are not feasible, *flow conditioners*[78] of various types can be installed upstream. This applies to *all* types of flowmeters, not just the obstruction types. However, some are less sensitive to flow disturbances than others, so one should always inquire about this feature.

Since flow rate is proportional to $\sqrt{\Delta p}$, a 10:1 change in Δp corresponds to only about a 3:1 change in flow rate. Since a given Δp-measuring instrument becomes quite inaccurate below about 10 percent of its full-scale reading, this nonlinearity typical of all obstruction meters (other than the laminar-flow element) restricts the accurate range of flow measurement to about 3:1. That is, a meter of this type cannot be used accurately below about 30 percent of its maximum flow rating. (Refinements in differential-pressure transmitters, including built-in

---

[76]R. W. Miller, op. cit., chap. 10.

[77]R. W. Miller, op. cit., chap. 8.

[78]D. M. Feener, "Applying Flow Conditioning," *Control Engineering,* Feb. 1999, pp. 107–110; Vortab Co., San Marcos, CA, 800-854-9959 (www.vortab.com).

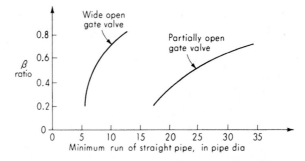

Figure 7.30
Effect of upstream disturbances.

microprocessors programmed to correct systematic errors peculiar to each unit,[79] can improve the low-range accuracy and increase the range to about 20:1.) The square-root nonlinearity also causes difficulties in pulsating-flow measurement,[80] where the average flow rate (the rate to be measured) has a fluctuating component superimposed on it. Let us consider, as a simple example, a flow Q where

$$Q = Q_{av} + Q_p \sin \omega t \qquad Q_p < Q_{av} \qquad \text{(7.50)}$$

and a flowmeter such that $\Delta p = KQ^2$. The Δp presented as input to the pressure-measuring system is then

$$\Delta p = K(Q_{av}^2 + 2Q_{av}Q_p \sin \omega t + Q_p^2 \sin^2 \omega t) \qquad \text{(7.51)}$$

If the Δp instrument has a low-pass filtering characteristic, it will tend to read the average value of Δp. This is seen to be

$$\Delta p_{av} = K\left(Q_{av}^2 + \frac{Q_p^2}{2}\right) \qquad \text{(7.52)}$$

Thus if we take a measured Δp_{av} and compute the corresponding Q_{av} from it, using $Q_{av} = \sqrt{\Delta p_{av}/K}$, we get a flow rate *higher* than actually existed. A further difficulty caused by the nonlinearity occurs when flow rate must be integrated to get total flow during a given time interval. Then the square root of the Δp signal must be taken before integration or this compensation must be included in the integrating device.

---

[79]J. E. Corley, "Intelligent Transmitters," *In Tech,* pp. 49–52, July 1984; J. O. Gray, "Process Sensors in CIM Schemes," *Sensors,* June 1987.

[80]A. K. Oppenheim and E. G. Chilton, "Pulsating Flow Measurement—A Literature Survey," *Trans. ASME,* p. 231, February 1955; T. Isobe and H. Hattori, "A New Flowmeter for Pulsating Gas Flow," *ISA J.,* p. 38, December 1959; H. J. Sauer, Jr., P. D. Smith, and L. V. Field, "Metering Pulsating Flow in Orifice Installations," *Inst. Tech.,* pp. 41–44, March 1969; R. W. Miller, "Flow Measurement Engineering Handbook," pp. 5-46 to 5-54.

The orifice has the largest permanent pressure loss of any of the obstruction meters (other than the laminar-flow element). This is one of its disadvantages since it represents a power loss[81] that must be replaced by whatever pumping machinery is causing the flow. The permanent pressure loss is given approximately by $\Delta p(1 - \beta^2)$, where Δp is the differential pressure used for flow measurement. Thus for the usual range of β (0.2 to 0.7), the permanent pressure loss ranges from 0.96 Δp to 0.51 Δp. The actual power loss, in fact, may be quite small since the Δp recommended[82] for conventional flowmetering of liquids and gases is typically 100 in of water.

Orifice discharge coefficients are quite sensitive to the condition of the upstream edge of the hole. The standard orifice design requires that this edge be very sharp and that the orifice plate be sufficiently thin relative to its diameter. Wear (rounding) of this sharp edge by long use or abrasive particles can cause significant changes in the discharge coefficient. Flows that contain suspended solids also cause difficulty since the solids tend to collect behind the "dam" formed by the orifice plate and cause irregular flow. Often this problem can be solved by use of an "eccentric" orifice in which the hole is at the bottom of the pipe rather than on the centerline. This allows the solids to be swept through continuously. Liquids containing traces of vapor or gas may be metered if the orifice is installed in a vertical run of pipe with the flow upward. Gases containing traces of liquid may be similarly handled except that the flow should be downward.

When compressible fluids are metered, Eq. (7.49) is no longer correct since the density changes from station 1 to 2. Assuming an isentropic process between these two states and using the thermodynamic steady-flow energy equation, an equation for the mass flow rate G in lbm/s can be derived:[83]

$$G = 0.5250208 \cdot \frac{CY_1 F_a d_2^2}{\sqrt{1 - \beta^4}} \cdot \sqrt{(p_1 - p_2)\rho_1} \tag{7.53}$$

where, d_2 is in in, $(p_1 - p_2)$ is in psi, and density is in lbm/ft^3. Also, the constant C is *not* the same as C_d in Eq. (7.49). In Eq. (7.49), C_d is defined in terms of a *laboratory calibration* that takes into account *all* sources of deviation from theory, but is correct only for conditions existing at calibration. Equation (7.53) uses correction factors for compressible flow and thermal expansion, and presumes that the constant C will be *computed* from available empirical formulas. The name "discharge coefficient" is used for both these symbols, so one must be careful when using this term to make clear the intended meaning.

Equation (7.53) can actually be used for *both* liquids and gases; for liquids, set $Y_1 = 1.0$. For liquids, the density is mainly a function of temperature; however,

---

[81]W. M. Reese, Jr., "Factor the Energy Costs of Flow Metering into Your Decisions," *In Tech,* pp. 36–38, July 1980; R. W. Miller, op. cit., pp. 6-29 to 6-37.

[82]R. W. Miller, op. cit., pp. 9–50.

[83]R. W. Miller, op. cit., pp. 9–18.

Miller gives corrections that might be needed at very high pressure. For gases, the perfect gas law allows estimation of density for known pressure and temperature. When necessary, deviations from the perfect gas law can be corrected using a *compressibility factor;* again Miller gives the needed relations. Thermal expansion of the pipe and orifice diameters when temperature changes from the design conditions is taken into account by F_a, which is 1.0 when there is *no* temperature change. Usually, the *orifice* expansion dominates, and Miller gives a graph with F_a values for different materials and a range of temperatures. For example, for 300-series stainless steel and a 700F° temperature change, $F_a = 1.012$. The *adiabatic gas expansion factor* Y_1 is given by

$$
Y_1 = \left\{ \frac{\left(\frac{p_2}{p_1}\right)^{\frac{2}{k}} \cdot (1 - \beta^4) \cdot \left(\frac{k}{k-1}\right) \cdot \left[1 - \left(\frac{p_2}{p_1}\right)^{\frac{k-1}{k}}\right]}{\left[1 - \left(\beta^4 \cdot \left(\frac{p_2}{p_1}\right)^{\frac{2}{k}}\right)\right] \cdot \left(1 - \frac{p_2}{p_1}\right)} \right\}^{\frac{1}{2}}
$$ (7.54)

For example, if $p_2/p_1 = 0.95$, $k = 1.4$, and $\beta = 0.65$, then $Y_1 = 0.9657$. Equation (7.53) holds for any of the "obstruction" meters (orifice, flow nozzle, venturi); we must of course use the correct discharge coefficient C for the particular device and pressure tap location. For flow nozzles and venturis, Eq. (7.54) gives accurate values. However, for orifices, the actual situation deviates from the theoretical assumptions, and Y_1 must be found by relations based on experiment (see Miller).

When computing flow rates from measured pressure drops, dimensions, and material properties, we need a value for the discharge coefficient, but this quantity *varies with flow rate,* so an iterative procedure may be necessary, that is, we assume a tentative value for C (guided of course by charts such as Miller's) and compute a tentative value for flow rate. This flow rate then gives a *better* value for C, and we continue the process until we get no further significant change in the computed flow rate. If you have available math software that includes an "equation solver," you can avoid the iteration since *formulas* (based on "curve fits" of many experiments) are available relating C to β, pipe Reynolds number, and/or pipe diameter (for example, Miller, Table 9.1). Let us illustrate the procedure with a simple example.

Suppose in Eq. (7.53) we are using a square-edge orifice with corner taps to meter air flow and that $k = 1.400$, $d_1 = 4.000$ in, $d_2 = 2.600$ in, $p_2/p_1 = 0.9500$, $\rho_1 = 0.0853$ lbm/ft^3, $p_1 - p_2 = 1.500$ psi, and $F_a = 1.002$. Equation (7.54) gives $Y_1 = 0.9657$, but recall that, for orifices, we cannot use this theoretical value. Miller (op. cit., p. 9-18) gives the accepted empirical formula for our example as:

$$
Y_{1,\,\text{orifice}} = 1.00 - (0.41 + 0.35\beta^4)\frac{\left(1 - \dfrac{p_2}{p_1}\right)}{k} = 0.9831
$$ (7.55)

Equation (7.53) then becomes

$$G = \frac{0.5250208(C \cdot 0.9831 \cdot 1.002 \cdot 2.600^2)}{\sqrt{1 - 0.65^4}} \cdot \sqrt{1.500 \cdot 0.0853}$$

$$= 1.690C \qquad \text{lbm/sec}$$

We now need the empirical equation relating C to Reynolds number R_D. Miller (op. cit., p. 9-11) gives this as

$$C = C_\infty + \frac{b}{R_D^n} \qquad (7.56)$$

where C_∞ is the discharge coefficient at infinite Reynolds number (C tends to "converge" at high Reynolds numbers). For corner taps, Miller gives for our example

$$C_\infty = 0.5959 + 0.0312\beta^{2.1} - 0.184\beta^8 = 0.6027$$

$$b = 91.706\beta^{2.5} = 31.24 \qquad n = 0.75 \qquad (7.57)$$

While we could use the standard definition of Reynolds number found in fluid mechanics texts, Miller (op. cit., C-7) gives a more convenient form for this application:

$$R_D = \frac{15.27888G}{d_1 \mu}$$

where μ is the absolute viscosity in lbm/ft · s. Taking the viscosity as 1.28×10^{-5} for the purposes of this example, gives us $R_D = 2.99 \times 10^5 G$. We can then express C as a function of G:

$$C = 0.6027 + \frac{31.24}{(2.99 \times 10^5 G)^{0.75}}$$

Since $G = 1.690C$,

$$C^{1.75} - 0.6027C^{0.75} - 0.001648 = 0$$

Using the MATHCAD equation solver, we find that $C = 0.6051$, and thus the mass flow rate is 1.023 lbm/s and $R_D = 306,877$. As with any other calculated result, we should next compute the estimated *uncertainty* of the calculated flow rate. This would be done in the usual way with a *root-sum-square* combination of the individual uncertainties. The uncertainty of our C value is needed here, and Miller has a table (op. cit., Table 9.59) that covers all the different flowmeters and conditions. For corner, flange, and D-D/2 taps, $10^4 < R_D < 10^7$, and $0.6 < \beta < 0.75$ (our conditions), Miller gives the uncertainty as $\pm\beta$ percent, which is ± 0.65 percent in our case.

While most orifice applications involve essentially steady flow, dynamic response is sometimes of interest. The dynamic response of the differential-pressure sensor must be adequate, of course; the methods of Chap. 6 answer such questions. An analysis[84] is available that uses a nonlinear differential equation relating mass flow rate and the Δp signal and includes lumped fluid inertial and resistance effects. A linearized version defines a first-order dynamic system in which flow rate lags Δp with a time constant τ given by

$$\tau = \frac{C_d^2 A_0^2 \rho l}{w_o A_p} \tag{7.58}$$

where

$C_d \triangleq$ discharge coefficient

$A_o \triangleq$ orifice area

$\rho \triangleq$ fluid density

$l \triangleq$ length between pressure taps

$w_o \triangleq$ fluid mass flow rate at linearization operating point

$A_p \triangleq$ pipe area

The *flow nozzle, venturi tube,* and *Dall flow tube* (Fig. 7.31) all operate on the same principle as the orifice and are described by equations similar to Eq. (7.53) (see Miller for details). (The Dall flow tube belongs to a class of modified venturis of proprietary design called "low-loss" tubes, intended to save pumping costs by having an especially low *permanent* pressure loss, while retaining other good properties.) Because of their less abrupt flow-area changes, nozzles and venturis have higher discharge coefficients than orifices, as high as 0.99 in some cases. Venturis can have very low permanent pressure loss, so are popular for high-flow situations such as municipal water systems, where large savings in pumping costs are possible. The world's largest flowmeter,[85] a venturi of proprietary design, has a 15-ft diameter, is 52 ft long, and meters 800 million gallons of water per day for the Southern Nevada Water Authority. Venturis are of course also used for a wide range of flow rates for liquids and gases, one source[86] has standard models from 1-in to l0-ft diameters. A detailed discussion of a proprietary venturi design with many desirable features is available from a manufacturer.[87]

Flow nozzles are more expensive than orifices but cheaper than venturis, and are often used for high-velocity steam flows, being more dimensionally stable at high temperature and velocity than an orifice. To fairly compare the various obstruction meters, we should keep constant the accuracy of the pressure measurement, which requires that the measured differential pressure is the same for each meter. If we do this, the flow nozzle must have a smaller β ratio than the orifice, and losses increase with smaller β; thus, the nozzle will have a permanent pressure loss

[84]R. T. Lakey, Jr., and B. S. Shiralkar, "Transient Flow Measurements with Sharp-Edged Orifices," *ASME Paper* 71-FE-30, 1971.

[85]*Flow Control,* April/May 2000, pp. 10–12 (www.pfsflowproducts.com).

[86]BIF, Chagrin Falls, OH, 440-543-5885 (www.bifwater.com).

[87]Primary Flow Signal, Warwick, RI, 401-463-9199 (www.pfsflowproducts.com).

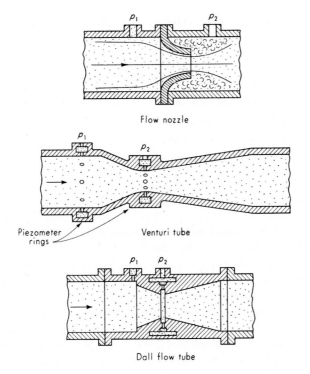

Figure 7.31
Variable-pressure-drop flowmeters.

about the same as the orifice. The venturi would also require a smaller β for the given Δp, but because of its streamlined form, its losses are low and nearly independent of β (about 10% of the measured Δp, for $0.2 < \beta < 0.8$).

Orifices, flow nozzles, and venturis can be "homemade" (using recommended dimensions given by sources such as Miller), or can be purchased as standard or special units from various vendors.[88] Manufacturers[89] of differential pressure sensors will also make up complete flowmeters, adding the needed orifice, nozzle, or venturi, which they may themselves make or purchase "outside."

Laminar-flow elements differ from the metering devices discussed above in that they are specifically designed to operate in the laminar-flow regime. Pipe flows generally are considered laminar if Reynolds number N_R is less than 2,000; however, in laminar-flow elements, often considerably lower values are designed for to ensure laminar conditions. The simplest form of laminar-flow element is merely a

[88]Superior Products, Inc., Torrance, CA, 800-593-0012 (www.orificeplates.com). (Standard and special orifices, nozzles, and venturis.)

[89]ABB (Fischer & Porter), Warminster, PA, 800-829-6001 (www.abb.com/automation).

length of small-diameter (capillary) tubing.[90] For $N_R < 2{,}000$, the Hagen-Poiseuille viscous-flow relation gives for incompressible fluids

$$Q = \frac{\pi D^4}{128 \mu L} \Delta p \qquad (7.59)$$

where $Q \triangleq$ volume flow rate, m^3/s

$D \triangleq$ tube inside diameter, m

$\mu \triangleq$ fluid viscosity, $N \cdot s/m^2$

$L \triangleq$ tube length between pressure taps, m

$\Delta p \triangleq$ pressure drop, Pa

One usually designs for $N_R \lesssim 1{,}000$ in such a device. Extremely small flows can be measured in this way; a 3-ft length of 0.004-in-diameter tubing measuring Δp with a 2-in water inclined manometer gives a threshold sensitivity of about 0.000175 in^3/h when hydrogen is flowing.[91]

A single capillary tube is capable of handling only small flow rates at laminar Reynolds numbers. To increase the capacity of laminar-flow elements, many capillaries in parallel (or their equivalent) may be employed. One variation[92] uses a large tube (about 1-in diameter) packed with small spheres. The passages between the spheres give the same effect as many capillary tubes. This particular instrument is designed to give a Reynolds number of 20 or less. A 5-in length of tube gives a Δp of 20 in of water for a 2 cm^3/min flow rate of air at 14.7 lb/in^2 absolute and 70°F. Another approach[93] uses a "honeycomb" element (see Fig. 7.32) with triangular members a few thousandths of an inch on a side and a few inches long. These devices have been used mainly to measure flow of low-pressure air, and standard models are available in ranges from 0.1 to 2,000 ft^3/min at pressure drops of 4 to 8 in of water. All the above laminar elements have the advantages accruing from a linear (rather than square-root) relation between flow rate and pressure drop; these are principally a large accurate range of as much as 100:1 (compared with 3:1 or 4:1 for square-root devices), accurate measurement of average flow rates in pulsating flow, and ease of integrating Δp signals to compute total flow. The laminar elements also can measure reversed flows with no difficulty. They are usually less sensitive to upstream and downstream flow disturbances than the other devices discussed. Their disadvantages include clogging from dirty fluids, high cost, large size, and high pressure loss (*all* the measured Δp is lost).

Laminar flow elements are widely used in pulsating-flow conditions as found in intake and exhaust manifolds of internal combustion engines, reciprocating air

---

[90]L. M. Polentz, "Capillary Flowmetering," *Instrum. Contr. Syst.*, p. 648, April 1961.

[91]Ibid.

[92]A. R. Hughes, "New Laminar Flowmeter," *Instrum. Contr. Syst.*, p. 98, April 1962.

[93]Meriam Instrument Co., Cleveland, OH, 216-281-1100 (www.meriam.com), *Tech Note* 2A.

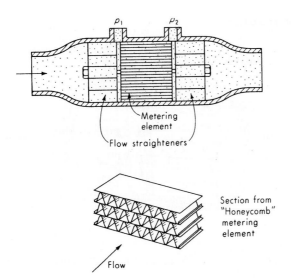

Figure 7.32
Laminar-flow element.

compressors, etc. An excellent paper[94] explores these issues in terms of the *dynamic* response characteristics of the instrument and also treats some nonlinear effects that influence both steady flow and dynamic measurements. A partial-differential-equation model leads to frequency response curves that allow prediction of dynamic errors in terms of dimensions and material properties. The curves are plotted from rather complex equations involving Bessell functions. When I first saw these curves, I was struck by their close resemblance to the amplitude-ratio and phase-angle curves of a simple first-order (lumped model) dynamic system. When I performed this analysis on a single capillary tube,[95] using only lumped fluid resistance and inertance in the model, the frequency response results were quite close to those of the paper. My first-order system time constant was $\tau = (\rho D^2)/(24\mu)$, which gave curves very close to those in the paper, obtained after a much more complex analysis. For air at 14.7 psia and 70°F, and tube diameter D in inches, the time constant becomes $0.0179D^2$ s. Note that the output Δp of the flowmeter *leads* the input flow rate in our first-order model, in contrast to the *lag* of most instruments. This opens the possibility of choosing the *lag* time constant of the Δp sensor to *match* the lead of the laminar element, giving a "perfect" (zero-order) dynamic response for the combination. (I have not seen this reported in the literature or tried it myself.)

---

[94]C. R. Stone and S. D. Wright, "Non-Linear and Unsteady Flow Analysis of Flow in a Viscous Flowmeter," *Trans. Inst. M C,* vol. 16, no. 3, 1994, pp. 128–141.

[95]E. O. Doebelin, "System Dynamics," Marcel Dekker, 1998, New York, p. 226.

Meriam accutube

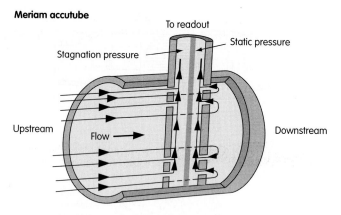

Figure 7.33
Averaging pitot tube.
(Courtesy of Meriam Instrument Co., Cleveland, OH, 216-281-1100, www.meriam.com.)

Averaging Pitot Tubes

The pitot-static tube described earlier for point velocity measurements can be calibrated for volume flow rate when it is located at a specific point in a specific duct; however, such single-point flow-rate measurements become inaccurate when velocity profiles differ from calibration conditions. Division of the total flow area into several equal-area annuli, measurement of local velocity at the center of area of each annulus by moving the pitot tube to each such point, and summation of these individual flow rates to obtain the total are possible for laboratory applications but impractical for on-line process monitoring or control. By using a fixed probe with several spatially dispersed sensing ports, whose readings are averaged in a plenum to produce a single "impact" pressure and a single "static" pressure, a practical volume flowmeter is obtained which produces a Δp signal related to volume flow rate by a square-root relation similar to that of the "obstruction" meters. Figure 7.33 shows one version available in pipe sizes ranging from 0.5- to 240-in diameter, for liquid, gas, or steam service. Advantages of averaging pitot tubes[96] include the possibility of installation in operating pipelines without shutdown (by using a sealable-drill technique), compatibility with common $\sqrt{\Delta p}$-type instrumentation, and low permanent pressure losses (compared with those of orifices) leading to claimed[97] savings of $1,000/year for a single steam line.

An alternative to multiple, averaged pitot tubes is multiple, averaged thermal sensors using the constant-temperature principle of Fig. 7.19. A commercial

---

[96]J. B. Schnake, "Emerging Flow Technology Boosts Accuracy and Savings," *Intech,* Jan. 1998, pp. 52–56; D. Thomas and R. D'Angelo, "Know Your Pitot," *Flowcontrol,* Sept. 2000, pp. 30–33.

[97]"Steam Measurement," *Appl. Note* 93-00-07-04 (580), Honeywell Process Control Div., Ft. Washington, PA, 1980.

version[98] for gas mass flow uses specially designed, rugged probes which include both the flow sensor and a temperature-compensating resistance thermometer in the R_3 location. The flow sensor thus provides a temperature-compensated mass velocity (ρV) signal, which is linearized before being summed with those from the other sensors in a mass flowmetering array. Good accuracy is obtained by basing the individual probe locations on a preliminary study of the duct-velocity profile with a traversing probe. Typically a duct 2×2 ft uses four probes while a 6×8 ft duct might require eight.

Constant-Pressure-Drop, Variable-Area Meters (Rotameters)[99]

A rotameter consists of a vertical tube with tapered bore in which a "float" assumes a vertical position corresponding to each flow rate through the tube (see Fig. 7.34). For a given flow rate, the float remains stationary since the vertical forces of differential pressure, gravity, viscosity, and buoyancy are balanced. This balance is self-maintaining since the meter flow area (annular area between the float and tube) varies continuously with vertical displacement; thus the device may be thought of as an orifice of adjustable area. The downward force (gravity minus buoyancy) is constant, and so the upward force (mainly the pressure drop times the float cross-section area) must be constant also. Since the float area is constant, the pressure drop must be constant. For a *fixed* flow area, Δp varies with the square of flow rate, and so to keep Δp *constant* for differing flow rates, the area must vary. The tapered tube provides this variable area. The float position is the output of the meter and can be made essentially linear with flow rate by making the tube area vary linearly with the vertical distance. Rotameters thus have an accurate range of about 10:1, considerably better than square-root-type elements. Accuracy is typically ± 2 percent full scale (± 1 percent if calibrated) with repeatability about 0.25 percent of reading. Assuming incompressible flow and the above described simplified model, we can derive the result

$$Q = \frac{C_d(A_t - A_f)}{\sqrt{1 - [(A_t - A_f)/A_t]^2}} \sqrt{2gV_f \frac{w_f - w_{ff}}{A_f w_{ff}}} \qquad (7.60)$$

where
$Q \triangleq$ volume flow rate, ft³/s
$C_d \triangleq$ discharge coefficient
$A_t \triangleq$ area of tube, ft²
$A_f \triangleq$ area of float, ft²
$g \triangleq$ local gravity, ft/s²
$V_f \triangleq$ volume of float, ft³
$w_f \triangleq$ specific weight of float, lbf/ft³
$w_{ff} \triangleq$ specific weight of flowing fluid, lbf/ft³

---

[98]J. Kurz, T. Ramey, and J. Willis, "The Book of Eva (*pamphlet*)," 1987; J. Kurz, "Automatic Isokinetic Stack Sampling," Kurz Instruments, Monterey, CA, 1987, 800-424-7356 (www.kurz-instruments.com).

[99]V. P. Head, "Coefficients of Float-Type Variable-Area Flowmeters," *Trans. ASME*, pp. 851–862, August 1954.

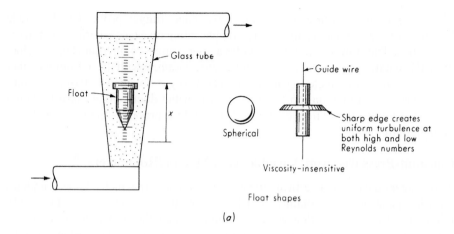

Glass tube

Float

Spherical

Guide wire

Sharp edge creates
uniform turbulence at
both high and low
Reynolds numbers

Viscosity-insensitive

Float shapes

(a)

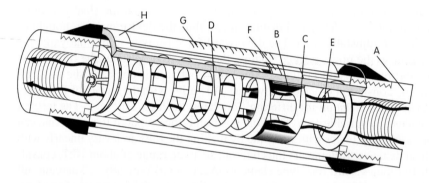

(b)

OPERATING THEORY

Enclosed within a high pressure casing (A), a high strength magnet (B) in tandem with the sharp-edged annular orifice disk (C), is pressed towards the zero flow rate position by a linear rate compression spring (D). A tapered metering pin (E) is positioned concentrically within the annular orifice disk and provides a variable-area opening that increases by the square of linear displacement of the orifice disk. Fluid flow creates a pressure differential across the orifice disk, pressing the magnet/orifice disk duo against the compression spring. Flow rate is read by aligning the magnetically coupled follower (F) with the graduated scale (G) located within the environmentally sealed window (H). The variable-area orifice design provides pressure differentials and orifice displacements that are linearly proportional to fluid flow rate.

Figure 7.34
Rotameter.

If the variation of C_d with float position is slight and if $[(A_t - A_f)/A_t]^2$ is always much less than 1, then Eq. (7.60) has the form

$$Q = K(A_t - A_f) \qquad (7.61)$$

And if the tube is shaped so that A_t varies linearly with float position x, then $Q = K_1 + K_2 x$, a linear relation. The floats of rotameters may be made of various materials to obtain the desired density difference [$w_f - w_{ff}$ in Eq. (7.60)] for metering a particular liquid or gas. Some float shapes, such as spheres, require no guiding in the tube; others are kept central by guide wires or by internal ribs in the tube. Floats shaped to induce turbulence can give viscosity insensitivity over a 100:1 range. The tubes often are made of high-strength glass to allow direct observation of the float position. Where greater strength is required, metal tubes can be used and the float position detected magnetically through the metal wall. If a pneumatic or an electrical signal related to the flow rate is desired, the float motion can be measured with a suitable displacement transducer. Flow rates beyond the range of the largest rotameter may be measured by combining an orifice plate and a rotameter in a bypass arrangement.[100]

Figure 7.34b[101] shows a version that relies on a spring (rather than a gravity) force, and thus need not be oriented vertically. Units are available with visual or electrical output for liquids (0.05 to 150 gpm) and air (2 to 1300 scfm), up to 6000 psi pressure and 600°F. Accuracy in the middle third of the range is about ±2.5 percent.

Turbine Meters

If a turbine wheel is placed in a pipe containing a flowing fluid, its rotary speed depends on the flow rate of the fluid. By reducing bearing friction and keeping other losses to a minimum, one can design a turbine whose speed varies linearly with flow rate; thus a speed measurement allows a flow-rate measurement. The speed can be measured simply and with great accuracy by counting the rate at which turbine blades pass a given point, using a magnetic proximity pickup to produce voltage pulses. By feeding these pulses to an electronic pulse-rate meter, one can measure flow rate; by accumulating the total number of pulses during a timed interval, the total flow is obtained. These measurements can be made very accurately because of their digital nature. If an analog voltage signal is desired, the pulses can be fed to a frequency-to-voltage converter. Figure 7.35 shows a flowmetering system of this type.

The simplest analytical model of this meter assumes one-dimensional flow and a rotor that feels *no* retarding torques, such as bearing friction, fluid viscosity, or inertial effects when accelerating. Since the fluid thus feels *nothing* as it proceeds through the meter, it continues in its original straight path, and we can calculate the rotor angular speed by requiring that the rotor just "get out of the way" of the fluid. With this assumption, we state that as the fluid moves from location a to b in Fig. 7.35b, the rotor *must* move from c to b, and we can write

$$\frac{r\omega \, dt}{V dt} = \tan \beta \qquad \omega = \frac{V \tan \beta}{r} \tag{7.62}$$

[100]"Guide to By-Pass Rotameter Application," *Tech. Bull.* T-023, Brooks Instrument Div., Hatfield, PA, 215-362-3700 (www.brooksinstrument.com).

[101]Lake Monitors, Inc., Milwaukee, WI, 800-850-6110 (www.lakemonitors.com).

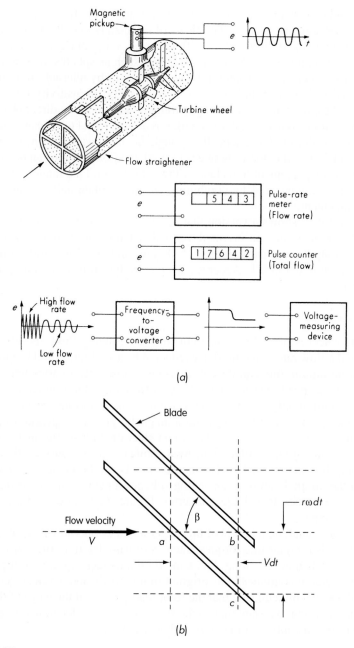

Figure 7.35
Turbine flowmeter.

For one-dimensional flow, the flow velocity V and volume flow rate Q are proportional, so we predict the rotor speed to be proportional to flow rate. Our simplified model works with a "point value" (r) for the rotor blade radius. More detailed

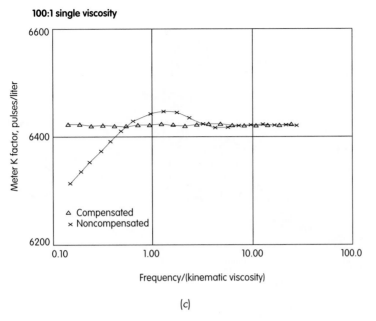

100:1 single viscosity

Meter K factor, pulses/liter

Frequency/(kinematic viscosity)

(c)

Figure 7.35
(Concluded)

studies[102] use the average of the squares of the inner and outer blade radii and include effects that we neglect, to achieve more correct but also more complex results. These more detailed studies and complicated results are perhaps of more interest to meter designers than users, so we now concentrate more on application information.

The ideal meter of our simple result would give a rotor speed (and thus an electrical pulse rate) exactly proportional to the volume flow rate of the fluid. Various real-world effects (mainly viscosity) make a real meter deviate from such perfection and require *calibration* to achieve practical utility. To measure volume flow rate with a turbine meter, we need to know the meter's "*K* factor" (pulses/liter), that is, for each pulse from the pickup, how much volume has passed through the meter. Ideally, the *K* factor would be a fixed constant, but a real meter, for a given liquid and temperature (this fixes the viscosity) has a *K* factor that varies with flow rate. Figure 7.35*c*[103] shows a common way of displaying the variation of *K* with flow rate. The curve labeled "noncompensated" is typical of the basic instrument. For

---

[102]W. F. Z. Lee and H. J. Evans, "Density Effect and Reynolds Number Effect on Gas Turbine Flowmeters," *J. Basic Engineering,* Dec. 1965, p. 1044; M. Rubin, R. W. Miller, and W. G. Fox, "Driving Torques in a Theoretical Model of a Turbine Meter," *J. Basic Engineering,* June 1965, pp. 413–420; W. F. Z. Lee et al., "Gas Turbine Flowmeter Measurement of Pulsating Flow," *J. Engineering Power,* Oct. 1975, pp. 531–539; L. A. Salami, *Trans Inst M C,* vol. 7, no. 4, July–Sept 1985; W. M. Jungowski and M. H. Weiss, "Effects of Flow Pulsation on a Single Rotor Turbine Meter," *J. Fluid Engineering,* vol. 118, March 1996, pp. 198–201; D. Wadlow, "Turbine Flowmeters," *Sensors,* Oct. 1999, pp. 16–3l, Nov. 1999, pp. 16–38.
[103]Flow Technology, Inc., Phoeniz, AZ, 800-528-4225 (www.ftimeters.com).

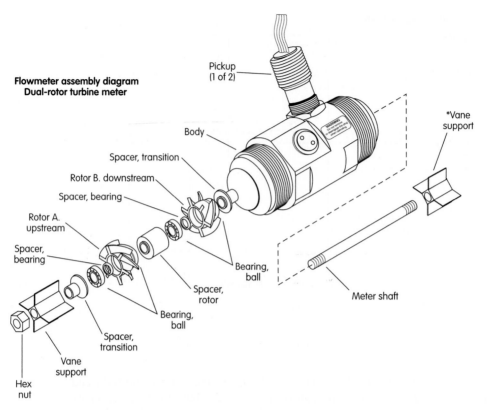

Flowmeter assembly diagram
Dual-rotor turbine meter

Pickup
(1 of 2)

*Vane
support

Body

Spacer, transition

Rotor B. downstream

Spacer, bearing

Rotor A.
upstream

Spacer,
bearing

Bearing,
ball

Spacer,
rotor

Meter shaft

Bearing,
ball

Spacer,
transition

Vane
support

Hex
nut

Figure 7.36
Dual-rotor (counter-rotating) turbine meter. *(Courtesy of Flowdata, Richardson, TX, 800-833-2448, www.flowdata.com.)*

low flow rates, viscous effects cause K to vary; K becomes essentially constant for a range (typically about 10:1) of higher flow rates. The K factor stays constant for even higher rates but this region is not useable because of high pressure drop and reduced bearing life.

By including a temperature sensor and a microprocessor loaded with data to allow calculation of viscosity and density from the measured temperature, the instrument (labeled "compensated" in Fig. 7.35c) can correct for these effects and thus extend the "flat" range of the K curve to get a "turndown" (also called *range-ability*) of 100:1. The density measurement also allows calculation of *mass* flow rate, often preferred to volume. If the meter is used with up to three different liquids, a "universal viscosity curve calibration" is available to deal with this. The referenced instruments also use a frequency-measuring technique that overcomes the time lags common to conventional frequency-to-voltage converters. By measuring the *time interval* between passage of two blades, the rotor speed can be updated as rapidly as every 7.5 ms. All the above mentioned compensating calculations are also performed at this same update rate, allowing measurement of rapidly changing flows.

Improved behavior over the classical design is also available in the *dual-rotor* instrument;[104] one version of which uses *counter-rotating* rotors as shown in Fig. 7.36. The basic instrument has only one pickup. Addition of the second pickup improves the turndown, provides diagnostic information on meter condition, and reduces sensitivity to flow turbulence. Another meter,[105] widely used for natural gas sales transactions, where high accuracy has a direct economic benefit, has two rotors that turn in the *same* direction. The upstream (main) rotor's pulse output is calibrated to indicate 110 percent of the true flow, while the downstream (sensing) rotor indicates 10 percent. True flow is obtained by subtracting the two readings. Because of the unique design of the meter, if, say, bearing friction of the main rotor increases from the calibration value and slows the main rotor by 2 percent, the sensing rotor *also* changes speed by 2 percent, so the subtraction still gives an accurate reading. In addition to *correcting* for many sources of error, this meter also reports how much conditions have changed from the original calibration, alerting the user to meter or flow-conditioning problems that might need attention, even though the flow rate is being accurately measured. This diagnostic feature uses the *ratio* of the two rotor speeds. When this changes from the original value, we know that *something* is causing a calibration change.

Commercial turbine meters are available with full-scale flow rates ranging from about 0.01 to 30,000 gal/min for liquids and 0.01 to 15,000 ft^3/min for air. Nonlinearity within the design range (often about 10:1) can be as good as 0.05 percent in the larger sizes. The output voltage of the magnetic pickups is of the order of 10 mV rms at the low end of the flow range and 100 mV at the high. Pressure drop across the meter varies with the square of flow rate and is about 3 to 10 lb/in^2 at full flow. Turbine meters behave essentially as first-order dynamic systems for small changes about an operating point.[106] However, the time constant (typically 2 to 10 ms at maximum flow) is inversely proportional to the operating point flow rate. If a frequency-to-voltage converter is used to get an analog voltage output, however, its response speed may not be negligible since the operating frequencies of turbine meters are of the order of 100 to 2,000 Hz. That is, the frequency-to-voltage converter requires low-pass filtering, which rejects frequencies somewhat below the turbine operating frequency and is thus limited in transient response also. For example, if the turbine is putting out 500 Hz, the low-pass filter will have to cut off at *least* at 500 Hz and probably considerably lower. If a first-order filter is designed to attenuate 20 dB at 500 Hz, it will have a time constant of 0.0032 s, thus adding this much to the lag of the overall system. Recall that some[107] F/V converters do *not* use low-pass filters and update the frequency reading for every single input pulse,

---

[104]W. F. Z. Lee, D. C. Blakeslee and R. V. White, "A Self-Correcting and Self-Checking Gas Turbine Meter," *J. Fluid Engineering,* June 1982, pp. 143–149.

[105]Auto-Adjust Turbo-Meter, Equimeter Inc., DuBois, PA, 800-375-8875 (www.equimeter.com).

[106]J. Grey, "Transient Response of the Turbine Flowmeter," *Jet Propul.,* p. 98, February 1956;
G. H. Stevens, "Dynamic Calibration of Turbine Flowmeters by Means of Frequency Response Tests," *NASA* TM X-1736. 1969.

[107]Model FV-801, Ono Sokki, Inc., Addison, IL, 630-627-9700 (www.onosokki.com).

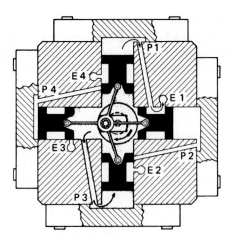

Figure 7.37
Piston-type positive-displacement flowmeter.
(Courtesy Fluidyne Instrumentation, Oakland, CA.)

allowing for faster response (as in the Flow Technology turbine meter discussed earlier).

Positive-Displacement Meters[108]

These meters are actually positive-displacement fluid motors in which friction and inertia have been reduced to a minimum. The flow of a fluid through volume chambers of definite size causes rotation of an output shaft. Figure 7.37 shows a self-porting four-piston design with fluid entering at $P3$ and leaving at $E1$, causing a clockwise shaft rotation which sequentially brings pairs ($P4$-$E2$, $P1$-$E3$, $P2$-$E4$, $P3$-$E1$) of inlet/outlet ports into operation through the three-way valve action of each piston. Such meters exhibit little sensitivity to viscosity (accuracy actually improves with increased viscosity, because of decreased leakage) and can give high accuracy over wide flow ranges (up to 1,000:1, see Fig. 7.38b). Since the meters have a maximum allowable pressure drop [typically 20 to 40 lb/in^2 differential (140 to 290 kPa)], high viscosity reduces the maximum allowable flow rate for a given meter (Fig. 7.38b). As a result of precision clearances in the instrument, fluid filtration at the 10-μm level is recommended. Meter construction materials also limit the fluids that can be metered. To eliminate friction-producing fluid seals, the meter shaft is coupled to the speed sensor magnetically through a nonmagnetic wall. This "soft" coupling can introduce errors for unsteady flows. Since the crankshaft kinematics produces an epicycloidal speed variation for a steady flow rate, special analog tachometers that compensate for this fluctuation are available.

[108]R. C. Baker and M. V. Morris, "Positive-Displacement Meters for Liquids," *Trans Inst M C,* vol. 7, no. 4, July–Sept. 1985.

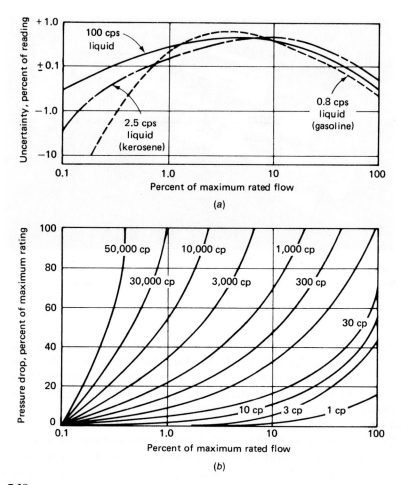

Figure 7.38
Performance of piston-type positive-displacement flowmeter.
(Courtesy Fluidyne Instrumentation, Oakland, CA.)

Digital (pulse-rate) output is also available that uses photo-optic techniques or rotary differential-transformer methods. The latter method does not use the soft magnetic coupling and thus follows unsteady crankshaft motions faithfully. Both methods provide direction sensing, so that reverse flows actually subtract pulses, to give true net flow in totalizing systems using up-down counters. For larger flow rates and for food/beverage applications requiring a cleanable, "pocketless" design, a meter based on helical rotors (Fig. 7.39) is available with performance similar to that of the four-piston design. An advantage shared by positive-displacement meters in general is their insensitivity to distorted inlet/outlet flow profiles; thus flow straighteners and/or long runs of straight pipe upstream/downstream of the meter usually are not required. Dynamic response specifications are not usually quoted;

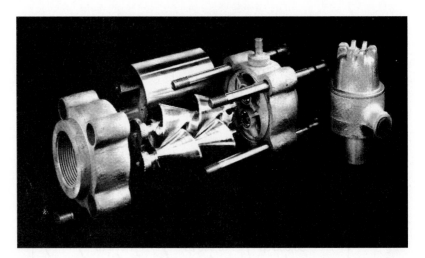

Figure 7.39
Helix-type positive-displacement flowmeter.
(Courtesy Fluidyne Instrumentation, Oakland, CA.)

however, the basic principle implies fast response since the output rotation is (ideally) "kinematically coupled" to the fluid flow rate. Deviations from ideal behavior which lead to response lags include fluid compressibility and meter leakage. Also, because of meter inertia and friction, rapid flow changes can cause the meter pressure drop to exceed the maximum allowable, causing metering errors or actual damage.

An interesting variation[108] on the positive-displacement concept eliminates both the leakage present in all such meters and the loading effect of the meter on the measured flow process. Both effects are chargeable to the meter pressure drop, and would disappear if it were zero. This is accomplished by using an active feedback scheme in which the meter shaft actually is driven by an electric servomotor whose drive voltage is obtained from the output of a differential-pressure transducer connected across the meter. Whenever this differential pressure is not zero, the motor either speeds up or slows down until the pressure is again zero; thus the meter speed "tracks" the impressed flow rate without extracting any energy from the flow.

Metering Pumps

A variable-displacement positive-displacement pump, if properly designed, can serve both to *cause* a flow rate and to *measure* it simultaneously. The principle again is merely that a positive-displacement machine, except for leakage and compressibility, delivers a definite flow rate of fluid at a given speed. In many pumps of this kind, the operating speed is fixed and the flow rate is varied by changing pump

---

[108]Laboratory Equipment Corp., Mooresville, IN, model PLV.

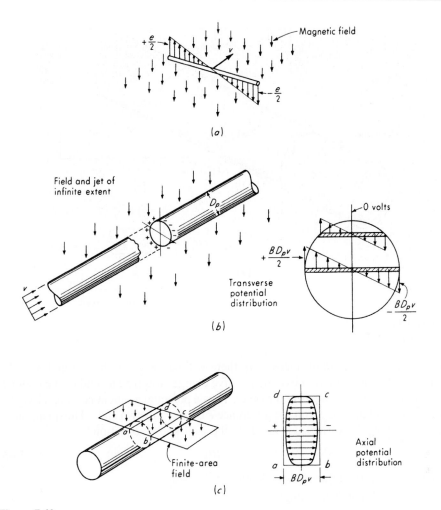

Figure 7.40
Electromagnetic flowmeter.

displacement, usually with some form of mechanical linkage. Since these pumps are used often in automatic control systems, many are designed to accept pneumatic or electrical input signals which adjust the pump displacement in a linear fashion. The flow rate of such a system can be set with an accuracy of the order of 1 percent.

Electromagnetic Flowmeters

The electromagnetic flowmeter[109] is an application of the principle of induction, shown in Fig. 7.40a. If a conductor of length l moves with a transverse velocity v

---

[109]J. A. Shercliff, "The Theory of Electromagnetic Flow Measurement," Cambridge University Press, New York, 1962.

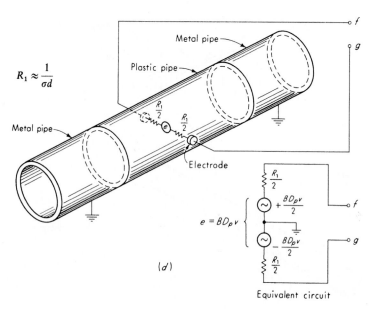

$$R_1 \approx \frac{1}{\sigma d}$$

$e = BD_p v$

(d)

Equivalent circuit

Figure 7.40
(Concluded)

across a magnetic field of intensity *B,* there will be forces on the charged particles of the conductor that will move the positive charges toward one end of the conductor and the negative charges to the other. Thus a potential gradient is set up along the conductor, and there is a voltage difference *e* between its two ends. The quantitative relation among the variables is given by the well-known equation

$$e = Blv \qquad (7.63)$$

where $B \triangleq$ field flux density, Wb/m^2 = V · s/m^2

$l \triangleq$ conductor length, m

$v \triangleq$ conductor velocity, m/s

If the ends of the conductor are connected to some external circuit that is stationary with respect to the magnetic field, the induced voltage, in general, will cause a current *i* to flow. This current flows through the moving conductor, which has a resistance *R,* causing an *iR* drop, so that the terminal voltage of the moving conductor becomes $e - iR$.

We consider now a cylindrical jet of conductive fluid with a uniform velocity profile, traversing a magnetic field as in Fig. 7.40*b.* In a liquid conductor the positive and negative ions are forced to opposite sides of the jet, giving a potential distribution as shown. The maximum voltage difference is found across the ends of a horizontal diameter and is $BD_p v$ in magnitude. In a practical situation, the magnetic field is of limited extent, as in Fig. 7.40*c;* thus no voltage is induced in that part of the jet outside the field. Since these parts of the fluid are, however, still conductive paths, they tend partially to "short-circuit" the voltages induced in the section

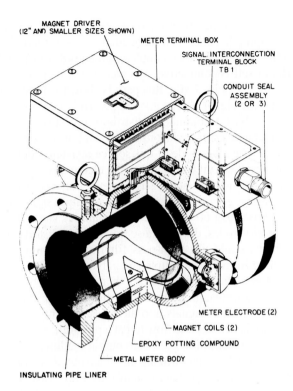

MAGNET DRIVER
(I2" AND SMALLER SIZES SHOWN)

METER TERMINAL BOX

SIGNAL INTERCONNECTION
TERMINAL BLOCK
TB 1

CONDUIT SEAL
ASSEMBLY
(2 OR 3)

METER ELECTRODE (2)

MAGNET COILS (2)

EPOXY POTTING COMPOUND

METAL METER BODY

INSULATING PIPE LINER

Figure 7.41
Construction details of magnetic flowmeter.

exposed to the field; thus the voltage is reduced from the value $BD_p v$. If the field is sufficiently long, this effect will be slight at the center of the field length. A length of 3 diameters usually is sufficient.[110]

In a practical flowmeter (see Fig. 7.41[111]), the "jet" is contained within a stationary pipe. The pipe must be nonmagnetic to allow the field to penetrate the fluid and usually is nonconductive (plastic, for instance), so that it does not provide a short-circuit path between the positive and negative induced potentials at the fluid surface. This nonconductive pipe has two electrodes placed at the points of maximum potential difference. These electrodes then supply a signal voltage to external indicating or recording apparatus. While Fig. 7.40 shows a uniform velocity profile, it has been proved mathematically[112] that e corresponds to the average velocity of

---

[110]T. C. Hutcheon, "Electrical Characteristics of the Magnetic Flow Detector Head," *Instrum. Eng.,* p. 1, April 1964.

[111]*Tech. Bull.* 10D-14, Fischer and Porter Corp., Warminster, PA, 1977.

[112]A. Kolin, "An Alternating Field Induction Flowmeter of High Sensitivity," *Rev. Sci. Instrum.,* vol. 16, p. 109, May 1945; N. C. Wegner, "Effect of Velocity Profile Distortion in Circular Transverse-Field Electromagnetic Flowmeters," *NASA* TN D-6454, 1971.

any profile which is symmetric about the pipe centerline. Because it is impractical to make the entire piping installation nonconductive, a short length (the flowmeter itself) of nonconductive pipe must be coupled into an ordinary metal-pipe installation. Since the fluid itself is conductive, this means that there is a conductive path between the electrodes. In Fig. 7.40d, this path is shown split into two equal parts $R_1/2$ and containing the signal source $e = BD_p v$. This resistance is not simple to calculate since it involves a continuous distribution of resistance over complex bodies. However, it can be estimated from theory, and once a device is built, it can be directly measured. The magnitude of this source resistance determines the loading effect of any external circuit connected to the electrodes.

The magnetic field used in such a flowmeter conceivably could be either constant[113] or alternating, giving rise to a dc or an ac output signal, respectively. For many years, ac systems (50 or 60 Hz) were most common, since they reduced polarization effects at the electrodes, did not cause flow-profile distortion from magneto-hydrodynamic effects, could use high-pass filtering to eliminate slow, spurious voltage drifts resulting from thermocouple and galvanic actions, and allowed use of high-gain ac amplifiers whose drift was less than that of dc types of comparable gain. These advantages outweighed the major disadvantage that the powerful ac field coils induced spurious ac signals into the measurement circuit. To cancel this error, one had to periodically stop the flow to get a pipe-full, zero-velocity condition and then adjust a balance control to get a zero output reading.

About 1975, industrial meters utilizing an "interrupted dc" field became available, and currently the market is shared mainly by these two types. In the interrupted dc meter, a dc field is switched in a square-wave fashion between the working value and zero, at about 3 to 6 Hz. When the field is zero, any instrument output that appears is considered to be an error; thus by storing this zero error and subtracting it from the total instrument output obtained when the field is next applied, one implements an "automatic zero" feature which corrects for zero errors several times a second. Additional advantages include power savings of up to 75 percent and simpler wiring practices. A disadvantage in some applications is the slower response time of about 7 s (60-Hz systems have about 2 s).

For a 60-Hz ac system with tapwater flowing at 100 gal/min in a 3-in-diameter pipe, e is about 3 mV rms. The resistance between the electrodes is given by theory[114] as approximately $1/(\sigma d)$, where $\sigma \triangleq$ fluid conductivity and $d \triangleq$ electrode diameter. For tapwater, $\sigma \approx 200 \ \mu S/cm$; so if $d \approx 0.64$ cm, there is a resistance of about 7,800 Ω as the internal resistance of the voltage source producing e, which requires the sensor amplifier to have an input impedance that is large relative to this value. Standard magnetic flowmeters accept fluids with conductivities as low as about 5 $\mu S/cm$, with special systems going down to about 0.1 $\mu S/cm$. Gasoline, with $10^{-8} \ \mu S/cm$, is definitely not measurable; alcohol, with 0.2 $\mu S/cm$, is just

[113]H. M. Hammac, "An Application of the Electromagnetic Flowmeter for Analyzing Dynamic Flow Oscillations," *NASA* TM X-53570, 1967.

[114]V. P. Head, "Electromagnetic Flowmeter Primary Elements," *Trans. ASME,* p. 662, Dec. 1959.

barely measurable; mercury (a liquid metal), with 10^{10} μS/cm, presents no conductivity problem. While commercially available systems are limited to the 0.1 μS/cm value, research devices which work with dielectric fluids such as liquid hydrogen have been demonstrated.[115]

A specialized but important application area which has received much attention is that of blood flow[116] in the vessels of living specimens. Miniaturized sensors allow measurements on vessels as small as 1 mm. To obtain dynamic flow capability, ac or switched dc ("square-wave") systems with 200 to 1,000 Hz frequency are used.

While ac and switched dc systems predominate, "pure" dc types have been employed in metering liquid metals, such as mercury. Here, no polarization problem exists. Also, an insulating pipe liner is not needed, since the conductivity of the liquid metal is very good relative to that of an ordinary metal (stainless-steel) pipe. This means that a metal pipe is not very effective as a "short circuit" for the voltage induced in a liquid-metal flow. Also, no special electrodes are necessary, the output voltage being tapped off the metal pipe itself at the points of maximum potential difference. Small, inexpensive pure dc meters using a permanent magnet field and special sintered silver-silver chloride electrodes, which greatly ease the classical electrode instability problems of dc meters, have appeared.[117] Originally developed for blood flow, they are usable with many liquids with conductivity greater than a few microsiemens per centimeter. Because of the permanent-magnet field and dc operation, the electronics can be very simple, giving a fast-response instrument of modest stability and accuracy useful in various research applications.

Since contact between the electrode and the liquid sometimes presents problems, one manufacturer[118] has developed a unit (useable to 0.05 μS/cm) where capacitive electrodes are buried inside a ceramic flow tube, giving a perfectly smooth and crevice-free surface. Another recent development[119] is the ability to measure flow accurately when the pipe is not full, as is common in many wastewater systems, which are designed to handle a wide range of flows and are thus often partially filled.

Some general features of electromagnetic flowmeters include the lack of any flow obstruction; ability to measure reverse flows; insensitivity to viscosity, density, and flow disturbances as long as the velocity profile is symmetrical; wide linear range; and rapid response to flow changes (instantaneous for a "pure" dc system; limited by the field frequency in an ac or switched dc system).

---

[115]V. Cushing, D. Reily, and G. Edmunds, "Development of an Electromagnetic Induction Flowmeter for Cryogenic Fluids," NASA Lewis Res. Ctr. contract NASW-381, *Final Rept.,* 1964.

[116]R. S. C. Cobbold, "Transducers for Biomedical Measurements," chap. 8, Wiley-Interscience, New York, 1974.

[117]"In Vivo Metric Systems," *Tech. Bull.,* no 1, Healdsburg, CA, Dec. 1980.

[118]Krohne Inc., Peabody, MA, 800-356-9464 (www.krohne.com).

[119]J. Flood, "Single-Sensor Measurement of Flow in Filled or Partially Filled Process Pipes," *Sensors,* Sept. 1997, pp. 54–57; B. Doney, "EMF Flow Measurement in Partially Filled Pipes," *Sensors,* Oct. 1999, pp. 65–68.

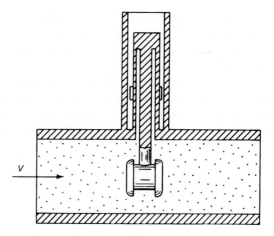

Figure 7.42
Drag-force flowmeter.

Drag-Force Flowmeters

A body immersed in a flowing fluid is subjected to a drag force F_d given by

$$F_d = \frac{C_d A \rho V^2}{2} \tag{7.64}$$

where $C_d \triangleq$ drag coefficient
$A \triangleq$ cross-section area, ft^2
$\rho \triangleq$ fluid mass density, slug/ft^3
$V \triangleq$ fluid velocity, ft/s

For sufficiently high Reynolds number and a properly shaped body, the drag coefficient is reasonably constant. Therefore, for a given density, F_d is proportional to V^2 and thus to the square of volume flow rate. The drag force can be measured by attaching the drag-producing body to a cantilever beam with bonded strain gages. A commercial instrument[120] uses the hollow-tube arrangement of Fig. 7.42 to isolate the gages from the flowing fluid. Units are available for pipe sizes from 0.5- to 60-in diameter. The drag body for the 0.5-in size is about 0.40-in diameter, has a drag coefficient of about 4.5, and produces about 2 lb of drag force at maximum flow. Drag bodies for larger pipe sizes take up a much smaller fraction of the pipe area and have a drag coefficient of about 1.5; maximum force is about 10 lb. The strain tube has wall thickness of about 0.010 to 0.030 in at the gage location, depending on the model. A full bridge of 350-Ω strain gages is designed for about 1000 $\mu\epsilon$. Liquids, gases, and steam can be metered over wide ranges of temperature, pressure, and flow rate. If the drag body is made symmetrical, reversing flows

---

[120]Target Flowmeter, Venture Measurement Co. LLC, Spartanburg, SC, 800-778-9251 (www.aaliant.com).

can be measured. Relative to most other flowmeters, dynamic response is quite fast; natural frequencies of 70 to 200 Hz are possible.

Ultrasonic Flowmeters

Small-magnitude pressure disturbances are propagated through a fluid at a definite velocity (the speed of sound) *relative to the fluid.* If the fluid also has a velocity, then the *absolute* velocity of pressure-disturbance propagation is the algebraic sum of the two. Since flow rate is related to fluid velocity, this effect may be used in several ways as the operating principle of an "ultrasonic" flowmeter.[121] The term *ultrasonic* refers to the fact that, in practice, the pressure disturbances usually are short bursts of sine waves whose frequency is above the range audible to human hearing, about 20,000 Hz. A typical frequency might be 10 MHz.

The two major methods ("transit time" and "doppler") of implementing the above phenomenon depend on the existence of transmitters and receivers of acoustic energy. A common approach is to utilize piezoelectric crystal transducers for both functions. In a transmitter, electrical energy in the form of a short burst of high-frequency voltage is applied to a crystal, causing it to vibrate. If the crystal is in contact with the fluid, the vibration will be communicated to the fluid and propagated through it. The receiver crystal is exposed to these pressure fluctuations and responds by vibrating. The vibratory motion produces an electrical signal in proportion, according to the usual action of piezoelectric displacement transducers. The need, in most instruments, for a fairly narrow, well-defined acoustic beam explains the use of high (10 MHz) frequencies. For a crystal to be an efficient transmitter of acoustic energy, its diameter D must be large compared with the wavelength λ of the oscillation. The conical beam projected from a circular crystal has a half-angle α given by $\sin \alpha = 1.2\lambda/D$; thus the desired small angles also require a small λ/D ratio. Since compactness requires reasonably small (≈ 1 cm) values of D, we need to use wavelengths on the order of 1 mm. Water, for example, has $\lambda = 1.5 \times 10^6/f$ mm ($f \triangleq$ frequency, Hz), so the need for megahertz frequencies is apparent.

Figure 7.43*a* shows the most direct application of these principles. With zero flow velocity the transit time t_0 of pulses from the transmitter to the receiver is given by

$$t_0 = \frac{L}{c} \tag{7.65}$$

where $L \triangleq$ distance between transmitter and receiver and $c \triangleq$ acoustic velocity in fluid. For example, in water, $c \approx 5,000$ ft/s, and so if $L = 1$ ft, $t_0 = 0.0002$ s. If the fluid is moving at a velocity V, the transit time t becomes

$$t = \frac{L}{c + V} = L\left(\frac{1}{c} - \frac{V}{c^2} + \frac{V^2}{c^3} - \cdots\right) \approx \frac{L}{c}\left(1 - \frac{V}{c}\right) \tag{7.66}$$

---

[121]L. C. Lynnworth, "Ultrasonic Flowmeters," chap. 5 in W. P. Mason and R. N Thurston (eds.), "Physical Acoustics," Academic, NY, 1979; D. Clayton, "Ultrasonic Flowmeters Get the Nod," *Control Engineering,* Sept. 1998, pp. 64–69.

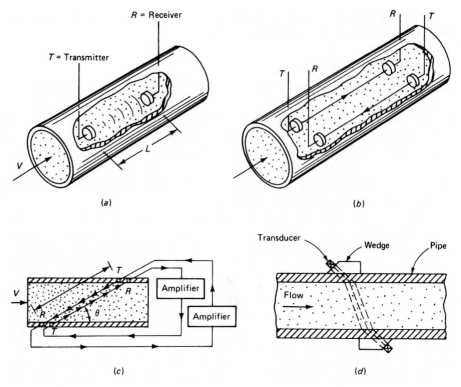

Figure 7.43
Ultrasonic transit-time flowmeters.

and if we define $\Delta t \triangleq t_0 - t$, then

$$\Delta t \approx \frac{LV}{c^2} \tag{7.67}$$

Thus, if c and L are known, measurement of Δt allows calculation of V. While L may be taken as constant, c varies, for example, with temperature, and since c appears as c^2, the error caused may be significant. Also, Δt is quite small since V is a small fraction of c. For example, if $V = 10$ ft/s, $L = 1$ ft, and $c = 5,000$ ft/s, then $\Delta t = 0.4$ μs, a very short increment of time to measure accurately. Since the measurement of t_0 is not directly provided for in this arrangement, the modification of Fig. 7.43b may be preferable. If t_1 is the transit time with the flow and t_2 is the transit time against the flow, we get

$$\Delta t \triangleq t_2 - t_1 = \frac{2VL}{c^2 - V^2} \approx \frac{2VL}{c^2} \tag{7.68}$$

This Δt is twice as large as before and also is a time increment that physically exists and may be measured directly. However, the dependence on c^2 is still a drawback.

A 4 Path Chordal LEFM

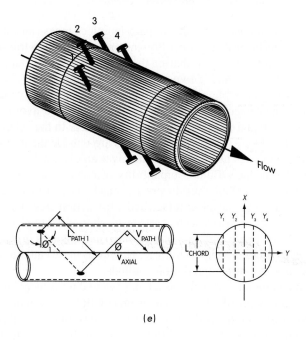

(e)

Figure 7.43
(Concluded)

In Fig. 7.43c two self-excited oscillating systems are created by using the received pulses to trigger the transmitted pulses in a feedback arrangement. The pulse repetition frequency in the forward propagating loop is $1/t_1$ while that in the backward loop is $1/t_2$. The frequency difference is $\Delta f \triangleq 1/t_1 - 1/t_2$, and since $t_1 = L/(c + V\cos\theta)$ and $t_2 = L/(c - V\cos\theta)$, we get

$$\Delta f = \frac{2V\cos\theta}{L} \tag{7.69}$$

which is independent of c and thus not subject to errors due to changes in c. This technique was used for many years but has been largely replaced (due to advances in digital-time-measurement accuracy and resolution) by methods based on *direct* measurement of the two transit times and the following data processing:

$$V = \frac{L}{2} \cdot \frac{t_2 - t_1}{t_2 t_1} \quad \text{proof:} \quad \frac{(c + V - c + V)L}{(c - V) \cdot (c + V) \cdot \dfrac{L^2}{(c - V) \cdot (c + V)}} \cdot \frac{L}{2} = \frac{2VL}{L^2} \cdot \frac{L}{2} = V \tag{7.70}$$

We see that this method also does not require knowledge of the acoustic velocity c. The details of time measurement vary with the manufacturer and are usually proprietary, however *averaging* of several measurements before reporting a flow rate is usually necessary. At least one meter[122] uses *cross-correlation* to measure the time intervals.

To cut costs and reduce errors due to path-length changes caused by buildup of deposits on the transducer faces, most flowmeters now use two transducers, rather than four. This is achieved by timesharing the single pair of transducers so it is used alternately for upstream and downstream propagation. High-speed electronic switching makes this possible, and since there is now only one value of L (rather than L_1 and L_2 when unequal deposit layers build up), we avoid the bias error ($\Delta t \neq 0$ when $V = 0$) and independence of c is retained. In the "clamp-on" type[123] of ultrasonic meter (Fig. 7.43d), the transducers are outside the pipe, which eliminates the fouling problems just described and gives an extremely convenient installation devoid of transducer/fluid compatibility problems. Generally, the existing pipe can be employed, and the transducers are attached by mechanical clamping or adhesive bonds, leaving the pipe intact and obstruction-free. Clamp-on meters exhibit their own problem areas, such as "acoustic short circuiting" (receiving transducer feels both liquid-path and pipe-path signals) and changes in beam path due to clamp slippage, temperature expansion, etc. However, these can often be overcome by proper design or use of compensation methods.

A problem common to both clamp-on and "wetted transducer" meters is sensitivity to flow velocity profile. Unlike electromagnetic meters, which read the correct average velocity as long as the flow profile is axisymmetric, ultrasonic meters of the type described above will give different readings for axisymmetric profiles of different shape but identical average velocity [Eq. (7.69) assumed a *uniform* profile]. The reason is that the pulse transit time is related to the integral of fluid velocity along the path. Thus 1 cm of path near the wall is weighted equally with 1 cm near the centerline, even though the contributions of the annular flow *areas* of these two regions would not be the same. If we defined a "meter coefficient" to be 1.0 for a uniform profile, this would drop to 0.75 for laminar flow ($N_R < 2,000$) and vary between 0.93 and 0.96 for turbulent flows with $4,000 < N_R < 10^7$. Rather than using the "tilted diameter" signal path assumed so far in our discussion, if one utilizes a *chordal* path, it has been shown[124] that the meter coefficient is nearly the same for both laminar and turbulent flow. Another approach is to employ several signal paths and average their results. Also, if flow conditions are reproducible, computed corrections always can be applied, even for a meter with a single diametral path. Figure 7.43e[125] shows a four-path chordal arrangement used in a transit-time flowmeter that also infers temperature from measured values of c and

[122]Digital Correlation Transit Time Flowmeter, Polysonics Inc., Houston, TX, 281-879-3700 (www.polysonicsinc.com).

[123]L. C. Lynnworth, "Clamp-On Ultrasonic Flowmeters: Limitations and Remedies," *Inst. Tech.,* pp. 37–44, September 1975; J. Baumoll, Letter to Editor, *Inst. Tech.,* p. 4, November 1975.

[124]L. C. Lynnworth, "Ultrasonic Flowmeters," op. cit., p. 430.

[125]H. Estrada, "General Principles of LEFM Time-of-Flight Ultrasonic Flow Measurements"; "Identifying and Bounding the Uncertainties in LEFM Flow Measurements," March 2000, Caldon, Inc., Pittsburgh, PA, 412-341-9920 (www.caldon.net).

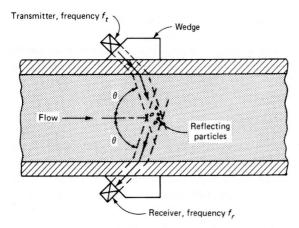

Transmitter, frequency f_t

Wedge

θ

Flow

θ

Reflecting particles

Receiver, frequency f_r

Figure 7.44
Clamp-on Doppler ultrasonic flowmeter.

pressure. Knowing the temperature and what fluid is being metered, the density and thus *mass* flow rate can be obtained. The referenced papers show how the improved accuracy of this design has allowed nuclear power plants to increase their power level by 1.7 percent and reduce the occurrence of safety-related overpower events. Such apparently small percentage changes can mean millions of dollars per year for a typical plant. The flowmeters are used for the feedwater flow, which must be accurately measured to calculate the power level.

The other main category of commercial ultrasonic meters uses the doppler principle. Whereas the "transit time" meters discussed above require relatively clean fluid to minimize signal attenuation and dispersion, doppler meters will not work at all unless sufficient reflecting particles and/or air bubbles are present.[126] For example,[127] in an 8-in pipe, 10 percent by volume of 40-μm reflectors or 0.2 percent of 100-μm reflectors would be needed. Doppler meters usually employ a clamp-on configuration; Fig. 7.44 shows one of several possible arrangements. Here the transmitter propagates a continuous-wave (CW) ultrasonic (0.5 to 10 MHz) signal into the fluid (assumed to have a uniform-profile velocity V), whereupon particles (assumed to be moving at velocity V) reflect some of the energy to the receiver. Analysis[128] of a wetted-transducer design gives

$$\Delta f = f_t - f_r = \frac{2f_t \cos \theta}{c} V \qquad (7.71)$$

[126]R. A. Moss, "Doppler Dilemma," *Flow Control,* Oct. 2000, pp. 44–51.

[127]J. M. Waller, Guidelines for Applying Doppler Acoustic Flowmeters, *In Tech,* pp. 55–57, October 1980.
T. R. Schmidt, What You Should Know about Clamp-on Ultrasonic Flowmeters, *In Tech,* pp. 59–62, May 1981.

[128]R. S. C. Cobbold, op. cit., p. 281.

The apparent dependence on c is misleading for the clamp-on design of Fig. 7.44, since it can be shown[129] (1) that changes in c cause compensating changes in cos θ and (2) that for such a design, one should interpret θ to be the transducer wedge angle and c to be the propagation velocity for the *wedge* material, not the fluid. This makes temperature effects manageable and contributes to the practicality of the doppler flowmeter.

In Fig. 7.44, the signal seen by the receiver is actually a complex signal, not a simple sine wave at frequency f_r. The two frequencies f_t and f_r get "mixed" in such a way that the received signal contains beat frequencies that are their sum and difference. A Fast Fourier Transform (FFT) analysis of this complex signal extracts its frequency content, and it is possible to "pick out" the peak corresponding to the difference frequency, which is the measure of flow velocity, as seen in Eq. (7.71). A refinement[130] on this basic method uses *two* different transmit frequencies. This allows a special FFT processing method that is effective in rejecting noise signals, such as those produced by variable-frequency electric drives often used to pump fluids. With regard to flow profile effects, the doppler meter "interrogates" only the sensing volume defined by the intersection of the transmitted and reflected beams; thus it measures an average velocity over this space (not over a tilted diametral path from wall to wall, as in a transit-time meter). This "point" velocity measurement can be calibrated, of course, to give total volume flow rate; however, the flow profile must be reproducible. (Also, any shift in the position of the sensing volume can cause errors even if the flow profile is reproducible.) If our goal is to measure "point" velocities rather than gross flow rates, this "problem" of the doppler meter actually becomes an advantage. A pulsed (rather than CW) system[131] using range-gating techniques actually has been utilized to get velocity profiles in blood vessels.

Because of the availability of meters for both clean and dirty fluids, little or no obstruction or pressure drop, convenient installation for clamp-on types, and good rangeability, ultrasonic meters are finding increasing application.[132] Ultrasonic meters have been used mainly for liquids but are now being applied to gases and steam (the present market is about 70 percent liquid and 30 percent gas; clamp-on is currently used only for liquids). Measurement of flare gas in refineries is a common application. One unit[133] uses a correlation transit-time technique together with proprietary methods for measuring average molecular weight to provide velocity (0.1 to 275 ft/s) and volume and mass flow rates (requires pressure and temperature sensors). A steam meter from the same source measures volume flow rate of wet steam and mass flow rate of saturated or superheated steam, again with help from pressure and temperature sensors and steam tables. High-temperature

---

[129]L. C. Lynnworth, "Ultrasonic Flowmeters," op. cit., p. 444.

[130]Polysonics Inc., Hydra SX40, Houston, TX, 281-879-3700 (www.polysonicsinc.com).

[131]R. S. C. Cobbold, op. cit., p. 286.

[132]"Engineer/Users Guidebook to Doppler Flow Measurement in Liquids," Polysonics, Houston, TX, 1986; C. Robinson, "Obstructionless Flow Meters," *In Tech,* pp. 33–36, December 1986.

[133]Panametrics, Inc., Waltham, MA, 800-833-9438 (www.panametrics.com).

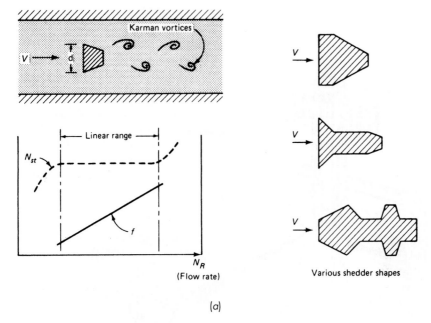

(a)

Figure 7.45
(a) Vortex-shedding flowmeter principles; (b) vortex-shedding flowmeter details.
(*Courtesy Neptune Eastech, Edison, NJ.*)

applications may require special techniques.[134] The American Gas Association has studied and now approves the use of multipath ultrasonic meters for custody transfer of natural gas, a large market that requires high accuracy.[135] Our emphasis has been on flow in pipes, but we want to at least mention an application to measure "external" flow; anemometers[136] or wind-velocity measurement. These are transit-time devices that can measure wind speed and direction in two or three dimensions.

Vortex-Shedding Flowmeters

The phenomenon of vortex shedding ("Karman vortex street") downstream of an immersed solid body of "blunt" shape when a steady flow impinges upstream is well known in fluid mechanics and is the basis of the vortex-shedding flowmeter.[137]

---

[134]L. Lynnworth, "High-Temperature Flow Measurement with Wetted and Clamp-On Ultrasonic Sensors," *Sensors,* Oct. 2000, pp. 36–52.

[135]"Measurement of Gas by Multipath Ultrasonic Meters," Transmission Measurement Committee Rept. No. 9, American Gas Association (www.aga.org), June 1998; J. Yoder, "Ultrasonics Reverberate through Flowmeter Market," *In Tech,* July 2000, pp. 44–47.

[136]R. Lockyer, "A Sonic Anemometer for General Meteorology," *Sensors,* May 1996, pp. 12–18; Vaisala Inc., Handar Business Unit, Sunnyvale, CA, 408-734-9640 (www.handar.com).

[137]E. Ribolini, "Intelligent and Mass Vortex Flowmeters," *In Tech,* Feb. 1996, pp. 31–37; J. Olin, "Generation-Next Mass Flowmeter Arrives," *In Tech,* Jan. 1999, pp. 58–61.

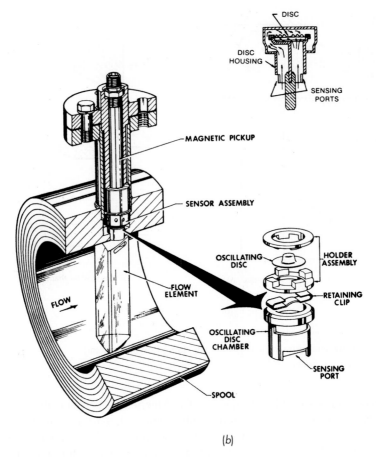

(b)

Figure 7.45
(Concluded)

When the pipe Reynolds number N_R exceeds about 10,000, vortex shedding is reliable, and the shedding frequency f is given by

$$f = \frac{N_{st}V}{d}$$

where
     $V \triangleq$ fluid velocity

     $d \triangleq$ characteristic dimension of shedding body

     $N_{st} \triangleq$ Strouhal number, an experimentally determined number, nearly constant in the useful flowmetering range (for example, 0.21 for cylinders)

By proper design of the shedding body shape, N_{st} can be kept nearly constant over a wide range of N_R (and thus flow rate), making f proportional to V and thus giving a "digital" flowmetering principle based on counting the vortex shedding rate (see Fig. 7.45*a*). Various shedder shapes and frequency sensing schemes have been developed by several manufacturers. The vortices cause alternating forces or local pressures on the shedder; piezoelectric and strain-gage methods can be employed to detect these. Hot-film thermal anemometer sensors buried in the shedder can detect the periodic flow-velocity fluctuations. The interruption of ultrasonic beams by the passing vortices can be used to count them. Vortex-induced differential pressures will cause oscillation of a small caged ball whose motion can be detected with a magnetic proximity pickup. Figure 7.45*b* shows yet another scheme which senses differential pressure with an elastic diaphragm.

A wide variety of liquids and gases and steam may be metered. Linear ranges of 15:1 are common, with 200:1 sometimes possible. Vortex frequencies at maximum flow rate are the order of 200 to 500 Hz, and the frequency responds to changing flow rate within about 1 cycle. If a frequency-to-voltage converter is used to realize an analog output voltage, then its low-pass filter will determine overall system response, which can be quite fast at high flow rates but must become slower at low flow rates (because the filter time constant must be set longer to give acceptable ripple at the lowest frequency, typically 5 to 10 Hz). When the fastest response is needed, a design[138] using ultrasonic sensing and a small triangular shedder (blunt face at the *back*) is available. A high-frequency ultrasonic beam traverses the pipe at the vortex location and is modulated by the vortex-shedding frequency. Filtering extracts the modulation envelope, which is the shedding frequency. With the blunt face at the back, the vortices are shed from the sharp corners of the triangle, which reduces the "time jitter" from vortex to vortex, compared with other shedder shapes. The small shedder size also gives higher frequencies (up to about 6000 Hz) for a given flow velocity. If we then use a "zero-time-constant" type of *F/V* converter (such as the Ono Sokki), we can update our velocity measurement for *each* pulse, giving, for example, one reading every 0.001 s for a frequency of 1000 Hz. If the fastest response is not needed, we can get higher accuracy by *averaging* several pulses, a technique commonly used in all vortex meters. As with the ultrasonic flowmeters, desirable characteristics of vortex-shedding meters have allowed them to take over a portion of the flowmeter market from the more traditional types of instruments.

Miscellaneous Topics

We briefly mention here three topics that cannot be treated in depth because of space limitations but which are felt worthy of more than just a bibliographic entry. *Two-phase flow*[139] occurs in a number of practical applications and has been of

---

[138]J-TEC Associates, Inc., Cedar Rapids, IA, 800-959-0872 (www.j-tecassociates.com); R. D. Joy, "Ultrasonic Vortex Meter Is a Wide-Range Gas Flowmeter," *Measurements & Control,* Oct. 1984; "Air Flow Measurement for Engine Control," *SAE* 760018, Jan. 1982.

[139]G. F. Hewitt, "Measurement of Two Phase Flow Parameters," Academic, New York, 1978; P. P. Kremlevsky, "Flow Rate Measurement in Multi-Phase Flows," Begell House, New York, 2000.

particular interest for steam/water and air/water mixtures in steam power plants.[140] Severe environments and lack of well-established calibration standards make instrument development particularly difficult. Few commercial instruments are available, and special designs or adaptations of standard devices usually are necessary.

Metering of *cryogenic fluids* such as liquid oxygen[141] and hydrogen (used in rocket engines), liquefied natural gas, etc., presents problems of extreme low temperatures and explosion hazards. Use of safer liquid nitrogen as a surrogate fluid for calibration purposes[142] has been established as a valid procedure and is the basis of a calibration facility which provides a mass flow uncertainty of about ± 0.2 percent over a flow range of 20 to 210 gal/min at temperatures of 70 to 90 K.

The use of *cross-correlation* as a basic flowmetering principle is being researched and may lead to a family of commercial instruments, since it can be implemented in several ways[143] by utilizing "tagging signals" already present in the flow or intentionally introduced. Briefly, the signals from two sensors separated axially in the pipe by a distance L and responding to the tagging signals are cross-correlated by using a range of delays. Since the tagging signals are swept from one sensor to the other at flow velocity V, the cross-correlation function should exhibit a peak at a delay equal to L/V seconds, which allows calculation of V since L is known. Turbulent eddies naturally present and detected by doppler ultrasonic methods, or thermal pulses intentionally injected and then sensed by thermocouples are only two of many possible schemes being considered. The practicality of the method hinges strongly on the availability of low-cost correlators based on integrated-circuit and microprocessor technology.

A thermal "transit-time" flowmeter[144] intended for low flows such as those found in chromatography offers 9 models with ranges from 0.02 to 0.2 mL/min to 10 to 100 mL/min. In Fig. 7.46a and b, we see the physical construction and functional block diagram, where entering liquid is "injected" with a heat pulse of about 0.2-s duration. The liquid sweeps this pulse downstream at flow velocity V where it is detected by a thermistor. The time between the upstream and downstream heat pulses, together with the known distance traveled, allows calculation of the velocity and thus the flow rate. Second-derivative analog-signal processing (Fig. 7.46d) is used to sharpen the sensed pulse (Fig. 7.46c) to allow more accurate time measurement.

[140]"Development of Instruments for Two-Phase Flow Measurements," *Rept.* ANCR-1181, Aerojet Nuclear Co., Idaho Falls, ID, 1974.

[141]D. B. Mann, "ASRDI Oxygen Technology Survey," vol. 6, "Flow Measurement Instrumentation," *NASA* SP-3084, 1974.

[142]J. A. Brennan et al., "An Evaluation of Selected Angular Momentum, Vortex Shedding and Orifice Cryogenic Flowmeters," *NBS Tech. Note* 650, Cryogenics Div., Boulder, CO, 1974.

[143]M. S. Beck, "Correlation in Instruments: Cross Correlation Flowmeters," *J. Phys. E: Sci. Instrum.,* vol. 14, no. 1, 1981.

[144]M-Tek Div. of Vacuum Research Corp., Pittsburgh, PA, 800-245-5101 (www.vacuumresearchcorp.com); T. E. Miller and H. Small, "Thermal Pulse Time-of-Flight Liquid Flow Meter," *Analytical Chemistry,* 1982, vol. 54, pp. 907–910.

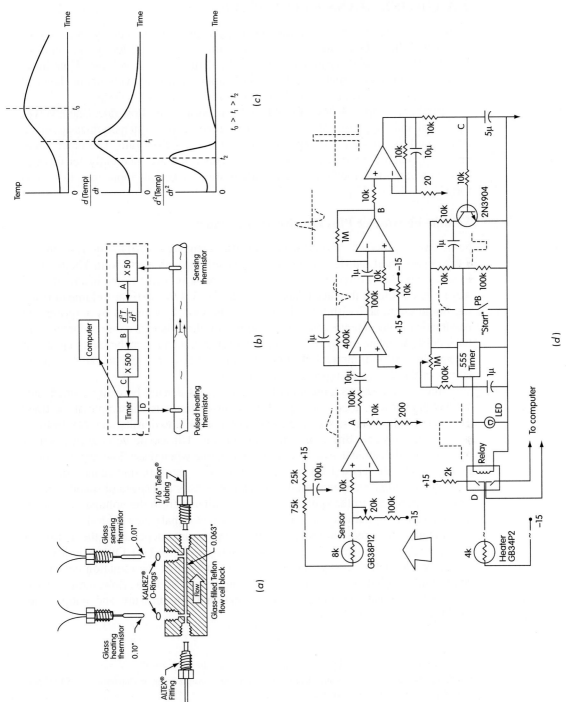

Figure 7.46
Thermal transit-time flowmeter.

7.3 GROSS MASS FLOW RATE

In many applications of flow measurement, mass flow rate is actually more significant than volume flow rate. As an example, the range capability of an aircraft or liquid-fuel rocket is determined by the *mass* of fuel, not the volume. Thus flowmeters used in fueling such vehicles should indicate mass, not volume. In chemical process industries also, mass flow rate is often the significant quantity.

Two general approaches are employed to measure mass flow rate. One involves the use of some type of volume flowmeter, some means of density measurement, and some type of simple computer to compute mass flow rate. The other, more basic, approach is to find flowmetering concepts that are inherently sensitive to mass flow rate. Both methods are currently finding successful application in various forms.

Volume Flowmeter Plus Density Measurement

Some basic methods[145] of fluid-density measurement are shown in Fig. 7.47. In Fig. 7.47a a portion of the flowing liquid is bypassed through a still well. The buoyant force on the float is directly related to density and may be measured in a number of ways, such as the strain-gage beam shown. In Fig. 7.47b a definite volume of flowing liquid contained within the U tube is continuously weighed by a spring and pneumatic displacement transducer. Flexible couplings isolate external forces from the U tube. A pneumatic force-balance feedback system also can be used to measure the weight. This minimizes deflection and thus reduces errors due to variable spring effects of flexible couplings and flexure pivots.

Figure 7.47c shows a method of measuring gas density by using a small centrifugal blower (run at constant speed) to pump continuously a sample of the flow. The pressure drop across such a blower is proportional to density and may be measured with a suitable differential-pressure pickup. Ultrasonic volume flowmeters often use an ultrasonic density-measuring technique when mass flow rate is wanted. In Fig. 7.47d the crystal transducer serves as an acoustic-impedance detector. Acoustic impedance depends on the product of density and speed of sound. Since a signal proportional to the speed of sound is available from the volume flowmeter, division of this signal into the acoustic-impedance signal gives a density signal.[146] The attenuation of radiation from a radioisotope source depends on the density of the material through which the radiation passes (see Fig. 7.47e). Over a limited (but generally adequate) density range, the output current of the radiation detector is nearly linear with density, for a given flowing fluid.[147] For gas flow, indirect measurement of density by means of computation from pressure and temperature signals (Fig. 7.47f) is also common.[148]

[145]D. Capano, "The Ways and Means of Density," *In Tech,* Nov. 2000, pp. 86–87.

[146]L. C. Lynnworth et al., "Ultrasonic Mass Flowmeter for Army Aircraft Engine Diagnostics," *USAAMRDL Tech. Rept.* 72-66, Fort Eustis, VA, 1973.

[147]Thermo Measure Tech, Round Rock, TX, 800-736-0801 (www.tnksi.com).

[148]"Gas Mass Flow Computation," *Ref.* A0-09-01, Honeywell Inc., Fort Washington, PA, 1977.

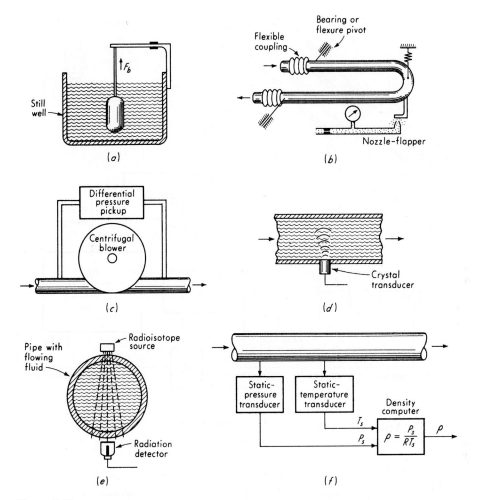

Figure 7.47
Fluid-density measurement.

Densitometers based on the effect of density on the natural frequency of a vibrating system are available for both liquids and gases and are in wide use.[149] Figure 7.47g shows a U-tube device, with process liquid flowing continuously through the tube. The first bending mode of vibration of the U tube is well approximated by standard cantilever-beam formulas. The beam stiffness is unaffected by the internal liquid, but the mass per unit length varies directly with the liquid density ρ_f, giving a density-dependent natural frequency f_n of

$$f_n = \frac{0.5605\sqrt{EI}}{\sqrt{\pi L^2}}\sqrt{\frac{1}{\rho_f R_i^2 + \rho_t(R_o^2 - R_i^2)}} \tag{7.72}$$

[149]Automation Products, Inc., Houston, TX, 800-231-2062 (www.dynotrolusa.com).

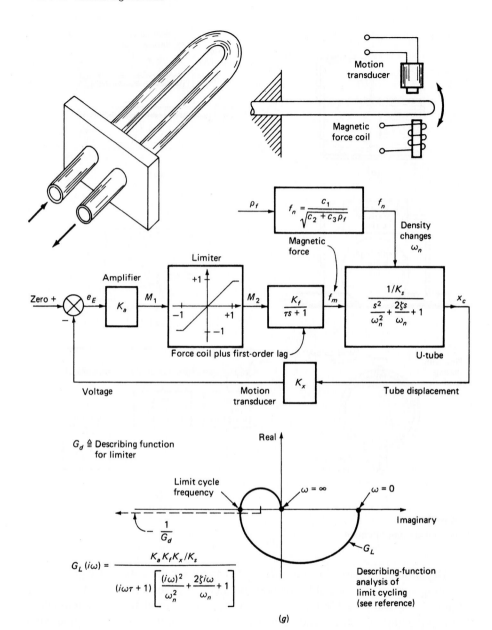

$$f_n = \frac{c_1}{\sqrt{c_2 + c_3 \rho_f}}$$

Density changes ω_n

Magnetic force

Amplifier

Limiter

Force coil plus first-order lag

$\dfrac{K_f}{\tau s + 1}$

$\dfrac{1/K_s}{\dfrac{s^2}{\omega_n^2} + \dfrac{2\zeta s}{\omega_n} + 1}$

U-tube

Voltage

Motion transducer

K_x

Tube displacement

$G_d \triangleq$ Describing function for limiter

Real

Limit cycle frequency

$\omega = \infty$

$\omega = 0$

Imaginary

$-\dfrac{1}{G_d}$

G_L

Describing-function analysis of limit cycling (see reference)

$$G_L(i\omega) = \frac{K_a K_f K_x / K_s}{(i\omega\tau + 1)\left[\dfrac{(i\omega)^2}{\omega_n^2} + \dfrac{2\zeta i\omega}{\omega_n} + 1\right]}$$

(g)

Figure 7.47
(Concluded)

where

$L \triangleq$ tube length

$I \triangleq$ tube-area moment of inertia (bending)

$\rho_t \triangleq$ tube-material mass density

$R_i, R_o \triangleq$ tube inside and outside radii

An intentionally unstable, amplitude-limited feedback system maintains continuous oscillation at a limit-cycling frequency.[150] This frequency is very close to the tube's density-dependent natural frequency and "tracks" it as the natural frequency changes with density. Standard electronic frequency-measuring schemes read out the density with high resolution and accuracy, while microprocessors can be easily used (the frequency counters are "digital") to linearize the frequency-density relation and/or provide temperature compensation from temperature-sensor signals. Design criteria include a desire for both high frequency (for good resolution over short counting intervals) and high sensitivity of frequency to density. For given E, L, R_o, ρ_f, and ρ_t, the natural frequency f_n exhibits a maximum with respect to R_i, which can be found by calculus. For $L = 5.50$ in, $R_o = 0.30$ in, water flowing, and a stainless-steel tube, $R_i = 0.2293$ in, maximizes f_n at about 500 Hz. Maximization of sensitivity $(df_n/d\rho_f)$ becomes analytically difficult, so we use numerical differentiation and find $R_i = 0.291$; and since frequency for this R_i was still close to 500 Hz, we choose $R_i = 0.291$. Stress analysis for 1,000 lb/in^2 working pressure showed an adequate safety factor. To check the final design, a simulation was run for the feedback system of Fig. 7.47g, with $K_x = K_f = K_s = 1.0$, $K_a = 1,000$, $\tau = 1.0$ s, $\zeta = 0.05$, and density varies over a ± 25 percent range. The limit-cycling frequency was found to change from 467 to 537 Hz as the density varied over its complete range.

In computing mass flow rate from volume flowmeter and densitometer (density-measuring device) signals, the necessary form of computer varies somewhat, depending on the type of flowmeter. For "head" meters (those producing a differential pressure or electrical signal proportional to ρV_{av}^2, such as an orifice) the computer multiplies the ρ signal by the ρV_{av}^2 to form $\rho^2 V^2$ and then takes the square root of this to get ρV, which is proportional to the mass flow rate. For "velocity" flowmeters, such as the turbine and electromagnetic types, the available signal is proportional to V_{av}; thus the computer simply must multiply this by the ρ signal, and the square-root operation is unnecessary. When "differential pressure" types of volume flowmeters are used to measure mass flow rate of gases or steam, sensors for differential pressure, absolute upstream pressure, and temperature are needed. One unit,[151] which uses an averaging pitot tube, incorporates all these sensors into the tube structure so that only a single pipe penetration is needed. An electronics module mounted immediately outside the pipe carries out the needed calculations nine times per second, and provides ± 1.3 percent (of mass flow *reading*) accuracy over a flow range of up to 8:1. Associated software includes a database of fluid properties for 110 fluids that are needed for density, compressibility, and viscosity calculations. Only two wires are needed to communicate with a central system since digital data using frequency-shift-keying (FSK) can be superimposed on the same wires carrying the 4 to 20 mA analog mass-flow signal.

---

[150]E. O. Doebelin, "Control System Principles and Design," chap. 8, Wiley, New York, 1985.

[151]Mass ProBar, Rosemount Inc., Chanhassen, MN, 800-999-9307 (www.rosemount.com).

Direct Mass Flowmeters

While the above indirect methods of mass-flow-rate measurement often are satisfactory and are in wide use, it is possible to find flowmetering concepts that are
more directly sensitive to mass flow rate. These may have advantages with respect
to accuracy, simplicity, cost, weight, space, etc., in certain applications. We discuss
briefly some of the more common principles in terms of the practical hardware
through which they have been realized.

A principle widely employed in aircraft fuel-flow measurement depends on the
moment-of-momentum law of turbomachines. Fluid mechanics shows that, for one-
dimensional, incompressible, lossless flow through a turbine or an impeller wheel,
the torque T exerted by an impeller wheel on the fluid (minus sign) or on a turbine
wheel by the fluid (plus sign) is given by

$$T = G(V_{ti}r_i - V_{to}r_o) \tag{7.73}$$

where $G \triangleq$ mass flow rate through wheel, slug/s (kg/s)
$V_{ti} \triangleq$ tangential velocity at inlet, ft/s (m/s)
$V_{to} \triangleq$ tangential velocity at outlet, ft/s (m/s)
$r_i \triangleq$ radius at inlet, ft (m)
$r_o \triangleq$ radius at outlet, ft (m)
$T \triangleq$ torque, ft $\cdot$ lbf(N $\cdot$ m)

Consider now the system of Fig. 7.48. The flow to be measured is directed through
an impeller wheel which is motor-driven at constant speed. If the incoming flow has
no rotational component ($V_{ti} = 0$); and if the axial length of the impeller is enough
to make $V_{to} = r\omega$, the driving torque necessary on the impeller is

$$T = r^2 \omega G \tag{7.74}$$

Since r and ω are constant, the torque T (which could be measured in several ways)
is a direct and linear measure of mass flow rate G. However, for $G = 0$, torque will

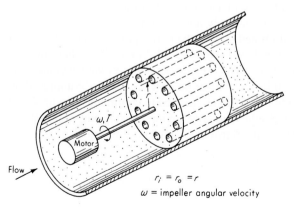

Figure 7.48
Angular-momentum element.

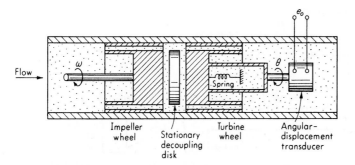

Figure 7.49
Angular-momentum mass flowmeter.

not be zero, because of frictional effects; furthermore, viscosity changes also would cause this zero-flow torque to vary. A variation on this approach is to drive the impeller at constant *torque* (with some sort of slip clutch). Then, impeller *speed* is a measure of mass flow rate according to

$$\omega = \frac{T/r^2}{G} \tag{7.75}$$

The speed ω is now nonlinear with G, but may be easier to measure than torque. If a magnetic proximity pickup is used for speed measurement, the time duration t between pulses is inversely related to ω; thus measurements of t would be linear with G.

A further variation is shown in Fig. 7.49. A constant-speed motor-driven impeller again imparts angular momentum to the fluid; however, no torque or speed measurements are made on this wheel. Closeby, downstream, a second ("turbine") wheel is held from turning by a spring restraint. For the impeller, $V_{to} = r\omega$; furthermore, this becomes V_{ti} for the turbine. Since the turbine cannot rotate, if it is long enough axially, the angular momentum is removed and V_{to} for the turbine is zero. Then the torque on the turbine is

$$T = r^2 \omega G \tag{7.76}$$

If the spring restraint is linear, the deflection θ is a direct and linear measure of G and can be transduced to an electrical signal in a number of ways. The decoupling disk reduces the viscous coupling between the impeller and turbine so that, at zero flow rate, a minimum of viscous torque is exerted on the turbine wheel.

Figure 7.50[152] shows a version of Fig. 7.48 which embodies several advantageous features and is the basis of successful commercial instruments. Again a spring measures torque, but now both ends of the spring rotate at the same speed, and spring deflection is inferred from the time interval between pulses produced by two magnetic pickoffs. This makes the Δt signal independent of impeller speed ω, since a smaller ω gives a smaller torque (and spring deflection), but this is exactly

---

[152]Eldec Corp., Lynnwood, WA, 425-743-8507 (www.eldec.com).

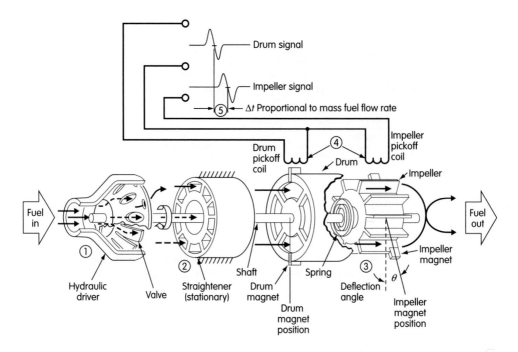

Drum signal

Impeller signal

Δt Proportional to mass fuel flow rate

MEASURING MASS FUEL FLOW: ① Fuel first enters the hydraulic driver, which provides the torque to rotate the shaft drum and impeller. ② The fuel then passes through a stationary straightener and into the impeller. ③ The mass of the fuel flowing through the rotating impeller causes it to deflect proportionally against the spring. ④ Impeller deflection relative to the drum is measured by pulses generated by magnets (attached to the drum and the impeller) rotating past two pickoff coils. ⑤ The time between pulses, caused by the angular displacement of the impeller relative to the drum, is directly proportional to the mass flow rate of the fuel.

Figure 7.50
Flow-driven angular-momentum meter.

compensated by the increase in Δt resulting from ω itself being lower. This independence of ω allows use of a drive employing a fluid turbine and thus eliminating the constant-speed electric motor drive completely. While the fluid drive need not maintain an exact speed, a simple internal valving scheme maintains a nominal speed for greater accuracy.

A number of other concepts using angular-momentum principles have been suggested. They include the vibrating gyroscope meter, the Coriolis meter, the rotating gyroscope meter, the twin-turbine meter, and the S-tube meter.[153] A Coriolis-type meter which has been successfully reduced to commercial practice[154] is shown in Fig. 7.51. Fluid flowing at mass flow rate G kilograms per second passes through

[153]C. M. Halsell, "Mass Flowmeters," *ISA J.,* p. 49, June 1960; J. Haffner, A. Stone, and W. K. Genthe, "Novel Mass Flowmeter," *Contr. Eng.,* p. 69, October 1962.

[154]K. O. Plache, "Coriolis/Gyroscopic Flow Meter," *Mech. Eng.,* pp. 36–41, March 1979; Micro Motion Inc., Boulder, CO, 800-760-8119 (www.micromotion.com).

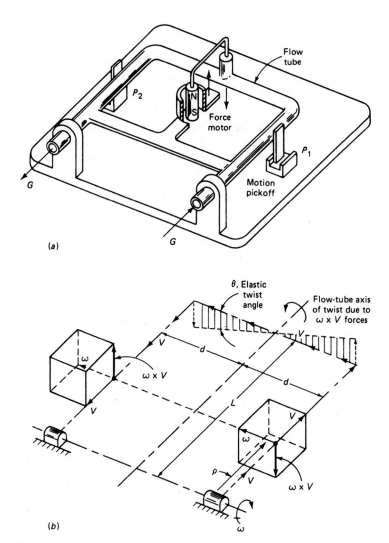

Figure 7.51
Coriolis-type mass flowmeter.

a C-shaped pipe cantilevered from two supporting brackets. The pipe is maintained
in steady sinusoidal bending vibration (at its natural frequency, 50 to 80 Hz, as a
cantilever beam) by an electromagnetic feedback system. This is a self-excited drive
system which always runs at beam natural frequency (and thus minimum power),
even though this frequency changes when fluid density changes. This is accomplished by deriving the force-motor drive signal from a velocity sensing coil wound
on the same form as the driver coil (they share the same permanent-magnet core).
Amplitude is stabilized by feedback which compares sense-coil voltage (velocity)

with a fixed reference signal. The mechanical "tuning fork" configuration minimizes the vibratory force into the frame.

Coriolis-type meters require that the fluid experience an angular velocity ω whose vector is perpendicular to the fluid velocity V. In this example, ω is an oscillatory motion produced by the C tube bending about its cantilever supports. For the simplified analysis of Fig. 7.51b, ω is treated as a rigid-body rotation about a fixed axis, and fluid flow is represented by a single velocity V rather than by a velocity profile. The absolute acceleration $\ddot{r}$ of a point located by a vector ρ from the origin (located by a vector R from a fixed reference point) of a coordinate system rotating with vector angular velocity ω is given by[155]

$$\ddot{\mathbf{r}} = \ddot{\mathbf{R}} + \omega \times (\omega \times \rho) + \dot{\omega} \times \rho + \ddot{\rho}_r + 2\omega \times \dot{\rho}_r \qquad (7.77)$$

For our example $\mathbf{R} \equiv 0$, so $\ddot{\mathbf{R}} = 0$, and $\dot{\rho}_r = \mathbf{V}$. The flowmeter motion pickoffs sense the twist angle θ; thus we are interested in only those inertia forces which cause twist. In Eq. (7.77) the only term of this type is the Coriolis acceleration $2\omega \times \mathbf{V}$ (see Fig. 7.51b).

An element of fluid mass dM at location ρ produces an inertia force of magnitude $(dM)(2\omega \times \mathbf{V})$ and direction opposite to $\omega \times \mathbf{V}$. Since $\mathbf{V}$ changes sign from the right to the left leg of the C tube, a pair of right/left mass elements produces an inertia torque dT:

$$dT = 2(2\omega \times \mathbf{V})(dM)d = 2(2\omega \times \mathbf{V})\left(\frac{G}{V}d\rho\right)d \qquad (7.78)$$

$$T = \int_0^L dT = 4\omega Gd \int_0^L d\rho = 4Ld\omega G \qquad (7.79)$$

The angular velocity ω oscillates sinusoidally, so the torque T is sinusoidal also. It acts as a driving torque, tending to twist the C tube; and since the C-tube torsional natural frequency is well above this driving frequency, the torsional spring/mass system acts essentially as a spring of stiffness K_s, allowing calculation of the twist angle θ from

$$\theta = \frac{4Ld\omega}{K_s}G \qquad (7.80)$$

The motion pickoffs P_1 and P_2 (both optical and magnetic types have been utilized) are located near the pipe neutral position and produce a switching action when the pipe passes the pickoff location; "proportional" sensors are not used. Because of twist θ, one of the pickoffs will be triggered a time interval Δt later than the other. If the average angular velocity over this Δt is ω_{av}, then

$$\theta = \frac{(L\omega_{av})\Delta t}{2d} \approx \frac{L\omega\,\Delta t}{2d} \qquad (7.81)$$

[155]G. W. Housner and D. E. Hudson, "Applied Mechanics–Dynamics," p. 31, Van Nostrand, New York, 1950.

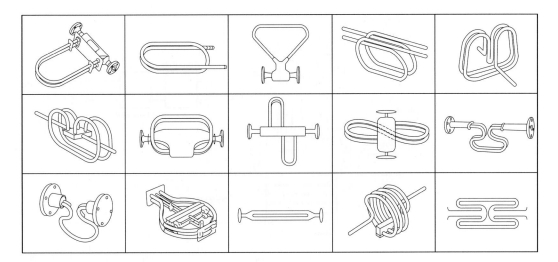

Figure 7.52
Various configurations for Coriolis mass flowmeters.

where the instantaneous value $\omega \approx \omega_{av}$ because the motion is sensed over a small fraction of the total cycle. Combining (7.80) and (7.81), we get

$$G = \frac{K_s}{8d^2}\,\Delta t \qquad (7.82)$$

which shows that Δt is a linear measure of mass flow rate. In the actual system, the Δt measurement is implemented[156] as a pulse-width modulation scheme using a gated oscillator feeding an up/down counter. Also total flow over any time interval is obtained easily by digital integration. This type of meter is obstructionless; is essentially insensitive to viscosity, pressure, and temperature; and can handle clean liquids, mixtures, foams and slurries, and liquids with entrained gases. Since Δt is measured once per bending cycle, the meter can respond rapidly to changing flows; however, averaging over several cycles generally is used to improve accuracy for measurements of average flow rate.

The U-tube Coriolis arrangement is only one of many possible configurations that utilize the same basic principle.[157] Figure 7.52 (Direct Measurement Corp., Longmont, CO, 1997) shows various forms, some of which are used in commercial instruments for both liquids and gases. The single straight-tube form[158] has a low pressure drop, is self-draining, and is easily cleaned. Some unusual applications of Coriolis technology include the *net oil computer,*[159] used to measure crude oil and brine emulsions in oil-field operations. Here the usual mass flow signal is combined

[156]K. O. Plache, op. cit.

[157]T. Andrews, "Coriolis Flowmeters: Old Principle, New Technology," *In Tech,* Jan. 2000, pp. 68–69; "Coriolis Gas Metering Moving into Mainstream," *Control Engineering,* Nov. 1998, p. 7; G. Cabak and S. Hough, "Vibration Testing Confirms the Design of Flowmeters," *Sound and Vibration,* Sept. 1992, pp. 6–10.

[158]www.micromotion.com; www.khrone.com.

[159]www.micromotion.com.

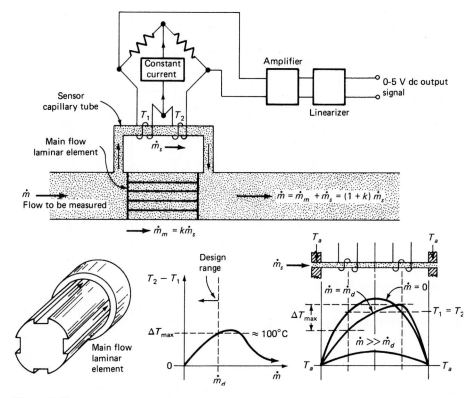

Figure 7.53

with a density signal (from the natural frequency of the Coriolis U-tube) and a temperature signal. Using known properties of water and salt solutions, the proportions of oil and water in the flow can be determined.

Mass flowmeters based on thermal principles may take a number of forms. Figure 7.53[160] shows, simplified, one type which combines thermal principles with a bypass concept, using laminar flow to provide a family of gas flowmeters covering the range from 10 cm³/min full scale to 200 ft³/min. These flowmeters are also available combined with a flow-control valve to form a mass-flow feedback controller, widely used in semiconductor manufacturing processes and elsewhere. In a bypass flowmetering scheme (usable, in principle, with *any* type of flowmeter), one actually measures only a *fraction* of the total flow, obtaining the value of the total flow by multiplication by a flow-ratio factor, assumed fixed and known. Often a bypass method is used to allow metering of a large flow rate with a smaller, inexpensive meter. In the present thermal meters, the bypass concept allows use of a single, small, low-power sensing tube as the basic element for a whole family of meters, up to those for very high flows. To ensure a fixed ratio between the main flow and the small bypass flow, both flow passages are designed to have laminar flow over the entire range of the meter. Since both elements feel the same pressure

---

[160]Sierra Instruments, Monterey, CA, 831-373-0200 (www.seirrainstruments.com).

drop at all times, the main flow rate is always proportional to the sensed bypass rate, making the total flow (what we want to measure) also proportional. In the referenced instruments, the ratio of total flow to bypassed flow ranges from 1 (smallest meter) to 100,000 (largest meter).

The sensor tube has a diameter of 0.031 in and a length 50 to 100 times this, giving laminar flow for the entire measuring range, with the maximum sensor-tube flow always being about 20 cm^3/min. The usually much larger main flow requires a larger flow area while still maintaining laminar conditions. Thin, rectangular flow channels milled into the periphery of a cylinder (see Fig. 7.53) are used here. Two electric windings on the outer surface of the sensor tube act as both heaters and resistance-temperature detectors and provide a constant-heat input to the fluid at all flow rates. For zero flow rate ($\dot{m}_s = 0$) the system is "thermally symmetric," giving the symmetric temperature profile shown in Fig. 7.53, and making $\Delta T \triangleq T_1 - T_2$ equal to zero. The two windings are in adjacent legs of a bridge circuit, and bridge resistances are selected to give a balanced bridge (no output voltage) for the zero-flow condition. At this no-flow condition, all the electric heating is being used to balance the heat losses from the tube to its surroundings, which are assumed to remain at a fixed ambient temperature T_a. If we now allow a steady fluid mass flow rate $\dot{m}_s$ from left to right and wait for the new thermal equilibrium, the new temperature profile will be *unsymmetric,* with T_1 dropping a large amount but T_2 staying nearly constant. Temperature T_1 drops because the mass flow of hot fluid leaving this region carries energy away at a rate $\dot{m}_s c_p T_1$, requiring an increase in the local heat-transfer rate (from tube to fluid) to support this energy loss and maintain equilibrium. To allow this heat-transfer rate to increase, the ΔT between tube and fluid must increase, which requires a drop in T_1 proportional to $\dot{m}_s$. Temperature T_2 changes little when flow occurs since this region *both* receives and loses heat (at about the same rate) from the flowing fluid. Although the desired mass-flow signal is associated mainly with T_1, use of the $T_2 - T_1$ signal available from the bridge circuit has the advantages of giving a zero output for zero flow rate and making the instrument less sensitive to ambient temperature, since T_1 and T_2 are affected about equally by changes in T_a. For very large flow rates (*beyond* the instrument's design range), the cooling effect of the fluid makes both T_1 and T_2 much lower and also more nearly equal, since a fluid particle passes so rapidly from region 1 to region 2 that little temperature change can occur.

In the referenced commercial instruments, calibration of each instrument against standards plus use of a built-in five-break-point electronic linearizer gives accuracy of ± 1 percent of full scale, including nonlinearity. Step response has a 2 percent settling time of 1 to 3 s. The meter pressure drop at maximum flow ranges from 0.08 to 1.5 lb/in^2, depending on meter size.

A variation[161] on the above principle uses a feedback technique to control three heaters so as to maintain a *fixed* temperature profile as flow rate changes. The

---

[161]L. D. Hinkle and C. F. Mariano, "Toward Understanding the Fundamental Mechanisms and Properties of the Thermal Mass Flow Controller," *J. Vac. Sci. Technol.,* vol. 9(3), May/June 1991; J. J. Sullivan, "A Guide to the Selection of MKS Flow Controllers and Control Valves for Semiconductor Processing," MKS Instruments Inc., Andover, MA, 800-227-8766 (www.mksinst.com).

output signal is the heater voltage required to maintain the profile. This design is claimed to improve linearity and response speed.

For larger flow rates, *immersible* thermal flowmeters[162] based on the sensitivity of hot-wire anemometers to mass flow rate are available in two forms. *In-line* units have a single sensor in a flow tube up to 8 inches in diameter. *Insertion* units have either a single sensor at the end of a long tube or multiple sensors distributed along a 1-in tube for getting average flow rates in large ducts.

Thermal mass flowmeters are widely used in semiconductor and MEMS manufacture, but one study[163] showed that 70 percent of units replaced as faulty were actually OK! The article makes a case for better understanding of details of meter operation and gives guidelines for diagnosing problems. The dominance of thermal meters in the semiconductor industry has been challenged by new designs based on pressure-drop across a special flow-resistance element.[164] Using two pressure transducers and one temperature sensor, and incorporating detailed calibration information for the actual gas used, these units claim better accuracy and much faster response.

PROBLEMS

7.1 Water flows in a 1-in-diameter pipe at 10 ft/s. If a pitot-static tube of 0.5-in diameter is inserted, what velocity will be indicated? Assume one-dimensional frictionless flow. Find the pitot-static-tube diameter needed to reduce the above error to 1 percent.

7.2 For the system of Fig. 7.14, what diameter sphere is needed to obtain a 1-lb force from a 50 mi/h wind when atmospheric pressure is 14.7 lb/in^2 and temperature is 70°F? If the 50 mi/h wind is assumed to be the full-scale value and if 0.05-in full-scale deflection of the differential transformer is desired, what must be the flexure-plate spring constant? If the total moving mass is assumed to be due to the spherical shell alone and if the shell is made of $\frac{1}{16}$-in-thick aluminum, estimate the usable frequency range of this instrument.

7.3 If, in the system shown in Fig. 7.19, the amplifier lag is not neglected, so that $(i/e_e)(D) = K_a/[(\tau_1 D + 1)(\tau_2 D + 1)]$, find the value of K_a that will put the feedback system just on the margin of instability (use the Routh criterion). If $\tau = 0.001$ s and $\tau_1 = \tau_2 = 0.000001$, what is the maximum allowable value of $K_b K_a K_e/I_m$ for marginal stability? If a value of $K_b K_a K_e/I_m$ about one-fifth of that giving marginal stability can be used and if Eq. (7.43) is assumed applicable under these conditions, what percentage improvement of τ_{ct} as compared with τ may be achieved?

[162]J. G. Olin, "An Engineering Tutorial: Thermal Mass Flowmeters," *In Tech,* Aug. 1993, pp. 37–41; "Flow Monitors for Continuous Emissions Monitoring," 1992; "Reducing Cost of Ownership with Thermal Mass Flow Meters," Sierra Instruments, www.seirrainstruments.com.

[163]V. Comello, "If It Ain't Broke, Don't Replace It," *R&D,* Sept. 1997, p. 81.

[164]D. T. Mudd et al., Pressure-Based MFC's Improve Gas Control, Semiconductor International, Mar. 2002, pp. 75–81. Fugasity Corp., Amherst, NH, 660-880-8635 (www.fugasity.com).

7.4 The frequency response of a hot-wire anemometer system also determines its ability to resolve *spacewise* variations in velocity. That is, at a given instant of time, if the velocity pattern in a flow were "frozen," the velocity component in a certain direction would be different at various stations along the line of travel of the gross flow. When this velocity structure is swept past a hot wire by the average velocity, it requires adequate frequency response to resolve the spacewise velocity variations. Consider a simple spacewise variation in which velocity deviation from average is given by $v = v_0 \sin (2\pi x/\lambda)$, where λ is the wavelength in inches and x is displacement in inches along the flow direction. If the average flow velocity is V_0 in/s, find an expression for the smallest wavelength λ_{min} that can be resolved by a system whose flat frequency response extends to f_0 hertz. Plot λ_{min} versus V_0 for a system with f_0 of 10,000 Hz.

7.5 Analyze the error in flow-rate measurement caused by thermal expansion of an orifice plate.

7.6 A capillary-tube laminar-flow element is needed to measure water flow of 0.01 in³/min at 70°F. A flowmeter pressure drop of 3 in of water is desired. If the element is designed for a Reynolds number of 500, what length and diameter of tubing are needed?

7.7 Using the simplified model discussed in the text, derive Eq. (7.60).

7.8 Using the assumptions discussed following Eq. (7.60), one can write for the weight flow rate

$$W = K_1(A_t - A_f) \sqrt{w_{ff}(w_f - w_{ff})} \qquad \text{lbf/s}$$

where K_1 includes all the other constants in Eq. (7.60). To make the weight-flow indication relatively insensitive to changes in fluid density w_{ff}, the float density w_f should be twice the density of the flowing fluid. Show the truth of this statement. *Hint:* Set $\partial W/\partial w_{ff} = 0$.

7.9 A wind-velocity-measuring sphere (as in Fig. 7.14) is mounted on a buoy (as in Fig. P7.1) and has a diameter of 18 cm, a mass of 0.5 kg, $C_d = 0.567$, and the force transducer spring constant $= 3,600$ N/m. Assume there is no wind, but the buoy is rocking about a single axis with $\theta = 0.2 \sin 2t$ radians (t in seconds). This motion will cause a force transducer reading even though there is no wind. Analyze to find the wind measurement error caused by this spurious input. Then suggest several methods for reducing and/or compensating for this kind of error.

7.10 Outline the procedure you would use to design a drag-force flowmeter of the type shown in Fig. 7.42. The given specifications include static sensitivity, dynamic response, flow-velocity range to be covered, allowable size, and fluid-density range to be covered.

7.11 Perform a dynamic analysis on the system of Fig. 7.47*a* to obtain the transfer function relating fluid density as an input to strain-gage bridge voltage as an output. What is the effect of changes in liquid level in the still well on the output signal? What is the effect of thermal (volume) expansion of the float?

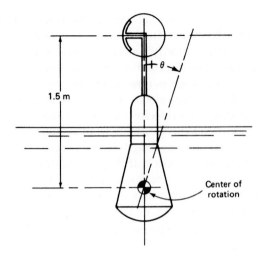

Figure P7.1

If the entire assembly is aboard a vehicle that is accelerating vertically, explain the effect on the output.

7.12 Figure P7.2 shows a laminar-flow element and associated differential-pressure transducer. We wish to study the dynamic response of this flow measurement system. Model the laminar element as a lumped liquid inertance and resistance (no fluid compliance) and treat the pressure transducer/tubing system as heavily damped, slow-acting. Find $[(p_1 - p_2)/Q](D)$.

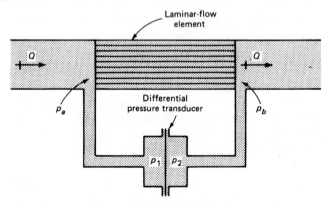

Figure P7.2

7.13 Perform a static analysis on the system of Fig. 7.47b to get a relation between fluid density as an input and nozzle-flapper pressure as output. List modifying and/or interfering inputs for this instrument.

7.14 Modify the system of Fig. 7.47*b,* using a feedback principle (null-balance system) to keep vertical deflection nearly zero for all densities. A bellows may be used to provide the rebalancing force.

7.15 For the system of Fig. 7.49, suppose that the full-scale flow rate is 10 lbm/s, an impeller speed of 100 r/min is used, and $r = 1.0$ in. If full-scale transducer rotation is to be 20°, what torsional spring constant is required? If this spring constant must be increased to improve dynamic response, what design changes are possible to achieve this?

7.16 Intuitively, what would you expect the dynamic response of the system of Fig. 7.49 to be? What design parameters would have a major influence on this response and in what way? Devise an experimental technique for subjecting this instrument to an approximate step input.

7.17 Derive Eq. (7.72). Then use calculus to maximize f_n with respect to R_i, and check the numerical values given. Use numerical differentiation to maximize $df_n/d\rho_f$ with respect to R_i, and check given results. Use standard thick-wall-cylinder stress formulas to check strength. Use SIMULINK or other available simulation methods to study the limit-cycling behavior.

BIBLIOGRAPHY

1. "Flow: Its Measurement and Control in Science and Industry," vol. 2, ISA, Research Triangle Park, NC, 1981.

2. "Flow: Its Measurement and Control in Science and Industry," vol. 1, ISA, Research Triangle Park, NC, 1971.

3. J. A. Shercliff, "The Theory of Electromagnetic Flow Measurement," Cambridge University Press, New York, 1962.

4. B. E. Richards (ed.), "Measurement of Unsteady Fluid Dynamic Phenomena," CRC Press, Boca Raton, FL, 1977.

5. S. Blechman, "Techniques for Measuring Low Flows," *Instrum. Contr. Syst.,* p. 82, October 1963.

6. R. B. Crawford, "A Broad Look at Cryogenic Flow Measurement," *ISA J.,* p. 65, June 1963.

7. T. Isobe and H Hattori, "A New Flowmeter for Pulsating Gas Flow," *ISA J.,* p. 38, December 1959.

8. A. K. Oppenheim and E. G. Chilton, "Pulsating Flow Measurement—A Literature Survey," *ASME Trans.,* p. 231, February 1955.

9. E. L. Upp, "Flowmeters for High-Pressure Gas," *Instrum. Contr. Syst.,* p. 151, March 1965.

10. N. F. Cheremisinoff, "Applied Fluid Flow Measurement," Marcel Dekker, New York, 1979.

11. M. R. Neuman et al. (eds.), "Physical Sensors for Biomedical Applications," CRC Press, Boca Raton, FL, 1980.

12. A. Noordergraaf, "Circulatory System Dynamics," Academic, New York, 1978.

13. G. P. Katys, "Continuous Measurement of Unsteady Flow," Macmillan, New York, 1964.

14. R. R. Dowden, "Fluid Flow Measurement, A Bibliography," British Hydromechanics Res. Assoc. Cranfield, Bedford, England, 1971.

15. R. P. Benedict, "Fundamentals of Temperature, Pressure, and Flow Measurements," Wiley, New York, 1969.

16. L. C. Lynnworth, "Ultrasonic Measurements for Process Control," Academic Press, New York, 1989.

17. H. H. Bruun, "Hot-Wire Anemometry," Oxford University Press, New York, 1995.

18. W. A. Wakehan, A. Nagashima, and J. V. Sengers, "Experimental Thermodynamics," vol. III, "Measurement of the Transport Properties of Fluids," CRC Press, Boca Raton, FL, 1991.

19. J. G. Webster (ed.), "The Measurement, Instrumentation, and Sensors Handbook," chaps. 21, 28, 29, and 76, CRC Press, Boca Raton, FL, 1999.

20. R. W. Miller, "Flow Measurement Engineering Handbook," 3rd ed., McGraw-Hill, New York, 1996.

21. T. B. Morrow et al., "Operational Factors that Affect Orifice Meter Accuracy," *Flow Control,* Jan. 2003, pp. 32–40.

CHAPTER 8

Temperature and Heat-Flux Measurement

8.1 STANDARDS AND CALIBRATION

The International Measuring System sets up independent standards for only four fundamental quantities: length, time, mass, and temperature. Standards for all other quantities are basically derived from these. We discussed previously the standards of length, time, and mass; we consider now the temperature standard.[1] It should be noted first that temperature is fundamentally different in nature from length, time, and mass. That is, if two bodies of like length are "combined," the total length is twice the original; the same is true for two time intervals or two masses. However, the combination of two bodies of the same temperature results in exactly the same temperature. Thus the idea of a standard unit of mass, length, or time that can be divided or multiplied indefinitely to generate any arbitrary magnitude of these quantities cannot be carried over to the concept of temperature. Also, even though statistical mechanics relates temperature to the mean kinetic energies of molecules, these kinetic energies (which are dependent on only mass, length, and time standards for their description) are not measurable at present. Thus an *independent* temperature standard is required.

The fundamental meaning of temperature, just as for all basic concepts of physics, is not given easily. For most purposes the zeroth law of thermodynamics gives a useful concept: For two bodies to be said to have the same temperature, they must be in thermal equilibrium; that is, when thermal communication is possible between them, no change in the thermodynamic coordinates of either occurs. The zeroth law says that when two bodies are each in thermal equilibrium with a third body, they are in thermal equilibrium with each other. Then, by definition, the bodies are all at the same temperature. Thus if we can set up a reproducible means of establishing a range of temperatures, unknown temperatures of other bodies may

---

[1]R. P. Benedict, "Fundamentals of Temperature, Pressure, and Flow Measurements," 2d ed., Wiley, New York, 1977.

677

be compared with the standard by subjecting any type of "thermometer" successively to the standard and to the unknown temperatures and allowing equilibrium to occur in each case. That is, the thermometer is calibrated against the standard and afterward may be used to read unknown temperatures.

In choosing the means of defining the standard temperature scale, conceivably we could employ any of the many physical properties of materials that vary reproducibly with temperature. For instance, the length of a metal rod varies with temperature. To define a temperature scale numerically, we must choose a reference temperature and state a rule for defining the difference between the reference and other temperatures. (Mass, length, and time measurements do *not* require universal agreement on a reference point at which each quantity is assumed to have a particular numerical value. Every centimeter, for example, in a meter is the same as every other centimeter.)

Suppose we take a copper rod 1 m long, place it in an ice-water bath which we have taken as our reference temperature source, and measure its length. Let us choose to call the ice-bath temperature 0°. We are now free to define any rule we wish to fix the numerical value to be assigned to all lower and higher temperatures. Suppose we decide that each additional 0.01 mm of expansion will correspond to $+1.0°$ on our temperature scale and each 0.01 mm of contraction to $-1.0°$. If the expansion phenomenon were reproducible, such a temperature scale would, in principle, be perfectly acceptable as long as everyone adhered to it. Would it be correct to say that each degree of temperature on this scale was "equal" to every other degree? That depends on what we mean by "equal." If "equal" means that each degree causes the same amount of expansion of the copper rod, then all degrees are equal. If, instead, we consider the expansion of, say, iron rods, then equal amounts of expansion would not, in general, be caused by a 1° (copper scale) change from -6 to $-5°$ as by a 1° change from 100 to 101°. Or, suppose we based our scale on conduction heat transfer in silver, for example. If a temperature difference from 100 to 200° causes a given heat-transfer rate, will the same rate be caused by a temperature difference from -50 to $+50°$? The answer is, in general, no.

The point of the above discussion is that, while our arbitrarily defined temperature scale is, in principle, as good as any other such scale based on some material property, its graduations have no particular significance with regard to physical laws *other* than the one used in the definition. We measure temperature for some *reason,* such as computing thermal expansion, heat-transfer rate, electrical conductivity, gas pressure, etc. The forms of the equations employed to make such calculations depend on the nature of the standard used to define temperature. A temperature scale that gives a simple form to thermal-expansion equations may give complex forms to all other physical relations involving temperature. Since this difficulty is common to *all* standards based on the properties of a particular substance, a way of defining a temperature scale independent of *any* substance is desirable.

The thermodynamic temperature scale[2] proposed by Lord Kelvin in 1848 provides the theoretical base for a temperature scale independent of any material

---

[2]F. W. Sears, "Thermodynamics, Kinetic Theory and Statistical Mechanics," p. 116, Addison-Wesley, Reading, MA, 1950.

property and is based on the Carnot cycle. Here a perfectly reversible heat engine transfers heat from a reservoir of infinite capacity at temperature T_2 to another such reservoir at T_1. If the heat taken from reservoir 2 is Q_2 and that supplied to reservoir 1 is Q_1, for a Carnot cycle $Q_2/Q_1 = T_2/T_1$; this may be taken as a *definition* of temperature ratio. If, also, a number is selected to describe the temperature of a chosen fixed point, then the temperature scale is completely defined. At present, the fixed point is taken as the triple point (the state at which solid, liquid, and vapor phases are in equilibrium) of water because this is the most reproducible state known. The number assigned to this point is 273.16 K since this makes the temperature interval from the ice point (273.15 K) to the steam point equal to 100 K. This would thus coincide with the previously established centigrade (now called Celsius) scale as a matter of convenience.[3]

While the Kelvin absolute thermodynamic scale is ideal in the sense that it is independent of any material properties, it is not physically realizable since it depends on an ideal Carnot cycle. Fortunately it can be shown[4] that a temperature scale defined by a constant-volume or constant-pressure gas thermometer using an ideal gas is *identical* to the thermodynamic scale. A constant-volume gas thermometer keeps a fixed mass of gas at constant volume and measures the pressure changes caused by temperature changes. The perfect-gas law then gives the fact that temperature ratios are identical to pressure ratios. The constant-pressure thermometer keeps mass and pressure constant and measures volume changes caused by temperature changes. Again, the perfect-gas law says that temperature ratios are identical to volume ratios. These ratios are identical to those of the thermodynamic scale; thus if the same fixed point (the triple point of water) is selected for the reference point, the two scales are numerically identical. However, now there is the problem that the ideal gas is a mathematical model, not a real substance, and therefore the gas thermometers described above cannot actually be built and operated.

To obtain a physically realizable temperature scale, *real* gases must be utilized in the gas thermometers; the readings must be corrected, as well as possible, for deviation from ideal-gas behavior; and then the resulting values are accepted as a definition of the temperature scale. The corrections for non-ideal-gas behavior are obtained for a constant-volume gas thermometer as follows: The thermometer is filled with a certain mass of gas, and mercury is added until the desired volume is achieved (see Fig. 8.1). Suppose that this is done with the system at the ice-point temperature. The gas pressure is measured; let us call it p_{i1}. Then the system is raised to the steam-point temperature, causing volume expansion. By adding more mercury, however, the volume can be returned to the original value. The pressure will be higher now; we call it p_{s1}. For an ideal gas, the ratio of the steam-point and ice-point temperatures also would be given by the pressure ratio p_{s1}/p_{i1}. If we repeat this experiment but use a different mass of gas, thus giving different ice-point and

---

[3]Ibid., p. 8.
[4]Ibid., p. 116.

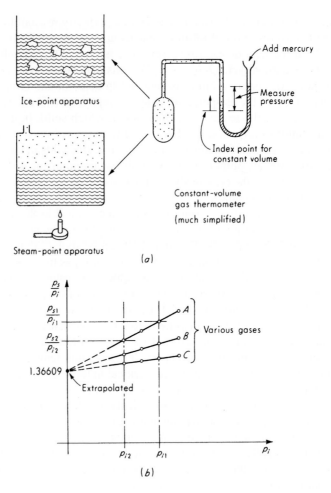

Figure 8.1
Gas-thermometer temperature scale.

steam-point pressures p_{i2} and p_{s2}, we find that $p_{s1}/p_{i1} \neq p_{s2}/p_{i2}$. This is a manifestation of the nonideal behavior of the gas; an ideal gas would have $p_{s1}/p_{i1} = p_{s2}/p_{i2}$.

Real gases approach ideal-gas behavior if their pressure is reduced to zero; thus we repeat the above experiment with successively smaller masses of gas, generating the curve A of Fig. 8.1b. Since we cannot use zero mass of gas, the zero-pressure point on this curve must be obtained by extrapolation. This zero-pressure point is taken as the true value of the pressure ratio corresponding to the steam-point/ ice-point temperature ratio. If this experiment is repeated with *different* gases (B, C in Fig. 8.1b), all the curves intersect at the same point, showing that the procedure is independent of the type of gas used. Actual results give the numerical value $p_s/p_i = 1.36609 \pm 0.00004$. If we take $T_s/T_i = p_s/p_i$, the choice of a numerical value for any chosen reference point (such as calling $T_i = 273.15$ K) completely fixes the entire temperature scale. Such a scale, unfortunately, is not practical for day-to-day temperature measurements since the procedures involved are extremely tedious and time-consuming. Also, gas thermometers actually have a lower precision and repeatability than some other temperature-measuring devices, such as resistance thermometers. This situation led to the acceptance in 1927 of the International Practical Temperature Scale (IPTS) which, with revisions in 1948, 1954, 1960, 1968, and 1990, is the temperature standard today.

The International Practical Temperature Scale is set up to conform as closely as practical with the thermodynamic scale. At the triple point of water, the two scales are in exact agreement, by definition. Five other primary fixed points are used. These are the boiling points of liquid oxygen ($-182.962°$C) and water ($100°$C) and the freezing points of zinc ($419.58°$C), silver ($961.93°$C), and gold ($1064.43°$C). Various secondary fixed points also are established, with the lowest being the triple point of hydrogen ($-259.34°$C), which also is the lowest value defined on the scale (the gold point $1064.43°$C is the highest defined fixed point). In addition to the fixed points, the International Practical Temperature Scale also specifies certain instruments, equations, and procedures to be utilized to interpolate between the fixed points. From -259.34 to $630.74°$C, a platinum resistance thermometer is the interpolating instrument; however, since one equation will not serve the entire range, equations for several subranges are defined. Each equation contains certain constants whose values must be obtained from the readings of that particular thermometer at specific fixed points. Once these constants have been found, the equation allows us to calculate temperature values from thermometer readings at any point within the given subrange. A similar procedure is employed between 630.74 and $1064.43°$C, except now the interpolating instrument is a platinum/10% rhodium and platinum thermocouple.

Above the gold point, the International Practical Temperature Scale *is* defined and uses a narrow-band radiation pyrometer ("optical" pyrometer) and the Planck equation to establish temperatures. The formula is

$$\frac{J_t}{J_{\text{Au}}} = \frac{e^{1.4388/\lambda(t_{\text{Au}}+273.15)} - 1}{e^{1.4388/\lambda(t+273.15)} - 1} \tag{8.1}$$

where $t_{\text{Au}} \triangleq$ gold-point temperature, $1064.43°$C
         $\lambda \triangleq$ effective wavelength of pyrometer, cm

The quantity J_t/J_{Au} is measurable with the pyrometer and is the ratio of the spectral radiance of a blackbody at temperature t to one at temperature t_{Au}. Since λ can be determined for a given pyrometer, Eq. (8.1) allows calculation of t when J_t/J_{Au} has been measured. In principle, this method can be applied to arbitrarily high temperatures, but in practice few reliable results above $4000°$C are known.

The highest meaningful temperatures, existing in the interior of stars and for short times in atomic explosions, are inferred from kinetic theory to be in the range 10^7 to 10^9 K. Definition of temperature, much less measurement, is difficult at these extremes, although spectroscopic methods have given useful results. At the other extreme, temperatures of 10^{-6} K have been produced by using the concept of nuclear cooling. Magnetic susceptibility of certain materials has been employed to measure temperatures in the extremely low ranges.

The question of the accuracy of temperature standards may be considered from two viewpoints. First, how closely can the International Practical Temperature Scale be reproduced; second, how closely does it agree with the thermodynamic absolute scale? The highest reproducibility of the International Practical Temperature Scale occurs at the triple point of water, which can be realized with a precision of a few ten-thousandths of a degree, giving an accuracy of about 1 ppm. For either lower or higher temperatures the accuracy falls off. Figure 8.2a[5] summarizes these data and shows calibration uncertainties for various instruments, with more recent data in Fig. 8.2b.[6] The question of agreement among the various empirical scales (such as the International Practical Temperature Scale) and the absolute thermodynamic scale involves the fact that, in general, the thermodynamic scale is considerably less reproducible than the empirical scales. For example, the steam-point temperature is reproducible to 0.0005° with a platinum resistance thermometer, but to only 0.02° with a gas thermometer. The disagreement between the International Practical Temperature Scale and the absolute thermodynamic scale has been estimated in Celsius degrees as[7]

$$\frac{t}{100}\left(\frac{t}{100} - 1\right)[0.04106 - 7.363(10^{-5})t] \qquad 0° < t < 444.6°C \qquad \textbf{(8.2)}$$

where t is the Celsius temperature. The error is seen to be zero at $t = 0$ and 100°C and has a maximum value of about 0.14C° near $t = 400$°C.

The IPTS is being evaluated continuously,[8] and new versions appear periodically. A detailed discussion of the most recent (1990) version (ITS-90) is available in various sources.[9] We should note that although the current scale is defined to only 0.65 K, measurements below this value are regularly made.[10]

Calibration of a given temperature-measuring device generally is accomplished by subjecting it to some established fixed-point environment, such as the melting and boiling points of standard substances, or by comparing its readings with those of some more accurate (secondary standard) temperature sensor which itself has been calibrated. The latter generally is accomplished by placing the two devices in intimate thermal contact in a constant-temperature-controlled bath. By varying the temperature of the bath over the desired range (allowing equilibrium at each point), the necessary corrections are determined. When calibrating a sensor against

---

[5]"Accuracy in Measurements and Calibrations," *NBS Tech. Note* 262, 1965.

[6]*NBS Spec. Publ.* 250, 1987.

[7]"Temperature, Its Measurement and Control in Science and Industry," vol. 2, p. 93, Reinhold, New York, 1955.

[8]R. P. Hudson, "Measurement of Temperature," *Rev. Sci. Instrum.*, vol. 51, no. 7, pp. 871–881, July 1980.

[9]"Manual on the Use of Thermocouples in Temperature Measurement," 4th ed., *ASTM,* Philadelphia, PA, 1993, appendix II; *NIST Tech. Note* 1265, August 1990.

[10]R. P. Hudson, op. cit.

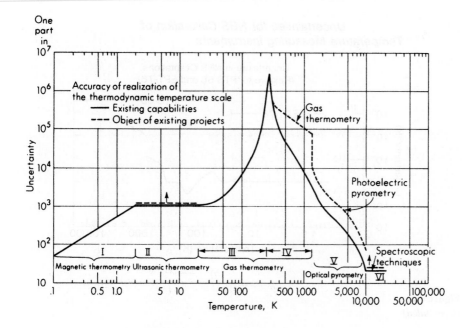

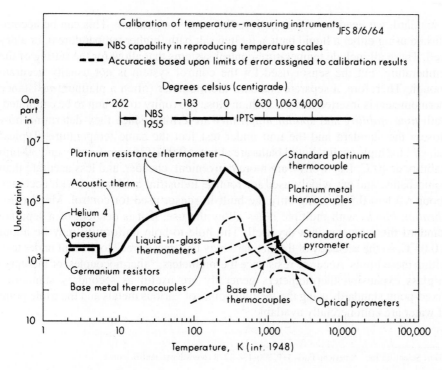

(a)

Figure 8.2
Temperature standards.

Uncertainties for NBS Calibration of Temperature Measuring Instruments

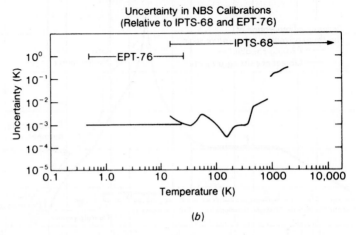

(b)

Figure 8.2
(Concluded)

a standard, we need to subject both to the same temperature. This can be accomplished using either a liquid bath, a molten salt bath (higher temperatures), or a dry well. These all include temperature-control systems that allow for the setting of the temperature, but the sensor used for the control system is not usually accurate enough. Therefore, a separate standard thermometer (often a platinum resistance thermometer) is inserted into the bath, in close proximity to the unit to be calibrated. Bath *time stability* and *spatial uniformity* are critical since they determine how closely the standard and the unit under test feel the same temperature. Typical values[11] for high-quality liquid baths are about ±0.001°C for uniformity and 30-min stability at 40°C. Dry wells[12] are more convenient, cheaper, and less accurate than liquid baths, and are widely used for routine industrial calibration, with accuracy about ±0.1 to 0.5°C, using only the built-in sensor used for control. Metal *equilibration blocks* with multiple holes allow the use, if it is desired, of a separate standard thermometer in a dry well. The hole-to-hole uniformity may be about ±0.05°C, so the accuracy of the standard may be "transferred" to the unit under test, within these limits. Accurate resistance thermometers,[13] thermocouples, or mercury-in-glass expansion thermometers generally are useful as secondary standards. Fixed-point standards using the melting points of various metals and the triple point of water are commercially available.[14]

---

[11]Hart Scientific Inc., American Fork, UT, 800-438-4278 (www.hartscientific.com).

[12]Hart Scientific Inc.

[13]D. I. Curtis, "Temperature Calibration and Interpolation Methods for Platinum Resistance Thermometers," *Rept.* 68023F, Rosemount Inc., Minneapolis, MN, 1980.

[14]Hart Scientific Inc.

8.2 THERMAL-EXPANSION METHODS

A number of practically important temperature-sensing devices utilize the phenomenon of thermal expansion in one way or another. The expansion of solids is
employed mainly in bimetallic elements by utilizing the differential expansion
of bonded strips of two metals. Liquid expansion at essentially constant pressure
is used in the common liquid-in-glass thermometers. Restrained expansion of
liquids, gases, or vapors results in a pressure rise, which is the basis of pressure
thermometers.

Bimetallic Thermometers

If two strips of metals A and B with different thermal-expansion coefficients α_A and
α_B but at the same temperature (Fig. 8.3a) are firmly bonded together, a temperature

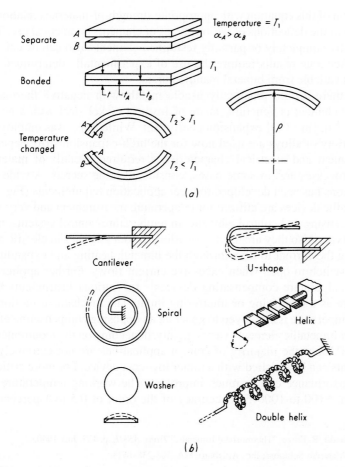

Figure 8.3
Bimetallic sensors.

change causes a differential expansion and the strip, if unrestrained, will deflect into a uniform circular arc. Analysis[15] gives the relation

$$\rho = \frac{t\{3(1 + m)^2 + (1 + mn)[m^2 + 1/(mn)]\}}{6(\alpha_A - \alpha_B)(T_2 - T_1)(1 + m)^2} \tag{8.3}$$

where
$\rho \triangleq$ radius of curvature
$t \triangleq$ total strip thickness, 0.0005 in $< t < 0.125$ in in practice
$n \triangleq$ elastic modulus ratio, E_B/E_A
$m \triangleq$ thickness ratio t_B/t_A
$T_2 - T_1 \triangleq$ temperature rise

In most practical cases, $t_B/t_A \approx 1$ and $n + 1/n \approx 2$, giving

$$\rho \approx \frac{2t}{3(\alpha_A - \alpha_B)(T_2 - T_1)} \tag{8.4}$$

Combination of this equation with appropriate strength-of-materials relations allows calculation of the deflections of various types of elements in practical use. The force developed by completely or partially restrained elements also can be calculated in this way. Accurate results require the use of experimentally determined factors[16] which are available from bimetal manufacturers.

Since there are no practically usable metals with negative thermal expansion, the B element is generally made of Invar, a nickel steel with a nearly zero $[1.7 \times 10^{-6}$ in/(in $\cdot$ C$°$)] expansion coefficient. While brass was employed originally, a variety of alloys are used now for the high-expansion strip, depending on the mechanical and electrical characteristics required. Details of materials and bonding processes are, in some cases, considered trade secrets. A wide range of configurations has been developed to meet application requirements (Fig. 8.3b).

Bimetallic devices are utilized for temperature measurement and very widely as combined sensing and control elements in temperature-control systems, mainly of the on-off type. Also they are used as overload cutout switches in electric apparatus by allowing the current to flow through the bimetal, heating and expanding it, and causing a switch to open when excessive current flows. Further applications are found as temperature-compensating devices[17] for various instruments that have temperature as a modifying or interfering input. The mechanical motion proportional to temperature is employed to generate an opposing compensating effect. The accuracy of bimetallic elements varies greatly, depending on the requirements of the application. Since the majority of control applications are not extremely critical, requirements can be satisfied with a rather low-cost device. For more critical applications, performance be much improved. The working temperature range is about from -100 to $1000°$F. Inaccuracy of the order of 0.5 to 1 percent of scale

[15]S. G. Eskin and J. R. Fritze, "Thermostatic Bimetals," *Trans. ASME,* p. 433, July 1940.

[16]Engineered Materials Solutions, Inc., Attleboro, MA, 508-236-1623.

[17]R. Gitlin, "How Temperature Effects Instrument Accuracy," *Contr. Eng.,* April, May, June 1955.

range may be expected in bimetal thermometers of high quality. Bimetal elements have been combined with conductive plastic potentiometers for automotive sensing applications.[18]

Liquid-in-Glass Thermometers[19]

The well-known liquid-in-glass thermometer is adaptable to a wide range of applications by varying the materials of construction and/or configuration. Mercury is the most common liquid utilized at intermediate and high temperatures; its freezing point of $-38°F$ limits its lower range. The upper limit is in the region of $1000°F$ and requires use of special glasses and an inert-gas fill in the capillary space above the mercury. Compression of the gas helps to prevent separation of the mercury thread and raises the liquid boiling point. For low temperatures, alcohol is usable to $-80°F$, toluol to $-130°F$, pentane to $-330°F$, and a mixture of propane and propylene giving the lower limit of $-360°F$.

Thermometers are commonly made in two types: total immersion and partial immersion. Total-immersion thermometers are calibrated to read correctly when the liquid column is completely immersed in the measured fluid. Since this may obscure the reading, a small portion of the column may be allowed to protrude with little error. Partial-immersion thermometers are calibrated to read correctly when immersed a definite amount and with the exposed portion at a definite temperature. They are inherently less accurate than full-immersion types. If the exposed portion is at a temperature different from that at calibration, then a correction must be applied. Corrections for full- and partial-immersion thermometers used at conditions other than those intended are determined most accurately by the use of a special "faden" thermometer,[20] designed to measure the average temperature of the emergent stem. If such a thermometer is not available, the correction may be estimated by suspending a small auxiliary thermometer close to the stem of the thermometer to be corrected, as in Fig. 8.4. This auxiliary thermometer estimates the mean temperature of the emergent stem. When a partial-immersion thermometer is employed at correct immersion but with a surrounding air temperature different from that of its original calibration condition, the correction may be calculated from (for mercury-in-glass)

$$\text{Correction} = 0.00009n(t_{cal} - t_{act}) \qquad F° \qquad (8.5)$$

where $\quad n \triangleq$ number of scale degrees equivalent to emergent stem length, $F°$

$\quad t_{cal} \triangleq$ air temperature at calibration, $F°$

$\quad t_{act} \triangleq$ actual air temperature at use, $F°$ (from auxiliary thermometer)

[18]P. J. Sacchetti and D. R. Phillips, "A Ratiometric Temperature Sensor for High Temperature Applications," *SAE Paper* 80024, 1980.

[19]J. F. Swindells, "Calibration of Liquid-in-Glass Thermometers," *Natl. Bur. Std. (U.S.), Circ.* 600, 1959; T. D. McGee, "Principles and Methods of Temperature Measurement," Wiley, New York, 1988, chap. 6.

[20]J. F. Swindells, ibid.

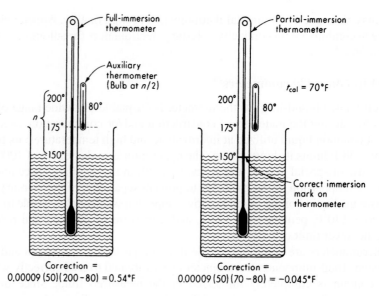

Full-immersion thermometer

Auxiliary thermometer (Bulb at $n/2$)

$200°$ $80°$

n { $175°$

$150°$

Partial-immersion thermometer

$t_{cal} = 70°F$

$200°$ $80°$

$175°$

$150°$

Correct immersion mark on thermometer

Correction =
0.00009 (50)(200 − 80) = 0.54°F

Correction =
0.00009 (50) (70 − 80) = −0.045°F

Figure 8.4
Full- and partial-immersion thermometers.

When a total-immersion thermometer is utilized at partial immersion, the same formula may be used except that $t_{cal} - t_{act}$ is replaced by (main-thermometer reading) − (auxiliary-thermometer reading). For Celsius thermometers the constant 0.00009 becomes 0.00016.

The accuracy obtainable with liquid-in-glass thermometers depends on instrument quality, temperature range, and type of immersion. For full-immersion thermometers the best instruments, when calibrated, are capable of errors as small as 0.4F° (range −328 to 32°F), 0.05F° (range −69 to 32°F), 0.04F° (range 32 to 212°F), 0.4F° (range 212 to 600°F), and 0.8F° (range 600 to 950°F). Errors in partial-immersion types may be several times larger, even after corrections have been applied for air-temperature variations. All the above figures refer to the ultimate performance attainable with the best instruments and great care in application. Errors in routine day-to-day measurements may be much larger.

Pressure Thermometers[21]

Pressure thermometers consist of a sensitive bulb, an interconnecting capillary tube, and a pressure-measuring device such as a Bourdon tube, bellows, or diaphragm (Fig. 8.5). When the system is completely filled with a liquid (mercury and xylene are common) under an initial pressure, the compressibility of the liquid is often small enough relative to the pressure gage $\Delta V/\Delta p$ that the measurement is

[21]D. M. Considine (ed.), "Process Instruments and Controls Handbook," p. 68, McGraw-Hill, New York, 1957.

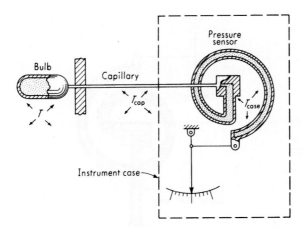

Figure 8.5
Pressure thermometer.

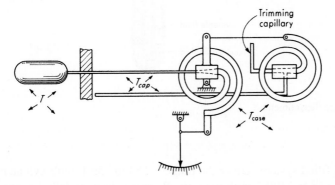

Figure 8.6
Compensation methods.

essentially one of volume change. For gas or vapor systems, the reverse is true, and the basic effect is one of pressure change at constant volume.

Capillary tubes as long as 200 ft may be used for remote measurement. Temperature variations along the capillary and at the pressure-sensing device generally require compensation, except in the vapor-pressure type, where pressure depends on only the temperature at the liquid's free surface, located at the bulb. A common compensation scheme using an auxiliary pressure sensor and capillary is shown in Fig. 8.6. The motion of the compensating system is due to the interfering effects only and is subtracted from the total motion of the main system, resulting in an output dependent on only bulb temperature. The "trimming" capillary (which may be lengthened or shortened) allows the volume to be changed to attain accurate case compensation by experimental test. Bimetal elements also are used to obtain case and partial capillary compensation.

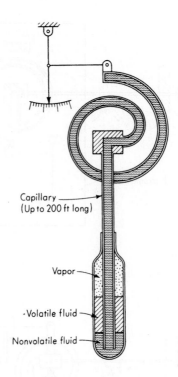

Figure 8.7
Vapor-pressure thermometer.

Liquid-filled systems cover the range −150 to 750°F with xylene or a similar liquid and −38 to 1100°F with mercury. Response is essentially linear over ranges up to about 300°F with xylene and 1000°F with mercury. Elevation differences between the bulb and pressure sensor different from those at calibration may cause slight errors. Gas-filled systems operate over the range −400 to 1200°F with linear ranges as great as 1000°F; errors due to capillary temperature variations usually are small enough not to justify compensation. Case compensation is accomplished with bimetal elements. Vapor-pressure systems are usable in the range −40 to 600°F. The calibration is strongly nonlinear; special linearizing linkages are needed if linear output is required. Characteristics of the system vary, depending on whether the bulb is hotter than, colder than, or equal in temperature to the rest of the system, since this determines where liquid and vapor will exist. The most versatile arrangement is shown in Fig. 8.7, where the volatile-liquid surface is *always* in the bulb. Capillary and case corrections are not needed in such a device since the vapor pressure of a liquid depends on only the temperature of its free surface. Commonly used volatile liquids include ethane (vapor pressure changes from 20 to 600 lb/in² gage for a temperature change from −100 to + 80°F), ethyl chloride (0 to 600 lb/in² gage for 40 to 350°F), and chlorobenzene (0 to 60 lb/in² gage for 275 to 400°F). The accuracy of pressure thermometers under the best conditions is of the order

±0.5 percent of the scale range. Adverse environmental conditions may increase this error considerably.

A unique gas thermometer[22] uses a bulb filled with activated carbon powder and inert gas (the 0.007″ ID capillary has *only* gas). The carbon powder has a huge surface area (125 *acres*/lb!) on which large volumes of gas can be adsorbed. Once adsorbed, the gas molecules are no longer "flying around" and causing fluid pressure, so the bulb can be charged (at, say, room temperature) with a lot of gas *without* causing a high initial pressure. When the temperature of the bulb changes, some of the adsorbed gas is "freed" to exert pressure (the amount of gas that can be held in adsorbtion is a function of temperature). This principle allows lower bulb pressure for a given temperature range, giving lower stress and longer life. A compact helical Bourdon tube immersed in silicone damping oil gives sufficient deflection to allow direct drive of the pointer (no gears). Units with ranges from −320 to 1200°F are available.

8.3 THERMOELECTRIC SENSORS (THERMOCOUPLES)[23]

If two wires of different materials *A* and *B* are connected in a circuit as shown in Fig. 8.8a, with one junction at temperature T_1 and the other at T_2, then an infinite-resistance voltmeter detects an electromotive force *E*, or if an ammeter is connected, a current *I* is measured. The magnitude of the voltage *E* depends on the materials and temperatures. The current *I* is simply *E* divided by the total resistance of the circuit, including the ammeter resistance.

There is some confusion in the literature (including earlier editions of this text!) as to the *location* of the thermoelectric voltage in the circuit, with some authors implying that it occurs at the *junction* of the two metals. This was shown *not* to be the case some time ago, but was overlooked by some, including widely accepted "committee publications" of engineering societies. While this misconception has no effect on many routine applications, it obscures some serious problems, and the *correct* interpretation allows a more systematic analysis of any thermoelectric circuit. What *is* the correct interpretation?[24] The thermoelectric emf is actually an effect *distributed* along the length of each *single* metal wire and exists even if the wire is not connected to anything. Its magnitude E_σ depends on a material property called the *absolute Seebeck coefficient* σ, and the distribution of temperature along the wire. Since the wire that produces the Seebeck emf also has ordinary electrical resistance *R*, if current *I* is allowed to flow, the net voltage will be reduced by an amount *IR*. We can thus model each wire segment as a Seebeck emf source in series with a specific resistance. The Seebeck coefficient is *defined* by the relation

[22]Dresser Industries Instrument Div., Stratford, CT, 203-378-8281 (www.dresserinstruments.com).

[23]"Manual on the Use of Thermocouples in Temperature Measurement," 4th ed., *ASTM MNL* 12, Philadelphia, PA, 1993; R. Bently, "Theory and Practice of Thermoelectric Thermometry," CSIRO Div. of Applied Physics, Sydney, Australia, 1995.

[24]*ASTM MNL* 12, R. P. Reed, chap. 2.

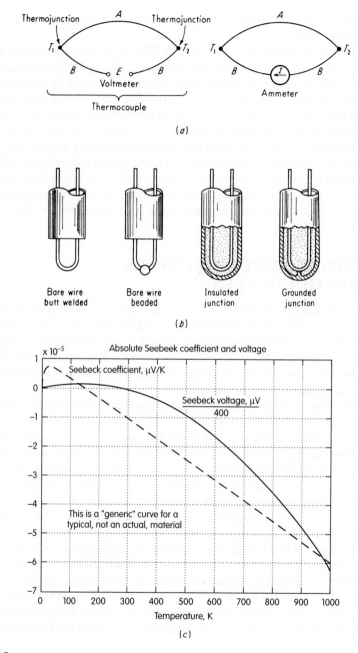

Figure 8.8
Basic thermocouple and junction types.

$$\sigma(T) \triangleq \frac{dE_\sigma}{dT} \qquad \text{and thus} \qquad E_\sigma(T) = \int \sigma(T)dT + C \qquad \textbf{(8.6)}$$

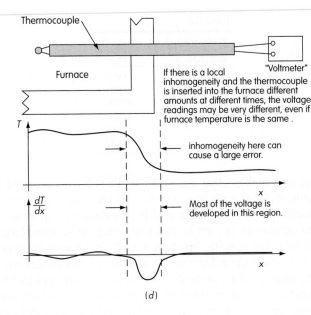

Thermocouple

Furnace

"Voltmeter"

If there is a local inhomogeneity and the thermocouple is inserted into the furnace different amounts at different times, the voltage readings may be very different, even if furnace temperature is the same .

inhomogeneity here can cause a large error.

Most of the voltage is developed in this region.

(d)

Figure 8.8
(Concluded)

The constant of integration C can be made zero by choosing to reference the Seebeck emf to 0 K, as shown in the experimentally measured data of Fig 8.8c, and leading to the equations

$$E_\sigma = \int_0^{T_2} \sigma(T)dT - \int_0^{T_1} \sigma(T)dT = \int_{T_1}^{T_2} \sigma(T)dT = E_\sigma(T_2) - E_\sigma(T_1) \quad (8.7)$$

The net voltage is thus easily obtained from two points on the curve of Fig. 8.8c by subtraction.

All the above relations *assume* that the material is *homogeneous* (σ is a function only of T); however, this is never exactly true, and in some cases the inhomogeneity is enough to cause large errors. The inhomogeneity can occur anywhere along the length of the wires and thus the *false* focus on the *junction* could obscure the real source of error, so this is a major contribution of the "new" viewpoint. Thermocouple materials *as received* from qualified suppliers *are* usually sufficiently homogeneous for most applications; however, fabrication processes, contaminating environments, handling stresses, and excessive temperatures can cause large inhomogeneity and measurement error. Mathematically, inhomogeneity can be modeled by letting $\sigma = \sigma(T, x)$, that is, the Seebeck coefficient depends on both temperature and location x in the wire. We then have

$$E_\sigma(x, T) = \int_{x_1}^{x_2} \sigma(x, T) \cdot \frac{dT}{dx} \cdot dx \quad (8.8)$$

Table 8.1

Temperature, °C	Voltage, mV
−20	−0.995
−10	−0.501
0	0.000
10	0.507
20	1.019
30	1.537

This expression also makes clear that changes in voltage occur only where the temperature *gradient dT/dx* is not zero. Often, long sections of a thermocouple have only a small gradient, and most of the emf is generated in a short section of high gradient. If inhomogeneity occurs in this high gradient section, large errors are possible (Fig. 8.8*d*). The specific function $\sigma(x, T)$ is a useful concept but cannot actually be known since the effects of the sources of inhomogeneity mentioned above cannot be quantified theoretically. It *is* possible however to devise experimental tests[25] that reveal the presence of inhomogeneity, allowing one to discard defective thermocouples. Note that removing a suspected unit and carefully calibrating it in a metrology lab will *not* correct errors due to inhomogeneity since the lab environment does not duplicate the use environment. An *in situ*[26] calibration, when possible, *will* correct such errors, but only if the "use" environment (such as immersion length, clamping forces, etc.) is not changed.

Having now warned you about the subtleties of inhomogeneity errors, we will concentrate on those many situations where this is *not* a problem. While the basic operation of thermocouples is best considered in terms of the *absolute* Seebeck coefficient and voltage of each wire, practical thermocouples always consist of two (or more) materials. The simplest and most common situation is just two materials, and such pairs are described by their *relative* Seebeck coefficient, which is just the difference of their absolute coefficients. That is, in Fig. 8.8*a*

$$E = \int_{T_1}^{T_2} \sigma_A \, dT - \int_{T_1}^{T_2} \sigma_B \, dT = \int_{T_1}^{T_2} (\sigma_A - \sigma_B) \, dT = \int_{T_1}^{T_2} \sigma_{AB} \, dT \qquad \textbf{(8.9)}$$

The relative Seebeck emf E is what is listed as a function of temperature in standard tables supplied by manufacturers of thermocouple wire. These tables refer to the situation where one of the thermojunctions (called the *reference junction*) is at a *known* temperature, usually the ice point, 0°C. Note that in *every thermocouple* the temperature of one of the junctions *must* be known in some way before we can find the temperature of the other junction from voltage measurements. Table 8.1 shows

---

[25]C. G. Burkett Jr. and W. A. Bauserman Jr., "A Computer-Controlled Apparatus for Seebeck Inhomogeneity Testing of Sheathed Thermocouples"; R. P. Reed and W. A. Bauserman, "Measurement of Thermoelectric Inhomogeneity of Thermocouples, LAR-15198," NASA Langely Res. Ctr., Oct. 1994.

[26]R. Bentley, op. cit., pp. 123–124.

a few values from a table for the *iron-constantan* (type J) thermocouple. In computerized data systems, we may prefer *equations* rather than tables when converting voltages and temperatures. Such equations are obtained by curve fitting the data found in the experimentally measured tables. For the iron-constantan (Type J) thermocouple the equations[27] are

$$E = 5.04e - 2T + 3.05e - 5T^2 - 8.57e - 8T^3 + \cdots +$$
$$1.56e - 23T^8 \quad E \text{ in mV, } T \text{ in } ^\circ C \tag{8.10}$$

$$T = 1.95e1E - 1.23E^2 - 1.08E^3 + \cdots -$$
$$8.38e - 5E^8 \quad E \text{ in mV, } T \text{ in } ^\circ C \tag{8.11}$$

The reference gives the complete equation with the constants expressed to many decimal places.

If the reference junction is at 0°C and the *IR* drop in the wires leading to the "voltmeter" is negligible, then the measured voltage E is simply looked up in the table (may require interpolation, which *can* be done linearly) or substituted into the equation, to find the measuring-junction temperature. While milliameters (which do allow *IR* drops) are sometimes used to read thermocouples, this is now quite rare. Most often, we use a potentiometer (theoretically draws *no* current) or an amplifier with such a high input resistance (relative to the thermocouple's) that the current and thus the *IR* drop is negligible. Since ice baths may not be practical for many industrial applications, the reference junction will often *not* be at 0°C. An *isothermal block* is often used and can accommodate multiple channels of thermocouples if many different temperatures must be measured. A typical unit might have 15 channels, 14 of which are for thermocouples and one for a semiconductor temperature sensor (see Sect. 8.5) that measures the temperature of the isothermal block. This is a block of metal (often aluminum) which is a good thermal conductor, surrounded by good thermal insulation, and has fastened to it the screw terminal strip where the thermocouple wires are attached. The temperature of the block is allowed to drift with room temperature, but the block is designed so that its temperature is spatially very uniform so that all 14 thermocouple reference junctions (the screw terminal locations) will have the same temperature at any time. The semiconductor sensor does *not* require a reference junction and can thus take the block's temperature "all by itself."

Once we know the reference junction temperature (by whatever means), we can easily use the table or equations to compute the unknown measuring junction temperature. From Eq. (8.9), it is clear that the integral from 0 to T_{meas} is just the sum of the integrals from 0 to T_{ref} and from T_{ref} to T_{meas}. Thus, using our simple Table 8.1 as the data source, if $T_{ref} = 10°C$ and $E = 1.030$ mV, then we add the voltage (0.507) corresponding to T_{ref} to 1.030 to get 1.537 mV. Looking in the table for this value, we find it at 30°C, which is T_{meas}. (These numbers have obviously been chosen for our convenience.) Note that if we (as in most computerized systems) use equations rather than table look up, we need *both* Eq. (8.10) and (8.11).

---

[27]*ASTM MNL* 12, chap. 10.

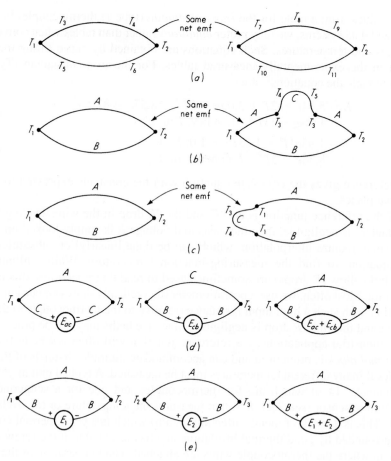

Figure 8.9
Thermocouple laws.

In analyzing various types of thermocouple circuits that arise in practice, certain "laws of thermocouples" known for many years are useful. These laws can all be derived from the basic relations (8.7) and (8.9), but we here simply state them without proof and illustrate typical applications in Fig. 8.9. *They apply only to homogeneous elements!*

1. The thermal emf of a thermocouple with junctions at T_1 and T_2 is totally unaffected by temperature elsewhere in the circuit if the two metals used are each homogeneous (Fig. 8.9a).

2. If a third homogeneous metal C is inserted into either A or B (see Fig. 8.9b), as long as the two new thermojunctions are at like temperatures, the net emf of the circuit is unchanged irrespective of the temperature of C away from the junctions.

3. If metal C is inserted between A and B at one of the junctions, the temperature of C at any point away from the AC and BC junctions is immaterial. As long as the junctions AC and BC are both at the temperature T_1, the net emf is the same as if C were not there (Fig. 8.9c).

4. If the thermal emf of metals A and C is E_{AC} and that of metals B and C is E_{CB}, then the thermal emf of metals A and B is $E_{AC} + E_{CB}$ (Fig 8.9d).

5. If a thermocouple produces emf E_1 when its junctions are at T_1 and T_2, and E_2 when at T_2 and T_3, then it will produce $E_1 + E_2$ when the junctions are at T_1 and T_3 (Fig. 8.9e).

These laws are of great importance in the practical application of thermo-couples. The first states that the lead wires connecting the two junctions may be safely exposed to an unknown and/or a varying temperature environment without affecting the voltage produced. Laws 2 and 3 make it possible to insert a voltage-measuring device into the circuit to actually measure the emf, rather than just talk-ing about its existence. That is, the metal C represents the internal circuit (usually all copper in precise instruments) between the instrument binding posts. The instru-ment can be connected in two ways, as shown in Fig. 8.9b and c. Law 3 also shows that thermocouple junctions may be soldered or brazed (thereby introducing a third metal) without affecting the readings. Law 4 shows that all possible pairs of metals need not be calibrated since the individual metals can each be paired with *one* stan-dard (platinum is used) and calibrated. Any other combinations then can be *calcu-lated;* calibration is not necessary.

In considering the fifth law, we should note that, in using a thermocouple to measure an unknown temperature, the temperature of one of the thermojunctions (called the reference junction) must be known by some independent means. A volt-age measurement then allows us to get the temperature of the other (measuring) junction from calibration tables.

Thermocouple circuits can get more complex than the basic schemes shown so far. Some of the complications come from *unintentional* thermojunctions as a result of apparatus construction that unavoidably joins dissimilar metals. Figure 8.9f[28] shows how the metals used in binding posts, solder, amplifier sockets, and inte-grated circuit amplifiers create a system with a total of 20 thermojunctions, only two of which are the intentional Chromel/constantan temperature sensor. The uninten-tional junctions are of course not within the control of the thermocouple user, we rely on the electronics designers to minimize their bad effects and give us guidance on warm-up times that will reduce their magnitude. All such circuits can be analyzed using the concept of the absolute Seebeck coefficient of each material. Figure 8.9g gives an example with three temperature zones and four materials. We see that if the two junctions at T_2 *are* actually at the same temperature, and similarly for T_3, then the voltage E_{meas} measured at the amplifier will be exactly the same as would be expected from the iron/constantan thermocouple itself, and need only be adjusted for the reference temperature T_2, which would of course have to be known. If the junctions assumed to both be at T_2 in Fig. 8.9g have *different* temperatures,

---

[28]D. H. Sheingold (ed.), "Transducer Interfacing Handbook," Analog Devices, Norwood, MA, 1980, p. 9.

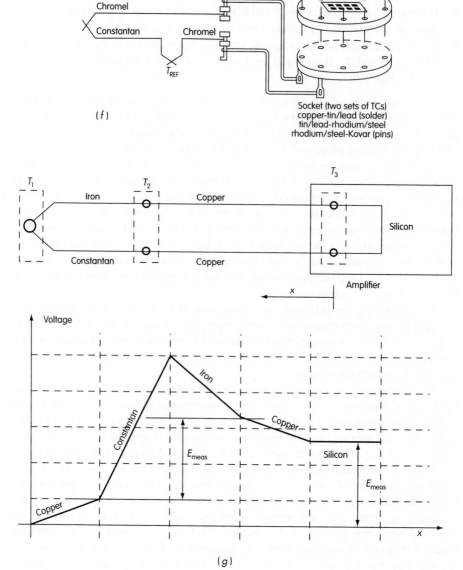

Figure 8.9
(Concluded)

then the graph clearly would show that the two sections labeled "copper" do *not* cancel each other, leading to an error. The same would hold true for the junctions assumed at T_3.

Thermoelectric temperature measurement depends only on the Seebeck effect, which we have discussed earlier. There are, however, two other effects, the *Peltier* and the *Thompson* which we want to at least mention. If current is allowed to flow in a thermocouple circuit (usually *not* the case in temperature-sensing devices), thermoelectric energy conversion processes will cause heating or cooling effects. The Peltier effect is concentrated at the junctions, and the Thompson effect is distributed along the wires. These effects will raise or lower the wire temperatures from the values that would exist if no current was flowing. Regardless, the Seebeck voltage will be correct *for those new temperatures.* Fortunately, for the metals used in thermoelectric temperature sensing, these effects are negligibly small *even if current is allowed to flow.* The Peltier effect *is* used for small-scale thermoelectric cooling but only by employing *semiconductor* materials; the metals used in temperature sensing have Peltier coefficients that are much too small and thermal conductivities that are much too large.

Common Thermocouples

Thermocouples formed by welding, soldering, or merely pressing the two materials together give identical voltages. If current is allowed to flow, the currents may be different since the contact resistance differs for the various joining methods. Welding (either gas or electric) is used most widely although both silver solder and soft solder (low temperatures only) are used in copper/constantan couples. Special capacitor-discharge welding devices (particularly needed for very-fine-wire thermocouples) are available. Ready-made thermocouple pairs are, of course, available in a wide range of materials and wire sizes.[29]

While many materials exhibit the thermoelectric effect to some degree, only a small number of pairs are in wide use. They are platinum/rhodium, Chromel/Alumel, copper/constantan, and iron/constantan. Each pair exhibits a combination of properties that suit it to a particular class of applications. Since the thermoelectric effect is somewhat nonlinear, the sensitivity varies with temperature. The maximum sensitivity of any of the above pairs is about 60 μV/C° for copper/constantan at 350°C. Platinum/platinum-rhodium is the least sensitive: about 6 μV/C° between 0 and 100°C.

The accuracy of the common thermocouples may be stated in two different ways. If you utilize standard thermocouple wire (which is *not* individually calibrated by the manufacturer) and make up a thermocouple to be used without calibration, you are relying on the wire manufacturer's quality control to limit deviations from the published calibration tables. These tables give the *average* characteristics, not those of a particular batch of wire. Platinum/platinum-rhodium is the most accurate; error is of the order of $\pm$ 0.25 percent of reading. Copper/constantan gives $\pm$0.5 percent or $\pm$1.5F° (whichever is larger) between -75 and 200°F and $\pm$0.75 percent between 200 and 700°F. Chromel/Alumel gives $\pm$5F° (32 to 660°F) and $\pm$0.75 percent (660 to 2300°F). Iron/constantan has $\pm$66 μV below 500°F and $\pm$1.0 percent from 500 to 1500°F. If higher accuracies

---

[29]Omega Engineering Inc., Stamford, CT, 888-826-6342 (www.omega.com).

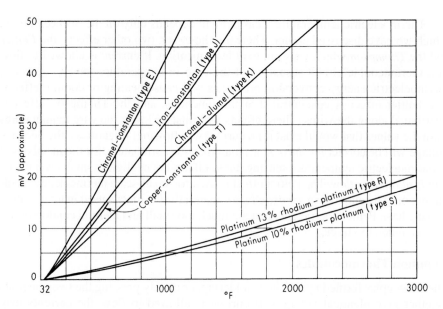

Figure 8.10
Thermocouple temperature/voltage curves.

are needed, the individual thermocouple may be calibrated. An indication of the achievable accuracy is available from National Bureau of Standards (now NIST) listings[30] of the results they will guarantee. At the actual calibration points, the error ranges from 0.05 to 0.5C°. Interpolated points are less accurate: 0.1 to 1.0C°, except platinum/platinum-rhodium at 1450°F, 2.0 to 3.0C°. The realization of this potential accuracy in applying such calibrated thermocouples to practical temperature measurement is, of course, dependent on the application conditions and is rarely possible.

Platinum/platinum-rhodium thermocouples are employed mainly in the range 0 to 1500°C. The main features of this combination are its chemical inertness and stability at high temperatures in oxidizing atmospheres. Reducing atmospheres cause rapid deterioration at high temperatures as the thermocouple metals are contaminated by absorbing small quantities of other metals from nearby objects (such as protecting tubes). This difficulty, causing loss of calibration, is unfortunately common to most thermocouple materials above 1000°C.

Chromel ($Ni_{90}Cr_{10}$)/Alumel ($Ni_{94}Mn_3Al_2Si_1$) couples are useful over the range −200 to +1300°C. Their main application, however, is from about 700 to 1200°C in nonreducing atmospheres. The temperature/voltage characteristic is quite linear for this combination (see Fig. 8.10).

Copper/constantan ($Cu_{57}Ni_{43}$) is used at temperatures as low as −200°C; its upper limit is about 350°C because of the oxidation of copper above this range. Iron/constantan is the most widely utilized thermocouple for industrial applications

---

[30]"Thermocouple Calibration," *Instrum. Contr. Syst.,* p. 1663, September 1961.

and covers the range -150 to $+1000°C$. It is usable in oxidizing atmospheres to about 760°C and reducing atmospheres to 1000°C.

Problems with type K thermocouples in some applications have led to the development of type N (Nicrosil/Nisil) thermocouples.[31] In severe high-temperature environments, thermocouples using the mineral-insulated, metal-sheathed (MIMS) construction are common. Here the thermocouple wires are enclosed within a protective metal tube, which is filled with magnesium oxide powder as insulation. The tube is then swaged to reduce the diameter and compress the assembly. In some cases, type K units of this construction suffer errors due to oxidation, contamination, magnetic transformations, and nuclear irradiation. Some of these errors are of the inhomogeneity type discussed earlier. By using type N wire and carefully selecting alloys for the sheath tube, improved performance has been obtained.

Thermocouple manufacturers have a wealth of experience concerning the application of thermocouples to diverse temperature-measuring problems and should be consulted if special types of problems are foreseen in a particular case.

Reference-Junction Considerations

For the most precise work, reference junctions should be kept in a triple-point-of-water apparatus[32] whose temperature is $0.01 \pm 0.0005C°$. Such accuracy is rarely needed, and an ice bath is used much more commonly. A carefully made ice bath is reproducible to about $0.001C°$, but a poorly made one may have an error of $1C°$.[33] Figure 8.11 shows one method of constructing an ice-bath reference junction. The main sources of error are insufficient immersion length and an excessive amount of water in the bottom of the flask. Automatic ice baths that use the Peltier cooling effect as the refrigerator, rather than relying on externally supplied ice (which must be continually replenished), are available with an accuracy of $0.05C°$.[34] These systems use the expansion of freezing water in a sealed bellows as the temperature-sensing element that signals the Peltier refrigerator when to turn on or off by displacing a microswitch.

Since low-power heating is obtained more easily than low-power cooling, some reference junctions are designed to operate at a fixed temperature higher than any expected ambient. A feedback system operates an electric heating element to maintain a constant and known temperature in an enclosure containing the reference junctions. Since the reference junction is not at 32°F, the thermocouple-circuit net voltage must be corrected by adding the reference-junction voltage before the measuring-junction temperature can be found. This correction is, however, a constant.

---

[31]C. P. Furniss, "Improved Temperature Measurement in Heat Treatment Furnaces Using Special Sensors," JMS Southeast, Inc., Statesville, NC, 800-873-1835.

[32]Hart Scientific Inc.

[33]C. L. Feldman, "Automatic Ice-Point Thermocouple Reference Junction," *Instrum. Contr. Syst.,* p. 101, January 1965.

[34]Ibid.

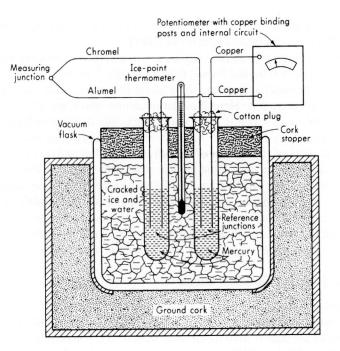

Figure 8.11
Ice-bath reference junction.

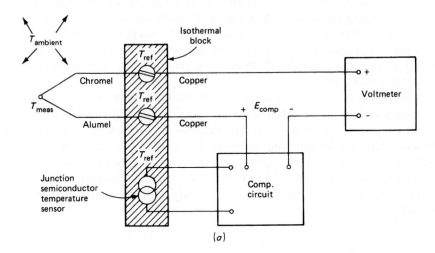

Figure 8.12
Isothermal block reference-junction technique.

Figure 8.12 sketches a reference-junction technique widely utilized (in various versions) for digital thermometer instruments, data loggers, and data acquisition

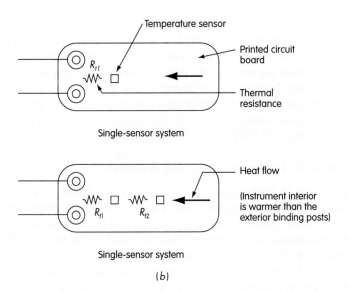

Figure 8.12
(Concluded)

systems. Wires from the measuring junction are screwed directly to an isothermal block terminal strip. The temperature of this block (which has *no* active temperature control) drifts with ambient temperature; but because of careful thermal design, at all times it is *uniform* over its length (within about $\pm 0.05°C$). This reference temperature is measured by independent means [often a junction semiconductor sensor (Sec. 8.5)], and compensation circuitry[35] develops a voltage E_{comp} which is combined with that from the measuring junction so that the net voltage presented to the voltmeter represents T_{meas}. In multichannel instruments with microprocessor computing power, the isothermal block can accept many thermocouple pairs (of *mixed* types, if desired) since the T_{ref} sensor now sends its temperature data to the computer, which computes the needed voltage correction for each different thermocouple and applies it digitally as each channel is scanned. Such a "software" correction scheme, of course, must be intermittent (limiting response speed) because of the sampled nature of digital computation, while the "hardware" technique of Fig. 8.12a is continuous and does not significantly degrade sensor dynamics.

When it is not feasible to use an isothermal block, a built-in temperature sensor located close to the instrument binding posts (where the thermocouple reference junctions will be created) may be used. This sensor is of necessity located inside the instrument housing and is thus somewhat remote from the binding posts. When the instrument is thoroughly warmed up and the ambient temperature is steady, the sensor does a good job of measuring the binding post temperature since there is little heat flow between them. During the warm-up period or when the ambient temperature is changing, there *is* heat flowing between the two locations and the

---

[35]D. H. Sheingold, op. cit., p. 122.

sensor reports a wrong value for the reference junction temperature. One manufacturer[36] deals with this problem by using *two* sensors inside the housing (Fig 8.12*b*). The temperature *difference* of the two sensors is a measure of the heat flow, which is assumed to be about the same for the path between the binding posts and the nearest sensor. Assuming the circuit board thermal resistances R_{t1} and R_{t2} are known and fixed, we can compute the binding post temperature, and thus the proper reference-junction compensation voltage, from the two sensor signals. Tests on these units[37] have shown a decrease in error from 1.76 percent to 0.5 percent.

Special Materials, Configurations, and Techniques

Increasing interest in high-temperature processes in jet and rocket engines and nuclear reactors has led to requirements for reliable temperature sensors in the range 2000 to 4500°F. New thermocouples developed for these applications include rhodium-iridium/rhodium,[38] tungsten/rhenium,[39] and boron/graphite.[40] Rhodium-iridium is usable to about 4000°F under proper conditions and has a sensitivity of the order of 6 μV/C°. Various alloys of tungsten and rhenium may be utilized up to 5000°F under favorable conditions and have about the same sensitivity at the highest temperatures as rhodium-iridium. Boron/graphite has a high sensitivity (about 40 μV/C°) and is usable for short times up to 4500°F.

An alternative solution to high-temperature problems may be found in various cooling schemes. Two such[41] in actual use are shown in Fig. 8.13. In Fig. 8.13*a* the hot-gas flow whose temperature is to be measured impinges on a small tube carrying cooling water, causing a temperature rise of about 100°F. If heat-transfer coefficients are known, measurements of water flow rate, temperature, and temperature rise allow calculation of the hot-gas temperature. Figure 8.13*b* shows another approach in which the hot gas is aspirated through a heat exchanger, cooling it to about 1000°F. Knowledge of heat-transfer characteristics and flow rates again allows calculation of the hot-gas temperature. Such methods have been used in the range 5000 to 8000°F.

Figure 8.14 shows in simplified fashion the principle of a pulse-cooling technique[42] which allows use of Chromel/Alumel thermocouples (melting point 2550°F) to measure temperatures up to 7000°F. (Further developments[43] have since

[36]Action Instruments Inc., San Diego, CA, 800-767-5726 (www.actionio.com).

[37]R. Brewer, "Thermocouples' Cold-Junction Blues No Longer Play," In Tech, July 1999, pp. 60–63.

[38]P. D. Freeze, "Review of Recent Developments of High-Temperature Thermocouples," *ASME Paper* 63-WA-212, 1963.

[39]Ibid.

[40]*Bull.* BGT-1, 1963, Astro Industries Inc., Santa Barbara, CA.

[41]*NASA, SP*-5015, p. 128, 1964.

[42]A. F. Wormser and R. A. Pfuntner, "Pulse Technique Extends Range of Chromel-Alumel to 7000°F," *Instrum. Contr. Syst.,* p. 101, May 1964.

[43]D. Kretschmer, J. Odgers, and A. F. Schlader, The Pulsed Thermocouple for Gas Turbine Applications, *ASME Paper* 76-GT-1, 1976; G. E. Glawe, H. A. Will, and L. N. Krause, "A New Approach to the Pulsed Thermocouple," *NASA* TM X-71883, 1976.

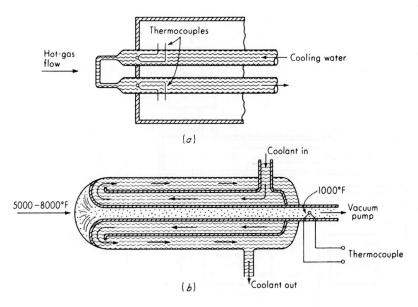

Figure 8.13
Cooled thermocouples.

appeared.) The measuring junction is kept at a low temperature by a cooling air flow. When this flow is shut off by a solenoid valve, the thermocouple starts to heat up, following the first-order equation

$$\tau \frac{dT_{tc}}{dt} + T_{tc} = T_{gas} \tag{8.12}$$

where $\tau \triangleq$ thermocouple time constant (assumed known)

          $T_{tc} \triangleq$ thermocouple temperature

          $T_{gas} \triangleq$ hot-gas temperature

This equation shows that T_{gas} can be *computed* any time after the cooling is shut off if dT_{tc}/dt is known. A voltage proportional to dT_{tc}/dt can be obtained by use of the differentiating circuit shown and, when it is summed with a voltage proportional to T_{tc}, provides a signal proportional to T_{gas}. Theoretically this signal is available immediately after the cooling is shut off; in practice, however, the cooling is left off a finite interval during which the value of T_{gas} is recorded. The cooling is turned on again before the thermocouple is overheated. (A variation on this scheme uses a high-speed actuator to insert and then retract the probe from the high-temperature region.) In the actual system of footnote 42 (Wormser and Pfuntner), additional computing elements also compute the numerical value of τ; thus this need not be known beforehand.

Still another approach to high-temperature gas measurements (particularly suited to dusty combustion gases because of its nonfouling characteristics) is the

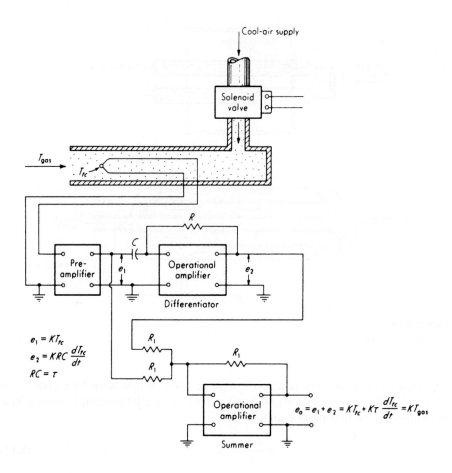

$$e_1 = KT_{tc}$$

$$e_2 = KRC \frac{dT_{tc}}{dt}$$

$$RC = \tau$$

$$e_o = e_1 + e_2 = KT_{tc} + K\tau \frac{dT_{tc}}{dt} = KT_{gas}$$

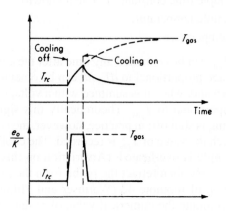

Figure 8.14
Pulsed-thermocouple technique.

venturi pneumatic pyrometer of Fig. 8.15.[44] Here the hot gas at temperature T_h is aspirated through a water-cooled tube provided with two venturi flow-metering sections, one at the hot end and the other at the cold. The "cold" end (maximum temperature about 1600°C) has a platinum resistance thermometer (thermocouples also have been employed) to measure T_c, the temperature of the cooled gas. Two differential-pressure transducers measure ΔP_h and ΔP_c at the venturis. The two venturis carry the same mass flow rate, and the pressure drops are small enough to use incompressible flow relations, making $\Delta P_c/\Delta P_h \approx \rho_h/\rho_c$. Also, since $P_c \approx P_h$, the ideal-gas law gives $T_h/T_c \approx \rho_c/\rho_h$, which leads finally to

$$T_h = KT_c \frac{\Delta P_h}{\Delta P_c} \qquad (8.13)$$

where K is a calibration factor to account for various deviations from theory. Equation (8.13) is readily implemented in an analog multiplier/divider once the platinum thermometer signal is linearized, which gives a linear output of 40 μV/°C. Temperatures T_h up to about 2500°C can be measured with accuracy about ± 2 percent of reading.

The measurement of rapidly changing internal and surface temperatures of solid bodies may be accomplished with arrangements such as those in Fig. 8.16. The main requirements in such applications are that the thermojunction be of minimum size and be precisely located and that any materials placed into the wall have thermal properties identical with those of the wall, so that temperature distributions are not distorted. In Fig. 8.16a[45] the thermojunction is formed by drawing an abrasive tool, such as a file or an emery cloth, across the end of the sensing tip. This action flows metal from one thermocouple element to the other since the 0.0002-in mica insulation is easily bridged over, thus forming numerous microscopic hot-weld thermojunctions. Subsequent erosion or abrasive action forms new thermal junctions continuously as the tip wears away. Such thermocouples have time constants as small as 10^{-5} s and are available in materials usable to 5000°F and 10,000 lb/in^2 pressure. In Fig. 8.16b[46] two thermojunctions are formed by plating a thin rhodium film over the end of a coaxial pair of thermocouple metals. Since the rhodium/metal A and rhodium metal B junctions are at the same temperature, the third metal (rhodium) has no effect. The plating is performed by vacuum evaporation and results in a rhodium layer 10^{-4} to 10^{-5} in thick. Theoretical calculations indicate the time constant of such a probe is of the order of 0.3 μs.

When the surface whose temperature is to be measured is suitable as one member of a thermocouple pair, the *intrinsic thermocouple* of Fig. 8.17a may be used for fast response measurements. If this is not possible but the surface is an electric conductor, then the arrangement of Fig. 8.17b can be employed. The

[44]*Data Sheet* 39, Land Instruments Inc., Tullytown, PA; J. Chedaille and Y. Braud, "Measurements in Flames," pp. 35–48, Crane, Russak & Co., New York, 1972.

[45]Nanmac Corp., Framingham, MA, 508-872-4811 (www.nanmac.com).

[46]D. Bendersky, "A Special Thermocouple for Measuring Transient Temperatures," *Mech. Eng.,* p. 117, February 1953.

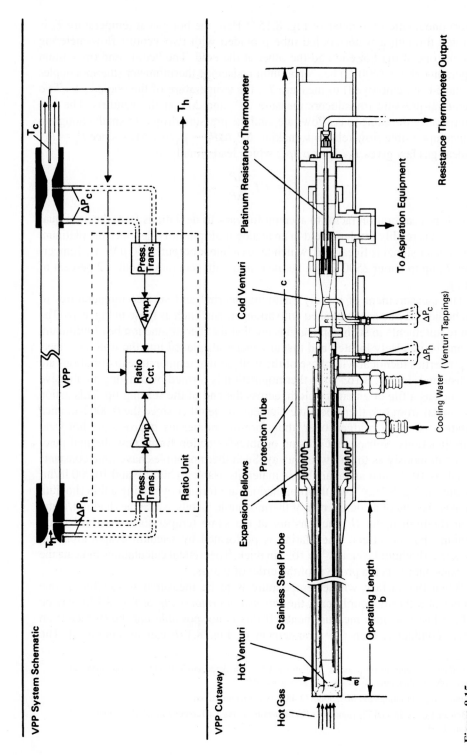

Figure 8.15
Venturi pneumatic pyrometer.

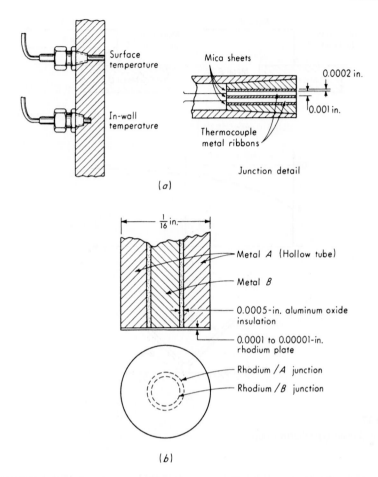

Figure 8.16
High-speed thermocouples.

dynamic response for such systems has been studied[47] by using analytical solutions of partial-differential-equation models of various types, numerical methods, and experimental tests. For the system of Fig. 8.17a, the response to a step change in surface temperature is given in Fig. 8.17c, where

$$a \triangleq \frac{1}{1 + \sqrt{(k\rho c)_w/(k\rho c)_s}} \qquad t_{95\%} \triangleq \frac{85\rho_s c_s R^2}{k_s}\left(\frac{k_w}{k_s}\right)^{1.08} \qquad \textbf{(8.14)}$$

where $R \triangleq$ wire radius, $k \triangleq$ thermal conductivity, $\rho \triangleq$ density, $c \triangleq$ specific heat, subscript $s \triangleq$ surface, and subscript $w \triangleq$ wire. For a constantan wire with $R = 0.001$

[47]N. R. Keltner, "Heat Transfer in Intrinsic Thermocouples—Application to Transient Temperature Measurement Errors," *SC*-RR-72 0719, Sandia Corp., Albuquerque, NM, 1973.

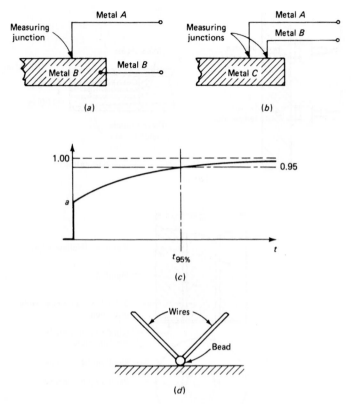

Figure 8.17
Intrinsic and bead-type thermocouples.

in and an iron surface, $a = 0.63$ and $t_{95\%} = 0.00092$ s. Figure 8.17d shows a related but different situation, in which an "ordinary" bead-welded thermocouple is fastened to the surface by either welding or adhesives. Response characteristics of this configuration also are available.[48] Thin-foil (0.0002 to 0.0005 in thick) thermocouples manufactured and attached like strain gages also are available[49] for surface-temperature measurement.

An actively compensated surface temperature instrument[50] uses a differential thermocouple to sense heat flow away from the measured surface and control a built-in heater to null this heat loss. When the heat loss is zero, the contact thermocouple gives a reading close to the undisturbed surface temperature.

[48]K. Wally, "The Transient Response of Beaded Thermocouples Mounted on the Surface of a Solid," *ISA Trans.,* vol. 17, no. 1, 1978.

[49]RdF Corp., Hudson, NH, 800-445-8367 (www.rdfcorp.com).

[50]Isotech Model 944, Leico Industries, New York, NY, 212-765-5290; B. D. Foulis, "Surface Temperature Measurement Using Contact Thermometry," *Isotech J. Thermometry,* vol. 6, no. 2, 1995.

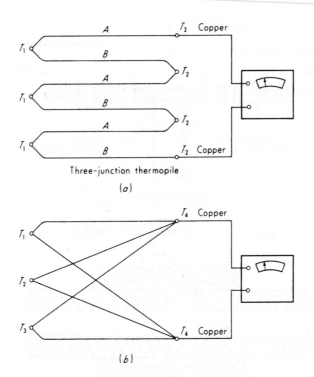

Figure 8.18
Multiple-junction thermocouples and flow couple.

Thermocouples in common use are made from wires ranging from about 0.020 to 0.1 in in diameter; the larger diameters are required for long life in severe environments. Since speed of response, conduction and radiation errors, and precision of junction location all are improved by the use of smaller wire, very-fine-wire thermocouples are utilized in special applications requiring these attributes and where lack of ruggedness is not a serious drawback. Such couples are available ready-made[51] in most common materials and wire sizes from 0.0005 to 0.015 in diameter. The time constant of an iron/constantan couple of 0.0005-in-diameter wire for a step change from 200 to 100°F in still water is about 0.001 s.

Several thermocouples may be connected in series or parallel to achieve useful functions (Fig. 8.18). The series connection with all measuring junctions at one temperature and all reference junctions at another is used mainly as a means of increasing sensitivity.[52] Such an arrangement is called a thermopile and for n thermocouples gives an output n times as great as a single couple. A typical Chromel/constantan thermopile has 25 couples and produces about 1 mV/F°.

[51]Omega Engineering Inc., Stamford, CT; Baldwin-Lima-Hamilton, Waltham, MA, *Tech. Data* 4336-1.

[52]R. A. Schnurr, "Thermopiles Aid in Measuring Heat Rejection," *Gen. Motors Eng. J.,* p. 8, April–May–June 1963.

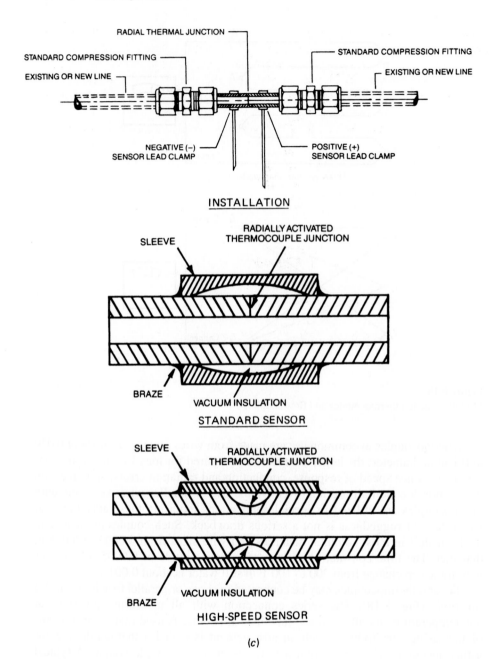

Figure 8.18
(Concluded)

Common potentiometers can resolve 1 μV, thus making such an arrangement sensitive to 0.001F°. The parallel combination generates the same voltage as a single couple if all measuring and reference junctions are at the same temperature. If the measuring junctions are at different temperatures and the thermocouples are all the same resistance, the voltage measured is the average of the individual voltages. The temperature corresponding to this voltage is the average temperature only if the thermocouples are linear over the temperature range being measured. A 20-junction thermopile with the two sets of junctions built into two standard pipe fittings is available[53] for sensitive measurements of differential temperatures in flowing fluids.

By inserting a short section of tubing made of butt-welded thermocouple materials into a pipeline, the temperature of the flowing fluid may be measured with durability, fast response, and high accuracy in a nonintrusive way. Ready-made sensors in a variety of materials and configurations are commercially available[54] (Fig. 8.18c).

8.4 ELECTRICAL-RESISTANCE SENSORS

The electrical resistance of various materials changes in a reproducible manner with temperature, thus forming the basis of a temperature-sensing method. Materials in actual use fall into two main classes: conductors (metals) and semiconductors. Conducting materials historically came first and traditionally have been called resistance thermometers. [More recently the terminology *resistance temperature detector* (RTD) has come into use.] Semiconductor types appeared later and have been given the generic name *thermistor.* Any of the various established techniques of resistance measurement may be employed to measure the resistance of these devices, with both bridge and "ohmmeter" methods being common.

Conductive Sensors (Resistance Thermometers)

The variation of resistance R with temperature T for most metallic materials can be represented by an equation of the form[55]

$$R = R_0(1 + a_1 T + a_2 T^2 + \cdots + a_n T^n) \tag{8.15}$$

where R_0 is the resistance at temperature $T = 0$ (see Fig. 8.19). The number of terms necessary depends on the material, the accuracy required, and the temperature range to be covered. Platinum, nickel, and copper are the most commonly used and generally require, respectively, two, three, and three of the a constants for a highly accurate representation. Tungsten and nickel alloys are also in use. Only constant a_1 may often be used since quite respectable linearity may be achieved over limited ranges. Platinum,[56] for instance, is linear within ± 0.4 percent over the ranges -300 to

[53]Delta-T Co., Santa Clara, CA.

[54]Flowcouple, Paul Beckman Co., Elkins Park, PA, 215-663-0250.

[55]T. D. McGee, op. cit., chap. 8.

[56]"Platinum Resistance Temperature Sensors," *Bull.* 9612, Rosemount Inc., Chanhassen, MN, 800-999-9307 (www.rosemount.com).

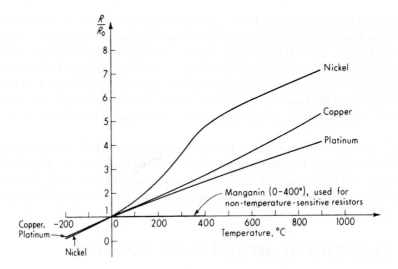

Figure 8.19
Resistance/temperature curves.

$-100°F$ and -100 to $+300°F$, ±0.3 percent from 0 to 300°F, ±0.25 percent from -300 to $-200°F$, ±0.2 percent from 0 to 200°F, and ±1.2 percent from 500 to 1500°F. Standard platinum resistance thermometers (SPRT) are used between 13.8 K and 962°C to realize the ITS-90 scale between the fixed points defined by freezing and melting point cells. Since the needs of industrial measurement and calibration vary, a range of platinum thermometers with different levels of accuracy are available.[57]

Sensing elements are made in a number of different forms (see Fig. 8.20). For measurement of fluid temperatures, the winding of resistance wire may be encased in a stainless-steel bulb to protect it from corrosive liquids or gases. Open-type pick-ups expose the resistance winding directly to the fluid (which must be noncorrosive) and give faster response. Various flat grid windings are available for measuring surface temperatures of solids. These may be clamped, welded, or cemented onto the surface. Thin deposited films of platinum also are used in place of wire windings. Surface-temperature transducers affixed to bodies may exhibit spurious output due to interfering strain inputs.[58] These strains may be due to loading of the structure or differential expansion. Careful design (Fig. 8.20b) can minimize strain effects. A small, low-cost monolithic RTD sensor[59] with on-chip signal conditioning using op-amp circuitry requires only application of 5-volt power to measure temperature in the -50 to 150°C range, with output of 22.5 mV/C° and 1% accuracy and linearity. Micro-size RTDs have been embedded in wafers used in

[57]Hart Scientific Inc.

[58]A. B. Kaufman, "Bonded-Wire Temperature Sensors," *Instrum. Contr. Syst.,* p. 103, May 1963.

[59]AD22100, Analog Devices, Wilmington, MA, 781-262-5645 (www.analog.com).

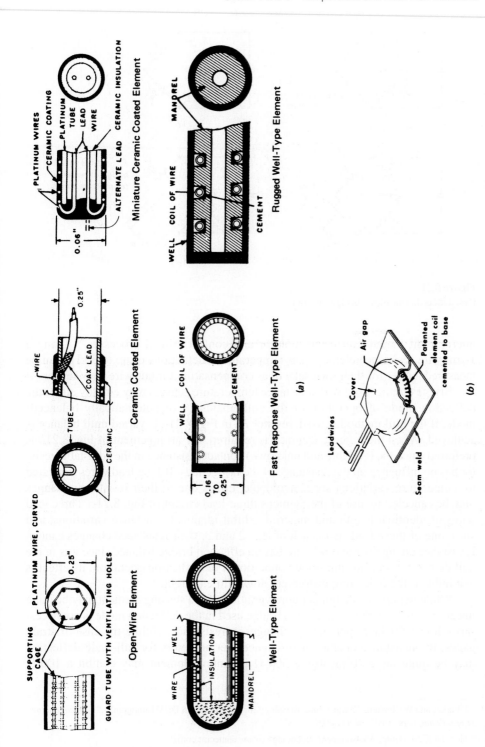

Figure 8.20
Resistance temperature detector (RTD) construction. (*Courtesy Rosemount Engineering, Minneapolis, MN.*)

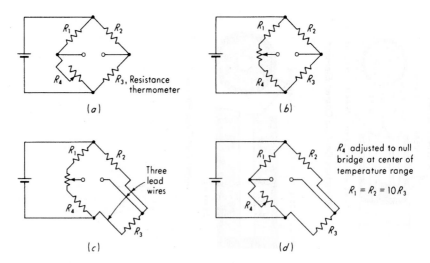

Figure 8.21
Resistance-thermometer bridge circuits.

microcircuit manufacturing to monitor and control critical processes.[60] Using a 1-μm wire resistance element, gas temperature fluctuations of up to 3 kHz can be measured, either for their own sake or as compensation for hot-wire flow sensors.[61]

Bridge circuits used with resistance temperature sensors may employ either the deflection mode of operation or the null (manually or automatically balanced) mode. If the null method is used, resistor R_4 in Fig. 8.21a is varied until balance is achieved. When the highest accuracy is required, the arrangement of Fig. 8.21b is preferred since the (variable and unknown) contact resistance in the adjustable resistor has no influence on the resistance of the bridge legs. If long lead wires subjected to temperature variations are unavoidable, errors due to their resistance changes may be canceled by use of the Siemens three-lead circuit of Fig. 8.21c. Three lead wires of identical length and material exhibit identical resistance variations, and since one of these leads is in each of legs 2 and 3, their resistance changes cancel. Resistance change in the third wire has no effect on bridge balance since it is in the null detector circuit for null-mode operation. For deflection operation, its effect is negligible if the indicating instrument draws little current.

While the resistance/temperature variation of the sensing element may be quite linear, the output voltage signal of a bridge used in the deflection mode is not necessarily linear for large percentage changes in resistance. Unlike the case of strain gages, the resistance change of resistance thermometers for full-scale deflection may be quite large. Typically, a 500-Ω platinum element may exhibit a 100-Ω

[60]J. Parker and W. Renken, "Temperature Metrology for CD Control in DUV Lithography," *Semiconductor International,* Sept. 1997, pp. 111–116.

[61]Model 90C20, Dantec Measurement Technology (www.dantecmt.com).

change over its design range. For a bridge with four equal arms, this would cause severe nonlinearity; however, by making the fixed arms R_1, R_2 of considerably higher resistance (say 10:1) than R_3 and R_4 and by balancing the bridge at the middle of the temperature range (rather than at one end), good linearity may be achieved (Fig. 8.21d). Typically,[62] a platinum element covering a range from 0 to 100°C using the 10:1 resistance ratio mentioned above gives a nonlinearity of only 0.5C°. For nickel elements, whose resistance/temperature variation is quite nonlinear, this nonlinearity and the bridge nonlinearity can be made to nearly cancel by proper design[63] since the two effects are of opposite directions. Commercially available[64] linear bridge modules requiring only ac power input and connection to a 100-Ω platinum probe (the most common resistance value) provide total accuracy (including linearity) of ± 0.1 percent of span for standard spans of -100 to $+500$°F, 0 to $+1000$°F, -100 to $+200$°C, 0 to $+200$°C, and 0 to $+500$°C. A linear dc output of 1 mV/degree with 0 mV at 0°F or 0°C can be sent directly to data acquisition systems, reducing their computation load since no software linearization is required. Two such bridges also may be directly connected to measure differential temperature between two probes.

Resistance-thermometer bridges may be excited with either ac or dc voltages. The direct or rms alternating current through the thermometer is usually in the range 2 to 20 mA. This current causes an I^2R heating which raises the temperature of the thermometer above its surroundings, causing the so-called self-heating error. The magnitude of this error depends also on heat-transfer conditions and usually is quite small. A 450-Ω platinum element of open construction carrying 25-mA current has a self-heating error of 0.2F° when immersed in liquid oxygen. Actually, by using an unsymmetric pulse type of excitation voltage whose rms (heating) value is small compared with its peak value, quite large instantaneous currents (and thus large peak output voltages) may be obtained without significant self-heating. Such pulse-excitation voltages (Fig. 8.22) can be obtained by commutating a dc source; this also allows timesharing of the bridge among several resistance sensors. As much as a 5-V full-scale bridge output signal can be obtained from resistance sensors by using this technique or others.[65]

While bridge techniques are the classical method for conditioning RTD sensors, the four-wire "ohmmeter" technique of Fig. 8.23 is widely employed in digital thermometers and data acquisition systems, where sensor nonlinearity is corrected in the computer software. Since a precise *current* source (usually a few milliamperes) is utilized, resistance changes in these two lead wires have no effect on sensor current, while high voltmeter input impedance (typically 200 MΩ) makes current in its two lead wires (and thus these lead-wire resistance errors) negligible.

[62]"Temperature Recording from Platinum Resistance Sensors," Brush Instruments, Cleveland, Ohio.

[63]D. R. Mack, "Linearizing the Output of Resistance Temperature Gages," *Exp. Mech.*, p. 122, April 1961.

[64]Linear Bridge Model 414L, Rosemount Inc., Minneapolis, MN.

[65]*Bull.* 9612, Rosemount Inc.

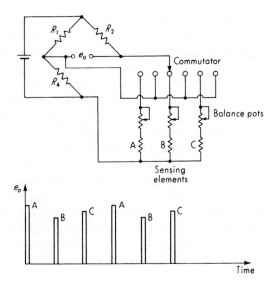

Figure 8.22
Pulse-excitation technique.

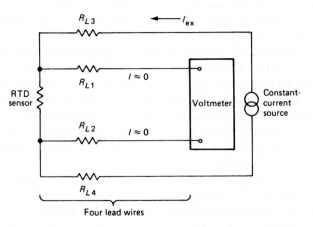

Figure 8.23
Four-wire ohmmeter technique.

Resistance-thermometer elements range in resistance from about 10 to as high as 25,000 Ω. Higher-resistance elements are less affected by lead-wire and contact resistance variations, and since they generally produce large voltage signals, spurious thermoelectric emf's due to joining of dissimilar metals also are usually negligible. Platinum is used from -450 to $1850°F$, copper from -320 to $500°F$, nickel from -320 to $800°F$, and tungsten from -450 to $2000°F$. Average temperatures may be measured by using resistance thermometers, as in Fig. 8.24a, while differential

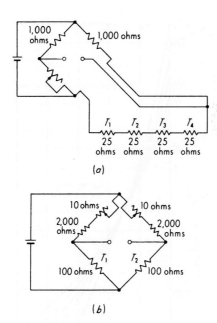

Figure 8.24
Average- and differential-temperature sensing.

temperature is sensed by the arrangement of Fig. 8.24b.[66] Differential-temperature measurements to an accuracy of 0.05C° have been accomplished in a nuclear-reactor-coolant heat-rise application.[67]

Bulk Semiconductor Sensors (Thermistors)[68]

The earlier types of bulk semiconductor resistance temperature sensors were made of manganese, nickel, and cobalt oxides which were milled, mixed in proper proportions with binders, pressed into the desired shape, and sintered. These were given the name *thermistor* and are in wide use today. Compared with conductor-type sensors (which have a small positive temperature coefficient), thermistors have a very large negative coefficient. While some conductors (copper, platinum, tungsten) are quite linear, thermistors are very nonlinear. Their resistance/temperature relation is generally of the form

$$R = R_0 e^{\beta(1/T - 1/T_0)} \tag{8.16}$$

where $R \triangleq$ resistance at temperature T, Ω

$R_0 \triangleq$ resistance at temperature T_0, Ω

---

[66]"Temperature Recording from Platinum Resistance Sensors," op. cit.

[67]B. G. Kitchen, "Precise Measurement of Process Temperature Differences," *ISA J.*, p. 39, February 1959.

[68]H. B. Sachse, "Semiconductor Temperature Sensors," Wiley, New York, 1975; G. Lavenuta, "Negative Temperature Coefficient Thermistors," *Sensors*, May–Sept. 1997 (5 parts).

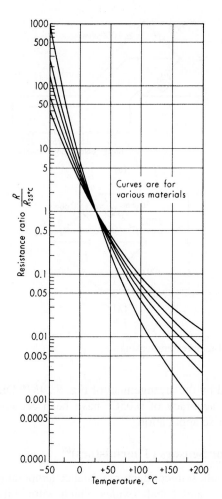

Figure 8.25
Thermistor resistance temperature curves.

$$\beta \triangleq \text{constant, characteristic of material, K}$$
$$e \triangleq \text{base of natural log}$$
$$T, T_0 \triangleq \text{absolute temperatures, K}$$

The reference temperature T_0 is generally taken as 298 K (25°C) while the constant β is of the order of 4,000. Computing $(dR/dT)/R$, we find the temperature coefficient of resistance to be given by $-\beta/T^2 \ \Omega/(\Omega \cdot C°)$. If β is taken as 4,000, the temperature coefficient at room temperature (25°C) is -0.045, compared with $+0.0036$ for platinum. While the exact resistance/temperature relation varies somewhat with the particular material used and the configuration of the resistance element, Fig. 8.25 shows the general type of curve to be expected.

Thermistors are available commercially in the form of beads, flakes, rods, and disks. However, most temperature-sensing applications will employ ready-made

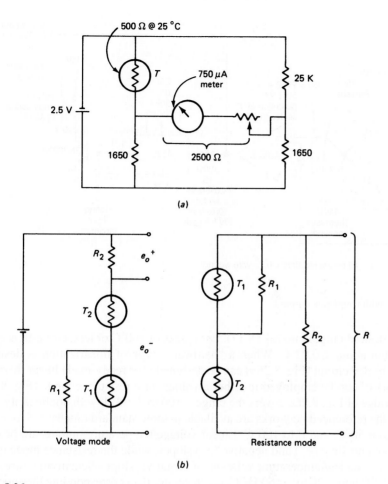

(a)

(b)

Figure 8.26
Thermistor linearization networks.

probes. Resistance at 25°C can vary over a wide range, from 500 Ω to several meg-ohms. The usable temperature range is from about −200 to +1000°C; however, a single thermistor cannot be used over such a large range. When the thermistor is utilized with a computerized data system, absolute temperature T (in kelvins) can be computed from measured resistance R (in ohms) from[69]

$$\frac{1}{T} = A + B \ln R + C (\ln R)^3 \qquad (8.17)$$

Constants A, B, and C are gotten by solving three simultaneous equations obtained by substituting three pairs of known R, T points (at the low, middle, and high ends

[69]"Practical Temperature Measurements," *Appl. Note* 290, p. 20, Hewlett-Packard Corp., Palo Alto, CA, 1980.

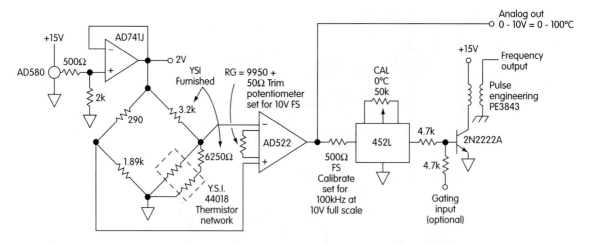

All bridge resistors = 0.1% wire wound

Figure 8.27
Linearized thermistor network with frequency output.

of the desired range) into Eq. (8.17). For ranges of 100°C or less, curve-fit accuracy is within about ±0.02°C. When a "hardware" type of linearization is desired, a linear bridge circuit (Fig. 8.26a) can be designed,[70] or ready-made linear thermistor networks[71] can be employed in either a voltage or a resistance mode (Fig. 8.26b). The bridge of Fig. 8.26a covers the range −100 to +100°F with nonlinearity ±1°F, while the referenced networks are available in three standard ranges: −5 to +45°C, −30 to +50°C, and 0 to 100°C. Linear voltage/temperature outputs are provided with both positive (e_o^+) and negative (e_o^-) slopes, while the resistance mode gives a linear resistance/temperature relation of negative slope. Sensitivities are about ±6 mV/°C and −20 to −150 Ω/°C, much greater than corresponding thermocouple or RTD values. Maximum nonlinearities are the order of ±0.06 to ±0.5°C. Units can be packaged in various ways; one form is a 9 × 9 × 3 mm epoxy molded unit. Figure 8.27[72] shows a linearized thermistor network used in a bridge circuit, with output to a voltage-to-frequency converter, to give a temperature-proportional frequency output with 0.001°C resolution. This gives noise-tolerant signal transmission and easy digital readout on a frequency counter.

Other bulk semiconductor temperature sensors include carbon resistors and silicon[73] and germanium[74] crystal elements. Carbon resistors are merely the

[70]"Thermistors," *Appl. Note* AN-1A, Fenwal Electronics, Pawtucket, RI, 401-727-1300 (www.fenwal.com).

[71]Linear Thermistor Networks, Fenwal Electronics; YSI 4800 LC, YSI, Dayton, OH, 800-747-5367 (www.ysi.com).

[72]D. H. Sheingold (ed.), op. cit., p. 150.

[73]J. R. Pies, "A New Semiconductor for Temperature Measuring," *ISA J.*, p. 50, August 1959.

[74]J. S. Blakemore, "Germanium for Low-Temp Resistance Thermometry," *Instrum. Contr. Syst.*, p. 94, May 1962.

commercial carbon-composition elements commonly used as resistance elements in radios and other electronic circuitry. The 0.1 to 1 W rated resistors with room-temperature resistance of 2 to 150 Ω are used widely for the measurement of cryogenic temperatures in the range 1 to 20 K. From about 20 K downward, these elements exhibit a large increase in resistance with a decrease in temperature given by the relation[75]

$$\log R + \frac{K}{\log R} = A + \frac{B}{T} \qquad (8.18)$$

where R is the resistance at Kelvin temperature T and A, B, and K are constants determined by calibration of the individual resistor. Reproducibility of the order of 0.2 percent is obtained in the range 1 to 20 K.

Silicon, with varying amounts of boron impurities, can be designed to have either a positive[76] or negative temperature coefficient over a particular temperature range. The resistance/temperature relation is quite nonlinear. A typical element shows a resistance change (from the nominal value at 25°C) of -80 percent at -150°C to $+180$ percent at $+200$°C. The temperature coefficient near room temperature is of the order of $+0.7$ percent/C°. Germanium, doped with arsenic, gallium, or antimony, is employed for cryogenic temperatures, where it exhibits a large decrease in resistance with increasing temperature. The relation is quite nonlinear but very reproducible, giving precise measurements within 0.001 to 0.0001 K near 4 K when adequate care is taken in technique. Commercially available elements cover a range from about 0.5 to 100 K, a typical unit changing resistance from 7,000 Ω at 2 K to 6 Ω at 60 K. A wide variety of sensors (including capacitance types useful in strong magnetic fields) plus measurement and control electronics are available from firms specializing in cryogenic measurements.[77]

8.5 JUNCTION SEMICONDUCTOR SENSORS

Junction semiconductor devices such as diodes[78] and transistors[79] exhibit temperature sensitivities which can be exploited as temperature sensors. Here we give details on a commercially available transistor-based device[80] (Fig. 8.28) that is

---

[75]L. G. Rubin, "Temperature," *Electron. Progr.,* p. 1, Raytheon Corp., Autumn 1963.

[76]W. Tarpley, "Temperature-Resistance Characteristics of TSP102 Silicon Thermistor," *Bull.* CA-195, Texas Instruments, Dallas, 1978; "Silicon Temperature Sensors," *Tech. Publ.* 9398 042 60041, Amperex Elect. Corp., Slatersville, RI, 1989.

[77]Lakeshore Cryotronics, Westerville, OH, 614-891-2244 (www.lakeshore.com); J. G. Weisend II, Handbook of Cryogenic Engineering, Taylor and Francis, Philadelphia, PA, 1998, chap. 4.

[78]R. W. Treharne and J. A. Riley, "A Linear-Response Diode Temperature Sensor," *Inst. Tech.,* pp. 59–61, June 1978; R. A. Pease, "Using Semiconductor Sensors for Linear Thermometers," *Inst. & Cont. Syst.,* pp. 80–81, June 1972; H. B. Sachse, op. cit.; M. F. Estes and D. Zimmer, Jr., "New Temperature Probe Locates Circuit Hot Spots," *Hewlett-Packard J.,* pp. 30–32, March 1981.

[79]H. B. Sachse, op. cit.

[80]"Two-Terminal IC Temperature Transducer AD590," *Publ.* C426C-5, Analog Devices, Norwood, MA, 1979.

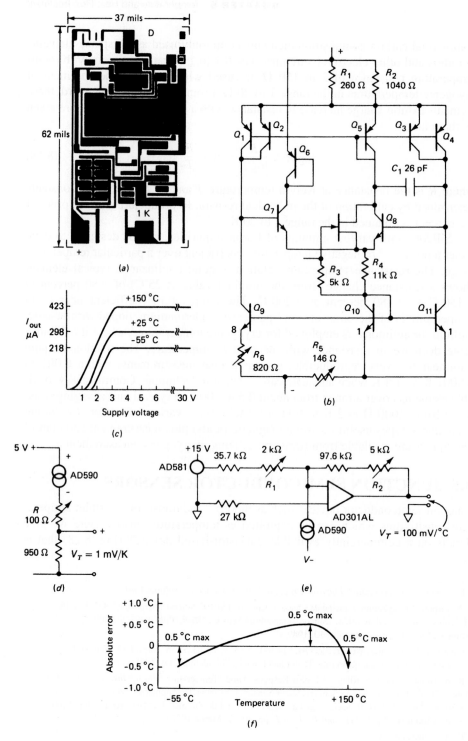

Figure 8.28
Junction semiconductor temperature sensor.

produced as an integrated-circuit (IC) chip and available in chip form, in small metal can or ceramic flat-pack packages, or in ready-made probes. The metallization diagram (Fig. 8.28a) of the chip shows the small size of the basic sensor, while Fig. 8.28b shows internal details. It is a fundamental property of silicon transistors that if two identical transistors are operated at a constant ratio r of collector-current densities, then the difference in their base-emitter voltages is $(kT/q) \ln r$. Since Boltzman's constant k and the charge q of an electron are constants, the resulting voltage is directly proportional to absolute temperature T. In the referenced device, transistors Q_8 and Q_{11} produce the voltage, resistors R_5 and R_6 convert it to a current, and Q_{10} (whose collector current tracks those in Q_9 and Q_{11}) supplies all the bias and substrate leakage current for the rest of the circuit, forcing the total current (flowing between plus and minus power supply terminals) to be proportional to temperature. Resistors R_5 and R_6 are laser-trimmed on the wafer to calibrate each device to give $1\ \mu A/°C$ at 25°C.

The simplest application circuit would require only an unregulated supply of 4 to 30 V, a microammeter, and the device, all connected in series. (Figure 8.28c shows the insensitivity to supply-voltage drifts; between 5 and 15 V, the effect is only 0.2 $\mu A/V$.) For such a circuit and a design range of -55 to $+155°C$, the accuracy (including nonlinearity) of the medium-grade device (five accuracy/price grades are available) would be about $\pm 4.2°C$. This basic device accuracy can be improved by adding "trim" circuitry and calibrating at known temperatures. Figure 8.28d shows a simple one-temperature trim circuit (which also gives a voltage, rather than current output) that improves accuracy to $\pm 1.5°C$. The two-temperature trim circuit of Fig. 8.28e (which also provides some amplification) gives the curve of Fig. 8.28f (accuracy $\pm 0.5°C$). (The AD581 is a precision reference voltage supply; AD301AL is an op-amp.) Just as in other electric sensors, self-heating effects are present and should be checked. For example, the metal-can package at 25°C in a stirred oil bath has a 0.06°C rise, while in still air this would be 0.72°C. Note that if the trim circuit techniques mentioned above are employed with calibration heat-transfer conditions the same as actual application conditions, self-heating causes no error.

The current-source nature of the device allows measurement of average temperature by using parallel connection of multiple sensors, and selection of minimum temperature by utilizing series connection (Fig. 8.29a). The op-amp circuit of Fig. 8.29b gives a sensitive indication of differential temperature, while Fig. 8.29c shows application to a thermocouple reference-junction compensator of the "hardware" type (AD580 is a precision reference voltage). Junction semiconductor sensors also are widely utilized in "software" reference-junction compensation. Here the sensor measures reference-junction temperature and sends a proportional voltage to the data acquisition system computer, where measuring-junction temperature is corrected digitally.

The main advantages of junction semiconductor temperature sensors are linearity, simple external circuitry, and good sensitivity. However, their upper range is restricted to about 200°C by the damage limits of silicon transistors.

726

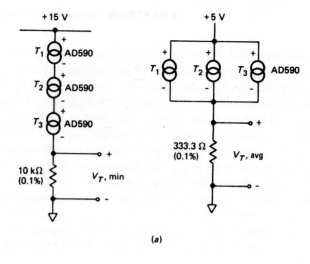

(a)

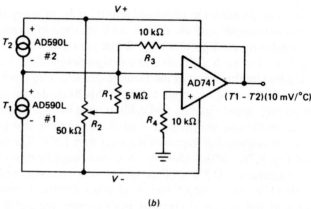

(b)

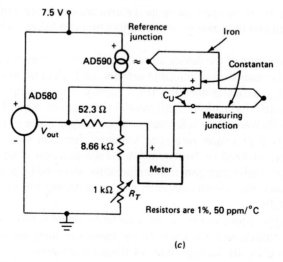

(c)

Figure 8.29
Junction semiconductor sensor applications.

8.6 DIGITAL THERMOMETERS

In most applications, the nonlinearity of thermocouples and RTDs is sufficiently great that we must use calibration tables (or curve-fit formulas derived from them) to convert voltage or resistance measurements to the corresponding temperatures. When many measurements are to be made, this is inconvenient, time-consuming, and prone to error. Also, with thermocouples (TCs), the reference-junction temperature must be accounted for. Because of these considerations (which are *not* present in most other sensors), the fact that temperature is industry's most measured quantity, and the fact that TCs and RTDs are popular, special-purpose digital voltmeters called *digital thermometers* have been developed and found a ready market.

Thermometers for RTDs generally use four-wire ohmmeter (rather than bridge) resistance measurement methods and provide the needed current source excitation (two- and three-wire sensors can be accommodated, but at reduced accuracy). Linearization methods for both TC and RTD thermometers can be strictly analog, nonmicroprocessor digital, or microprocessor digital (the most common). Simple units provide linearization for only a single type of TC (or RTD), while microprocessor linearization techniques allow switch selection from among six types and Fahrenheit or Celsius readout in more sophisticated instruments. Reference-junction compensation in TC thermometers usually uses an isothermal block whose temperature is monitored by a junction semiconductor sensor, which sends these data to the same microprocessor that does the linearization. Use of dual-slope analog/digital conversion for low noise and high resolution (RTDs: 0.01° low range, 0.1° high range; TCs: 0.1°) plus microprocessor computing lags, limits sampling rates to two or three readings per second. *Instrument* accuracies are consistent with the quoted resolutions; however, accuracy of the actual *temperature* measurement will be considerably less because of deviations of the individual TC or RTD from published tables and the usual problems of sensing any physical variable precisely. Sharing of a single digital thermometer among many TCs or RTDs is possible by using scanner accessories. Since digital information is already available, interfacing to printers, computers, etc., is also convenient.

8.7 RADIATION METHODS[81]

All the temperature-measuring methods discussed up to this point require that the "thermometer" be brought into physical contact with the body whose temperature is to be measured. Also, except for the pulsed thermocouple of Fig. 8.14, the temperature sensor generally is intended to assume the same temperature as the body being measured. This means that the thermometer must be capable of withstanding this temperature, which in the case of very hot bodies presents real problems, since the thermometer may actually melt at the high temperature required. Also, for bodies that are moving, a noncontacting means of temperature sensing is most convenient.

---

[81]R. Vanzetti, "Practical Applications of Infrared Techniques," Wiley, New York, 1972; J. C. Richmond and D. P. Dewitt, "Applications of Radiation Thermometry," *ASTM* TP 895, 1985.

Furthermore, if we wish to determine the temperature variations over the surface of an object, a noncontacting device can readily be "scanned" over the surface. Radiation thermometers discussed in this section are widely used in manufacturing control systems in many industries. Footnote 81 (Richmond and Dewitt) gives a particularly useful discussion of the selection and application of a radiation thermometer for a computer-based system.[82]

To solve problems of the type mentioned above, a variety of instruments based in one way or another on the sensing of radiation have been devised. These might, in general, be called radiometers; however, common usage employs terms such as *radiation pyrometer, radiation thermometer, optical pyrometer,* etc., to describe a particular type of instrument. Since this terminology is not standardized, you must inquire into the basic operating principle of a given instrument to be sure what its characteristics are, rather than relying on the name given to it. Instruments can be usefully classified into those that are intended to measure temperature at a "point" and those that give a temperature "map" over an extended area (imagers). Imagers can be further divided into "scanning" and "staring" types.

Other important applications of infrared radiation include night vision, missile guidance, and infrared spectroscopy. In missile guidance (the Sidewinder[83] missile is an outstanding example), the missile is designed to "home" on the infrared radiation emitted by the hot jet exhaust of the target aircraft's engine. Infrared spectroscopy[84] is used for the analysis of gases, liquids, and solids, identifying and determining the concentration of molecules or molecular groups.

Radiation Fundamentals

Radiation-type temperature sensors operate with electromagnetic radiation whose wavelengths lie in the visible and infrared portions of the spectrum. The visible spectrum is quite narrow: 0.3 to 0.72 μm. The infrared spectrum generally is defined as the range from 0.72 to about 1,000 μm. Bordering the visible spectrum on the low-wavelength side are the ultraviolet rays, while microwaves border the infrared spectrum on the high side. Radiation-type temperature sensing devices utilize mainly some part of the range 0.3 to 40 μm.

Physical bodies (solids, liquids, gases) may emit electromagnetic radiation or subatomic particles for a number of reasons. As far as temperature sensing is concerned, we need be concerned with only that part of the radiation caused solely by temperature. Every body above absolute zero in temperature emits radiation dependent on its temperature. The ideal thermal radiator is called a *blackbody*. Such

---

[82]R. A. Holman, "Mold Temperature Measurement for Glass-Pressing Processes," *ASTM* STP 895, pp. 67–73, 1985.

[83]R. Westrum, "Sidewinder: Creative Missile Development at China Lake," Naval Inst. Press, Annapolis, MD, 1999. A fascinating story of how creative engineers/physicists working at a remote base in the Mojave Desert overcame bureaucracy to develop the world's best air-to-air missile.

[84]P. R. Griffiths and J. A. de Haseth, "Fourier Transform Infrared Spectroscopy," Wiley, New York, 1986; G. J. Dixon, "Compact FT-IR Spectrometers Address Real-World Problems," *Laser Focus World,* Feb. 1998, pp. 111–118.

a body would absorb completely any radiation falling on it, and for a given temperature, emit the maximum amount of thermal radiation possible. The law governing this ideal type of radiation is Planck's law, which states that

$$W_\lambda = \frac{C_1}{\lambda^5(e^{C_2/(\lambda T)} - 1)} \tag{8.19}$$

where $W_\lambda \triangleq$ hemispherical spectral radiant intensity, W/(cm$^2 \cdot \mu$m)
$C_1 \triangleq$ 37,413, W $\cdot \mu$m^4/cm^2
$C_2 \triangleq$ 14,388, μm $\cdot$ K
$\lambda \triangleq$ wavelength of radiation, μm
$T \triangleq$ absolute temperature of blackbody, K

The quantity W_λ is the amount of radiation emitted from a flat surface into a hemisphere, per unit wavelength, at the wavelength λ. Equation (8.19) thus gives the distribution of radiant intensity with wavelength. That is, a blackbody at a certain temperature emits *some* radiation per unit wavelength at every wavelength from zero to infinity, but not the same amount at each wavelength. Figure 8.30*a* shows the curves obtained from Eq. (8.19) by fixing T at various values and plotting W_λ versus λ. The curves exhibit peaks at particular wavelengths, and the peaks occur at longer wavelengths as the temperature decreases. The area under each curve is the total emitted power and increases rapidly with temperature. Equations giving the peak wavelength λ_p and the total power W_t are

$$\lambda_p = \frac{2,891}{T} \quad \mu\text{m} \tag{8.20}$$

and
$$W_t = 5.67 \times 10^{-12} T^4 \quad \text{W/cm}^2 \tag{8.21}$$

Figure 8.30*b* shows the wavelength range over which 90 percent of the total power is found for various temperatures. Note that lower temperatures require measurement out to longer wavelengths. In Fig. 8.30*a*, note also, say, for the 600 K and 650 K curves, that the vertical distance between the two curves is much greater at short wavelengths, such as 2 μm, than at long wavelengths, such as 12 μm. This means that the radiation changes much more, for a given temperature change, at short wavelengths, making instruments that operate at such wavelengths more sensitive.

While the concept of a blackbody is a mathematical abstraction, real physical bodies can be constructed to approximate closely blackbody behavior. Such radiation sources are needed for calibration of radiation thermometers and generally take the form of a blackened conical cavity of about 15° cone angle. The temperature is adjustable, automatically controlled for constancy, and measured by some accurate sensor such as a platinum resistance thermometer. A family of calibrators[85] covers the range −20 to 3000°C with emissivity as good as 0.999, +0.0005, −0.0001,

---

[85]M300 Series, Mikron Instrument Co., Oakland, NJ, 800-631-0176 (www.mikroninst.com); J. Merchant, "Blackbody Calibration Sources Function as Standards," *Laser Focus World*, Apr. 1995, pp. 105–110.

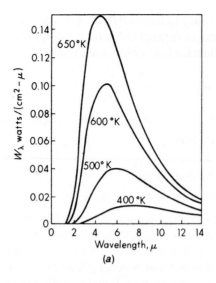

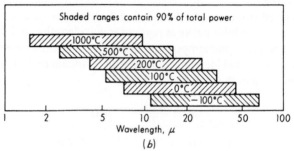

Figure 8.30
Blackbody radiation.

compared to the perfect blackbody which has 1.000. Accuracies vary with the model, but are typically about ±0.25% of reading, ±1°C. Model M380 uses metal freezing-points to establish certain discrete temperatures with uncertainty in the range 0.05 to 0.50°C. While it is possible to construct a nearly perfect blackbody, the bodies whose temperatures are to be measured with some radiation-type instrument often deviate considerably from such ideal conditions. The deviation from blackbody radiation is expressed in terms of the emittance of the measured body.

Several types of emittance have been defined to suit particular applications. The most fundamental form of emittance is the *hemispherical spectral emittance* $\varepsilon_{\lambda, T}$. Let us call $W_{\lambda a}$ the *actual* hemispherical spectral radiant intensity of a real body at temperature T, and let us assume it can be measured (by using optical bandpass filters). Then $\varepsilon_{\lambda, T}$ is defined as

$$\varepsilon_{\lambda, T} \triangleq \frac{W_{\lambda a}}{W_{\lambda}} \tag{8.22}$$

where W_λ is the blackbody intensity at temperature T. Thus emittance is dimensionless and always less than 1.0 for real bodies. In the most general case, it varies with both λ and T. With the definition of Eq. (8.22), the radiation from a real body may be written as

$$W_{\lambda a} = \frac{C_1 \varepsilon_{\lambda, T}}{\lambda^5 (e^{C_2/(\lambda T)} - 1)} \tag{8.23}$$

Similarly, the total power W_{ta} of an actual body is given by

$$W_{ta} = C_1 \int_0^\infty \frac{\varepsilon_{\lambda, T}}{\lambda^5 (e^{C_2/(\lambda T)} - 1)} \, d\lambda \tag{8.24}$$

and if we assume that W_{ta} can be measured experimentally, we may define the *hemispherical total emittance* $\varepsilon_{t, T}$ by

$$\varepsilon_{t, T} \triangleq \frac{W_{ta}}{W_t} \tag{8.25}$$

where W_t is the blackbody total power at temperature T. Thus if $\varepsilon_{t, T}$ is known, the total power of a real body is given by

$$W_{ta} = 5.67 \times 10^{-12} \, \varepsilon_{t, T} \, T^4 \qquad \text{W/cm}^2$$

If a body has $\varepsilon_{\lambda, T}$ equal to a constant for all λ and at a given T, it is called a *graybody*. In this case we see that $\varepsilon_{\lambda, T} \equiv \varepsilon_{t, T}$. Also the curves of $W_{\lambda a}$ versus λ have exactly the same shape as for W_λ. Since many radiation thermometers operate in a restricted band of wavelengths, the *hemispherical band emittance* $\varepsilon_{b, T}$ has been defined by

$$\varepsilon_{b, T} \triangleq \frac{\displaystyle\int_{\lambda_a}^{\lambda_b} \{ \varepsilon_{\lambda, T} / [\lambda^5 (e^{C_2/(\lambda T)} - 1)] \} \, d\tau}{\displaystyle\int_{\lambda_a}^{\lambda_b} \{ 1 / [\lambda^5 (e^{C_2/(\lambda T)} - 1)] \} \, d\tau} \tag{8.26}$$

This is seen to be just the ratio of the total powers, actual and blackbody, within the wavelength interval λ_a to λ_b for bodies at temperature T. If the actual power can be measured directly, $\varepsilon_{b, T}$ can be found without knowing $\varepsilon_{\lambda, T}$. For a graybody, $\varepsilon_{b, T} \equiv \varepsilon_{\lambda, T}$. When we shortly discuss complete instruments for radiation thermometry, we will find some are designed to measure at one specific wavelength, others for a specific *band* of wavelengths, and still others for "all" wavelengths. This wavelength "tailoring" shapes the instrument's readings to suit a particular class of applications and is accomplished by the designer's choice of detector, lens material, and optical filters. It also means that any emittance value that we might apply to correct for nonblackbody conditions must be the appropriate *type* of emittance for a particular instrument.

Sometimes it is possible to *coat* the target surface with a "black" material of known and high emissivity without interfering with its other functions. One application[86] used this method to facilitate temperature imaging of hypersonic nose cones and wing structures in wind tunnel tests. An extensive database of emittance values for black, white, reflective, and transmissive materials is available.[87]

If a radiation thermometer has been calibrated against a blackbody source, knowledge of the appropriate emittance value allows correction of its readings for nonblackbody measurements. Unfortunately, emittances are not simple material properties such as densities, but rather depend on size, shape, surface roughness, angle of viewing, etc. This leads to uncertainty in the numerical values of emittances, which is one of the main problems in radiation temperature measurement. Real-world objects are characterized by emittance ε, reflectance r, and transmittance t, which are related by

$$\varepsilon = 1 - r - t \qquad (8.27)$$

When r and/or t is not zero, we have nonblackbody behavior, and measurement errors as in Fig. 8.31a are possible. Commercial radiation thermometers usually include an emittance adjustment with a range from about 0.2 to 1.0. Thus if material emittance is known, it can be compensated for. The most reliable technique for determining emittance for such purposes requires that at some time we be able to measure specimen temperature (simultaneous with the radiation measurement) by some independent means, such as a thermocouple. Since in this method we are using the *instrument itself* to determine the emittance, there is no question that we are using the correct *type* of emittance. The radiometer emittance dial is adjusted until the temperatures indicated by the two instruments agree. If there are no changes in conditions, this emittance setting can be used for any future radiation temperature measurements without, of course, requiring the presence of the thermocouple. Since emittance can vary with temperature, such an emittance calibration may be necessary over the desired temperature range. In such a case, a calibration in which the ε dial is left fixed at some arbitrary value, and a table of indicated versus true temperatures is developed, might be more useful for correction purposes. One commercial instrument[88] measures (and corrects for) target emittance by impinging a laser beam on the target and measuring the fraction of energy reflected.

Plastic film manufacture[89] makes wide use of radiation temperature measurement and is a good example for emittance considerations based on Eq. (8.27). Reflectance r is about 0.04 for all plastics in the infrared range and is independent of film thickness. Transmittance t is given for cellulose acetate film (as measured by a spectrophotometer) in Fig. 8.32. Note that in the range 8 to 10 μm, film of any thickness between 0.001 and 0.1 in behaves almost as a blackbody; thus we should design our radiometer for sensitivity to only this range.

---

[86]H. Kaplan, "Black Coatings Are Critical in Optical Design," *Photonics Spectra,* Jan. 1997, pp. 48, 49.

[87]Stellar Optics Research International Corp. (SORIC), Thornhill, Ont., Canada, 905-731-6088; S. McCall, "Engine Black Paint," *Photonics Spectra,* Jan. 1996, p. 12.

[88]The Pyrometer Instrument Co., Northvale, NJ, 201-768-2000 (www.pyrometer.com).

[89]Plastic Film Measurement, TN100, Ircon Inc., Niles, IL, 847-967-5151 (www.ircon.com).

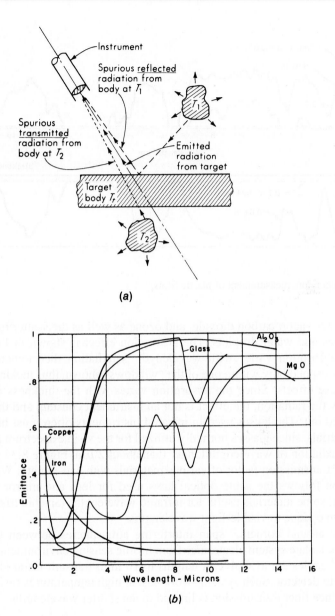

(a)

(b)

Figure 8.31
Measurement problems with reflective/translucent targets; emittance of real materials.

Another source of error is the losses of energy in transmitting the radiation from the measured object to the radiation detector. Generally the optical path consists of some gas (often atmospheric air) and various windows, lenses, or mirrors used to focus the radiation or protect sensitive elements from the environment. In atmospheric air the attenuation of radiation is due mainly to the resonance-absorption

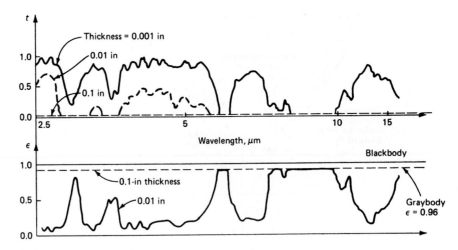

Figure 8.32
Radiation temperature measurement of plastic films.

bands of water vapor, carbon dioxide, and ozone as well as the scattering effect of dust particles and water droplets. The absorption effect is shown in Fig. 8.33.[90] Since the absorption varies with wavelength, a radiation thermometer can be designed to respond only within one of the "windows" shown, thus making it insensitive to these effects. Since the absorption varies with the thickness of the gas traversed by the radiation, the effect is not an instrument constant and thus cannot be calibrated out. The lenses used in infrared instruments often must be made of special materials, since glasses normally utilized for the visible spectrum are almost opaque to radiation of wavelength longer than about 2 μm. Figure 8.34 shows the variation of transmission factor of various materials with wavelength. While infrared radiation follows the same optical laws used for lens and mirror design as visible light, some materials useful for infrared wavelengths (arsenic trisulfide, for example) are opaque to visible-light wavelengths.

To "see around corners," span interfering atmospheres between target and detector, and isolate system electronics from hostile sensing environments, flexible fiber optic[91] cables are available in lengths to 10 m to transmit infrared radiation from target to detector. Such systems work best at high temperatures (using silicon detectors) since fiber transmission is limited to the shorter wavelengths.

Radiation Detectors: Thermal and Photon

In all radiation thermometers (other than the disappearing-filament optical pyrometer), the radiation from the measured body is focused on some sort of *radiation*

[90]Ircon Inc.

[91]O. W. Uguccini and F. G. Pollack, "High-Resolution Surface Temperature Measurements on Rotating Turbine Blades with an Infrared Pyrometer," *NASA* TN D-8213, 1976.

Transmission for atmosphere at 80°F, and several relative humidities

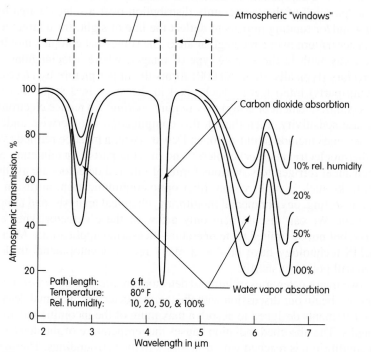

Figure 8.33
Atmospheric absorption.

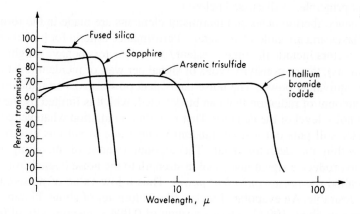

Figure 8.34
Spectral transmission of optical material.

detector,[92] which produces an electrical signal in response to the incoming radiant flux. Detectors are made as single elements and also as arrays. A single-element detector can be used to measure temperature "at a point" (really the *average* over

---

[92]J. D. Vincent, "Fundamentals of Infrared Detector Operation and Testing," Wiley, New York, 1990.

some area) or as a component in a *scanned* imaging instrument. Scanned imagers develop a "picture" of the temperature distribution over a selected area by using oscillating and/or rotating mirrors to deflect the line of sight, from target to detector, in a raster pattern over a rectangular area. The somewhat complex moving parts are done away with in the *staring* type of imager, where a rectangular array of microdetectors (typically about 80,000) forms the target picture by *electronically* (no moving parts) interrogating each pixel in the array in a sequential manner. Detectors (single-element or array) may require *cooling*[93] to reduce electronic noise and increase sensitivity, even though this complication is otherwise undesirable. Cooling schemes include liquid-nitrogen Dewars (which must be refilled every few hours), thermoelectric (Peltier) coolers[94] (no moving parts), or miniature Stirling-type mechanical refrigerators. Various forms of infrared detectors are available for different tasks such as night vision, fiber optic communication, intrusion alarms, agricultural sorting, environmental monitoring, chemical analysis, and temperature measurement. We can discuss here only a few of these detector types and will emphasize, but not be limited to, temperature-measuring applications. Microcircuit and MEMS technology have produced many recent developments in detectors, which we will point out shortly.

The two major categories of infrared detectors are *thermal detectors* and *photon detectors;* we begin our discussion with the thermal detectors. Thermal detectors are blackened elements designed to absorb a maximum of the incoming radiation at all wavelengths. The absorbed radiation causes the temperature of the detector to rise until an equilibrium is reached with heat losses to the surroundings. Thermal detectors usually measure this temperature, using a resistance thermometer, thermistor, or thermocouple (thermopile) principle. (*Pyroelectric* thermal detectors operate on a different principle, as discussed below.)

Resistance-thermometer and thermistor elements are made in the form of thin films or flakes and are called *bolometers*. Performance criteria for both thermal and photon detectors include the time constant (most detectors behave roughly as first-order systems), the responsivity (volts of signal per watt of incident radiation), and the noise-equivalent power. The noise-equivalent power gives an indication of the smallest amount of radiation that can be detected, which is limited by the inherent electrical noise level of the detector. That is, with no radiation whatever coming in, the detector still puts out a small random voltage due to various electrical noise sources within the detector itself. The amount (watts) of incoming radiation required to produce a signal just equal in strength to the noise (signal-to-noise ratio of 1) is called the *noise-equivalent power.* Thus a low value of noise-equivalent power is desirable. An evaporated-nickel-film bolometer of about 35 mm² area has a resistance of about 100 Ω, a time constant of 0.004 s, responsivity of 0.4 V/W, and noise-equivalent power of 3×10^{-9} W. A thermistor bolometer of 0.5 mm² area

[93]H. W. Messenger, "Microcooler Reduces Sizes of Thermal-Imaging Radiometers," *Laser Focus World,* July 1991, pp. 53–54; P. Finney, "Quantum Effect Infrared Sensing," *Sensors,* Dec. 1995, pp. 32–50.

[94]A. W. Allen, "Thermoelectrically Cooled IR Detectors Beat the Heat," *Laser Focus World,* Mar. 1997, pp. s15–s19.

might have 3-MΩ resistance, a time constant of 0.004 to 0.030 s, responsivity of 700 to 1,200 V/W, and noise-equivalent power of 2×10^{-10} W. Thermopiles of an area from 0.4 to 10 mm^2 have resistance of the order of 10 to 100 Ω, time constants in the range 0.005 to 0.3 s, responsivity of 3 to 90 V/W, and noise-equivalent power of 2×10^{-11} to 7×10^{-10}.

MEMS technology has produced thermopiles[95] with about 50 junctions in a 2×2 mm area, for use in point temperature-measuring instruments. For staring-type imaging systems, uncooled microbolometer arrays,[96] with 320×240 pixels in a sealed vacuum case, are achieved using integrated circuit and micro-machining techniques. Each 50×50 μm pixel has a vanadium oxide surface (emittance ≈ 0.8) that absorbs the incoming radiation, causing the surface to heat up. Vanadium oxide is a semiconductor material used as a thermistor, so its electrical resistance decreases with temperature, and typical thermistor circuitry is used to develop an output voltage. To keep the heat from quickly leaking off by conduction, the pixel is supported on thin silicon legs, designed to have a high thermal resistance (about 8×10^6 °C/W).

A class of thermal detector with quite different properties is the *pyroelectric,* useful over the spectral range of 0.001 (soft X rays) to 1,000 μm (far infrared). While not used much for temperature measurement, they have application in fire sensors, intruder alarms, gas analysis, and laser diagnostics. Because these are important applications, we will present some details even though they are not directly related to temperature sensing. These devices can have a wide spectral response (similar to thermopiles and bolometers), but do not suffer from the slow response speed characteristic of other thermal detectors. The basis of the improved speed lies in the fact that the electrical output depends on the time rate of change in detector temperature rather than on the detector temperature itself. Materials that exhibit the pyroelectric effect include lithium tantalate crystals, ceramic barium titanate, and polyvinyl fluoride plastic films.[97] Figure 8.35 shows spectral response for coated and uncoated lithium tantalate sensors.[98] Black absorbing coatings, which increase spectral uniformity and overall sensitivity, are used often with infrared radiation; however, they reduce speed of response by introducing a -10 dB/decade rolloff in amplitude ratio at high frequencies, with a typical breakpoint being at about 2,000 Hz.[99] When coatings are not employed, very fast

[95]W. Schmidt, "New Manufacturing Technology Improves Thermopile Sensors," *Laser Focus World,* Aug. 1995, pp. 77–79; Thin Film Thermopile Detectors, Meggitt Avionics, Manchester, NH, 800-842-4291 (www.meggittavi.com).

[96]J. Kreider et al., "Uncooled Infrared Arrays Sense Image Scenes," *Laser Focus World,* Aug. 1997, pp. 139–150; Thermacam PM 595, FLIR Systems, N. Billerica, MA, 800-464-6372 (www.flir.com).

[97]J. Cohen and S. Edelman, "Polymeric Pyroelectric Sensors for Fire Protection," *AFAPL Rept.* TR-74-16, Wright Patterson Air Force Base, OH, 1975.

[98]Molectron Detector Inc., Portland, OR, 800-366-4340 (www.molectron.com).

[99]W. M. Doyle, "A User's Guide to Pyroelectric Detection," *Electro-Opt. Syst. Des.,* pp. 12–17, November 1978; E. H. Putley, "The Pyroelectric Detector," chap. 6 in R. K. Willarson and A. C. Beer (eds.), "Semiconductors and Semimetals," vol. 5, Academic, New York, 1970; J. Cooper, "A Fast Response Pyroelectric Thermal Detector," *J. Sci. Instrum.,* vol. 39, pp. 467–472, 1962.

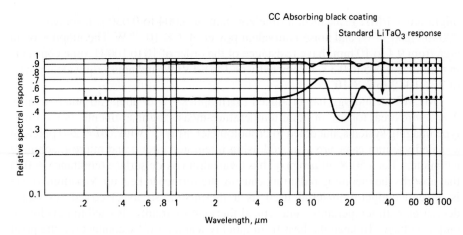

Figure 8.35
Wavelength response of pyroelectric detectors.

response (rise times on the order of 170 ps)[100] are possible since the electrical output can be shown[101] to respond to the instantaneous volume-averaged detector temperature and thus does not have to "wait" for the high surface temperature created by a fast radiation pulse to diffuse throughout the detector mass (a relatively *slow* thermal process).

In Fig. 8.36*a*, a pyroelectric crystal is shown with thin metal film electrodes, one of which faces the radiation (edge-type electrodes also are possible). In some cases face electrodes, while sometimes thin enough to be essentially transparent to the impinging radiation, may significantly effect spectral response and detector speed. Since pyroelectric materials are dielectrics, the electrode/crystal sandwich is a capacitor, in addition to being a radiation-sensitive charge generator. Pyroelectric materials exhibit a permanent electric polarization (charge) which is sensitive to temperature. Often a "poling" process which involves application of an external electric field at room or elevated temperature to obtain maximum alignment of the material's domains is employed to develop the strongest effect. Pyroelectric materials all exhibit a Curie temperature (610°C for lithium tantalate), above which the effect is lost and repolarization would be required.

Since connection of any measuring circuit to the sensor electrodes allows a flow of charge which ultimately neutralizes the polarization effect, the sensor, unlike thermopiles and bolometers, is not able to directly measure steady radiation (however, addition of a chopper system allows this capability). You may note that this behavior is analogous to that of the piezoelectric effect discussed in Chap. 4;

[100]W. B. Tiffany, "Commercial Applications of Pyroelectric Detectors," *Electro-Opt. Syst. Des.,* pp. 11–14, November 1976.

[101]P. A. Schlosser, "Investigation of the Radiation Imaging Properties of Pyroelectric Detector Arrays," Ph.D. diss., Nuclear Eng. Dept., The Ohio State University, Columbus, 1972.

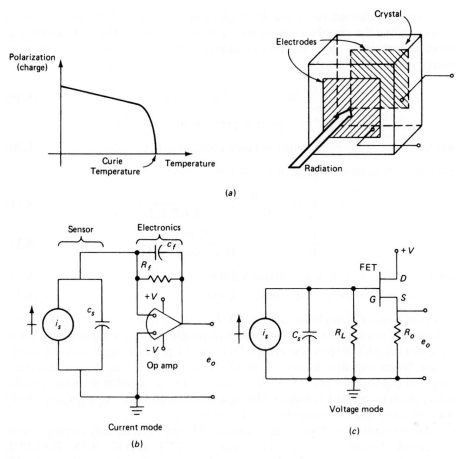

Figure 8.36
Pyroelectric detector and electronics.

and, in fact, one of the earliest pyroelectric materials studied (barium titanate) is employed also in piezoelectric applications. Both pyroelectric and piezoelectric behavior require the material to be ferroelectric; it must exhibit permanent electrical polarization. This piezoelectric response can cause spurious readings in pyroelectric detectors whose housings are subject to mechanical or acoustical excitation and even for nonvibrating housings when rapidly changing radiation input (such as short laser pulses) excites detector vibration.

For the linear portion of the polarization/temperature curve, charge q is proportional to volume-averaged detector temperature T, which gives sensor current i_s as

$$i_s = K_p \frac{dT}{dt} \tag{8.28}$$

where K_p is a sensitivity constant of the particular device. By modeling the thermal system as detector thermal capacitance C_t connected to a constant-temperature heat sink by a thermal resistance R_t, neglecting the coating dynamics mentioned earlier, calling the absorbed radiant power W (watts), and treating all variables as small perturbations away from an equilibrium operating point, conservation of energy gives

$$W\, dt - \frac{T - 0}{R_t}\, dt = C_t\, dT \qquad \textbf{(8.29)}$$

$$(\tau_t D + 1)T = R_t W$$

where $\qquad \tau_t \triangleq$ thermal time constant $\triangleq R_t C_t \qquad \textbf{(8.30)}$

For the "current mode" circuit of Fig. 8.36b, we have

$$i_s = K_p D\left(\frac{R_t W}{\tau_t D + 1}\right) = -\frac{e_o}{R_f/(R_f C_f D + 1)} \qquad \textbf{(8.31)}$$

$$\frac{e_o}{W}(D) = -\frac{K_e D}{(\tau_t D + 1)(\tau_e D + 1)} \qquad \textbf{(8.32)}$$

where $\quad K_e \triangleq K_p R_t R_f \triangleq$ sensitivity, V/(W/s) $\qquad \textbf{(8.33)}$

$\qquad \tau_e \triangleq$ electrical time constant $\triangleq R_f C_f \qquad \textbf{(8.34)}$

A typical system[102] might have $K_e = 50$, $\tau_t = 0.1$ s, and $\tau_e = 10^{-5}$ s. While the resistance R_f is an "intentional" component (typically $10^8\ \Omega$) the capacitance C_f may represent parasitic effects (the order of 0.01 to 0.1 pF) associated with the wiring layout. When a smaller R_f (say 10^5 to $10^6\ \Omega$) is used to extend the flat frequency range, an intentional C_f of 0.5 to 5.0 pF may need to be wired in to reduce oscillations caused by op-amp dynamics. Note that sensor capacitance C_s (typically 25 pF) does not enter into the system transfer function.

An alternative ("voltage mode") form of electronics commonly employed with pyroelectric detectors uses a field-effect transistor (FET) as in Fig. 8.36c. For a FET, point G (the gate) may be treated as essentially an open circuit, while the source (S) to drain (D) current, and thus e_o, is proportional to the voltage at G. Analysis gives

$$e_G = i_s\left(\frac{R_L}{R_L C_s D + 1}\right) = \frac{K_p R_t R_L D W}{(\tau_e D + 1)(\tau_t D + 1)} \qquad \textbf{(8.35)}$$

And if $e_o = K_g e_G$ (with K_g typically 0.7 V/V), we get

$$\frac{e_o}{W}(D) = \frac{K_e D}{(\tau_t D + 1)(\tau_e D + 1)} \qquad \textbf{(8.36)}$$

where $\quad K_e \triangleq K_g K_p R_t R_L \triangleq$ sensitivity, V/(W/s) $\qquad \textbf{(8.37)}$

$\qquad \tau_e \triangleq$ electrical time constant $\triangleq R_L C_s \qquad \textbf{(8.38)}$

Typical[103] values might be $R_L = 10^{11}\ \Omega$, $R_o = 10^4\ \Omega$, $C_s = 30$ pF, $\tau_t = 1$ s, and $K_e = 20{,}000$.

---

[102]PI-30 Detector/Op Amp, Molectron Detector Inc.

[103]Eltec Instruments Model 406, Cat. 8-78, 8-79, Daytona Beach, FL.

Equations (8.32) and (8.36) clearly show that the current-mode electronics exhibits the same form of transfer function as the voltage-mode, although each mode can be designed with widely varying numerical values. For the values given as typical above, Fig. 8.37 shows the frequency response (recall that an additional −10 dB/decade rolloff would be present at high frequency if an absorbing coating were present). The voltage-mode system of this example shows no range of flat response, although, of course, it could be obtained easily by changes in numerical values. Actually, a flat response is not always necessary or desirable. For example, to obtain response to *steady* radiation (which neither voltage mode nor current mode directly provides), a chopper is utilized to convert the steady radiation flux to a periodic variation. Since the chopping is at a fixed frequency, such an instrument need only have adequate amplitude ratio at that frequency, but there is no requirement for flat amplitude ratio. Also, some applications require measurement of the total energy of a radiation pulse. This requires integration, with respect to time, of the input signal W (which is power), and this is accomplished if the pulse-frequency content lies mainly in the downward-sloping (−20 dB/decade) region of the frequency response. Finally, if an overall system with flat amplitude ratio is desired in which the detector and preamp provide a −20 dB/decade slope, then the preamp can be followed with a dynamic compensator (approximate differentiator) to achieve this.

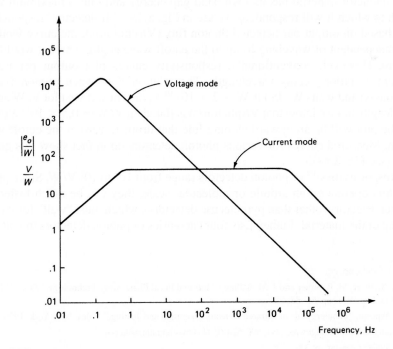

Figure 8.37
Pyroelectric-detector frequency response.

Our brief treatment of pyroelectric detectors has not included any discussion of electrical noise, always important in practical design studies. Useful data on this area may be found in the literature.[104]

We now turn to the *photon* (also called *quantum*) detectors, which are generally based on semiconductor materials. These materials exhibit *band gaps,* levels of electron energy that are not allowed, that is, some electrons will be found in a low-energy *valence band,* some in a high-energy *conduction band,* but none in the energy gap between the two. To provide an electrical signal, some electrons must be in the conduction band. When the material is exposed to incident photon flux (radiant energy), the individual photons may add enough energy to some of the electrons that they "jump" across the band gap, from the valence band to the conduction band, creating an electrical signal. This general effect is used by detectors in several ways, leading to classes of photon detectors called *photoconductive, photovoltaic, metal-insulator-semiconductor,* and *Schottky barrier* detectors.[105] Photons carry an energy E (electron-volts, eV) that is related to wavelength λ (μm) by the formula[106] $E = 1.24/\lambda$; thus, a photon of wavelength 5.6 μm has an energy of 0.22 eV. Photons of longer wavelengths will have *less* energy. To kick an electron from the valence band to the conduction band, a photon must have energy greater than the material's band-gap energy.

Indium antimonide (InSb) has a band-gap energy of 0.22 eV; thus, this material will not be sensitive to wavelengths greater than about 5.6 μm, its *cutoff wavelength.* Photons with longer wavelengths pass through the material undetected. Each semiconductor material has its own band-gap energy and thus a maximum wavelength to which it will respond, as we see in Fig. 8.38.[107] If detector "responsivity" were based on output per detected photon flux (V/(photon/s), the curve would be flat (independent of wavelength) up to the cutoff wavelength, where it would drop to zero. However, conventionally, responsivity curves plot output per *watt* of incident radiation versus wavelength. The relation[108] between photon flux N_λ (photons/s) and watts W (J/s) is $W = 2 \times 10^{-19} N_\lambda/\lambda$. Thus, if we plot V/W against wavelength and we know that V/(photon/s) is flat, then $V/W = 0.5 \times 10^{19} \lambda$ (V/N_λ) and the plot will be an upward sloping line that drops to zero at the cutoff wavelength. Measured curves for various photon detectors do in fact show this general trend (see Fig. 8.38).

Responsivities[109] of photon detectors range from 10^3 to 10^6 V/W. Since photon detectors operate on an atomic or molecular scale, they can be much faster than thermal detectors (other than pyroelectric detectors), which must "wait" for the bulk heating of the material. Finney lists four categories of photon detectors in wide use:

---

[104]W. M. Doyle, op. cit.

[105]D. A. Scribner, M. R. Kruer and J. M. Killiany, "Infrared Focal Plane Array Technology," *Proc. IEEE,* vol. 79(1), Jan. 1991, pp. 66–85.

[106]J. D. Vincent, "Fundamentals of Infrared Detector Operation and Testing," Wiley, New York, 1990, p. 33.

[107]Hamamatsu Corp. Bridgewater, NJ, 800-524-0504 (www.hamamatsu.com).

[108]J. D. Vincent, op. cit., p. 32.

[109]P. Finney, "Quantum Effect Infrared Sensing," *Sensors,* Dec. 1995, pp. 32–50.

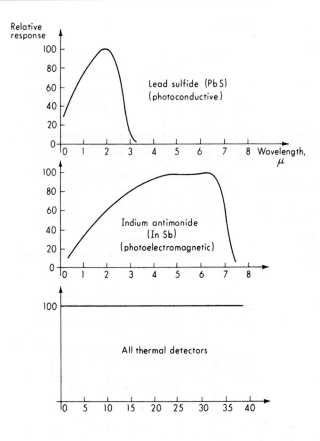

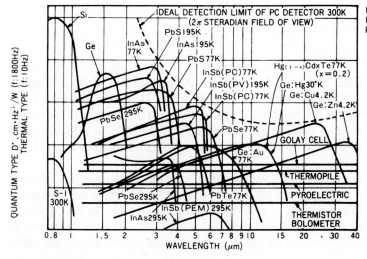

Figure 8.38

Radiation-detector spectral sensitivity.

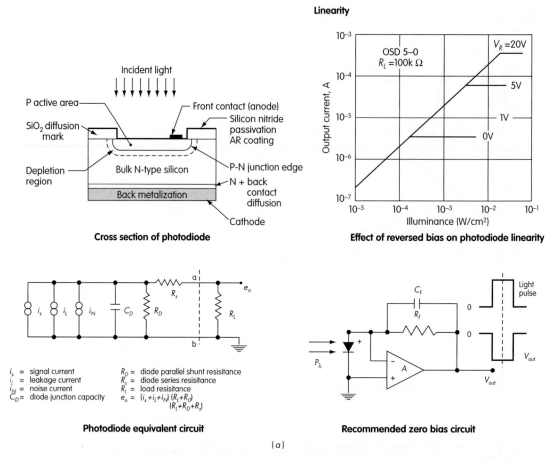

Linearity

Cross section of photodiode

Effect of reversed bias on photodiode linearity

Photodiode equivalent circuit

i_s = signal current
i_L = leakage current
i_N = noise current
C_D = diode junction capacity

R_D = diode parallel shunt resistance
R_s = diode series resistance
R_L = load resistance
$e_o = (i_s + i_L + i_N) \dfrac{(R_L + R_D)}{(R_L + R_D + R_s)}$

Recommended zero bias circuit

(a)

Figure 8.39
Basic detector circuits.

lead salts (PbS, PbSe, and PbTe), indium compounds (InSb and InAs), various formulations of HgCdTe, and platinum silicide (PtSi). He quotes response times ranging from 100 ns to 5 ms. Most imaging systems must be fast enough to provide 30 frames/s, the standard video rate, and utilize InSb, HgCdTe, or PtSi. Temperature resolution is about $0.1°C$ when detector cooling to liquid nitrogen levels (77 K) is used.

The scope of this text prevents us from presenting more details on the construction, electronic circuitry, etc. for all the detectors but a few physical features of selected devices will at least give you some useful familiarity. We first look at a *photodiode,* one of the simpler photon detectors, which however is representative of the class. Figure 8.39a[110] shows the physical construction, equivalent circuit, effect

---

[110]Centro Vision Inc., Newbury Park, CA, 800-700-2088 (www.centrovision.com).

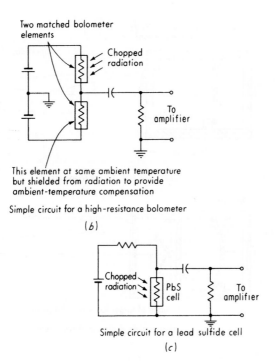

Two matched bolometer elements

Chopped radiation

To amplifier

This element at same ambient temperature but shielded from radiation to provide ambient-temperature compensation

Simple circuit for a high-resistance bolometer

(b)

Chopped radiation

PbS cell

To amplifier

Simple circuit for a lead sulfide cell

(c)

Figure 8.39
(Concluded)

of bias voltage on the linear range, and a simple but practical op-amp circuit. The photodiode shown is a *silicon* device, intended mainly for the visible range of wavelengths, but sometimes used for the "near infrared" since its cutoff wavelength is about 1100 nm. Photon detectors of the other types mentioned earlier would share many general features, so our discussion has wide applicability, though not in every detail. In the equivalent circuit model, note the presence of components of total current due to *noise* and *leakage* in addition to the desired signal current related to incoming photon flux. The device can be used *without* the external load resistor R_L; this would be called "photovoltaic" operation. Incoming photon flux would provide a proportional signal current that would flow through the internal diode resistance R_D, producing an output voltage measurable at terminals *a* and *b* with a high-impedance voltmeter (diode series resistance R_s would then have no effect). The circuit can be modified by installing an external *bias voltage* to control the linear range of operation, as shown in the graph for bias voltages from 0 to 20V. Finally, most practical circuits use op-amps, giving the *photoconductive* mode of operation (Fig. 8.39a showing the version with zero bias voltage). The diode's capacitance and resistance then become of no consequence since they are effectively "short-circuited" at the op-amp input terminals. The diode current is thus forced to pass through the op-amp feedback elements (R_f and C_f), external components that can be selected for desired performance of the complete circuit. Figure 8.39b and c show some other simple circuits for detectors.

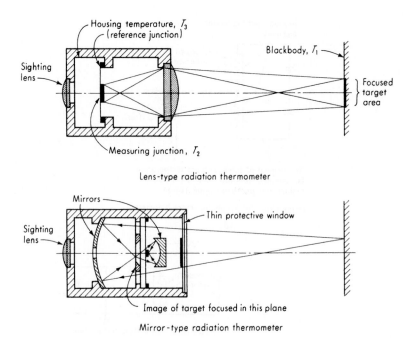

Figure 8.40
Lens- and mirror-type radiation thermometers.

Some type of circuit must be employed to realize a usable electrical signal (generally a voltage) from a radiation detector. Thermopile devices generally work with an uninterrupted stream of radiation and require no circuitry other than the usual reference junction, which is commonly left at ambient temperature. Bolometers and photoconductive cells often employ a chopper to interrupt the radiation at a fixed rate of the order of several hundred hertz. This leads to an ac-type electric signal and allows use of high-gain ac amplifiers. Typical circuits for such arrangements are shown in Fig. 8.39*b* and *c*.

Unchopped (DC) Broadband Radiation Thermometers

We begin our study of complete radiation-sensing instruments by consideration of the simplest type. These instruments[111] employ a blackened thermopile as detector and focus the radiation by means of either lenses or mirrors. Figure 8.40 shows the basic construction of this class of instruments. The reference of footnote 111 gives a very complete analysis of such devices.

Basically, for a given source temperature T_1, the incoming radiation heats the measuring junction until conduction, convection, and radiation losses just balance

---

[111]T. R. Harrison, "Radiation Pyrometry and Its Underlying Principles of Radiant Heat Transfer," Wiley, New York, 1960.

the heat input. The measuring-junction temperature usually is less than 40°C above its surroundings even if the source is incandescent. An oversimplified analysis gives

$$\text{Heat loss} = \text{radiant heat input}$$

$$K_1(T_2 - T_3) = K_2 T_1^4 \tag{8.39}$$

If the thermocouple voltage is proportional to $T_2 - T_3$, the voltage output should be proportional to T_1^4. Figure 8.41 shows an actual calibration curve of such an instrument together with the ideal relationship. For high temperatures the agreement is quite close. The temperatures T_2 and T_3 are both influenced by the environmental temperature; thus compensation generally must be provided for this, particularly if the instrument is intended to measure low temperatures. This compensation may include a thermostatically controlled housing temperature.

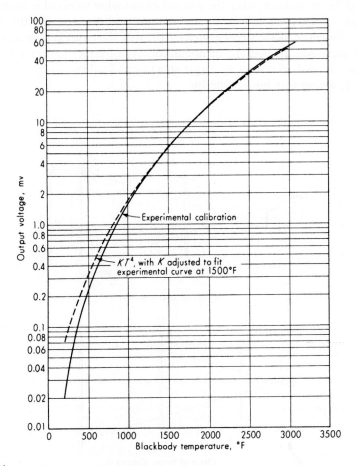

Figure 8.41
Theoretical and experimental calibration curves.

The thermopiles used may have from 1 or 2 to 20 or 30 junctions. A small number of junctions has less mass and thus faster response, but lower sensitivity limits application to high temperatures. Response is roughly that of a first-order system, with time constants ranging from about 0.1 (high-temperature systems) to 2 s (low-temperature systems). Instruments of this class can measure temperatures as low as 0°F; that is, actually the thermopile is cooler than the ambient temperature and reads a negative voltage. Theoretically there is no upper limit to the temperatures that can be measured in this way.

Conceivably, an instrument of the above type could be constructed with no focusing means, i.e., no lens or mirror. A simple diaphragm with a circular aperture (Fig. 8.42) would define the target from which radiation is received. To define smaller target areas for a given target distance, a smaller aperture could be utilized; however, there would be a proportionate loss in incoming radiation and thus sensitivity. The reading of such an instrument is independent of the distance between the target and the instrument, since the amount of radiation received is limited by the solid angle of the cone defined by the aperture and detector, and this is always the same. However, as the target distance increases, the target area necessary to fill the cone increases. If, because of nonuniformity of target temperature or small target size, we wish to restrict the target area to small values, a very small aperture is required, giving very low sensitivity.

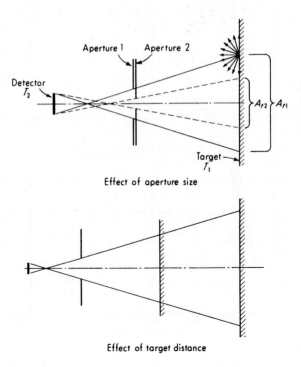

Effect of aperture size

Effect of target distance

Figure 8.42
Effect of aperture size and target distance.

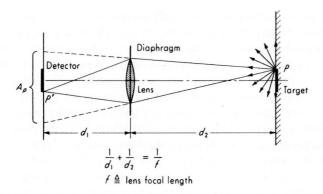

$$\frac{1}{d_1} + \frac{1}{d_2} = \frac{1}{f}$$

$f \triangleq$ lens focal length

Figure 8.43
Advantages of focusing.

The basic purpose of lens or mirror systems is to overcome this restriction, thus allowing the resolution of small targets without loss of sensitivity. Thus, in Fig. 8.43, radiation emanating from the point P on the target is focused on the corresponding point P' of the detector. If a simple diaphragm, rather than a focused lens, had been used, the same amount of energy would be spread out over the area A_p, with the detector receiving only a fraction of the radiation. Commercial lens- or mirror-type instruments generally have parameters such that targets 2 ft or more distant are adequately focused with a fixed-focus lens, and the instrument calibration is independent of distance as long as the target fills the field of view. Minimum target size to fill the field of view is of the order of one-twentieth of the target distance for common instruments. For targets closer than 2 ft, focusing is necessary and may affect the calibration, depending on instrument construction and closeness of target. Target diameters of 0.1 to 0.3 in at target distances of 4 to 12 in are available. Since the focal length of a lens depends on the index of refraction, which in turn varies with wavelength, not all wavelengths are focused at the same point. In particular, if one focuses a lens visually (using visible light), the longer infrared wavelengths, which contribute a large portion of the total energy at lower temperatures, will be out of focus. Such an instrument may have to be focused by adjusting for maximum thermopile output rather than for sharpest visual definition. Another effect of lenses is selective transmission, as shown in Fig. 8.34. The use of mirrors rather than lenses is an attempt to alleviate some of these problems. However, instruments of both types are employed widely with success.

While perfectly practical, instruments of the above type are not used as much as they have been in the past with the exception of a miniature version that takes advantage of microcircuit thermopiles not available until recently. Marketed as the "infrared thermocouple,"[112] these small (typically 0.5-in diameter and 2 in long)

---

[112]"The IRt/c Book," Exergen Corp., Watertown, MA, 800-422-3006 (www.exergen.com). This 129-page "book" is much more than a product catalog. It discusses many important practical topics in infrared temperature sensing and includes 95 "technical notes" detailing myriad application problems and their solutions using the product; F. Pompei, "A Case for the Unpowered IR Thermocouple," *Sensors,* June 1995.

inexpensive, noncontact sensors are widely used in temperature measurement and control. The basic concept (U.S. patent #5,229,612, 1993) uses microcircuit thermopiles with a number of junctions sufficient to provide an unamplified output voltage that is equal to what a *single* thermocouple of that type (type J, K, R, and S are available) would produce if it were at *target* temperature. This allows the output voltage to be directly connected to standard instruments designed to read out thermocouple temperatures for measurement and/or control purposes, that is, the thermopile itself is *not* at target temperature and is in fact just as "cool" as the older, larger instruments of Fig. 8.40. However, the thermopile acts "as if" it were a single thermocouple in contact with the target. This behavior holds only for a restricted range of target temperatures, but this is not a problem in most *control* applications since the controller is striving to keep the target temperature very near a desired value.

For strictly measurement applications, a calibration table relating temperature to voltage allows the thermopile to be used over a much larger range. For example, a unit which is useable over the range -50 to 550°F could be configured to act as a "single thermocouple" for target temperatures with about an 80° range *anywhere* within its "useable" range. Resolution, typically 0.0001C°, is very good because no noise-producing electronics are involved.

In addition, the response of the microcircuit thermopile can be quite fast, typically 0.1 s. While the thermopile itself has the broad wavelength capability typical of thermal detectors, built-in optical filters allow tailoring of the spectral response to meet application needs. A high-emissivity model intended for nonmetal targets uses a 6.5- to 14-μm range while a low-emissivity unit useful for metals uses a 0.1- to 5-μm range. By using a germanium lens, a 2- to 20-μm range is achieved. Unlike more complex radiation thermometers, these devices do not include an emissivity "dial" or adjustment, but several clever design features minimize errors due to unknown emissivity. For example, nonmetal targets typically change emissivity at about -2 percent/100F°, so this correction is built into the device. Also, emissivity of a given target may be included in the device calibration if the user applies a "microscanner" available from the manufacturer. This device uses a conical cavity to create a local blackbody condition on the target surface, allowing accurate temperature measurement, which is then transferred to the unit being calibrated (see Fig. 8.44). Once this calibration is accomplished, the microscanner is no longer needed.

Chopped (AC) Broadband Radiation Thermometers

A number of advantages accrue when the radiation coming from the target to the detector is interrupted periodically (chopped) at a fixed frequency; therefore many infrared systems employ this technique. When high sensitivity is needed, amplification is required, and high-gain ac amplifiers are easier to construct than their dc counterparts. This is usually the main reason for using choppers. Additional benefits related to ambient-temperature compensation and reference-source comparison also may be obtained. Systems employing thermal (broadband) and photon

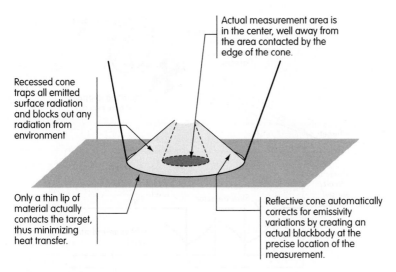

Actual measurement area is
in the center, well away from
the area contacted by the
edge of the cone.

Recessed cone
traps all emitted
surface radiation
and blocks out any
radiation from
environment

Only a thin lip of
material actually
contacts the target,
thus minimizing
heat transfer.

Reflective cone automatically
corrects for emissivity
variations by creating an
actual blackbody at the
precise location of the
measurement.

Figure 8.44
Microscanner calibrator for infrared thermocouple sensor.

(restricted-band) detectors and choppers are in common use. Here we consider those utilizing thermal detectors.

The time constants of adequately sensitive thermopile detectors are generally too long to allow efficient use of chopping; thus the faster bolometers, usually of the thermistor type, are employed. Figure 8.45 shows the basic elements of a blackened-chopper radiometer. A mirror focuses the target radiation on the detector; however, this beam is interrupted periodically by the chopper rotating at constant speed. Thus the detector alternately "sees" radiation from the target and radiation from the chopper's blackened surface. For high target temperatures, sufficient accuracy may be achieved by leaving the chopper temperature at ambient. Higher accuracy, particularly at low target temperatures, is obtained by thermostatically controlling the chopper temperature.

An arrangement similar to that of Fig. 8.39b is used for the detector circuit. The output voltage of the detector circuit is essentially as shown in Fig. 8.45. By amplifying this in an ac amplifier, the mean value (which is subject to drift) is discarded, and only the difference between the target and chopper radiation levels is amplified. If the chopper radiation level is considered as a known reference value, the target radiation and thus its temperature may be inferred. To provide a dc output signal related to target temperature and suitable for recording or control purposes, the ac amplifier is followed by a phase-sensitive demodulator and filter circuit. The necessary synchronizing signal for the demodulator may be generated by placing a magnetic proximity pickup near the chopper blades. While the response time of the detector itself may be of the order of a millisecond, the chopper frequency and necessary demodulator filter time constant greatly reduce the overall system speed. High chopper speeds allow faster overall response, but reduce sensitivity if the detector time constant is too large, since the detector does not have time to reach equilibrium during the time when either the target or the chopper is in view.

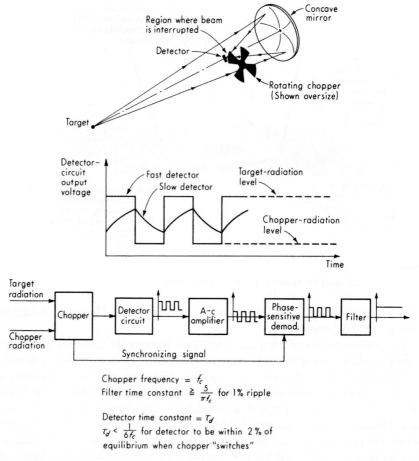

Figure 8.45
Blackened-chopper system.

Chopped (AC) Selective-Band (Photon) Radiation Thermometers

The use of photon detectors allows faster response speeds and may reduce sensitivity to ambient temperature. Since such detectors respond *directly* to the incident photon flux, rather than detecting a temperature change, the disturbing influence of ambient temperature changes is restricted mainly to changes in the responsivity of the detector. When radiation is expressed in terms of photon flux rather than watts, the formulas are somewhat modified. Equation (8.19) becomes

$$N_\lambda = \frac{2\pi c}{\lambda^4(e^{C_2/(\lambda T)} - 1)} \tag{8.40}$$

where $N_\lambda \triangleq$ hemispherical spectral photon flux, photons/(cm$^2 \cdot$ s $\cdot \mu$m)

$c \triangleq$ speed of light, 3×10^{10} cm/s

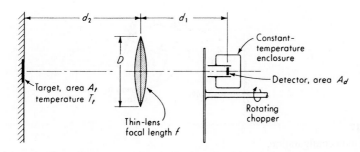

Figure 8.46
Photon-detector system.

The peak of the photon-flux curves occurs at a different wavelength from that of the radiant intensity. It is given by

$$\lambda_{p,p} = \frac{3,669}{T} \quad \mu m \tag{8.41}$$

The total photon flux for all wavelengths is

$$N_t = 1.52 \times 10^{11}\, T^3 \quad \text{photons/(cm}^2 \cdot \text{s)} \tag{8.42}$$

Figure 8.46 shows the basic arrangement of an instrument[113] using a photon detector and a chopper. Basic optics gives

$$\frac{1}{d_1} + \frac{1}{d_2} = \frac{1}{f} \tag{8.43}$$

and

$$A_t = \frac{A_d d_2^2}{d_1^2} \tag{8.44}$$

Combining these gives

$$A_t = \frac{A_d (d_2 - f)^2}{f^2} \tag{8.45}$$

Now if $d_2 \gg f$, we get approximately

$$A_t = \frac{A_d d_2^2}{f^2} \tag{8.46}$$

If the detector is a square with side L_d, the resolved target will be a square of side $L_d\, d_2/f$. For example, a 1-mm^2 detector used with a lens with a focal length of 75 mm requires a target of size $d_2/75$ to fill exactly the field of view.

For the general configuration of Fig. 8.47, basic radiation laws give

$$\frac{\text{Radiation incident on } dA_2}{dA_2} = \frac{\cos \theta_1 \cos \theta_2}{\pi r^2} (\text{radiation emitted from } dA_1) \tag{8.47}$$

---

[113]"Infrared Thermometry," Infrared Industries, Santa Barbara, CA.

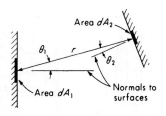

Figure 8.47
Basic radiation configuration.

We can apply this to the configuration of Fig. 8.46 to find the radiation received over the area of the lens from the target. If d_2 is large compared with the size of target and lens (usually true), then $\cos \theta_1 \approx \cos \theta_2 \approx 1$, and dA_1 and dA_2 may be replaced by A_1 and A_2 in Eq. (8.47). To simplify the analysis, let us assume that the detector has uniform spectral response from $\lambda = 0$ to $\lambda = 6.9$ and zero response beyond and that the target is a graybody with emittance ε. The radiation emitted by the target is then

$$A_t \varepsilon \int_0^{6.9} N_\lambda \, d\lambda = A_t \varepsilon E(T_t) \qquad \text{photons/s} \tag{8.48}$$

where

$$E(T_t) \triangleq \int_0^{6.9} N_\lambda \, d\lambda \qquad \text{a function of } T_t$$

The radiation received at the lens is

$$\frac{A_t \varepsilon E(T_t)(\pi D^2/4)}{\pi d_2^2} = \frac{A_t \varepsilon E(T_t) D^2}{4 d_2^2} \tag{8.49}$$

If the lens transmits $100 K_{tr}$ percent of the flux incident on it (perfect transmission has $K_{tr} = 1.0$) and if the system is focused so that the target image just fills the detector area, then the photon flux density at the detector is given by

$$N_d \triangleq \frac{K_{tr} A_t \varepsilon E(T_t) D^2}{4 d_2^2 A_d} \qquad \text{photons/(cm}^2 \cdot \text{s)} \tag{8.50}$$

For the indium antimonide photoelectromagnetic detector, the output voltage is proportional to N_d. Since for a focused target $A_t/(A_d d_2^2) = 1/f^2$, Eq. (8.50) can be written as

$$N_d = \frac{K_{tr} D^2}{4 f^2} \varepsilon E(T_t) \tag{8.51}$$

The output voltage of the overall instrument is $K_{dr} K_a N_d$, where $K_{dr} \triangleq$ detector responsivity, V/[photons/(cm$^2 \cdot$ s)], and K_a is the amplifier gain, V/V. Thus

$$\text{Instrument output voltage} \triangleq e_o = \frac{K_{dr} K_a K_{tr} D^2}{4 f^2} \varepsilon E(T_t) \tag{8.52}$$

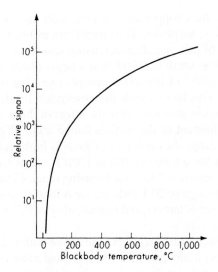

Figure 8.48
Indium antimonide system-response curve.

Note that the first factor in Eq. (8.52) is a constant of the instrument while the second $[\varepsilon E(T_t)]$ is a function of target temperature and emittance only. Also, as long as the target is focused, the reading is independent of the distance from the target. The variation in $E(T_t)$ with T_t for a blackbody target can be found by experimental calibration of the overall instrument. Its general shape is shown in Fig. 8.48. Since Eq. (8.42) shows that the total photon flux varies as T^3 and since the detector output is roughly proportional to total flux, the instrument output signal varies approximately as T^3; thus, for a graybody

$$e_o \approx K\varepsilon T^3 \qquad (8.53)$$

Since the instruments are calibrated against blackbody sources, for nonblackbodies a value of ε must be known in order to find T. If the value of ε used is in error, an error in T will result. Because of the third power law, however, errors in ε do not cause proportionate errors in T. Rather

$$\frac{T_{\text{actual}}}{T_{\text{assumed}}} = \left(\frac{\varepsilon_{\text{assumed}}}{\varepsilon_{\text{actual}}}\right)^{1/3} \qquad (8.54)$$

Thus if we assume $\varepsilon = 0.8$ when it is really 0.6, the temperature error is only 10 percent.

An instrument[114] of the above general class which uses mirror optics has a range from 100 to 2000°F (8000°F with calibrated aperture), focusing range 4 ft to infinity (18 to 60 in optional), 0.5° field of view, output signal 10 mV full scale, and an overall time constant (in chopped mode) of 2, 20, or 200 ms. For the study of

---

[114]Infrared Industries, Santa Barbara, CA.

very rapid transients, the chopper may be turned off and the system operated with just the detector and ac amplifier. Thus transients as brief as 10 μs may be measured. Other models of this manufacturer, using lens optics and lead sulfide detectors, have target sizes as small as 0.009 in at a target distance of 2.8 in.

Another instrument[115] of this class accepts radiation only in the wavelength band 4.8 to 5.6 μm. This band avoids the absorption bands of atmospheric water vapor and carbon dioxide, thus removing the effect of these variables on instrument response. The measurement of the surface temperature of glass is facilitated also since in this spectral range the emittance of glass is high and independent of thickness. This instrument has a range of 100 to 1000°F, target size equal to its distance divided by 57, time constant 0.2 to 0.5 s, focusing range 17 in to infinity, calibration accuracy 2 percent of range or 20°F (whichever is larger), resolution 0.25 percent of range or 3°F (whichever is larger), and repeatability of 0.5 percent of range or 10°F (whichever is larger).

While photon detectors are themselves selective with respect to wavelength, the use of optical bandpass filters provides a further (and more controllable) selectivity which is employed widely, as discussed in the previous section. Also, all types of radiation thermometers are inherently nonlinear because of the T^4 relation between temperature and radiant flux and the voltage/flux nonlinearities of the various detectors. The total instrument nonlinearity can be compensated electronically by either analog or digital computer means, and such linearization is widely available.

Automatic Null-Balance Radiation Thermometers

Figure 8.49 shows an interesting variation of the chopped, selective band type of instrument. Here the rotating chopper exposes the detector alternately (150 Hz) to radiant flux from the target and from a feedback-controlled reference source [an infrared light-emitting diode (LED)]. Thus the detector output signal is a square wave whose mean value is the average of the two flux levels and whose amplitude is their difference. The difference signal is extracted and amplified by an ac amplifier (acting as a high-pass filter), and the amplitude of this ac error signal is converted to a dc level (the feedback-system error signal) by a synchronous rectifier (phase-sensitive demodulator). Whenever the reference-source radiant flux is not equal to that from the measured target, the error signal drives the reference flux in the direction to null the error. By using sufficiently high gain in the feedback system, the reference flux almost perfectly tracks that from the measured target, and thus measurement of the LED drive signal is equivalent to measurement of the target flux. Such a system is largely unaffected by drifts in detector sensitivity because the detector is located in the forward path of the feedback system. The stability requirements, of course, are transferred from the detector to the LED reference source, but careful selection and aging give excellent stability here. The LED response is sufficiently fast that overall system dynamics is determined by the low-pass filter required in the demodulator, just as in conventional (nonfeedback)

---

[115]Ircon, Inc., Chicago.

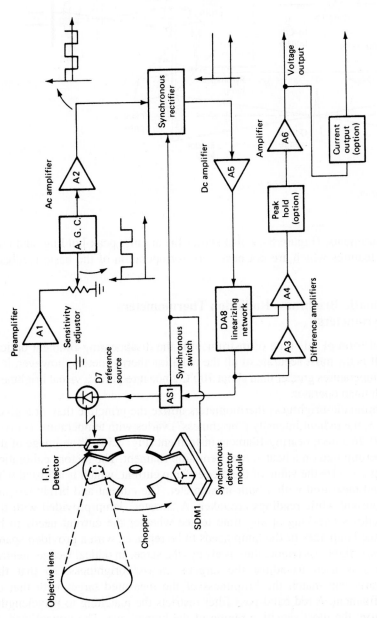

Figure 8.49

Automatic-null-balance radiation thermometer. *(Courtesy Williamson Corp., Concord, MA.)*

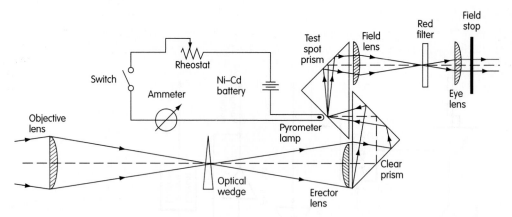

Figure 8.50
Disappearing-filament optical pyrometer.

chopper instruments. (Figure 8.49 also shows linearizing, peak-holding, and current output features which are not necessary for operation of the basic feedback principle.)

Monochromatic-Brightness Radiation Thermometers (Optical Pyrometers)

The classical form of this type of instrument is the disappearing-filament optical pyrometer. It is the most accurate of all the radiation thermometers; however, it is limited to temperatures greater than about 700°C since it requires a visual brightness match by a human operator.

Monochromatic-brightness thermometers utilize the principle that, at a given wavelength λ, the radiant intensity ("brightness") varies with temperature as given by Eq. (8.19). In a disappearing-filament instrument (Fig. 8.50),[116] an image of the target is superimposed on a heated tungsten-lamp filament. The factory calibration of each lamp provides the value of the proper lamp current, which is adjusted with a rheostat and measured with a built-in ammeter. The current and brightness are then held constant while readings are taken. A "master" lamp provided with the instrument allows checking at any time to see whether the current needs to be adjusted as the lamp ages or the lamp needs to be replaced with a provided spare. Since the *lamp* brightness temperature is always the same, an optical wedge (neutral density filter) is used to adjust the target's *viewed* brightness, so that the human operator can match the brightness of the measured target with that of the viewed filament. A red band-pass filter restricts the matching to wavelengths near 0.655 μm, the most sensitive region of the human eye. The optical wedge, mounted on a rotatable dial enscribed with temperature values, is adjusted until the

---

[116]Pyrometer Instrument Co., Northvale, NJ, 800-468-7976 (www.pyrometer.com).

lamp image disappears in the superimposed target image, whereupon the target temperature is read from the dial. If the target is not a blackbody, tables are provided to adjust for known emittance (ε) values. A special model designed for molten iron and steel in the open ($\varepsilon = 0.4$) can be read directly.

When the brightnesses of the target and lamp are equal, we can write

$$\frac{\varepsilon_{\lambda_e} C_1}{\lambda_e^5 (e^{C_2/(\lambda_e T_t)} - 1)} = \frac{C_1}{\lambda_e^5 (\varepsilon^{C_2/(\lambda_e T_L)} - 1)} \tag{8.55}$$

where
$\varepsilon_{\lambda_e} \triangleq$ emittance of target at wavelength λ_e
$\lambda_e \triangleq$ effective wavelength of filter, usually 0.65 μm
$T_t \triangleq$ target temperature
$T_L \triangleq$ lamp brightness temperature

For T less than about 4000°C, the terms with $e^{C_2/(\lambda_e T)}$ are much greater than 1, which allows Eq. (8.55) to be simplified to

$$\frac{e_{\lambda_e}}{e^{C_2/(\lambda_e T_t)}} = \frac{1}{e^{C_2/(\lambda_e T_L)}} \tag{8.56}$$

Then
$$\varepsilon_{\lambda_e} = e^{-(C_2/\lambda_e)(1/T_L - 1/T_t)} \tag{8.57}$$

and finally
$$\frac{1}{T_t} - \frac{1}{T_L} = \frac{\lambda_e \ln \varepsilon_{\lambda_e}}{C_2} \tag{8.58}$$

If the target is a blackbody ($\varepsilon_{\lambda_e} = 1.0$), there is no error since $\ln \varepsilon_{\lambda_e} = 0$ and $T_t = T_L$. If ε_{λ_e} is not 1.0 but is known, Eq. (8.58) allows calculation of the needed correction. The errors caused by inexact knowledge of ε_{λ_e} for a particular target are not as great for an optical pyrometer as for an instrument sensitive to a wide band of wavelengths. The percentage error is given by

$$\frac{dT_t}{T_t} = -\frac{\lambda_e T_t}{C_2} \frac{d\varepsilon_{\lambda_e}}{\varepsilon_{\lambda_e}} \tag{8.59}$$

Thus, for a target at 1000 K, a 10 percent error in ε_{λ_e} results in only a 0.45 percent error in T_t. For radiation thermometers *in general,* Eq. (8.59) shows that operation at *short* wavelengths results in less temperature error for a given emittance error. The use of a monochromatic red filter aids the operator in matching the brightness of target and of lamp since color effects are eliminated. Also, the target emittance need be known only at one wavelength. If ε_{λ_e} is exactly known, temperatures can be measured with optical pyrometers with errors on the order of 3C° at 1000°C, 6C° at 2000°C, and 40C° at 4000°C. With special optical systems, targets as small as 0.001 in can be measured at distances of 5 or 6 in.

Because of its manual null-balance principle, the optical pyrometer is not usable for continuous-recording or automatic-control applications. To overcome this drawback, automatic brightness pyrometers[117] have been developed.

---

[117]S. Ackerman and J. S. Lord, "Automatic Brightness Pyrometer Uses a Photomultiplier 'Eye,'" *ISA J.,* p. 48, December 1960; J. S. Lord, "Brightness Pyrometry," *Instrum. Contr. Syst.,* p. 109, February 1965.

Two-Color Radiation Thermometers

Since errors due to inaccurate values of emittance are a problem in all radiation-type temperature measurements, considerable attention has been paid to possible schemes for alleviating this difficulty. Although no universal solution has been found, the two-color concept[118] has met with some practical success. The basic concept requires that W_λ be determined at two different wavelengths and then the ratio of these two W_λ's be taken as a measure of temperature. For the usual conditions of practical application, the terms $e^{C_2/(\lambda T)}$ are much greater than 1.0, and we may write with close approximation

$$W_{\lambda 1} = \frac{\varepsilon_{\lambda_1} C_1}{\lambda_1^5 e^{C_2/(\lambda_1 T)}} \tag{8.60}$$

$$W_{\lambda 2} = \frac{\varepsilon_{\lambda_2} C_1}{\lambda_2^5 e^{C_2/(\lambda_2 T)}}$$

Then

$$\frac{W_{\lambda 1}}{W_{\lambda 2}} = \frac{\varepsilon_{\lambda_1}}{\varepsilon_{\lambda_2}} \left(\frac{\lambda_2}{\lambda_1}\right)^5 e^{(C_2/T)(1/\lambda_2 - 1/\lambda_1)} \tag{8.61}$$

For a graybody, $\varepsilon_{\lambda_1} = \varepsilon_{\lambda_2}$; thus

$$\frac{W_{\lambda 1}}{W_{\lambda 2}} = \left(\frac{\lambda_2}{\lambda_1}\right)^5 e^{(C_2/T)(1/\lambda_2 - 1/\lambda_1)} \tag{8.62}$$

and we see that the ratio $W_{\lambda 1}/W_{\lambda 2}$ is independent of emittance as long as it is numerically the same at λ_1 and λ_2. In one commercial instrument[119] the two filters ($\lambda_1 = 0.71 \pm 0.02$, $\lambda_2 = 0.81 \pm 0.02 \ \mu$m) are mounted on a rotating wheel so that the incoming radiation passes alternately through each on its way to a silicon photon detector. Special electronic circuitry performs operations equivalent to taking the ratio of $W_{\lambda 1}$ and $W_{\lambda 2}$. Instruments covering the range 1500 to 4000°F are available as standard models. A dual solid-state detector is available[120] that allows construction of relatively simple two-color instruments (Fig. 8.51).

Blackbody-Tipped Fiber-Optic Radiation Thermometer

Earlier we mentioned the use of optical fibers to transmit radiation for temperature measurements. In this section a distinctly different use of fiber optics is the basis of a unique, new type of radiation thermometer. Originally developed at NBS, it is now commercially available[121] and is intended mainly for the range of 500 to 2000°C in hot gases such as in gas-turbine combustors. The sensing probe is made from a single-crystal aluminum oxide (sapphire) optical fiber 0.25 to 2 mm in diameter and 5 to 100 cm long. A cup-shaped blackbody cavity is formed at the sensing

[118]T. P. Murray and V. G. Shaw, "Two-Color Pyrometry in the Steel Industry," *ISA J.*, p. 36, December 1958.

[119]A. S. Anderson, "The Dual Wavelength Radiometer," Williamson Corp., Concord, MA, 1980.

[120]J16 Si Series, Judson Infrared Montgomeryville, PA.

[121]R. R. Dils, "An Introduction to Optical Fiber Thermometry (TDS 1001)," Accufiber Corp., Vancouver, WA, 1983. Currently (2003) marketed by Luxtron, Santa Clara, CA, 800-627-1666 (www.luxtron.com).

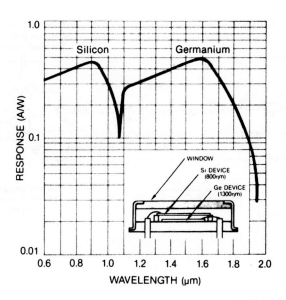

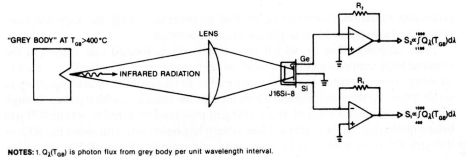

NOTES: 1. $Q_\lambda(T_{GB})$ is photon flux from grey body per unit wavelength interval.

2. Ratio $\frac{S_2}{S_1}$ can be used to determine "Grey body" temperature T_{GB} regardless of emissivity or absolute signal levels.

Figure 8.51
Dual detector for two-color pyrometer.

tip by vacuum-sputtering an opaque 2-μm-thick film of platinum or iridium which precisely defines the sensing region. The cup-shaped film may have a squared or angled end and use length/diameter ratios from about 2 to 80, with the larger ratios more closely approximating blackbody conditions but the smaller ones giving better spatial resolution. A protective sapphire film is also formed over the metal sensing film. When the probe is immersed in a hot gas, the sensing film receives heat by convection, conduction, and radiation and its temperature rapidly (due to large area and small mass) approaches the gas temperature. Just as in other probes, of course, conductive and radiative heat *losses* prevent the sensing film from exactly reaching the gas temperature; however, the present probe exhibits significant advantages since the conductivity of sapphire is 14 times smaller than that of typical thermocouple metals and the fiber's optical transparency also reduces radiation errors. Whatever temperature the sensing film achieves, it acts as a blackbody cavity,

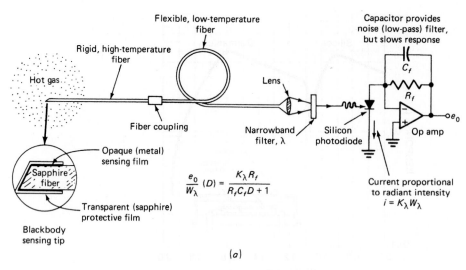

Flexible, low-temperature
fiber

Rigid, high-temperature
fiber

Hot gas

Lens

Capacitor provides
noise (low-pass) filter,
but slows response

C_f

R_f

oe_0

Op amp

Fiber coupling

Narrowband
filter, λ

Silicon
photodiode

Opaque (metal)
sensing film

Sapphire
fiber

Transparent (sapphire)
protective film

Blackbody
sensing tip

$$\frac{e_0}{W_\lambda}(D) = \frac{K_\lambda R_f}{R_f C_f D + 1}$$

Current proportional
to radiant intensity
$i = K_\lambda W_\lambda$

(*a*)

Figure 8.52(*a*)
Blackbody-tipped fiber-optic radiation thermometer.

projecting radiation characteristic of that temperature along the sapphire fiber toward a silicon photodiode or photomultiplier detector.

The sapphire fiber on which the sensing tip is formed can withstand the required high temperatures but is rigid and expensive, so it is optically coupled to an inexpensive, flexible, low-temperature fiber to transmit the optical signal up to about 100 m. At the receiving end a lens focuses the radiation, which passes through a narrowband optical filter of about 100-μm passband centered at 600 or 800 μm before impinging on the detector. (The system has been used with either the 600- or 800-μm filters or with both in a two-color ratio mode.) Figure 8.52a shows the complete system schematically. Since the detectors used produce an output signal which is very linear with the radiant power input [W_λ of Eq. (8.19)], a calibration method using ratios of detector outputs and based on Eq. (8.19) may be used.[122] Suppose we expose the instrument to a known calibration temperature T_0 and record an instrument output voltage E_0. If we now expose the instrument to an *unknown* temperature T, we will get some instrument reading E, and we wish to be able to calculate the corresponding temperature. We will *assume* the instrument follows Eq. (8.19) and will use a number for λ that is the *peak* wavelength of the narrowband-filter response. This is approximate since Eq. (8.19) presumes "filters" of infinitesimal passband whereas the real filters pass wavelengths about $\pm$ 15 percent either side of the peak. All these approximations must be *checked* by comparing their predictions with accepted standards. Using the excellent approximation $e^{C_2/(\lambda T)} - 1 \approx e^{C_2/(\lambda T)}$ in Eq. (8.19), we get

$$T = \left(\frac{1}{T_0} + \frac{\lambda}{C_2} \ln \frac{E_0}{E}\right)^{-1}$$

[122]K. G. Kreider, "Fiber Optic Thermometry," *ASTM* STP 895, pp. 151–161, 1985.

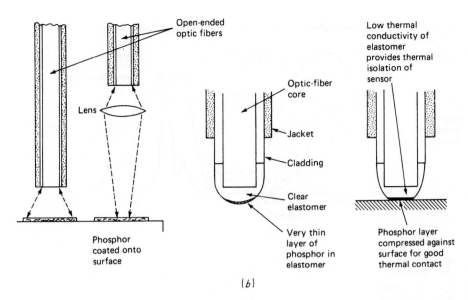

Open-ended
optic fibers

Low thermal
conductivity of
elastomer
provides thermal
isolation of
sensor

Optic-fiber
core

Lens

Jacket

Cladding

Clear
elastomer

Phosphor
coated onto
surface

Very thin
layer of
phosphor in
elastomer

Phosphor layer
compressed against
surface for good
thermal contact

(b)

Figure 8.52(b)
Application of fluoroptic temperature sensing.

Kreider used such a technique with $T_0 = 1320$ K and found errors of less than 5 K over the range 1000 to 1600 K, using a platinum/platinum-rhodium thermocouple standard accurate to 0.25 percent.

Fluoroptic Temperature Measurement

A commercially available fiber-optic temperature sensor is the fluoroptic[123] system. Here a material (magnesium fluorogermanate activated with tetravalent manganese) exhibiting a long, exponentially decaying fluorescence either is applied to the surface whose temperature is to be measured or is incorporated into the tip of a probe which will be pressed against a surface (Fig. 8.52b). A xenon flash lamp source, filtered as shown in Fig. 8.52c, sends a short pulse of "blue" light out along the optical fiber toward the fluorescent film, which then emits a longer-wavelength (red) fluorescence that propagates back along the fiber through another filter into an optical detector. The decay time of the fluorescence is a reproducible function of the temperature of the sensing film (Fig. 8.52c) and is measured with conventional electronic timing circuitry. Good accuracy is easily achieved since the time intervals are relatively long compared to the clock period (0.1 to 1.0 μs) of standard electronic timers. About 10 readings (flashes) per second are possible, with single-flash precision being about $\pm$ 0.5°C, while averaging 20 or 30 flashes reduces this figure to

[123]K. A. Wickersheim and M. H. Sun, "Fiberoptic Thermometry and Its Applications," *J. Microwave Power,* pp. 85–94, 1987; Luxtron, Santa Clara, CA, 800-627-1666 (www.luxtron.com).

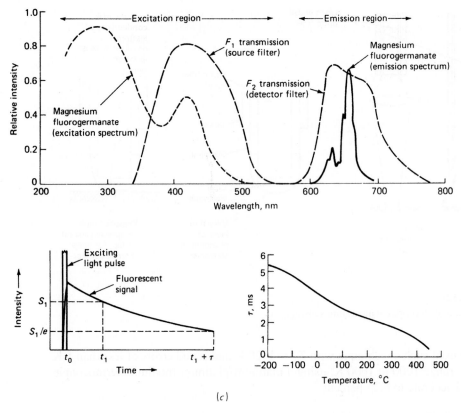

(c)

Figure 8.52(c)
Principles of fluoroptic temperature sensing.

±0.1°C. The calibration accuracy is ± 0.2°C at the calibration temperature and ±0.5°C at 50° either side of calibration temperature.

Since the probe and sensor are totally nonelectrical, the fluoroptic system is well suited to measurements in areas of high electrical noise. A 0.001-in-diameter microprobe allows measurements in tiny devices such as integrated circuits.

Infrared Imaging Systems

While infrared instruments which measure a "point" target meet many application needs, sometimes the temperature *distribution* along a selected line or over a selected area is more useful. *Line scanners*[124] use a single detector and a rotating or oscillating mirror to scan the detector's line of sight over the target surface, giving the temperature distribution along the scanned line. Such instruments are particularly appropriate for moving targets such as webs of plastic film or huge rotating cement kilns. The scanner gives the temperature profile across the moving web of film during its manufacture, allowing better control of the process. The temperature profile of the cement kiln reveals local hot spots which indicate a need for repair of its refractory lining.

When a two-dimensional image of a selected area is needed, various forms of infrared *camera* are available. Most of these systems have the function of providing a television-like visual display in which the colors represent various temperature levels of the surface of some two-dimensional object (target) on which the infrared camera is focused. While these colors are accurately related to infrared energy levels emitted from the target, recall that all infrared temperature sensing requires knowledge of the emittance of the target surface to convert the detector signal to degrees of temperature.

Both staring and scanning systems are available, but staring systems are preferred in many cases since they do away with complex moving parts and can often gather more of the available radiation. That is, a *scanning system* with a single detector element can only spend a short time at any point of the target surface since it must move the line of sight over the entire target quickly enough to complete a scan in about 1/30 s (the standard television frame rate is 30 frames/s). In a *staring system,* the electrons produced by the incoming photons can be "accumulated" over a time interval (called the *integration time*) by the detector array since the electronic readout uses a charge-coupled-device (CCD) that stores the charge on each pixel's tiny capacitor. This charge accumulation can occupy the entire $\frac{1}{30}$ s available between frames for *every* pixel in the array.

While there is some variation from company to company, it is quite common to use an uncooled microbolometer array for lower cost portable systems and a platinum silicide array cooled to about 77 K (with a miniature Stirling-cycle refrigerator) for the higher performance systems. Typical[125] specifications for the microbolometer instrument with a 320 × 240 pixel array include a resolution of 0.1C°, a spectral range of 7.5 to 13 μm, spatial resolution of 1.3 mrad (0.0013 times the range), and temperature ranges from − 40 to 2000°C. Microbolometer instruments do not require cooling to cryogenic levels but do need temperature stabilization, usually by means of a thermoelectric (Peltier) heating/cooling system. A cooled platinum silicide unit from the same vendor has similar specifications except for the spectral range, which is 3.6 to 5 μm.

In some systems, emissivity measurement and correction have been implemented by using computer-aided thermal-image storage and processing.[126] In small-target applications such as microelectronic devices, emissivity measurement of each individual pixel (spatial resolution of 0.038 mm) has been achieved. The procedure requires the device to be heated sequentially to two *known* low-level temperatures. (The device temperature here is spatially *uniform* since the device's electric power is *not* applied.) Actually, a "radiation reading" at *one* known temperature is sufficient to calculate emissivity. Two readings can improve statistical reliability and

[125]FLIR Systems, N. Billerica, MA, 800-464-6372 (www.flir.com).

[126]H. Kaplan, "Thermal Imaging Systems for Measuring Temperature Distribution," *ASTM* STP 895, pp. 96–98, 1985.

Infrascope optical schematic diagram

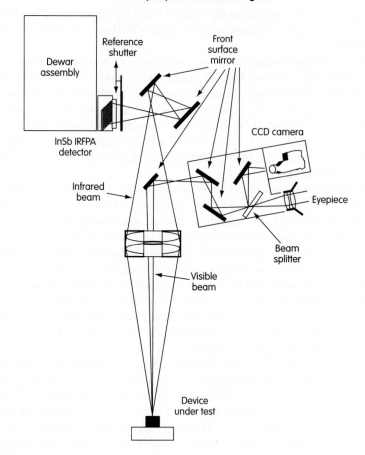

Figure 8.53
Staring infrared microscope.

detect emissivity trends with temperature, which can be extrapolated to the higher temperatures existing when the final measurement is made (and corrected for emissivity) with the device's power applied so as to reveal its temperature distribution under normal operation. By comparing images of each device as it "comes off the production line" with computer-stored images of *acceptable* devices, this technique provides a valuable nondestructive quality-control test.

Infrared radiation (using proper lens or mirror materials) can be focused just as in the more familiar visible telescopes and microscopes, with infrared microscopes being available as commercial products. Single point, scanning, and staring versions allow measurement of tiny targets,[127] often microelectronic devices. The staring instrument referenced (Fig. 8.53) uses liquid nitrogen cooling (77 K) with an indium antimonide (InSb) detector array (1.5 to 5.5 μm) to achieve a temperature

---

[127]Barnes Engineering Div., Shelton, CT, 203-926-1777.

resolution of 0.02C°, a spatial resolution of 4 μm, and a temperature range of 20 to 300°C. Using a CCD camera in the *visible* spectral range, the infrared thermal image is overlaid on the visible one to facilitate the location of points of interest. Each pixel is emissivity corrected, as is the "Narcissus effect," where the cold detector sees its own reflection on the surface of a highly reflective target.

As microelectronic device features have become smaller and smaller, a fundamental infrared limitation has been encountered. For *any* "light" signal, the spatial resolution is limited by the wavelength λ, according to Lord Rayleigh's criterion:

$$\text{Resolution} = 0.61\lambda/(NA)$$

where NA is the numerical aperture of the microscope ($NA \approx 0.5$/f# of optics). For optics with an f# of, say, 2.0, the resolution is about 2.44λ; so for infrared radiation of, say, 2 μm wavelength, we would be unable to resolve features smaller than about 5 μm size. Features of thermal interest are already smaller than this, and the trend has always been toward smaller and smaller sizes. To measure thermal aspects of these tiny targets, the method of *fluorescent microthermal imaging*[128] has been developed, giving a resolution of about 0.3 μm. Here a thin film of fluorescent material is deposited on the target surface and illuminated with ultraviolet light, giving a fluorescent response that contains the desired thermal information.

8.8 TEMPERATURE-MEASURING PROBLEMS IN FLOWING FLUIDS

In attempting to measure the static temperature of flowing fluids (particularly high-speed gas flows), we encounter certain types of problems irrespective of the particular sensor being used. These have to do mainly with errors caused by heat transfer between the probe and its environment and the problem of measuring static temperature of a high-velocity flow with a stationary probe.

Conduction Error[129]

Let us consider the so-called conduction error first. Figure 8.54*a* shows a common situation. A probe has been inserted into a duct or other flow passage and is supported at a wall. In general, the wall will be hotter or colder than the flowing fluid; thus there will be heat transfer, and this leads to a probe temperature different from that of the fluid. We analyze a simplified model of this arrangement to find what measures can be taken to reduce and/or correct the error to be expected in such a case. Figure 8.54*b* shows the simplified model to be employed in analyzing this situation. A slender rod extends a distance L from the wall. We assume the rod

[128]D. L. Barton and P. Tangyunyong, "Fluorescent Microthermal Imaging," Sandia National Laboratories, Albuquerque, NM, 1997, 505-844-7085.

[129]R. P. Benedict, "Fundamentals of Temperature, Pressure, and Flow Measurements," Wiley, New York, 1977, chap. 12; F. D. Werner et al., "Stem Conduction Error for Temperature Sensors," *Bull.* 9622, Rosemount Inc., Minneapolis, MN.

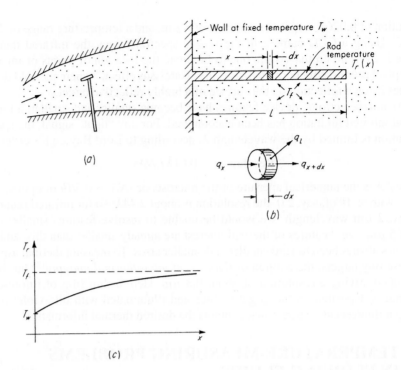

Figure 8.54
Probe configuration and conduction-error analysis.

temperature T_r is a function of x only; it does not vary with time or over the rod cross section at a given x. A fluid of constant and uniform temperature T_f completely surrounds the rod and exchanges heat with it by convection. For a steady-state situation

Heat in at x = (heat out at $x + dx$) + (heat loss at surface)

$$q_x = q_{x + dx} + q_l$$

One-dimensional conduction heat transfer gives

$$q_x = -kA \frac{dT_r}{dx} \tag{8.63}$$

where $k \triangleq$ thermal conductivity of rod and $A \triangleq$ cross-sectional area. Then

$$q_{x + dx} = q_x + \frac{d}{dx}(q_x)\, dx = -kA \frac{dT_r}{dx} + \frac{d}{dx}\left(-kA \frac{dT_r}{dx}\right) dx \tag{8.64}$$

Now if k and A are assumed constant,

$$q_{x + dx} = -kA \frac{dT_r}{dx} - kA \frac{d^2 T_r}{dx^2}\, dx \tag{8.65}$$

We assume the heat loss by convection at the surface to be given by

$$q_l = h(C\, dx)(T_r - T_f) \qquad (8.66)$$

where $h \triangleq$ film coefficient of heat transfer

$C \triangleq$ "circumference" of rod (it need not be circular)

$C\, dx =$ surface area

So we have

$$\frac{d^2 T_r}{dx^2} - \frac{hC}{kA} T_r = -\frac{hC}{kA} T_f \qquad (8.67)$$

If we take h and C as being constants, then Eq. (8.67) is a linear differential equation with constant coefficients and is readily solved for T_r as a function of x. We need two boundary conditions to accomplish this. Clearly, $T_r = T_w$ at $x = 0$ is one such condition. The simplest assumption at $x = L$ is an insulated end; this gives $dT_r/dx = 0$ at $x = L$. Even if the end is not insulated, if L is quite large, we see intuitively that the variation of T_r with x must be as in Fig. 8.54c; thus $dT_r/dx \approx 0$ for $x = L$. Using these two boundary conditions with Eq. (8.67) gives

$$T_r = T_f - (T_f - T_w)\left[\frac{e^{-mL}}{2\cosh mL}e^{mx} + \frac{e^{mL}}{2\cosh mL}e^{-mx}\right] \qquad (8.68)$$

where

$$m \triangleq \sqrt{\frac{hC}{kA}} \qquad (8.69)$$

Since generally the temperature-sensing element (thermocouple bead, thermistor, etc.) is located at $x = L$, we evaluate Eq. (8.68) there to get

$$\text{Temperature error} = T_r - T_f = \frac{T_w - T_f}{\cosh mL} \qquad (8.70)$$

Equation (8.70) may be used in two ways: to indicate how to design a probe support to minimize the error and to allow us to calculate and correct for whatever error there might be. It is clear that the error is reduced if T_w is close to T_f. Insulating or actively controlling the temperature of the wall encourages this. The term $\cosh mL$ will be large if m and L are large. Thus the probe should be immersed (L is called the *immersion length*) as far as practical (see Fig. 8.55[130]). We see that to make m large, h should be large (high rate of convection heat transfer) and k should be small (the probe support made of insulating material). The term C/A depends on the shape of the rod. For the usual circular cross section, $C/A = 2/r$, where r is the rod radius. Thus we see that the rod should be of small cross section to reduce error.

[130]F. D. Werner et al., ibid.

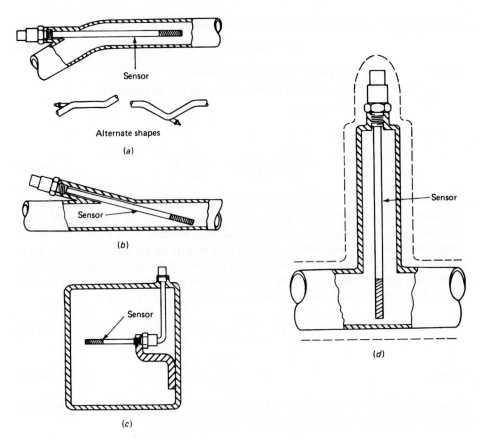

Sensor

Alternate shapes

(a)

Sensor

Sensor

(b)

Sensor

(c)

Sensor

(d)

Figure 8.55
Schemes to reduce conduction error.

If the boundary condition at $x = L$ is changed to a more realistic (and complicated) one in which there is convection heat transfer with a film coefficient h_e at the end, then at $x = L$

$$T_r - T_f = \frac{T_w - T_f}{\cosh mL + [h_e/(mk)]\ \sinh mL} \tag{8.71}$$

Since the error predicted by (8.71) is less than that of (8.70), the use of the simpler relation is conservative.

Radiation Error[131]

Additional error is caused by radiant-heat exchange between the temperature probe and its surroundings. This occurs simultaneously with the previously studied

---
[131]R. P. Benedict, op. cit.

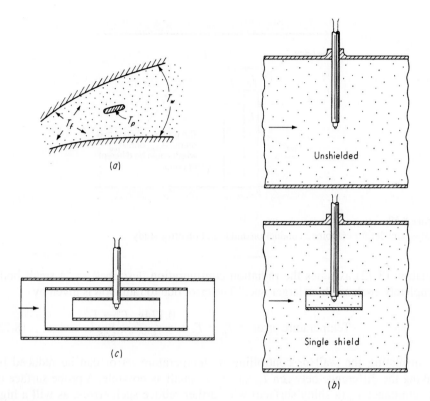

Figure 8.56
Radiation-error analysis and shielding.

conduction losses, but here we first consider it separately for simplicity. We also assume radiation exchange only between the probe and the surrounding walls, neglecting radiation of the gas itself or the absorption by the gas of radiation passing through it. Neglecting conduction losses, we may consider the probe as in Fig. 8.56a. For steady-state conditions

$$\text{Heat convected to probe} = \text{net heat radiated to wall}$$

$$hA_s(T_f - T_p) = 0.174\varepsilon_p A_s \left[\left(\frac{T_p}{100} \right)^4 - \left(\frac{T_w}{100} \right)^4 \right] \tag{8.72}$$

where
$h \triangleq$ film coefficient at probe surface, Btu/(h · ft² · °F)

$A_s \triangleq$ probe surface area

$\varepsilon_p \triangleq$ emittance of probe surface

$T_p \triangleq$ probe absolute temperature, °R

$T_w \triangleq$ wall absolute temperature, °R

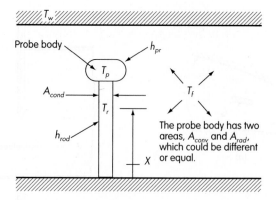

Figure 8.57
Analysis model for combined conduction and radiation error study.

Equation (8.72) assumes the radiation configuration described as "a small body completely enclosed by a larger one." The error due to radiation is given by

$$\text{Temperature error} = T_p - T_f = \frac{0.174\varepsilon_p}{h}\frac{T_w^4 - T_p^4}{10^8} \tag{8.73}$$

By insulating the wall or controlling its temperature, error can be reduced by making the difference between T_w and T_p as small as possible. A probe surface of low emittance ε_p (a shiny surface) will further reduce such errors, as will a high value of heat-transfer coefficient h. To obtain a high value of h when the fluid velocity is low, the aspirated type of probe may be utilized. Here a high local velocity is induced at the probe by connecting a vacuum pump to the probe tubing. Equation (8.73) may be used also to calculate corrections if numerical values of the needed quantities are available.

   We treated conduction and radiation errors *separately* since the error can then be given by simple formulas (available from the referenced literature) that reveal the effect of parameters most clearly. Conduction and radiation often *coexist,* and since the radiation effect is nonlinear and some boundary conditions are not compatible, superposition of the individual solutions is not really correct. After some thought, I came up with an analysis of the combined situation which I believe is reasonable. Figure 8.57 shows my analysis model for the combined case. Equation (8.67) still holds, but now we use h_{rod} for the rod's h and h_{pr} for the probe "body." Boundary conditions are $T_r(0) = T_w$ and $T_r(L) = T_p$. These lead to the partial solution

$$T_r = C_1 e^{mx} + C_2 e^{-mx} + T_f \qquad C_2 \triangleq \frac{T_p - T_f - (T_w - T_f)e^{mL}}{e^{-mL} - e^{mL}} \qquad C_1 \triangleq T_w - T_f - C_2$$
$$\tag{8.74a}$$

Even though T_p in practice would be a known measured value, we at this point treat it as an *unknown.* Otherwise, we could with no further analysis compute T_f from the known values of system parameters, T_w and T_p, with *no* consideration of the radiation effect. To include the radiation, we write a heat balance for the probe body as

$$h_{pr} A_{conv} (T_f - T_p) - kA_{cond} \frac{dT_r}{dx}\bigg|_{x=L} = \frac{0.174 \varepsilon_p A_{rad}}{10^8} (T_p^4 - T_w^4) \quad \textbf{(8.74b)}$$

where

$$\frac{dT_r}{dx}\bigg|_{x=L} = mC_1 e^{ml} - mC_2 e^{-mL}$$

The solution can be given in terms of formulas but is too complicated to use for studying the effects of parameters on the error. I therefore wrote a MATHCAD script for it that allows easy exploration of numerical values. If you do the same and try some numbers, note that if you set the radiation to zero, the values for T_f will not match those obtained using Eq. (8.70) since the boundary condition at $x = L$ is taken differently in the two models. My new analysis is probably more correct since real probes do have a "probe body" at $x = L$, not just a "rod."

Probes with some form or another of radiation shield are employed widely to reduce radiation errors. Figure 8.56b shows a probe with a single shield. The principle of all radiation shields is to interpose between the probe and the wall a body (the shield) whose temperature is closer to the fluid temperature than is the wall. Thus the probe "sees" the shield rather than the wall, and if the temperature of the shield is close to fluid temperature, the probe radiation heat loss will be small. For pure radiation heat transfer, it is easy to show that interposing a single screen between a body and its surroundings will reduce the heat loss to one-half the former value, since the screen comes to a temperature $T_{screen}^4 = (T_{body}^4 + T_{surroundings}^4)/2$. For n screens the heat loss is reduced to $1/(n + 1)$ of the unscreened value. All these results are for the simplest case in which the screens and surroundings completely enclose the body. For actual probe shields, various geometric and emittance factors complicate the situation. Also, additional heating of the shield by convection raises its temperature and reduces probe error. Experimental tests[132] with concentric circular cylinder shields (Fig. 8.56c) have shown the following:

1. A significant decrease in error may be achieved by adding more shields, at least up to about four.

2. Little is gained by increasing the length/diameter ratio beyond 4:1 in attempting to reduce the unshielded angles at the open ends.

3. For multiple shields the spacing between shields must be sufficiently large to prevent excessive conduction heat transfer between shields and to allow high enough flow velocity for good convection from gas to shields. A double shield with only $\frac{1}{32}$-in spacing acted almost as a single shield.

To illustrate the need for shielding, an unshielded probe exposed to $1800°F$ gas flow at $270 \text{ lbm}/(\text{ft}^2 \cdot \text{min})$ may exhibit an error of $160F°$. A suitable quadruple shield can reduce this to about $20F°$.

[132]W. J. King, "Measurement of High Temperatures in High-Velocity Gas Streams," *Trans. ASME*, p. 421, July 1943.

Another shielding technique employs an electrically heated shield with an additional temperature sensor fastened to the shield. The heat input to the shield is adjusted until the probe sensor and shield sensor register identical temperatures. At this point, probe and shield should both be at the fluid temperature, with the heat loss from the shield to the cooler wall being replaced by the shield heater.

Velocity Effects[133]

It is often necessary to determine the static temperature of a flowing gas, since its physical properties depend on this temperature. To measure this temperature directly with a probe, however, requires that the probe be stationary with respect to the fluid; thus it must be moving at the same velocity as the fluid. Since usually this is impractical, various indirect methods of measuring static temperatures are in use. If we can measure static pressure and density, sound velocity, or index of refraction, formulas allow calculation of static temperature. Experimental techniques based on each of these principles have been developed. However, for routine measurements a different approach is employed. This involves placing a stationary probe in the stream and calculating the static temperature from the readings of this probe, by using suitable corrections. Ideally, if a perfect gas is decelerated from free-stream velocity to zero velocity adiabatically (not necessarily isentropically), the temperature rises from the free-stream static temperature T_{stat} to the so-called stagnation or total temperature T_{stag}, where

$$\frac{T_{\text{stag}}}{T_{\text{stat}}} = 1 + \frac{\gamma - 1}{2} N_m^2 \tag{8.75}$$

where
$$\gamma \triangleq c_p/c_v, \text{ ratio of specific heats}$$
$$N_m \triangleq \text{Mach number}$$
$$T_{\text{stag}}, T_{\text{stat}} \triangleq \text{absolute temperatures}$$

This result holds for both subsonic and supersonic flows because the shock wave that forms ahead of a probe in supersonic flow affects only the entropy, not the total enthalpy of the gas. For air, Eq. (8.75) becomes

$$\frac{T_{\text{stag}}}{T_{\text{stat}}} = 1 + 0.2N_m^2 \tag{8.76}$$

and if $N_m < 0.22$, T_{stag} is within 1 percent of T_{stat}. Thus for sufficiently low velocities, a stationary probe can be utilized to read static temperature directly. For higher Mach numbers, T_{stat} can be calculated from Eq. (8.75) if γ and N_m are known. Measurement of Mach numbers with a pitot-static tube is discussed in Chap. 7.

Unfortunately, real temperature probes do not attain the theoretical stagnation temperature predicted by Eq. (8.75). Even if the conduction and radiation errors

---

[133]R. P. Benedict, op. cit., chap. 11; T. M. Stickney et al., "Rosemount Total Temperature Sensors," *Tech. Rept.* 5755, Rosemount Inc., Minneapolis, MN, 1975 (now B. F. Goodrich Aircraft Sensors Div., Burnsville, MN, 952-892-4253, www.bfg-sensors.com).

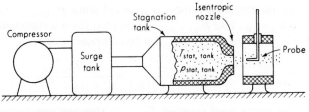

$T_{\text{stat, tank}}$ kept near room temperature to minimize conduction and radiation errors

Figure 8.58
Recovery-factor calibration setup.

discussed earlier in this section are corrected for, there remain further deviations of the actual situation from the assumed ideal. Correction for these effects generally is accomplished by experimental calibration to determine the recovery factor r of the particular probe. This is defined by

$$r \triangleq \frac{T_{\text{stag, ind}} - T_{\text{stat}}}{T_{\text{stag}} - T_{\text{stat}}} \tag{8.77}$$

where $T_{\text{stag, ind}} \triangleq$ temperature actually indicated by probe. If r is assumed to be a known number for a given probe, then combination of Eqs. (8.75) and (8.77) gives

$$T_{\text{stat}} = \frac{T_{\text{stag, ind}}}{1 + r\,[(\gamma - 1)/2]\,N_m^2} \tag{8.78}$$

A probe that measures T_{stag} exactly would have a recovery factor of 1.0 while one that measures T_{stat} exactly would have $r = 0$.

A possible apparatus[134] for determination of r is shown in Fig. 8.58. The flow velocity in the stagnation chamber is $\frac{1}{100}$ of the nozzle flow velocity; thus measurement of tank temperature and pressure is carried out accurately under essentially zero-velocity conditions. By careful design to minimize friction and heat transfer, the nozzle can be made to provide an almost-perfect isentropic expansion. The validity of this assumption has been checked experimentally. For an isentropic process from tank to nozzle,

$$T_{\text{stag, nozzle}} = T_{\text{stat, tank}} \tag{8.79}$$

and

$$p_{\text{stag, nozzle}} = p_{\text{stat, tank}} \tag{8.80}$$

In a free jet

$$p_{\text{stat, nozzle}} = p_{\text{atmosphere}} \tag{8.81}$$

Now the nozzle Mach number N_m can be computed from the standard pitot-tube formulas since $p_{\text{stat, nozzle}}$ and $p_{\text{stag, nozzle}}$ are both known. However, the actual use of

---

[134]H. C. Hottel and A. Kalitinsky, "Temperature Measurements in High-Velocity Air Streams," *J. Appl. Mech.*, p. A-25, March 1945.

a pitot tube (with its attendant errors) is avoided since $p_{\text{stag, nozzle}}$ is obtained by measurement of $p_{\text{stat, tank}}$, and $p_{\text{stat, nozzle}}$ is obtained from a barometer reading of $p_{\text{atmosphere}}$. Once N_m is known, $T_{\text{stat, nozzle}}$ can be computed from

$$T_{\text{stat, nozzle}} = \frac{T_{\text{stag, nozzle}}}{1 + [(\gamma - 1)/2]\,N_m^2} = \frac{T_{\text{stat, tank}}}{1 + [(\gamma - 1)/2]\,N_m^2} \tag{8.82}$$

The reading of the probe itself supplies $T_{\text{stag, ind}}$. Thus we can compute r from its definition

$$r = \frac{T_{\text{stag, ind}} - T_{\text{stat, nozzle}}}{T_{\text{stag, nozzle}} - T_{\text{stat, nozzle}}} \tag{8.83}$$

For bare thermocouple sensors, the numerical value of r usually lies in the range 0.6 to 0.9, depending on the form of the junction (butt-welded, twisted, or spherical bead) and the orientation (wire parallel to flow or transverse to flow). To get a high value of r and one relatively independent of flow conditions such as velocity magnitude and direction, sensors (usually thermocouples or resistance thermometers) are built into probes that have been specifically designed to approach ideal stagnation conditions. Figure 8.59 shows two examples[135] of such probes. Desirable characteristics of probes include the following:

1. Low heat capacity in the sensing element for a fast response.
2. Conduction loss of lead wires minimized by exposing enough length of the lead to the stagnation temperature.
3. Radiation shield of low thermal conductivity and low surface emittance.
4. Vent holes provided to replenish continuously the fluid in the stagnation chamber; otherwise, it would be cooled by conduction and radiation. This flow must be kept small enough, however, that stagnation conditions are essentially preserved. The increased convection coefficient caused by the flow speeds the response and reduces the radiation error.
5. Blunt shape causes formation of a normal shock wave in supersonic flow. This increases the temperature in the boundary layer and reduces the heat loss from the probe. The shock wave also reduces the influence of misalignment.

In addition to the laboratory-type probes of Fig. 8.59, permanently installed stagnation-temperature sensors for rugged environments, such as control-system sensing in aircraft gas-turbine engine inlets, are needed. Added to the problems discussed above, such probes must survive impacts by birds and ice balls and include provisions for deicing and inertial separation of water droplets. Figure 8.60[136] shows characteristics of sensors of this type which use platinum resistance elements.

[135]R. W. Ladenburg et al., "Physical Measurements in Gas Dynamics and Combustion," p. 186, Princeton University Press, Princeton, NJ, 1954.

[136]Product Data Sheet 2186, Rosemount Inc., (now B. F. Goodrich Aircraft Sensors Div.).

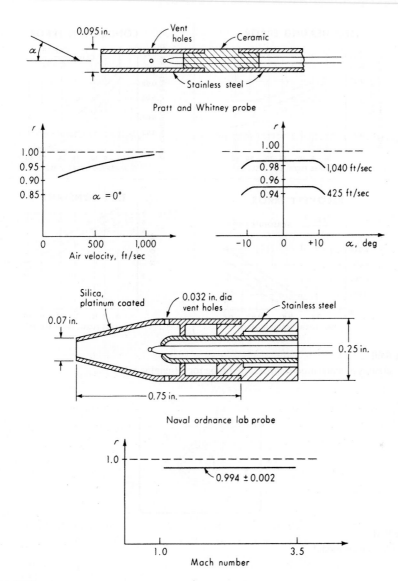

Figure 8.59
Stagnation-temperature probes.

8.9 DYNAMIC RESPONSE OF TEMPERATURE SENSORS

Since the conversion from sensing-element temperature to thermal expansion, thermoelectric voltage, or electrical resistance is essentially instantaneous, the dynamic characteristics of temperature sensors are related to the heat-transfer and -storage parameters that cause the sensing-element temperature to lag that of the measured

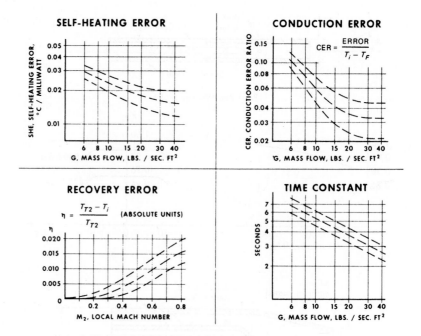

Figure 8.60
Characteristics of stagnation-temperature probe for jet-engine inlet.

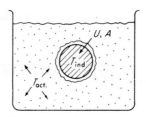

Figure 8.61
First-order sensor model.

medium. When a fluid-immersed sensing element is used "bare" (not in a protective well), often the model of Fig. 8.61 is adequate (see Fig. 8.17 for sensors attached to solids). Here heat losses are neglected, resistance to heat transfer is lumped in a single element, and energy storage is lumped in a single element. Conservation of energy gives

$$UA(T_{\text{act}} - T_{\text{ind}})\, dt = MC\, dT_{\text{ind}} \tag{8.84}$$

where $U \triangleq$ overall heat-transfer coefficient

$A \triangleq$ heat-transfer area

$T_{\text{act}} \triangleq$ actual temperature of surrounding fluid

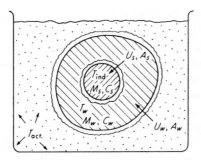

Figure 8.62
Second-order sensor model.

$T_{\text{ind}} \triangleq$ temperature indicated by sensor

$M \triangleq$ mass of sensing element

$C \triangleq$ specific heat of sensing element

This leads to

$$\frac{T_{\text{ind}}}{T_{\text{act}}}(D) = \frac{1}{\tau D + 1} \qquad (8.85)$$

where

$$\tau \triangleq \frac{MC}{UA} \qquad (8.86)$$

Clearly, speed of response may be increased by decreasing M and C and/or increasing U and A. Since U, in general, depends on the surrounding fluid and its velocity, τ is not a constant for a given sensor, but rather varies with how it is employed.

Since temperature sensors often are enclosed in protective wells[137] or sheaths, a thermal model taking into account heat-transfer resistance and energy storage in the well is of practical interest. Figure 8.62 shows such a configuration. Analysis gives

$$\frac{T_{\text{ind}}}{T_{\text{act}}}(D) = \frac{1}{\tau_w \tau_s D^2 + [\tau_w + \tau_s + M_s C_s/(U_w A_w)]D + 1} \qquad (8.87)$$

where $\tau_w \triangleq M_w C_w/(U_w A_w)$, time constant of well alone

$\tau_s \triangleq M_s C_s/(U_s A_s)$, time constant of sensor alone

We see that the addition of a well changes the form of response to second-order and increases the lag. The term $M_s C_s/(U_w A_w)$ is called the *coupling term* between the well and the sensor. If it is small compared with $\tau_w + \tau_s$, we have approximately

$$\frac{T_{\text{ind}}}{T_{\text{act}}}(D) = \frac{1}{\tau_w \tau_s D^2 + (\tau_w + \tau_s)D + 1} = \frac{1}{\tau_w D + 1}\frac{1}{\tau_s D + 1} \qquad (8.88)$$

which is just a cascade combination of the sensor and well individual dynamics.

---

[137]D. W. Richmond, "Selecting Thermowells for Accuracy and Endurance," *InTech.*, pp. 59–63, February 1980; J. A. Masek, "Guide to Thermowells," Omega Engineering, Stamford, CT, 1981 (www.omega.com).

The accuracy of the theoretical model can be increased by increasing the number of "lumps" of heat-transfer resistance and energy storage employed, the ultimate limit being an infinite number corresponding to a distributed-parameter (partial differential equation) rather than a lumped-parameter (ordinary differential equation) approach. When temperature sensors are employed as measuring devices in feedback-control systems, usually they are allowed to contribute no more than 30° phase lag at the frequency where the entire open-loop lag is 180°. Under such conditions, usually they are modeled adequately as simple first-order systems. The best[138] time constant to utilize for such a model is determined by an experimental ramp-input test and is numerically the steady-state time lag observed in such a test.

When greater accuracy is needed in utilizing the results of experimental tests to determine sensor dynamics, a model[139] using three time constants and a dead time may be employed. The transfer function is then

$$\frac{T_{\text{ind}}}{T_{\text{act}}}(D) = \frac{e^{-\tau_{dt}D}}{(\tau_1 D + 1)(\tau_2 D + 1)(\tau_3 D + 1)} \tag{8.89}$$

Numerical values of τ_1, τ_2, τ_3, and τ_{dt} may be obtained from step-function response tests. For example, a thermocouple used in a heavy-duty stainless-steel well had

$$\frac{T_{\text{ind}}}{T_{\text{act}}}(D) = \frac{e^{-2.6D}}{(21.6D + 1)(2.9D + 1)(2.1D + 1)} \tag{8.90}$$

where the time constants are in seconds.

When accurate numerical values are needed, generally experimental tests are required to determine temperature probe dynamics. For simple bare thermocouples, however, extensive research and testing have provided semiempirical formulas that allow calculation of the time constant with fair accuracy. One such relation[140] useful for temperatures from 160 to 1600°F, wire diameter 0.016 to 0.051 in, mass velocity 3 to 50 lbm/(ft$^2 \cdot$ s), and static pressure of 1 atm is

$$\tau = \frac{3{,}500\rho c d^{1.25}\, G^{-15.8/\sqrt{T}}}{T} \qquad \text{s} \tag{8.91}$$

where $\rho \triangleq$ average density of two thermocouple materials, lbm/ft^3

$c \triangleq$ average specific heat of two thermocouple materials, Btu/(lbm $\cdot$ F°)

$d \triangleq$ wire diameter, in

$G \triangleq$ flow mass velocity, lbm/(ft$^2 \cdot$ s)

$T \triangleq$ stagnation temperature, °R

---

[138]G. A. Coon, "Response of Temperature-Sensing-Element Analogs," *Trans. ASME,* p. 1857, November 1957.

[139]J. R. Louis and W. E. Hartman, "The Determination and Compensation of Temperature-Sensor Transfer Functions," *ASME Paper* 64-WA/AUT-13, 1964.

[140]R. J. Moffat, "How to Specify Thermocouple Response," *ISA J.,* p. 219, June 1957.

Within the above restrictions, this formula will predict τ for butt-welded junctions within about 10 percent. Another such result[141] based on tests for a Mach-number range of 0.1 to 0.9 and a Reynolds-number range of 250 to 30,000, gives

$$\tau = \frac{4.05\rho c d^{1.50}\{1 + [(\gamma - 1)/2]N_m^2\}^{0.25}}{p^{0.5}N_m^{0.5}T^{0.18}} \tag{8.92}$$

where $\gamma \triangleq$ ratio of specific heats

$N_m \triangleq$ Mach number

$p \triangleq$ static pressure, atm

This reference also presents a comprehensive analysis of conduction and radiation errors and the effects of differences in the thermal properties of the two metals used in a thermocouple.

Dynamic response tests of temperature sensors in the laboratory are not always accurate predictors of behavior when the sensor is installed in the process environment. Response degradations may be present in new sensors (if improperly installed) or may develop over time as a result of cracking of insulating cements or relocation of insulating powders used in probe construction. If actual installed response is critical to process safety and/or performance, *in situ* dynamic testing can provide the desired confidence. For both thermocouple and RTD sensors, the loop current step response (LCSR) technique[142] developed for nuclear power plant testing provides an *in situ* test method if the sensor lead wires may be disconnected for 15 to 60 min. The sensor circuit is subjected to a step increase (or decrease) of current, causing a heating (or cooling) effect, whose response can be analyzed to extract sensor time-constant information.

Dynamic Compensation of Temperature Sensors

When environmental conditions require a rugged temperature sensor, the mass may be so high as to cause a sluggish response. It may be possible to obtain an improved overall measuring-system response by cascading an appropriate dynamic compensation device with the sensor. Such schemes have been applied in practice with considerable success and may be implemented in a number of ways.[143] An *RC* network such as that shown for hot-wire anemometer compensation in Chap. 7 may be used if the sensor is essentially first-order. Second-order sensors also may be compensated,[144] operational-amplifier networks provide a convenient means of

---

[141]M. D. Scadron and I. Warshawsky, "Experimental Determination of Time Constants and Nusselt Numbers for Bare-Wire Thermocouples in High-Velocity Air Streams and Analytic Approximation of Conduction and Radiation Errors," *NACA, Tech. Note* 2599, 1952.

[142]T. W. Kerlin, H. M. Hashemian, and K. M. Peterson, "Time Response of Temperature Sensors," *ISA Trans.,* vol. 20, no. 1, 1981.

[143]C. E. Shepard and I. Warshawsky, "Electrical Techniques for Compensation of Thermal Time Lag of Thermocouples and Resistance Thermometer Elements," *NACA, Tech. Note* 2703, 1952; Louis and Hartman, op. cit.

[144]J. R. Louis and W. E. Hartman, op. cit.

implementation. Since the compensation is correct only for specific sensor dynamics, changes in numerical values or form of transfer function caused by changes in operating conditions can lead to loss of compensation. In general, increased speed of response is traded for overall sensitivity in such compensation schemes. This can be made up by additional amplification, but only up to a point; then the inherent noise level prevents further improvement. However, improvement of 100:1 or more is often possible. Dual-sensor thermocouple probes using two different wire sizes and the ratio of their two FFT-computed frequency spectra have shown some promise in compensation under varying flow conditions.[145]

8.10 HEAT-FLUX SENSORS

Requirements for measurement of local convective, radiative, or total heat-transfer rates have led to the development of several types of heat-flux sensors.[146] Here we review briefly the operating principles and characteristics of the most common types.

Slug-Type (Calorimeter) Sensors

In Fig. 8.63a a slug of metal is buried in (but insulated from) the surface across which the heat-transfer rate is to be measured. Neglecting losses through the insulation and the thermocouple wires, we may write

$$\text{Heat transferred in} = \text{energy stored}$$

$$Aq\,dt = Mc\,dT \tag{8.93}$$

where
$A \triangleq$ surface area of slug, cm^2
$q \triangleq$ local heat-transfer rate, W/cm^2
$M \triangleq$ mass of slug, kg
$c \triangleq$ specific heat of slug, $W \cdot s/(kg \cdot {}^\circ C)$
$T \triangleq$ slug temperature, $^\circ C$

Then

$$q = \frac{Mc}{A}\frac{dT}{dt} \tag{8.94}$$

and thus q may be determined by measuring dT/dt if Mc/A is known. Since the thermocouple reads T rather than dT/dt, a graphical, numerical, or electrical differentiation must be carried out to get q. For greater accuracy, the heat losses may be taken into account by modifying Eq. (8.94) to give

$$q = \frac{Mc}{A}\frac{dT}{dt} + K_l\,\Delta T \tag{8.95}$$

---

[145]D. L. Elmore, W. W. Robinson, and W. B. Watkins, "Dynamic Gas Temperature Measurement System," *NASA* CR-168267, 1983.

[146]T. E. Diller, "Advances in Heat Flux Measurements," vol. 23, Academic Press, New York, 1993, pp. 279–368. This excellent review has 306 references.

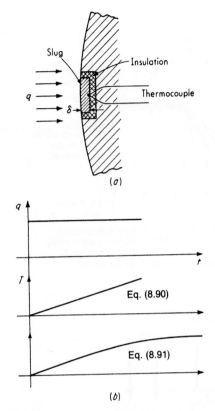

Figure 8.63
Slug-type and sandwich-type heat-flux sensors.

where $K_l \triangleq$ loss coefficient, W/(cm^2 · °C)

$\Delta T \triangleq$ temperature difference between slug and casing (usually taken as temperature rise of slug by assuming constant casing temperature)

The numerical values of Mc/A and K_l for a given sensor are determined by calibration. Equation (8.94) predicts that, for a constant q, T increases linearly with time and without limit. Actually, the unavoidable heat losses eventually make dT/dt approach zero, as shown by the more correct Eq. (8.95).

If a slug-type sensor is used to measure an essentially steady q and if losses are negligible, the slug temperature rises linearly and the slope is easily measured. Overheating of the slug may require that the heat source be "turned off" before damage occurs. The slug will now cool and the cooling rate can be used to estimate the heat loss, again by measuring the slope of the T versus t curve. Diller (p. 309) states that the cooling rate should be less than about 5 percent of the measured heating rate if the losses in Eq. (8.95) are to be neglected. He also discusses other considerations such as the fact that if a heat flux is by convection (with film

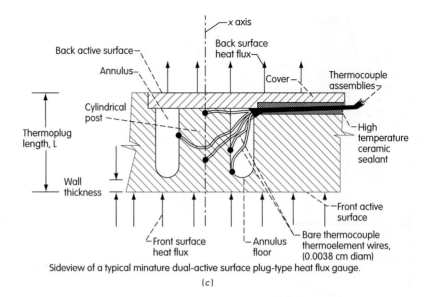

Sideview of a typical minature dual-active surface plug-type heat flux gauge.

(c)

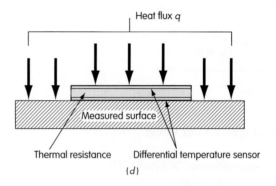

(d)

Figure 8.63
(Continued)

coefficient h) from a fluid stream at a constant temperature, the slug temperature changes exponentially with time constant Mc/hA. (This of course would *not* be a constant q; as the slug heats up, the temperature difference driving the heat flow gets smaller.) For this convection case, our assumption of a uniform temperature throughout the slug can be checked with a Biot number requirement, $hL/k < 0.1$, where L is the slug length and k is its conductivity.

While the above discussion reveals the basic principle of the slug-type sensor, a more correct and complex analysis[147] may be necessary in some cases. Here the

[147]C. H. Liebert, "Miniature Plug-Type Heat Flux Gages," *NASA TM* 106483, 1994; "Miniature High Temperature Heat Flux Gages," *NASA TM* 105403, 1992; "Heat Flux Measurements," *NASA TM* 101428, 1989.

rate of energy storage in the slug is the difference between the front-face heat transfer rate and the heat loss from the rear face. Assuming one-dimensional heat flow along the *x*-axis of the slug (of length *L* and conductivity *k*) we can write

Front surface heat flux = energy storage rate + back surface heat flux

$$q = \int_0^L \left(\rho c \cdot \frac{\partial T}{\partial t} \right) dx + k \left(\frac{\partial T}{\partial x} \right)_{x=L} \tag{8.96}$$

We again see the need to differentiate *T* with respect to time, but now we also need the spatial gradient $\partial T / \partial x$ and the time derivative must be known as a function of *x*. Figure 8.63*c* (Liebert, footnote 147) shows that we now need *several* thermocouples to get the needed information. The gage shown is used to measure *both* the front and back surface heat flux. If only the front surface flux is needed, the slug is made shorter so that it does not touch the cover plate, but rather has an air gap there. The terms in Eq. (8.96) are evaluated numerically from the measured data, using least-squares curve fits where needed. For greater accuracy, the variation of material properties with temperature was also taken into account.

Slug-type gages are usually fabricated by the user for a specific application (usually fast-transient measurements) and are less used than the "sandwich" type that we next discuss and that are available ready-made from various manufacturers. Figure 8.63*d* shows the concept, where a thin sheet of material serving as a thermal resistance R_t is sandwiched between two temperature sensors intended to measure the ΔT across the resistance, giving the heat flux *q* as $R_t \Delta T$. One version[148] uses thick-film microcircuit fabrication to produce the resistance element and a thermopile ΔT sensor with 1600 thermocouples/cm². Attachment is with two-sided, thermally conductive tape or the user's choice of adhesive. The 5 × 5 cm device is 0.6 mm thick, has a thermal resistance of 3C°/W, sensitivity of 25 mV/(W/cm²), rise time of about 0.06 s, and maximum temperature of 150°C. It also includes a single thermocouple that measures the face temperature. The thermal resistance of such devices should be as small as possible to minimize the local disturbance to heat-transfer conditions. This small thermal resistance makes the measured ΔT very small, but the multijunction thermopile overcomes this sensitivity problem. The front surface is coated with Zynolyte (a proprietary black coating) over a colloidal graphite prime to give an emissivity of 0.94, guaranteeing that the incoming radiant-heat flux will be almost entirely absorbed and measured, rather than being partially reflected.

Using a similar concept but *thin*-film methods, Vatell Corp. offers a series of Heat Flux Microsensors[149] that use the thermopile ΔT sensor but a thin-film platinum RTD for measurement of the front surface temperature. The 0.25-in diameter units can be flush mounted in a surface or attached to the end of a probe, and have response times of about 6 μs, sensitivity of 8 to 150 μV/(W/cm²), maximum face

---

[148]Episensor, Vatell Corp., Christiansburg, VA, 540-961-3576 (www.g3.net/vatell).

[149]J. M. Hager et al., "In-Situ Calibration of a Heat Flux Microsensor Using Surface Temperature Measurements," Vatell Corp., 1994.

(e)

Figure 8.63
(Concluded)

temperatures from 300 to 850°C, and minimum detectable heat fluxes of 0.01 to 0.25 W/cm². Figure 8.63*e* (Vatell and Virginia Tech Mechanical Engineering Department) shows several of these sensors embedded in a 3-fold scaled-up model of a gas turbine blade used to measure convective heat flux and thus the heat transfer coefficient at several points on the blade.

Steady-State or Asymptotic Sensors (Gardon Gage)

Figure 8.64 shows the essential features of this type of sensor, which was first proposed by R. Gardon.[150] A thin constantan disk is connected at its edges to a large copper heat sink, while a very thin (<0.005-in-diameter) copper wire is fastened at the center of the disk. This forms a differential thermocouple between the disk center and its edges. When the disk is exposed to a constant heat flux, an equilibrium temperature difference is rapidly established which is proportional to the heat flux. Since the thermocouple signal is now directly proportional to the heat flux, no differentiating process (such as is required in a slug-type sensor) is necessary. Furthermore, loss corrections generally are not needed, nor is a thermocouple

[150]R. Gardon, "An Instrument for the Direct Measurement of Intense Thermal Radiation," *Rev. Sci. Instrum.*, p. 366, May 1953.

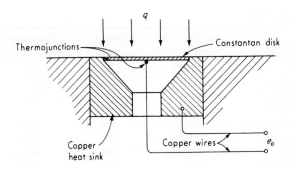

Figure 8.64
Gardon gage.

reference junction required. Instrument response[151] is approximately of the first-order type; thus

$$\frac{e_o}{q}(D) = \frac{K}{\tau D + 1} \tag{8.97}$$

where

$$K \triangleq \frac{d^2 K_e}{16 \delta k} \tag{8.98}$$

$$\tau \triangleq \frac{\rho c d^2}{16 k} \tag{8.99}$$

and $d \triangleq$ diameter of disk

$\delta \triangleq$ thickness of disk

$K_e \triangleq$ thermocouple sensitivity, mV/F°

$k \triangleq$ thermal conductivity of disk

$c \triangleq$ specific heat of disk

For copper/constantan, numerical values are

$$\frac{e_o}{q}(D) = \frac{0.0308(d^2/\delta)}{(5.96d^2)D + 1} \tag{8.100}$$

where d and δ are in inches, e_o is in millivolts, q is in Btu/(s · ft²), and $5.96d^2$ is in seconds. See Table 8.2 for other materials. Typical commercial units[152] are available for full-scale heat fluxes of 5 to 5000 W/cm², produce 10-mV full-scale output, and have time constants as fast as 0.0015 s.

Convenient foil-type heat-flow sensors which may be cemented directly to the measured surface and which use a differential thermopile to measure the ΔT across a polyamide plastic film are available.[153] The hot and cold junctions of the

---

[151]*Rept. HTL*-ER-4, op. cit.

[152]Vatell Corp.

[153]Micro-Foil Heat Flow Sensors, RdF Corp.

Table 8.2

Material	Specific heat, Btu/(lbm · °F)	Density, lbm/ft³	Conductivity, (Btu/h)/(ft · °F)
Iron	0.1065	491	41.5
Copper	0.0921	559	232
Chromel	0.107	545	11.1
Alumel	0.125	537	17.2
Constantan	0.094	556	12.6
Platinum	0.0324	1339	41.0
Nisil	0.12	543	16.6
Nicrosil	0.11	532	8.67

All these properties vary with temperature, so this table is intended only for estimates.

thermopile (Chromel/constantan or copper/constantan) are attached to opposite sides of the plastic (0.0005 to 0.01 in thick) to form a "sandwich." When the entire package (0.003 to 0.012 in thick, 0.3 × 0.5 in area) is cemented to the surface, any heat transferred must pass through the plastic film, causing a proportional ΔT, which is measured by the thermopile. If desired, a single thermocouple for measuring surface temperature itself may be included in the package. Maximum operating temperature is 500°F. In addition to the usual specifications, the manufacturer provides data on sensor thermal capacitance and resistance, allowing estimation of sensor disturbance effects.

Application Considerations

The introduction of the sensor into the wall locally alters the thermal properties of the wall and causes the measured heat flux to differ from that which would occur if the sensor were not present.[154] Thus it is desirable to match, insofar as feasible, the thermal properties of sensor and wall. For a Gardon gage, there is a radial temperature gradient over the disk which, if excessive, causes a variation in local convection coefficient and thus an error. By sacrificing sensitivity (and then recovering it by external amplication, if necessary) the temperature gradient and associated error may be reduced. When only the radiation component of the total flux is desired, the front of the sensor is covered with a thermally isolated sapphire window which passes the radiation flux but blocks the convective flux. The Diller reference (footnote 146) gives a very complete discussion of these and other design and application problems.

---

[154]D. R. Hornbaker and D. L. Rall, "Thermal Perturbations Caused by Heat-Flux Transducers and Their Effect on the Accuracy of Heating-Rate Measurements," *ISA Trans.*, p. 123, April 1964.

PROBLEMS

8.1 An Invar/brass cantilever bimetal (see Fig. 8.3b) has $t = 0.05$ in, a length of 2 in, a width of 0.5 in, and $t_A = t_B$ and $n + 1/n \approx 2$. Estimate the end deflection for temperature changes of 30 and 60°C. If the end is held fixed, estimate the force developed for temperature changes of 30 and 60°C.

8.2 A thermometer is to have a sensitivity of 10 in/°C when mercury is used in the neighborhood of room temperature. Obtain an expression relating capillary cross-section area and bulb volume to meet this requirement. Obtain expressions for the time constant if the bulb is spherical and, if it is cylindrical with a length equal to 5 diameters. Also find the ratio of these two time constants.

8.3 Sketch and explain the operation of a bimetallic compensator to replace the auxiliary pressure sensor in Fig. 8.6. Can both case and capillary compensation be obtained? Explain.

8.4 Analyze the system of Fig. 8.13a to obtain a steady-state relation between hot-gas temperature as an input and thermocouple voltage as an output.

8.5 Develop equations to estimate the dynamic response of the system of Fig. 8.13a.

8.6 Repeat Prob. 8.4 for the system of Fig. 8.13b.

8.7 In Fig. 8.14, if $\tau = 2.8$ s, $T_{gas} = 5000°F$, and the thermocouple damage limit is 2000°F, how long can the cooling be left off if the steady-state cooled thermocouple temperature is 500°F?

8.8 A resistance-thermometer circuit as in Fig. 8.21d has $R_1 = R_2 = 10,000\ \Omega$ and is to cover the temperature range 0 to 400°C. The thermometer element is platinum with a resistance of 1,000 Ω at 200°C. Plot the curve of bridge output voltage (open circuit) versus input temperature, using the data of Fig. 8.19. See Chap. 10 for bridge-circuit equations. Bridge excitation is 20 V.

8.9 A 500-Ω resistance thermometer carries 5-mA current. Its surface area is 0.5 in², and it is immersed in stagnant air, so that the heat-transfer coefficient is $U = 1.5$ Btu/(h · ft² · F°). Find its self-heating error. What would be the error in water with $U = 100$ Btu/(h · ft² · F°)?

8.10 A pulse-excited resistance thermometer (see Fig. 8.22) has an excitation voltage in the form of a rectangular pulse of 100-V height and 0.1-s duration. The pulse is on for 0.1 s and off for 0.9 s in a repetitive cycle. Compute the ratio of peak/rms voltage for this pulse. What average heating power would this voltage pulse produce in a 500-Ω resistor?

8.11 We want to analyze a proposed design for an isothermal block to be used for thermocouple reference junctions. The metal part is aluminum, 2 cm thick, 7 cm wide, and 30 cm long. It is surrounded on all sides by polyethylene-foam insulation, 2 cm thick. Find out how much the aluminum temperature varies from end to end when the two foam ends are exposed to different air temperatures.

8.12 Estimate the percentage of total power found above 10 μm for blackbody radiation at $T = 400\ K$.

8.13 Explain the disadvantage of a large time constant in a thermal-radiation detector using a chopper.

8.14 Derive Eq. (8.68).

8.15 Consider a subsonic air flow in a duct. Pitot-static-tube measurements give a static pressure of 100 lb/in^2 absolute and a stagnation pressure of 129.1 lb/in^2 absolute. A temperature probe with a recovery factor $r = 0.80$ extends from a 100°F wall a distance of 1 ft into the flow. The probe thermocouple reads 400°F. The probe support has a radius of 0.02 ft and a thermal conductivity of 100 Btu/(h · ft · F°). The end of the probe may be assumed insulated, and the surface convection coefficient is 10 Btu/(h · ft^2 · F°). Radiation effects are negligible. Calculate the static temperature of the flow.

8.16 Derive Eq. (8.87).

8.17 Make and analyze a third-order model of a temperature sensor analogous to the second-order model of Fig. 8.62.

8.18 A butt-welded 0.03-in bare-wire copper/constantan thermocouple is used to measure the temperature (near 100°F) of atmospheric air flowing at 100 ft/s. Estimate the time constant, using both formulas available.

8.19 In Eq. (8.96), explain how the measured data would be used to evaluate the two terms to the right of the equal sign.

8.20 Compare the sensitivity and response speed of Gardon gages of like diameter and thickness but made of (*a*) constantan disk, copper heat sink; (*b*) copper disk, constantan heat sink; (*c*) iron disk, constantan heat sink; (*d*) constantan disk, iron heat sink.

8.21 Sketch and explain a test setup for evaluating the step-function response of temperature sensors exposed to air flows of different velocities.

8.22 Sketch and explain a test setup for static calibration of heat-flux sensors.

8.23 Sketch and explain a test setup for step-function testing of heat-flux sensors.

8.24 Perform a linearized dynamic analysis of the venturi pneumatic pyrometer (Fig. 8.15).

8.25 Explain the operation of the circuits of Fig. 8.29*a*.

8.26 Derive Eq. (8.35).

8.27 Combine Eqs. (8.74*a*) and (8.74*b*) to solve for T_f, assuming that T_w, T_p, and all the physical parameters are known. Then write a computer program (a spreadsheet-type software such as MATHCAD is acceptable) to implement this solution. Include also in this program, for comparison, the T_f solutions from Eqs. (8.70) and (8.72). Then make up some reasonable numbers for all the needed values and compare the results from the three solutions for a variety of situations.

BIBLIOGRAPHY

1. R. J. Thorn and G. H. Winslow, "Recent Developments in Optical Pyrometry," *ASME Paper* 63-WA-224, 1963.

2. R. H. Tourin, "Recent Developments in Gas Pyrometry by Spectroscopic Methods," *ASME Paper* 63-WA-252, 1963.

3. P. S. Schmidt, "Spectroscopic Temperature Measurements," *Inst. Tech.*, pp. 35–38, December 1975.

4. G. Conn and D. Avery, "Infrared Methods," Academic, New York, 1960.

5. D. Greenshields, "Spectrometric Measurements of Gas Temperatures in Arc-Heated Jets and Tunnels," *NASA, Tech. Note* D-1960, 1963.

6. J. Grey, "Thermodynamic Methods of High-Temperature Measurement," *ISA Trans.*, vol. 4, no. 2, 1965.

7. B. Bernard, "Flame Temperature Measurements," *Instrum. Contr. Syst.*, p. 113, May 1965.

8. A. G. Gaydon, "The Spectroscopy of Flame," Wiley, New York, 1957.

9. T. R. Billeter, "Using Microwave Techniques for High Temperature Measurement," *Inst. & Cont. Syst.*, pp. 107–109, February 1972.

10. "Bibliography of Temperature Measurement 1953–1969," *NBS* SP-373, 1972.

11. S. S. Fam, L. C. Lynnworth, and E. H. Carnevale, "Ultrasonic Thermometry," *Inst. & Cont. Syst.*, pp. 107–110, October 1969.

12. G. D. Nutter, "Recent Advances and Trends in Radiation Thermometry," *ASME Paper* 71-WA/Temp-3, 1971.

13. T. J. Quinn, "Temperature," Academic, New York, 1983.

14. T. D. McGee, "Principles and Methods of Temperature Measurement," Wiley, New York, 1988.

15. R. I. Solonkhin and N. H. Afgan, "Measurement Techniques in Heat and Mass Transfer," Hemisphere, New York, 1985.

16. D. P. Dewitt and G. D. Nutter, "Theory and Practice of Radiation Thermometry," Wiley, New York, 1988.

17. J. G. Webster (ed.), "The Measurement, Instrumentation, and Sensors Handbook," chap. 6, CRC Press, Boca Raton, FL, 1999.

9 CHAPTER

Miscellaneous Measurements

9.1 TIME, FREQUENCY, AND PHASE-ANGLE MEASUREMENT

The fundamental standard of time is discussed in Sec. 4.2. The United States Frequency Standard is a cesium atomic fountain clock operated by NIST in Boulder, CO,[1] and has an uncertainty of 1.7 parts in 10^{15}. Time and frequency signals with varying degrees of accuracy are radio broadcast by NIST and are available to any laboratory with an appropriate receiver. Time code signals are also available from the Global Positioning System, as discussed in Sec. 4.16. A comprehensive history of time and frequency measurement is available from NIST.[2] Useful technical notes are also available from manufacturers of time- and/or frequency-measurement instruments.[3]

Perhaps the most convenient and widely utilized instrument for accurate measurement of frequency and time interval is the *electronic counter-timer*. Figure 9.1 gives a block diagram showing the basic operation of such devices. The instrument's time and frequency standard is a piezoelectric crystal oscillator[4] which generates a voltage whose frequency is very stable since the crystal is kept in a temperature-controlled oven. A typical frequency is 10^7 Hz, while the drift in frequency may be of the order of 3 parts in 10^7 per week. Over time, this gradual drift can cause errors; thus highly accurate measurements require periodic recalibration of the oscillator against a suitable standard such as the radio-broadcast signals. Instruments with various clock frequencies and stabilities are available. Quartz crystal clocks (oscillators) use various approaches to deal with the effects of temperature on the clock frequency, which of course affects all the timing and

---

[1]"NIST Launches New Primary Frequency Standard," *Cal Lab,* Jan.–Feb. 2000, pp. 14–15; M. A. Lombardi, "Traceability in Time and Frequency Metrology," *Cal Lab,* Sept.–Oct. 1999, pp. 33–40.

[2]"From Sundials to Atomic Clocks: Understanding Time and Frequency," *NIST MN* 155, 1999 ed.

[3]"The Science of Timekeeping," *AN* 1289, Dec. 2000, Agilent Technologies, 800-452-4844.

[4]V. E. Bottom, "Introduction to Quartz Crystal Unit Design," Van Nostrand, New York, 1982.

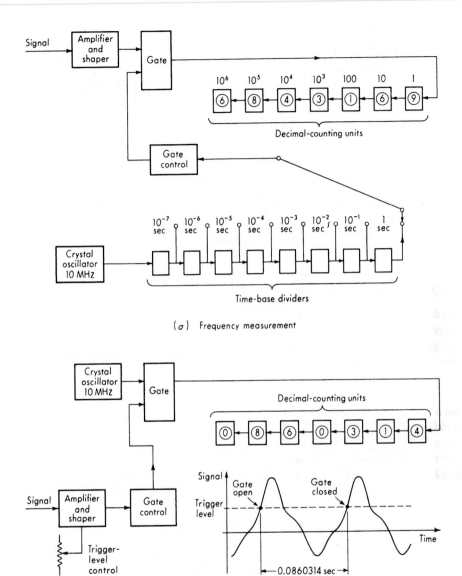

Figure 9.1
Basic counter applications.

frequency measurements. If the crystal is allowed to drift with room temperature (called *RTXO*), a frequency accuracy of about 3 parts per million (ppm) for 0 to 50°C is typical. Crystals can be temperature compensated (TCXO), providing an accuracy of 0.5 ppm. Using a feedback-controlled oven to stabilize the crystal temperature (OCXO), a frequency accuracy of about 0.003 to 0.2 ppm is achieved,

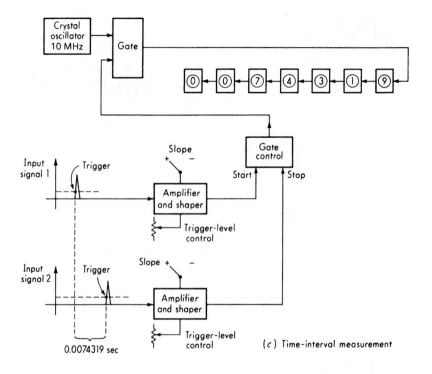

(c) Time-interval measurement

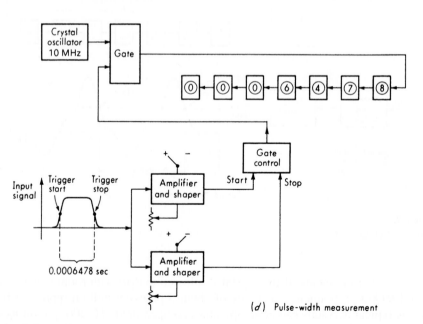

(d) Pulse-width measurement

Figure 9.1
(Continued)

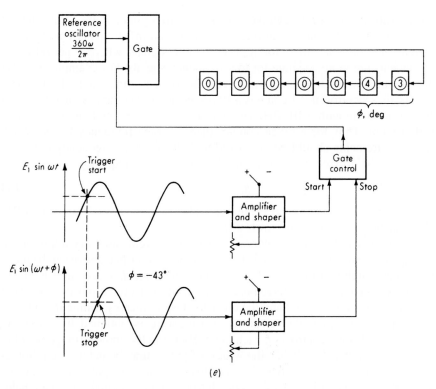

Figure 9.1
(Concluded)

depending on the quality of the oven. In addition to temperature errors there is also long-term drift. Using signals from the Global Positioning System, quartz oscillators can be calibrated, as needed, to give an accuracy of about 1 part in 10^{12} per day. Resolution of time interval measurements for the best instruments can be as good as 50 ps for a "single shot" event or 1 ps when averaging is used.

In Fig. 9.1*a* the instrument is set up for frequency measurement of a signal whose frequency is 6,843,169 Hz. This is accomplished by allowing the signal (suitably "shaped" to define each cycle more precisely) to go through a gating circuit to the decimal-counting units for a precisely timed interval. This interval may be selected in 10 : 1 steps from 10^{-7} to 1 s. Thus in the 1-s interval used in Fig. 9.1*a* the counters accumulate 6,843,169 pulses. This mode of operation is also called EPUT (events per unit time). To measure the frequency *ratio* of two signals *A* and *B,* use the scheme of Fig. 9.1*a,* connecting signal *A* as the "signal" but using signal *B* in place of the crystal oscillator (most counters provide for such an "external clock").

Sometimes it is more desirable to measure the period (rather than the frequency) of a signal. Figure 9.1*b* shows the arrangement used for this measurement. The trigger-level control is adjusted so that triggering occurs on the steepest part of the signal waveform to reduce error. There is usually provision for triggering on either a positive slope or a negative slope, as desired. Since there is an inherent potential error of ± 1 count in turning the gate on and off, frequency-mode measurements are more accurate for high-frequency signals whereas period-mode measurements are more accurate for low-frequency signals. For example, a 10-Hz signal measured in the frequency mode with the usual 1-s time interval gives only 10 counts; thus an error of ± 1 count is a 10 percent error. The same signal measured in the period mode with a 10-MHz counter gives 10^6 counts and an error of only 0.0001 percent. Thus for a given counter there is some frequency below which period measurements should be employed and above which frequency measurements should be utilized. For a 1-s sampling period, this frequency f_0 is given by

$$f_0 = \sqrt{f_c} \tag{9.1}$$

where $f_c \triangleq$ frequency of crystal oscillator ("clock" frequency). Thus a 10-MHz clock-frequency counter has $f_0 = 3,160$ Hz.

Measurements of the time interval between two events are very important in many experimental studies. The basic building blocks described above can be interconnected in a slightly different fashion, as in Fig. 9.1c and d, to accomplish this. In Fig. 9.1c two separate events have been transduced to electrical pulses; one event pulse is used to open the gate, and the other to close it, thereby timing the interval between them. Considerable versatility in triggering is obtained by providing trigger-level and slope controls on each input. By using the above arrangement but only one input signal (Fig. 9.1d), the widths of pulses may be determined. Sometimes additional circuits are provided to send to an oscilloscope pulses that show the exact point on the incoming signals at which triggering is initiated. These are helpful in adjusting the trigger-level and slope controls and in interpreting the resulting information.

Often measurement of the phase angle between two sinusoidal signals of the same frequency f_s is required. A general-purpose digital counter-timer can be employed for such measurements, as shown in Fig. 9.1e. To use this method, the amplitude of the two signals must be made equal and the triggering point of the two channels adjusted to be the same. Then the phase angle can be read directly with a resolution of $1°$ for the setup shown, or $0.1°$ if the reference frequency is set at $3,600 f_s$.

To prevent false counts as a result of the unavoidable noise on input signals, counter input circuits use a Schmitt trigger type of circuit with a built-in hysteresis effect (see Fig. 9.2). When frequency is measured, to cause one count in the counting register, a signal must cross *both* the upper and the lower hysteresis levels. [For period or time-interval measurement the gate is opened (or closed) when the signal crosses a selected hysteresis level (either upper or lower).] The difference between the two hysteresis levels is called the *counter sensitivity,* and typically this might be 150 mV peak to peak. This is really a *threshold* sensitivity since it fixes the amplitude of the smallest countable signal. But note that the optimum value is not necessarily the lowest, because of the false triggering problem mentioned earlier. The major error sources[5] in counter-timers are categorized as:

---

[5]"Fundamentals of the Electronic Counters," *Appl. Note* 200, Hewlett-Packard Corp., Palo Alto, CA, 1978.

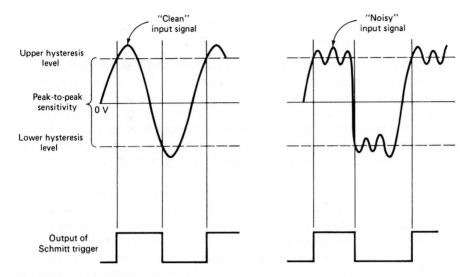

Figure 9.2
Use of counter hysteresis to reject noise.

1.   The ± 1 count ambiguity, a random error
2.   The time-base (clock) error, resulting from temperature and line-voltage changes, aging, and short-term instability
3.   The trigger error, a random error due to noise that causes early or late crossing of the hysteresis level
4.   Channel-mismatch error, a systematic error present in two-channel measurements when input-circuit rise times and/or propagation delays are not identical for the two channels.

Frequency measurements are subject only to the ± 1 count and time-base errors. Period measurements have these plus the trigger error, while time-interval measurements suffer from all four. A high-performance instrument utilized within 1 week of time-base calibration and within the usual line-voltage (10 percent) and temperature (0 to 50°C) limits might have a time-base error of 1 part in 10^8. Channel-mismatch error might be 1 ns and could be reduced by calibration of this effect. Analysis[6] of the trigger error gives the standard deviation of the timing error as

$$\text{rms trigger error} = \sqrt{\frac{N_c^2 + N_{sa}^2}{\dot{V}_a^2} + \frac{N_c^2 + N_{sb}^2}{\dot{V}_b^2}} \qquad (9.2)$$

where $N_c \triangleq$ counter noise, V rms (typically 0.2 to 2.0 mV)
$\quad\quad\quad\ N_{sa} \triangleq$ start-signal noise, V rms
$\quad\quad\quad\ N_{sb} \triangleq$ stop-signal noise, V rms

---

[6]Ibid.; "Understanding Frequency Counter Specifications," *Appl. Note* 200-4, Hewlett-Packard Corp.

$$\dot{V}_a \triangleq \text{slew rate of start signal at trigger point, V/s}$$
$$\dot{V}_b \triangleq \text{slew rate of stop signal at trigger point, V/s}$$

The importance of using timing pulses with high slew rates is clear from this result.

Since even simple counter-timers allow time or period measurements of very high *resolution* (six or seven digits may be displayed), consideration of the above errors is very important, since total accuracy may turn out to be *much* less than resolution. This total error includes not only the contributions of the counter, but also any uncertainties in the sensors that transduce some physical event to the voltage which the timer accepts as input (for example, see Fig. 4.70). These sensor-related errors may completely overshadow those of the timer.

While the description of Fig. 9.1 explains the operation of most basic counter-timers, much more sophisticated instruments that employ microprocessor technology to extend capabilities, automate procedures, and provide calculating features are available.[7] The referenced instrument still uses a 10-MHz crystal clock as the basic timing source: however, an interpolation scheme increases time resolution from the 100 ns of the clock itself to about 1 ns for the instrument and removes the ± 1 count uncertainty always present in "ordinary" counters. Automatic trigger-setting circuits allow implementation of many useful features, such as 10 to 90 percent rise- (or fall-) time measurements and slew-rate measurements. Rise time is found by setting trigger levels at the 10 and 90 percent levels (this is automatic) and making a time-interval measurement, which is stored in a temporary register. Then the two trigger levels are measured automatically, and the microprocessor subtracts these and divides by the rise time to obtain and display slew rate in volts per second. Many other automatic features such as nonambiguous phase-angle measurement and various statistical calculations may be selected by push-button.

Another common method of phase-angle measurement involves cross-plotting the two sinusoidal signals against each other, by using an *XY* plotter for very low frequencies and an oscilloscope for high frequencies. The cross plot can be shown to be an ellipse, and suitable measurements on this ellipse give the phase angle (see Fig. 9.3*a*). We have

$$e_i = E_i \sin \omega t$$

and

$$e_o = E_o \sin(\omega t + \phi)$$

If we set $t = 0$, $e_i = 0$, and $e_o = E_o \sin \phi$, then

$$\sin \phi = \frac{e_o|_{e_i=0}}{E_o} \tag{9.3}$$

Since $e_o|_{e_i=0}$ has two values (plus and minus), the quadrant of ϕ is ambiguous; however, usually this can be resolved by visual observation of the two sine waves plotted against time (say on a dual-beam oscilloscope) or from knowledge of the system characteristics. The direction of travel of the "spot" as it plots the ellipse also

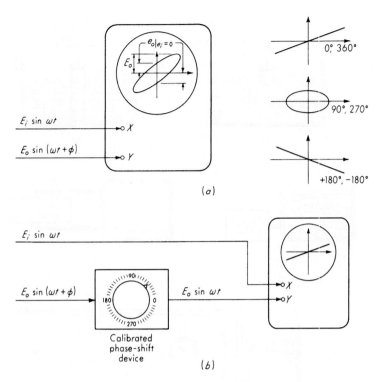

Figure 9.3
Phase angle from Lissajous figure.

resolves this difficulty, but may be hard to detect at high frequencies. An alternative method employing the same basic principle but a null technique is shown in Fig. 9.3b. Here the calibrated phase-shift circuit is adjusted until the ellipse degenerates into a straight line (0° phase shift). Then the phase angle ϕ is read directly from the phase-shifter dial. When the "sinusoidal" signals are noisy and/or distorted, special phase meters and tracking filters[8] may be necessary for accurate measurement. The FFT signal/system analyzers discussed in Chap. 10 can handle such problems also.

9.2 LIQUID LEVEL

Measurement and/or control of liquid level[9] in tanks is an important function in many industrial processes and in more exotic applications, such as the operation and fueling of large liquid-fuel rocket motors. Figure 9.4 illustrates some of the more common methods of accomplishing this measurement.

---

[8]E. O. Doebelin, "System Modeling and Response," p. 237, Wiley, New York, 1980.

[9]J. McIntyre, "Sorting out Liquid Level Sensors," *I & CS*, Feb. 1992, pp. 31–34; G. Voss, "The Principles of Level Measurement," *Sensors*, Oct. 2000, pp. 55–64.

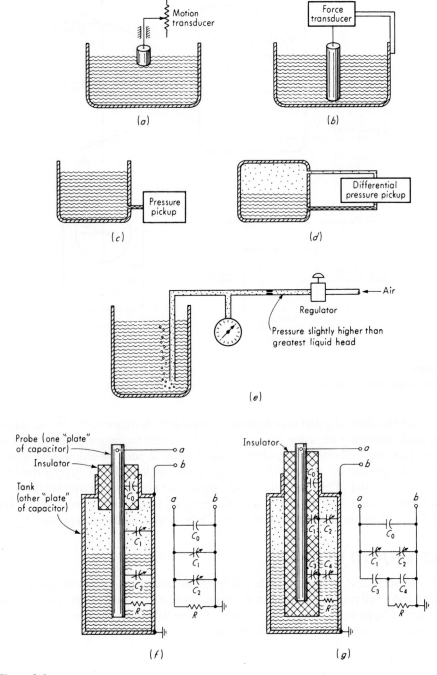

Figure 9.4
Liquid-level measurement.

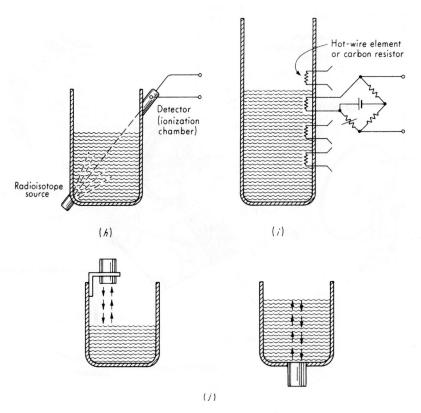

Figure 9.4
(Continued)

The simple float of Fig. 9.4*a* can be coupled to some suitable motion transducer to produce an electrical signal proportional to the liquid level. Figure 9.4*b* shows a "displacer" which has negligible motion and measures the liquid level in terms of buoyant force by means of a force transducer. A unit[10] intended for monitoring oil levels in engines, transmissions, and gearboxes uses a hollow cylindrical displacer restrained by springs. A full-scale change in the level causes a motion of about 0.070 in, which is measured by a Hall effect sensor to give a voltage change ranging from 1 V (empty) to 4 V (full). Note that a displacer (in contrast to a float) will not change its reading if the liquid level changes because of temperature change only. This is desirable since, when temperature changes (common in engines and gearboxes), the *quantity* of fluid has not changed, so a level sensor should *not* *indicate* a change. Since hydrostatic pressure is related directly to liquid level, the pressure-sensing schemes of Fig. 9.4*c* and *d* allow measurement of the liquid level in open and pressure vessels, respectively. In the "bubbler" or purge system of Fig. 9.4*e,* the gas pressure downstream of the flow restriction is the same as the

---

[10]Lucas Schaevitz Sensors, Hampton, VA, 800-745-8008 (www.schaevit.com).

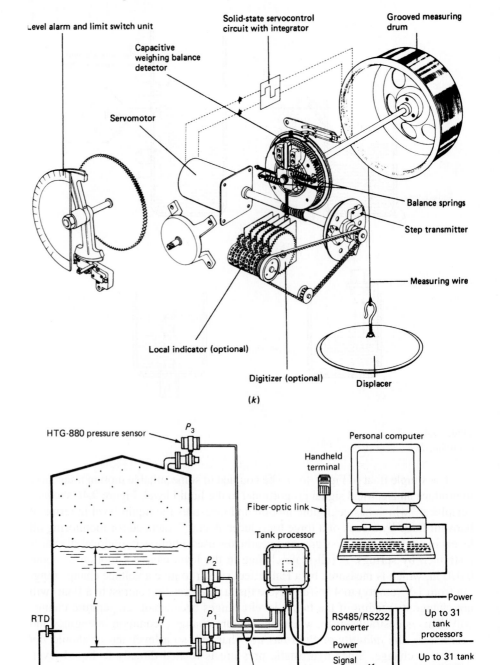

Level alarm and limit switch unit

Solid-state servocontrol circuit with integrator

Grooved measuring drum

Capacitive weighing balance detector

Servomotor

Balance springs

Step transmitter

Measuring wire

Local indicator (optional)

Digitizer (optional)

Displacer

(k)

HTG-880 pressure sensor

P_3

Handheld terminal

Personal computer

Fiber-optic link

Tank processor

P_2

L

H

P_1

RTD

Intrinsically safe wiring

RS485/RS232 converter

Power

Signal

Up to 31 tank processors

Power

Up to 31 tank processors

Up to 31 tank processors

(l)

Figure 9.4
(Continued)

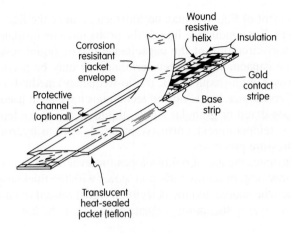

Construction of Metritape sensor

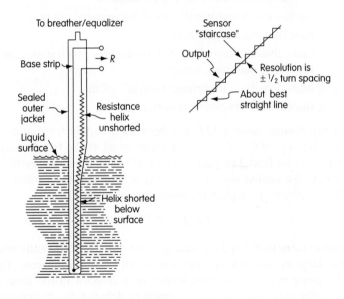

Metritape sensor operation

Figure 9.4
(Concluded)

hydrostatic head above the bubble-tube end. The flow of gas is quite small; a bottle of nitrogen used as a source of pressurized gas may last six months or more.

Capacitance variation[11] has been employed in various ways for level sensing. For essentially nonconducting liquids (conductivity less than 0.1 μS/cm^3), the

_____

[11]Magnetrol Intl., Downers Grove, IL, 630-969-4000 (www.magnetrol.com).

bare-probe arrangement of Fig. 9.4*f* may be satisfactory since the liquid resistance *R* is sufficiently high. For conductive liquids the probe must be insulated as in Fig. 9.4*g,* to prevent short-circuiting of the capacitance by the liquid resistance. The measurement of the capacitance between terminals *ab* may be accomplished in several ways. However, high-frequency ac (radio-frequency) methods offer significant advantages. Capacitance level-sensing techniques have been used with many common liquids, powdered or granular solids, liquid metals (high temperatures), liquefied gases (low temperatures), corrosive materials such as hydrofluoric acid, and in very-high-pressure processes.

Figure 9.4*h* illustrates the use of radioisotopes for level measurement. Since the absorption of beta-ray or gamma-ray radiation varies with the thickness of absorbing material between the source and the detector, a signal related to tank level may be developed. For analyzing such arrangements we may use the law

$$I = I_o e^{-\mu\rho x} \tag{9.4}$$

where $I \triangleq$ intensity of radiation falling on detector

$I_0 \triangleq$ intensity at detector with absorbing material not present

$e \triangleq$ base of natural logarithms

$\mu \triangleq$ mass absorption coefficient (constant for given source and absorbing material), cm^2/g

$\rho \triangleq$ mass density of absorbing material, g/cm^3

$x \triangleq$ thickness of absorbing material, cm

The gamma-ray source, cesium 137, has been used widely for liquid-level measurements and has a μ of 0.077 cm^2/g for water or oil, 0.074 for aluminum, 0.072 for steel, and 0.103 for lead. For gaging a tank of water, then, if a vertical radiation path (rather than the angled one of Fig. 9.4*h*) is assumed, the variation of *I* with liquid height *h* is given by (neglecting absorption of air path)

$$I = I_0 e^{-0.077h} \tag{9.5}$$

This exponential relation of *I* and *h* is nearly linear only for sufficiently small values of *h*. For *h* as large as, say, 100 cm, the nonlinearity is quite apparent. This can be overcome by using either a radiation source or a detector in the form of a strip oriented vertically rather than a "point" source or detector. Such arrangements are nonlinear for small values of $\mu\rho x_{max}$. Therefore point-to-point configurations are indicated for small ranges, whereas larger ranges require the more complex strip-to-point type. For a strip source (or detector), the strength (sensitivity for a detector) can be "tailored" to vary in just the right way with position along the strip to give a linear tank-level/detector-signal relation.

Figure 9.4*i* shows the method of using hot-wire or carbon resistor elements for the measurement of liquid level in discrete increments. The basic concept is that the heat-transfer coefficient at the surface of the resistance element changes radically when the liquid surface passes it. This changes its equilibrium temperature and thus its resistance, causing a change in bridge output voltage. By locating resistance elements at known height intervals, the tank level may be measured in discrete increments. Such arrangements have been used in filling fuel tanks of large rocket engines with cryogenic liquid fuels.

Ultrasonic "range-finding" techniques, discussed in Chap. 4, can be applied to liquid-level sensing, as in Fig. 9.4*j*. The "radar" liquid level sensors use similar principles but with *microwave* energy rather than acoustic. Two versions, *noncontact* and *guided wave,* are available. Details[12] of the radar devices and a comparison with capacitance-level control are available from one of the manufacturers.

An interesting variation on the classic float of Fig. 9.4*a* is found in the servo-positioned level gage of Fig. 9.4*k,*[13] widely employed in deep tanks and underground caverns to 500-ft depth. The thin displacer has density (relative to liquid in tank) such that it would sink if it were not supported by the measuring wire, which applies just enough lifting force to support the displacer at a design immersion of 1 to 2 mm. The lifting force is maintained at this value (as the tank level rises and falls) by the servodrive, whose error signal comes from the elastic force sensor which deflects from neutral in proportion to the difference between lifting force and the design value set by adjustment of the balance springs. A capacitance displacement pickoff transduces elastic deflection to an electrical error signal for the servomotor. Competing (nonservo) gages use a mechanical "constant-force" spring to supply the lifting force, being otherwise similar except that displacer immersion is much larger, typically 10 mm. These simpler and cheaper all-mechanical devices suffer, however, from several error sources which are minimized in the servotype gage: increased friction, variation in constant-force spring lift, greater sensitivity to liquid-density change because of greater immersion, and greater cable-weight effect as a result of using a heavier tape (rather than a light wire) to suspend the displacer. Comparison of overall error for a 20-m-deep tank shows ± 13.5 mm for mechanical gages and ± 2.9 mm for servogages. Since 1 mm in a 100-ft-diameter tank is 4.6 bbl of product (at, say, $40 per barrel), the economic importance of accuracy in a system used for custody transfer of product becomes obvious.

Hydrostatic tank gaging (HTG), an advanced version of Fig. 9.4*c* and *d,* is becoming attractive[14, 15] for large tanks where the measurements of density and total contained mass are required. Figure 9.4*l* shows a typical system using three vibrating-wire ("digital") pressure sensors and a platinum resistance (RTD) temperature sensor. A data table of tank volume versus level, and the height *H,* must also be available for entry into the system computer. Fluid statics gives

$$\rho = \frac{P_1 - P_2}{gH} \qquad L = \frac{P_1 - P_3}{\rho g} \qquad \text{Volume} = \int_0^L A(L)dL \qquad \text{Mass} = \rho V \quad \textbf{(9.6)}$$

---

[12]"Guided Wave Radar, A New Era in Process Level Measurement," *Bull.* 57-201.0; "Capacitance Level Control—the End of an Era?" *Bull.* 57-202.0; Magnetrol International, Downers Grove, IL (www.magnetrol.com).

[13]Enraf-Nonius Service Corp., Bohemia, NY.

[14]D. Blume and J. Myers, "Smart Digital Pressure Transmitters for Innovative Tank Gaging," *In Tech,* pp. 45–48, December 1986; C. Robinson, "Hydrostatic Tank Gaging," *In Tech,* pp. 37–40, February 1988.

[15]*Prod. Spec.* PSS 1-4A2 A, Foxboro Co., Foxboro, Mass.

where g is the local gravitational acceleration. Temperature data are used as a reference point for volume and to convert data to standard-temperature values. On a 50,000-bbl tank, improved accuracy of mass measurement ($\pm$ 0.1 percent compared to $\pm$ 0.3 percent for level-gage methods) could save $4000 worth of $1.00 per gallon product in a custody-transfer operation, helping to justify the relatively high cost (about $10,000 per tank in 1990) of the new system.

Figure 9.4m[16] shows a level-measurement sensor widely used in marine vessel applications and with "problem liquids" such as iron-ore slurries, frothing copper ore, or raw municipal sewage. A helical winding of resistance wire and a longitudinal gold stripe are separated by a small air gap, but are progressively brought into contact as liquid pressure collapses an insulating plastic sheath that surrounds the assembly. An output voltage proportional to the level is developed (analogous to a potentiometer displacement sensor). It takes about 10 cm of water pressure to collapse the tube, so one can't measure the last few inches in an emptying tank unless a small sump is provided. This bias error is reasonably constant, so it can be largely compensated by calibration.

9.3 HUMIDITY[17]

Knowledge of the amount of water vapor in a gas is important to the operation and/or automatic control of many industrial processes. This information may be gathered and presented in a number of ways, depending on the needs of the particular process and the measuring instrumentation utilized. In common use are the relative humidity (ratio of water partial pressure to saturation pressure), dew-point temperature, mixing ratio or specific humidity (mass of water per unit mass of dry gas), and volume ratio (parts of water vapor per million parts of air).

The ultimate standard for calibration of humidity-measuring devices is the NIST standard gravimetric hygrometer. This is a strictly laboratory apparatus in which the water vapor in an air sample is absorbed by suitable chemicals and then weighed very carefully. It directly determines the mixing ratio in grams per kilogram, covers the range 0.30 to 20.0, and has an uncertainty (systematic error plus 3 standard deviations) of about 0.1 percent of the reading. For less critical calibrations, NIST uses its two-pressure humidity generator.[18] This equipment generates air/water mixtures in the dew-point range $-70°C$ (uncertainty $\pm1.2C°$) to $+25°C$ ($\pm0.1C°$), relative-humidity range 10 to 98 percent at temperatures ranging from $-55°C$ (relative-humidity uncertainty ±2.5 percent) to $+40°C$ (relative-humidity

---

[16]Consilium Metritape, Littleton, MA, 978-486-9800; "Gauging Problem Liquids with a Resistance-Tape Level Sensor," *Sensors,* Aug. 1989, pp. 13–21; C. Randall, "Four Case Histories: Liquid Level Sensing Under Adverse Conditions," *Sensors,* Dec. 1991.

[17]A. Wexler, "A Study of the National Humidity and Moisture Measurement System," *NBS* IR75-933, *Natl. Bur. Std. (U.S.),* 1975; S. Soloman, "Sensors Handbook," chap. 39, McGraw-Hill, New York, 1998; P. R. Wiederhold, "Water Vapor Measurement," Marcel Dekker, New York, 1997.

[18]A. Wexler and R. D. Daniels, "Pressure-Humidity Apparatus," *J. Res. Natl. Bur. Std.,* vol. 48, p. 269, 1952; Thunder Scientific, Albuquerque, NM, 800-872-7728 (www.thunderscientific.com).

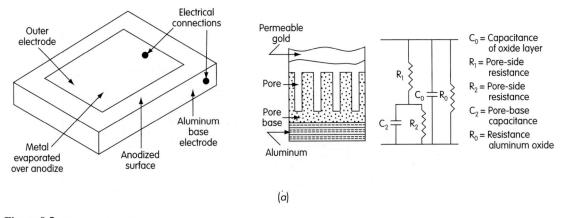

(a)

Figure 9.5
Aluminum oxide and chilled-mirror humidity sensors.

uncertainty 0.5 percent), mixing ratio in the range 0.0013 g/kg (uncertainty ± 0.0003 g/kg) to 20 g/kg (uncertainty $\pm$ 0.5 percent), and volume ratio 2 ppm (± 0.5 ppm) to 30,000 ppm ($\pm$ 0.5 percent).[19]

Classically, relative humidity has been found from psychrometric charts and the temperature readings of two thermometers. One, the dry-bulb thermometer, reads the ordinary air temperature; the other, the wet-bulb, is intended to read the temperature of adiabatic saturation. The latter measurement requires that the bulb be kept wet and a suitably high (about 1,000 ft/min) air velocity be maintained over the wet bulb. Automated instruments are available.[20]

For continuous recording and/or control of relative humidity, various electrical sensors are available. The *Dunmore* and *Pope* cells discussed in the fourth edition of this text have been largely replaced by thin-film or bulk *polymer* sensors. All relative humidity sensors require also a measurement of the (dry-bulb) temperature if a value of *absolute* moisture, such as vapor pressure or dew point, is to be calculated. The temperature and relative humidity sensors should be closely integrated, which is possible in some microscale devices.[21] Sensors for water *vapor pressure,* a measure of *absolute* moisture, often use the aluminum oxide technology[22] shown in Fig. 9.5a.[23] Water vapor is transported through the gold layer and equilibrates on the pore walls of the oxide layer. The number of water molecules absorbed on the oxide structure determines the conductivity of the pore walls, which changes the impedance measured between the device terminals. This impedance ranges from about 50 kΩ to 2 MΩ, depending on water vapor pressure, and is measured using 1-V,

[19]"Humidity Calibration Service," *Instrum. Contr. Syst.,* p. 123, November 1964.

[20]Thunder Scientific model 5A-IMP.

[21]E. Nowak, "Moisture Measurement with Capacitive Polymer Humidity Sensors," *Sensors,* Oct. 1996, pp. 86–91.

[22]P. R. Wiederhold, "Humidity Measurements," *Inst. Tech.,* pp. 85–91, June 1975.

[23]Panametrics, Waltham, MA, 800-833-9438 (www.panametrics.com).

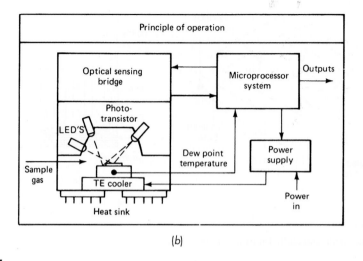

(b)

Figure 9.5

77-Hz AC excitation. These sensors can measure water content of both gases and (non-aqueous) liquids.

Dew-point temperature can be determined by noting the temperature of a polished metal surface (mirror) when the first traces of condensation ("fogging") appear. Commercial devices are available in which this operation has been completely automated by means of a feedback system. The mirror is cooled thermo-electrically (Peltier cooling), with the cooler's electrical input coming from an amplifier whose input is the error signal of the feedback system. Light-emitting diodes (LEDs) provide light to the mirror surface, where it is reflected to photo-transistors that generate a feedback signal which is compared with a set reference voltage to generate the system error signal. The reference voltage is set at a value that just barely maintains condensation (fogging) on the mirror. Any deviation from this level of fogging causes an increase (or a decrease) in light reflected to the phototransistors and a proportional change in the cooling effort. Thus the mirror surface is always kept at the dew-point temperature, which is sensed with an embedded platinum RTD. For dew points below 0°C, the system tracks the frost point. Figure 9.5b[24] shows a microprocessor-controlled version of this instrument which uses a dual-mirror system to correct continuously for mirror contamination, which causes errors and requires periodic shutdown and/or cleaning in systems without this feature. The compensating mirror is controlled to stay "dry" at a precise temperature differential above the "wet" mirror, and both mirrors experience essentially the same contamination; so the dry mirror can be employed as a refer-ence channel (phototransistor changes also are corrected by this scheme). Dew-point accuracy of $\pm$ 0.2°C over the range $-$ 75 to $+$ 60°C is achieved. Addition of ambient temperature and/or pressure sensors allows microprocessor calculation of

[24]*Bull.* 3–300, EG&G Environmental Equipment. Now Edge Tech, Milford, MA, 800-276-3729 (www.edgetech.com).

relative humidity, parts per million, grains, or pounds. The accuracy of this instrument leads to its wide use as a calibration standard for other sensors.[25] Recent versions include the *cryogenic* and the *cycling mirror* types.[26]

Moisture is one of the most difficult quantities to measure, and our brief treatment has left out many details. The text by Wiederhold is highly recommended for those who wish to pursue further study of this subject.

9.4 CHEMICAL COMPOSITION

In years past, the chemical composition of materials was necessarily determined by taking a sample to a laboratory and performing the required chemical tests, usually with somewhat tedious procedures. Today, many important measurements of this type are made on a relatively continuous and automatic basis without the need for a human operator. The need for such measuring systems is due largely to the desire to automatically control product quality directly in terms of its chemical composition, rather than inferring it from measurements of temperature, pressure, flow rate, etc. Even in manually controlled situations, the desire to increase production rates while also maintaining or improving quality leads to a need for rapid analysis methods. Rapid and accurate analysis techniques are very useful in research and development problems. These needs of industry have led to the development of a wide variety of instruments for measuring various aspects of chemical composition and related quantities.

Some examples of measurements of the above type include analysis of products of combustion, monitoring of the composition of dissolved gases in oil-well drilling mud, detection of alcohol contaminant in a heavy-hydrocarbon liquid stream, detection of explosive solvents in the atmosphere within a uranium-extraction kettle, measurement of pH of industrial-waste effluent to control river pollution, determination of alloying constituents in metals, outgassing of materials under high vacuum, use of rocketborne instruments for analyzing atmospheric gases at high altitudes, air-pollution studies, and analysis of anesthetic gases in blood. Most of the analyzers used for such measurements are rather complex systems, rather than simple sensors, and a discussion of these methods adequate for their selection and use is beyond the scope of this text. The references[27] serve this purpose for the interested reader.

There are, however, some simple sensors which do perform useful "chemical analysis," and now we describe one such from the important area of combustion control. Both industrial furnaces and automotive engine-control systems make use

---

[25]J. Tennermann, "The Chilled Mirror Dew Point Hygrometer as a Measurement Standard," *Sensors,* July 1999, pp. 49–54.

[26]P. R. Wiederhold, "The Principles of Chilled Mirror Hygrometry," *Sensors,* July 2000, pp. 46–51.

[27]D. M. Considine (ed.), "Process Instruments and Controls Handbook," 5th ed., McGraw-Hill, New York, 1999; J. G. Webster, "The Measurement, Instrumentation, and Sensors Handbook," CRC Press, 1999, Boca Raton, FL, chaps. 70–73; J. Janata, "Principles of Chemical Sensors," Plenum Press, New York, 1989.

of oxygen sensors of the zirconium oxide ceramic type.[28] The sensor consists of a closed-end hollow tube (similar in size and shape to a thermometer well) made of zirconium oxide ceramic with porous platinum electrodes coated on both the inner and outer surfaces. Above about 500°C the sensor becomes permeable to oxygen ions, and a solid electrolyte effect produces a voltage e_o millivolts between the two electrodes, given by the Nernst equation

$$e_o = 0.0215T \ln \frac{O_{2r}}{O_{2m}} \qquad (9.7)$$

where T is sensor temperature in kelvins, O_{2r} is the oxygen concentration in percent of a reference gas (often atmospheric air, $O_{2r} = 20.9$ percent) at the sensor inside surface, and O_{2m} is the oxygen concentration of the measured gas at the sensor outside surface. Sensors normally include a built-in RTD or platinum/rhodium thermocouple since T must be known to relate e_o to O_{2m}. Figure 9.6 graphs Eq. (9.7); accuracy is typically ± 0.1 percent of oxygen when the concentration is 2 percent, and the response time is about 5 s.

The area of chemical sensors is one of active development, and many useful devices[29] are now available, including some which approximate the human sense of smell.[30] Our treatment has been very brief, but the quoted book and paper references (27 to 30) will get the interested reader started on a more comprehensive study.

9.5 CURRENT AND POWER MEASUREMENT

Chapter 12 covers *voltage*-indicating and -recording devices and it might seem natural to include there a discussion of the measurement of electrical current also. However, while there are many familiar voltage-measuring instruments such as meters, potentiometers, strip-chart recorders, oscilloscopes, etc., there is only one "instrument" dedicated to current measurement, the analog or digital *ammeter.* (The digital ammeter is in fact a *voltmeter* with a built-in current sensor.) In this section, we thus concentrate on current *sensors,* which usually transduce a current into a voltage, which is then measured by one of the voltage-sensing instruments covered in Chap. 12. Ammeters are useful only for *steady* currents, while most sensors can also deal with dynamic currents.

---

[28]D. S. Howarth and R. V. Wilhelm, Jr., "A Zirconia-Based Lean Air-Fuel Ratio Sensor," *SAE Paper* 780212, 1978; J. A. Brothers and F. Hirschfeld, "Some New Applications for Zirconia Sensors," *Mech. Eng.,* pp. 35–37, November 1980; "Oxygen Monitoring Systems," App. Eng. Handbook, Dynatron, Inc., Wallingford, CT.

[29]P. Stupay, "Design and Performance Criteria of a Gas Detection System," *Semiconductor International,* July 1997, pp. 239–248; B. R. Kinkade, "Bringing Nondispersive IR Spectroscopic Gas Sensors to the Mass Market," *Sensors,* Sept. 2000, pp. 83–88; Figaro Gas Sensors, Glenview, IL, 847-832-1701 (www.figarosensor.com).

[30]J. Li, "The Cyranose Chemical Vapor Analyzer," *Sensors,* Aug. 2000, pp. 56–60; E. J. Staples, "A New Electronic Nose," *Sensors,* May 1999, pp. 33–40.

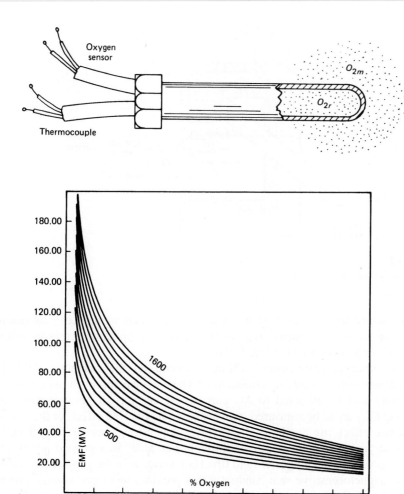

Figure 9.6
Zirconium-type oxygen sensor.

The simplest current sensor is just a *resistor* since it transduces the current through it to a voltage across it, that is, we simply insert our current-sensing resistor such that the current must pass through it. This insertion of course *disturbs* the circuit under observation, so the resistor's ohm value must be small relative to the circuit's equivalent resistance at the insertion point (see Fig. 3.30). Also, for dynamic measurements, we must recall that *real* resistors always have some parasitic inductance and capacitance, so the conversion from current to voltage will not exactly follow the simple $e = iR$ relation. While we may sometimes use "ordinary" resistors, special current-sensing resistors of low ohm value and small inductance

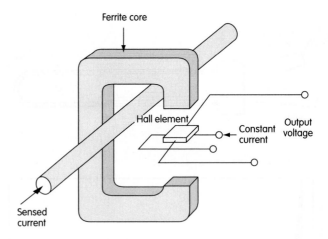

Figure 9.7
Current sensor based on Hall effect.

and capacitance are available[31] as are special integrated circuit devices[32] for reading the resistor voltage. For measuring small currents, the op-amp current-to-voltage converter[33] may be useful (see Fig. 8.39a).

Most other current-sensing schemes depend on the magnetic field that surrounds any current-carrying conductor.[34] The classical approach uses a *current transformer* and is restricted to AC currents. If currents of *general* waveform (including DC) are to be measured, sensors based on the Hall effect, or the magnetoresistance effect, are available. In a typical "open-loop" sensor (Fig. 9.7), the current-carrying conductor is passed through the "hole" in a gapped ferrite core used to concentrate the magnetic field (the Hall device is in the gap). A Hall generator is a magnetosensitive semiconductor that provides an output voltage proportional to the product of *its* current (which is held constant) and the component of magnetic field that is perpendicular to its surface. Since this field is proportional to the current being measured, the device output voltage is proportional to the sensed current. A typical device[35] has a range of 0 to 350 A and a frequency response from 0 to 1000 Hz, with other models offering lower current ranges and faster response. "Closed-loop" Hall sensors use a feedback nulling scheme[36] to improve linearity and reduce temperature dependence.

---

[31]Isotek Corp., Swansea, MA, 508-673-2900; Caddock Electronics, Roseburg, OR, 541-496-0700; IRC, Boone, NC, 704-264-8861.

[32]LT1787, Linear Technology, Milpitas, CA, 800-454-3276 (www.linear-tech.com).

[33]J. Phalan, "A Ten-Decade Logarithmic Current-to-Voltage Converter," *Sensors,* Feb. 1998, pp. 51–54.

[34]E. Ramsden, "5 Ways of Monitoring Electrical Current," *Sensors,* July 1999, pp. 26–37.

[35]Bell Technologies, Inc., Orlando, FL, 407-678-6900 (www.belltechinc.com).

[36]E. Ramsden, op. cit.

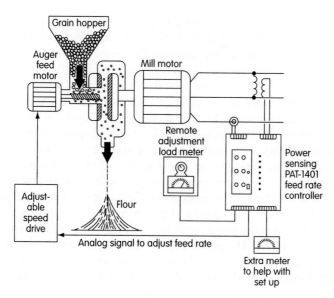

Figure 9.8

Power measurement/control in grinding application.

Magnetoresistive current sensors[37] use elements that change resistance in response to a magnetic field. To get an output voltage proportional to the sensed current, the elements usually are connected in a Wheatstone bridge arrangement, and a feedback technique (as used in closed-loop Hall sensors) gives good linearity and temperature insensitivity. Frequency response to about 50 kHz is possible.

Turning now to the measurement of *electrical power,* recall that the definition of instantaneous power taken from a source and supplied to a device is $P \triangleq e \cdot i$, where e and i are the instantaneous voltage across and current through the device, respectively. If the algebraic sign of P is negative, the device is *returning* power to the source. The *net energy* "used up" by the device over some time interval is the integral of the instantaneous power over that interval. The measurement (using the familiar electromechanical watt-hour meter) of AC power and energy supplied to our homes is used to calculate our utility bill. Power measurement is also used for other purposes. Figure 9.8[38] shows the grinding of materials such as grains, cereals, chemicals, and fibers. Such machines have an optimum power level, which can be maintained by measuring the drive motor power and using a feedback-control system to adjust the material feed rate such that the actual power equals the desired value. Similar applications exist for various machining operations, sensing of dull tools, viscosity indication in mixers, etc.

---

[37]H. Lemme and A. P. Friedrich, "The Universal Current Sensor," *Sensors,* May 2000, pp. 82–91; www.fwbell.com.

[38]Load Controls Inc., Sturbridge, MA, 888-600-3247 (www.loadcontrols.com).

Hall-effect devices are nicely suited to power sensing because they can inherently perform the *multiplication* needed, that is, the Hall output is the product of *its* current and the sensed current. By applying the motor *voltage* to the Hall current terminals, the Hall current (which is kept *constant* for a strictly current sensor) now becomes proportional to the voltage used in our power equation. The Hall output then becomes proportional to the product of voltage and current for the device whose power we are measuring. Electric motors are obviously an important area for power measurement and the three-phase AC induction motor is perhaps the most widely used in industry. Here the current and voltage waveforms are essentially sinusoidal, and the concept of *power factor* applies. If the load is balanced, we can measure the motor-consumed power by sensing the voltage across two of the phases and the current at the third.[39] For unbalanced loads, all three voltages and all three currents must be sensed. The consumed power for each phase is calculated from $P = EI \cos \theta$, where θ is the phase angle between the voltage and current and $\cos \theta$ is called the *power factor*. The phase angle is measured from the zero crossings of the voltage and current waveforms. Demodulation and filtering of the AC signals from the Hall devices is needed to get the desired "DC" output signal proportional to the power. This filtering reduces the response speed to power transients, but a 90-percent response time of about 0.02 s for a step change is claimed for the referenced instrument.

Some motor controls (SCR, PWM, etc.) produce voltage and current waveforms of much higher frequency content than the "smooth" AC waveforms just discussed. Also, power measurements on devices other than motors may involve high frequencies. Specialized power meters[40] for ranges from DC to 300 kHz are available.

9.6 USING "OBSERVERS" TO MEASURE INACCESSIBLE VARIABLES IN A PHYSICAL SYSTEM

Sometimes a physical quantity that we wish to measure is inaccessible or not cost/effective to measure directly. A commercial sensor for the variable may not be available or perhaps the environment is too severe to allow a sensor to survive for long. In such and similar situations, the concept of the *observer* may be useful. The basic idea of the observer probably comes from the technical area usually called *modern control theory*.[41] One of the tools of modern control theory is *state variable feedback*. Here, in order to control some variable associated with a system, we strive to measure *all* the state variables of that system and use these as feedback signals, each with its own gain, in an attempt to design a control system with desirable

---

[39]"Application Notes: Power Sensors and Load Controls," Load Controls Inc.

[40]WT1000 Series, Yokogawa Electric Corp., Newnon, GA, 770-253-7000 (www.yca.com); Valhalla Scientific, San Diego, CA, 800-548-9806 (www.valhallascientific.com).

[41]B. C. Kuo, "Automatic Control Systems," 6th ed., Prentice Hall, Englewood Cliffs, NJ, 1991, pp. 241–243; W. J. Palm, "Modeling, Analysis, and Control of Dynamic Systems," Wiley, New York, 1983, pp. 647–654.

behavior with respect to accuracy, stability, speed, etc. We are rarely able to directly measure all the state variables, and the observer concept was invented to somehow supply values of the *unmeasurable* variables, so that we would have available a full set of state variable "measurements" for use in our control-system design. While the observer concept is most often found in the context of a control system, this is not necessary; we could use it for strictly measurement purposes.

In its simplest form, the observer is nothing but a mathematical model of the physical system in which the inaccessible variable (or variables) is found. That is, to implement an observer, we must first have a model (set of equations) describing the system. As in any other modeling application, results will be better if our model is an *accurate* representation of the physical system. It is also necessary that we be able to *directly* (not with an observer) measure at least one of the physical system's *state variables*. Recall that the state variables of a system can be defined as those variables that would appear at the outputs of integrators in a simulation of the system. We also need to be able to measure directly the "command" input to the system. In some cases, observer theory also allows us to "measure" disturbing *inputs* to systems that may be difficult or impossible to measure conventionally. For example, in a motion-control system, there may be disturbing torques acting on the load that are impossible to measure directly, but we would like to use such a measurement to implement a feedforward-control mode to improve the system response.

Assuming that the command input to the real system is directly measurable with a suitable sensor (usually producing a voltage proportional to the physical input variable), this voltage can then be sent into our system mathematical model, generally implemented in a digital computer, much like the SIMULINK models you have used in simulation studies. The sensor voltage must of course be digitized in an A/D converter before the computer will accept it. If our measurement of the input, and our computerized mathematical model, are both accurate, then *any* variable accessible in the model can be "measured" there even if it is not physically accessible in the real system. If our mathematical models were highly accurate, this simplest form of the observer would be practical. Unfortunately, our models are never perfect, sometimes quite inaccurate, and subject to error as the physical system itself changes because of environmental influences such as temperature or wear. When such is the case, the simple form of observer described above may be ineffective.

A more practical form of observer still uses the system model as described above but adds feedback signals, designed to force the variables in our model to conform more closely to the physical system variables even when our model is somewhat inaccurate. Each of these feedback terms has a "gain" value, which we must properly adjust to get the best observer behavior. Since we have a feedback system that includes the physical system, the mathematical model, and the observer feedback configuration, this entire assemblage may now become unstable, just like any other feedback system, when the observer feedback terms are not properly designed. As with any other theoretical concept, the observer can sometimes be made to work well and at other times fail. The rest of these notes shows a few simple observer applications that will make the concept more clear, especially if you have never encountered it before.

The first example is a simple spring, mass, and damper system driven by an external force (Fig. 9.9). We have no trouble measuring the force input and the mass's displacement, but are for some reason unable or unwilling to measure directly the mass's velocity, which signal we need for some practical purpose, perhaps as a stabilizing signal in a motion-control system. We will study the use of an observer in this situation by simulating both the physical system and the observer in SIMULINK. When the mathematical model of the physical system, which is used in the observer, conforms exactly with the physical system's actual behavior and we are also able to exactly measure the physical input force f_i, then the observer will contain exact values of all the system variables, which we can then measure *from the observer,* rather than from the physical system. In fact, while our initial stated goal was to get values of the mass's velocity without directly measuring it, we will see that we can also get from the observer the mass's acceleration, the spring force, and the damper force since these all appear in the observer.

A more realistic simulation allows the observer's model of the physical system to be *incorrect* in some way, such as having wrong numerical values for *M, B,* and/or K_s. Now we need to add to the observer's structure the feedback signals mentioned earlier and also set the gains of these feedback paths at values which optimize the observer's behavior, that is, the observer must be "designed." Since the observer is now a feedback system, the various tools for feedback-system design can be applied. In our simple examples, we will find that "trial and error" using simulation is a feasible design tool. Figure 9.10 shows the SIMULINK diagram for the case where the model is perfect and we also measure the input force perfectly. Note that there is no need here to measure the state variables. There also is clearly no need to actually run this simulation since it is obvious that all the signals in the simulation of the physical system will be perfectly reproduced in the observer. In Fig. 9.11, we see the more realistic situation where the model has some incorrect numerical values. We want to first show that if we *do not* use the feedback paths in the observer, its signals will not be accurate "measurements" of the "inaccessible" variables. Be sure to note that our simulation still assumes that we measure the input force and mass displacement (state variable x_o) with perfect accuracy. These assumptions could of course also be relaxed and the effect studied with simulation.

In Fig. 9.11, the "feedback paths" that we have mentioned earlier and that are usually needed in practical applications of observers are formed by first getting a

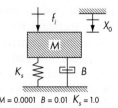

$M = 0.0001 \quad B = 0.01 \quad K_s = 1.0$

Figure 9.9
Candidate system for velocity measurement using an observer.

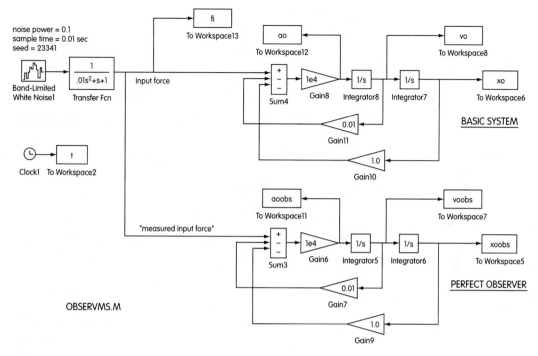

Figure 9.10
Simulation of system with a perfect observer.

signal that is the difference between the physically measured state variable (in our case, the mass's displacement x_o) and the observer's estimate $x_{o,\,obs}$ of that same variable. The goal of the feedback is to force this difference to be very small, ideally zero, by multiplying this "error" by feedback gains and applying signals to the input of the observer "summers," which produce the state variables. That is, if $x_{o,\,obs}$ is not sufficiently close to x_o, then we change the observer's acceleration so as to drive $x_{o,\,obs}$ in the proper direction. Note that this correction is somewhat indirect since the acceleration must pass through **Integrator9**, **Sum6**, and **Integrator10** before an effect is felt on $x_{o,\,obs}$. Also, an *additional* correction is applied to **Sum6** from the feedback channel that has gain **Kx**. If the gain of these feedback paths can be made large without causing instability, we may be able to make the observer's displacement signal closely match that of the physical signal even if the observer's system model is not perfect. If we can make the displacements of the physical system and the observer model match closely, then the observer may also give good estimates of the velocity and even the acceleration. If so, we can then measure these quantities with good accuracy in the observer, rather than in the physical system, where they may be inaccessible.

When forming the observer as a model of the actual system, we must force this model into state-variable form where *all* the state variables are explicitly shown, using a set of *first-order* differential equations. In the present example, we use *two first-order equations* as the observer model, rather than the single second-order equation we used to simulate the basic system. (We could, of course, also have used

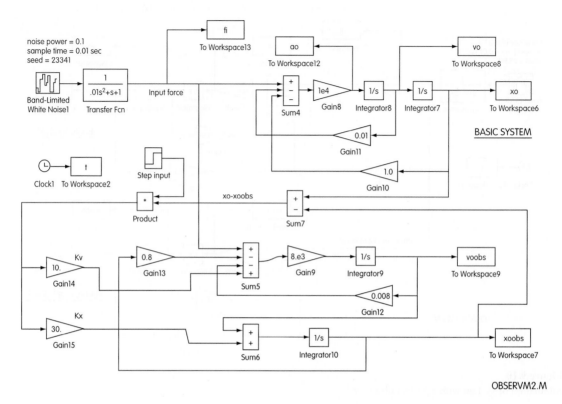

noise power = 0.1
sample time = 0.01 sec
seed = 23341

fi
To Workspace13

ao
To Workspace12

vo
To Workspace8

Band-Limited
White Noise1

$\dfrac{1}{.01s^2+s+1}$

Transfer Fcn

Input force

Sum4

1e4
Gain8

1/s
Integrator8

1/s
Integrator7

xo
To Workspace6

BASIC SYSTEM

0.01
Gain11

1.0
Gain10

Clock1 To Workspace2

t

Step input

xo-xoobs

Sum7

Product

Kv

10.
Gain14

0.8
Gain13

Sum5

8.e3
Gain9

1/s
Integrator9

voobs
To Workspace9

0.008
Gain12

Kx

30.
Gain15

Sum6

1/s
Integrator10

xoobs
To Workspace7

OBSERVM2.M

Figure 9.11
Simulation with an imperfect (more realistic) observer.

two first-order equations to model the basic system, but this is not necessary or usually desirable unless we prefer the state-variable way of modeling physical systems.) Each of the observer's first-order equation models includes an observer *gain;* in this case, we have the two gains called **Kv** and **Kx**. These gains must be "designed" to get good behavior from the observer. Analytical tools are available for this, but, as mentioned earlier, we will use simulation trial and error.

To be realistic, the observer model has intentionally been given *incorrect* numerical values for *M, B,* and K_s. Each value has been set 20 percent too low. During the first 1.0 s, the feedback gains are set to 0.0, thus disabling the observer feedback path and showing how the observer works for a mismatched model and no corrective feedback effects. For time *greater* than 1.0 s, the feedback paths are enabled, with specific values for **Kx** and **Kv**. By simply trying a few combinations of values for **Kx** and **Kv**, we soon discover that a wide range of values gives good performance in this example. In fact, we also find that *both* observer "loops" and their associated gains are not really needed. We can set one of the gains to zero and get good performance by adjusting the other. Figure 9.11 uses the values **Kv** = 10 and **Kx** = 30 to get the good results of Fig. 9.12. There we have set both gains to zero for the first 1.0 s of simulation and then suddenly "turned on" the observer by jumping the gains to 10 and 30 for the next 1.0 s. By plotting our graphs for the time

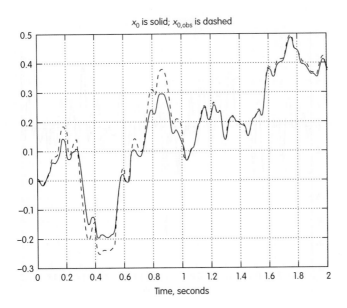

x_0 is solid; $x_{0,obs}$ is dashed

Time, seconds

Figure 9.12
Observer is "turned on" at $t = 1.0$ seconds.

period 0 to 2.0 s, we can see at a glance how the use of the feedback terms greatly improves the accuracy of the observer *even when the observer's system model is 20 percent inaccurate.* We can compare the observer's values with the real-system values for any of the signals available in the observer. These include the displacement, velocity, and acceleration of the mass, and also the spring force and damper force.

We see that after time $= 1.0$, the observer's "measurement" of the displacement is much more accurate since the feedback correction with gains of 10 and 30 is now in effect. It is, of course, foolish to measure the displacement in the observer since we have a *direct* measurement of this from the physical system. What we really wanted is the velocity, which is *not* measured in the physical system but *is* available as an estimate in the observer. Figure 9.13 shows this result. The acceleration, spring force, and damper force are also available from the observer; if you duplicate this simulation, you will see that they are also accurate.

Our next example deals with the thermal process shown in Fig. 9.14. Cans of food product travel through a convection heating oven to be thermally processed. We are able to measure air temperature T_{air} with a thermocouple and can-surface temperature T_s with a noncontacting, radiation temperature sensor. However, our main interest is in the *internal* temperature of the can contents, which is inaccessible. If we can provide a good thermal model of the system, we should be able to implement an observer to estimate the internal temperature T_c from measurements of T_{air} and T_s. Figure 9.14 shows a simple thermal model that will be suitable for our example. Heat transfers by convection, from the air temperature to the can surface temperature, through a thermal resistance R_{ts}. At the same time, heat transfers from

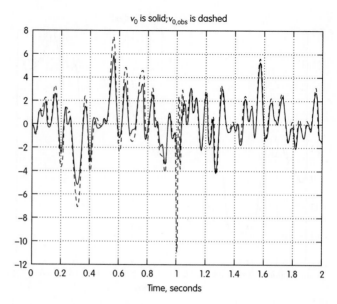

Figure 9.13
Velocity measurement using observer.

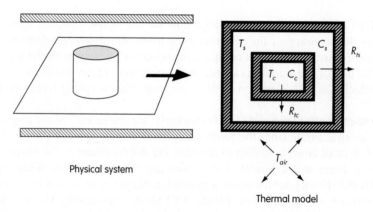

Physical system

Thermal model

Figure 9.14
Candidate system for observer measurement of inaccessible temperature.

the can surface to the fluid contents through a thermal resistance R_{tc}, raising the temperature of the thermal capacitance C_c. The difference of the two heat transfers raises the temperature of the thermal capacitance C_s. System equations are:

$$\frac{T_{air} - T_s}{R_{ts}} - \frac{T_s - T_c}{R_{tc}} = C_s \cdot \frac{dT_s}{dT} \tag{9.8}$$

$$\frac{T_s - T_c}{R_{tc}} = C_c \cdot \frac{dT_c}{dt} \tag{9.9}$$

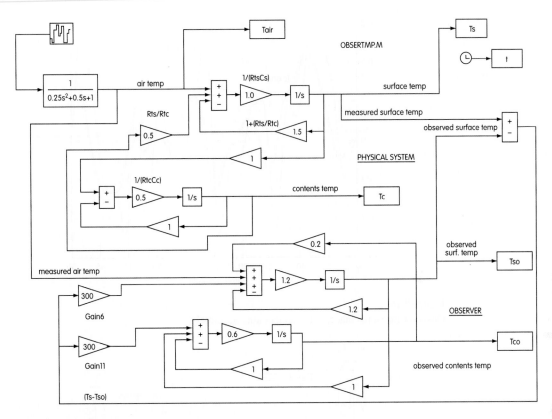

Figure 9.15
Simulation for thermal system observer study.

The simulation diagram for this system and the associated observer is shown above in Fig. 9.15.

To be realistic, we have again mismatched the physical system parameters and the numbers used in the observer's model. Both feedback paths depend on the observer "error" signal (Ts-Tso), so we form it from the directly measured Ts and the observer estimate Tso. As usual, the concept behind the observer requires that we provide corrective inputs to the summers that produce the derivatives of the two state variables in such a way that when (Ts-Tso) is not zero, the corrective inputs tend to drive this error toward zero. The corrective term Gain6(Ts-Tso) tends to drive Tso in the proper direction; if Tso is, say, larger than Ts, then the corrective term will be negative, driving Tso to become smaller. If Tso is too large, the reverse happens. For the corrective term Gain11(Ts-Tso), which is applied to the Tco summer, we are still trying to drive (Ts-Tso) toward zero, but it is now somewhat indirect since this corrective term works first on Tco, not Tso. To get the correct algebraic sign on this term, we can look at Eq. (9.8) to see that the effect of T_c on dT_c/dt is just *opposite* that of T_s. Note that we *do not* have a signal (Tc-Tco) to work on any error in Tco *directly*. Rather, Tco is brought closer to Tc only in an attempt to minimize (Ts-Tso). The feedback gains Gain6 and

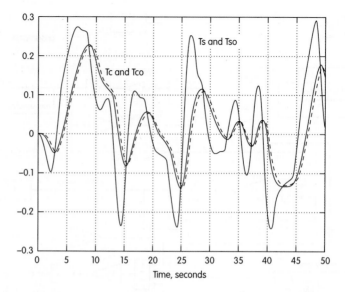

Figure 9.16
Observer measures inaccessible temperature accurately.

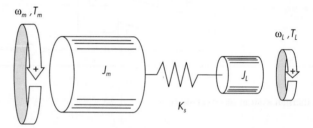

Figure 9.17
Motor and load for a motion control system.

`Gain11` must be "designed" to get the best estimates of both `Ts` and `Tc`, particularly `Tc`, since we can measure `Ts` directly, without the observer. In our example, these two gains were adjusted by simulation trial and error, but standard tools of feedback-system design could of course be used to set them in a more "scientific" manner. Figure 9.16 shows the performance attained with the gain values shown in Fig. 9.15. We see that both temperatures are accurately "measured" from the observer.

Our final example is from the field of electromechanical motion-control systems and was actually constructed and tested as a practical machine control.[42] Leonhard shows how an observer mechanized in a microprocessor greatly improved the performance of a rotary speed-control system using a DC motor. Figure 9.17

---

[42]W. Leonhard, "Control of Electrical Drives," Springer, New York, 1984, pp. 292–295.

shows the motor inertia J_m, the load inertia J_L, and the connecting shaft modeled as a torsion spring K_s. The motor inertia is actually larger than that of the load, because the load inertia (actually quite large) has been "reflected" to the motor shaft through a large gear ratio, common in many systems. The motor inertia feels the motor magnetic torque T_m, and the load feels a disturbing load torque T_L. Motor and load speeds are ω_m and ω_L, respectively. Leonhard states that the measurement of the load speed may be difficult or expensive in practice, and suggests the use of an observer for this measurement. The motor torque is, of course, under our control, but the load torque is not and is rarely susceptible to direct measurement. Since the load torque is one source of error in a speed-control system, we would like to estimate it and then use this estimate as a feedforward signal in our speed-control system. Finally, the shaft connecting motor and load has little inherent damping (we model it above as a pure spring), giving rise to a bad resonance effect with the two inertias. Leonhard suggests that a "synthetic" damping could be achieved if we could lay our hands on the speed *difference* between the two shafts. A negative control signal proportional to this difference, and applied to the motor's amplifier, produces a component of magnetic torque that acts exactly like a viscous damper connected between the two inertias. Leonhard proposes to implement all these control actions by using a suitably designed observer.

Newton's law applied to each inertia in turn gets us the system equation set

$$T_m - K_s(\omega_m - \omega_L) = J_m \cdot \frac{d\omega_m}{dt} \qquad \textbf{(9.10)}$$

$$T_L - K_s(\omega_L - \omega_m) = J_L \cdot \frac{d\omega_L}{dt} \qquad \textbf{(9.11)}$$

In an electromechanical motion-control system such as this, there is usually no need to measure the system input, which is needed in the observer, with a separate sensor, as we have done in the two previous examples. This is true because the input signal to the physical system *is already a voltage,* suitable for the observer's input (after digitizing it). The input to the motor comes from an amplifier that feeds the motor's armature, and if this amplifier is of the transconductance type, it produces a motor armature current, and thus a torque, that is directly proportional to the amplifier input. In the simulation diagram of Fig. 9.18, we therefore show the input to both the physical system and the observer as the motor torque **TM**, even though in an actual application, we would be using the amplifier voltage. If the amplifier were an ordinary voltage amplifier, the input would be taken as the amplifier input voltage, but now the observer's model of the physical system would have to include any dynamics between amplifier input and motor torque. The only signal that needs to be measured with a sensor is the motor speed.

The estimation of the load torque input is a capability not explained earlier. The concept is most directly perceived if we first assume that the observer model is perfect. If this were the case, then no corrective feedback terms would be needed on the summer/integrator paths for **wm** and **wL** *if* there were no disturbing torque. With a disturbing torque present, the two speed estimates will *not* be correct even for a

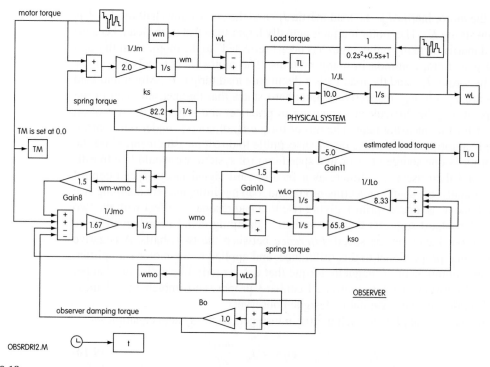

Figure 9.18
Simulation to study estimation of an unknown disturbing torque.

perfect model, and we thus need to somehow provide a corrective effort for this situation. This is done by forming the error signal (**wm-wmo**), multiplying it by a gain, and feeding this signal into the summer/integrator path whose output is **wLo**. That is, this signal is our estimate **TLo** of the load torque, which is needed to implement the Newton's law equation for **wL** in the observer. The observer will now adjust this signal to "make up the deficiency" in our estimate of **wL**. Of course, changes in **wLo** will causes changes in **wmo** since the spring couples the two inertias. Thus, adjustments in the estimated load torque effect *both* angular velocities. This single feedback correction will give estimates of **TL**, **wL**, and **wm**, but these estimates will probably be inaccurate if our model is not perfect. We thus add two more feedback correction paths, one entering the summer/integrator which produces **wmo**, and one entering the summer/integrator path for **wLo**. Each of the three paths uses the same error signal (**wm-wmo**) and of course needs to have its gain "designed" to a proper value.

To do our first study, in Fig. 9.18, we set the motor torque and observer feedback gains 8 and 10 (**Gain8** and **Gain10**, respectively) to zero. We want to use this configuration to see whether the observer can estimate the load torque "without any help" from the two other feedback paths. The physical system is modeled with no damping at all, and when we model the observer this way and try to set **Gain11** to get good estimates of the load torque, we find that the *system* goes *unstable*. This

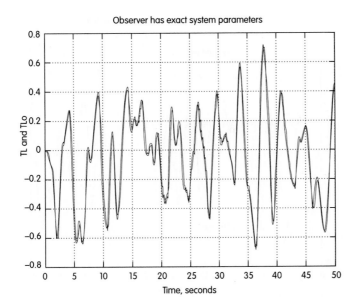

Figure 9.19
Load torque accurately estimated when observer knows correct system parameters.

situation brings out another feature of observer design. Just as we can set the feedback gains as we wish, we can also add other *dynamic* behaviors to the observer. For example, we have been calling the dynamics in the feedback path "gains," and, so far, they have been exactly that. They *could*, however, be "dynamic gains," that is, transfer functions giving various control modes, not just a proportional mode. I found in the current example that if I added to the observer a viscous damping term proportional to the difference in angular velocities, the whole system remained stable when I adjusted `Gain11`, and in fact I got good estimates not only of the load torque but also of the two angular velocities.

Figure 9.19 shows the performance for the numbers given in Fig. 9.18. We see that the load torque is accurately estimated. The observer model here has *exactly correct values* for the inertias and the spring constant. The load and motor angular velocities are also accurately tracked (graph not shown). The system natural frequency of 5 Hz is not apparent in any of these graphs. It could be seen if we looked at the velocity *difference* at a suitable scale.

We now want to be more realistic and allow the observer's model of the physical system to be somewhat inaccurate, as would be the case in a real application. In the simulation diagram, let us make the observer model's inertias 20 percent too large and spring constant 20 percent too small. This will make a significant change in the natural frequency. Activating all three feedback paths and setting their gains, we obtain the results shown in Fig. 9.20.

In the Leonhard reference, he does not use the "damping" feature in his observer; this was something that I tried and found to be useful. Leonhard does not evaluate his observer as a separate device used for measurement but only as an

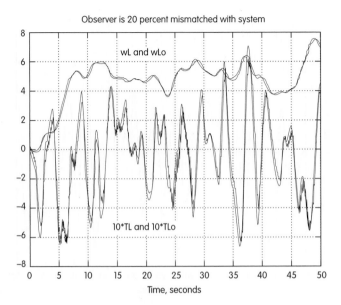

Figure 9.20
Observer remains accurate when system parameters are not perfectly known.

embedded system within the overall closed-loop motion-control system. In his implementation, he *does* use a damping term exactly like mine, but he uses it only as a stabilizing derivative-control mode ("synthetic damping") in the control amplifier of the physical system, not in the observer itself. Since the physical system, motor speed sensor, and observer are all coupled together as one overall system, the damping term may effect his system in a similar way to what we observed in our example. In any case, when we used the observer as a "stand-alone" measurement tool, the damping term certainly was useful in improving observer behavior.

This concludes our brief introduction to the use of observers in measurement systems. If you had not been aware of this idea earlier, you should now have enough basic concepts to try it in future situations where it may be useful. As mentioned earlier, it is not a panacea that always solves the practical problem. However, when more direct measurement approaches are not feasible, it may be worth trying. In all our examples, we did not try to use existing design tools for feedback systems as a means of designing the observer. Rather, we appealed to simple trial-and-error simulation. If one does a lot of observer work, it would perhaps be more efficient to initially use "scientific" design tools and only go to simulation for "fine tuning," just as we do in control-system design when observers are *not* used.

9.7 SENSOR FUSION (COMPLEMENTARY FILTERING)

When measuring a particular variable, a single type of sensor for that variable may not be able to meet all the required performance specifications. If several alternative sensors for making the particular measurement are available, we may sometimes *combine* several sensors into a measurement system that utilizes the best qualities of each individual device. That is, if one sensor type meets all but a few of the requirements, perhaps one of the alternative types will "fill in the gap" in performance. The second sensor makes up for the specifications that the first sensor lacks, and vice versa. Thus, the two sensors *complement* each other, giving rise to the name *complementary filtering*. Other names you will encounter for this same concept are *aiding* and *sensor fusion*. A more advanced version of a similar idea is called *Kalman filtering*. Our first example comes from the application area of aerospace vehicle systems,[43] in particular, the sensing of vehicle altitude above sea level. Merhav supplies the background for our specific application and also extends the concept to more complicated situations, including the Kalman filtering version.

The measurement of aircraft altitude has for many years been accomplished using a static pressure measurement since the altitude above sea level is related in a known way (see a section on "standard atmosphere" in a fluid mechanics book) to static pressure. The relation is nonlinear but is well known. In the upcoming discussion the altitude changes studied are small enough that we can neglect any nonlinear effects. The static pressure, and thus the "barometric altitude," is measured with any suitable pressure sensor, which receives an input pressure from a length of tubing coming from the static pressure tap on the aircraft's pitot tube. (All aircraft and missiles have a pitot tube for measuring altitude and airspeed.) The "good" feature of this barometric altimeter is that there is no significant zero drift with time. The "bad" features are that the pneumatic response is rather slow and there is significant noise due to air turbulence. If our flight vehicle requires a fast, noise-free altitude signal, the barometric approach may not be adequate. An alternative scheme for measuring altitude uses an accelerometer oriented to measure vertical acceleration; we then integrate this signal twice to get an altitude signal. This approach provides a fast response and is relatively noise-free. However, the slightest bias error in the accelerometer and/or integrators will cause an ever-increasing zero drift, which is totally unacceptable. Also, for level flight at any altitude, if the altitude signal accumulated at the output of the second integrator should be lost for any reason, there is no way to recover a value for the altitude. We see that the two schemes for measuring altitude are complementary; where one is bad the other is good. Thus, the *baroinertial altimeter*[44] is used in many practical flight vehicles.

In this application, the basic concept used to combine the two altitude sensor signals is not complicated; it uses well-known low-pass and high-pass filters, combined in a clever way. If a time-varying signal is applied to both a low-pass $[1/(\tau s + 1)]$ and a high-pass $[\tau s/(\tau s + 1)]$ filter, and if the two filter output signals are summed, the summed output signal is exactly equal to the input signal. In the altimeter system, we send the accelerometer/integrator altitude signal to the

[43]S. Merhav, "Aerospace Sensor Systems and Applications," Springer, New York, 1996, chap. 9.

[44]Ibid., p. 402; M. Kayton and W. R. Fried, "Avionics Navigation Systems," Wiley, New York, 1969, p. 317; M. Kayton and W. R. Fried, 2nd ed., 1997, p. 373.

high-pass filter because we want to wipe out the low-frequency drift part of the signal. The noisy altitude signal from the static pressure sensor is sent to the low-pass filter to attenuate the high-frequency noise portion of that signal. We now want to first study the simplest version of such a scheme.

Consider a system using the low-pass and high-pass filters as just described above. Let the input signal to the low-pass filter be a triangular wave of frequency 0.1 Hz and amplitude 100.0 ft of altitude, contaminated with some random noise. This signal will represent a slowly varying aircraft altitude, as measured by a pressure transducer attached to the pitot-static tube's static tap. The $\pm$ 100-ft altitude change is of course taking place around some operating altitude, say 5000 ft, and the random part of the signal models the atmospheric turbulence unavoidably picked up by the pitot tube.

Suppose the doubly integrated accelerometer signal has a constant error of 80 ft. Figure 9.21 shows a SIMULINK diagram for a sensor-fusion system combining the barometric and inertial altitude signals, using the low- and high-pass filters. The low-pass filter time constant τ is chosen so as to attenuate the random atmospheric noise; in this case, $\tau = 1.0$ s works well. The high-pass time constant must be the same, so we do not have design freedom to force the high-pass filter to "wash out" the bias more rapidly. Figure 9.22 shows how our system, given a little time, ignores the inertial bias error and attenuates the atmospheric noise.

The system just studied is not quite the same as the real baroinertial altimeter. The main difference is that any constant bias error in the accelerometer is integrated twice since the desired output is vertical *position,* not acceleration. Integrating a constant twice gives a parabolic drift, which will *not* be wiped out by our simple high-pass filter. We need a "stronger" high-pass filter. We can show mathematically

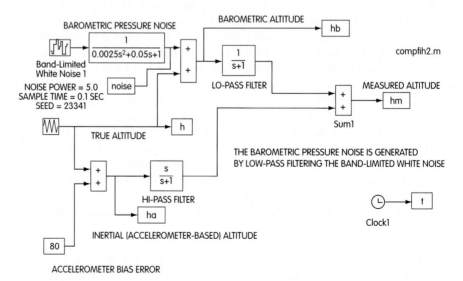

Figure 9.21
Sensor fusion of barometric and inertial altitude signals.

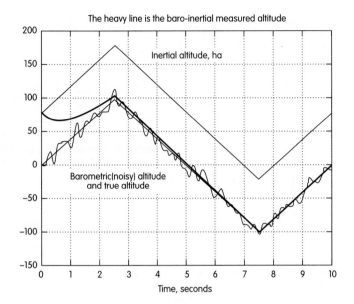

Figure 9.22
Sensor fusion washes out inertial bias and reduces atmospheric noise.

that a constant bias error will be eliminated by an s in the filter numerator, a ramp type of drift requires an s^2, and a parabolic drift requires an s^3. The high-pass filter denominator will now have to be a cubic polynomial; so that the filter is physically realizable. When this is done, the low-pass filter must also be modified, so that it has a quadratic numerator and the same cubic denominator. Figure 9.23 shows the system with the new filters, and Fig. 9.24 shows the resulting behavior.

Absolute Angle Measurement

The baroinertial altimeter is not an off-the-shelf product that one can buy "ready-made." Rather, it would need to be designed and assembled using available components by the electronics or controls group of the aircraft/missile manufacturer. An off-the-shelf product that *can* be purchased ready for use and that uses the same complementary-filtering principles discussed above is the Watson solid-state vertical reference, Model ADS-C132-1A.[45] This device can be mounted in, for example, an aerospace vehicle, automobile, or ship, on an antenna, etc. to measure the absolute angular position and angular rate about a chosen single axis. (Multiple devices can be packaged to measure motions about three perpendicular axes if such data is needed.)

Such sensors are replacing the classical "spinning wheel" gyroscopes discussed in Chap. 4 for certain applications. The two basic sensors used are a microelectromechanical (MEMS) rate gyro (Watson ARS-C122-1A) using piezoelectric tuning

---

[45]Watson Industries, 3041 Melby Road, Eau Claire, WI 54703, 800-222-4976.

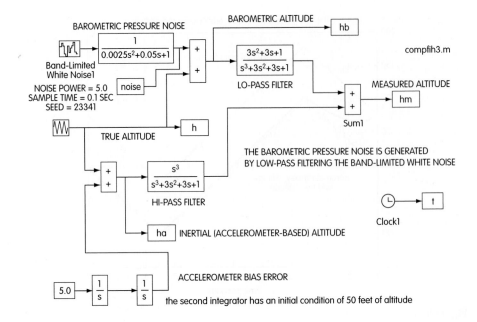

Figure 9.23
More realistic baroinertial altimeter.

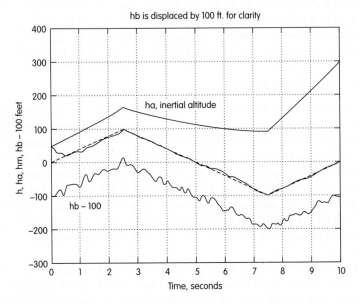

Figure 9.24
Response of realistic baroinertial altimeter.

forks (*no* spinning wheel) and a Lucas Accustar Electronic Clinometer.[46] The Lucas clinometer measures tilt angle relative to gravity vertical by immersing two circular sector capacitance plates in a dielectric liquid. Angular tilting causes one pair of plates to increase capacitance and the other to decrease. These capacitance changes cause a frequency change Δf in an oscillator, which Δf is then converted to a pulse-width-modulated (PWM) signal. By low-pass filtering the PWM signal, a DC voltage proportional to tilt angle is obtained. A range of about $\pm 60°$ and resolution of about $0.001°$ are typical performance values.

The Watson "solid-state" rate gyro gives a DC voltage output proportional to angular velocity, with a flat frequency response to about 50 Hz. Op-amp analog integration would give us angular position, but the bias error in the rate gyro, when integrated, quickly causes an unacceptable, ever-increasing drift of the position signal. The Lucas clinometer does *not* suffer from a drift problem (no integration is involved) and can thus be used to correct for the gyro drift problem. It *cannot,* however, be used by itself for angle measurements in applications (like measuring vehicle motions) that require a fast response since it is a first-order instrument with about a 0.30-s time constant. This makes its frequency response flat to only about 0.50 Hz, much too slow for many applications. The two sensors are thus good candidates for a complementary-filtering application, giving both angular position and angular velocity data over a 50-Hz bandwidth with negligible drift.

While the configuration of the separate high-pass and low-pass filters used earlier for the baroinertial altimeter discussion is most useful for *explaining* the basic concept of complementary filtering, the *practical implementation* uses instead a feedback type of configuration that produces identical differential equations and transfer functions. Also, in our study of the baroinertial altimeter, we treated both sensors as dynamically perfect zero-order systems. For our study of the Watson vertical reference, we want to use the feedback configuration employed in the actual device and to also use the realistic sensor models. Figure 9.25 shows the system block diagram, which can also be used to set up the SIMULINK simulation of Fig. 9.26. The sensor sensitivities must both be equal; we take them as 1.0. The rate gyro has a 50-Hz natural frequency and 0.5 damping ratio. The complementary filter has only two adjustments; we take them as $\omega_n = 0.2$ rad/s and $\zeta = 0.5$. The major effect is that of ω_n; larger values correct bias effects more quickly but filter noise effects less effectively. The gyro bias error is a constant 0.005 rad/s and the clinometer noise is a small random signal. The input angle is taken as zero for the first 50 s to see how the system "fights out" the gyro bias and attenuates the clinometer noise. At 50 s, the input angle steps up to 1.0 radian, so we can see the response to sudden changes. Figure 9.27 shows that the noise filtering works continuously, while the bias rejection takes about 10 s to recover from this sudden large transient. This recovery could be speeded up by setting ω_n to a larger value, but noise filtering would be less effective. If the actual noise were at a higher frequency and/or smaller amplitude than for our example, this increase in ω_n might be acceptable.

---

[46]Lucas Control Systems Schaevitz Sensors, 1000 Lucas Way, Hampton, VA 23666, 800-745-8008 (www.schaevitz.com).

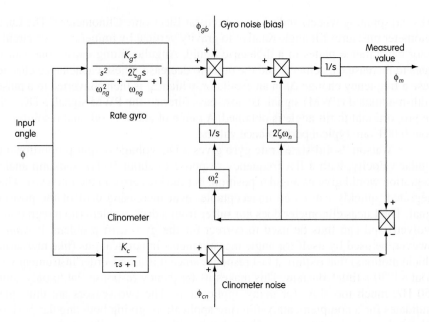

Figure 9.25
Sensor fusion of MEMS rate gyro and clinometer in Watson vertical reference instrument.

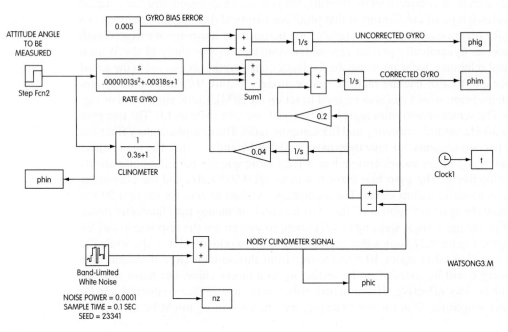

Figure 9.26
SIMULINK simulation for Watson vertical reference.

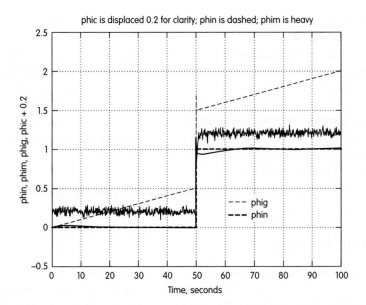

phic is displaced 0.2 for clarity; phin is dashed; phim is heavy

Figure 9.27
Sensor fusion removes gyro drift and reduces clinometer noise.

Our simple examples have shown the basic principle of sensor fusion. More complex applications have been made and can be found in the referenced literature.

PROBLEMS

9.1 Derive Eq. (9.1).

9.2 Prove that a plot of $A_o \sin(\omega t + \phi)$ against $A_i \sin \omega t$ is an ellipse.

9.3 Analyze the system of Fig. 9.4a to obtain the transfer function relating liquid level h_i to float motion x_o. Neglect dry-friction effects. If the liquid level increases very slowly and the float motion is subject to a dry-friction force F_f, develop a formula to estimate the maximum steady-state error.

9.4 Assume the force transducer in Fig. 9.4b is of the elastic deflection type, and obtain the transfer function relating liquid level h_i to force-transducer deflection x_o.

9.5 Discuss the effect of liquid density changes on the accuracy of liquid-level measurement in the systems of Fig. 9.4a and b.

9.6 For the system of Fig. 9.4a, discuss the effect on static and dynamic behavior of using a float that is a body of revolution but *not* a cylinder.

9.7 Repeat Prob. 9.6 for the system of Fig. 9.4b.

9.8 Discuss interfering and/or modifying inputs for the system of Fig. 9.4d. Assume the pressure pickup itself to be insensitive to such inputs.

9.9 Repeat Prob. 9.8 for the system of Fig. 9.4e.

9.10 For the system of Fig. 9.4*k,* make a study of the error caused by the weight of the measuring wire.

9.11 For the baroinertial altimeter of Fig. 9.23, prove that a high-pass filter must have in the numerator of its transfer function the term KD^m if the filter is to "wash out" bias terms that are constant ($m = 1$), ramp type ($m = 2$), and parabolic ($m = 3$). This, of course, is true for *any* high-pass filter, not just those used in the baro-inertial altimeter.

9.12 In Fig. 9.21, derive the differential equation relating **hm** to the three inputs **h**, pressure noise signal, and inertial altitude bias signal.

9.13 Repeat Prob. 9.12 for Fig. 9.23.

9.14 Use Fig. 9.25 to obtain the differential equation relating ϕ_m to ϕ, ϕ_{gb}, and ϕ_{cn}.

BIBLIOGRAPHY

1. P. Kartaschoff, "Frequency and Time," Academic, New York, 1978.

2. W. A. Olsen, "An Integrated Hot Wire-Stillwell Liquid Level Sensor System for Liquid Hydrogen and Other Cryogenic Fluids," *NASA, Tech. Note* D-2074, 1963.

3. "Liquid Hydrogen Level Sensors," *Instrum. Contr. Syst.,* p. 129, May 1964.

4. D. D. Kana, "A Resistive Wheatstone Bridge Liquid Wave Height Transducer," *NASA,* CR-56551, 1964.

5. L. Siegel, "Nuclear and Capacitance Techniques for Level Measurement," *Instrum. Contr. Syst.,* p. 129, July 1964.

6. C. L. Pleasance, "Accurate Volume Measurement of Large Tanks," *ISA J.,* p. 56, May 1961.

7. H. Hellivig, "Frequency Standards and Clocks: A Tutorial Introduction," *NBS, Tech. Note* 616, 1972.

8. P. R. Wiederhold, "Water Vapor Measurement," Marcel Dekker, New York, 1997.

9. R. E. Ruskin (ed.), "Principles and Methods of Measuring Humidity in Gases," vol. 1 of "Humidity and Moisture," Reinhold, New York, 1965.

10. "Dew Point Hygrometer Error Analysis," *Appl. Data* 3-051; EG&G Environmental Equipment. Now Edge Tech (www.edgetech.com).

11. P. L. Mariam, "Measuring Level in Hostile or Corrosive Environments," *Inst. Tech.,* pp. 45–47, April 1979.

12. K. Soleyn, "The Theory and Operation of Optical Chilled Mirror Hygrometers for Humidity Calibration," *Cal Lab,* July–August–September 2002, pp. 31–38.

PART 3

Manipulation, Transmission, and Recording of Data

CHAPTER 10

Manipulating, Computing, and Compensating Devices

The information or data generated by a basic measuring device generally require "processing" or "conditioning" of one sort or another before they are presented to the observer as an indication or a record. Devices for accomplishing these operations may be specific to a certain class of measuring sensors, or they may be quite general-purpose. In this chapter we consider those devices most often needed in building up measurement systems.

10.1 BRIDGE CIRCUITS

Bridge circuits of various types are employed widely for the measurement of resistance, capacitance, and inductance. Since we have seen that many transducers convert some physical variable to a resistance, a capacitance, or an inductance change, bridge circuits are of considerable interest. While capacitance and inductance bridges are important, the simpler resistance bridge is in widest use, and so we concentrate on it here. Adequate technical literature on all types of bridge circuits is readily available.[1]

Figure 10.1 shows a purely resistive (Wheatstone) bridge in its simplest form. The excitation voltage E_{ex} may be either ac or dc; here we consider only dc. In measurement applications, one or more of the legs of the bridge is a resistive transducer such as a strain gage, resistance thermometer, or thermistor. The basic principle of the bridge may be applied in two different ways: the null method and the deflection method. Let us assume that the resistances have been adjusted so that the bridge is balanced; that is, $e_{AC} = 0$. (It is easily shown that this requires $R_1/R_4 = R_2/R_3$.) Now we let one of the resistors, say R_1, change its resistance. This will unbalance the bridge, and a voltage will appear across AC, causing a meter

---

[1]E. Frank, "Electrical Measurement Analysis," chaps. 10 and 13, McGraw-Hill, New York, 1959.

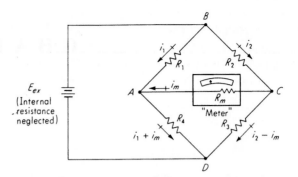

Figure 10.1
Basic Wheatstone bridge.

reading. The meter reading is an indication of the change in R_1 and actually can be utilized to compute this change. This method of measuring the resistance change is called the *deflection method,* since the meter deflection indicates the resistance change. In the *null method,* one of the resistors is adjustable manually. Thus if R_1 changes, causing a meter deflection, R_2 can be adjusted manually until its effect just cancels that of R_1 and the bridge is returned to its balanced condition. The adjustment of R_2 is guided by the meter reading; R_2 is adjusted so that the meter returns to its null or zero position. In this case the numerical value of the change in R_1 is related directly to the change in R_2 required to effect balance.

Both the null and deflection methods are employed in practice In the deflection method, a calibrated meter is needed, and if the excitation E_{ex} changes, an error is introduced, since the meter reading is changed by changes in E_{ex}. With the null method, a calibrated variable resistor is needed, and since there is no meter deflection when the final reading is made, no error is caused by changes in E_{ex}. The deflection method gives an output voltage across terminals AC that almost instantaneously follows the variations of R_1. This output voltage can be applied to an oscilloscope (rather than the meter shown in Fig. 10.1), and thus measurements of rapid dynamic phenomena are possible. The null method, however, requires that the balancing resistor be adjusted to null the meter before a reading can be taken. This adjustment takes considerable time if done manually; even when an instrument servomechanism makes the adjustment automatically, the time required is much longer than is allowable for measuring many rapidly changing variables. Thus the choice of the null or the deflection method in a given case depends on the speed of response, drift, etc., required by the particular application.

In order to obtain quantitative relations governing the operation of the bridge circuit, a circuit analysis is necessary. The following information is desired:

1. What relation exists among the resistances when the bridge is balanced ($e_{AC} = 0$)? The answer has been given as $R_1/R_4 = R_2/R_3$.
2. What is the sensitivity of the bridge? That is, how much does the output voltage e_{AC} change per unit change of resistance in one of the legs?
3. What is the effect of the meter internal resistance on the measurement?

We consider the question of bridge sensitivity first for the case where the "meter" has a very high internal resistance R_m, compared with the bridge resistances. If this is the case, the meter current i_m will be negligible compared with the currents in the legs. Often this situation is closely approximated in practice since most voltage-measuring instruments (digital voltmeters, oscilloscopes, chart recorders, etc.) have input amplifiers with 1-MΩ or more input resistance.

For $i_m = 0$ we have

$$i_1 = \frac{E_{ex}}{R_1 + R_4} \tag{10.1}$$

$$i_2 = \frac{E_{ex}}{R_2 + R_3} \tag{10.2}$$

$$e_{AB} = \text{voltage rise from } A \text{ to } B = i_1 R_1 = \frac{R_1}{R_1 + R_4} E_{ex} \tag{10.3}$$

$$e_{CB} = \frac{R_2}{R_2 + R_3} E_{ex} \tag{10.4}$$

and finally

$$e_{AC} = e_{AB} + e_{BC} = e_{AB} - e_{CB} = \left(\frac{R_1}{R_1 + R_4} - \frac{R_2}{R_2 + R_3} \right) E_{ex} \tag{10.5}$$

Thus we see that the output voltage is a linear function of the bridge excitation E_{ex} but, in general, a *nonlinear* function of resistances R_1, R_2, R_3, and R_4. If the bridge is balanced initially and then R_1, say, begins to change, the output voltage signal will *not* be directly proportional to the change in R_1. For certain practically important special cases, however, perfect linearity is possible. The best example is found in many strain-gage transducers in which, at the balanced condition, $R_1 = R_2 = R_3 = R_4 = R$. Also, the resistance changes are such that $+\Delta R_1 = -\Delta R_2 = +\Delta R_3 = -\Delta R_4$. Then we may write

$$e_{AC} = \left[\frac{R_1 + \Delta R_1}{(R_1 + \Delta R_1) + (R_4 + \Delta R_4)} - \frac{R_2 + \Delta R_2}{(R_2 + \Delta R_2) + (R_3 + \Delta R_3)} \right] E_{ex} \tag{10.6}$$

$$e_{AC} = \frac{\Delta R_1}{R} E_{ex} \tag{10.7}$$

Clearly, Eq. (10.7) shows a strictly linear relationship between e_{AC} and ΔR_1. Strict linearity also is obtained for R_2 and R_3 fixed and $\Delta R_1 = -\Delta R_4$.

Even when the above symmetry does not exist, the bridge response is very nearly linear as long as the ΔR's are small percentages of the R's. In strain gages, for example, the ΔR's rarely exceed 1 percent of the R's. Since the case of small ΔR's is of practical interest, we work out an expression for bridge sensitivity that is a good approximation for such a situation. From Eq. (10.5), $e_{AC} = f(R_1, R_2, R_3, R_4)$, and thus for small changes from the null condition we may write

$$\Delta e_{AC} = e_{AC} \approx \frac{\partial e_{AC}}{\partial R_1} \Delta R_1 + \frac{\partial e_{AC}}{\partial R_2} \Delta R_2 + \frac{\partial e_{AC}}{\partial R_3} \Delta R_3 + \frac{\partial e_{AC}}{\partial R_4} \Delta R_4 \tag{10.8}$$

Now,

$$\frac{\partial e_{AC}}{\partial R_1} = E_{ex} \frac{R_4}{(R_1 + R_4)^2} \qquad V/\Omega \tag{10.9}$$

$$\frac{\partial e_{AC}}{\partial R_2} = - E_{ex} \frac{R_3}{(R_2 + R_3)^2} \tag{10.10}$$

$$\frac{\partial e_{AC}}{\partial R_3} = E_{ex} \frac{R_2}{(R_2 + R_3)^2} \tag{10.11}$$

$$\frac{\partial e_{AC}}{\partial R_4} = - E_{ex} \frac{R_1}{(R_1 + R_4)^2} \tag{10.12}$$

The partial derivatives are taken as constants; thus Eq. (10.8) shows a linear relation between e_{AC} and the ΔR's.

We explained above, in a qualitative fashion, that if the meter resistance is "high enough," terminals AC may be thought of as an open circuit (no current i_m). It would be useful to have a more quantitative method of deciding whether the meter resistance was "high enough" and, if it were not, how to correct for it. We do this now.

By using Thévenin's theorem, the bridge circuit and the "meter" that loads it may be represented as in Fig. 10.2. Since we have been calling the bridge output voltage under assumed open-circuit conditions e_{AC}, this becomes the E_o of Fig. 3.22. Let us call the bridge output under the actual loaded condition e_{ACL}. Immediately we can write

$$i_m = \frac{e_{AC}}{R_{\text{total}}} = E_{ex} \frac{R_1/(R_1 + R_4) - R_2/(R_2 + R_3)}{R_m + R_1 R_4/(R_1 + R_4) + R_2 R_3/(R_2 + R_3)} \tag{10.13}$$

Knowing i_m, we can compute the actual voltage e_{ACL} across the meter under the condition where the meter draws current, since the voltage across the meter will be the product of the current i_m and the meter resistance R_m. Carrying this out and simplifying, we get

$$e_{ACL} = \frac{E_{ex}(R_1 R_3 - R_2 R_4)}{(R_1 + R_4)(R_2 + R_3) + [(R_1 + R_4)R_2 R_3 + R_1 R_4(R_2 + R_3)]/R_m} \tag{10.14}$$

Now

$$e_{AC} = \frac{E_{ex}(R_1 R_3 - R_2 R_4)}{(R_1 + R_4)(R_2 + R_3)} \tag{10.15}$$

and if we wish to display the effect of the meter resistance on the bridge output voltage, we can form the ratio of e_{ACL} to e_{AC}. After some manipulation, this can be shown to be

$$\frac{e_{ACL}}{e_{AC}} = \frac{1}{1 + (1/R_m)[R_2 R_3/(R_2 + R_3) + R_1 R_4/(R_1 + R_4)]} \tag{10.16}$$

Now we have a quantitative way of assessing the effect of the meter resistance R_m on the bridge output. We see that if $R_m = \infty$, then $e_{ACL} = e_{AC}$, as expected. If R_m is not infinite, there will be a *reduction* in the output signal, and the magnitude of this

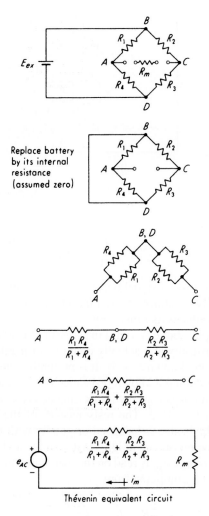

Replace battery
by its internal
resistance
(assumed zero)

Thévenin equivalent circuit

Figure 10.2
Thévenin analysis of bridge.

reduction depends on the relative values of R_m and the bridge "equivalent resistance" R_e, which is defined as

$$R_e \triangleq \frac{R_2 R_3}{R_2 + R_3} + \frac{R_1 R_4}{R_1 + R_4} \tag{10.17}$$

In terms of R_e, Eq. (10.16) becomes

$$\frac{e_{ACL}}{e_{AC}} = \frac{1}{1 + R_e/R_m} \tag{10.18}$$

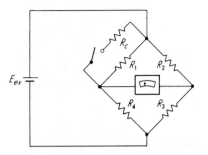

Figure 10.3
Shunt calibration method.

Thus, if $R_m = 10R_e$,

$$\frac{e_{ACL}}{e_{AC}} = \frac{1}{1.1} = 0.91 \tag{10.19}$$

and there is a 9 percent loss in signal because of the noninfinite meter resistance. This type of loss usually is referred to as a *loading effect,* that is, the meter "loads down" the bridge and reduces its sensitivity.

The theory developed above is useful in assessing the effects of various parameters on the bridge sensitivity and actually could be utilized to compute the sensitivity if all quantities were known exactly. It is preferable, however, to calibrate the bridge directly by introducing a known resistance change and noting the effect on the bridge output. Often this known resistance change is introduced by means of the arrangement shown in Fig. 10.3. The resistance R_c of the calibrating resistor is known accurately. If the bridge is balanced originally with the switch open, when the switch is closed, the resistance in leg 1 will change and the bridge will be unbalanced. The output voltage e_{AC} is read on the meter, and the resistance change ΔR that caused this voltage is computed from

$$\Delta R = R_1 - \frac{R_1 R_c}{R_1 + R_c} \tag{10.20}$$

The bridge sensitivity is then

$$S \triangleq \frac{e_{AC}}{\Delta R} \quad \text{V/}\Omega \tag{10.21}$$

This procedure gives an overall calibration since the values of all the resistors and the battery voltage are taken into account. Note that such a shunt calibration is *not* a complete calibration when the bridge is part of, say, a strain gage pressure transducer. A complete or "end-to-end" calibration requires that we apply known *pressures* to the transducer and note the bridge output voltage.

Figure 10.1 shows a bridge circuit with the bare essentials. Often additional features are necessary or desirable for the convenience of the user. Figure 10.4 shows a versatile arrangement providing the following capabilities:

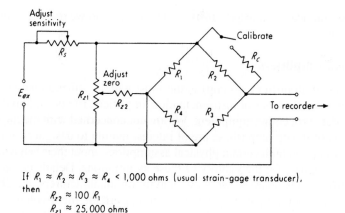

If $R_1 \approx R_2 \approx R_3 \approx R_4 <$ 1,000 ohms (usual strain-gage transducer),
then
  $R_{z2} \approx 100\, R_1$
  $R_{z1} \approx 25,000$ ohms

Figure 10.4
Bridge with sensitivity, balance, and calibration features.

1. Variation of overall sensitivity without the need to change E_{ex}
2. Provision for adjusting the output voltage to be precisely zero when the measured physical quantity is zero, even if the legs are not exactly matched
3. Shunt-resistor calibration

Commercial transducers also may include additional temperature-sensitive resistors to achieve temperature compensation (see Sec. 5.3).

When a bridge uses only a *single* sensor (strain gage, RTD, etc.) in one leg and the other three legs are bridge-completion resistors, the sensor is often remote from the bridge, connected to it by (sometimes long) lead wires. These lead wires are in series with the sensor, so any change in their resistance cannot be distinguished from sensor resistance changes, which leads to errors. This problem is solved by using a three-wire connection[2] to the readout device (see Fig. 8.21c). An alternative[3] to the Wheatstone bridge, which uses constant-current excitation in a novel way, has been invented and applied at several NASA facilities.

10.2 AMPLIFIERS

Since the electrical signals produced by most transducers are at a low voltage and/or power level, often it is necessary to amplify them before they are suitable for transmission, further analog or digital processing, indication, or recording. While our discussion is aimed mainly at *users* (rather than designers) of amplifiers, the use of operational amplifiers in the construction of some simple "homemade" devices

---

[2]TT-612, Measurements Group, Raleigh, NC (www.measurementsgroup.com); C. P. Wright, "Applied Measurement Engineering," Prentice-Hall, Englewood Cliffs, NJ, 1995, pp. 145–146.

[3]K. F. Anderson, "A Successor to the Wheatstone Bridge: NASA's Anderson Loop," Cal Lab, Sept.–Oct. 1998, pp. 23–32.

has become practical for nonspecialists in electronics, and we present some material along these lines.

Operational Amplifiers[4]

The operational amplifier (op-amp) is the most widely utilized analog electronic subassembly; it is the basis of instrumentation amplifiers, filters, and myriad analog and digital data processing equipment. We are not concerned with the internal electronic details of the op-amp (these are of interest mainly to designers of op-amps); we simply accept on faith certain physical assumptions about their behavior, relying on op-amp designers for the validity of these assumptions. Details on op-amps are available from many sources; we draw heavily on manufacturers' handbooks and catalogs.[5] For the op-amp of Fig. 10.5a, e_A and e_B are the input voltages, i_A and i_B are called bias currents, Z_D is the differential input impedance, A is the open-loop gain (amplification), Z_o is the output impedance, V_{os} is the offset voltage, and $\pm V_s$ are the power-supply voltages. In establishing the basic operation of devices that employ op-amps, usually we utilize an ideal op-amp model which neglects certain real-world characteristics in the interest of simple analysis and understanding. Once this basic operation is established, the deviations from perfection can be studied by using a more correct (and more complicated) model. These ideal and real-world models are compared usefully as follows:

Characteristic	Ideal value	Typical real-world value
Open-loop gain A	∞	100,000 V/V
Offset voltage V_{os}	0	± 1 mV @ 25°C
Bias currents i_A, i_B	0	10^{-6} to 10^{-14} A
Input impedance Z_D	∞	10^5 to 10^{11} Ω
Output impedance Z_o	0	1 to 10 Ω

Our ideal model also assumes instantaneous response (flat frequency response from zero to infinite frequency). Some op-amps are designed for low-frequency use only, others extend well into the megahertz range; so usually we can select a model whose dynamic behavior is close to the ideal, at least for the frequency range of interest. The op-amp simplified model of Fig. 10.5b assumes the ideal values for all the parameters and conventionally does not show the power supplies $\pm V_s$.

Assumption of the ideal behavior makes op-amp circuit analysis relatively simple, even for nonspecialists in electronics. Our first example, the *voltage-follower* of Fig. 10.5c, gives

$$e_o = (e_A - e_B - V_{os})A = (e_i - e_o - 0)\infty \tag{10.22}$$

$$\frac{e_o}{\infty} = 0 = e_i - e_o \qquad e_o = e_i \tag{10.23}$$

---

[4]G. B. Clayton, "Operational Amplifiers," 2d ed. Butterworth, London, 1979; W. G. Jung, "IC Op-Amp Cookbook," 3d ed., Prentice Hall (SAMS), Englewood Cliffs, NJ, 1995.

[5]"Linear Design Seminar" (1995), "Amplifier Application Guide" (1992), Analog Devices, Wilmington, MA, 781-262-5643 (www.analog.com).

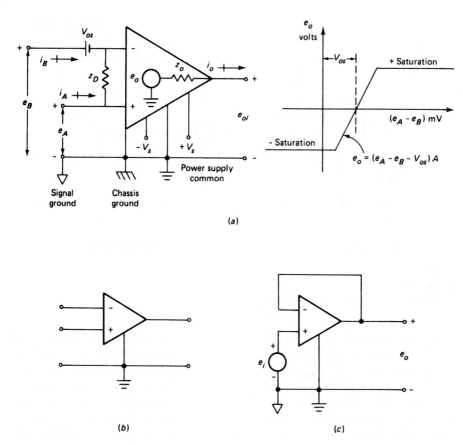

Figure 10.5
Op-amp terminology.

At first glance, a device that merely duplicates at its output a voltage which is applied to its input may not seem very useful. But note that the source (say a transducer) supplying e_i now works into an "infinite" impedance, and so no current or power is withdrawn from it. Yet e_o can be impressed across a load resistance (say a meter) which *does* draw current and consume power. Thus, while there is no voltage amplification, there is power amplification; such a device is sometimes called a *buffer amplifier.*

The output capabilities of a real op-amp are, of course, finite; typically, maximum e_o for proper operation might be specified at ± 10 V with ± 10-mA maximum current capability. Thus if we connected a resistive load of 1,000 Ω across e_o, both maximum voltage and current could be achieved. A 10,000-Ω load would reach 10 V at 1 mA, and a 100-Ω load would reach 10 mA at 1 V. Most op-amps are

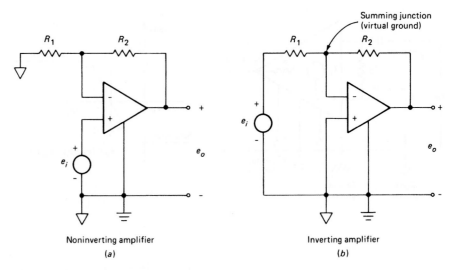

Figure 10.6
Noninverting and inverting op-amp configurations.

protected from short-circuits, so 0 Ω across e_o causes no damage; the output current merely saturates at its limiting value. To estimate the deviation of real-world behavior from the ideal, substitute the typical values from the table above into Eq. (10.22) to get $e_o = 0.9999e_i - 0.0009999$, clearly a very good approximation when $e_i \gg 1$ mV. When the output terminals at e_{ol} are open-circuit, the ideal output e_o and the "loaded" output voltage e_{ol} are identical, irrespective of the value of Z_o, since no output current flows. When e_{ol} is connected across a specific load impedance Z_l, then $e_{ol} = e_o - i_o Z_o$, causing another deviation from ideal behavior. For example, if e_i is 1.0 V, $Z_l = 1{,}000$ Ω, and $Z_o = 1$ Ω, then our earlier equation becomes

$$e_{ol} = (0.9999)(1.0) - 0.0009999 - \frac{0.9999 - 0.0009999}{1000 + 1} \, 1.0 = 0.9979 \text{ V} \quad \textbf{(10.24)}$$

which again compares favorably with the ideal value of 1.0.

When voltage amplification is desired, the inverting and noninverting configurations of Fig. 10.6 are basic. When the ideal model is employed, analysis gives for the noninverting amplifier

$$e_o = \left(\frac{R_2}{R_1} + 1 \right) e_i \qquad \textbf{(10.25)}$$

and for the inverting amplifier

$$e_o = -\frac{R_2}{R_1} e_i \qquad \textbf{(10.26)}$$

Clearly, in both cases, e_o / e_i amplification (called *signal gain*) is determined entirely by the resistance ratio R_2/R_1. For the inverting amplifier, the input resistance

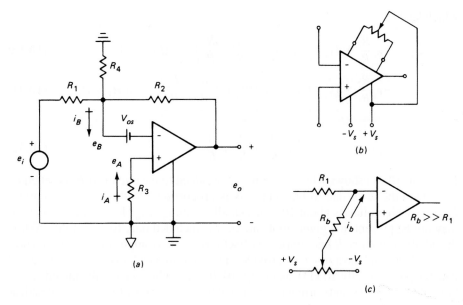

Figure 10.7
Offset-voltage and bias-current considerations.

presented to the source e_i is R_1, because e_i is effectively across R_1 (the summing junction is essentially at the same potential as the system ground). Analysis of the more correct model gives very nearly this same result. For the noninverting amplifier, input resistance is ideally infinite ($i_A \equiv 0$). However, analysis[6] of the more correct model gives

$$R_{in} = R_D\left(1 + A\frac{R_1}{R_1 + R_2}\frac{R_L}{R_o + R_L}\right) \tag{10.27}$$

which can easily be $1{,}000R_D$, a very large value. In the noninverting case, since $e_A - e_B$ always must equal zero in the ideal model, the junction of R_1 and R_2 must be at voltage e_i at all times; in the inverting case, this junction (now called the *summing junction*) must be at 0 V since e_A is grounded, and for this reason it is called a *virtual ground*.

In Fig. 10.7a, the inverting configuration (generalized by addition of R_3 and R_4) is analyzed for voltage and current offset errors:

$$\frac{e_i - e_B}{R_1} + \frac{e_o - e_B}{R_2} - \frac{e_B}{R_4} - i_B = 0 \tag{10.28}$$

$$e_B = e_A + V_{os} \tag{10.29}$$

[6]G. B. Clayton, op. cit., p. 22.

which lead to

$$e_o = -e_i \frac{R_2}{R_1} + (V_{os} + i_B R)\left(1 + \frac{R_2}{R_1} + \frac{R_2}{R_4}\right) \qquad \text{if } R_3 = 0 \qquad \textbf{(10.30)}$$

$$e_o = -e_i \frac{R_2}{R_1} + \underbrace{[V_{os} + (i_B - i_A)R]\left(1 + \frac{R_2}{R_1} + \frac{R_2}{R_4}\right)}_{\text{error}} \qquad \text{if } R_3 = R \qquad \textbf{(10.31)}$$

$$\frac{1}{R} \triangleq \frac{1}{R_1} + \frac{1}{R_2} + \frac{1}{R_4} \qquad \textbf{(10.32)}$$

Equation (10.31) shows why sometimes R_3 is added to the circuit: If R_3 is set equal to R, then the error due to bias currents is proportional to the *difference* current $i_B - i_A$, which often is about 10 times smaller than i_A or i_B. Resistance R_4 may or may not be present in a given application; it is included only for generality. While the addition of $R_3 = R$ may improve offset errors, many applications require additional efforts, and most op-amps provide a pair of terminals for connecting a manual balance pot (typically 10 kΩ), as in Fig. 10.7b. This adjustment nulls the effects of both offset voltages and currents, but the balance is gradually lost with time and/or temperature and must be retuned periodically.

When the offset is due mainly to bias current ($i_B R_1 > 4$ or 5 mV), the current injection bias circuit of Fig. 10.7c is preferred to the balance pot, since it causes less unbalance in the op-amp input stage. Some applications require use of both bias schemes.

Another basic configuration is the differential amplifier[7] of Fig. 10.8a. The input signal source in our example is a balanced grounded type such as found in a strain-gage bridge force transducer (Fig. 10.8b). At bridge null, e_x and e_y are both 5 V, and the amplifier output should be zero. When force application causes bridge unbalance, e_x goes to, say, 5.01 V and e_y to 4.99 V, and we want to amplify the 0.02-V difference. In the general circuit of Fig. 10.8a the original 5-V signal (common to both input paths) is called the *common-mode voltage* e_{CM}, while e_1 and e_2 represent the changes in e_x and e_y, respectively. Superimposing our earlier results for the inverting and noninverting configurations (allowed because of system linearity), we can write

$$e_o = -\frac{R_3}{R_1}(e_{CM} + e_1) + e_A\left(1 + \frac{R_3}{R_1}\right) \qquad \textbf{(10.33)}$$

$$e_A = (e_{CM} + e_2)\frac{R_4}{R_2 + R_4} \qquad \textbf{(10.34)}$$

which gives

$$e_o = -e_1 \frac{R_3}{R_1} + e_2 \frac{R_4}{R_2} \frac{R_3/R_1 + 1}{R_4/R_2 + 1} + e_{CM}\left(\frac{R_4}{R_2} \frac{R_3/R_1 + 1}{R_4/R_2 + 1} - \frac{R_3}{R_1}\right) \qquad \textbf{(10.35)}$$

---

[7]R. B. Northrup, "Introduction to Instrumentation and Measurements," CRC Press, Boca Raton, FL, 1997, pp. 31–40.

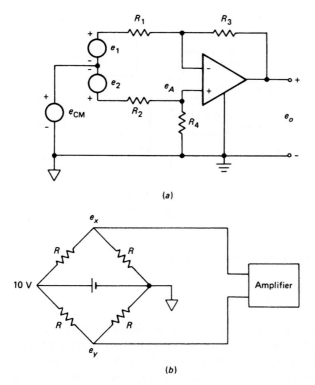

Figure 10.8
Differential input op-amp configuration.

We see now that to eliminate common-mode errors due to e_{CM}, we must be careful to make $R_4/R_2 = R_3/R_1$, since then we achieve our original goal, that is,

$$e_o = \frac{R_3}{R_1}(e_2 - e_1) \tag{10.36}$$

Usually this balancing is done by making $R_1 = R_2$ and $R_3 = R_4$, where R_1 and R_2 must include not only the amplifier's resistors but also the source resistance (in our example, the strain-gage bridge resistance) of e_1 and e_2.

Equation (10.35) assumes the plus and minus amplifier input channels to have perfectly matched gains. A real amplifier deviates from this, and the degree of deviation is specified by the *common-mode rejection ratio* (CMRR):

$$\text{CMRR} \triangleq \frac{(e_{CM})\,(\text{differential gain})}{e_o \text{ due to } e_{CM}} \tag{10.37}$$

For a CMRR of 80 dB (10,000), a differential gain [see Eq. (10.36)] of $R_3/R_1 = 100$, and an e_{CM} of 1 V, we would get a spurious output of 0.01 V even if $R_4/R_2 \equiv R_3/R_1$. Thus we get common-mode errors from *both* resistance mismatch and inherent amplifier-channel mismatch. These two effects may add or subtract, and sometimes

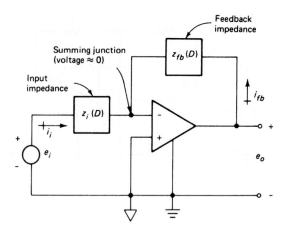

Figure 10.9
Op-amp generalized computing configuration.

intentional resistor mismatch is used to cancel amplifier common-mode error. Common-mode voltage e_{CM} is not limited to steady or intentionally applied effects like the bridge excitation of Fig. 10.8*b;* rather it is *any* signal that is applied equally to both differential input paths. Often it is a spurious noise voltage picked up by the circuitry from its surroundings. In any case, use of amplifiers with high common-mode rejection ratio and careful attention to symmetry in the lead-wire and amplifier impedances are helpful in reducing errors. The differential input resistance is $R_1 + R_2$, while the effective common-mode input resistances at e_1 and e_2 are $R_1 + R_3$ and $R_2 + R_4$, respectively; thus large R values are required for high input resistance. When these get too large, offset errors due to bias current and loss of common-mode rejection ratio at high frequencies (because of stray capacitance) limit use of this circuit; however, the *instrumentation amplifier* of the next section is available to meet such needs.

The inverting-type circuit with resistances replaced by generalized impedances $Z_i(D)$ and $Z_{fb}(D)$ is the basis of a wide variety of analog computing, filtering, and control devices (see Fig. 10.9):

$$i_i = \frac{e_i - 0}{Z_i(D)} = -i_{fb} = -\frac{e_o - 0}{Z_{fb}(D)} \tag{10.38}$$

$$\frac{e_o}{e_i}(D) = -\frac{Z_{fb}(D)}{Z_i(D)} \tag{10.39}$$

By proper choice of Z_{fb} and Z_i and by using combinations of resistors and capacitors (inductors are usually undesirable, not necessary, and rarely used), many practical e_o/e_i transfer functions may be synthesized (some are covered later in this chapter).

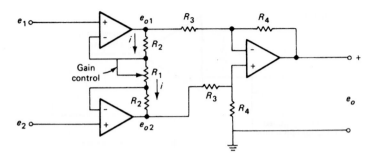

Figure 10.10
Instrumentation amplifier.

Instrumentation Amplifiers

Obviously, many kinds of amplifiers are utilized in instrumentation; however, a particular class and configuration conventionally is given the name *instrumentation amplifier,* mainly to distinguish it from the simpler op-amp circuits. Various designs[8] of instrumentation amplifier are in use; Fig. 10.10 shows a basic type[9] that utilizes three op-amps. You can construct instrumentation amplifiers, using op-amps as building blocks, or purchase them ready-made. Whereas op-amps are basic building blocks adaptable to a wide variety of unrelated uses, instrumentation amplifiers are "committed" devices of essentially fixed configuration intended for rather specific purposes. Their major characteristics include high common-mode rejection ratio and input impedance, low noise and drift, moderate bandwidth, and a limited range of gain (usually 1 to 1,000, programmable by a single resistor). In Fig. 10.10, since the op-amp inputs carry no current, a single current i flows from e_{o1} to e_{o2} through R_2, R_1, and R_2, giving

$$i = \frac{e_{o1} - e_1}{R_2} = \frac{e_1 - e_2}{R_1} = \frac{e_2 - e_{o2}}{R_2} \tag{10.40}$$

which leads to

$$e_{o1} - e_{o2} = (e_1 - e_2)\left(1 + 2\frac{R_2}{R_1}\right) \tag{10.41}$$

and finally

$$e_o = (e_2 - e_1)\frac{R_4}{R_3}\left(1 + \frac{2R_2}{R_1}\right) \tag{10.42}$$

Differential and common-mode input resistance are infinite ideally, with actual values of 10^9 Ω being readily available. A high common-mode rejection ratio (90 dB) is achieved even when source resistances are unbalanced by, say, 1,000 Ω.

---

[8]Ibid., p. III-33; "Isolation and Instrumentation Amplifiers: Application, Theory, Selection," Analog Devices; "Isolation and Instrumentation Amplifiers: Designer's Guide," Analog Devices; R. Morrison, Analog Devices; "Instrumentation Fundamentals and Applications," Wiley, New York, 1984.

[9]G. B. Clayton, op. cit., p. 144.

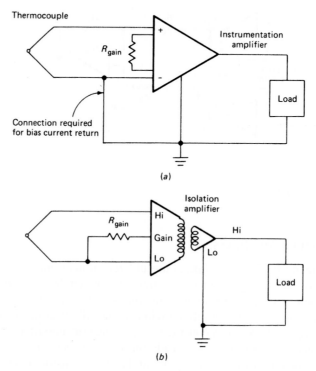

Figure 10.11
Instrumentation/isolation-amplifier comparison.

Isolation amplifiers[10] are a special subclass of instrumentation amplifiers intended for the most demanding applications, where low-level signals ride on top of high common-mode voltages; possibilities exist of troublesome ground disturbances and ground loops; processing circuitry must be protected from faults and power transients; interference from motors, power lines, etc., is heavy; and/or patient protection is important in biomedical applications. "Ordinary" instrumentation amplifiers require a return path for the bias currents. If this is not provided, these currents will charge stray capacitances, causing the output to drift excessively or to saturate. Therefore, when "floating" sources such as thermocouples are amplified, a connection to amplifier ground (Fig. 10.11*a*) must be provided; sometimes this leads to excessive noise problems. An isolation amplifier (Fig. 10.11*b*) does not require such a ground connection (the signal is now *isolated* from ground), and rejection of interfering noise may be much improved. A *two-port isolator* has its signal input circuit isolated from signal output and power input circuits. The ultimate in isolation is provided by *three-port isolators* which have, in addition, isolation between signal output and power input. To achieve isolation, two techniques

---

are employed: transformer coupling and optical coupling. To allow signal transmission from input to output without a conductive connection, the transformer method uses a modulation/demodulation (carrier amplifier) scheme, in which the ac signals are transferred from input to output across a transformer. Opto-isolators transduce input voltage to proportional light intensity by using LEDs. This light is transduced back to output voltage by light-sensitive diodes. Details of operation and comparison of the two methods are available in the literature.[11] Transformer isolated and "ordinary" instrumentation amplifiers may be compared[12] as follows:

Specification	Isolation amplifier	Instrumentation amplifier
Common-mode rejection ratio 5,000-Ω source unbalance, dc to 100 Hz	115 dB	80 dB
Common-mode voltage range	$\pm 2,500$ V dc ($\pm 7,500$ V peak)	± 10 V
Differential input voltage range	240 V rms ($\pm 6,500$ V peak)	± 10 V
Input-to-ground leakage	10^{11} Ω shunted by 10 pF	Feedback generated depends on linear circuit operation
Bias-current configuration	Single current; amplifier needs only two input conductors	Two currents; third wire needed for bias return
Small-signal bandwidth	dc to 2 kHz	dc to 1.5 MHz
Gain nonlinearity	0.05%	0.01%
Gain vs. temperature	$\pm 0.01\%/\,^{\circ}$C	$\pm 0.0015\%/\,^{\circ}$C
Offset vs. temperature	$\pm 300\ \mu$V$/\,^{\circ}$C	$\pm 150\ \mu$V$/\,^{\circ}$C

Figure 10.12[13] shows use of three isolation amplifiers (of a type that utilizes external modulator/demodulator drive signals) in a three-channel data acquisition system.

Transconductance and Transimpedance Amplifiers

Most amplifiers used in measurement systems accept a voltage as the input signal and produce a proportional voltage at their output terminals. In some applications, we require or prefer a *current* as the input or output signal. When the input is a voltage and the output is a current, the amplifier is called *transconductance* whereas the reverse situation is called *transimpedance*. Note that in an "ordinary" (voltage in, voltage out) amplifier, there generally *is* a current existing at each location, but our *purpose* is that the output voltage be some constant times the input voltage, that is, the transfer function gain has dimensions of volt/volt.

---

[11]B. Morong, "Isolator Stretches the Bandwidth of Two-Transformer Designs," *Electronics,* July 3, 1980.

[12]Isolation and Instrumentation Amplifiers: Designer's Guide, op. cit.

[13]Ibid.

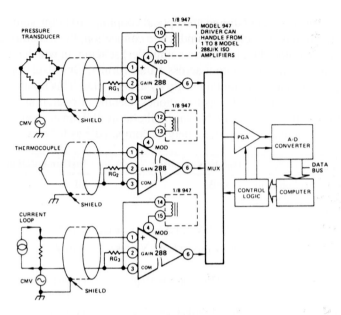

3 Channel Isolated Data Acquisition System

Figure 10.12
Isolation-amplifier application.

In a transconductance amplifier the dimensions of the transfer function are amps/volt, that is, when we apply a voltage input, we expect an output *current* proportional to that voltage. In an "ordinary" amplifier, there *will* be an output current, but it is determined by the output *circuit,* not the input voltage. Producing an output current that is somehow *independent* of the circuit in which it flows seems to be "magical," but it is the magic of feedback. That is, the transconductance amplifier is a feedback-control system in which the controlled variable is the output current. This current is measured by the amplifier, and the amplifier output *voltage* is raised or lowered to maintain the current at the value commanded by the input voltage. Of course, no feedback system is perfect so there are some practical limits on how well the current follows the voltage command; however, the concept works well enough that it is regularly used, mainly to *speed up* the response of circuits or electromechanical devices to commands. For example, the circuit of an electro-hydraulic servovalve has significant inductance, which slows the buildup of current, magnetic force, and thus valve spool motion when the input voltage is changed. By using a transconductance amplifier to supply the servovalve coils, the inductive lag is largely suppressed, giving a useful speed increase. Some amplifiers[14] can be converted from volt/volt to amp/volt operation by simply throwing a switch.

---

[14]Model PA-138, Labworks Inc., Costa Mesa, CA, 714-549-1981 (www.labworks-inc.com).

Transimpedance amplifiers are used when a sensor produces a current signal, rather than the usual voltage. This need is easily accommodated by standard op-amp techniques, as shown in the photodiode application of Fig. 8.39a.

Noise Problems, Shielding, and Grounding[15]

In the progression from single-amplifier op-amp circuits to instrumentation and finally isolation amplifiers, capability for successfully dealing with noisy environments significantly improves, at the expense of greater cost, complexity, and size. Clearly, you should use any available methods to minimize noise effects so that the simplest and cheapest type of amplifier which provides satisfactory service can be utilized. Here we cover only briefly some of the main considerations, leaving details to the listed references. We begin by stating that transducers of low impedance generally cause less severe noise problems and should be employed in preference to high-impedance devices, when possible.

Gradual changes in amplifier offset voltages and bias currents resulting from temperature, time, and line voltage sometimes are called drift, rather than noise, but we include them here for completeness. If the temperature and/or line voltage can be kept more nearly constant, errors caused by these disturbances are, of course, reduced. Drift performance of amplifiers varies greatly from model to model; thus a close check of specifications allows a proper choice. Remember that any junction of dissimilar metals in a circuit is an unintentional thermocouple capable of introducing temperature-related errors. [An *integrated-circuit* (IC) amplifier, its socket, and a pair of binding posts connected to one intentional thermocouple can have 18 *unintentional* thermojunctions in series.] If frequency response to dc is not required, use of ac coupling can greatly reduce all drift effects. In digital data systems, often *automatic zero* techniques are utilized to compensate drift. Here, since data are already intermittent because of the sampling required in digital systems, the input can be shorted periodically (made zero) and the resulting output (which must be the total drift error) temporarily put in memory. Then the next data reading can be corrected for drift by adding the memorized error.

Inductive pickup, electrostatic pickup, and ground loops can cause large error voltages, often concentrated at the power-line frequency. The changing magnetic field that surrounds any conductor carrying alternating current can link signal wires and produce large interfering voltages by inductive pickup. Because of the stray capacitance present between any adjacent conductors, a varying electric field can couple noise voltages to a signal circuit electrostatically. In solving (or preventing) such noise problems, start at the source by taking all measures possible to remove equipment such as power lines, motors, transformers, fluorescent lamps, relays, etc., from the near neighborhood of sensitive signal circuits, since the listed devices

---

[15]R. Morrison, "Grounding and Shielding Techniques in Instrumentation," Wiley, New York, 1977; D. H. Sheingold (ed.), "Transducer Interfacing Handbook," chap. 3, Analog Devices, 1980; "Elimination of Noise in Low-Level Circuits," Gould Inc., Cleveland, OH; "Noise Control in Strain Gage Measurements," *TN-501*, Measurements Group, Raleigh, NC, 1980.

produce both inductive and electrostatic interference. If further measures are necessary, the use of an enclosing conductive shield (Fig. 10.13) can decrease electrostatic pickup. The shield functions by capturing charges that otherwise would reach
the signal conductors. Once captured, these charges must be drained off to a satisfactory ground, or else they would be coupled to the signal conductors through the
shield-to-cable capacitance.

Shielding of the above type is *not* practical for magnetically induced noise since
magnetic shielding at low (60 Hz) frequencies requires thick (2.5 mm) shields of
ferromagnetic metals. Rather, inductive noise is minimized by *twisting* the two
signal conductors so that the "loop area" available for inducing error voltages is
reduced and the mutual inductances between the noise source and each wire are
balanced, to give a canceling effect. Commercially available cable provides twisted
conductors, wrapped foil shields, and a grounding drain wire to meet the needs of
both electrostatic and inductive noise reduction. Flexing of a cable can generate
noise as a result of rubbing friction within the cable itself (triboelectric effect), and
so cables should be properly secured. When cable flexing is unavoidable, a
construction employing a conductive tape between the primary insulation and the
metal-foil shield reduces triboelectric noise considerably.

A *ground loop* is created by connecting a signal circuit to more than one
ground. If the multiple "ground" points were truly at identical potentials, no problem would arise. However, the conductor that serves as ground (piece of heavy
wire, metal chassis of instrument, ground plane of printed-circuit board, etc.) generally carries intentional or unintentional currents and has some resistance; thus two
points some distance apart will *not* have identical potentials. If this potential difference produces current flows through the shield and/or signal circuit, then large

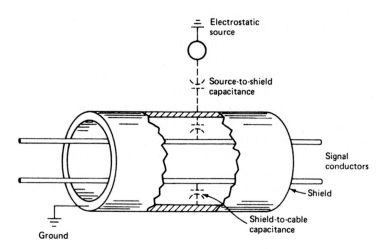

Figure 10.13
Electrostatic shielding.

noise voltages may occur. In Fig. 10.14a, two ground loops, one through the shield and the other through a signal wire, are caused by improper grounding practices. Current in the signal wire directly causes error voltages (because of wire resistance), while shield current couples voltages into the signal circuit through shield-to-cable capacitance. In Fig. 10.14b, the shield ground loop is broken simply by grounding the shield at only one point (the signal source), while use of a floating-input (isolated) amplifier breaks the other loop, thus greatly reducing noise pickup. If the amplifier has a floating internal shield ("guard shield" or "guard") surrounding its input section, then the cable shield is still grounded only at the signal source; but the amplifier end of the cable is connected to this guard, effectively extending it to the signal source. A final technique for noise control is filtering (see Sec. 10.3), but usually this can be utilized only if signal and noise occupy different frequency ranges.

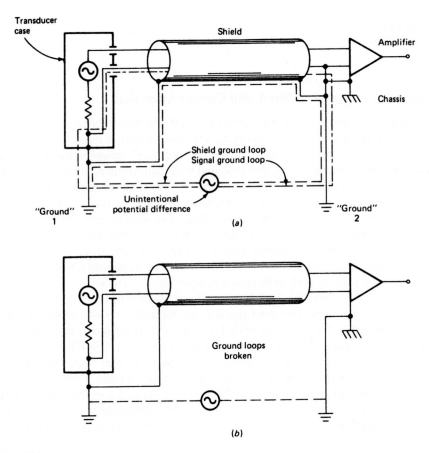

Figure 10.14
Ground-loop problems.

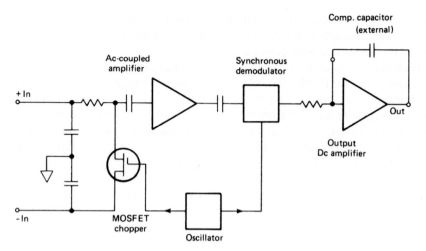

Figure 10.15
Chopper amplifier.

Chopper, Chopper-Stabilized, and Carrier Amplifiers

In earlier generations of op-amps, chopper and chopper-stabilized designs were necessary when low drift was important, and these amplifier types were employed widely. Presently, the drift performance of "ordinary" op-amps has been much improved, and their simplicity, low cost, and wide bandwidth make them preferable to chopper-based designs in most cases. However, some extreme low-noise and low-drift applications still are best handled with chopper-type instruments; thus manufacturers continue to produce them, though in smaller quantities than earlier. A floating-input data amplifier design[16] based on a combination of a chopper-stabilized op-amp and an isolation amplifier with an op-amp front end has 200,000 gain, 100 nV/°C temperature drift, 5 μV/year time drift, 160-dB CMRR, and input noise of 1 μV peak to peak in its design bandwidth of a few hertz.

Figure 10.15 shows the configuration of a chopper-type op-amp[17] designed for noninverting operation. The input signal is fed through a resistor to the MOSFET chopper (switch). When the MOSFET is off (high resistance), the error signal appears at the input to the ac-coupled amplifier. When the MOSFET transistor is on (low resistance), the input to this amplifier is reduced to near zero. The difference between the on and off voltages at the amplifier is a square wave of amplitude slightly less than the error voltage. (A 3,500-Hz oscillator drives the chopper on and off and provides a synchronizing signal for the demodulator.) The ac-coupled amplifier amplifies the square-wave error signal; its output is ac-coupled to a

---

[16]D. H. Sheingold, "Transducer Interfacing Handbook," op. cit., pp. 226–228.

[17]Model 261, Analog Devices.

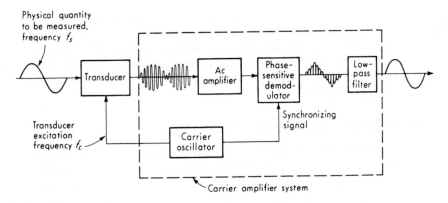

Figure 10.16
Carrier amplifier system.

synchronous demodulator which reconstructs the input signal. The drift of the input stage is not present in the demodulated signal, since it was not chopped by the input network. An output dc op-amp, connected as a low-pass filter with some additional gain, completes the system. For such an arrangement, offset and drift, referred to the *amplifier input,* are equal to the output dc amplifier stage input drift and offset divided by the ac-coupled amplifier gain. Typically, output-stage drift might be 100 μV/°C and ac gain is 1,000, which gives a system drift of 0.1 μV/°C, a very low value. As in any modulation/demodulation/filtering scheme, frequency response is limited by the modulation (chopping) frequency and the necessary low-pass output filter. Noninverting amplifiers built with this op-amp can have flat (-3 dB) frequency response to about 100 Hz, extremely high gain (100,000), and low noise (0.4 μV peak to peak, 0.01 to 1 Hz).

In a chopper-stabilized[18] (rather than chopper) op-amp, the input signal low-frequency components are diverted through a low-pass filter, amplified with a ("drift-free") chopper amplifier (gain $\approx$ 1,000), and then further amplified in an "ordinary" dc amplifier (gain $\approx$ 50,000). The unfiltered input signal is summed with the chopper amplifier output at the input of the final dc stage; thus low-frequency components go through both the chopper and dc amplifiers (total gain $\approx$ 5 $\times$ 10^7) while high-frequency go through only the dc (gain $\approx$ 50,000). Such a system combines the low-drift behavior of the chopper amplifier with the wide bandwidth of the dc type, overcoming the frequency-response limitations of pure chopper amplifiers (response to several hundred kilohertz is possible).

When a transducer *requires* ac excitation (LVDTs are perhaps the most common example), the carrier amplifier system of Fig. 10.16 is still employed widely. When *either* ac or dc excitation is possible (strain-gage transducers, for example), the simplicity and bandwidth of dc amplifiers make dc excitation the

---

[18]E. O. Doebelin, "Measurement Systems," Rev. Ed., pp. 619–621, McGraw-Hill, New York, 1975.

preferred choice in most cases. However, there still are some applications[19] in which the high gain, low drift, and good interference rejection of carrier systems are needed, so they are still available from several manufacturers. The final low-pass filter (needed to smooth the demodulator output), of course, sets the frequency-response limit of the entire system. A typical 5,000-Hz carrier system gives flat response to about 1,000 Hz.

Charge Amplifiers and Impedance Converters

Wide use of piezoelectric accelerometers, pressure pickups, and load cells has led to the development of an amplifier type that offers some advantages over the usual voltage amplifier in certain applications. Such a *charge amplifier* is shown connected to a piezoelectric transducer in Fig. 10.17. The idealized form is shown in Fig. 10.17*a*, where we note that a FET-input operational amplifier is used with a capacitor C_f in the feedback path. Assuming, as usual, that the input voltage e_{ai} and current i_a of the operational amplifier are small enough to take as zero, we get

$$K_q Dx_i = -C_f De_o \qquad (10.43)$$

$$e_o = -\frac{K_q x_i}{C_f} \qquad (10.44)$$

Equation (10.44) indicates that e_o would be instantaneously and linearly related to displacement x_i without the usual loss of steady-state response associated with piezoelectric transducers and voltage amplifiers. Unfortunately, this advantage is not realizable since a system constructed as in Fig. 10.17*a* would, because of nonzero op-amp bias current, exhibit a steady charging of C_f by the bias current until the amplifier saturated. To overcome this problem, in the practical circuit of Fig. 10.17*b* a feedback resistance R_f is included to prevent this small current from developing a significant charge on C_f. Analysis of this new circuit gives

$$\frac{e_o}{x_i}(D) = \frac{K\tau D}{\tau D + 1} \qquad (10.45)$$

where $K \triangleq \dfrac{K_q}{C_f},$ V/in

$$\tau \triangleq R_f C_f, \qquad \text{s} \qquad (10.46)$$

Equation (10.45) is of identical form to the transfer function of a piezoelectric transducer and a *voltage* amplifier and exhibits the same loss of static and low-frequency response. The advantages of the charge amplifier are found in Eq. (10.46). We note that both the sensitivity K and the time constant τ are now independent of the capacitance of the crystal itself and the connecting cable, whereas with a voltage amplifier neither of these advantages is obtained. Thus long cables

---

[19]"Carrier Frequency Amplifiers and DC Amplifiers, A Comparison," Hottinger Baldwin, VD 83001a, Framingham, MA, 1987.

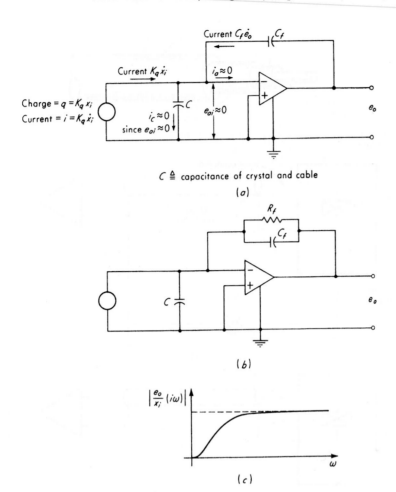

$C \triangleq$ capacitance of crystal and cable

(a)

(b)

(c)

Figure 10.17
Charge amplifier.

(often several hundred feet in practical setups) do not result in a reduced sensitivity or a variation in frequency response. These advantages and others[20] are sufficient to make the charge amplifier of practical interest in many systems. Disadvantages[21] that may arise in certain applications include a possibly poorer signal-to-noise ratio and a reduction in natural frequency of the transducer because of a loss of stiffness, caused by what amounts to a short-circuit across the crystal.

When utilized with quartz-crystal transducers,[22] the value of C_f is from 10 to 100,000 pF and R_f is 10^{10} to 10^{14} Ω. For $C_f = 100,000$ pF and $R_f = 10^{14}$ Ω,

---

[20]D. Pennington, "Charge Amplifier Applications," Endevco Corp., Pasadena, CA, April 1964.

[21]Wilcoxon Research, Bethesda, MD, *Res. Bull.* 5.

[22]Kistler Instrument Corp., Clarence, NY, *Tech Notes* 133762 and 130662.

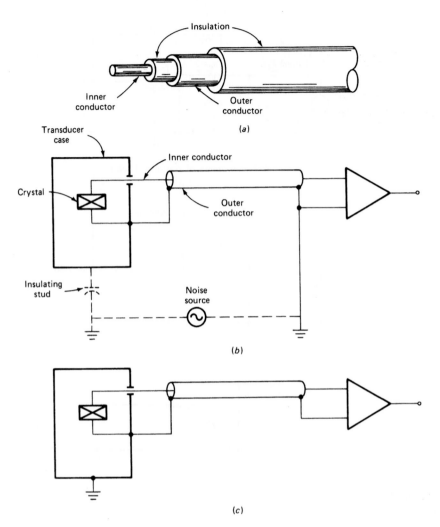

Figure 10.18
Piezoelectric transducer shielding techniques.

$\tau = 10^6$ s, showing that practically dc response, allowing static calibration and measurement, is possible under these conditions. For ceramic-type transducers, C_f is from 10 to 1,000 pF and R_f from 10^8 to 10^{10} Ω, making the maximum τ about 10 s and static measurements thus usually impractical.[23] Rather than the twisted, shielded pair recommended earlier for general-purpose low-level signal cables, usually piezoelectric transducers employ a miniature coaxial cable (Fig. 10.18a).

[23]Ibid.

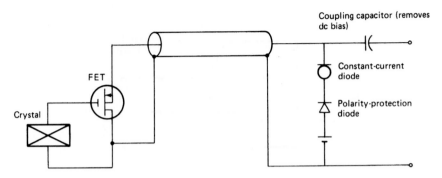

Figure 10.19
Piezoelectric transducer impedance converter.

Good noise rejection can be achieved by several means.[24] Figure 10.18*b* shows use of an insulated transducer mounting (to break the ground loop) with a case-grounded transducer and single-ended grounded amplifier. For accelerometers, the insulated mounting (more compliant than "steel to steel") may not always be acceptable because of dynamic-response loss. Then a case-grounded transducer with grounded mounting may be necessary, and this requires a floating or differential charge amplifier if a noisy environment is present (Fig. 10.18*c*).

Another approach to piezoelectric signal conditioning is the *impedance converter* of Fig. 10.19. Here a field-effect transistor is connected directly across the crystal, presenting a very high impedance, but providing a large output (10 V) at a low impedance ($<100 \ \Omega$). A single coaxial cable carries both power (from a 28-V battery) and readout signal. The FET device is small enough to be built into the transducer case, and such transducers are common where their lower allowable temperature is acceptable. No charge amplifier is needed; the output can go directly to conventional voltage amplifiers.

Concluding Remarks

The versatility of electronics ensures that most reasonable amplifier requirements can be met, though exotic conditions may require complex and expensive systems. Thus sensors, rather than electronics, usually limit overall system performance. The 5- or 10-mA current output of most instrumentation amplifiers is adequate except when high-current loads, such as galvanometers or motors, must be driven. Such needs are met simply by adding a high-power output stage to the amplifier. Many amplifiers also include built-in filters with switch-selectable bandwidths, convenient for noise suppression.[25]

---

[24]J. Wilson, "Noise Suppression and Prevention in Piezoelectric Transducer Systems," *Sound & Vib.,* pp. 22–25, April 1979.

[25]Ectron Corp., San Diego, CA, 800-732-8159 (www.ectron.com).

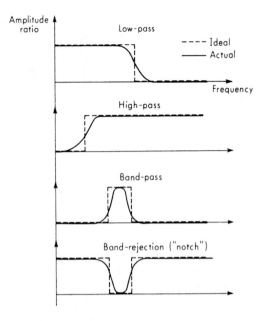

Figure 10.20
Basic filter characteristics.

10.3 FILTERS

The use of frequency-selective filters to pass the desired signals and reject spurious ones has been discussed. Figure 10.20 summarizes the most common frequency characteristics used. Filters may take many physical forms; however, the electrical form is most common and highly developed with regard to both theory and practical realization. By use of analogies, the material on electrical filters may suggest the configurations of mechanical, hydraulic, acoustical, etc., systems that will provide the desired filtering action in specific problems.

Low-Pass Filters

The simplest low-pass filters commonly in use are shown in several different physical forms in Fig. 10.21. They all have identical transfer functions given by

$$\frac{e_o}{e_i}(D) = \frac{x_o}{x_i}(D) = \frac{p_o}{p_i}(D) = \frac{T_o}{T_i}(D) = \frac{1}{\tau D + 1} \qquad \textbf{(10.47)}$$

Since these are all simple first-order systems, the attenuation is quite gradual with frequency, 6 dB/octave. This does not give a very sharp distinction between the frequencies that are passed and those that are rejected. By adding more "stages" (see Fig. 10.22*a*) the sharpness of cutoff may be increased. The use of inductance elements (Fig. 10.22*b*) also may lead to better filtering action. When a filter is inserted into a system, it is necessary to take into account possible loading effects by use of appropriate impedance analysis.

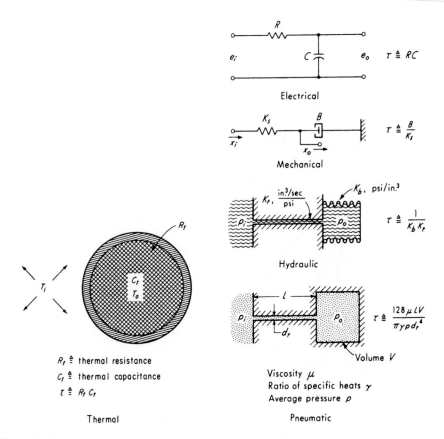

Figure 10.21
Low-pass filters.

All the filters shown thus far are *passive* filters, since all the output energy must be taken from the input. Today many electrical filters are *active* devices based on op-amp technology. Whereas passive filters have very low noise, require no power supplies, and have a wide dynamic range, active filters[26] are much more adjustable and versatile, can cover very wide frequency ranges, have very high input and very low output impedances (which makes cascading and interconnection simple), and can be configured for simple switching from low-pass to high-pass and combination for band-pass or band-reject behavior.

Figure 10.23*a* shows an active second-order low-pass filter, while Fig. 10.23*b* illustrates a "state-variable filter" with electrically adjustable parameters.[27] The

[26]"The Application of Filters to Analog and Digital Signal Processing," Rockland Systems Corp., Rockleigh, NJ, 1980; D. E. Johnson, J. R. Johnson, and H. P. Moore, "A Handbook of Active Filters," Prentice-Hall, Englewood Cliffs, NJ, 1980.

[27]D. H. Sheingold (ed.), "Nonlinear Circuits Handbook," pp. 138–141, Analog Devices, 1976.

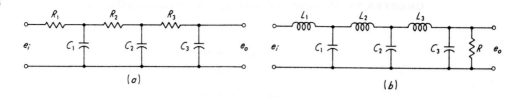

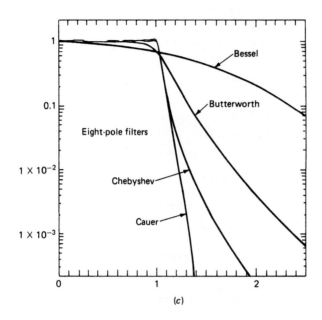

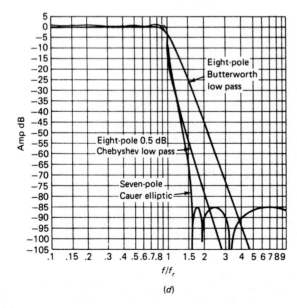

Figure 10.22
Sharper-cutoff low-pass filters.

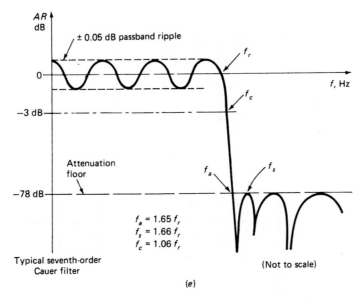

± 0.05 dB passband ripple

f_r

f_c

f, Hz

−3 dB

Attenuation floor

f_a

f_s

−78 dB

$f_a = 1.65 f_r$
$f_s = 1.66 f_r$
$f_c = 1.06 f_r$

Typical seventh-order
Cauer filter

(Not to scale)

(e)

Figure 10.22
(Concluded)

state-variable filter provides three simultaneous outputs: a low-pass, a high-pass, and a bandpass:

$$\frac{e_{lp}}{e_i}(D) = -\frac{1}{D^2/\omega_n^2 + 2\zeta D/\omega_n + 1} \tag{10.48}$$

$$\frac{e_{hp}}{e_i}(D) = -\frac{(100R^2C^2/V_c^2)D^2}{D^2/\omega_n^2 + 2\zeta D/\omega_n + 1} \tag{10.49}$$

$$\frac{e_{bp}}{e_i}(D) = -\frac{(20\zeta RC/V_c)D}{D^2/\omega_n^2 + 2\zeta D/\omega_n + 1} \tag{10.50}$$

These three outputs can be utilized individually or summed in various ways to obtain additional filtering effects. For example, if the low-pass and high-pass filters are summed (with appropriate coefficients), we can get a notch filter effect. The use of the multipliers (not necessary in an "ordinary" state-variable filter) allows adjustment of ω_n by simply applying the appropriate voltage at V_c. If digital control were desired, multiplying digital/analog converters could replace the analog multipliers.

Perhaps the most stringent requirements for sharp cutoff in low-pass filters are found in the antialiasing filters used in digital data systems. Here sophisticated active filters can provide 115 dB/octave rolloff, and pairs of filters [needed for two-channel (system) analyzers] are closely matched in phase ($\pm 1°$) and amplitude ratio (± 0.1 dB). Several types of filters useful for antialiasing and other purposes have been developed to meet different requirements and are in common use.[28] When a

[28]"Active Filter Products Design and Selection Guide"; "Series 900 Tuneable Lowpass Filter Instruments," Frequency Devices Inc., Haverhill, MA.

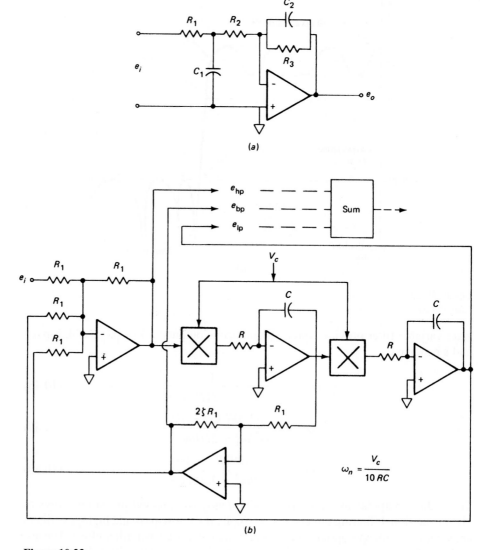

Figure 10.23
Active low-pass filter and voltage-controlled state-variable filter.

linear variation of phase shift with frequency within the passband is desired (to minimize signal-shape distortion, see Fig. 3.106), the Bessel type of filter is best; however, its sharpness of cutoff will be the worst of the types we show here. An eight-pole unit might typically[29] have a transfer function

$$\frac{e_o}{e_i}(D) = \prod_{i=1}^{4} \left(\frac{D^2}{\omega_{ni}^2} + \frac{2\zeta_i D}{\omega_{ni}} + 1 \right)^{-1} \tag{10.51}$$

[29]Ibid.

where $\omega_c \triangleq$ cutoff frequency, amplitude ratio down 3 db

$$\omega_1 = 1.778\omega_c \qquad \zeta_1 = 1.976$$
$$\omega_2 = 1.832\omega_c \qquad \zeta_2 = 1.786$$
$$\omega_3 = 1.953\omega_c \qquad \zeta_3 = 1.297$$
$$\omega_4 = 2.189\omega_c \qquad \zeta_4 = 0.816$$

This filter would attenuate 48 dB/octave beyond ω_c, and its phase shift would be linear (accurate to four digits) to slightly beyond ω_c, that is, $\phi = -3.179\omega/\omega_c$ rad for $\omega \leq \omega_c$. When the sharpness of cutoff near ω_c is more important than the linear phase, the Butterworth design[30] may be preferable. Butterworth filters have no peak in their amplitude ratio but do have overshoot for step inputs. MATLAB/SIMULINK and other math software provide convenient modules for many common types of filters. For example, the MATLAB statement: `[b,a]=butter(8,100*2*pi,'s')` will return the proper polynomial numerator (b) and denominator (a) for an eighth-order low-pass filter with 100-Hz cutoff (-3 db) frequency. If you want to use this filter in a SIMULINK block diagram, provide a transfer-function module and enter the numerator as b and denominator as a. If you want to see the numerical values for the natural frequency and damping ratio of all 4 complex poles, just enter `damp(a)`.

Even though the attenuation at very high frequencies is the same 48 dB/octave as for the Bessel, near ω_c the rolloff is much better (see Fig. 10.22c). Also below ω_c the amplitude ratio is flatter, but the phase is quite nonlinear (ϕ is $-152°$ at $\omega_c/2$ and $-360°$ at ω_c).

As antialiasing filters ahead of A/D converters in digital systems, Cauer elliptic filters are quite popular. The transfer function for a seventh-order unit might typically be

$$\frac{e_o}{e_i}(D) = \frac{\prod_{j=1}^{3} (D^2/\omega_{mj}^2 + 1)}{(D/\omega_{n1} + 1)\prod_{i=2}^{4} (D^2/\omega_{ni}^2 + 2\zeta_i D/\omega_{ni} + 1)} \qquad \textbf{(10.52)}$$

where

$$\omega_{m1} = 1.695\omega_c \qquad \omega_{n1} = 0.428\omega_c \qquad \zeta_2 = 0.562$$
$$\omega_{m2} = 2.040\omega_c \qquad \omega_{n2} = 0.634\omega_c \qquad \zeta_3 = 0.230$$
$$\omega_{m3} = 3.508\omega_c \qquad \omega_{n3} = 0.901\omega_c \qquad \zeta_4 = 0.0625$$
$$\omega_{n4} = 1.037\omega_c$$

The amplitude-ratio curve drops off very sharply to low values near ω_c but does not go smoothly to zero at high frequency; rather it shows some "bumps" and an *attenuation floor.* Once the resolution (number of bits) of the A/D converter and the highest signal frequency f to be accurately measured have been chosen, one can estimate the filter characteristics needed. One approach is to set the attenuation floor at a level equal to one-half the size of the LSB of the A/D converter, since signals this small are just beyond the resolution of the system. For example, a 12-bit

---

[30]Ibid.

converter has a resolution of 1/4096, making the LSB about -72 dB, so we would design the attenuation floor to be -78 dB. The frequency f_a at which the attenuation floor is first reached is then treated as the highest frequency present in the input to the A/D converter. So, by using the Shannon sampling theorem, our sample rate would be chosen as twice this frequency. Cauer filters have some ripple (typically ± 0.6 percent) in the passband-amplitude ratio (see Fig. 10.22e); the frequency at the last ripple before starting the sharp rolloff is called f_r. The phase is quite linear with frequency to about $f/f_c = 0.5$ but gets rather nonlinear beyond (ϕ is $-141°$ at $f_c/2$ and $-371°$ at f_c). For the above typical data, a 12-bit digital system to handle 100.-Hz analog data would use a rate of about 330 samples per second. If filters with less abrupt cutoff were used instead, much faster sampling rates would be needed to preserve the specifications listed above, requiring faster sampling hardware and more digital memory. Further discussion of these questions can be found in the literature.[31] When waveform distortion (due to nonlinear phase, amplitude ripple, and/or amplitude attenuation) is accepted as a price for sharp cutoff, it may be possible to recover accuracy in the *final* data by including digital correction calculations for these aberrations in the system software. This requires that the distortions due to the filter not be excessive and be known and repeatable.

High-Pass Filters

Figure 10.24 shows the simplest passive high-pass filters, which all have the transfer function

$$\frac{e_o}{e_i}(D) = \frac{x_o}{x_i}(D) = \frac{x_o/K_{px}}{p_i}(D) = \frac{\tau D}{\tau D + 1}$$

Again, the attenuation is quite gradual, and more complex passive or active configurations are needed to obtain a more sharply defined cutoff.

Bandpass Filters

By cascading a low-pass and a high-pass filter, we can obtain the bandpass characteristic (Fig. 10.25). To sharpen the rejection on either side of the passband, we can simply use the sharper low- and high-pass sections mentioned above or an active bandpass filter, such as that of Fig. 10.23b. Very sharply tuned bandpass filters (used in some signal and system analyzers) are available by heterodyning or Fourier filtering.[32]

Band-Rejection Filters

A common application of a band-rejection filter is found in the input circuits of self-balancing potentiometer and *XY* recorders. These instruments are subject to

---

[31]S. Smith and S. McGinn, "The Design of High-Performance Data Acquisition Systems," *Sound & Vib.*, pp. 18–24, Nov. 1987; *Appl. Notes* 503 and 504, Neff Inst. Corp., Monrovia, CA.

[32]E. O. Doebelin, "System Modeling and Response," pp. 232–243, Wiley, New York, 1980.

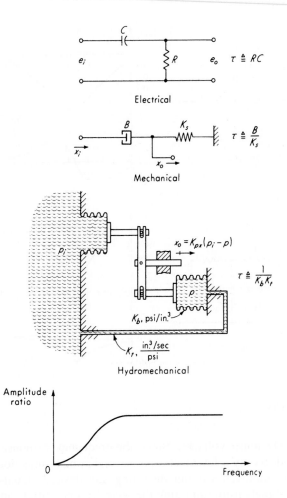

Figure 10.24
High-pass filters.

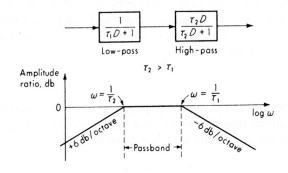

Figure 10.25
Bandpass filter.

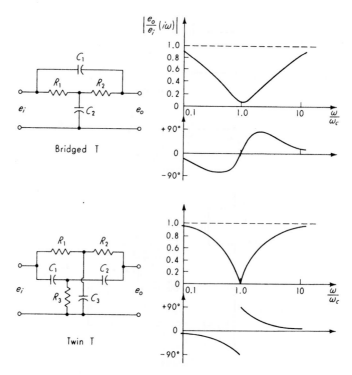

Figure 10.26
Band-rejection filters.

interfering 60 Hz noise voltages. Since the frequency response of the overall recorder is good to only a few cycles per second, a band-rejection filter tuned to 60 Hz may be employed without distorting any desired signals. Such a filter prevents noise signals from saturating the recorder's amplifiers and distorting the proper amplification of the desired signals.

Passive networks commonly utilized for rejection of a band of frequencies include the bridged-T and the twin-T networks (Fig. 10.26). While the bridged-T does not completely reject any frequency, the twin-T can be so designed. Equations and charts for designing these filters are available.[33] One approach to active band-rejection filtering is shown in Fig. 10.23b.

Digital Filters

As part of the general trend to replace analog electronics by digital wherever this is feasible and economic, digital filtering[34] is now quite common. Digital filtering can essentially duplicate any of the classical filter functions of Fig. 10.20 and can

[33]J. E. Gibson and F. B. Tuteur, "Control System Components," p. 43, McGraw-Hill, New York, 1958.

[34]R. E. Bogner and A. G. Constantinides (eds.), "Introduction to Digital Filtering," Wiley, New York, 1975.

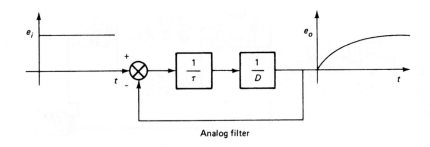

Analog filter

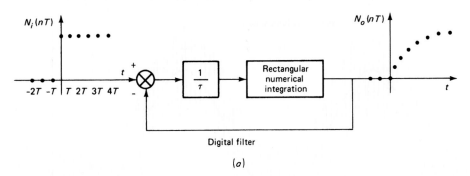

Digital filter

(a)

Figure 10.27
Digital low-pass filter.

produce certain useful effects not possible in the analog domain. It also has the usual digital benefits of accuracy, stability, and adjustability by software (rather than hardware) changes. Digital filtering is an algorithm by which a sampled signal (or sequence of numbers), acting as an input, is transformed to a second sequence of numbers called the output. The algorithm may correspond to low-pass, high-pass, or other forms of filtering action. The output sequence can be employed in further digital processing or can be converted from digital to analog, producing a filtered version of the original analog signal. Since entire books are devoted to the theory of digital signal processing/filtering, here we give only a brief discussion.

A digital version of the simplest low-pass filter [Eq. (10.47)] can be found by converting the analog system's differential equation to a difference equation by using, say, rectangular numerical integration, as in Fig. 10.27:

$$N_o(nT) = N_o[(n-1)T] + \frac{T}{\tau}\{N_i[(n-1)T] - N_o[(n-1)T]\}$$

Here the sampling interval of the digital system is T seconds, and $N_o(nT)$ and $N_i(nT)$ are number sequences for the input and output, evaluated only at discrete-time points such as T, $2T$, $3T$, etc. Readers unfamiliar with difference equations may wish to do a sample calculation with the above equation (use $T = 1$ s, $\tau = 5$ s) to verify the step-response behavior of Fig. 10.27a. For digital filters the sampling frequency $1/T$ must be 2.5 or more times the highest frequency in the analog data to prevent aliasing problems.

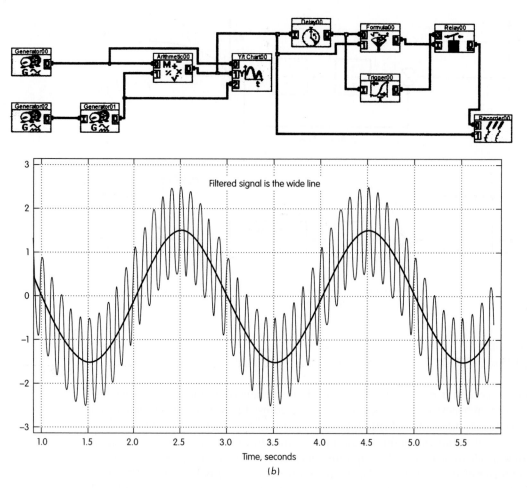

Figure 10.27
(Concluded)

Most data-acquisition software, such as DASYLAB discussed in detail in Chap. 13, provides ready-made digital filters that implement most of the classical filtering functions useful in data processing, relieving the user of any need for digital-filter design expertise. Also, specialized filtering functions *not* available as ready-made icons may sometimes be implemented using combinations of available operations, as shown in the following DASYLAB example. A simple *notch filter*[35] has been used to remove a spurious signal caused by vibration of the supporting boom of a sonic anemometer used for wind-speed studies. To remove the spurious

---

[35]*WPL Application Note No.* 6, Applied Technologies, Inc., Longmont, CO, 303-684-8722 (www.apptech.com, 1990).

signal at frequency f_{noise}, we simply compute the *average* of sample data points separated in time by one-half the period $1/(2 f_{noise})$ of the offending signal. This averaging procedure cancels the noise signal since a delay of a half period gives points that are 180° out of phase. In our example, the sampling rate is 1000 Hz, the desired signal is at 0.5 Hz, and the noise is at 10 Hz, so the required delay is 0.05 s. DASYLAB has a module that delays signals by a specific number of blocks, so if we set our block size at 50, the delay will be 0.05 s. The delayed and undelayed signals are sent to a FORMULA module set up to add the two signals and divide by two. Starting from time zero, the averaged signal would not be correct until 0.05 s has passed, so we use TRIGGER and RELAY modules to exclude it until the delayed signals start to appear. (This last feature is not really necessary in a practical system since the first 50 ("wrong") points are easily ignored.) The DASYLAB diagram and graphs of the noisy and filtered signals are shown in Fig. 10.27*b*. If the noise frequency deviates from the assumed value, the filtering will of course not be as perfect as shown but can still be effective.

Other special-purpose digital filters include the *median filter*[36] and the *zero-phase*[37] filter. The median filter can operate with real-time data and provides a noise-smoothing effect *without* "rounding the corners" of step-like signal components. The zero-phase filter can be used only for post-processing but has the unusual feature of providing the desired amplitude ratio with *no phase shift or time delay whatsoever.*

A Hydraulic Bandpass Filter for an Oceanographic Transducer

While filtering with electrical networks is most common, in some instances other physical forms offer advantages. This section illustrates this idea with an example of a filter that has been successfully constructed and used.[38]

The Scripps Institution of Oceanography at La Jolla, California, employs pressure transducers in its studies of ocean-wave phenomena. A particular study required measurements of waves whose frequencies are lower than those of ordinary gravity waves (which one observes visually) and higher than those due to tides. These waves of intermediate frequency are of rather low amplitude relative to those due to tides and to gravity and thus are difficult to measure with a pressure pickup which treats all frequencies about equally. The bandpass filter and pressure pickup of Fig. 10.28 solves this problem since it is "tuned" to the frequency range of interest, which is about 0.001 Hz. Such low frequencies are very difficult to handle with electrical circuits, but the hydraulic filter shown gives very good results with quite simple and reliable components.

---

[36]MATLAB medfilt1; J. E. Maisel, "Median Filter Combines Both Lowpass and Highpass Characteristics," *Personal Engineering & Instrumentation News,* Feb. 1993, pp. 61–65.

[37]MATLAB filtfilt.

[38]F. E. Snodgrass, "Shore-Based Recorder of Low-Frequency Ocean Waves," *Trans. Am. Geophys. Union,* p. 109, Feb. 1958.

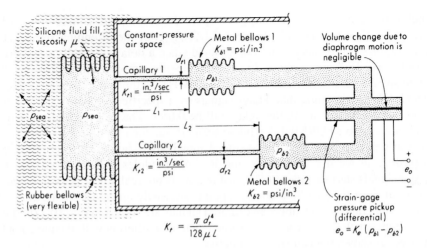

Figure 10.28
Hydraulic bandpass filter.

In use, the pressure transducer is located underwater, often buried in a foot of sand for temperature insulation, with a "snorkel" tube extending up through the sand to sense water pressure. This pressure is directly related to the height of the waves passing overhead; thus a record of pressure-transducer output voltage is a record of wave activity. Analysis of the system gives

$$\frac{e_o}{p_{\text{sea}}}(D) = \frac{K_e(\tau_2 - \tau_1)\,D}{(\tau_1 D + 1)(\tau_2 D + 1)} \tag{10.53}$$

where $K_e \triangleq$ sensitivity of differential pressure pickup [mV/(lb/in^2)] and

$$\tau_1 \triangleq \frac{1}{K_{t1} K_{b1}} \qquad \text{s} \tag{10.54}$$

$$\tau_2 \triangleq \frac{1}{K_{t2} K_{b2}} \qquad \text{s} \tag{10.55}$$

We can easily show that the frequency ω_p of peak response is given by

$$\omega_p = \sqrt{\frac{1}{\tau_1 \tau_2}} \qquad \text{rad/s} \tag{10.56}$$

The amplitude ratio M_p at this frequency is

$$M_p = \frac{K_e(\tau_2 - \tau_1)}{\tau_2 + \tau_1} \tag{10.57}$$

and the phase angle at ω_p is zero.

Mechanical Filters for Accelerometers

While electrical filters at the *output* of piezoelectric accelerometers are in common use, mechanical filters at the *input* can protect the accelerometer from shock,

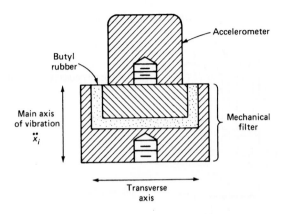

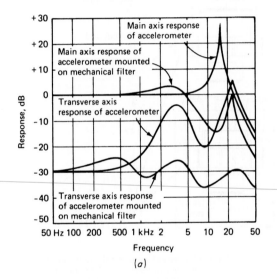

Figure 10.29

Mechanical filter for accelerometer.

prevent electrical saturation resulting from large, high-frequency acceleration when only low-frequency motions are of interest, and provide electrical isolation to break ground loops and thus reduce electrical noise pickup. Figure 10.29[39] shows the construction (utilizing butyl rubber as both the spring element and electrical insulation) and response characteristics of such filters. Other manufacturers offer mechanically filtered piezoelectric and piezoresistive accelerometers.[40] The piezoresistive unit actually uses an existing accelerometer design with an external shock isolator

[39]Model UA 0559, B&K Instruments, Marlboro, MA.

[40]Endevco Models 7255A and 7270AM6; A. S. Chu, "Built-In Mechanical Filter in a Shock Accelerometer," Endevco Corp., San Juan Capistrano, CA, 800-982-6732 (www.endevco.com).

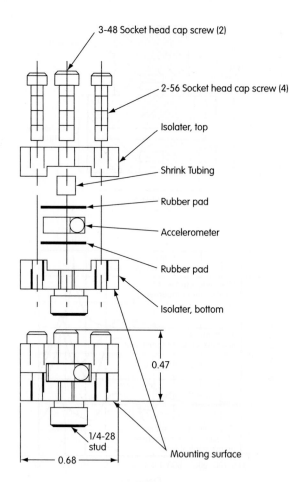

Exploded view of improved SNL mechanical
isolater with acceptable performance to
60,000 g for DC-10kHz.

(*b*)

Figure 10.29
(Concluded)

consisting of two thin sheets of special rubber sandwiched between the outside housing of the accelerometer and two rigid plates; the whole assembly is screwed together as shown in Fig. 10.29*b*.[41]

---

[41]V. I. Bateman, F. A. Brown, and M. A. Nusser, "High Shock, High Frequency Characteristics of a Mechanical Isolator for a Piezoresistive Accelerometer, the ENDEVCO 7270AM6," *Report SAND2000-1528,* July 2000, Sandia National Laboratories, Albuquerque, NM.

Filtering by Statistical Averaging

All the filters mentioned above are of the frequency-selective type and, of course, require that the desired and spurious signals occupy different portions of the frequency spectrum. When signal and noise contain the same frequencies, such filters are useless. A basically different scheme may be employed usefully under such circumstances if the following is true:

1. The noise is random.
2. The desired signal can be caused to repeat itself.

If these two conditions are fulfilled, it should be clear that if one adds up the ordinates of several samples of the total signal at like values of abscissa (time), the desired signal will reinforce itself while the random noise will gradually cancel itself. This will occur even if the frequency content of signal and noise occur in the same part of the frequency spectrum. It can be shown that the signal-to-noise ratio improves in proportion to the square root of the number of samples utilized. Thus theoretically the noise can be eliminated to any desired degree by adding a sufficiently large number of signals. In practice, various factors prevent realization of theoretically optimum performance. *Lock-in amplifiers* use a similar principle to extract minute signals buried in random noise and surrounded by large-amplitude, discrete interfering frequencies.[42]

10.4 INTEGRATION AND DIFFERENTIATION

Often in measurement systems it is necessary to obtain integrals and/or derivatives of signals with respect to time. Depending on the physical nature of the signal, various devices may be most appropriate. Generally accurate differentiation is harder to accomplish than integration, since differentiation tends to accentuate noise (which is usually high frequency) whereas integration tends to smooth noise. Thus second and higher integrals may be found easily while derivatives present real difficulties.

Integration

Mechanical (ball-and-disk) and electromechanical (integrating motor and velocity servo) devices for integration are used only in specialized applications and are described in the reference.[43] Algorithms for digital (numerical) integration are covered in numerical analysis texts, but we rarely need to consult these since most data-acquisition software (such as the DASYLAB used in this text) provide ready-made modules. When, say, a second or third integral of a signal is needed, we merely cascade two or three of the basic integrator modules. *Electronic-analog integration* is still a useful operation in many measurement systems and uses the classical op-amp integrator. This device was the heart of the, now obsolete, analog

---

[42]*Tech. Notes,* IAN 23 to 49, Ithaco Co. Ithaca, NY.

[43]E. O. Doebelin, "Measurement Systems," 4th ed., McGraw-Hill, New York, 1990, pp. 790–792.

computer, which has been largely replaced by digital-simulation software such as SIMULINK used in this text.

Figure 10.30 shows the op-amp type of integrator. By use of high-quality chopper-stabilized op-amps, an integrator of quite low drift can be constructed in this way. Accuracies of the order of 0.1 percent for short-term operation and 1 percent over 14 h are typical of high-quality electronic integrators of this type. If higher integrals are desired, such units may be cascaded; however, drift becomes more troublesome. (See Prob. 10.30 for possible solutions to drift problems.) In addition to providing a closer approximation to true integration than the passive networks discussed in the following paragraph, the presence of the amplifier (with its own power supply) means that power can be supplied to the device following the integrator without taking any significant power from the device supplying the integrator. That is, op-amp circuits generally can have a high input impedance and low output impedance.

All the low pass filters of Fig. 10.21 may be utilized as *approximate integrators* for input signals within a restricted frequency range. This can be shown as follows:

$$\frac{e_o}{e_i}(i\omega) = \frac{1}{i\omega\tau + 1} \tag{10.58}$$

Now if $\omega\tau \gg 1$,

$$\frac{e_o}{e_i}(i\omega) \approx \frac{1}{i\omega\tau} \tag{10.59}$$

and thus

$$\frac{e_o}{e_i}(D) \approx \frac{1}{\tau D} \tag{10.60}$$

$$e_o \approx \frac{1}{\tau}\int e_i\,dt \tag{10.61}$$

Thus, if the frequency spectrum of the input signal is such that $\omega\tau \gg 1$ for all significant frequencies, a good approximation to the desired integrating action is obtained. For a given τ, the approximation improves as ω increases. It appears as if any ω can be accommodated by choosing τ sufficiently large. However, large τ

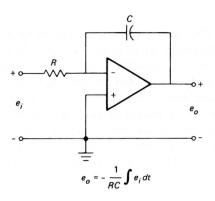

Figure 10.30
Electronic integrator.

decreases the magnitude of the output; thus this can be carried only as far as the noise level of the system permits.

A form of "digital integration" that does not involve numerical algorithms is available from sensor signals that are in the form of a pulse rate proportional to the measured quantity. Flow meters of the turbine and vortex-shedding type are good examples of this since we often want to integrate flow rate to get total flow over a timed interval. (If some sensor does not *inherently* produce a pulse rate, we can convert a "time-varying DC" signal to a pulse rate using a voltage-to-frequency converter.) Once we have a pulse-rate signal, the integral is obtained in an accurate and drift-free form by simply applying the signal to an electronic counter set to *accumulate* the total number of pulses occurring over the desired time interval.

Differentiation

Because differentiation is basically problematic due to its noise accentuation, we should always consider alternatives. For example, if you have a displacement signal and want also a velocity signal, consider velocity *transducers* such as the moving-coil pickup or the rate gyro. These solutions should be evaluated against the differentiating circuits discussed below.

All the high-pass filters of Fig. 10.24 may be employed as *approximate differentiators* for input signals within a restricted frequency range, as shown by the following analysis:

$$\frac{e_o}{e_i}(i\omega) = \frac{i\omega\tau}{i\omega\tau + 1} \tag{10.62}$$

Now if $\omega\tau \ll 1$,

$$\frac{e_o}{e_i}(i\omega) \approx i\omega\tau \tag{10.63}$$

$$\frac{e_o}{e_i}(D) \approx \tau D \tag{10.64}$$

$$e_o \approx \tau \frac{de_i}{dt} \tag{10.65}$$

We note here that for a given τ, the approximation improves for lower values of ω. Again τ may be reduced to extend accurate differentiation to higher frequencies. However, small τ reduces sensitivity; thus noise level is limiting, just as in the approximate integrators.

Use of op-amps results in both approximate and "exact" differentiators of improved performance relative to the passive high-pass filters discussed above. Figure 10.31 shows some of these circuits. In Fig. 10.31*a,* analysis of this "exact" differentiator gives

$$\frac{e_o}{e_i}(D) = -RCD \tag{10.66}$$

This circuit is rarely useful because the ever-present noise (generally of high frequency relative to the desired signal) will completely swamp the desired signal

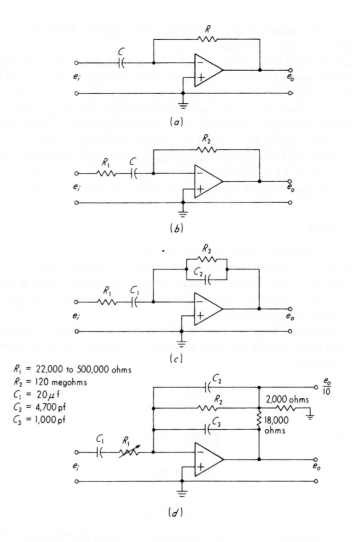

Figure 10.31
Electronic differentiators.

at the output. All exact differentiators must suffer from this problem. It can be alleviated only by shifting to approximate differentiators that include low-pass filters to take out the effects of high-frequency noise. Figure 10.31*b* shows a common scheme which, when analyzed, gives

$$\frac{e_o}{e_i}(D) = -\frac{R_2\,CD}{R_1\,CD + 1} \tag{10.67}$$

This gives an accurate derivative for frequencies such that $R_1 C \omega \ll 1$ and amplifies high-frequency noise only by an amount R_2/R_1. To attenuate noise, we must use a second-order type of low-pass filter, such as given by the circuit of Fig. 10.31c. Analysis yields

$$\frac{e_o}{e_i}(D) = -\frac{R_2 C_1 D}{(R_2 C_2 D + 1)(R_1 C_1 D + 1)} \tag{10.68}$$

Figure 10.31d shows an actual circuit[44] designed for measuring the rate of charging or discharging of batteries and using a solid-state operational amplifier. Analysis gives

$$\frac{e_o}{e_i}(D) = -\frac{10 R_2 C_1 D}{(R_1 C_1 D + 1)[R_2(10 C_3 + C_2)D + 1} \tag{10.69}$$

The output is read on a meter which may be connected to e_o or $e_o/10$, depending on the size of the output. For the numerical values given and $R_1 = 22,000$,

$$\frac{e_o}{e_i}(D) = -\frac{24,000D}{(0.44D + 1)(1.764D + 1)} \tag{10.70}$$

We note that for $De_i = 10$ mV/min the output is 4 V. If, say, $\frac{1}{2}$ mV of 60-Hz noise is present at the input, then the output noise is only 41 mV, which is about 1 percent of the desired output.

As a simple design exercise, let us now develop an analog differentiator for a motion-measurement application. Suppose we have a damped pendulum whose angular oscillation is measured with a conductive-plastic potentiometer. (I have such an apparatus that I use for demonstrations in one of my instrumentation classes, so realistic data is available. The pot used there is a 1000-Ω, 1.25-W, single-turn unit excited with 25 V.) The pendulum is moved about one radian away from its rest position and released; the resulting oscillation is at about 1 Hz and lasts for about 3 s. We want to also get a velocity signal but a tachometer generator is much too noisy for this application. The angular-displacement voltage displayed on a thermal-writing recorder appears to the eye to be perfectly smooth and noise free. We realize of course that conductive-plastic pots *do* have some surface roughness that may show up when we differentiate, but the visually smooth displacement signal encourages us to try to differentiate electrically.

We have shown differentiating circuits that have various degrees of accuracy and complexity, but it is usually wise to try the simpler ones first and go to the more complex ones only if necessary. We therefore start with the simplest passive circuit, the high-pass filter of Fig. 10.24. Equation (10.63) shows that the circuit is accurate for "sufficiently low" frequencies, so our first task is to estimate the *highest* frequency that our circuit might see. In our application, this is relatively easy to do since the displacement signal is a damped sine wave of known frequency (about 1 Hz). We could compute the frequency spectrum of the actual transient wave form,

---

[44]"The Lightning Empiricist," Philbrick Researches Inc., Boston, MA, Oct. 1963.

but this is probably overkill when the signal is so simple. We will just take, conservatively, the highest frequency to be about 2 Hz. Next, the requirement that $\omega\tau \ll 1.0$ must be made *"specific,"* that is, *how much* less than 1.0 do we want it to be? Let us take $\omega\tau = 0.01$, a quite stringent accuracy requirement. With our largest frequency as being 2.0 Hz, the required time constant is 0.000796 s, which is also the *sensitivity* of our differentiator [volts/(volt/sec)]. Note the *trade-off*; if we opt for more accuracy ($\omega_{max}\tau < 0.01$), then we must also accept less sensitivity.

There is an infinite number of combinations of R and C that will give our desired τ value, but we can use loading considerations to narrow the choice since the differentiator circuit will be a load on our 1000-Ω pot. We saw in the text section on potentiometers that the load resistance should be at least 10 times the pot resistance to ensure nonlinearity less than 1.5 percent. Let us conservatively take $R = 30,000\ \Omega$, which makes $C = 0.0265\ \mu F$. To be practical, we should recall that resistors and capacitors are *not* marketed in an infinite number of values. Rather, there are *standard* values and *accuracy* ranges (± 10 percent, ± 1 percent, etc.). *Adjustable R's and C's* are of course available, but they are larger, more expensive, and less reliable than fixed components, so we use them only when necessary. We thus choose the standard R and C values closest to our needs, 30 kΩ and 0.02 μF, respectively, giving $\tau = 0.0006$ rather than 0.000796. This smaller τ provides less sensitivity but greater accuracy ($\omega_{max}\tau = 0.00754$). There is probably no need to pay for the high accuracy components that are available; the 10-percent grade is really adequate since we can *measure* the actual R and C values once they are purchased. It seems as if we are "giving up" the τ value that we "really want," but the accuracy and sensitivity that we have chosen are really somewhat arbitrary, so we should not think of them as unchangeable. Finally, *all* measurement systems should be statically and dynamically *calibrated* before being used for critical measurements, so if our design calculations are a little rough, we will get accurate values later at calibration time.

The system as designed above was operated and gave a very clean displacement signal. However, the velocity signal, while usable, had noticeable random-appearing noise whose frequency content was visually estimated as much higher than the desired 1-Hz signal, and so low-pass filtering was considered. To design this filter in the most "scientific" way, we could do a spectrum analysis on the noise to document its frequency content, but again a simpler approach may be adequate since we *will* test the overall system before accepting the new design. We want the filter to attenuate the noise but not significantly degrade the quality of the differentiation, so its cutoff frequency should be a little above the highest frequency (2 Hz) that is to be accurately differentiated. We choose for the simplest low-pass filter (Fig. 10.21) a trial value of break-point frequency of 25 rad/s (4 Hz), which requires a time constant of 0.04 s. Again, loading considerations are used to pick a specific R value, but now we need to also consider loading with respect to the *recorder* input resistance of 1 MΩ. Our usual 10:1 rule for choosing impedance values in a chain of cascaded subsystems must now be relaxed since the recorder input resistance cannot be chosen. If we "split the difference," we can get about a 6:1 ratio for the final two interfaces by making the filter $R = 180$ kΩ, making $C = 0.22$ μF. This filtered differentiator was run and gave good results.

The above design was achieved rather quickly by assuming *no* loading between the various stages, which allowed each section to be designed as an isolated device. Applying the 10:1 impedance-ratio rule, the errors in this approximation should not be excessive. Fortunately, we can *check* our approximation by *analyzing* the complete system *without* assuming no loading. That is, analysis of any system of *known* components is always much easier than system design to meet specifications, particularly if there are many components. This approach is widely applied in measurement-system design, not just in this example. Figure 10.32*a* shows the overall system; the differentiator/filter/recorder combination shown in Fig. 10.32*b* is easily analyzed as a *complete system* and the results compared with our approximate ("no-loading") predictions:

$$\frac{e_o}{De_i}(D) = \frac{R_1 C_1}{(R_1 C_1 D + 1)}\frac{1}{(R_2 C_2 D + 1)} \qquad \text{assuming no loading} \qquad \textbf{(10.71)}$$

$$\frac{e_o}{De_i}(D) = \frac{\dfrac{R_1 R_3}{R_1 + R_2 + R_3}\cdot C_1}{R_1 R_2 C_1 C_2 \cdot \dfrac{R_3}{R_1 + R_2 + R_3}D^2 + \dfrac{(R_1 + R_2)\cdot R_3 C_2 + (R_2 + R_3)\cdot R_1 C_1}{R_1 + R_2 + R_3}D + 1} \qquad \begin{array}{l}\text{exact analysis}\\[6pt]\textbf{(10.72)}\end{array}$$

We could have derived Eq. (10.72) *before* any attempts to choose numerical values, but its complexity shows how difficult those choices would be. Having, however,

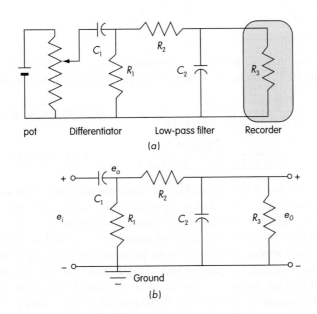

pot Differentiator Low-pass filter Recorder

(a)

(b)

Figure 10.32

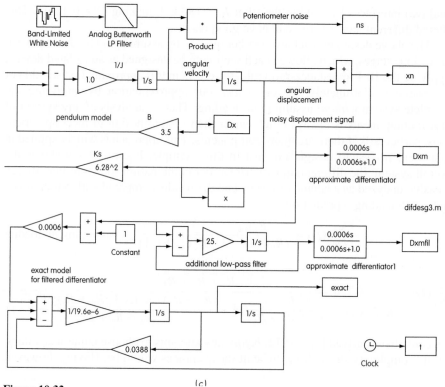

Figure 10.32
(Continued)

easily gotten reasonable estimates of numerical values, we can also easily substitute them into Eq. (10.72) to *check* whether our approximate design will work.

$$\frac{e_o}{De_i}(D) = \frac{0.0006}{24.0 \times 10^{-6}D^2 + 0.0406D + 1} \qquad \text{assuming no loading} \qquad \textbf{(10.73)}$$

$$\frac{e_o}{De_i}(D) = \frac{0.000495}{19.6 \times 10^{-6}D^2 + 0.0388D + 1} \qquad \text{exact analysis} \qquad \textbf{(10.74)}$$

Since readers do not have available the actual physical setup that I used, a simulation is now presented that is a close approximation to my apparatus. The pendulum is easily modeled as a rotational second-order system with inertia, damping, and a spring representing the gravitational torque on the pendulum (linear behavior is assumed; small angles). Since we want to access the velocity, I used the "expanded" model rather than a single-block transfer-function icon for the pendulum. Exact modeling of the pot *noise* is not possible without the laborious generation of a SIMULINK lookup table that uses thousands of points taken directly from a recording of the actual pot noise. An acceptable alternative is to use the band-limited random-signal source and a low-pass filter to generate a noise signal that

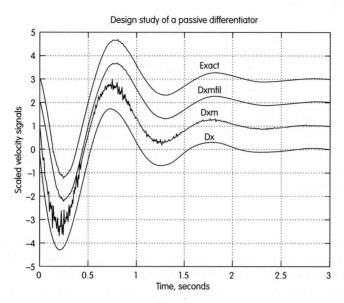

Figure 10.32
(Concluded)

"looks like" the actual noise. Pot noise is generated mainly by the variation of the wiper contact resistance as the wiper slides over the roughness of the conductive plastic resistance element. Observation of the actual noise signal shows that the noise is roughly proportional to slider velocity, so I incorporated this feature into my simulation. This proportionality is not unreasonable since a higher velocity gives a more rapid change of both contact resistance and noise voltage. The pot roughness is not really a random phenomenon although it appears so to casual visual inspection. Careful inspection of actual recorder graphs shows that the noise voltage repeats quite closely when we move over the same region of the resistance element. As a result of gradual wear, there would be some change after perhaps thousands of cycles of pot motion. Using these concepts, I adjusted the noise source and the filter until the simulated noise appeared to reasonably match the real signal.

Figure 10.32*c* shows the SIMULINK diagram for this study, which includes the approximate differentiator, the filtered version of this differentiator using the "no-loading" model, and finally the exact model of the filtered differentiator. Figure 10.32*d* shows the exact velocity signal Dx, the noise in the unfiltered unit's signal Dxm, the improvement realized by filtering Dxmfil, and the close agreement of the no-loading and exact circuit models for the filtered units. The filtered differentiator shows only a little noise (near the velocity peaks) and is a close replica of the true velocity. We used the design values for the *R*'s and *C*'s; in practice, we would have used the actual (measured) values. Finally, the transfer function of the actual circuit (Fig. 10.32*b*) could be measured in various ways to get the true dynamic behavior without the errors resulting from inexact assumptions used in the theoretical analysis. One efficient method would excite the input with a random signal and use a

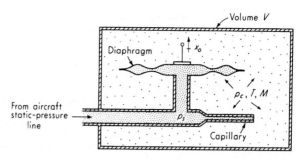

Figure 10.33
Rate-of-climb sensor.

two-channel spectrum analyzer to measure the amplitude-ratio and phase-shift curves. Some analyzers also provide *curve-fitting* software, which would give us the numerical values of coefficients in the system transfer function.

A final example of a differentiator using nonelectrical methods is the aircraft rate-of-climb indicator shown in Fig. 10.33. Since atmospheric pressure varies with altitude, a device that measures rate of change of atmospheric pressure can indicate rate of climbing or diving. While actual design requires a more critical study,[45] here we consider a simplified linear analysis to show the main features. Static pressure p_s corresponding to aircraft altitude is fed from the vehicle's static pressure probe to the input tube of the rate-of-climb indicator. Leakage through a capillary tube into the chamber of volume V occurs at a mass flow rate assumed to be $K_c(p_s - p_c)$ lbm/s. Air in the chamber follows the perfect-gas law $p_c V = MRT$. Motion of the output diaphragm is according to $x_o = K_d(p_s - p_c)$. By assuming K_c and T to be constant, analysis gives

$$\frac{x_o}{Dp_s}(D) = \frac{K}{\tau D + 1} \tag{10.75}$$

where

$$K \triangleq \frac{K_d V}{RTK_c} \qquad \text{in/[(lb/in}^2)/\text{s]} \tag{10.76}$$

$$\tau \triangleq \frac{V}{RTK_c} \qquad \text{s} \tag{10.77}$$

Thus x_o, which may be measured with any displacement transducer, is an indication of rate of change in p_s, and thereby a measure of rate of climb, if pressure is assumed to be a linear function of altitude. Since this is not exactly true, various compensating devices are needed in a practical instrument for this and other spurious effects.

[45]D. P. Johnson, "Aircraft Rate-of-Climb Indicators," *NACA, Rept.* 666, 1939.

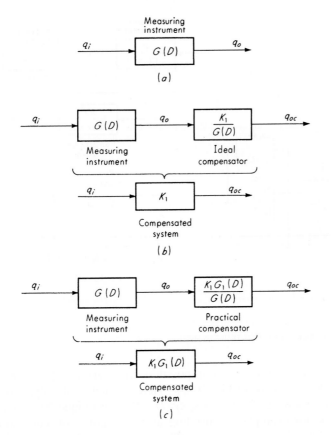

Figure 10.34
Generalized dynamic compensation.

10.5 DYNAMIC COMPENSATION

Sometimes it is not possible to obtain the desired behavior from a measuring device solely by adjusting its own parameters. To get fast response from a thermocouple, for example, very fine wire must be used. Perhaps the vibration and temperature environment might be so severe that such a fine-wire thermocouple would be destroyed before any readings could be obtained. For this and similar situations, dynamic compensation may provide a solution.

Figure 10.34 shows the general arrangement by which dynamic compensation may be employed. Ideally an instrument with transfer function $G(D)$ is cascaded with a compensator $K_1/G(D)$, and thus (if negligible loading is assumed) the overall system has *instantaneous response* since its transfer function is just the constant K_1. This result is, of course, too good to be true. The practical difficulty lies in the construction of the compensator $K_1/G(D)$, which generally is not realizable physically because of the need for perfect differentiating effects. While perfect compensation

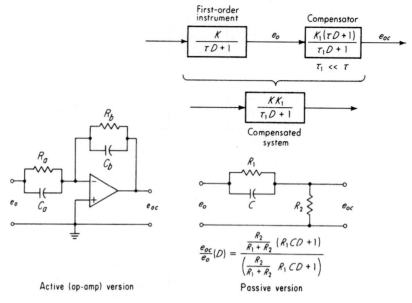

Figure 10.35
Dynamic compensation for first-order system.

for instantaneous response is *not* possible, very great improvements *may* be achieved with the scheme of Fig. 10.34c. Here the undesirable dynamics $G(D)$ are *replaced* with more desirable ones, $G_1(D)$. This technique has been used with good success, for example, in speeding up the response of temperature-sensing elements and hot-wire anemometers. These are basically first-order instruments; thus $G(D) = K/(\tau D + 1)$. While the compensator, in general, can take any suitable physical form, because most sensors produce an electrical output, most compensators in use are electric circuits. The compensator generally utilized for first-order systems takes the form shown in Fig. 10.35. Note that for the passive version any increase in speed of response ($\tau_1 \ll \tau$) is paid for by a loss of sensitivity in direct proportion, since $K_1 = R_2/(R_1 + R_2) = \tau_1/\tau$. If this loss of sensitivity is not tolerable, additional amplification is needed. Usually it is placed between the sensor and the compensator because then it will also serve to unload the two circuits from each other. [The active (op-amp) version can *combine* compensation and amplification in one stage.] While several stages of such compensation may be used sometimes, such staging cannot be carried beyond a certain point because of additional noise introduced by the amplifiers and accentuated by the compensators. However, a response speedup of the order of 100:1 is feasible and has been achieved for thermocouples and hot-wire anemometers.

Theoretically the concept of dynamic compensation is applicable to any-order system. A good example of a more complex application is found in the "equalization" of vibration shaker systems. Figure 10.36a shows the basic arrangement of a shaker system for sinusoidal vibration testing. Many tests involve "sweeping" the frequency of the oscillator through a certain range while maintaining the

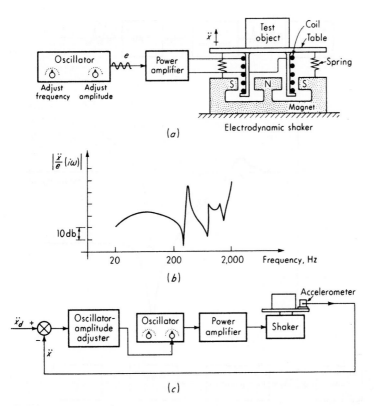

Figure 10.36
Vibration shaker systems.

acceleration amplitude constant at the test object. While it is not difficult to maintain constant the amplitude of oscillator output voltage e while sweeping through the frequency range, the acceleration $\ddot{x}$ will not be constant since the transfer function $(\ddot{x}/e)(i\omega)$ is not constant over this range. In fact, because of various resonances in the electromechanical shaker, test fixtures, and the test object itself, severely distorted frequency response is not uncommon (Fig. 10.36b). This difficulty can be overcome by use of a feedback scheme as in Fig. 10.36c. Here the actual acceleration $\ddot{x}$ is compared with the desired value $\ddot{x}_d$; if they differ, the amplitude of the oscillator is adjusted to obtain correspondence. This adjustment is performed automatically and continuously as the frequency range is swept.

When random vibration testing (see Fig. 10.37) rather than pure sinusoidal is desired, the above approach is not applicable directly since now the signal $\ddot{x}$ is random and the noise source cannot be adjusted in any simple fashion to force $\ddot{x}$ to have the desired frequency spectrum (the spectrum of e). One approach is to provide dynamic compensation such that the transfer function $(\ddot{x}/e)(i\omega)$ is flat over the desired frequency range. Then the spectrum at $\ddot{x}$ will be the same as that put in at e. The necessary dynamic compensation here is one which can put "peaks" where there are "notches" and notches where there are peaks in the curve of Fig. 10.36b.

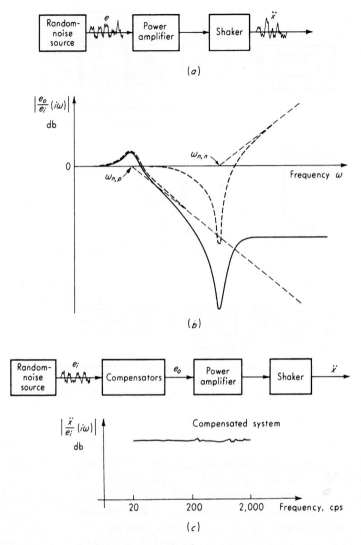

Figure 10.37
Dynamic compensation for vibration shaker.

Thus the overall curve can be made relatively flat. A compensator that will provide one peak and one notch has the form

$$\frac{e_o}{e_i}(D) = \frac{D^2/\omega_{n,n}^2 + 2\zeta_n D/\omega_{n,n} + 1}{D^2/\omega_{n,p}^2 + 2\zeta_p D/\omega_{n,p} + 1} \tag{10.78}$$

Figure 10.37*b* shows the frequency response of such a compensator in which the peak occurs at a lower frequency than the notch. The reverse is also possible, if needed. Since ζ_n and ζ_p are adjustable, the compensator can be "tailored" to cancel

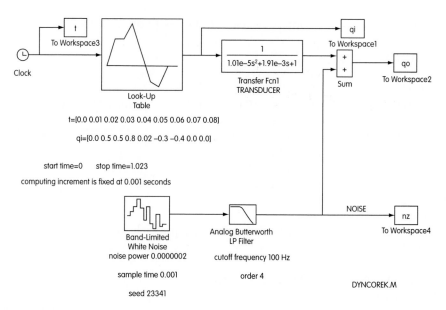

t=[0.0 0.01 0.02 0.03 0.04 0.05 0.06 0.07 0.08]

qi=[0.0 0.5 0.5 0.8 0.02 –0.3 –0.4 0.0 0.0]

start time=0   stop time=1.023

computing increment is fixed at 0.001 seconds

Figure 10.38
SIMULINK simulation to generate data for digital off-line dynamic compensation example.

exactly the undesired shaker-system dynamics. When several peaks and notches are present (as in Fig. 10.36b), several compensators are used; as many as 10 are not uncommon.

While the analog dynamic compensators described operate "online" in "real time," dynamic compensation also can be realized "offline" digitally, by using frequency-spectrum or equivalent time-domain methods.[46] We merely compute $Q_o(i\omega)$ from the measured $q_o(t)$ (using, say, fast Fourier transform methods), divide $Q_o(i\omega)$ by the (assumed known) $(Q_o/Q_i)(i\omega)$ of the instrument to get $Q_i(i\omega)$, and then inverse-transform to get $q_i(t)$. Off-line compensation of this kind is easily demonstrated with a simple MATLAB/SIMULINK example. In Fig. 10.38 we use SIMULINK to generate the response, to a specific transient, of a "too-slow" second-order transducer of known dynamics. This response (1024 time points) is then used as input to the following MATLAB m-file:

```
% program to correct an instrument for dynamic error
% you need to run the simulink (dyncorek) first
% first, let's plot the transducer input and output
```

[46]G. L. Schulz, "Method of Transfer Function Calculation and Distorted Data Correction," Air Force Weapons Laboratory, *Rept.* AFWL-TR-73-300, Kirtland AFB, NM, March 1975; M. R. Weinberger, "Flight Instrument and Telemetry Response and Its Inversion," *NASA* CR-1768, Sept. 1971; L. W. Bickle, "A Time Domain Deconvolution Technique for the Correction of Transient Measurements," *Rept.* SC-RR-71-0658, Sandia Labs, Albuquerque, NM, 1971; H. Grote, "Restoration of the Input Function in Measuring Instrumentation," U.S. Army Electronics Command, *Rept.* ECOM-2662, Ft. Monmouth, NJ, 1966; D. L. Elmore, W. W. Robinson, and W. B. Watkins, "Dynamic Gas Temperature Measurement System," *NASA* CR-168267, 1983.

```
% you can add appropriate printing statements below, if
you wish
plot(t,qi,t,qo,'--')
axis([0 .2 -.5 1.])
pause(5)
% now compute the direct Fast Fourier Transform of the
transducer output signal
qoftran=fft(qo, 1024);
qoftran=qoftran(1:512);
magqof=0.001.*abs(qoftran);
freqs=[0:1/1.023:511/1.023]';
plot(freqs,magqof)
pause(5)
% now compute the instrument's frequency response
w=2*pi*freqs;
s=w*1i;
tf=1./(1.01e-5.*s^2+1.91e-3.*s+1);
plot(w,abs(tf))
pause(5)
% now divide qo(iw) by qo/qi(iw)
Qiw=(0.001.*qoftran)./(tf);
% now compute the inverse Fourier transform of Qiw, to
recover the input signal in the time domain
qit=real(ifft(Qiw))*500*2*pi/pi;
tt=0:2*pi/(500*2*pi):2*pi/(500*2*pi/511);
plot(tt,qit);grid;xlabel('time, seconds')
axis([0 .2 -.5 1])
hold on
plot(t(1:512),qi(1:512),'r--')
```

If you run this demonstration, you will see that the correction is almost perfect. Some FFT system analyzers (Sec. 10.13) provide this kind of capability almost in real time. The effect of real-world imperfections can be studied by adding noise (see bottom of Fig. 10.38) and/or using *incorrect* sensor dynamics.

10.6 POSITIONING SYSTEMS

The need to position an object in response to a command arises in many measurement systems:

> Traversing systems for fluid-flow probes
> Alignment of optical components
> Scanning of laser beams (scanning vibrometer, etc.)
> Coordinate measuring machines
> Pen or stylus positioning in strip-chart recorders

Scanning-probe microscopes

Stage positioning in microdevice manufacture

Inertial-element positioning in servo accelerometers

Manual control using ordinary micrometers to position slides or stages is the simplest approach and is often adequate. For extra fine resolution, differential screws[47] and elastic flexure motion transformers[48] are available. If an apparatus is to be "automated," various forms of electrical or computer control may be needed. Motion control is an important topic with its own extensive theory and literature, and might be divided, based on *power level,* into "machine" (high power) and "instrument" (low power) categories. Both areas use open-loop or closed-loop (feedback) methods, depending on the application. Open-loop methods generally require only an *actuator* to produce the motion, while feedback schemes need also a motion *sensor.* Ready-made positioning systems are available from various vendors; at other times, we may need to build our own.

The actuators used are usually electromechanical "motors" of one sort or another. Translational motions can be produced by inherently translational actuators (linear DC motors, voice-coil actuators, or piezo translators) or by rotary motors with rotary-to-translation mechanisms (ball screws or rack/pinion). Rotary motors include galvanometers, DC brush or brushless motors, stepper motors, AC two-phase servomotors, and rotary piezodrives. Software for computer control of stepper or DC servomotors is available from some software vendors, allowing easy integration of data acquisition and motion control.[49]

For the finest resolution, *piezoactuators* provide control at micrometer and nanometer levels.[50] Maximum motion is about 100 μm, but one can "piggyback" a piezoactuator on a motorized micrometer to get both long range and fine resolution. The Inchworm,[51] a unique totally-piezo actuator, gives long range (typically 50 mm) and fine (as fine as 1 nm) resolution without the use of micrometer screws. The Picomotor[52] uses a piezoactuator to rotate a micrometer screw, giving 0.1 μm resolution and up to 2-in standard travel. *Stepping motors* can often be used in open-loop mode so long as the stepping rate and load torque are kept below critical values to prevent loss of steps.[53] The controller can sometimes be quite simple since the motor responds to a single control pulse by moving a fixed amount, say, $\frac{1}{2000}$ of a revolution. While the velocity will be "jumpy," we can accurately control the

[47]Differential Micrometer, Newport Corp., Irvine, CA, 800-222-6440 (www.newport.com).

[48]"Linearity of Elastic Actuators," 1995, A. E. Hatheway Inc., Pasadena, CA, 626-795-0514 (www.aehinc.com).

[49]"Motion Control Tutorial," catalog 2001, pp. 690–717, National Instruments (www.ni.com).

[50]Polytec PI, Auburn, MA (www.polytecpi.com); "The Nanopositioning Book," Queensgate Instruments, East Meadow, NY (www.nanopositioning.com).

[51]Burleigh Instruments, Fishers Park, NY (www.burleigh.com).

[52]New Focus Inc., Sunnyvale, CA, 408-980-8088 (www.newfocus.com); A. Tuganov, "Replacing Manual Actuators," *Lasers & Optronics,* Feb. 1994, pp. 15–16.

[53]E. O. Doebelin, "System Dynamics," Marcel Dekker, New York, 1998, pp. 300–306.

position by inputting a known number of pulses. The *average* velocity is controlled by using a desired pulse *rate*. For translational motion, a ball screw with a pitch of, say, 1 mm/rev would give 2 μm/step resolution for the motor just quoted.

Feedback systems using various forms of DC servomotors give smoother velocity and position control than open-loop steppers, but require more care in design and use because of the potential instability possible in any feedback system. Proper design, however, routinely gives reliable, high-performance systems. The classical actuator here is the brush-type servomotor using a permanent-magnet field and armature control. In the fourth edition of this text, a dynamic analysis of the *galvanometer* was presented as part of the discussion of *galvanometer oscillograph recorders*. In the years since that edition, this class of recorders has become largely obsolete, but the galvanometer itself has not, being used for many high-speed motion-control applications such as laser-beam scanning. The dynamic behavior of the galvanometer is essentially the same as most rotary and translational DC motors, so its analysis serves to illuminate the behavior of all these actuators. While the *brushless* versions of the rotary and translational (voice coil) motors require rotor-position sensors and electronic-commutation circuits, the *overall* dynamic behavior is again the same.[54]

Figure 10.39 shows the structure of a spring-restrained galvanometer intended for oscillatory motion rather than continuous rotation. A brush-type *motor* would of course *not* have the springs and *would* have brushes and a commutator. If we analyze the spring-restrained version, it is easy to later set the spring constant to zero. The commutator has no effect on the dynamic model so need not be included. Models for translational devices simply substitute the translational versions of the rotary components (spring, damper, etc.). We assume a fixed magnetic field provided by permanent magnets; wound fields are only necessary in high-power motors, not in our "instrument" applications. The figure is quite schematic; to get a nearly radial field, there is usually some iron (not shown) inside the coil frame. The voltage applied to the coil will usually come from an amplifier, which we model as a voltage source in series with the amplifier output resistance.

To get the system differential equations, we will need to apply Kirchoff's voltage loop law to the electric circuit and Newton's law to the rotating inertia. These two equations will be a coupled set of simultaneous equations that we will then reduce to a single equation in one unknown, allowing the definition of a useful transfer function relating input voltage to output motion. The coil is modeled with resistance, inductance, and the back emf effect present in all DC motor/generators. Kirchoff's law then gives

$$i_g(R_s + R_g) + L_g \frac{di_g}{dt} + HNlb \frac{d\theta_o}{dt} - e_s = 0 \qquad \textbf{(10.79)}$$

The details of the back emf term will be unfamiliar to most readers; they can be found in most introductory physics texts. The flux density (H teslas) is produced by the permanent magnet, while N is the number of turns in the coil, and *lb* represents

---

[54]Ibid., pp. 292–293.

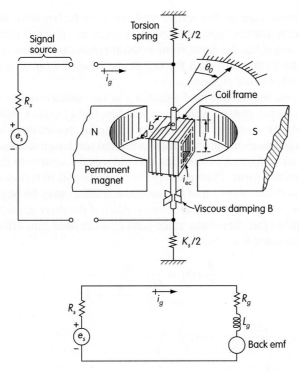

Figure 10.39
Analysis model for galvanometer rotary actuator.

the coil area exposed to the magnetic field. This equation has *two* unknowns, coil current i_g and coil rotation angle θ_o, so we need to now write the following Newton's law to get the required second equation:

$$HNlbi_g - \frac{(Hlb)^2}{R_f} \cdot \frac{d\theta_o}{dt} - K_s\theta_o - B \cdot \frac{d\theta_o}{dt} = J \cdot \frac{d^2\theta_o}{dt^2} \qquad (10.80)$$

The terms on the left side are all torques, and the two left-most terms will again be unfamiliar; however, details are available from physics texts since the back emf and torque effects are fundamental to basic DC motor/generator theory. The left-most term is the basic one in that it relates magnetic force and torque to dimensions and field strength. Here the *lb* term is now *not* coil area. Rather, the magnetic force on a *single* vertical conductor is Hli_g; thus, the *torque* will be that force times lever arm b. The second term from the left is a subtle one dependent on the nature of the coil *frame*. If the frame is made from a conducting material, it is in reality *a coil of one turn* and will thus have a voltage induced into it, according to our earlier back-emf formula. Dividing this voltage by the resistance R_f of the coil frame gives a current i_{ec} that flows in the frame. (Such currents are conventionally called *eddy currents.*) Since this current is in the same magnetic field as the coil *wires,* the frame will feel a magnetic torque that we calculate with the same formula as used for the coil. The algebraic sign of this torque will however be negative in our equation since such effects always *oppose* the motion (angular velocity) that causes them. Note that this torque has the same form as the mechanical viscous-damping torque and adds directly to that torque. If the coil frame is made from an insulating material, then this effect is not present.

We now have the needed two equations in two unknowns, so they could be solved for if the input voltage were given as a function of time. Combining the two equations leads to a single third-order differential equation, which frustrates attempts to get solutions in the letter form preferred for design studies. Accordingly, we now make an additional assumption that reduces the system to the familiar and useful second-order form. This assumption neglects coil inductance and is often valid since experience shows that the inductance effect may be negligible for the useful frequency range of practical devices. Also, if the driving amplifier is of the *transconductance* type, then inductance (and also the back-emf effect) is negated. Neglecting inductance leads to:

$$\frac{\theta_o}{e_s}(D) = \frac{K_\theta}{\dfrac{D^2}{\omega_n^2} + \dfrac{2\zeta D}{\omega_n} + 1} \tag{10.81}$$

where

$$K_\theta \triangleq \frac{HNlb}{K_s(R_s + R_g)}, \text{rad}/\text{V}$$

$$\omega_n \triangleq \sqrt{\frac{K_s}{J}}, \text{rad}/\text{s}$$

$$\zeta \triangleq \frac{B + \dfrac{(Hlb)^2}{R_f} + \dfrac{(HNlb)^2}{R_s + R_g}}{2\sqrt{K_s J}}$$

Note that damping is present even if B is zero; the two other terms in the numerator must have the same dimensions as B and are thus an electrical form of damping. For galvanometers that were used in oscillographs, this electrical damping alone was sufficient to give the optimum $\zeta = 0.65$ for ω_n less than about 400 Hz (higher natural frequencies required a mechanical damper). These relations give useful estimates of system performance in terms of basic physical parameters and are available *only* because we neglected inductance. Using them as design tools, we can establish tentative numerical values for the parameters. It is then easy to use an available inductance value to solve the more correct third-order model and see whether inductance is really negligible.

The spring effect was present in the galvanometers used in (now obsolete) oscillograph recorders; it is also currently used in optical-scanning galvanometers of the *resonant* and open-loop types.[55] Resonance is used as a design principle in a number of machines and instruments. By driving a lightly damped system sinusoidally at its natural frequency, we can get large motions with very small

---

[55] J. I. Montagu, "Galvanometric and Resonant Low Inertia Scanners," GSI Lumonics, Bedford, MA, 781-275-1300 (www.gsilumonics.com).

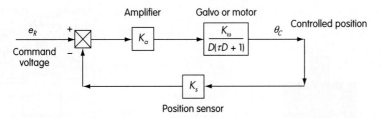

Figure 10.40
Instrument servomechanism using a DC motor or galvanometer.

driving force and power. Scanners of this type are of course limited to applications where the sinusoidal motion meets the application's needs. Open-loop scanners can reproduce arbitrary motion commands if their frequency content is below about 10 percent of the natural frequency. If a scanner must produce more accurate motions of an arbitrary nature, a feedback approach is more appropriate. Accordingly, we remove the spring and add a position sensor, summing junction, and amplifier to form an instrument servomechanism. With the spring removed, application of a constant input voltage results in a steady-state *speed* rather than position as follows:

$$\frac{\theta}{e_s}(D) = \frac{\dfrac{HNlb}{R_s + R_g}}{D\left(\dfrac{J}{\left[B + \dfrac{(Hlb)^2}{R_f} + \dfrac{(HNlb)^2}{R_s + R_g}\right]}D + 1\right)} = \frac{K_\omega}{D(\tau D + 1)} \qquad (10.82)$$

This transfer function has the same form as that of an armature-controlled DC motor.[56] The closed-loop galvanometer scanner and DC motor instrument servomechanism have the same system diagram (Fig. 10.40), which leads to the closed-loop transfer function:

$$\frac{\theta_C}{e_R}(D) = \frac{K_{cl}}{\dfrac{D^2}{\omega_n^2} + \dfrac{2\zeta D}{\omega_n} + 1} \qquad (10.83)$$

where

$$K_{cl} \triangleq \frac{1}{K_s}$$

$$\omega_n \triangleq \sqrt{\frac{K_a K_\omega K_s}{\tau}}$$

$$\zeta \triangleq \frac{1}{2\sqrt{\tau K_a K_\omega K_s}}$$

A fast response can be obtained by raising the loop gain $K_a K_\omega K_s$, but this may cause poor damping unless τ is very small (which requires a large B). Most such

[56]E. O. Doebelin, "System Dynamics," op. cit., pp. 484–489.

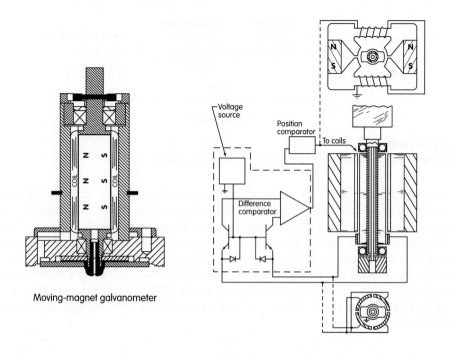

Moving-magnet galvanometer

Moving-iron galvanometer

Figure 10.41
Moving-magnet and moving-iron galvanometer actuators. *(Courtesy GSI Lumonics, www.gsilumonics.com.)*

servos *do not* use intentional mechanical dampers. Rather, they usually add some *derivative control* to the amplifier, which provides a form of electrical damping. The derivative control takes the form of a "lead-lag" term $[(\tau_L D + 1)/(\tau_G D + 1)]$ to provide some low-pass filtering for the differentiator. Inclusion of this effect will raise the system to the third order, which also then allows for consideration of stability, *not possible* with second-order models.

Some galvanometer scanners use transduction schemes other than the moving coil, that is, *moving magnet* and *moving iron*[57] (Fig. 10.41). While the basic torque-producing mechanism is different from that of the moving coil, and requires a separate analysis, the end result is a transfer function of a similar form. Closed-loop systems are usually used, with capacitance displacement sensors being a common form. All three technologies are in use, the choice being made on the usual criteria of cost and performance. The fastest moving-magnet galvos may have a small-signal step response as fast as 0.3 ms.[58] Miniature scanners based on MEMS technology are under development (GSI Lumonics).

[57]R. P. Aylward, "The Advances & Technologies of Galvanometer-Based Optical Scanners," Cambridge Technology, Inc., Cambridge, MA, 617-441-0600 (www.camtech.com); G. Marshall, "Optical Scanning," Marcel Dekker, New York, 1991; J. Montagu and A. Bukys, "Moving Magnet Galvanometer Scanners," *General Scanning Tech Note* 12/92, GSI Lumonics, Bedford, MA (www.gsilumonics.com).

[58]Cambridge Technology, Inc., Model 6800HP.

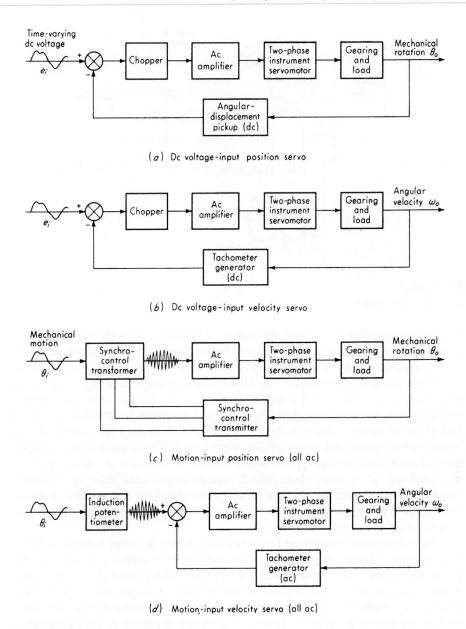

Figure 10.42
Instrument servomechanisms.

Of the many applications, we mention here a recent system[59] that uses two separate orthogonal mirror scanners to *xy* deflect a laser beam used to micromachine liquid-crystal displays. The workpieces are mounted on an *xy* slide, positioned

[59]R. Bann et al., "Micromachining System Accommodates Large Wafers," *Laser Focus World,* Jan. 2001, pp. 189–192.

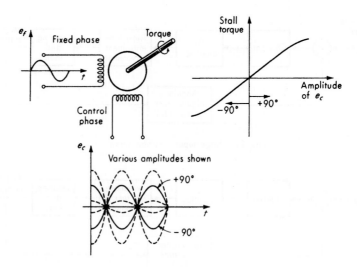

Figure 10.43
Two-phase servomotor.

by two servomechanisms under the mirrors. The slides move the workpiece to a desired location, and the galvos then quickly scan the beam over that subarea. This design was faster than an alternative that used slides to control the *entire* motion.

Some instrument servos (see Fig. 10.42) utilize ac amplifiers and two-phase ac instrument servomotors even if the input signal is dc. The use of ac amplifiers is based on their freedom from drift, high gain and reasonable cost, while the use of two-phase ac motors relates to their low friction (no brushes are needed as in a brush-type dc motor) and controllability. Figure 10.43 briefly summarizes the operating characteristics of this type of motor. One of the phases is of fixed amplitude. The amplitude of the other phase (which must be displaced in phase by ± 90 electrical degrees from the fixed phase) controls the direction and amount of torque developed. When the controlled phase reverses polarity (goes from $+90$ to $-90°$ or vice versa), the torque reverses.

The schematic and graphs of Fig. 10.44 show how a dc signal is converted to ac by a chopper and how a reversal in polarity of the dc error signal e_{aa} results in a $180°$ phase shift (from $+90$ to $-90°$ or vice versa) in the motor control-phase voltage e_{dd}, thereby causing the required reversal in the direction of torque. We see that whenever $e_i \neq e_p$, there will be an error voltage e_{aa} which is converted to ac and amplified so that it tends to drive the motor in a direction to change e_p until it equals e_i. If e_i is changing and if the amplifier gain is high enough, e_p (and therefore output motion θ_o) will "track" e_i with very little error.

This completes our discussion of positioning devices used in measurement systems. For the feedback-type systems, more details are available in the voluminous control-systems literature.[60] Because the various instrument servos we have

---

[60]E. O. Doebelin, "Control System Principles and Design," Wiley, New York, 1985.

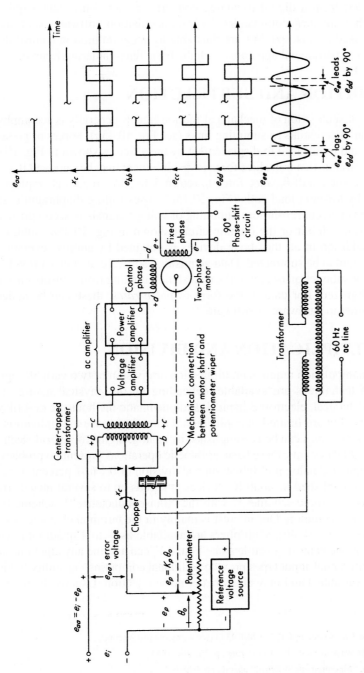

Figure 10.44
Position servo.

described respond to a voltage command and require little input current, they may be commanded from a digital-to-analog converter present on a data-acquisition board, which is in turn commanded by data-acquisition software, such as the DASYLAB used in this text. We are thus able to integrate motion control that is needed in our experimental apparatus with the basic data-gathering function.

10.7 ADDITION AND SUBTRACTION

The addition or subtraction of mechanical-motion signals generally is accomplished by use of gear differentials or summing links (see Fig. 10.45*a*). Forces or pressures are summed and transduced to displacement by the schemes shown in Fig. 10.45*b*. The spring restraints that transduce force to displacement may be removed if a feedback system using a null-balance force to return deflection to zero is employed. A totalizer for hydrostatic load cells (Fig. 10.45*c*[61]) uses rolling diaphragm seals to achieve 0.0005 percent resolution. Summing of voltage signals is accomplished by the simple series circuit or the op amp circuit shown in Fig. 10.45*d*. Subtraction rather than addition in all the above devices is obtained by simply reversing the sense of the input to be subtracted. Data-acquisition software of course provides for the addition or subtraction of any of the data signals coming from sensors once they have been digitized. For data in the form of *pulse rates,* these can be added or subtracted either in hardware or software.

10.8 MULTIPLICATION AND DIVISION

When data manipulation requires multiplication or division of two variable signals, a number of techniques are available, depending on the physical nature of the signals.[62] Mechanical, electromechanical, and pneumatic methods are described in the literature,[63] but are limited in application. General-purpose multiplication and division in data systems can be accomplished by analog, digital, or hybrid electronic means. For digital systems, these basic arithmetic operations present no problem for any size computer. Analog multiplication and division of 0.1 to 1 percent accuracy are readily accomplished in small IC devices at speeds up to several megahertz by using various electronic schemes,[64] with the "transconductance"[65] scheme being perhaps the most common. Our interest is mainly in the terminal characteristics of the devices (see Fig. 10.46). Multipliers are available as 1- to 4-quadrant devices; the 4-quadrant one is most versatile since V_x and V_y can assume any algebraic signs whatever. Differential-input types provide additional computing capability (additive constants or variables) and may allow better noise rejection. Usually the reference

[61]Emery Winslow Co., Seymour, CT, 203-881-9333 (www.emerywinslow.com).

[62]S. A. Davis, "31 Ways to Multiply," *Contr. Eng.,* p. 36, Nov. 1954.

[63]E. O. Doebelin, "Measurement Systems," op. cit., pp. 649–655.

[64]"Multiplier Application Guide," Analog Devices, 1980.

[65]Ibid.; D. H. Sheingold, "Nonlinear Circuits Handbook," op. cit.

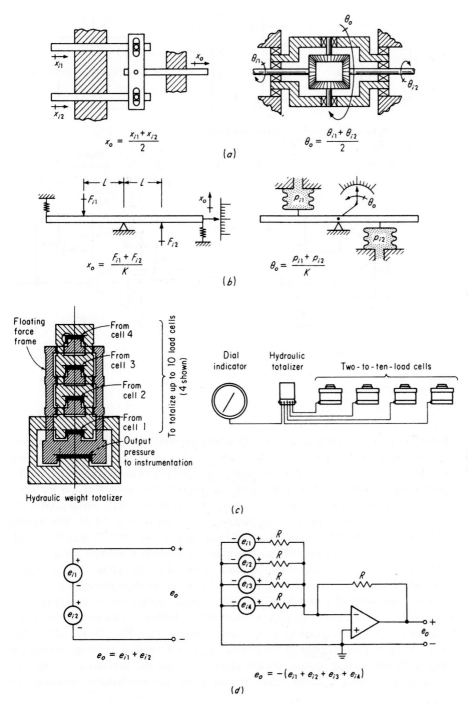

Figure 10.45
Addition and subtraction.

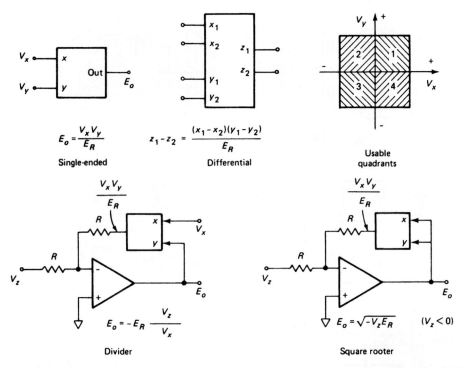

$$E_o = \frac{V_x V_y}{E_R}$$

Single-ended

$$z_1 - z_2 = \frac{(x_1 - x_2)(y_1 - y_2)}{E_R}$$

Differential

Usable quadrants

$$\frac{V_x V_y}{E_R}$$

$$E_o = -E_R \frac{V_z}{V_x}$$

Divider

$$\frac{V_x V_y}{E_R}$$

$$E_o = \sqrt{-V_z E_R} \qquad (V_z < 0)$$

Square rooter

Figure 10.46
Electronic multiplier/divider.

value E_R is 10 V, so that full-scale (± 10 V) inputs give full-scale (± 10 V) output. Division capability requires only the addition of a properly connected op-amp (included and switch-selectable in many multiplier/divider packages); but the denominator, of course, must remain of one polarity to avoid division by zero. Squaring is done easily by connecting the signal to both inputs, while taking the square root necessitates the op-amp feedback scheme of Fig. 10.46. For hybrid analog/digital applications, *multiplying digital/analog converters*[66] may be useful. Here a time-varying analog voltage becomes the reference voltage for a digital/ analog converter, which makes the D/A output equal to the product of the analog input and the instantaneous value of the digital input to the digital/analog converter.

A Wheatstone bridge may be utilized for multiplication if one of the signals can be transduced to a voltage (which is used as the bridge-excitation voltage) and the other can be transduced to a resistance change (see Fig. 10.47). Then the bridge output is proportional to the product $q_{i1}q_{i2}$ for small-percentage resistance changes. While analog and/or digital electronics are the preferred schemes in general-purpose data systems, there is still room for mechanical ingenuity in

---

[66]"Appl. Guide to CMOS Multiplying D/A Converters," Analog Devices, 1978.

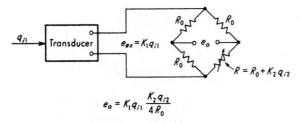

Figure 10.47
Wheatstone-bridge multiplier.

special-purpose devices, such as the Btu meter of Fig. 10.48. This instrument/
computer implements the relation

$$\text{Btu} = C \int \dot{M}(T_1 - T_2)\, dt \tag{10.84}$$

to find the energy change in heated (or cooled) liquid flows in piping systems and
requires subtraction, multiplication, and integration. Two pressure thermometers
measure T_1 and T_2; the subtraction is effected by connecting one Bourdon tube to
the pointer and the other tube to the scale, which produces a direct indication of
$T_1 - T_2$. These same rotations also position two notched cams, forming a gap
proportional to $T_1 - T_2$, which actuates a cam roller follower. The cam roller rotates
continuously around the cams at a speed proportional to flow rate; this rotation is
provided by either a positive-displacement or a turbine-type flowmeter (both are
shown in Fig. 10.48, but one would be employed in an actual system). Two revolution counters (odometers), one reading Btu and the other reading gallons, are driven
in intermittent fashion from a serrated drive wheel. As the rollers pass over and into
the recess generated by the cams, two cam levers attached to the rollers engage the
serrated circumference of the drive wheel. Thus each rotation of the cam roller
carrier produces a partial rotation of the drive wheel that is proportional to $T_1 - T_2$.
Integration is achieved simply by accumulating total rotations since "average
speed" is proportional to $\dot{M}(T_1 - T_2)$. The entire instrument requires no external
power sources and achieves an accuracy of ± 1.5 percent for ΔT greater than 50
percent of rated.

10.9 FUNCTION GENERATION AND LINEARIZATION

When we need to generate a specific nonlinear function of a mechanical-motion
signal, the use of cams, linkages, and noncircular gears allows great freedom since
almost any reasonable function can be approximated adequately by one or a combination of these methods (see Fig. 10.49). The use of instrument servos also allows
these methods to be employed with electrical signals, if, for some reason, all-
electronic methods are not feasible.

Nonlinear potentiometers are used widely in function generation. They are
constructed in basically the same manner as potentiometer displacement transducers
except that a specific *nonlinear* relation between θ_i and e_o is wanted, rather than the

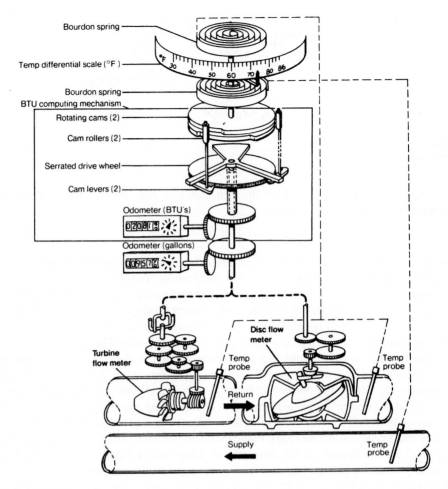

Figure 10.48
All-mechanical computing Btu meter.
(Courtesy Hersey Measurement Co., Spartanburg, SC.)

linear relation desired for a motion transducer (see Fig. 10.50). A wide variety of functions are possible by distributing the resistance winding in a proper nonlinear fashion on the mandrel. Techniques also have been developed for constructing nonlinear potentiometers by using conducting plastic or deposited-film (rather than wirewound) resistance elements. While functions of rather arbitrary form are available as special items, certain basic functions are used so commonly that they are obtainable ready-made as stock items. These include sine and cosine over 360°, sine or cosine over 360°, 180° sine, 90° sine, ±75° tangent, square function, and logarithmic function. The conformity of the voltage-output/rotation-input relation to the theoretical function is of the order of 0.3 to 2 percent of full scale, depending on the type of function and the instrument quality.

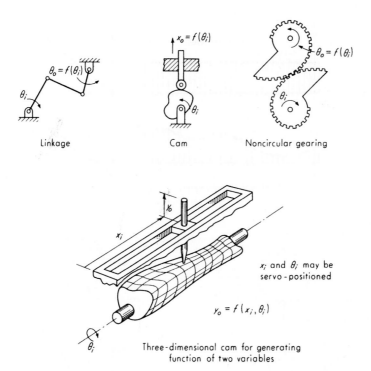

Figure 10.49
Mechanical function generation.

Potentiometer-type pressure transducers for flow rate, altitude, and airspeed applications are standard items. Use of a computer-controlled mechanical milling process shapes the conductive-film resistance element of each transducer to give the desired voltage/pressure relation. This milling is done on each individual transducer with pressure and excitation voltage applied, thus compensating for any individual mechanical differences from unit to unit. The flow-rate transducer accepts a Δp signal from orifice-type flow elements and includes a square root function to develop an output linear with flow rate. Airspeed transducers are also differential-pressure units that accept stagnation and static pressures from aircraft pitot tubes and provide the proper nonlinear function to give an output linear with airspeed. A 0- to 1,000-kn unit conforms to airspeed linearity within ± 10 kn and can resolve ±1.5 kn. Altitude transducers accept aircraft static pressure as input and linearize the U.S. standard atmosphere function with (for a 0- to 70,000-ft device) resolution of 10 ft.

When very accurate sine and/or cosine functions are needed (as in navigation and fire-control computers where resolution and composition of vectors must be performed), the use of resolvers rather than nonlinear potentiometers may be indicated. Resolvers are small ac rotating machines similar to synchros. In general, they have two stator windings and two rotor windings (see Fig. 10.51). If one of the

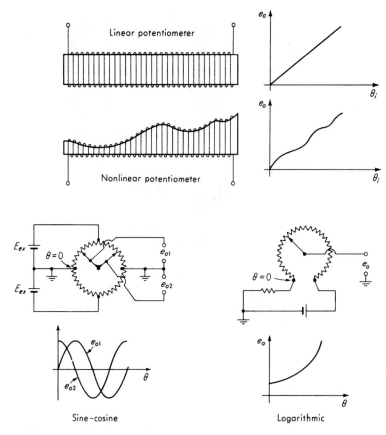

Figure 10.50
Nonlinear potentiometers.

stator windings is excited with an ac signal of constant amplitude (60 or 400 Hz is employed commonly) and the other is short-circuited, rotation of the rotor through an angle θ_i from a null position gives at the two rotor windings ac signals whose amplitudes are proportional to $\sin \theta_i$ and $\cos \theta_i$, respectively. Other important computing functions such as converting vehicle rotation angles to earth coordinates in navigation systems also can be performed by resolvers. A typical high-accuracy resolver has an excitation voltage of 26 V maximum at 400 Hz, open-circuit output voltage of 0 to 26 V, residual null voltage of 1 mV maximum, and a maximum deviation from the desired functional relation of 0.01 percent.

All the function generators shown so far involve moving parts and so are limited in speed. When the speed of all-electronic function generation is necessary or an electromechanical solution is unacceptable for other reasons, multiplier/ dividers, multifunction modules, or diode function generators are available.[67] A

---

[67]D. H. Sheingold, "Nonlinear Circuits Handbook," op. cit.

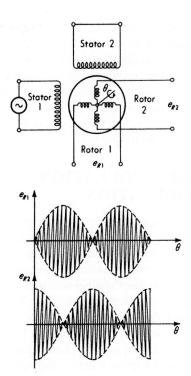

Figure 10.51
Resolver.

cascade of multipliers with outputs scaled and summed with op-amp summers can generate a power series with adjustable coefficients. For example, two multipliers and one summer can generate $y = a_1 x + a_3 x^3$, which can fit $\sin x$ within ± 0.6 percent over the range $-\pi/2$ to $\pi/2$. Linearizers for hot-wire anemometers often utilize this sort of power-series scheme with terms up to the fourth power. Multifunction modules[68] provide the relation

$$E_o = \frac{10}{E_{\text{ref}}} V_y \left(\frac{V_z}{V_x}\right)^m \tag{10.85}$$

where V_x, V_y, V_z are variable or constant input voltages (≥ 0), $E_{\text{ref}} \approx 9$ V, and m can be set in the range 0.2 to 5. This device, by itself or combined with others, provides very versatile function generation for linearization or other purposes. Diode-function generators approximate the function in a piecewise linear way and have been used to convert thermocouple voltages to their corresponding temperatures.[69]

Once sensor signals have been digitized, data-acquisition (or general-purpose mathematical) software can perform function generation of arbitrary complexity.

---

[68]Model 433, Analog Devices.
[69]E. O. Doebelin, "Measurement Systems," 4th ed., op. cit., pp. 821–822.

DASYLAB, for example, allows formulas of any complexity, involving any of the sensed signals. Alternatively, a table-lookup module can be used with discrete points, such as an experimental calibration curve. These digital methods are usually *slower* than all-electronic analog solutions but are adequate for most applications. If higher-speed digital calculations are essential, one can sometimes use a DSP *(digital-signal processing)* board as part of the data system. These special purpose boards can relieve the personal computer processor of those computations that need to be faster than the PC itself can handle.

10.10 AMPLITUDE MODULATION AND DEMODULATION

We have seen a number of examples of measurement systems in which interconversion between ac and dc signals was necessary and/or desirable. The dc-to-ac conversion is a form of amplitude modulation, whereas ac-to-dc conversion is called *demodulation* or *detection.* In measurement systems, usually the demodulation must be phase-sensitive, so that the algebraic sign of the original time-varying dc signal is preserved. Modulation/demodulation may be accomplished by assorted electronic schemes; the availability of small, accurate, and inexpensive IC analog multipliers makes the approach of Fig. 10.52 popular. There we show both modulation and synchronous demodulation, with an intermediate stage of ac amplification (since this is often the reason for using a modulation/demodulation scheme at all). The carrier oscillator can be either sine wave or square wave, and its frequency should be 5 to 10 times the highest signal frequency so that the low-pass filter can

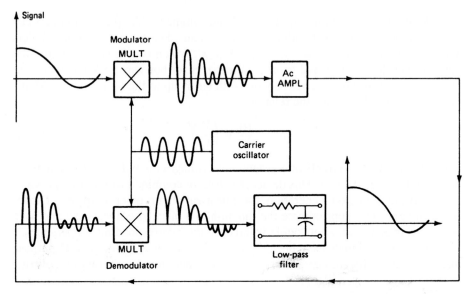

Figure 10.52
Multiplier-type modulator/demodulator system.

achieve a proper trade-off between ripple and speed of response. Many infrared temperature instruments employ a scheme of this sort, except that the modulation is done mechanically with a rotating chopper wheel and a proximity pickup on this wheel generates the synchronizing waveform for the demodulator.

10.11 VOLTAGE-TO-FREQUENCY AND FREQUENCY-TO-VOLTAGE CONVERTERS

The conversion of a dc voltage input to a periodic-wave output whose frequency is proportional to the dc input may serve several useful functions in measurement systems. Such devices are employed widely in FM/FM telemetry systems since the voltage-to-frequency (V/F) conversion process is a form of frequency modulation. Also they are used in the integrating digital voltmeter where a dc signal is converted to a periodic wave of proportional frequency. Then this wave is applied to an electronic counter for a fixed time interval, giving a reading proportional to the average dc voltage over the time interval. The recording of dc voltages on magnetic tape recorders also is accomplished through the use of frequency modulation. The inverse operation, frequency-to-voltage (F/V) conversion, accepts input signals of virtually any waveform and produces a time-varying DC output proportional to the rate at which the input signal crosses a fixed (but selectable) threshold, often zero volts. A major use is with sensors that produce pulse rates proportional to some physical variable: turbine flowmeters, vortex-shedding flowmeters, proximity pickups that count gear teeth on a rotating shaft, etc. We often read out such sensors with an electronic counter/timer set to give an average pulse-rate, say, every second. If, instead, we want a continuous recording on a strip-chart recorder that accepts voltage inputs, then an F/V converter can be used. Most such converters use a final low-pass filter to smooth the ripple, and this filter sets the dynamic response of the converter. For low-pulse rates, this filter must have a low cutoff frequency, giving the converter a slow response. Circuit details of V/F and F/V converters are available.[70] Some[71] F/V converters avoid the problem of slow filter response by using a digital approach (see Tachometer Encoder Methods, Sec. 4.4).

10.12 ANALOG-TO-DIGITAL AND DIGITAL-TO-ANALOG CONVERTERS; SAMPLE/HOLD AMPLIFIERS

Because most sensors have analog output while much data processing is accomplished with digital computers, devices for conversion between these two realms obviously play an important role. For motion inputs, the shaft-angle encoders of Chap. 4 provide analog-to-digital (A/D) conversion; here we concentrate on all-electronic devices whose analog input or output is a voltage.

---

[70]E. O. Doebelin, "Measurement Systems," 4th ed., op. cit., p. 824.

[71]Model FV-801, Ono Sokki, Addison, IL, 800-922-7174 (www.onosokki.net).

Binary bits (n)	(2^n)	Equivalent percent or fraction of range of least-significant bit*		Residual $1 - \sum\limits_{1}^{n}\left(\dfrac{1}{2^n}\right)$
		Percent	ppm	
1	2	50.0	500 000.	0.5
2	4	25.	250 000.	0.25
3	8	12.5	125 000.	0.125
4	16	6.25	62 500.	0.062 5
5	32	3.125	31 250.	0.031 25
6	64	1.562 5	15 625.	0.015 625
7	128	0.781 25	7 812.5	0.007 812 5
8	256	0.390 625	3 906.25	0.003 906 25
9	512	0.195 313	1 953.13	0.001 953 13
10	1 024	0.097 656	976.56	0.000 976 56
11	2 048	0.048 828	488.28	0.000 488 28
12	4 096	0.024 414	244.14	0.000 244 14
13	8 192	0.012 207	122.07	0.000 122 07
14	16 384	0.006 104	61.04	0.000 061 04
15	32 768	0.003 052	30.52	0.000 030 52
16	65 536	0.001 526	15.26	0.000 015 26
17	131 072	0.000 763	7.63	0.000 007 63
18	262 144	0.000 381	3.81	0.000 003 81
19	524 288	0.000 191	1.91	0.000 001 91
20	1 048 576	0.000 095	0.95	0.000 000 95
21	2 097 152	0.000 048	0.48	0.000 000 48
22	4 194 304	0.000 024	0.24	0.000 000 24
23	8 388 608	0.000 012	0.12	0.000 000 12
24	16 777 216	0.000 006	0.06	0.000 000 06

*May be limited by noise and other uncertainties in actual circuit.

Figure 10.53
Resolution of A/D and D/A converters.

We begin with digital-to-analog (D/A) converters since they are used as components in some A/D converters. The most fundamental property of any D/A or A/D converter is the number of bits for which it is designed since this is a basic limit on resolution (see Fig. 10.53).[72] Units with 8 to 12 bits are most common; however, resolution to about 18 bits is available, but special care must be taken in the application[73] (the least significant bit represents only 38 μV). Most D/A converters utilize a principle similar to that of Fig. 10.54, the so-called R-2R ladder network. Basic to the accuracy of such devices is the stability of the reference voltage and

---

[72]"Design Engineers' Handbook and Selection Guide, A/D and D/A Converter Modules," BR-1021, Analogic Corp., Wakefield, MA, 1980.

[73]"Designer's Guide to High Resolution Products," Analog Devices, 1981; "Analog-Digital Conversion Handbook," 3d ed., Analog Devices, 1986.

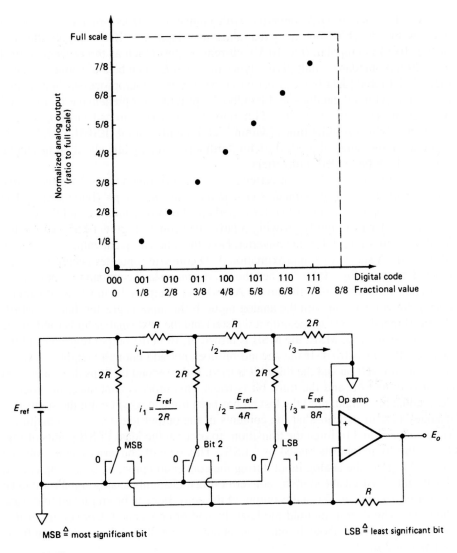

Figure 10.54
Digital-to-analog conversion (3 bits).

resistor values. Even with the best components, the highest-resolution converters need periodic calibration to remain within specification. To understand the operation of the 3-bit D/A circuit shown, note that the switches (called *current-steering switches*) are connected to "ground" irrespective of whether they are "on" (1) or "off" (0), because the op-amp negative input is a virtual ground. So we can redraw the circuit in this way and quickly see why the currents are as labeled. The op-amp merely sums the currents "steered" to it by the switches and produces an output voltage proportional to this sum.

A wide variety of D/A converters with a range of cost/performance and special features are available from many manufacturers: bipolar (± 10 V) outputs (the unit of Fig. 10.54 is unipolar, 0 to 10 V), current outputs (such as the process control 4- to 20-mA standard), multiplying types in which E_{ref} can be a dynamic variable, digitally buffered (data-bus-compatible) types for easy microprocessor interfacing, etc. Manufacturers' catalogs and handbooks provide comprehensive listings and discussions of specifications. In addition to resolution (mentioned earlier), here we mention only that settling times [within $\frac{1}{2}$ least-significant bit (LSB)] for full-scale input are in the range of 3 to 30 μs for "ordinary" units, while about 10 ns can be achieved by 8-bit "video" converters.

Let us turn now to A/D converters.[74] The most common types[75] are *successive approximation, flash, pipelined, sigma delta,* and *dual slope* (integrating). The successive-approximation converter is perhaps the most common, and the sigma-delta converter is a rapidly growing relative newcomer. Figure 10.55a shows the ideal behavior of a 3-bit A/D converter. Note the inherent quantization uncertainty of $\pm\frac{1}{2}$ LSB. A successive-approximation A/D converter operates according to the block diagram of Fig. 10.56.[76] When the "start conversion" command is applied, the D/A converter (which is built into the A/D device) outputs the most-significant bit (MSB) for comparison with the analog input. If the input is greater than the MSB, the MSB remains on (1 in the output register) and the next smaller bit is tried; if the input is less than the MSB, the MSB is turned off (0 in the output register) and the next smaller bit is tried. If the second bit does not add enough weight to exceed the input, it is left on and the third bit is tried. If the second bit "tips the scales" too far, it is turned off and the third bit is tried. This process continues, in order of descending bit weight, until the last bit has been tried, at which point the status line changes state to indicate that the contents of the output register now constitute a valid conversion. (To force the transitions to occur at the ideal $\frac{1}{2}$ LSB points of Fig. 10.55, the comparator is biased by $\frac{1}{2}$ LSB.) This type of converter cannot tolerate much change in the analog input during the conversion process, so a sample/hold device must be used ahead of the converter if fast-changing analog signals are to be converted accurately. The status output of the converter can be employed to release the sample/hold from its hold mode at the end of conversion. Converters of this type are available to about 16 bits (conversion time ≈ 30 μs); 8-bit units can be as fast as 1 or 2 μs.

Sigma-delta converters became popular during the late 1990s and are now in wide use. Their operating principle is quite complex, so we concentrate here on some of the features, leaving the electronic details for the literature.[77] These

---

[74]B. M. Gordon, "Linear Electronic Analog/Digital Conversion Architectures, Their Origins, Parameters, Limitations and Applications," *IEEE Trans.,* vol. CAS-25, no. 7, July 1978.

[75]B. Black, "Analog-to-Digital Converter Architectures and Choices for System Design," *Analog Dialogue,* vol. 33-8, 1999, Analog Devices, Wilmington, MA (www.analog.com).

[76]D. H. Sheingold (ed.), "Analog-Digital Conversion Notes," Analog Devices, 1977.

[77]O. Josefsson, "Using Sigma-Delta Converters, Ask the Applications Engineer," 1997, pp. 29–38; D S Patch, "Sigma-Delta Conversion Technology," Winter 1990, Analog Devices.

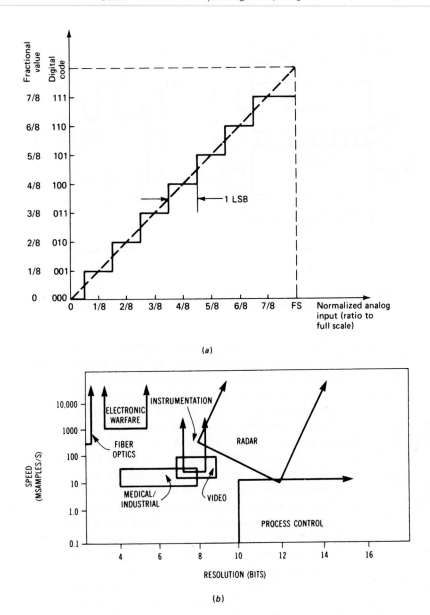

Figure 10.55
Analog-to-digital conversion.

converters use massive "over sampling"; the input signal is sampled many times (typically 256 times) faster than the signal would otherwise require and then decimation and digital filtering yield the desired resolution and effective sample rate. Originally developed for digital audio systems such as compact disks, they have

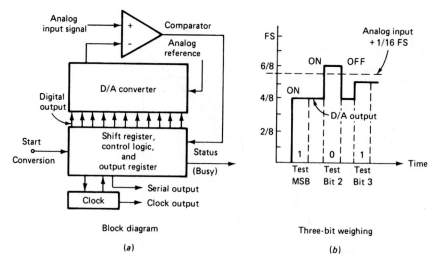

Figure 10.56
Successive-approximation A/D converter.

several advantages[78] in data-acquisition systems. Expensive, sharp cutoff, analog anti-aliasing prefilters required with other converter types may be replaced with simple analog, first-order, low-pass filters.[79] The digital low-pass filtering built into the converter provides a large dynamic range, flat amplitude ratio, and constant delay over the passband. Cost is low enough that we can often use a separate converter for each channel, avoiding the time-skew and other problems associated with multiplexing many data channels into one A/D.

Several variations of the dual-slope principle are utilized; Fig. 10.57[80] shows a basic implementation. Conversion here is indirect since the analog input is first converted to a time interval which is then digitized by using a counter. The analog input V_{in} is applied to an integrator, and a counter (counting clock pulses) is started at the same time. After a preset number of counts (and thus a fixed time T), the input is disconnected and a reference voltage of opposite polarity is applied to the integrator. At this switching instant, integrator output is proportional to the average value of V_{in} over time interval T. The integral of V_{ref} is an opposite-going ramp of slope $V_{ref}/(RC)$. The counter, reset to zero at time T, now counts until the integrator output crosses zero, Δt seconds later. Therefore Δt (and thus counter output) will be proportional to the average value of V_{in} over time interval T. In Fig. 10.57, V_{in} has been offset by V_{ref} and divided by 2, which allows a bipolar analog input to produce an offset binary output suitable as input for computer systems.

[78]A. Brower et al., "New Developments in Large-Scale Dynamic Data Acquisition Systems," *Sound & Vib.,* April 1998, pp. 18–22.

[79]J. C. Cole, "Acceleration Data Acquisition Using a Δ-Σ Converter," *Sensors,* July 1999, pp. 55–60.

[80]D. H. Sheingold (ed.), "Analog-Digital Conversion Notes."

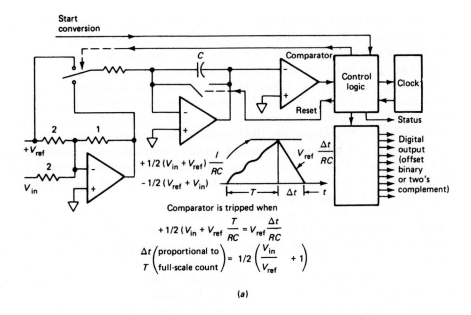

(a)

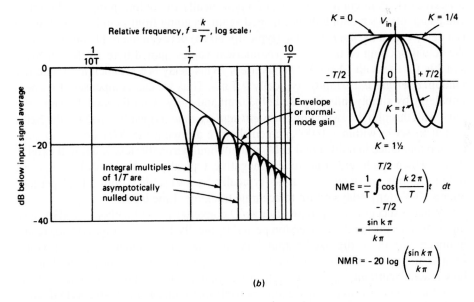

(b)

Figure 10.57
Dual-slope A/D converter.

Dual-slope converters have a number of advantages. Accuracy is unaffected by capacitor value or clock frequency, since these effect the up slope and down ramp equally. The integrating effect rejects high-frequency noise and averages changes in V_{in} during the integration period T. By choosing T as an integral multiple of the most prevalent noise signal's period (say, $T = \frac{1}{60}$ s for 60-Hz noise), theoretically perfect noise rejection at 60, 120, 180 Hz, etc., is obtained. Of course, this choice leads to the main disadvantage of dual-slope A/D converters: the slow conversion rate, usually less than 30 per second. A final advantage, widely implemented, is the easy inclusion of automatic-zero capability. Here, before each measurement cycle the input is short-circuited; thus any output produced by the integration process over T must be zero-drift error, which is subsequently subtracted from the next reading, to give continuous zero correction. (Successive-approximation A/D devices do not use automatic zero because of the speed penalty.) Integrating A/D converters are available to about 17-bit resolution. Conversion rates are usually 3 to 30 per second; however, rates up to about 250 per second are available, although this sacrifices the line-frequency noise rejection.

We conclude this section with a brief discussion of sample/hold amplifiers (SHAs),[81] often needed when successive-approximation A/D converters are employed with fast-changing inputs. Figure 10.58a shows the principle of the tracking-type[82] SHA we wish to discuss. When it is in tracking (sampling) mode, the switch is closed and the output follows the input, usually with a gain of +1. To engage the hold mold, the switch is opened, and ideally the output signal retains the last value it had when the hold command was given. This process continues until sampling (tracking) is again commanded, whereupon ideally the output jumps to the input value and follows it until hold is engaged again. In going from sample to hold, there is a delay called the *aperture time* (typically 50 ns) between the hold command and the actual opening of the switch. (Numerical values quoted here are for an SHA suitable for use with 14-bit A/D converters.) This delay is reproducible to within a random *aperture uncertainty* (typically 0.5 ns). Thus we can compensate for aperture delay by simply initiating the hold command early. After the switch opens, we must wait through the *switching transient settling time* (typically 1 μs) for the SHA output to settle within +0.003 percent before allowing the A/D device to start conversion. Once in hold, a real SHA will not hold perfectly steady, but will exhibit *droop,* which can be either positive or negative, typically 1 μV/s. In switching from hold to sample, an *acquisition time* (typically 5 μs to settle within ±0.01 percent after a 10-V step) delays return to accurate tracking of the input.

When a successive approximation A/D converter is used without an SHA, any input changes during the conversion period (typically 12 μs, see Fig. 10.59a) result in uncertainty in the digitized output. Take for example, a signal 10 sin $2\pi ft$ volts. Its maximum rate of change (*slew rate*) is $20\pi f$ volts per second. In deciding on how much uncertainty we can tolerate, a reasonable lower bound would be $\pm\frac{1}{2}$ LSB since the A/D converter's uncertainty is itself that great. For a 14-bit A/D device, $\pm\frac{1}{2}$ LSB is 610 μV (assuming 10 V full scale). Now we can calculate the highest allowable frequency of input such that the uncertainty of the digitized output is no

---

[81]J. V. Wait, "Sampled Data Reconstruction Errors," *Inst. & Cont. Syst.,* pp. 127–129, June 1970; L. W. Gardenhire, "Selecting Sample Rates," *ISA J.,* p. 59, April 1964.

[82]"Designers' Guide to High Resolution Products," Analog Devices.

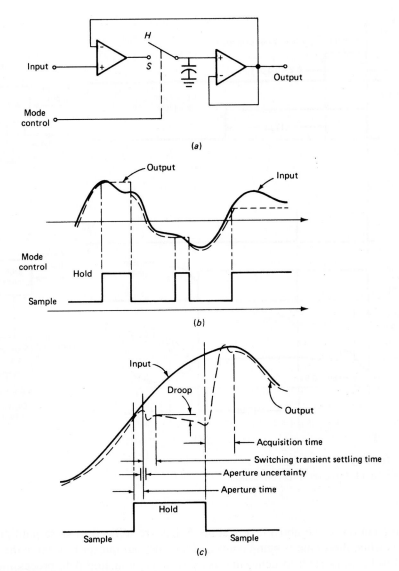

Figure 10.58
Sample-hold amplifier.

greater than $\pm\frac{1}{2}$ LSB as 0.8 Hz! This severe frequency limitation is greatly relaxed when an SHA with properties as given above is connected at the A/D input. The sequence of events is now as in Fig. 10.59*b*, where we see that the total time for a conversion is somewhat longer. However, the uncertainty calculation now uses the SHA aperture uncertainty of 0.5 ns (in place of the A/D conversion time of 12 μs), which gives the highest allowable frequency as 19 kHz, a 24,000:1 improvement.

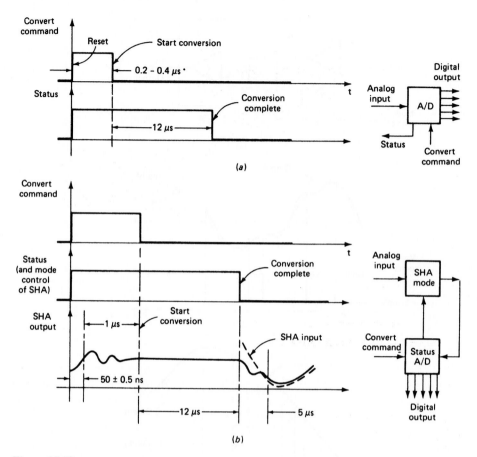

Figure 10.59
Effect of sample/hold on A/D conversion.

In addition to the above application, SHAs are utilized to store multiplexer outputs while the signal is being converted and the multiplexer is seeking the next signal to be converted; to determine peaks/valleys in analog data processing; to establish amplitudes in resolver-to-digital conversion; to facilitate analog computations involving signals obtained at different time instants; to hold converted data between updates in data distribution systems; and to perform "synchronous filtering." In synchronous filtering,[83] the source of the noise must be intermittent and predictable, such as the ignition spark of an oil-burner igniter. The same pulse that triggers the igniter can be used to put an SHA connected to the data signal into hold, protecting sensitive equipment from transient overload and slow recovery.

---

[83]D. H. Sheingold, "Transducer Interfacing Handbook," op. cit., p. 64.

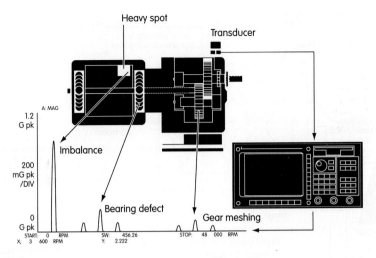

Figure 10.60
Vibration problem diagnosis using single-channel frequency-spectrum analysis.

10.13 SIGNAL AND SYSTEM ANALYZERS (SPECTRUM ANALYZERS)

In the analysis and design of many devices and systems, it is necessary to have accurate knowledge about the characteristics of the inputs to the system. Once a system has been built, often its performance is checked by studying its output. Equipment for carrying out such studies may be characterized as *signal-analysis equipment.* Closely related to this is the problem of experimentally defining the characteristics (transfer function, frequency response, etc.) of a physical system which may be too complex to analyze accurately by theory alone. Equipment for such experimental modeling[84] investigations might be called *system-analysis equipment* and generally utilizes coordinated simultaneous measurements of both the system input and output signals, together with suitable data processing to obtain conveniently the desired system characterization.

Perhaps the most widely used signal-analysis equipment is that which measures the frequency spectrum of a fluctuating physical quantity. The most common applications are in the field of sound and vibration, where the frequency spectrum of a sound pressure, stress, acceleration, etc., may be very useful in diagnosing faults in an operating machine or system. These faults can be traced to their origin by noting peaks in the frequency content at certain frequencies and then finding the machine parts that run at speeds which would produce such frequencies. Figure 10.60[85] shows a rotating machine with an attached accelerometer to measure its vibration. Frequency analysis of this vibration signal reveals peaks at frequencies that are

---

[84]E. O. Doebelin, "System Modeling and Response," op. cit.

[85]"Effective Machinery Measurements Using Dynamic Signal Analyzers," *Appl. Note* 243-1; "The Fundamentals of Signal Analysis," *Appl. Note* 243, Hewlett-Packard (Agilent Technologies).

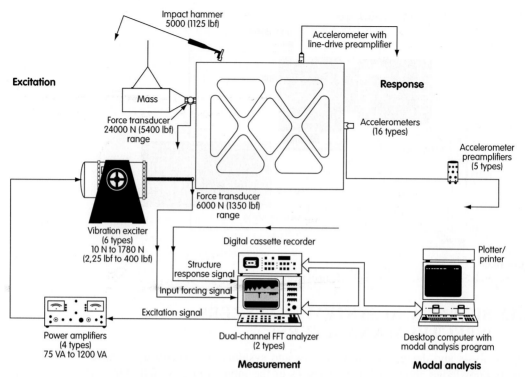

Figure 10.61
System dynamic analysis using two-channel spectrum analysis.

associated with unbalance, bearing defects, and gear meshing, allowing diagnosis of machine problems. Such studies are used both during the design phase of a new machine and also in preventive maintenance programs for machinery installed in factories or process plants. The spectrum of a new and properly operating machine can be stored in the analyzer's memory and compared with a current spectrum, raising alarm signals when certain spectrum features deviate excessively from normal. This technique is also used as a manufacturing quality-control tool to inspect products, say, electric motors, as they come off the production line. For *system* rather than *signal* analysis, we need a two-channel instrument to simultaneously measure input (say, force) and output (say, acceleration) to define a transfer function, as shown in Fig. 10.61.[86] When more than one pair of input/output relations is of interest, *multichannel analyzers* are available. The spatial variation ("mode shape") of resonant vibration at the various natural frequencies is also useful for guiding redesign efforts that redistribute material to achieve more optimum stress and deflection patterns and that alter the natural frequencies to avoid machine-excitation peaks. A number of commercially available computer-aided-design (CAD) systems

[86]"Structural Testing Using Modal Analysis," *BG* 0112-12, Bruel & Kjaer, Norcross, GA, 800-332-2040 (www.bkhome.com).

include, in their methodology, the integration of dynamic test data of this type into the overall design-build-test-redesign sequence. Thus powerful theoretical methods (such as finite elements) can be employed when their accuracy and cost are appropriate, while automated dynamic test methods provide confidence in situations where theory is not yet completely reliable. The CAD system ties the two approaches together with a user-friendly interface.

While analog and hybrid analyzers[87] designed for sinusoidal testing are still in use, most current signal and system analyzers utilize the digital fast Fourier transform (FFT) approach,[88] which accommodates test signals of all types—sinusoidal, pulse, and random. These methods can be implemented offline on any general-purpose digital computer (by using tape-recorded data and appropriate software) or in essentially real time (by using special-purpose computers marketed as signal/system analyzers and including all necessary auxiliary equipment for rapid and convenient use). Two main types of instruments are available: single-channel devices useful for spectrum analysis of single signals and multichannel units intended for system analysis that employ pairs of simultaneous input/output signals. The digital discrete Fourier transform process utilized by these instruments also can be thought of as a narrow-bandpass filtering operation, which makes the analyzers a form of digital filter. Such signal and system analysis can also be performed with specialized modules included in general-purpose data-acquisition software, such as the DASYLAB discussed in Chap. 13. (Section 13.4 gives a detailed example of an FFT application.) Several manufacturers offer self-contained (not PC-based) data-acquisition instruments/recorders[89] or digital oscilloscopes[90] that include the FFT capability which is the basis of signal and system analysis.

The characteristics of a recent machine[91] with 16-bit A/D and up to 4 channels gives some idea of the performance to be expected. Built-in amplifiers allow direct input from most measurement transducers, with AC or DC coupling, either differential or single-ended. Over sampling using a sigma-delta A/D allows simple antialiasing[92] (Fig. 10.62) with a fixed analog input filter and digital filtering in the A/D. Machines of this type usually take a fixed number of time points, say, 2048, and vary the sample rate to suit the selected frequency span. The upper frequency ranges from 102.4 kHz (1 channel) to 25.6 kHz (4 channels) with real-time performance of 25.6 kHz/channel. Twenty frequency spans (0.098 Hz to 51.2 kHz, 2 channels) can be selected, with up to 800 lines of frequency data in each span. (Zoom[93] operation allows finer frequency resolution within any selected span.) Note

---

[87]E. O. Doebelin, "System Modeling and Response," op. cit., sec. 6.1.

[88]Ibid., secs. 6.1–6.3.

[89]Model TA220-1200, Soltec, San Fernando, CA, 800-423-2344 (www.solteccorp.com).

[90]Model DL708E Digital Scope, Yokogawa, Newnan, GA, 770-253-7000 (www.yca.com).

[91]Agilent (Hewlett-Packard) 35670A, Product Overview.

[92]See Sect. 13.4. Also, T. L. Lago, "Digital Sampling According to Nyquist and Shannon," *Sound and Vibration,* Feb. 2002, pp. 20–22; R. Welaratna, "Effects of Sampling and Aliasing in the Conversion of Analog Sounds to Digital Format," *Sound and Vibration,* Dec. 2002, pp. 12–13.

[93]N. Thrane, Zoom-FFT, B&K Tech. Rev., no 2, 1980, Bruel & Kjaer.

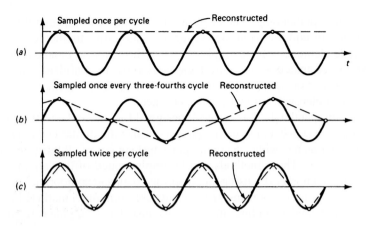

Figure 10.62
Aliasing caused by too sparse sampling.

that for 2048 time points in the FFT, there would be 1024 frequency points, but only the first 800 are plotted since the rest are too close to the aliasing limit ($N/2$) and are thus not reliable. As in any FFT calculation, if we want to study low frequencies, we must gather longer time samples; a 0- to 20-Hz span with 1024 lines would require $1024/20 = 51.2$ s of time data (again, only 800 of the 1024 lines would be plotted). Order tracking[94] for the study of machines rotating at variable speed is also available.

For system analysis, a suitable system input signal is needed, and the analyzer provides a selection[95] of time-varying voltages: random, burst random, periodic chirp, burst chirp, pink noise, fixed sine, swept sine, and arbitrary. If the system tested does not itself accept voltage as an input, the user must provide the necessary voltage-to-physical variable transducer, such as the vibration shaker of Fig. 10.61. Of course, the system output must *also* be transduced (measured) with a voltage-output instrument. The *results* of a system measurement are the amplitude ratio and phase angle curves of the system sinusoidal transfer function. The analyzer will curve fit the measured results with analytical transfer functions of your choice, using as many as 20 terms each in the numerator and denominator.

The gathering of the digital sample requires different times, depending on the selected frequency range. But once the sample is in memory, the calculation of the discrete Fourier transform by the FFT algorithm always requires the same time—a fraction of a second. Thus (at least for the lower frequency ranges), all the incoming analog data can be analyzed without "gaps," since the next sample is being gathered while the present sample is being analyzed. This is called *real-time operation,* and it is another significant advantage over earlier analog instruments, which took

[94]*Agilent Appl.,* Note 243-1.
[95]E. O. Doebelin, "System Modeling and Response," op. cit., Chap. 6.

much longer to obtain a single spectrum and so had no chance of detecting and displaying spectrum variations, as they occur, in time-variant signals. The basic result of the FFT calculation for an analog function sampled within the selected time window is its Fourier transform (magnitude and phase angle) in the form of discrete values at equally spaced frequencies in the selected range. In a dual-channel instrument analyzing the input and output signals of a "linear" physical system, the ratio of the two transforms, of course, gives the sinusoidal transfer function of the system. This is true irrespective of whether the input excitation is a swept-frequency sine wave, a single pulse, or a random signal of Gaussian or binary form, as long as the input-frequency content is adequate to "exercise" the system over the frequency range of interest. Actually, it has been found more accurate to compute the transfer function from the ratio of the cross spectrum of input and output signals and the power spectrum of the input. The cross spectrum is obtained by multiplying the output transform and the complex conjugate of the input transform, while the power spectrum is just the product of the input transform and its complex conjugate. Although this calculation gives results identical to those of the simple transform ratio when only a single record is analyzed, it gives improved accuracy when noise is present and averages of several records are used. The analyzer provides simple but versatile means for accumulating averages of various types.

In addition to the frequency-domain calculation of Fourier transform, power spectrum, cross spectrum, and transfer function, the analyzer provides the corresponding time-domain functions, such as autocorrelation function, cross-correlation function, and impulse response, all at the push of a button. Statistical functions such as probability density and cumulative distribution can be computed also. A built-in CRT display with full alphanumeric annotation displays all results. The results also can be read out for hard copy on external plotters or sent to external computers for further processing. A built-in microprocessor with simple front-panel programming allows selection of a wide variety of additional useful functions, such as time delays or advances, integration or differentiation, subtraction of background noise, correction for known transducer dynamics, recall of past setups for panel controls, etc. Addition of a special external minicomputer system gives computer-aided-design capabilities including animated-mode shape displays and finite-element analysis.

PROBLEMS

10.1 Derive the balanced-bridge relationship $R_1/R_4 = R_2/R_3$.

10.2 For a Wheatstone bridge, show that if $R_1 = R_2 = R_3 = R_4$ at balance, and if $\Delta R_2 = \Delta R_3 = 0$ and $\Delta R_1 = -\Delta R_4$, then the output voltage is a perfectly linear function of ΔR_1 no matter how large ΔR_1 gets.

10.3 Discuss qualitatively the effect on bridge operation, for both the null method and the deflection method, of the excitation voltage source having an internal resistance.

10.4 In the system of Fig. 10.4, what considerations determine the numerical value of R_s?

10.5 In a Wheatstone bridge, $R_1 = 3{,}000$, $R_4 = 4{,}000$, $R_2 = 6{,}000$, and $R_3 = 8{,}000 \ \Omega$ at balance. Find the open-circuit output voltage if $\Delta R_1 = 30$, $\Delta R_2 = -20$, $\Delta R_3 = 40$, and $\Delta R_4 = -50 \ \Omega$, and $E_{ex} = 50$ V. If the bridge output is connected to a meter of 20,000-Ω resistance, what will the output voltage now be?

10.6 Obtain $(e_o/e_i)(D)$ for the circuit of Fig. 10.23a.

10.7 Derive Eq. (10.45).

10.8 Why does a charge amplifier essentially amount to a short circuit across the crystal?

10.9 Derive the transfer functions of the circuits of (a) Fig. 10.22a, (b) Fig. 10.22b.

10.10 Derive the transfer function of the hydromechanical filter of Fig. 10.24.

10.11 In the circuit of Fig. 10.24, let e_i be supplied by a sinusoidal generator with an internal resistance of 1,000 Ω and an open-circuit voltage of 10 V peak to peak. Also let $C = 10 \ \mu\text{F}$ and $R = 10 \ \Omega$. If e_o is open-circuit, what voltage will actually appear at the e_o terminals for frequencies of 0, 100, 1,000, 10,000, and 100,000 Hz?

10.12 Derive Eqs. (10.48), (10.49), and (10.50).

10.13 Explain how a notch filter can be used in the feedback path of a high-gain feedback system to construct a bandpass filter.

10.14 Derive Eqs. (10.53) to (10.57). Show also that the phase angle at ω_p is zero.

10.15 Sketch the configuration of a hydraulic bandpass filter which has an amplitude-ratio attenuation of 40 dB/decade on either side of the passband. This is twice the attenuation rate of the system of Fig. 10.28. Only components of the type used in Fig. 10.28 are allowed. Short "transition" regions of slope $\pm$ 20 dB/decade are allowed between the flat response portion and $\pm$ 40 dB/decade portions. You must derive the transfer functions to prove your "invention" works as claimed.

10.16 Derive Eq. (10.66).

10.17 For the system of Fig. 10.31a, let $e_i = e_{\text{signal}} + e_{\text{noise}}$, where $e_{\text{signal}} = 10 \sin 20t$ and $e_{\text{noise}} = 0.1 \sin 377t$. What is the signal-to-noise ratio before and after the differentiation?

10.18 The input to a differentiator with transfer function $(e_o/e_i)(D) = D$ is a random signal with a constant mean-square spectral density of 0.001 V^2/Hz from 0 to 10,000 Hz and zero elsewhere. Calculate the rms voltage at the input and at the output of the differentiator.

10.19 Derive Eq. (10.68).

10.20 Derive Eq. (10.69).

10.21 Using the system of Eq. (10.70) and the e_i of Prob. 10.17, compute the signal-to-noise ratio at both the input and the output.

10.22 Derive Eq. (10.75).

10.23 Discuss sensitivity/response-speed trade-offs in the system of Fig. 10.33.

10.24 Derive the transfer functions of the compensating circuits of Fig. 10.35.

10.25 Design a compensating network to speed up by a factor of 10 the response of a thermocouple with a time constant of 1 s. Thermocouple resistance is 10 Ω and full-scale output is 5 mV. The amplifier/recorder available has maximum full-scale sensitivity of 0.1 mV and an input resistance of 100,000 Ω.

10.26 List and explain the action of all effects that tend to degrade the static accuracy of the system in Fig. 10.44.

10.27 Derive the equation of the op-amp summing circuit of Fig. 10.45d.

10.28 Derive the operating equation of the mechanical filter of Fig. 10.29.

10.29 Discuss the dynamic response of the system of Fig. 10.48.

10.30 The "standard" op-amp integrating circuit, as used in analog-computer applications, becomes almost impossible to employ in those measurement instrumentation applications for which the signal is a short-duration (a few milliseconds) pulse. Since a "computing loop," as found in all analog-computer applications where a differential equation is being solved, is not present, drift is not self-correcting. Also, if input signal e_i is, say, 5 V but lasts only 1 ms, then $\int e_i \, dt$ is only 0.005 V $\cdot$ s, and we need a very large value of $1/(RC)$ to get an output voltage of convenient (volt level) size. For a typical $1/(RC) = 1,000$, if e_i (which comes from some sensor and/or amplifier) differs from zero (before the pulse to be integrated occurs) by only a small amount, say 0.005 V, then e_o will "drift" at the rate of 5 V/s, causing saturation (and disabling of the integrator) in just a few seconds. An approximate integrator which solves this drift problem is shown in Fig. P10.1.

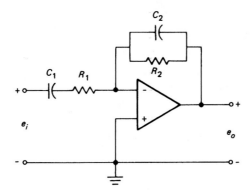

Figure P10.1

(*a*) Find $(e_o/e_i)(D)$.

(*b*) Take $R_1C_1 = R_2C_2$ and plot logarithmic frequency-response curves (dB and ϕ) for $(e_o/e_i)(i\omega)$. Superimpose on this graph the frequency response of a perfect integrator with $RC = R_1C_2$. Discuss the nature of the approximation for the approximate integrator. Why is drift no longer a serious problem?

(c) Take $R_1 = 10^6 \ \Omega$, $R_2 = 10^9 \ \Omega$, $C_1 = 10^{-6}$ F, and $C_2 = 10^{-9}$ F. Use SIMULINK (or other available digital simulation) to check performance of the approximate integrator if e_i is a rectangular pulse of height 5 V and duration 0.001 s. Would the integrator work well for short pulses of other waveforms? Why? What about "long-duration" pulses?

BIBLIOGRAPHY

1. A. Svoboda, "Computing Mechanisms and Linkages," McGraw-Hill, New York, 1948.

2. G. A. Korn, "Minicomputers for Engineers and Scientists," McGraw-Hill, New York, 1973.

3. D. M. Auslander and P. Sague, "Microprocessors for Measurement and Control," Osborne McGraw-Hill, Berkely, CA, 1981.

4. A. F. Arbel, "Analog Signal Processing and Instrumentation," Cambridge University Press, New York, 1980.

CHAPTER 11

Data Transmission and Instrument Connectivity

When the components of a measurement system are located more or less remotely from one another, it becomes necessary to transmit information among them by some sort of communication channel. Also in some cases, even though components are close together, transmission problems arise because of relative motion of one part of the system with respect to another. We examine briefly questions of this sort and some of the equipment commonly used to solve such problems.

The transmission of information is amenable to mathematical analysis totally dissociated from any hardware considerations, and there is a large body of technical literature on this subject. This science of communication has been extremely useful in putting the design of hardware on a rational basis, showing the trade-offs in competitive systems, and putting theoretical limits on what can be done. Its consideration, however, is beyond the scope of this text, and we restrict ourselves to rather qualitative, hardware-oriented discussions.

11.1 CABLE TRANSMISSION OF ANALOG VOLTAGE AND CURRENT SIGNALS

Perhaps the most common situation is that in which a simple cable is used to transmit an analog voltage signal from one location to another. The accurate analysis of a cable or transmission line involves the use of a distributed-parameter (partial differential equation) approach since the properties of resistance, inductance, and capacitance are not lumped or localized.[1] Figure 11.1a shows the model generally employed for such an analysis. An approximation suitable for low frequencies is the lumped network of Fig, 11.1b. The shunt conductance has been neglected here since, in practice, generally it is negligible. If the line is not too long and the

---

[1] H. H. Skilling, "Electric Transmission Lines," McGraw-Hill, New York, 1951.

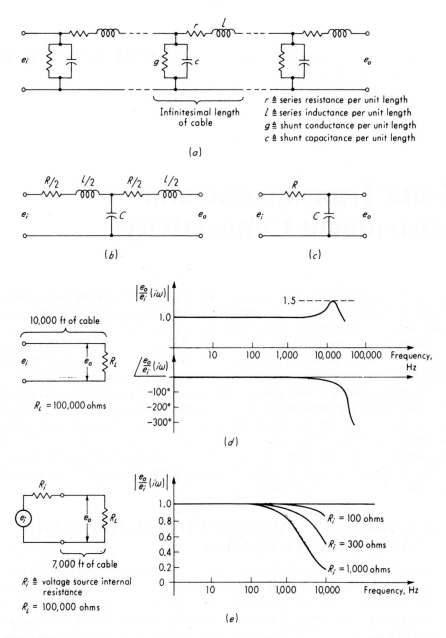

$r \triangleq$ series resistance per unit length
$l \triangleq$ series inductance per unit length
$g \triangleq$ shunt conductance per unit length
$c \triangleq$ shunt capacitance per unit length

Infinitesimal length
of cable

(a)

(b)   (c)

10,000 ft of cable

$R_L = 100,000$ ohms

(d)

$R_i \triangleq$ voltage source internal
resistance

$R_L = 100,000$ ohms

7,000 ft of cable

(e)

Figure 11.1
Cable models and response.

frequencies are not too high, the crude model of Fig. 11.1c may be adequate. The inductance has been totally neglected; R represents the total resistance of both conductors in the cable, while C represents the total capacitance between them.

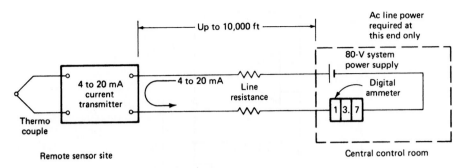

Figure 11.2
Current loop of 4 to 20 mA.

These can be numerically calculated from the values per foot of length given by cable manufacturers. Typically, resistance per foot might be of the order of 0.01 Ω while capacitance would be about 30 pF/ft. Thus a 1,000-ft length of two-conductor cable has $R = 20\ \Omega$ and $C = 30,000$ pF. The actual frequency response of a length of cable can be measured experimentally to get its exact characteristics. Figure 11.1d and e shows some typical results.[2] The use of cables up to 7,000 ft long to transmit low-level (± 10 mV full scale) data has been accomplished.[3] However, even short cables (less than 10 ft) can cause difficulties in high-impedance transducers such as the piezoelectric type. The use of charge amplifiers rather than voltage amplifiers may be helpful in such cases.

A hard-wired (cable) data-transmission technique widely utilized in the process industries is the so-called current loop (usually a 4- to 20-mA range). Transmitters are available which convert millivolt, thermocouple, RTD, frequency, slidewire potentiometer, or bridge-circuit inputs into a proportional output current. A zero input signal produces 4-mA current while full-scale input produces 20-mA. Such transmitters are available in several forms. The two-wire version of Fig. 11.2 is particularly convenient since the connection between the central control room and the remote sensor requires only two wires to transmit both power and signal. The voltage appearing across the transmitter output terminals (which will vary as output current changes) is actually the transmitter's power supply, but the transmitter operation is insensitive to changes in this voltage as long as it stays above some minimum, say 9 V. Thus for a system supply voltage of, say, 80 V, line resistance can be as high as $(80 - 9)/0.02 = 3,550\ \Omega$. The transmitter is a true current source, which makes the system relatively immune to induced noise voltages and line resistance changes (within the 3,550-Ω limit). While an actual transmitter is somewhat more complicated, the op-amp circuit of Fig. 11.3 shows in principle how such current-source behavior may be achieved.

---

[2]R. L. Smith, "Transmission of Low-Level Voltage Over Long Telephone Cables," *NASA, Tech. Note* D-1320, p. 14, January 1963.

[3]Ibid.

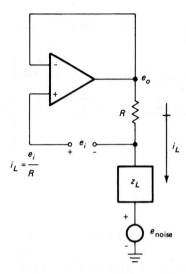

Figure 11.3
Basic op-amp current source.

Circuit details for the actual devices are available in the literature.[4] While these 4- to 20-mA devices might be considered "old-fashioned" in today's digital world, they still account for the majority of process variable transmitters used by the industry.[5] Part of their success is attributable to their compatibility with the HART (*highway-addressable remote transducer*) field-communications protocol[6] originally developed by Rosemount Inc., but now open to use by anyone. This is a widely used digital signal transmission scheme that superimposes a digital signal "on top of" the 4- to 20-mA analog signal and uses the same wires, with analog and digital signals separately available at all times. HART-compatible 4- to 20-mA transmitters (which include sensors for many process variables) are available from many manufacturers. The digital signal, which allows for *two-way* communication between transmitters, computers, and hand-held communicators (see Fig. 11.4), is used for many purposes, for example, adjusting transducer zero and span, PID control mode tuning, control system set point adjustment, and transducer identification. HART devices use *frequency-shift keying* (FSK) as the digital-signal transmission method. Here sinusoidal signals of two different frequencies represent digital 1's and 0's (1 cycle of 1200 Hz is a 1 and approximately 2 cycles of 2200 Hz is a 0). The presence of the digital signals does not interfere with the analog signal because the digital signals have an average value of zero, are of a low level relative to the analog signal, and are usually filtered by a 10-Hz low-pass filter that reduces the ripple to 0.01 percent of the full-scale analog signal. The HART system is *not* a

---

[4]E. O. Doebelin, "Measurement Systems," 4th ed., McGraw-Hill, New York, 1990, pp. 847, 848.

[5]D. Harrold, "4-20 mA Transmitter Alive and Kicking," *Control Engineering,* Oct. 1998, pp. 109–114; "Analog Transmitters Lead the Pack," *Control Engineering,* Jan. 1999, pp. 73–79.

[6]"HART Technical Overview," HART Communication Foundation, Austin, TX, 512-794-0369 (www.hartcomm.org).

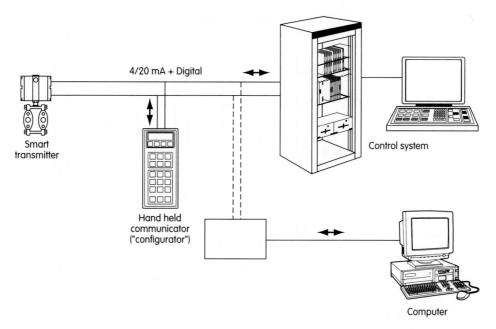

Figure 11.4
HART digital communication system.

high-speed digital channel; the maximum bit rate is 1200 bits/s, but this is adequate for many applications as attested by over 5 million installations. Digital-to-analog converters with 4- to 20-mA current output are available.

11.2 CABLE TRANSMISSION OF DIGITAL DATA

When data must be transmitted very long distances (100 mi is not unusual), analog signals tend to be corrupted by the response characteristics of the transmission line and the pickup of spurious noise voltages from a number of sources. Under such conditions, it may be desirable to convert the analog data to some digital form, transmit them in digital form, and then reconvert to analog form if desired. A time-honored example of digital transmission is the telegraph system, in which letters and numbers are represented by a system of coded pulses (dots and dashes).

In many cases the information can be transmitted over lines that were installed for other purposes. Electric power systems, for example, transmit information signals over their power transmission lines simultaneously with the transmission of 60-Hz power. It is simply necessary to keep the frequencies utilized sufficiently separated to allow easy filtering for elimination of unwanted signals. Telephone lines are also in wide use for data transmission, a familiar example being the modem-based systems used to connect personal computers to the Internet, with data rates on the order of 56 kBd (1 baud (Bd) = 1 bit/s) on "ordinary" phone lines.

Digital communications are, of course, also necessary over shorter distances. The 4- to 20-mA current transmitters discussed in Sec. 11.1 can be designed to

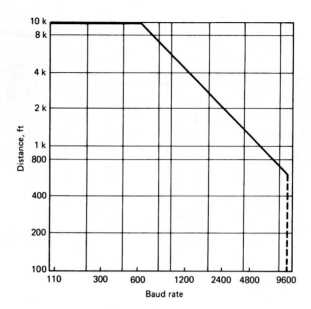

Figure 11.5
Distance/speed trade-off for current loop of 4 to 20 mA.

provide digital transmission at up to 9600 Bd and distances up to 10,000 ft (see Fig. 11.5[7]). The standard ASCII RS-232 transmission is limited to about 50 ft at rates up to 9600 Bd. Higher speed interfaces, such as IEEE-488 ("GBIB")[8] support rates up to about 2 Mbits/s, connecting up to 15 devices, all of which should be within about 3 m of the host computer. All high-speed interfaces require careful matching of the receiving instrument's input impedance to the characteristic imped-ance of the connecting cable[9] so as to avoid confusing reflections.

The high accuracy of digital data transmission as compared with analog is due to the fact that the size or precise shape of a pulse in a digital system is not particu-larly important. Rather, the system operates on the presence or absence of some sort of pulse. Thus even rather severe degradation of pulse shape by the transmission medium does not affect the accuracy of a digital system *at all* as long as the pres-ence or absence of a pulse can be detected.

11.3 FIBER-OPTIC DATA TRANSMISSION

The use of optical (rather than electrical) means of data transmission is of increas-ing importance. Methods of transmitting both analog and digital information are

[7]Publ. C603b-15-1/81, Analog Devices.

[8]R. B. Northrop, "Introduction to Instrumentation and Measurements," CRC Press, Boca Raton, FL, 1997, p. 429.

[9]Ibid., pp. 438–444.

available.[10] While communication over either an air path or a glass fiber is possible, we emphasize the fiber approach.[11] Basically, an electrically controllable light source (often an LED), an optical fiber, and a photodetector (often a silicon photodiode) are needed to make a complete system. Fiber optics offers wide bandwidths and high data rates while avoiding many of the interference and security constraints associated with traditional electronic communications. Glass cable can carry signals to or from instrumentation at high common-mode voltage levels without creating electrical paths between devices, that is, it provides high isolation. Optical fibers also are immune to electromagnetic interference from sources such as lightning or power switching, and provide good noise rejection. Glass cables produce no sparks; one application uses a fiber running inside a natural-gas pipeline. Data security is enhanced since tapping into fiber-optic lines is extremely difficult without causing fiber failure. Theoretically, fiber-optic cables can support data bandwidths up to the terahertz range. However, limitations of the terminal photonic devices (LEDs and photodiodes) presently limit use to low gigahertz range.[12]

11.4 RADIO TELEMETRY

The word *telemetry* means simply measurement at a distance and includes all forms of such systems, irrespective of the methods of transmission or physical nature of the hardware. When interconnecting wires are not possible or desirable, data may be transmitted by radio, using various detailed schemes.[13]

Radio telemetry probably received its greatest impetus from the requirements of aircraft and missile flight testing during and after World War II. Considerable standardization based on the requirements of such systems has been accomplished, and our discussion here reflects this emphasis. Figure 11.6 shows the FM/FM system of radio telemetry. The symbol "FM/FM" refers to the fact that two frequency-modulation processes are employed. In the first process, time-varying dc voltages are converted to proportional frequencies, by using voltage-to-frequency converters as described in Sec. 10.11. When employed in FM/FM telemetry systems, these converters generally are called *subcarrier oscillators*. Instead of being voltage-to-frequency converters, subcarrier oscillators may be inductance-controlled. The inductance of a variable-inductance transducer forms part of the oscillator circuit,

---

[10]"I. Math, Basic Optical Links," *Electro-Opt. Syst. Des.,* pp. 48–51, Sept. 1977.

[11]W. F. Trover, "Fiber Optic Communications for Data Acquisition and Control," *In Tech,* pp. 33–41, April 1981; B. G. Mahrenholz and R. R. Little, Jr., "A Fiber Optic Link for High-Speed Data Acquisition," *In Tech,* pp. 43–46, April 1981; "High Sensitivity–Low Speed Fiber Optic Transmitter and Receiver," PDS-408A, PDS-425, and AN-97, Burr-Brown Res. Corp., Tucson, AZ.

[12]R. B. Northrop, op. cit., pp. 444–454.

[13]C. M. Harris and C. E. Crede (eds.), "Shock and Vibration Handbook" vol. 1, pp. 19–76, McGraw-Hill, New York, 1961; M. H. Nichols and L. L. Rauch, "Radio Telemetry," Wiley, New York, 1956; F. Carden, "Telemetry System Design," Artech House, London, 1995; O. J. Strock and S. M. Rueger, "Telemetry System Architecture," *ISA,* Research Triangle Park, NC, 1995; P. L. Walter, "Shock and Vibration Measurements via Space Telemetry," *Sound & Vib.,* Sept. 1998, pp. 18–23.

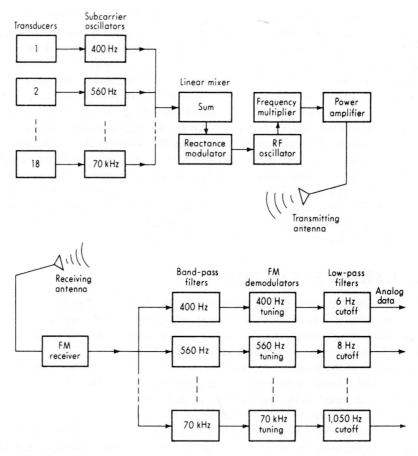

Figure 11.6
FM/FM radio telemetry.

and changes in inductance cause proportional changes in frequency from the center frequency.

The FM/FM system of Fig. 11.6 has 18 available channels; thus 18 different physical variables may be measured and transmitted simultaneously. Note that the center frequencies range from 400 to 70,000 Hz. Such low frequencies cannot be practically transmitted by radio propagation since they would require antennas of immense size, because the size of an antenna must be of the order of the wavelength to be transmitted [wavelength in meters = 3×10^8/(frequency in hertz)]. Thus an additional frequency modulation to boost all frequencies into the radio-frequency (RF) range is employed. Rather than utilize a separate radio-frequency transmitter for each of the 18 channels (which is wasteful of the crowded RF spectrum and requires much more equipment), the 18 channels are "mixed" (added) and sent out

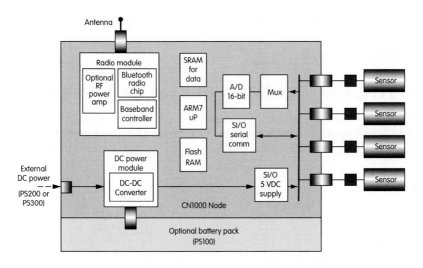

CN1000 architecture

Figure 11.10
BLUETOOTH wireless data transmission system.

occurs. To alleviate multipath problems, systems use *correlators* to select a solid signal path, *antenna diversity* (more than one antenna), and packet-radio methods to constantly resynchronize the radio.

Data-acquisition instruments using spread-spectrum wireless technology have appeared recently. The Fluke Wireless Data Logger[20] accepts analog input signals from up to 21 sensors and transmits the data to a wireless modem installed on a PC, with an indoor range of 800 ft and a line-of-sight range of $\frac{1}{4}$ mi, at data rates up to 38.4 kBd. Inexpensive wireless-data transmission is promised by two competing technologies: *Bluetooth* and *IEEE802.11b*. At the time of this writing (2003), neither system had a large installed base, but one or both are likely to be important before long. One sensor manufacturer[21] offers Bluetooth hardware and software using spread-spectrum frequency-hopping methods in the 2.4-GHz band, with a range of about 10 m. A 4-channel unit ($1 \times 4 \times 4$ in, Fig. 11.10)[21] has 16-bit resolution and provides 100 samples/s for each channel. Higher power versions transmit up to 100 m, and future models will support sampling rates above 5 kHz.

11.5 PNEUMATIC TRANSMISSION

Transmission of pressure signals in industrial pneumatic control systems is accomplished regularly over distances of several hundred feet. Pneumatic-transmission-line dynamics is analogous to that of electric cables but, of course, at a much lower

---

[20]Model 2625A/WL, Fluke Corp. (www.fluke.com).

[21]Crossbow Technology Inc., San Jose, CA (www.xbow.com); M. Dunbar, "Where Wireless Sensor Communications and Internet Meet," *Sensors,* Sept. 2000.

frequency. Adequate simplified models employing a dead time equal to the acoustic transmission time and either a first-order or second-order system are available and are discussed in Chap. 6.

11.6 SYNCHRO POSITION REPEATER SYSTEMS

Figure 11.11 illustrates synchro position repeater systems used for transmitting low-power mechanical motion over considerable distances with only a three-wire interconnecting cable. Whenever the two angles θ_i and θ_o are not identical, an

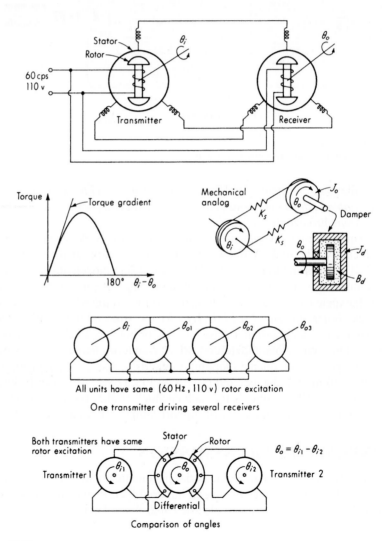

Figure 11.11
Torque-synchro angle transmission.

electromagnetic torque is exerted on the rotor of *each* machine, which tends to bring the shafts into alignment. Thus, if the transmitter shaft θ_i is turned, the receiver shaft θ_o will follow accurately as long as there is no appreciable torque load on the θ_o shaft. The accuracy of such systems depends on the torque gradient (torque per unit error angle) of the transmitter/receiver system. A typical value might be 0.35 in · oz/degree. The electrical system serves only to transmit power from the θ_i to the θ_o shaft; all the mechanical work taken out at θ_o must be provided as mechanical work at θ_i. The torque gradient is reduced as the resistance of the connecting cable is increased; thus long cables result in reduced accuracy. In a typical unit, 10-Ω resistance in each of the three wires results in a 50 percent loss of torque. The dynamic response of these systems is essentially second-order; a mechanical analog is shown in Fig. 11.11. Sometimes a damper is put on the θ_o shaft to reduce oscillations since little inherent damping is present. When one transmitter drives several receivers (all units of identical size), the torque available at each receiver is $2/(N + 1)$ times the torque for a single pair, where N is the number of receivers. The synchro differential shown in Fig. 11.12 is useful for comparing two rotations at a location remote from either. Its static and dynamic behavior is essentially the same as for a transmitter/receiver pair. When these open-loop synchro-systems cannot meet accuracy requirements, the instrument servos (feedback systems) of Fig. 10.42 may provide a solution.

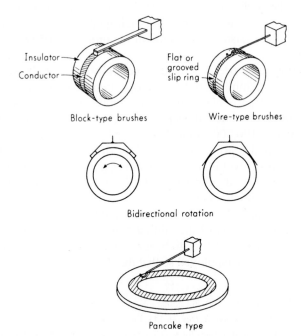

Figure 11.12
Slip-ring configurations.

11.7 SLIP RINGS AND ROTARY TRANSFORMERS

When transducers must be mounted on the rotating members of machines, some means must be provided to bring excitation power into the transducer and to take away the output signal. Some transducers (such as synchros) themselves are rotating "machines" in which such data and/or power transmission between a rotating and a stationary member is necessary. When only a small relative motion is involved, continuous flexible conductors (often in the form of light coil springs) can be employed. In some cases of limited rotation through a few revolutions, the connecting wires simply can be allowed to wind or unwind on the rotating shaft. Commercially available "twist capsules"[22] can provide up to several hundred circuits and up to ±320° of rotation. This manufacturer also is a source of slip rings, fiber-optic rotating joints, and rotating hydraulic and pneumatic joints. However, continuous high-speed rotation requires slip rings, radio telemetry, or some form of magnetic coupling between rotating and stationary parts.

Figure 11.12 shows the common forms of slip rings.[23] Rings are made of coin gold, silver, or other noble metals and alloys. Block-type brushes often are sintered silver graphite while wire-type brushes are alloys of platinum, gold, etc. An important consideration in slip rings used to transmit low-level instrumentation signals is the electrical noise produced at the sliding contact. One component of this noise is due to thermocouple action if the brush and ring are of different materials. The other main effect is a random variation of contact resistance from surface roughness, vibration, etc. If the contact carries current, a variation in contact resistance causes a noise voltage to appear at the contact. A high-quality miniature sliding slip ring may exhibit a contact-resistance variation of the order of 0.05 Ω peak to peak and 0.005 Ω rms.[24]

While slip rings have been operated successfully at about 100,000 r/min, applications above 10,000 r/min generally require extreme care because of heating and vibration problems. A particular slip-ring assembly[25] usable to 100,000 r/min and intended for strain-gage work had peak-to-peak noise voltage of 0.02 mV at 52,000 r/min and 0.40 mV at 100,000 r/min. This assembly used liquid cooling and lubrication of slip rings and bearings and gave a brush life of 30 h at 35,000 r/min. At 52,000 r/min the noise level in a typical strain-gage circuit gave a signal-to-noise ratio of about 150:1. Hard gold rings of $\frac{1}{4}$-in diameter were utilized with two cantilevered wire-tuft brushes per ring.

When slip rings are used with strain-gage circuits, particular care must be taken since the resistance variation of the sliding contact may be comparable with the small strain-gage resistance change to be measured. If possible, a full bridge on the rotating member should be employed so that the sliding contacts can be taken out of

---

[22]Electro-Tec Corp., Blacksburg, VA, 800-382-5366.

[23]A. J. Ferretti, "Slip Rings," *Electromech. Des.*, p. 145, July 1964.

[24]E. J. Devine, "Rolling Element Slip Rings for Vacuum Application," *NASA, Tech. Note* D-226l, p. 11, 1964.

[25]A. J. Ferretti, op. cit., p. 159.

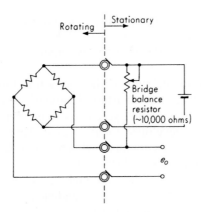

Figure 11.13
Bridge-circuit slip-ring configuration.

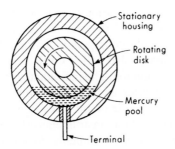

Figure 11.14
Mercury-pool slip ring.

the bridge circuit. This arrangement (Fig. 11.13) greatly reduces the effects of slip-ring resistance variations. For the most demanding applications, more complex schemes are available[26] to reduce noise to even lower levels.

A rotating disk dipping into a mercury pool (see Fig. 11.14) can perform the same function as a conventional slip ring. A commercially available device[27] is usable from 0 to 10,000 r/min; has a contact resistance of 0.005 Ω, contact-resistance variation of $\pm$ 0.00025 Ω for 0 to 600 rpm and no measurable resistance variation from 600 to 10,000 rpm; is compensated for self-generated thermoelectric voltages; and can be made with 2 to 160 terminals.

Ordinary sliding slip rings may not operate properly in the high-vacuum environment of space. Preliminary research[28] using a thrust-type ball bearing as the

---

[26]C. C. Perry and H. R. Lissner, "The Strain Gage Primer," p. 186, McGraw-Hill, New York, 1955; P. K. Stein, "Measurement Engineering" vol. 2, chap. 29, Stein Engineering Services, Inc., Phoenix, AZ, 1964.

[27]Meridian Laboratory, Middleton, WI, 800-837-6010 (www.meridianlab.com).

[28]E. J. Devine, op. cit.

signal-transfer mechanism indicates that rolling contact slip rings may provide a solution for such problems. A particular test at 2,000 r/min and vacuum of 2×10^{-9} torr gave operation for over 100 million revolutions at a resistance variation of $0.002 \ \Omega$ rms.

Another alternative to slip rings is the rotary transformer. With this device, signal and/or power voltages are transferred through an annular air gap between concentrically rotatable primary and secondary transformer coils. Figure 11.15[29] shows two data systems (developed for jet-engine testing) which employ rotary transformers. One system is used with slowly changing thermocouple signals and utilizes an 8-bit A/D converter to digitize the temperature data, which are then transmitted serially (one bit at a time) through the rotary transformer. The other system handles dynamic strain-gage data of bandwidth 150 to 3,000 Hz, and since no dc data are present, this information can be directly coupled to the transformer.

11.8 INSTRUMENT CONNECTIVITY

In addition to connecting *sensors* to various signal processing devices, we also often want to interconnect "stand-alone" instruments (digital multimeters, frequency counters, etc.) to each other and to computers. Various "buses"[30] are available for such purposes, the IEEE-488.2, also called GPIB for *general-purpose interface bus,*[31] being one of the most common. The individual instruments must of course be GBIB compatible, and this is often an available option on many standard units. A bus is usually considered to have a set of *unbroken* signal lines that carry digital data in parallel (rather than serial) fashion; instruments are connected by "tapping into" the bus at various locations. The instruments do not interfere with each other because signal transmission is *two* way. A computer or other controller coordinates the system, managing "who is talking and who is listening" at any given time.

GPIB data transfer rates range from 1.5 Mbytes/s for 15-m cables to 8 Mbytes/s for 2-m cables. Shielded 24-conductor cables with stackable, that is, hermaphrodite (both male and female), connectors are used, and up to 15 devices may be connected in linear or star configurations. Hardware and software for various Windows-based and Macintosh computers is available, so "intranets" within a facility and internet connections to remote facilities can be set up (Fig. 11.16[32]). The Ethernet networking interface, widely used for high-speed business and information technology data, can also be applied to measurement and control applications (Fig. 11.17[33]).

---

[29]D. J. Lesco, J. C. Sturman, and W. C. Nieberding, "On-the-Shaft Data Systems for Rotating Engine Components," *NASA* TM X-68112, 1972.

[30]R. B. Northrop, op. cit., pp. 428–438.

[31]"GPIB Instrument Control," National Instruments Catalog 2001, www.ni.com, pp. 718–817.

[32]Ibid., p. 767.

[33]Ibid., p. 860.

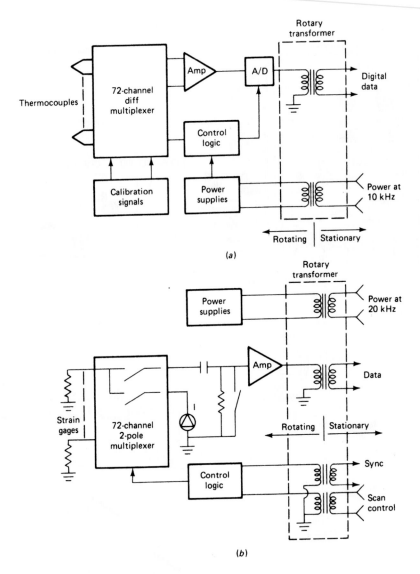

Figure 11.15
Rotary-transformer applications.

For simpler and lower-speed applications, one of the available *serial* (rather than parallel) data communications links[34] using only a few conductors may be appropriate, especially for very long distances. The original version, RS232C, was developed in the 1960s about 10 years before GPIB; later improved versions

[34]R. B. Northrop, op. cit., pp. 431–435.

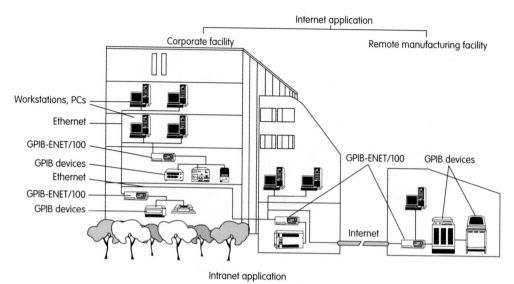

Figure 11.16
GPIB data transmission system.

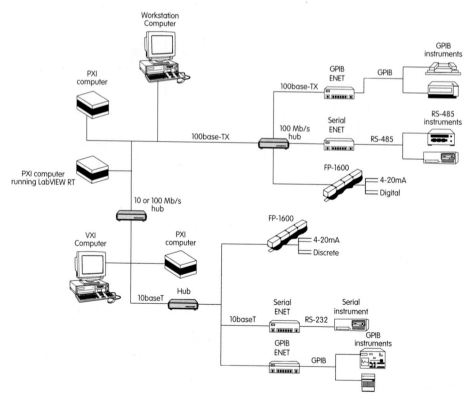

Figure 11.17
ETHERNET networking interface for data transmission.

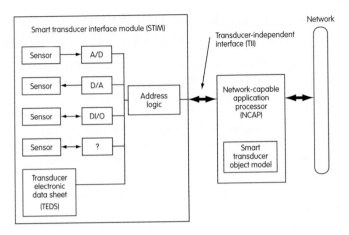

Figure 11.18
IEEE-1451.2 system for connecting smart transducers and computers.

include RS-423, RS-449, and RS-485. RS-232C is limited to about 20-kBd data rates for a 15-m cable. The later improved versions can handle 10 MBd on a 12-meter cable, reducing progressively to 80 kBd for 1200 m.

Another standard that addresses the problems of connecting smart transducers to computers is IEEE-1451.2.[35] Figure 11.18 shows this scheme made up of a network, a *network-capable application processor* (NCAP), a *transducer-independent interface* (TII), and a *smart transducer interface module* (STIM). The transducers themselves are considered to be part of the STIM; in fact, to provide the critical *self-identification* features, the transducer *must* be inseparable from the STIM electronics during normal use.[36] This allows the computer to identify each transducer through its *transducer electronic data sheet* (TEDS) by sending questions and receiving answers about things such as physical location, transducer characteristics, date of last calibration, etc. The computer can also *command* adjustments to the transducer, such as range and zero settings.[37] While this specific standard is based on sound technology, it faces competition from other standards, such as Foundation Fieldbus and Profibus.[38]

---

[35]R. N. Johnson, "IEEE-1451.2 Update," *Sensors,* Jan. 2000, pp. 17–27.

[36]B. A. O'Mara, "Designing an IEEE-1451.2-Compliant Transducer," *Sensors,* Aug. 2000, pp. 46–51.

[37]J. Moore, "Using the IEEE-1451.2 Correction Engine to Compensate a Multivariable Smart Pressure Transmitter," *Sensors,* Aug. 1999, pp. 47–51; L. E. Eccles, "IEEE-1451.2 Engineering Units Conversion Algorithm," *Sensors,* May 1999, pp. 107–112; G. Murphy, "Smart Transmitters Get Help from On-Board Microcontrollers and IEEE-1451.2," *Control Solutions,* Sept. 2000, pp. 71–77. J. Pearson, "Fast-Tracking Plug and Play," *Sensors,* Dec. 2002, p. 6; D. Potter, "Plug-and-Play Sensors," *Sensors,* Dec. 2002, pp. 14–22.

[38]N. Sheble, "IEEE-1451 Standard Loses Ground," *In Tech,* Aug. 2000, p. 168; W. Boyes, "The Fieldbus Wars Ended in a Draw," *Flood Control,* Feb. 2003, pp. 34–39.

11.9 DATA STORAGE WITH DELAYED PLAYBACK (AN ALTERNATIVE TO DATA TRANSMISSION)

When real-time data transmission is not required, difficult transmission problems can sometimes be avoided by using this technique. Obtaining a temperature/time history for the contents of a can of food product as it passes through the processing line is a good example. Miniature (1 × 2 in) data systems are available[39] that include a battery, sensor, signal conditioning, sample-timing clock, A/D converter, and 1,000-point digital memory, all hermetically sealed. Sealed in the food can, the system records the desired data as the can proceeds through the processing steps. Then the can is opened, and the data are read out for analysis. Units to measure temperature, pressure, or humidity are available, with adjustable sample rates from 1 Hz to once a day.

Similar loggers[40] are widely used to study the shock/vibration environment of various transport modes (truck, ship, aircraft, rail) by fastening the unit into a typical carton, which is then shipped in the usual manner. Triaxial accelerometers record each axis of motion, with on-board computers calculating total vector acceleration, package drop height, etc. Units as small as a hen's egg are available. If desired, a GPS (Global Positioning System) interface can be included that records the geographical location and time of occurrence of each measured event so a detailed history of the entire transportation route can be documented. After the trip, the data is uploaded to a PC, where specialized software provides complete analysis and graphing. Some units[41] provide a wireless-communication link using existing cell phone service.

PROBLEMS

11.1 Find a general expression for $(e_o/e_i)(D)$ for the system of Fig. 11.1b. If $R = 20 \ \Omega$, $C = 0.3 \ \mu\text{F}$, and $L = 0.2$ mH, plot the logarithmic frequency-response curves.

11.2 A synchro repeater system has one transmitter and five receivers. The torque gradient of a single pair of devices with very short cable connections is 0.5 in · oz/degree, and 10 percent of this is lost for each ohm of cable resistance. Each receiver drives a dial with 0.05 in · oz of friction. If the allowable error is 0.5° and cable resistance is 0.05 Ω/ft., find the maximum allowable cable length.

11.3 Explain the operation of the system of Fig. 11.3.

---

[39]Mesa Laboratories, Datatrace Div., Lakewood, CO, 800-525-1215.

[40]Dallas Instruments, Dallas, TX, 800-527-7071 (www.dallasinstruments.com).

[41]IST, Okemos, MI, 517-349-8487 (www.isthq.com).

BIBLIOGRAPHY

1. R. H. Cerni and L. E. Foster, "Instrumentation for Engineering Measurement," chap. 5, Wiley, N.Y., 1961.

2. R. J. Barber, "21 Ways to Pick Data Off Moving Objects," *Contr. Eng.,* p. 82, Oct. 1963; p. 61, Jan. 1964.

3. E. H. de Grey and J. G. Bayly, "Measuring through Vessel Walls," *ISA J.,* p. 82, May 1963.

4. M. H. Nichols and L. I. Rauch, "Radio Telemetry," Wiley, New York, 1956.

5. F. Carden, "Telemetry System Design," Artech House, London, 1995.

6. O. J. Strock and S. M. Rueger, "Telemetry System Architecture," *ISA,* Research Triangle Park, NC, 1995.

7. R. B. Northrop, "Introduction to Instrumentation and Measurements," CRC Press, Boca Raton, FL, 1997.

8. C. F. Coombs, Jr., "Electronic Instrument Handbook," 2nd ed., McGraw-Hill, New York, 1995.

12 CHAPTER

Voltage-Indicating and -Recording Devices

The majority of signals in measurement systems ultimately appear as voltages. Since voltage cannot be seen, it must be transduced to a form intelligible to a human observer. The form in which the data are presented is generally that of a pointer moving over a scale, a pen writing on a chart (including thermal writing on heat-sensitive paper and electron beams writing on cathode-ray tubes), visual presentation of a set of ordered digits, or printout of digital data by ink-jet or laser printers, or other technology. We consider the most common types of such indicating and/or recording devices.

12.1 STANDARDS AND CALIBRATION

Figure 12.1[1] gives information on the primary standards for voltage, resistance, capacitance, and inductance. For most routine purposes in engineering laboratories, the secondary standards (voltage-calibration sources, precision resistor decades, etc.) available from many manufacturers are adequate. Multifunction calibrators are available[2] that can perform most of the needed electrical calibrations.

12.2 ANALOG VOLTMETERS AND POTENTIOMETERS

While digital voltmeters are very popular, analog meters are still the preferred choice for certain applications.[3] The most widely employed meter movement for dc and (with rectifiers) ac measurement in electronics and instrumentation work is the

---

[1]"Accuracy in Measurements and Calibration," *NBS, Tech. Note* 262, 1965.

[2]Fluke Model 9100, Fluke Corp., Everett, WA (www.fluke.com).

[3]J. Harte, Jr., "Analog Panel Meters—Alive and Well!" *Inst. & Cont. Syst.*, pp. 19–23, July 1979; J. Hayes, "Digital or Moving-Coil Meters?" *Mach. Des.*, pp. 113–115, Sept. 5, 1974.

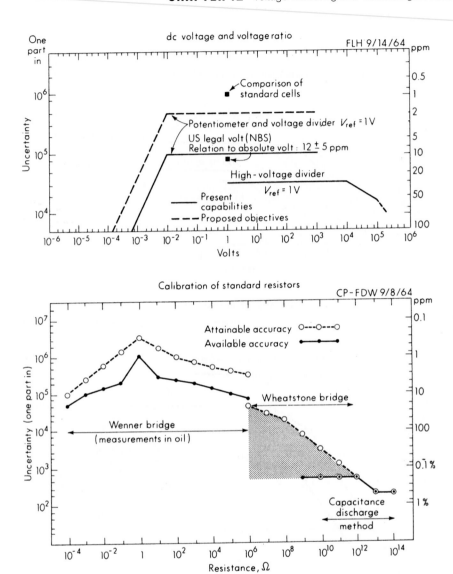

Figure 12.1
Electrical standards.

classical D'Arsonval movement (see Fig. 12.2). This basically current-sensitive device is used to measure voltage by maintaining circuit resistance constant by means of compensating techniques (see Fig. 2.17a). Relatively recent improvements on this basic configuration include taut-band suspension (rather than pivot-and-jewel bearings), individually calibrated scale divisions, and expanded-scale instruments. Taut-band suspension completely eliminates bearing friction, reduces inertia and temperature effects, increases ruggedness, and results in less loading on

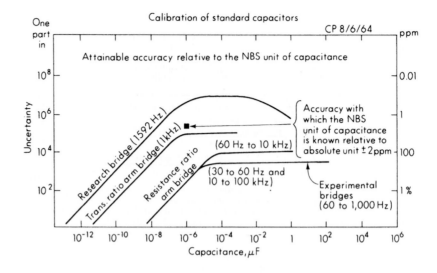

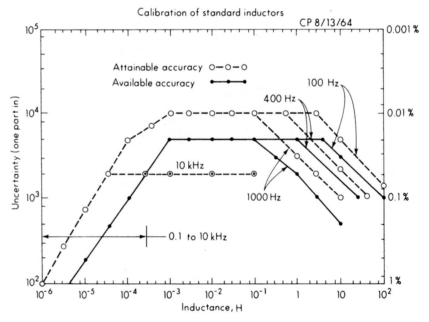

Figure 12.1
(Concluded)

the measured circuit since the reduced friction requires less power drain. The increased accuracy made possible by taut-band construction can be provided at reasonable cost through the use of automatic calibration systems, which print an individual scale for each and every instrument. Expanded-scale instruments use a precision voltage-suppression circuit to measure a small variation around a larger

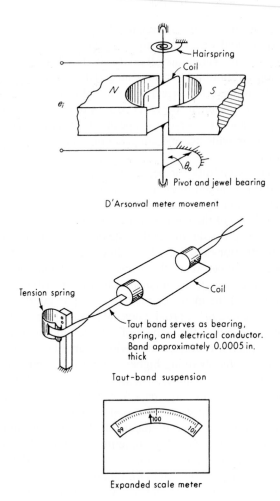

D'Arsonval meter movement

Taut-band suspension

Expanded scale meter

Figure 12.2
DC analog meters.

voltage. Thus if you need to measure a 100-V signal accurately, it is possible to do this with a meter whose scale goes from 99 to 101 V. Static inaccuracies of 0.1 percent are attainable in a rugged and portable instrument with these methods.

Transistorized voltmeters (see Fig. 3.28) still utilize the D'Arsonval meter movement, but precede it by amplifier circuits. These increase the input impedance and overall sensitivity. Such instruments generally accept a wide range of dc and ac input voltages and have static error of the order of 1 to 3 percent of full scale.

When ac (not necessarily sinusoidal) voltages are to be measured with a D'Arsonval movement, it is necessary to perform rectification. Depending on the circuitry used, a meter may be sensitive to the average, peak, or rms value of the input waveform. It is common practice to calibrate the scale of the meter to read the rms value no matter what quantity is fundamentally sensed. This procedure is accurate only if pure sinusoidal waveforms are being measured since, in this case only, the peak, average, and rms values are all related by fixed constants and thus can be included in the scale calibration. For nonsinusoidal waveforms, peak- or average-sensing meters will not read the correct rms value. In some cases, peak or average value is actually what is wanted; however, rms is desired most often. A true rms voltmeter is complex and expensive; thus peak- and average-sensing meters calibrated to read rms are in wide use and are generally satisfactory, except in the most critical applications.

Figure 12.3 shows circuits for peak, average, and rms meters. In the peak circuit, the capacitor is charged to the peak value of a periodic input voltage. This charge cannot leak off rapidly because of the one-way conduction of the diodes and the high input impedance of the voltmeter (transistorized voltmeters often use peak sensing). The voltage across the meter thus stays near the peak value of the input with only slight fluctuations resulting from diode reverse leakage and meter non-infinite impedance. The meter reads the *largest* peak, whether positive or negative. In the average-reading circuit, the input is full-wave-rectified, and the low-pass filtering characteristic of the meter movement is employed to extract the average value. The rms-reading circuit[4] approximates the required square-law parabola with a few straight-line segments in the fashion of a diode function generator. The average voltage on the capacitor is utilized to provide a variable bias on the diodes in the function generator, thereby obtaining higher accuracy than possible in a fixed-bias unit using the same number of diodes. The averaging required in obtaining an rms value is performed by the meter's low-pass filtering characteristic while the square-root operation is obtained simply by meter-scale distortion.

When highly accurate rms measurements of nonsinusoidal signals are required (random signals are a good example), sometimes methods based on the heating power of the waveform are employed since heating power is directly proportional to the mean-squared voltage. Voltmeters for random signals must be able to handle peaks that are large compared with the rms value. This is specified by the peak factor of the meter. Large peak factors (ratio of peak to rms value) are desirable; values up to about 10 are available.

When the most accurate measurements of dc voltage are needed, potentiometers rather than deflection meters are employed. The potentiometer is a null-balance instrument in which the unknown voltage is compared with an accurate reference voltage, which can be adjusted until the two are equal. Since, at the null point, no current flows, errors due to *IR* drops in lead wires are eliminated. Such *IR* drops are always present when a D'Arsonval-type meter is used to measure voltage directly. Figure 12.4*a* shows the basic potentiometer circuit. We see that a galvanometer (just a very sensitive D'Arsonval movement) is utilized as a null detector. It detects the presence or absence of current by deflecting whenever the unknown and reference voltages are unequal. However, it need not be calibrated since it must indicate only

---

[4]C. G. Wahrman, "A True RMS Instrument," *B & K Tech. Rev.,* B & K Instruments, Marlboro, MA, 1963.

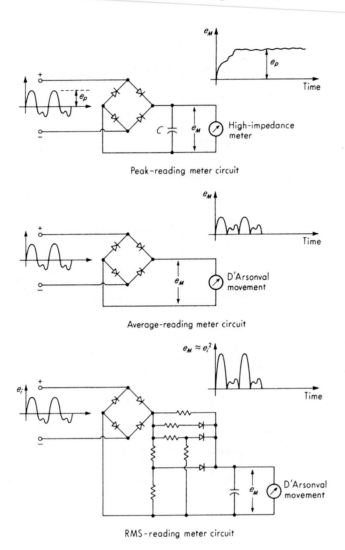

Figure 12.3
Peak, average, and rms circuits.

the presence of current, not its numerical value. The basic circuit of Fig. 12.4a is not practical since the accuracy of the reference voltage picked off the slidewire is directly influenced by changes in the dry-cell voltage. Since the dry cell supplies power to the slidewire, its voltage is bound to gradually drop off. This problem is solved in the practical circuit of Fig. 12.4b by inclusion of an additional component, the standard cell.

Figure 12.4c shows the Weston cadmium saturated standard cell which is a basic working standard of voltage. Its terminal voltage is 1.018636 V and is reproducible to the order of 0.1 to 0.6 ppm. Its accuracy in terms of the fundamental

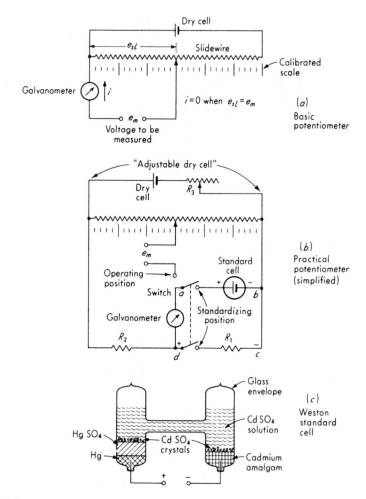

Figure 12.4
Manually balanced potentiometer.

mass, length, and time standards can be established to only about 10 ppm, however. Its temperature coefficient is $-40 \ \mu V/C°$; thus close temperature control obviously must be employed in the most exacting situations. Since its accuracy is destroyed if any appreciable current is drawn from it over a time interval, a standard cell cannot be substituted for the dry cell of Fig. 12.4a. So the standard cell must be utilized as an intermittent reference against which the slidewire excitation voltage can be checked whenever desired. The *unsaturated* Weston cell is employed in practical instruments since it is more portable. Its terminal voltage varies from one unit to another. However, its drift at constant temperature is only about -0.003 percent per year; thus it is perfectly adequate for most purposes. Its temperature coefficient is about $-10 \ \mu V/C°$.

The operation of the circuit of Fig. 12.4b is as follows: When the slide-wire scale on the potentiometer was calibrated originally at the factory, slidewire excitation-adjusting resistor R_3 was set at a fixed value and resistor R_1 was adjusted until, when loop $abcd$ was completed by the switch, no current flowed in the galvanometer. This means that the voltage drop across R_1 was just equal to the standard-cell voltage. Now the slidewire, R_1, and R_2 are all fixed and stable resistors; thus if the voltage across R_1 is at its calibration value, the slidewire excitation voltage also must be at its calibration value. Thus, whenever we wish to check the calibration (this is called standardization), we merely complete the loop $abcd$ momentarily (so as not to draw much current from the standard cell) and note whether the galvanometer deflects. If it does, we adjust the slidewire excitation with R_3 until deflection ceases. Then we are assured that the slidewire excitation is at its original calibration value. The resistor R_2 is merely a current-limiting resistor to prevent drawing large current from the standard cell through the slidewire path, which is fairly low-resistance.

Fairly common and inexpensive potentiometers which can be read to the nearest microvolt are available. More sophisticated instruments intended for the most accurate calibration work provide greater accuracy and sensitivity. One such unit measures in three ranges: 0 to 1.611110 V in steps of 1.0 μV, 0 to 0.1611110 in steps of 0.1 μV, and 0 to 0.01611110 in steps of 0.01 μV. The total parasitic thermoelectric voltage is less than 0.1 μV. The limit of error on the high range is ± 0.003 percent of reading ± 0.1 μV, while on the medium and low ranges it is ± 0.005 percent of reading ± 0.1 μV. These values approach the level of the National Standards achieved by the NIST, which are about 0.001 percent from 0.01 to 1,000 V. Today, manual potentiometers are used mainly in calibration laboratories. Most other precision-voltage measurements are made with digital voltmeters/multimeters or with "virtual" voltmeters based on digital data-acquisition systems. The potentiometer principle is still widely applied to "strip chart" (self-balancing potentiometer) *recorders,* where the slidewire motion required to balance the instrument becomes the pen motion of the recorder.

12.3 DIGITAL VOLTMETERS AND MULTIMETERS

While analog meters require no power supply, give a better visual indication of trends and changes, suffer less from electric noise and isolation problems, and are simple and inexpensive, digital meters offer higher accuracy and input impedance, unambiguous readings at greater viewing distances, smaller size, and a digital electrical output (for interfacing with external equipment) in addition to visual readout. The three major classes of digital meters are panel meters, bench-type meters, and systems meters. All employ some type of A/D converter [often the dual-slope integrating type (see Sec. 10.12)] and have a visible readout that displays the converter output. Usually panel meters are dedicated to a single function (and perhaps even a fixed range), while bench and system meters often are *multimeters;* that is, they can read ac and dc volts and amperes as well as resistance, over several ranges. The basic circuit is always dc volts; current is converted to volts by passing it through a

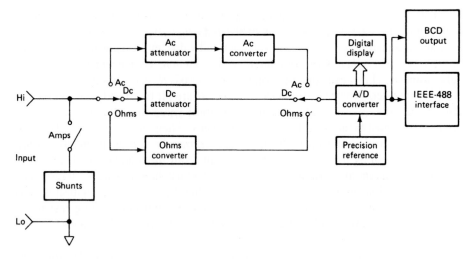

Figure 12.5
Digital multimeter configuration.

precision low-resistance shunt, while alternating is converted to direct current by employing rectifiers and filters. For ohms measurement, the meter includes a precision low-current source that is applied across the unknown resistor; again this gives a dc voltage which is digitized and read out as ohms (see Fig. 12.5). Bench meters are intended mainly for stand-alone operation and visual reading, while systems meters provide at least an electrical binary-coded decimal output (in "parallel" with the visual display) and perhaps sophisticated interconnection and control capability (such as the IEEE-488 interface of Fig. 12.5) or even microprocessor-based computing power.

Digital panel meters[5] are available in a very wide variety of special-purpose functions. Readouts range from basic 3-digit (999 counts, accuracy ± 0.1 percent of reading, ± 1 count) to high-precision $4\frac{3}{4}$-digit ones ($\pm 39{,}999$ counts, accuracy ± 0.005 percent of reading, ± 1 count). Units are available to accept inputs such as dc volts (microvolts to ± 20 V), ac volts (true rms measurement), line voltage, strain-gage bridges (meter provides bridge excitation), RTDs (meter provides sensor excitation), thermocouples of many types (meter provides cold-junction compensation and linearization), and frequency inputs such as pulse tachometers. A high-precision unit with input resistance of $10^9 \, \Omega$, ± 0.0025 percent resolution ($10 \, \mu$V), and ± 0.005 percent of reading ± 1 count accuracy uses dual-slope A/D conversion with automatic zero. The reading rate is 2.5 per second when free-running and 10 per second maximum when externally triggered. These meters can be obtained with TRI-STATE binary-coded decimal outputs. TRI-STATE outputs provide a "disconnected" state in addition to the usual digital HI and LO. This facilitates

[5]Nonlinear Systems, San Diego, CA (www.nonlinearsystems.com); Datel Systems, Mansfield, MA (www.datel.com).

interconnection to microcomputer data busses since any number of devices can be serviced by a single bus, one at a time, by "disconnecting" all but the two that are talking to each other.

Bench- and systems-type digital meters ($4\frac{1}{2}$ to $8\frac{1}{2}$ digits) are available from many sources; we quote some data from one manufacturer whose yearly catalog[6] and free "textbook"[7] include technical information sections that are of general utility. Figure 12.6[8] summarizes the capabilities of several instrument classes for voltage and current measurement. *Nanovoltmeters* (nVM) and *digital multimeters* (DMM) have input impedances of about 10 GΩ while *electrometers* (used to measure low current and charge, and very large resistance) may have input impedance of 200 TΩ. *Picoammeters* use special techniques (Fig. 12.7a[9]) to accurately measure small currents. *Source-measure units* (SMU) combine an electrometer, DMM, voltage source, and current source to create a versatile instrument for studying voltage/current relations in all kinds of electrical devices. They provide adjustable excitation (voltage or current source) and measure the resulting response. Figure 12.7b[10] shows how four such units are used to measure op-amp characteristics (most applications require *only* one unit). Systems-type DMMs are an alternative to PC-based data-acquisition systems or data loggers, providing multichannel voltage-, current-, resistance-, temperature-, and frequency-measurement capability. Data analysis features and sampling rates do not match PC-based systems but may be sufficient and cost/effective for many applications.

12.4 ELECTROMECHANICAL SERVOTYPE *XT* AND *XY* RECORDERS

Figure 12.8 shows the principle of servotype *XT* recorders employed for obtaining indication and simultaneous recording of a voltage $e_i(X)$ against time (T). The instrument servomechanism is designed so that displacement x_o tracks the voltage e_i accurately over the design frequency range. Variations on this general principle include use of ac or dc amplifiers, ac or dc motors, rotary or translational motors, various mechanical-drive arrangements (piano wire and pulleys, etc.), assorted writing schemes, and different displacement transducers (potentiometers, RVDTs, capacitance, etc.). Adjustable chart drive speeds (these establish the time base) were obtained originally from constant-speed synchronous motors and pushbutton mechanical change gears; however, now the step-motor drive of Fig. 12.8 is more common.

---

[6]2000/01 Catalog, Keithley Instruments, Inc., Cleveland, OH, 800-552-1115 (www.keithley.com).

[7]"Low Level Measurements," 5th ed., Keithley Instruments, Inc. (There are many practical tips on general methods and specific applications.)

[8]Ibid.

[9]Ibid.

[10]Keithley 236/237, "Source Measure Units Applications Overview," Keithley Instruments, Inc., p. 45.

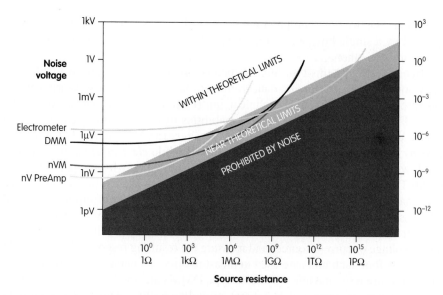

Choosing the right instrument. In DMMs, SourceMeter instruments, and data acquisition products, the limiting factor is usually the instrument's digital resolution or accuracy. In most applications, this is much better than what is needed and noise is not a factor. However, in low-current or high-resistance measurements, the limiting factor is usually current noise in the device, connections, or instrument. In low-voltage or resistance measurements, voltage noise is typically the limiting factor. As the graph indicates, nanovoltmeters have the lowest voltage noise; therefore, they provide the best measuring capability when measuring low resistances. Electrometers and sensitive SourceMeter instruments (SMUs) have the lowest current noise so they provide the best capability when measuring low currents or charge or high resistances.

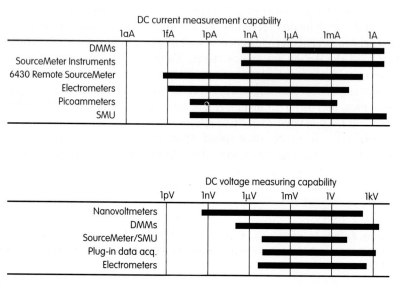

Figure 12.6
Capabilities of digital voltage and current instruments.

Voltage burden can cause errors at any current level

The voltage burden is the terminal voltage of an ammeter. An ideal ammeter will not alter the current flowing in a circuit when connected in place of a conductor. Thus, it must have zero resistance and therefore zero voltage burden.

Digital multimeters use the shunt ammeter technique shown in **Figure 1** to measure current. The meaurement method is to develop a voltage across a sensing resistor. The resistor is chosen such that 200mV corresponds to the maximum current reading on a selected range. The voltage burden specification is the 200mV developed across the sensing resistor.

Feedback picoammeters such as the 485 and Keithley electrometers use a technique in which the voltage burden is the input voltage of an op amp, as shown in **Figure 2**.

The output voltage of the op amp is precisely related to the input current. Since input voltage is output voltage divided by op amp gain (typically 100,000), the voltage burden is only microvolts. The maximum specified voltage burden of the 485 is only 0.2mV.

An example of the problems caused by high voltage burden is shown in **Figure 3**. In measuring the emitter current of a transistor, the DMM causes a very significant error (200mV out of 300mV) while the 485 voltage burden creates negligible error (0.2mV out of 300mV). Even though the basic measurement is well within the range of a DMM, the 485 makes a more accurate measurement since, due to its low voltage burden, the 485 is much closer to an ideal ammeter.

(a)

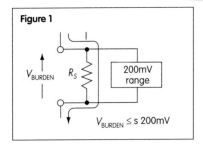

Figure 1

$V_{BURDEN} \leq$ s 200mV

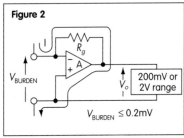

Figure 2

$V_{BURDEN} \leq$ 0.2mV

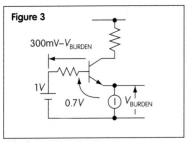

Figure 3

Ideal ammeter: $V_{BURDEN} = 0$
   $= 0\%$ error

485: $V_{BURDEN} \leq 0.2mV$
   $\leq 0.07\%$ error

DMM: $V_{BURDEN} \leq 200mV$
   $\leq 67\%$ error

Figure 12.7
(*a*) Picoammeter techniques. (*b*) One application of source/measure devices.

One subclass of this type of recorder utilizes rather wide charts (10 in) and is intended for slowly changing inputs (<1 Hz for full-scale travel). Throw-away fiber ink pens are employed most, but some recorders use heated styli and heat-sensitive paper. With suitable preamps, most any desired voltage range can be accommodated, with resolution to as little as a few microvolts being common. Multichannel operation (a separate servo for each pen) is possible either reduced-width side by side or full-width overlapping (if pens are staggered to avoid mechanical interference); the limit is about six pens. Staggered pens give an undesirable chart-displacement error between channels. Some recorders[11] compensate for this with a

---

[11]Soltec Corp., San Fernando, CA, 800-423-2344 (www.solteccorp.com).

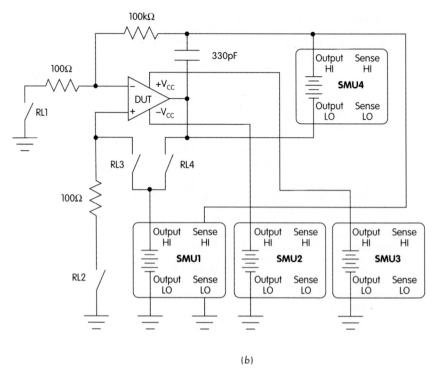

(b)

Figure 12.7
(Concluded)

microprocessor-implemented interchannel time delay. When inputs change very slowly, another approach to multichannel operation multiplexes the various input voltages into one servo and uses a numbered print wheel (or thermal dot matrix printhead) to print an identifying number or symbol for each channel of data (24 or 30 channels is a practical limit). Static accuracy of this entire subclass of recorders is about 0.1 percent full scale.

To cross-plot one variable against another, the *XY* configuration of Fig. 12.9 is available. Here the paper stands still (held down by vacuum or electrostatic attraction) while two independent servos move the pen horizontally and vertically. This type of plotter, because of its mechanical complexity and resulting maintenance problems, is becoming obsolete and is available from only a small number of manufacturers. The *xy* function is today mainly implemented with PC-based data-acquisition systems that use a CRT to display the graph, an ink-jet or laser printer for a hard copy, or a thermal-array recorder with an *xy* option. One exception is the digital-input plotters used to produce engineering drawings from CAD software, where the large size output often needed is not possible with the usual PC-type printer.

Higher-speed pen-type recorders[12] (usually called *direct-writing oscillographs*), which were good to about 40 Hz, are also becoming obsolete, as are the even faster

[12]E. O. Doebelin, "Measurement Systems," 4th ed., McGraw-Hill, New York, 1990, p. 871.

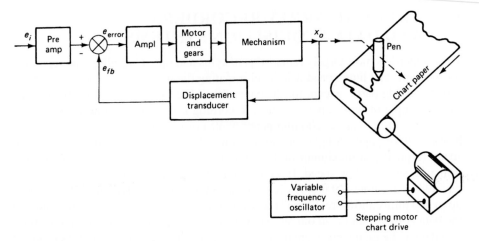

Figure 12.8
Servotype pen recorder.

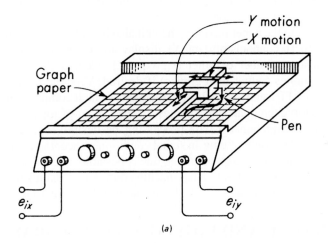

Figure 12.9

(8000 Hz) optical-writing[13] types. The pen-type units used ink pens or heated styli (with heat-sensitive paper) while the optical type used photo-sensitive paper to record the trace of a light beam deflected by a fast galvanometer. The high-speed recording provided by these instruments is now largely accomplished with high-speed digital sampling and storage, with the data then being read-out at a rate slow enough for a thermal-array recorder or PC-type printer.

[13]Ibid, p. 872.

12.5 THERMAL-ARRAY RECORDERS AND DATA ACQUISITION SYSTEMS

These recorders have largely replaced pen/stylus and optical oscillograph recorders for medium- and high-speed data by eliminating the messy ink and all the troublesome "moving parts" except for the paper drive. We use one manufacturer's unit[14] as an example of typical performance. Recording is done with a linear array (typically 8 dots/mm over the 200 mm paper width) of tiny heated elements writing on heat-sensitive paper. Single- or multichannel data is sampled and stored at high speed (typically, a maximum of 1 MHz), and then read out to the thermal-array printer at suitably low speed. If we use only a single channel, the 1-MHz sample rate allows accurate (10 points/cycle) measurement of 100-kHz data. Paper speeds up to 200 mm/s allow "real-time" recording for data up to about 20 Hz; beyond that the data is stored and read out at a slower rate to the printer. For continuous high-speed data, the total record length is limited by the available memory. This is one of the disadvantages relative to the older "pen-and-ink" recorders, which could record high-speed data for relatively long times. While recording "slow" data in real-time mode, it is possible to capture fast transients if a suitable trigger event is provided. Various forms of "add-on" memory are available for the thermal-array recorders, so long-term recording *is* possible.

In addition to the paper recorder, the unit also has a 6 × 8 in. color LCD display showing up to 16 channels of data. *XY* display and recording is possible between one (x) and any number of other (y_1, y_2, etc.) selected channels. Data can be presented in tables (rather than graphs) if desired. Versatile mathematical operations, including FFT and "waveform judgement" (which compares measured waveform with a stored template and activates a trigger if out-of-tolerance) are available. Should more comprehensive data processing be desired, data can be sent to a computer. A wide variety of plug-in signal conditioners accommodate most sensors. Long distance measuring using a phone line is possible by connecting to a modem. Automatic data transmission to a fax machine during data acquisition is also possible. The basic unit is about 5 × 6 × 12 in. and weighs about 15 lb.

12.6 ANALOG AND DIGITAL CATHODE-RAY OSCILLOSCOPES/DISPLAYS AND LIQUID-CRYSTAL FLAT-PANEL DISPLAYS

Oscilloscopes ("scopes") are used for the measurement and display of voltage signals whose frequencies can reach several gigahertz, but 50 MHz is more than adequate for most mechanical engineering applications. The classical version is the *analog* scope, while the more recent *digital storage* scope provides many features that make it popular, though it has not yet completely displaced the analog type

---

[14]Model TA220, Soltec Corp. (www.solteccorp.com).

(hundreds of thousands[15] of analog scopes are still sold each year). For most mechanical-engineering applications, digital scopes are preferred. Some very high frequency applications[16] are best served by analog scopes, mainly because the display locally brightens when the trace velocity slows or several traces are overlaid, as happens with some modulated and digital data. This local brightening aids in visual interpretation. Some digital scopes use special techniques[17] to mimic this analog advantage. Another approach[18] combines analog and digital scopes in one instrument. Liquid-crystal flat-panel displays provide functions similar to those found in cathode-ray units but in a more compact form (as is required in laptop computers). In addition to *XT* and *XY* traces of data signals, all these displays are also used for computer graphics of various types, such as solid models of parts designed by CAD software and stress distributions from stress-analysis software. Since digital scopes usually feature significant signal-processing capability, they may appear under different names, such as *waveform recorder* or *compact data-acquisition system.* The unifying feature here is that the input signals are digitized in an A/D converter, processed in various ways, visually displayed, and available for permanent recording or transmission to a network or computer. From this viewpoint, the thermal-array data system discussed in Sec. 12.5 could thus legitimately be called a digital-storage scope, though it is not marketed as such. As already mentioned, digital scopes offer a menu of useful signal-processing features, including cursors, that may be moved to selected points on the waveform, with digital read-out of the time and voltage values, and *dual* cursors that allow for the accurate measurement of voltage and time *differences.* The display of waveform features such as period, frequency, or rise time is also standard.

Figure 12.10 shows in simplified fashion the functional operation of a typical analog cathode-ray oscilloscope.[19] A focused narrow beam of electrons is projected from an electron gun through a set of horizontal and vertical deflection plates. Voltages applied to these plates create an electric field which deflects the electron beam and causes horizontal and vertical displacement of its point of impingement on the phosphorescent screen. By proper design this displacement can be made closely linear with deflection-plate voltage. The phosphorescent screen emits light which is visible to the eye and may be photographed for a permanent record.

The most common mode of operation is that in which a plot of the input signal against time is desired. This may be accomplished by driving the horizontal deflection plates with a ramp voltage, thus causing the spot to sweep from left to right at a constant speed. To ensure that the sweep and the input signal applied to the vertical

---

[15]T. Lecklider, "Scopes and Digitizers Masquerade As Each Other," *Evaluation Engineering,* Jan. 2000, pp. 42–48.

[16]M. Rowe, "DSO Displays: Almost as Good as Analog," *Test & Measurement World,* Feb. 2000, pp. 41–52.

[17]M. Rowe, op. cit.

[18]Fluke Inc. (www.fluke.com).

[19]R. van Erk, "Oscilloscopes: Functional Operation and Measuring Examples," McGraw-Hill, New York, 1978; C. F. Coombs, Jr. (ed.), "Electronic Instrument Handbook," 2nd ed., McGraw-Hill, New York, 1995, pp. 14.–14.61.

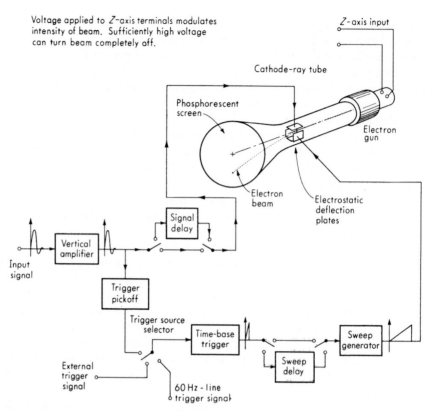

Figure 12.10
Cathode-ray oscilloscope.

deflection plates are synchronized properly, the triggering of the sweep can be initiated by energizing the trigger circuit from the leading edge of the input signal itself
(see Fig. 12.10). This results in a loss of the first instants of the input signal on the
screen, but generally this is not serious since only about 1 mm of deflection is
needed to cause triggering. In those cases where this loss is objectionable, oscilloscopes with signal delay (Fig. 12.11) are available. These delay the application of
the input signal to the vertical deflection plates so that the sweep starts *before* the
rise of the input signal on the screen. Thus the complete input signal is recorded. Most oscilloscopes also provide for triggering from either positive-going or
negative-going voltages (or both[20]), and the instant of triggering can be adjusted
from the minimum 1-mm level upward to any point on the input waveform. Triggering also can be controlled from external signals or the 60-Hz power-line signal.
When external trigger signals which are conveniently available occur somewhat

---

[20]C. Baker, "Digital Storage and Plug-In Versatility Distinguish New 10 MHz Oscilloscope," *Tekscope*,
vol. 12, no. 14, Tektronix Corp., Beaverton, OR, Dec. 1980.

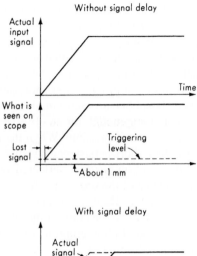

Figure 12.11
Signal delay.

before the input signal of interest, a sweep-delay feature may be useful; some instruments provide this capability. Digital scopes are particularly versatile with respect to triggering, such as providing *pre-* and *post* triggering, which allow for the capture of data both before and after a trigger event, respectively. By putting the digitized data into a circular buffer, a selected total number of samples can be apportioned into a batch of pretrigger data and a batch of posttrigger data, allowing us to see the *entire* trace of an event, rather than losing the early parts, as would be the case with many analog scopes. The circular buffer (say, 5000 total samples) *continuously* inputs the data stream so that it will not "miss" any details but pushes out the earliest data as new values come in, always holding 5000 points. If we program the scope for, say, 1000 pretrigger and 4000 posttrigger points, then when the trigger event occurs, the buffer is "frozen," and we can then display 1000 points of what was happening before the trigger and 4000 points of what happened after. The trigger event might be, for example, a proximity-pickup pulse from an engine crankshaft 35° before top dead center, or a data signal rising above 0.1 V. Since the deflection sensitivity of the cathode-ray tube itself is only of the order of 0.1 cm/V, oscilloscopes include amplifiers so that the instrument can directly handle input signals down to microvolts. Input amplifiers typically provide 1-MΩ input impedance, selectable single-ended or differential input, and selectable ac or dc coupling. Oscilloscopes are also useful for *XY* plotting. For such operation the horizontal deflection plates are merely disconnected from the sweep generator and connected to an amplifier identical to the vertical amplifier.

Cathode-ray tubes are obtainable with a number of different phosphors on the screen.[21] The choice of phosphor controls the intensity of light available for visual observation or photographic recording, as well as the persistence of the trace after the electron beam has moved on. Both long- and short-persistence phosphors are available. Long-persistence phosphors are useful in visual observation of transients since the entire trace is visible long enough for an observer to note its characteristics. Persistence for several seconds is possible. When the moving-film method of trace photography is used, a very-short-persistence phosphor is necessary to prevent blurring. (In this method the electron beam is deflected vertically only, while the film is moved horizontally in front of the screen at a fixed, known velocity.) Persistence of less than 1 μs is available. Dual-persistence phosphors provide either a long or short persistence, depending on the color of the filter utilized over the scope screen. The most common method of photographing oscilloscope traces uses a still camera and the 10-s Polaroid[22] film process. A common method of photographing transients employs a double exposure to record both the trace and the grid lines. With the grid-line illumination turned off and the camera shutter held open, the transient is triggered, thus recording its image on the film. Then the shutter is closed. Now the grid lines are turned on, and the shutter is snapped in the normal manner (say $\frac{1}{25}$ s at F:16) to superimpose the grid lines on the picture. This procedure is necessary since the illumination from the grid lines is so great that it would completely fog the film if left on during the long time that the shutter is left open to catch the transient. All these cumbersome photographic techniques are eliminated in digital scopes; giving one of their most important advantages over analog. Any displayed waveform, whether a "one-shot" transient or a continuous signal, is easily captured on the screen, with the same brightness for fast or slow signals. External or built-in printers quickly provide clear hard copy of graphical traces or tables of numerical values.

To obtain multichannel capability in oscilloscopes, several approaches[23] are taken. The dual-beam oscilloscope has two separate electron beams in one cathode-ray tube, with separate deflection plates and amplifiers for each beam. In some units both beams use the same sweep system; thus the two traces are plotted against the same time base. Completely independent beams allowing different time bases on each trace are available. The other approach (dual trace) uses a single-beam cathode-ray tube and a high-speed electronic switch to timeshare the beam among several input signals. Such multitrace systems are available to give up to four traces on a single screen. Multichannel capability is easily implemented in digital scopes since the data is stored in memory and can be read out to the display screen in whatever way we please, using any of the multichannel digitizing methods common in

---

[21]R. A. Bell, "CRT Phosphor Selection," *Inst. & Cont. Syst.*, pp. 86–90, March 1970.

[22]Polaroid Corp., Cambridge, MA.

[23]V. Lutheran and B. Floersch, "Dual-Beam: An Often Misunderstood Type of Oscilloscope," *EDN*, August 5, 1974.

data-acquisition systems. All the channels could share, through a multiplexer, a single A/D converter, with the usual time-skew problems and solutions. Alternatively, to avoid time skew, each channel could have its own A/D converter. In any case, to display a stationary pattern on the screen, the data for a given sweep is read out of memory in a repetitive cycle, fast enough to give a bright and steady picture with the screen's phosphor decay rate.

Versatility of operation was achieved in analog scopes by the use of *plug-in* units for a single mainframe. These include multitrace, op-amp (used to build versatile analog-signal processing such as integration, differentiation, or filtering), carrier amplifiers, spectrum analyzers, high-gain amplifiers, and special time bases. Digital scopes can duplicate many of these features with internal software, so plug-in units are not needed.

Most laboratory scopes have an 8×10 cm screen, but larger displays are useful for viewing by large groups, for presentation of many channels of data, etc. At one time, scopes with 21-in. screens were available, but today, large-screen displays may be better achieved using *virtual oscilloscopes,* that is, PC-based digital data-acquisition systems using software to emulate scope behavior, with display on the PC's monitor, which, even in large sizes, is much cheaper than a large-screen scope. *Projection* display systems can show the PC screen greatly enlarged on a standard "movie" screen for viewing by large groups.

Limited screen size and inherent nonlinearity of the CRT deflection system make the error of the basic analog oscilloscope (both voltage and time) about 2 to 5 percent. In dual-beam or dual-trace scopes, timing accuracy can be improved by applying an oscillatory signal of known frequency to one input as a reference. Voltage accuracy can be increased similarly in a dual-trace instrument by applying a reference square wave of accurately known amplitude to one input. Many digital scopes use 8-bit A/D converters to achieve higher sampling rates without requiring too much memory. This gives a resolution of about 0.4 percent, but input amplifier nonlinearity and noise errors give an overall voltage accuracy of 1 to 2 percent. The time accuracy is about the same. Resolution finer than 8 bits is available on some hardware scopes, and PC-based virtual scopes may also provide differential input, a feature usually *not* provided on hardware scopes. Differential input[24] is often required when the input signal comes directly (rather than through a pre-amp) from transducers that supply only millivolt-level signals. Differential input also helps improve accuracy in electrically noisy environments and/or when long cables are necessary. A 14-bit card[25] can provide 50 Msamples/s simultaneous sampling on 2 channels, with up to 1 billion points of acquisition memory.

Digital scopes have both *acquisition memory* and *display memory.*[26] The acquisition memory is just the "raw" sample points as they come from the A/D converter. These are sent to a microprocessor, where they may be manipulated in various ways

[24]"Differential Oscilloscope Measurements," Tektronix, Inc., Beaverton, OR, 800-835-9433 (www.tektronix.com).

[25]Gage Applied, Inc., Lachine, QC, Canada, 800-567-4243 (www.gage-applied.com).

[26]"XYZ's of Oscilloscopes," Tektronix, Inc.

before going to the display memory, which actually sends the final *waveform* points to the display. For example, several sample points may be used to generate a single waveform point, perhaps using some kind of averaging. Interpolation is also possible, though it is not regularly used, to get smoother, more accurate displays. Mathematically, a sin x/x interpolation recovers bandwidth-limited waveforms *perfectly* from the discrete sample points, but the computational burden makes it impractical.[27] Usually there are only two display choices: "dots" (one for each sample point) or "connect the dots" (linear interpolation with straight-line segments).

12.7 VIRTUAL INSTRUMENTS

Many of the measurement functions provided by "actual" (hardware) instruments are available with PC-based digital data-acquisition systems and associated software as *virtual instruments*. Virtual instruments that emulate, for example, DMMs, scopes, spectrum analyzers, arbitrary function generators, counters, data loggers, strip-chart recorders, and *xy* plotters are possible because, once the analog data is digitized in a suitable A/D conversion board, software can then mimic all these hardware functions. The PC monitor can display a graphic image that looks like the "real" instrument's panel with all the usual knobs and buttons for adjusting the instrument settings and showing the graphical or numerical readout. Chapter 13 shows how such systems can be applied to practical measurement situations, using the DASYLAB software as an example. Such virtual instruments are also widely used in process-control panels and aircraft cockpits. There, instead of a huge panel with many fixed meters, a single (or multiple) computer monitor can display only those instruments that the operator wants to observe at a particular time. Such systems can also be programmed to *force* critical instruments onto the display when alarm conditions exist. It is also possible to build a data system that combines virtual instruments with real instruments, if that is desired. While general-purpose software such as DASYLAB provides for many different virtual instruments, some types are so widely used that special-purpose hardware and software is available. A prime example of this is virtual oscilloscopes, where manufacturers[28] provide special industrially hardened PCs, digitizing boards, and comprehensive scope software to assemble a high-performance measurement system.

12.8 MAGNETIC TAPE AND DISK RECORDERS/REPRODUCERS

Magnetic tape recording has a long history in both consumer electronics (voice and music recording) and engineering instrumentation. When recording high-frequency engineering data, such as acoustic or vibration data, was needed for relatively long periods, tape recording was often the only choice.[29] Recent developments in

---

[27]C. F. Coombs, Jr., op. cit., p. 14.44.

[28]Gage Applied, Inc., (www.gage-applied.com).

[29]D. Banaszak, "Processing Vibration and Acoustic Data from a Digital Recorder," *Sound & Vib.,* Aug. 1996, pp. 14–16; B. Shipman, "Selecting the Right Data Recorder," *Sound & Vib.,* Oct. 1997, pp. 12–16.

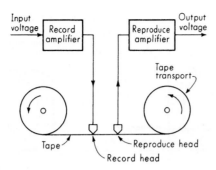

Figure 12.12
Tape recorder/reproducer.

high-speed and high-capacity hard disk drives, and ever-faster PCs, are threatening to take over from tape technology many engineering recording tasks.[30] One such system,[31] based on a fast Pentium processor, replaces the tape recorder, data-acquisition system, and monitoring oscilloscope that would usually make up a tape recording system. A complete analog "front end" includes individual differential-input, programmable-gain amplifiers, antialiasing filters, and simultaneous sampling (each channel has its own 16-bit sigma-delta A/D converter). With a 9-GByte hard drive, 8 channels of 1-kHz data can be recorded for 62 h; higher frequency data and/or more channels scale proportionately (16 channels of 80-kHz data, the fastest allowed, data runs for 23 min). When such systems require larger data-storage capacity and increased data reliability and security, *redundant arrays of independent disks* (RAID[32]) technology may be the solution. Depending on the application, such systems should be considered alternatives to tape recording and the most cost-effective approach chosen. We now give a brief discussion of tape technology.

The magnetic recorder/reproducer has a number of unique features derived mainly from its ability to record a voltage, store it for any time, and then reproduce it in electrical form essentially identical to its original occurrence. Recording methods include the direct, FM, pulse-duration modulation (PDM), and digital techniques.[33] We consider first the direct and FM modes of operation as applied to tape devices.

Figure 12.12 shows a functional diagram of a tape recorder/reproducer, and Fig. 12.13 shows a closeup of the record and reproduce heads. A current i proportional to the input voltage is passed through the winding on the record head, producing a

[30]J. J. Jachman, "Digital Data Acquisition Has Changed Our Jobs," *Test & Measurement World,* Feb. 15, 1997, pp. 20–22; "Data Acquisition System Provides Many Functions," *Sound & Vib.,* Apr. 2001, pp. 18–20.

[31]DataMAX Instrumentation Recorder, RC Electronics, Santa Barbara, CA, 805-685-7770 (www.rcelectronics.com).

[32]RC Electronics (www.rcelectronics.com); General Technics Inc., Ronkonkoma, NY, 800-487-2523 (www.gtweb.net/about.html).

[33]"Magnetic Tape Recording Technical Fundamentals," Bell & Howell Datatape Div., Pasadena, CA, 1979.

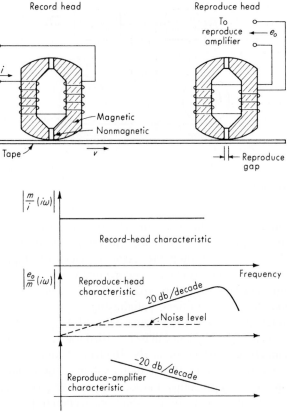

Figure 12.13
Record and reproduce heads

magnetic flux $\phi = K_\phi i$ at the recording gap. The tape (thin plastic coated with iron oxide particles) passes under the gap, and the oxide particles retain a state of permanent magnetization proportional to the flux existing at the instant the particle leaves the gap. (Actually the applied flux and induced magnetization are not proportional because of the nonlinearity of the magnetic-hysteresis curve. Effectively, however, a close linearity is obtained by a high-frequency bias technique.[34]) Thus, with a sinusoidal input signal

$$i = i_0 \sin 2\pi ft$$

and a tape speed of v inches per second, the intensity of magnetization along the tape varies sinusoidally with distance x according to

$$\text{Magnetization} \triangleq m = K_m K_\phi i_0 \sin\left(\frac{2\pi f}{v} x\right) \tag{12.1}$$

---

[34]Ibid.

where $m = K_m \phi$. The wavelength of the magnetization variation is then v/f inches. For example, a 60-Hz signal at a 60-in/s tape speed gives a wavelength of 1 in. If the tape with this signal on it is passed under the reproduce head, then a voltage proportional to the rate of change of flux bridging its gap will be generated in its coil. Note that since the output voltage depends on the rate of change of flux, if a direct current at the input had produced a constant tape magnetization, the reproduce head would have given *zero* output.

Thus the technique described above, the so-called direct recording process, can be used with varying input signals only, with about 50 Hz being the usual lower limit of frequency. Furthermore, since the reproducing head has a differentiating characteristic, the reproduce amplifier must have an integrating characteristic in order for the system output to be proportional to the input. An upper frequency limit also exists, because at sufficiently high frequencies, for a given reproduce gap and tape speed, one wavelength of magnetization will become equal to or less than the gap width. Then the average magnetization in the gap will be zero, and no output voltage will be generated. For example, at the fastest common tape speed, 240 in/s, and a gap width of 0.00008 in, this occurs at 3.0 MHz. Actually, the system is usable to only about half this frequency with reasonable accuracy. The frequency range of the direct recording process is approximately within the band 100 Hz to 2 MHz. The direct recording process does not give particularly high accuracy. Essentially this is limited by the signal-to-noise ratio, which is of the order of 25 dB (about 18:1). The rather high noise level is the result of minute defects in the tape surface coating to which the direct recording process is sensitive.

When more accurate recording and response to dc voltages was required, generally the FM system was employed in the past. Here the input signal is used to frequency-modulate a carrier, which is then recorded on the tape in the usual way. Now, however, only the *frequency* of the recorded trace is significant, and tape defects causing momentary amplitude errors are of little consequence. The frequency modulators employed here are similar in principle to those discussed under voltage-to-frequency converters and subcarrier oscillators in Chaps. 10 and 11. However, the frequency deviation for tape recorders is ±40 percent about the carrier frequency. The reproduce head reads the tape in the usual way and sends a signal to the FM demodulator and low-pass filter, where the original input signal is reconstructed. The signal-to-noise ratio of an FM recorder may be of the order of 40 to 50 dB (100:1 to 330:1), indicating the possibility of inaccuracies smaller than 1 percent. By using sufficiently high carrier frequencies (432 kHz), the flat (±1 dB) frequency response of FM recorders may go as high as 80,000 Hz at 120 in/s tape speed. To conserve tape when high-frequency response is not needed, generally a range of tape speeds is provided. When the tape speed is changed, the carrier frequency is altered in direct proportion. This makes the recorded wavelength of a given dc input signal the same, no matter what tape speed is being used, since ±40 percent full-scale frequency deviation is utilized in all cases. Signals may be recorded at one tape speed and played back at any of the others without change in magnitude, but with a compression or an expansion of the time scale. A common set of specifications might be as follows:

Tape speed, in/s	Carrier frequency, kHz	Flat frequency response ±0.5 dB, Hz	RMS signal-to-noise ratio
120	108	0–20,000	50
60	54	0–10,000	50
30	27	0–5,000	49
15	13.5	0–2,500	48
$7\frac{1}{2}$	6.75	0–1,250	47
$3\frac{3}{4}$	3.38	0–625	46
$1\frac{7}{8}$	1.68	0–312	45

Most instrumentation tape recorders today use *digital* technology rather than FM. We included this brief treatment of FM methods for historical interest and in recognition that some FM systems produced earlier might still be in use. Digital instrumentation tape recorders are mainly of two broad classes: those that accept data that is *already* in digital form and those designed to accept the analog signals that come directly from most sensors. The latter would thus include the usual amplification, multiplexing, sampling, and digitizing hardware that we have discussed earlier as part of other types of digital data-acquisition systems. Our discussion of digital-tape technology thus concentrates on the methods used to record data that is already in digital form. Mechanically, digital instrumentation recorders use the "helical scan" method employed in the common videotape (VCR) and digital audio tape (DAT) recorders, with similar economical tape cassettes. To get the tape and read/write heads to have the high *relative* velocity needed to resolve high-frequency data, the heads are placed on a rapidly rotating (4000 rpm[35]) drum at a skewed angle; that is, the moving tape is wrapped helically around the drum at a small angle. This arrangement allows for much slower tape speeds (16.3 mm/s) than the stationary head configuration (120 in/s = 3048 mm/s). The Sony recorders can record 2 channels of 20-kHz data for 46 min or up to 32 channels of 1.25-kHz data. More advanced machines[36] offer a frequency response to 160 kHz. Some recorders[37] provide a "card slot" that accepts three different storage media: PC flashcard, MO (magneto-optical disk), or AIT (Sony's Advanced Intelligent Tape) tape cartridge. Users can select the medium that is most compatible with their needs. Typically, flashcards hold about 256 MBytes, MO hold about 1.3 GBytes, and AIT hold about 25 GBytes. Typical digital recorders use 16-bit quantization and provide about a 80-dB dynamic range and signal-to-noise ratio, much better than FM recorders.

Wide utilization of digital data systems requires that digital data often be directly recorded on magnetic-tape or -disk systems.[38] Pulse-code modulation

---

[35]Sony Model PC200Ax Series, Sony, Lake Forest, CA, 1-888-910-7669 (www.sonypt.com).

[36]Sony SIR-1000 Series, TEAC RX Series, Teac America, Montebello, CA, 323-727-4853 (www.teacrecorders.com).

[37]Teac GX-1.

[38]"Magnetic Tape Recording Technical Fundamentals," op. cit.; "Parallel Mode High Density Digital Recording Technical Fundamentals," Bell & Howell Datatape Div., Pasadena, CA, 1979.

(PCM), using one of several coding schemes, is employed extensively to record the data, which are assumed to be in the binary form of 0s and 1s (a high voltage level and a low level). The actual magnetic-recording technique is similar to direct analog (rather than FM) in that the response does not extend to zero frequency. Codes of the return-to-zero-level (type RZ) (see Fig. 12.14) are wasteful of band-width since each bit requires *two* level transitions. Non-return-to-zero-level (NRZ-L) codes are more frugal of bandwidth, but present recording problems since a long string of 1s or 0s is essentially "dc," which cannot be recorded properly. The frequency content of the signal depends, of course, on both the sequence of 1s and 0s and the tape speed. A rule of thumb says that the number of level changes per unit time should be about 1.5 times the low-frequency cutoff of the recorder. Thus for a 100-Hz cutoff, we would require a combination of tape speed and level changes such that there would be (on the average) 150 level changes per second. Most digital record- ing is done with the basic NRZ-L coding scheme or improved versions of it. One such version separates the NRZ-L data-stream into 7-bit words; inverts bits 2, 3, 6, and 7; and adds a parity bit to make an 8-bit word. The parity bit is chosen so as to make the total number of 1s in the 8-bit word an odd count. Advantages of such a scheme include reduction of dc content and a guarantee that there will be at least one transition every 14 bits.

Since digital recorder heads (just as analog recorders) can have multiple tracks, data can be recorded in serial, parallel, or serial-parallel format. In serial format, frequently used in instrumentation applications, data are recorded in a continuous stream on a single track. In parallel format, common in computer applications, data are recorded simultaneously on several tracks and the data have a relationship across the width of the tape, such as using an 8-track head to record 8-bit words.

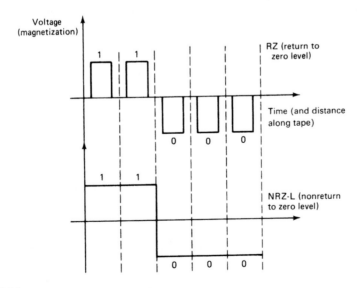

Figure 12.14
Digital recording formats.

Serial-parallel mode is a process in which two or more serial datastreams, longitudinally recorded, also have a relationship across the width of the tape. For example, the number 10011101 could be recorded on two tracks by using the code that alternate bits employ alternate tracks; thus one track would contain 1010, and the second would hold 0111. This technique is utilized in instrumentation recording at very high data rates.

BIBLIOGRAPHY

1. R. B. Northrop, "Introduction to Instrumentation and Measurements," CRC Press, New York, 1997.
2. M. B. Stout, "Basic Electrical Measurements," 2nd ed., Prentice-Hall, Englewood Cliffs, NJ, 1960.
3. "Low Level Measurements," 5th ed., Keithley Instruments, Inc., Cleveland, OH.
4. J. G. Webster (ed.), "The Measurement, Instrumentation and Sensors Handbook," CRC Press, New York, 1999, sec. VII.
5. I. F. Kinnard, "Applied Electrical Measurements," Wiley, New York, 1956.
6. J. Murphy, "Ten Points to Ponder in Picking an Oscilloscope," *IEEE Spectrum,* vol. 33 (7), pp. 69–77, 1996.
7. R. A. Witte, "Electronic Test Instruments, Theory and Application," Prentice-Hall, Englewood Cliffs, NJ, 1993.

CHAPTER 13

Data-Acquisition Systems for Personal Computers

In simple experiments, it may be most cost/effective to record data using voltmeters, oscilloscopes, oscillographs, strip-chart recorders, compact data loggers, or interconnected stand-alone instruments. When the experiments or other measurement functions get more complex and/or involve extensive data manipulation, a digital data-acquisition system using a personal computer and specialized software often is a better choice. If a personal computer with appropriate empty slots is available, you need to purchase suitable data-acquisition boards and software that provide the needed functions and are compatible with the hardware. Both hardware and software are available from a large number of sources and making the best choice involves careful consideration relative to cost and performance.

With respect to data-acquisition *software,* a major choice relates to ease of learning and use, as compared with versatility. If software is to provide for "all possible contingencies and applications," it tends to be more complex and harder to learn. Some software is designed to be used by programmers who specialize in producing custom data systems for particular applications and spend all their time in this kind of work. Here complex software with a long learning curve may be well justified, since once it is learned, it is not forgotten because of daily use, and its comprehensive capabilities mean that it will produce efficient systems for most any applications that might arise. Most engineers, however, have an *occasional* need to develop a data-acquisition system for a particular project and thus would welcome software which was designed for quick learning and use, so that it could be "picked up" quickly whenever needed.

In an undergraduate engineering curriculum, one might opt to prepare students to use either of the two types of software just briefly described, depending on how much curricular time was to be devoted to this topic. In this text, I have chosen to present the simpler, easier-to-learn type, which we have found to nicely meet the needs of our mechanical engineering students in the time available. Discussions with working engineers in industry who also use this software reveal that, even though it is simple to learn and use, it is capable of providing efficient systems for rather complex applications. Remember that "software efficiency" should include *both* the engineering time taken to develop the program and the speed with which the program runs. If our final system runs a little slower and perhaps does not provide every "bell and whistle," but took only a few hours or days to prepare, it may well be the most "efficient."

Many software products of the two general types just discussed are on the market, and while we do not want to endorse any specific ones, the purposes of this chapter are best served by showing *specific* rather than generic examples. In our measurement systems course we use DASYLAB,[1] a German product which was recently acquired by National Instruments,[2] a large U.S. company that markets LABVIEW, one of the most widely used comprehensive packages. Data-acquisition software may use graphical or command-line programming, or some combination. Comprehensive packages such as LABVIEW use a combination, while DASYLAB employs mostly graphical methods. One reason our students pick up DASYLAB so quickly is that they have earlier used the SIMULINK dynamic-system simulation language, which also builds its program by connecting icons with "wires." Newcomers to DASYLAB will find it very easy to learn even if they do not have such prior exposure.

13.1 ESSENTIAL FEATURES OF DATA-ACQUISITION BOARDS

While operations such as analog-to-digital conversion have been discussed in some detail elsewhere in this book, it is useful at this point to briefly highlight certain features. Digital data-acquisition software gets its input signals from data-acquisition boards installed in the computer. The choice and use of such boards involves a number of considerations that we only briefly mention here. A board will provide a definite number of *data channels,* and we require one channel for each measurement sensor. Channels can be connected either *single-ended* or *differential.* Typically, a card might provide 8 channels if you use them differentially or 16 if used single-ended. If our sensors provide low-level (say, millivolt) signals and perhaps use long cables, we may *have* to use differential inputs (reducing the total number of available channels) since this mode better suppresses noise.

The *resolution* of a board (number of bits) influences the smallest sensor voltage change that will be recognized. If the full-scale voltage for the A/D converter on the board is ± 10 V, a 12-bit card will divide this into $2^{12} = 4096$ subranges, making the least significant bit equal to 20 volts/4096 = 0.004883 V. Thus, a sensor voltage change smaller than about 5 mV may *not* result in any change in the digitized value. When we measure low-level (millivolt) signals, we need a card with on-board *amplification,* prior to the A/D conversion. Thus, if the amplifier has a gain

---

[1]DASYTEC, Amherst, NH, 800-731-5015.

[2]National Instruments, Austin, TX, 800-258-7022 (www.natinst.com).

of 1000 V/V, we can now resolve sensor voltage changes of about 5 μV with the 12-bit card described above. Some cards allow us to set the amplifier gain over a range of values to suit our sensor voltage ranges, which often are different from channel to channel. (We want to have each sensor "use up" the entire ± 10-V range of the A/D, in so far as possible, for accuracy.) *Programmable-gain* amplifiers let us change the gain "on the fly" to suit *each* channel since we often "time-share" a *single* amplifier among *all* the channels on a board. Programmable gain involves switching, which may slow down the sampling rate to allow the amplifier to "settle down" before actually taking a reading.

While digital-data systems have many advantages, a major disadvantage (relative to analog systems) is the need for *sampling,* that is, the sensors are only interrogated every, say, 0.001 s, so anything that happens in between is *completely lost.* We, of course, are well aware of this and carefully choose the rate sufficiently high to properly sample the *fastest-changing* sensor signals. If we plan to use our data *only* to study its *frequency spectrum* (FFT analysis), the slowest rate commonly used is about 2.56 samples per cycle of the *highest* frequency present in the data. More often, we want an accurate *time-domain* record and then 7 to 10 samples per cycle should be used. The sampling rate sometimes is set in the hardware (data-acquisition board) and sometimes by the software, such as DASYLAB.

Multifunction boards include not only A/D converters but also D/A converters, digital inputs and outputs, counter inputs, etc. If we use our system to not only gather measured data but also to *control* our apparatus, then D/A converters are often needed to send control signals from the computer to motors, heaters, etc., which can alter operating conditions. DASYLAB can utilize all these types of functions.

13.2 THE DASYLAB DATA-ACQUISITION AND -PROCESSING SOFTWARE

DASYLAB is a Windows-based application, so its operation will not be totally strange to most readers. It also relates somewhat to SIMULINK in that both build up complete systems using a large number of functional modules, interconnected by "wires." One of DASYLAB's most useful features is that it can be used in a "simulation" mode to try out new systems *without* the need to have any sensors or other laboratory apparatus connected to the computer. When using this simulation mode, the measurement sensors that would normally provide input to the A/D converter are simulated by DASYLAB-generated signals, using a module called `signal generator`. This module (sometimes in combination with other modules) can simulate most any kind of sensor signal (transient, periodic, random, or mixed) that you might want to represent true physical signals. Once these "sensor" signals have been generated, the *rest* of the DASYLAB system will be *exactly the same* for the "working" configuration as for the simulation mode. Thus, we can work out most of the "bugs" in a proposed new system *before* we go into the laboratory to actually take measurements.

Using the DASYLAB software that comes with each copy of this book, you can install a *simulation* version on your own computer. Then you can use this simulation mode to quickly learn how to use DASYLAB and/or try out your ideas for any data-acquisition and -processing system you might be interested in. The simulation version does *not* allow you to connect to data-acquisition boards and sensors to set up an *actual* (not simulation) system; you must purchase the appropriate DASY-LAB software (which includes detailed instruction manuals) for this purpose. While the *manuals* are not included with the enclosed simulation version, the upcoming discussions and examples in this text, together with the comprehensive on-line help (which *is* included in the simulation version) make it quick and easy to explore the capabilities of DASYLAB.

The DASYLAB Functional Modules

DASYLAB provides over a hundred functional modules, each of which is represented by a small icon. You select modules either from a *toolbar* of icons along the left side of the screen or from *pull-down menus* selected from a *menu bar* along the top of the screen (see Fig. 13.1). The menus contain *all* the modules, but the toolbar can hold only about 16 modules. If you do not like the default set of toolbar modules, you can customize it to suit your needs. Left clicking on a toolbar icon or a menu item moves a copy of that module onto the "worksheet" area of your screen. You can then drag it anyplace you like. A module can be *removed* from the screen by right double-clicking on it. Worksheets can be larger than the screen size; you can scroll both vertically and horizontally to provide space for large diagrams.

A particular data-acquisition and -processing system is created by selecting a suitable set of functions (icons) and properly connecting them with "wires." Icons display "ports" (terminals) that can be connected to other icons. Some icons have only an "in" port or only an "out" port; others have both kinds of ports. To connect one icon to another, click the cursor on, say, an output port and then move (by any path) the "plug" symbol that appears to the input port and click again. A "wire connection" of vertical and horizontal segments will appear, connecting the two modules. (An "autorouter" function cleverly "steers" the "wiring" to avoid crossing any modules. The autorouter can be disabled if you desire.) Once a "connecting wire" has been installed, you can *branch off* from it at any point to connect to other modules if that is needed; thus, the output of one module can be sent to many other modules. To create a branch, click once anywhere on the "wire" and then move the "plug tip" that appears to the desired input port, clicking again when you get there. To *delete* a wire, right click on it and the *section* to be deleted will change color. If that is the section you *want* to delete, right click again and it will disappear.

Before a module is connected to others, it can be positioned at will by clicking and dragging. *After* connections have been made, moving a module may (or may

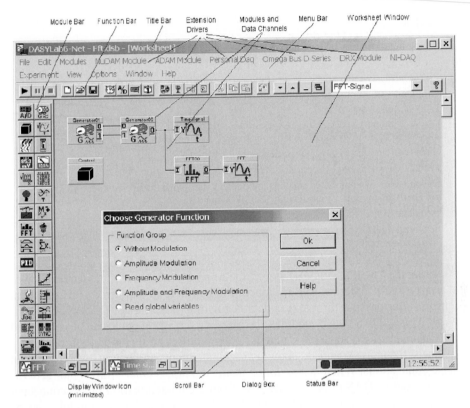

Figure 13.1
The DASYLAB worksheet screen.

not) result in disruption of some connections, due to the action of the autorouter. It is thus best to, as far as possible, position all the modules in a system before making the connections. One can, of course, recover from disruptions by deleting, moving, and/or reconnecting. In addition to the *module bar,* there is also a *function bar* and a *menu bar* at the top of the screen, as shown in Fig. 13.1. We will discuss these later, when it is more appropriate.

List and Brief Description of the Functional Modules

We next want to name and list all the modules and then briefly describe the ones most commonly used. The listing is by *groups;* modules in a group perform similar functions.

Groups	Modules
INPUT/OUTPUT	analog input analog output digital input digital output counter input signal generator
TRIGGER FUNCTION	combi trigger pre/post trigger start/stop trigger relay
CONTROL	PID control switch slider latch time delay stop TTL pulse generator
MATH	arithmetic trigonometry scaling differentiation/integration logical operations formula interpreter
STATISTICAL	statistical values (max min max position min position mean rms variance) min/max position in signal histogram regression counter
SIGNAL ANALYSIS	filter correlation FFT data window polar/cartesian
DATA REDUCTION	average block average separate merge/expand cut out time slice
DISPLAY	analog meter digital meter list display (tables) bar graph chart recorder y/t chart x/y chart histogram status display (lamp) write data
FILES	read data write data backup data
SPECIAL	new black box export/import (used with black box) action message time base signal adaptation

From the **INPUT/OUTPUT** group, the **analog input** module would be used in nearly every practical system since this is where signals from measurement sensors, after passing through the data-acquisition board, enter DASYLAB itself. As with many DASYLAB modules, the **analog input** module can be configured with as many as 16 separate channels. When you set up this module for say, five channels, the icon expands to show five separate inputs and outputs where you can make connections. When in *simulation mode,* the inputs come from *signal generators* internal to DASYLAB, rather than from a data-acquisition board connected to measurement sensors. Each channel of the **analog input** module can be assigned its own descriptive name and, if the board has a programmable-gain amplifier, its own voltage range; otherwise, all the channels have *the same* voltage range.

To use the **analog output** module, the board must have one or more D/A converters (multifunction boards typically have several). Analog output capability allows DASYLAB to be used for *control* purposes, not just for measurements. Control could be the *main* function of a DASYLAB system or it could be an *auxiliary* function used to automate or otherwise control various operations in a measurement system. For example, you might want DASYLAB to control a motor that positions a pitot tube, to make velocity measurements, at selected locations in a flow channel. A position sensor would tell DASYLAB (through an A/D converter) where the pitot tube was, and DASYLAB would compare this with a desired position to determine a position error. This error could be processed in DASYLAB's **PID control** module, and a command sent out through a D/A converter to a power amplifier driving the motor. When the pitot tube is driven to the correct position, DASYLAB could then take readings from the pitot tube's static and stagnation pressure sensors, and from this calculate the velocity.

The various **trigger** modules are used to start and stop the taking of data, much as in an oscilloscope. The **trigger** module "watches" the stream of digitized data coming from the board and when a signal crosses a selected trigger level, it sends a signal to a **relay** module. The **relay** module acts as a gate, which blocks the data stream until it receives a trigger signal, whereupon the data is accepted for storage and/or processing. Triggering is sometimes also controlled at the data-acquisition board, rather than in the software. The **stop** module can be used to control in several versatile ways how much data is collected. We can take a desired number of samples or blocks of data, or we can take data for a certain length of time in response to a pulse from some other module or a signal that is above (or below) a selected value. The **time delay** module has a number of uses. When we time-share a single A/D converter and amplifier among several data channels (the usual case), the channels may exhibit *time skew* since the channels are sampled *in sequence* rather than simultaneously. (There *are* ways to sample simultaneously but we cannot always use them.) If the time skew is excessive and we know the Δt between channels, we can bring the channels back into synchronism by passing the data streams through **time delay** modules set to make up for the time shift from channel to channel. This technique can also be used to correct for *sensors* that have different time shifts (phase angle shifts in their frequency-response curves).

Modules in the **MATH** group allow versatile processing of our data; their names (for example, **arithmetic**, **trigonometry**, and **logical operations**) usually indicate their capabilities. In a multichannel system, we can do calculations with a single channel or several channels. The **formula interpreter** module has up to 16 inputs [called **IN(0)**, **IN(1)**, **IN(2)**, ..., **IN(15)**] and 16 outputs. You can define an output to be almost any function of the inputs. The **scaling** module provides functions such as converting voltage to temperature for several types of thermocouples. Many sensors have linear calibration curves; the **scaling** module can use two points on such a curve to convert voltages into physical units. It also can perform the important *table lookup* function. Here we must create a reference file with a table of (x, y) pairs of points that document the relationship we want DASYLAB to use in getting y values from given x values.

Modules in the **STATISTICAL** group again have names that indicate familiar operations such as **average**, **standard deviation**, **min**, and **max**. The **position in signal** module is quite useful when we need to select, say, the first, last, or thirteenth data value from a stream of data. Once selected, the value can be used in further calculations if you desire. In the **SIGNAL ANALYSIS** group, the **filter** module provides low-pass, high-pass, band-pass, and band-reject capabilities, while the **FFT** module computes the frequency spectrum of time-varying signals. Sometimes we want to sample at a high rate but only *keep* a fraction of the total samples. An example is the *"burst mode"* of data acquisition, used to reduce the *time skew* between channels that we mentioned as a problem earlier. Here, we might sample 4 channels, at 1000 samples/s (250 samples/s for each channel) but only *keep* the values every 0.1 s. If we *sampled* every 0.1 s, there would be a time skew of 0.4 s between the first and the fourth channel, whereas the burst-mode approach gives a time skew of only 0.004 s. The **separate** module of the **DATA REDUCTION** group allows us to sample at any rate but then keep, say, only every 100th sample.

Modules in the **DISPLAY** group give us a choice of many ways to display the data or computed results. Again, most of the names (for example, **x/t chart**,

x/y chart, and **digital meter**) are self-explanatory. The **list display** module produces tables with a column for any variable we choose. The "index" for the table can be either the sample number or the time of the sample. Multichannel "strip-chart" recordings can be produced with either the **y/t chart** module or the **chart recorder** module, the **y/t chart** module being used for "fast" events and the chart recorder for "slow" events. The **status display** module provides "lamps" that "light up" when certain events occur.

The modules in the **FILES** group let you read from or write to files of various formats. It is also possible to transfer data, using the clipboard, directly between DASYLAB and other Windows applications that might be on your computer. In the **SPECIAL** group, the **black box** module lets you make up your own combinations of modules, which can then be reduced to a single icon for future use. You can in fact put black boxes *within* other black boxes. The **action** module provides for versatile control of DASYLAB functions and operations using the concept of *event-driven actions*. Here, a *sender* perceives an event and signals a *receiver* to perform some action. For example, a printing action could be enabled only when a certain signal exceeds a selected level. The **time base** module will often be used to create a "time column" in a list display (table) or when time values are needed in certain calculations. One use of the **signal adaptation** module is to synchronize data streams with different sampling rates, block sizes, or starting times. If you try to record a set of such signals on a chart recorder, it may refuse to record them. Sending one or more of the channels into a **signal adaptation** module *before* they get to the recorder will sometimes solve this problem.

13.3 DASYLAB SIMULATION EXAMPLE NUMBER ONE

We will be presenting three examples of simple but useful data-acquisition systems for specific practical applications using DASYLAB in simulation mode. They will be given as if the reader were actually developing the application on a computer, which is, of course, the best possible way to learn about DASYLAB's features and capabilities. Readers who do not have the software available can still get quite a good appreciation of its features and capability by following the text's presentation.

This first exercise includes some of the most common basic operations that occur in many practical data-acquisition systems. In the text below, left clicks are just called *clicks;* right clicks are called *right clicks.* While you are working your way through this exercise, feel free at any time to explore DASYLAB's context-sensitive **HELP SYSTEM**. It duplicates, and sometimes goes beyond, what could be found in the complete DASYLAB manual.

Simulating Sensor Signals and Recording Them versus Time

Log on to DASYLAB and begin constructing your system diagram by clicking on **FILE** and then **NEW** to create a blank worksheet screen. Then click on **generator** in the module bar along the screen's left edge; a dialog box appears that allows you to select different types of signal generators. Choose the default type (**without**

modulation). A **generator** module will now appear on your screen. (*Readers who do not have the software available for actual use should now look at Fig. 13.2; it shows the final complete worksheet, but you should be able to follow our steps as we build it piece by piece.*) Click on it and drag it to a position near the top left of your screen. Now right double-click on it and a **deletion** dialog box appears. Hit **OK** to delete the **generator** module. (Now you know how to delete *any* module.) Now click on **MODULES** in the menu bar at the screen top to display a menu of module types. Click successively on each item to explore *all* the available modules (recall that the module bar of icons can only show about 16 items). When you get to the menu item **CONTROL**, click on **generator** in the submenu to see how you can obtain the **generator** module from the modules *menu,* rather than by clicking on its icon in the module bar. If a module does not appear as an icon in the module bar, you can always find it somewhere in the menu. Now select **without modulation** to again display the **generator** module on your screen.

To set up the **generator** for a desired type of signal, double-click on its module to display a dialog box that allows choice of:

1. How many separate signals (channels) you want (as many as 16)
2. The type of signal (sine, square, etc.) for each channel
3. The frequency, amplitude, offset, and/or phase shift for each channel's signal

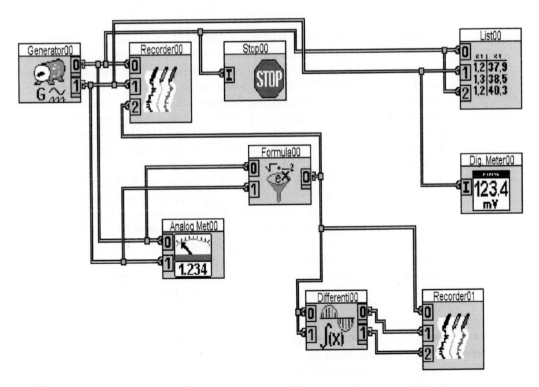

Figure 13.2
A simple example of a DASYLAB data acquisition/processing system.

We want *two* signals for this demo, so click on the + symbol at the right of the channel list and channels 0 and 1 will "light up." If a channel is *green,* you can select numbers for its frequency, etc. To change from *yellow* to *green,* just click on the *yellow;* then *that* channel's signal can be set up. Now set up channel 0 as a 0.117-Hz sine wave with amplitude 4.0 V and 0.0 offset and phase shift. Then set up channel 1 to be a 0.137-Hz triangular wave of amplitude 1.00 V and 0.0 offset and phase shift. Then hit **OK** to "install" these values.

Since we almost always want a chart recording of our experiment variables, let us next install a **chart recorder** module (you can use either the module bar or the module menu). When it appears, drag it to the right of your **generator** module and then double-click on it. For the **TIME AXIS**, select **CONTINUOUS** and **START AT LEFT**; for **ZOOMING OPTIONS**, select **X AND Y DIRECTION**; and for **DISPLAY TIME**, enter 21.0 s. Set up three channels (0, 1, and 2) and then click on **SCALING** to get another screen that allows you to set the "sensitivity" of each recorder channel. Set the three channels as follows (select **AUTO**, *not* **USER DEFINED**):

Channel	Display from	Display to
0	− 5.00	7.50
1	− 5.00	7.50
2	− 5.00	7.50

For **DISPLAY MODE**, select **DISPLAY CHANNEL** (x in box). We have set up three channels to accommodate another channel not yet present in our system. We *could* add recorder channel 2 *later,* but the **recorder** module tries to *expand,* and if space is not available ("bumping into other modules"), it *refuses* to expand. We can, of course, at that time rearrange the diagram, but prefer to avoid this by "planning ahead" now. Now, using the method explained earlier, "wire" the generator channels to the recorder inputs; generator 0 to recorder 0 and 2, and generator 1 to recorder 1. If you left channel 2 "empty" for later use, you would get an error message: **an input is open.** Click on **EXPERIMENT** in the menu bar and then on **EXPERIMENT SETUP**. In **GLOBAL SETTINGS**, set **SAMPLING RATE/CH** to 5.0 Hz, **BLOCK SIZE** to 1, and then click **OK**. When a **recorder** module is used, it is given a number, and an icon with this number appears near the bottom of the screen (you can drag it elsewhere if you wish). Find this icon (labeled **RECORDER 00**) and drag it into some clear space near the bottom of your worksheet. Then double click on it and the recorder "chart" will appear. This chart window has its own menu bar and function bar that allows zooming, unzooming, grid lines, etc. Move the cursor slowly over the function bar to identify the various functions available. Then click on the **GRID** tool (the eighth from the left) to get chart grid lines.

To start the experiment, you can either click on the **START** symbol (extreme left on the main function bar) or click on **EXPERIMENT** and then on **START**. You should then see the three chart traces building up as time goes by. To *pause* the collecting of data, click on the **PAUSE** symbol (just to the right of the **START** symbol). To resume data collection, click on **PAUSE** again; to stop collecting data, click on the **STOP** symbol (just to the right of **PAUSE**). If you do not click on **PAUSE** or **STOP**, the data collecting will go on "forever."

Stopping an Experiment at a Selected Time

Often we want our system to stop taking data after a certain time or a certain number of samples. The **stop** module can be used for this; let us add it to our system by clicking on its module-bar icon. Suppose we want data taking to stop after 101 samples have been collected in channel 0. Connect ("wire") channel 0 to the input (I) of the **stop** module. Then double-click on the **stop** module to get a dialog box. In **ACTION AFTER**, select **SAMPLES** in **ACTION**, select **stop** in **PARAMETER**, enter 101, and then hit **OK**. Now start the experiment again; it should stop itself after about 20 s of chart time (101 samples at 5 samples/s). If everything seems to be working fine, this might be a good time to start *saving* this file. As in any computer programming, it is wise to save the program periodically so that errors or crashes do not wipe out all your hard work. Go to **FILE**, then **SAVE AS**, and type in the filename *xxxxxxxx*.dsb, where the *x*'s are as much of your last name as will fit the space.

Chart Recorder Options

We next want to explore some of the manipulations possible with the chart recorder. Expand the chart window to full screen to show the entire menu bar and function bar. Move the cursor slowly along the function bar to see what each icon means. Also, click on each menu-bar item to examine the pull-down menus. Your screen should be showing the **MULTIPLE CHARTS** display (three separate traces and axis sets). Sometimes we prefer a one-chart display. Click on **ONE CHART** to see this; then return to **MULTIPLE CHARTS**. Now click on **GRID** to turn it on, off, and on again. Try **ZOOM** and **UN-ZOOM**, **CURSOR**, **ZOOM BETWEEN CURSOR**, **COLORS AND LINES**, and **FONT STYLE**. We often want to *annotate* our graphs. Click on **TEXT** in the menu bar and then on **EDIT NEW TEXT STRING**. Move the cursor to a "clear space" on the graph and click. A window appears where you can write a note; type in your full name and click **OK**. The note appears; you can drag it wherever you please. Now **PRINT** this graph.

Producing Tables or Lists

We often want *tables* of data, not just graphs. Click on the **LIST** icon to place this module on your screen. Double-click on it and configure it for three channels. Set each channel (remember the green and yellow business earlier) for eight digits with four decimals, and then click on **OPTIONS**. Select **NORMAL FORMAT**, **MEMORY PER CHANNEL 9999** samples, **SHOW TIME CHANNEL** (x), and **SAMPLE TIME** as **TYPE OF TIME CHANNEL**. Now, "wire" the three channels to the **LIST** module, and then double-click on the **LIST 00** icon (bottom of screen) to display the **LIST** window (a four-column table). Expand the window to make sure all four columns are visible. Then **START** the experiment to see the table values being entered; stop after about 30 samples and print this table. You can of course, as in other Windows applications, display *both* a graph and a table simultaneously, assuming there is sufficient screen space.

Analog and Digital Meters

Just as "real" meters are useful in many experiments, "virtual" meters are often employed as part of a digital data-acquisition system display. Digital meters are best for showing accurate values; analog meters are better for showing trends in the variable. Click on the **DIG. METER** icon to place a meter module on your screen. "Wire" it to channel 1 and double-click on it. Under **SETTINGS**, enter −0.5 for **LOWER LIMIT** and 0.5 for **UPPER LIMIT**. These limits can be used as *warnings* or *alarms* to an operator when the variable enters unsafe regions. Under **MODE**, select **LAST VALUE**. Then click on **COLORS**, and then in **COLOR CHANGE**, click on **UPPER**. Then click on the red color block to select red as the "warning color" when the upper limit is exceeded. Click **OK** and then **LOWER**, and again choose the red color. Now go back to the **DIGITAL METER** window and click on **OPTIONS**. Under **NUMERICAL VALUE**, set **DIGITS** at 8 and **DECIMALS** at 4, then click **OK**, **OK** to "back out" to the worksheet screen. Now **START** the experiment, with the digital meter displayed (you can make it large or small) and note how the meter's display readings changes color as the alarm limits are crossed.

Now place an **ANALOG METER** on your diagram and configure it for two channels (this can be done also with digital meters). Wire **GENERATOR** channel 0 and 1 to the meter module. Set **MODE** at **LAST VALUE** for each channel. Double-click on the **ANALOG MET00** icon (somewhere at bottom of screen) to display the two meters, grouped into a single panel. Start the experiment to see the meters operate. "Danger limits" can also be set and displayed on analog meters.

Some Simple Data-Processing Operations

Having shown some useful data-*display* methods available in DASYLAB, we now want to explore some *data processing*. The possibilities here are really limitless because we can *combine* the large selection of basic operations in an infinite variety of ways. We thus can only give a brief sampling of some basic operations. Click on **MODULES**, **MATHEMATICS**, and then **FORMULA MODULE** to place module **FORMULA 00** on your screen. We want to calculate the expression (channel 0) + 2 (Channel 1).[2] Double-click on the formula module now on your screen, and then, under **INPUT CHANNELS**, click on the + sign to set this to 2 (it *can* be set up for as many as 16 input channels). We want only one *output* channel, so use the channel bar (0 to 15) to activate only channel 0 (we *could* have as many as 16 outputs, each with its own formula). Then move the cursor to the "white space" labeled **FORMULA** and click in this space. Below the white space, click on **INPUT** and then on **IN(0)** to start writing the formula. Then click on **OPERATOR** and then on +. Then type 2.0 and click on **OPERATOR**, **x**, **CONSTANT**, **(**, **INPUT**, **IN(1)**, **CONSTANT**, **)**, **OPERATOR**, and **^**; then type 2. On your chart recorder module, *disconnect* recorder channel 2 from **GENERATOR** channel 0 and replace it with the output of **FORMULA 00**. Now display the chart recorder and start the experiment to see how your computed function behaves.

Integration and Differentiation

Sometimes we want to compute integrals and/or derivatives of sensor signals or computed results. We might, for example, integrate a mass flow meter signal (kg/s) to compute the total mass (kg) that passed through the meter over some timed interval, or we could differentiate a velocity signal to get acceleration. To explore this, now click on **MODULES**, **MATHEMATICS**, and then **differentiation/integration** to place this module on the screen. Configure output channel 0 as the derivative and channel 1 as the integral. Connect the output of the **FORMULA 00** module to *both* inputs since we want here to take the derivative and the integral of this computed quantity. Place a *new* chart recorder module on your screen to record the input and the two outputs of the DIFF/INT module. (By this time you probably know how to do these operations, so I am leaving out the details.) Your worksheet should now look like Fig. 13.2 or its equivalent. Run the experiment and observe the results; you may need to adjust some graph scaling to keep all the curves "on the paper." On the new chart recorder, write text notes to identify the input trace, the integral trace, and the derivative trace. Also print your full name on this graph.

13.4 DASYLAB SIMULATION EXAMPLE NUMBER TWO

This next exercise shows how DASYLAB computes and displays the frequency spectrum of a time-varying signal. Getting the frequency content of physical signals is one of the most useful data-analysis functions used by engineers and is provided in almost all data-acquisition and -processing software by using FFT methods. The DASYLAB FFT operation is quite simple to use. We again use the simulation mode to practice this skill. We will need to generate an appropriate time-varying signal and then send it to the **FFT** module. We need to be careful (both in simulation mode and of course in actual applications) to adjust our sampling rate so as to not *alias* the highest frequencies present in our time-varying signal. In a real-world application, we also usually need to use *analog* antialiasing (low-pass) filters on our sensor signals *before* they get to the A/D converter. Usually we need to apply a *window* to our digitized data before it is FFT analyzed. DASYLAB provides a selection of such windows. If you have ever used a *spectrum analyzer,* you may recall that these instruments usually provide the capability of *averaging* several data samples to get a more reliable frequency spectrum. DASYLAB also provides this capability. Recall some basic relationships for any FFT analysis:

1. If your data sample lasts T seconds, the lowest frequency that will be computed is $1/T$ Hz, the spacing between computed frequencies will also be $1/T,$ and the highest frequency will be given by $f_{\max} = (1/T) \cdot (N/2)$, where N is the number of samples taken of the time-varying signal. All FFT software works best if you choose N to be some power of 2, such as 1024 or 2048.

2. Frequency content is only computed at *discrete* frequencies, spaced at intervals of $1/T$ Hz. There is *no* information about frequency content available between these discrete values. If a true peak exists at 1.0 Hz and the nearest discrete computed frequencies are, say, 0.97 and 1.07 Hz, then the *indicated* peak will be at 0.97 or 1.07 (an error in frequency) and the *height* of the peak will also be in error. Also, if there are truly *two* sharp peaks very near each other and the frequency resolution is too coarse, then we may see one "broad" peak rather than two distinct narrow peaks. All these types of errors can be reduced by using a longer time sample (larger T), which increases the frequency resolution. If we maintain the same sampling rate, we pay the price of larger sample sizes and slower computation.

3. For a time-varying signal that contains frequencies up to f_{max} Hz, the sampling rate should be *at least* $2.56 \cdot f_{max}$ samples/s. In a practical application, f_{max} is usually set by our choice of the cut-off frequency of the antialiasing (low-pass) filters.

4. For a sampling rate of f_{samp} Hz, one-half this frequency is called the *Nyquist* frequency f_N. Any frequency content in the time-varying signal that is above the Nyquist frequency will cause *aliasing,* that is, there will be frequencies present in the computed spectrum that are not really present in the time-varying signal. The effects of aliasing can sometimes be detected fairly simply. If, for example, you sample at 100 Hz, then the Nyquist frequency would be 50 Hz. If, say, your signal had some frequency content at 65 Hz, you might get some computed frequency content at $50 - (65 - 50) = 35$ Hz. Thus, an actual frequency that is 15 Hz above the Nyquist frequency causes an aliased (incorrect) frequency at 15 Hz *below* the Nyquist frequency. A general rule for aliasing goes as follows: If the computed frequency spectrum shows a peak at a frequency f, this *could* be a true peak or it *could* be due to aliasing caused by frequencies $(f_{samp} \pm f)$, $(2f_{samp} \pm f)$, $(4f_{samp} \pm f)$, etc. This rule is general, so be on the lookout for "unexplained" frequency content; it might be due to aliasing. If we conscientiously use *antialiasing* (low-pass) filters as mentioned above, aliasing should not be a problem.

With the above background in mind, let us now start building our new DASYLAB application. Open a new DASYLAB worksheet and place a **generator** module at the upper left of your screen. Configure this generator for three channels (0, 1, and 2). Make channel 0 a 1.0-Hz *sine* wave of amplitude 2.0 (offset and phase shift are both 0.0). Make channel 1 a *noise* signal of amplitude 5.0 and 0.0 offset. Make channel 2 a *triangular* wave of frequency 11.5 Hz and amplitude 4.0 (the offset and phase are both 0.0). (Recall that a triangle wave has only the odd harmonics ($n = 1, 3, 5, \ldots$) and the amplitudes of the harmonics are given by $8A/(\pi n)^2$, where A is the amplitude of the triangle wave.) (*Readers who do not have the software available for actual use should now use Fig. 13.3 to follow the text's development.*) Now place a **filter** module to the right of the generator and configure it for one input. Set the filter for **LOW PASS**, 10.0 Hz, fourth **ORDER**, and **BUTTERWORTH** (see Chap. 10 for a brief discussion of filter types). Connect channel 1 of the generator to the filter input. We are low-pass filtering the

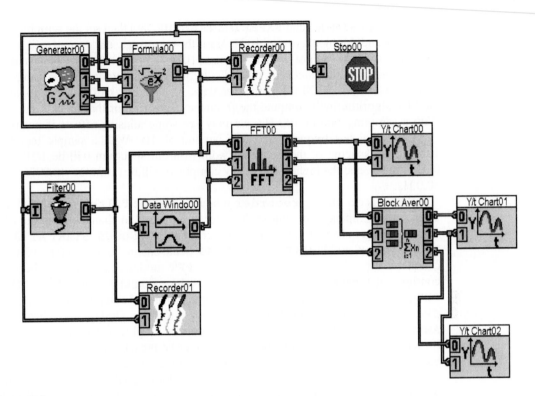

Figure 13.3
A DASYLAB application using frequency-spectrum (FFT) analysis.

random-noise signal to make it more physically realistic and to reduce its high-frequency content to prevent aliasing. Without filtering, it would have sharp pointed peaks, rather than the rounded peaks seen in real-world random variables.

Now place a **formula** module to the right of the filter and configure it for three inputs and one output. Connect channel 0 of the generator to channel 0 of the **formula** module, channel 2 of the generator to channel 2 of the **formula** module, and the filter output to channel 1 of the **formula** module. Then set up the **formula** module to perform the calculation **IN(0) + IN(1) + IN(2)**, that is, we want to sum the three signals to get the time-varying signal to be frequency analyzed. Place a **chart recorder** module to the right of the **formula** module, configure it for two inputs, and connect input 0 to channel 0 of the generator and input 1 to the **formula** output. Set the time axis as **CONTINUOUS**, **START AT LEFT**, **ZOOM DIRECTION X AND Y**, and **DISPLAY TIME** = 11.0 s. Set the scaling for both channels at −7.5 to +7.5, **AUTO**.

Place a **stop** module on your worksheet, connect it to channel 0 of the **generator** module and set it for:

ACTION AFTER	BLOCKS
ACTION	STOP
PARAMETER	10

Click on **EXPERIMENT** and **EXPERIMENT SETUP**, and then set the sampling rate at 100 Hz and the block size as 1024. With these settings, our system will gather 10 blocks of data, each of which will have 1024 samples in it. The FFT analysis always works with *blocks* of data. Each block will require (1024 samples)/ (100 samples/s) = 10.24 s to gather the data. As each block of data is completed, the FFT algorithm will compute the frequency spectrum of that block's samples. With a sampling rate of 100 Hz, we may see some aliasing effects if our time-varying signal has frequency content beyond 50 Hz. With a sample length of 10.24 s, the spacing of the frequency points on our spectrum will be 1/10.24 = 0.097656 Hz, and the highest frequency computed will be (1024/2) · (0.097656) = 50.0 Hz.

Next place a **chart recorder** module under the **filter** module, configure it for two inputs, and connect the filter input signal to channel 0 and the filter output signal to channel 1. Set this recorder for **CONTINUOUS, START AT LEFT, ZOOM X AND Y**, and 11.0-s **DISPLAY TIME**. Scale channel 0 for −3.0 to +3.0 and channel 1 for −7.5 to +7.5. Now place an **FFT** module on your worksheet and configure it for three inputs (this also gives three outputs). Select **AMPLITUDE SPECTRUM** as the **OPERATION** for channels 0 and 1, and **POWER DENSITY SPECTRUM**, usually called *power spectral density* (PSD), for channel 2. **AMPLITUDE SPECTRUM** will show the correct amplitude for any sine waves (or harmonics of periodic functions) that might be present in the analyzed signal, but it is not the most useful result for any random components. **POWER DENSITY SPECTRUM** gives the result most useful for random signals, and since most real-world signals have some random content, it is the most used in practical FFT analysis. *Both* displays, however, clearly show the presence of "peaks" and "valleys" in the spectrum, which is by itself often useful information.

Now place a **data window** module on your worksheet. Select **HANNING** as the window type and 1024 for **VECTOR LENGTH** (it should always be the same as the block size). Connect the output of the **formula** module (our time-varying signal) to channel 0 of the **FFT** and also to the input of the **DATA WINDOW**, that is, we want to do one spectrum analysis *with* a window (the recommended procedure) and one *without* to see their effect. For continuously varying data, such as ours, a window (often the Hanning) is usually suggested. (*Transient* data requires another treatment.[3]) If *no* window is used for continuous data, the variable suddenly "jumps up" from zero at the beginning of a block of data and suddenly "drops down" to zero at the end. These *sudden* changes are *not* part of the real signal's behavior and introduce into the computed spectrum spurious frequencies called *side lobes*. For example, if a *true* frequency peak occurred at, say, 15 Hz, the computed spectrum might also show some frequency content on either side of 15 Hz. These side lobes would be *errors* and not really representative of the signal's true spectrum. Windows such as Hanning's suppress these side lobes so that we are not misled about the frequency content of the signal. They accomplish this by replacing the sudden rise and fall by *gradual* changes for about the first and last 10 percent of the

[3]E. O. Doebelin, "System Dynamics," Marcel Dekker, New York, 1998, pp. 662–667.

block. Such windows make the width of computed peaks in the spectrum a bit wider than they really should be. Thus, if a spectrum truly has *two* sharp peaks very close to each other, they may show up in the computed spectrum as a *single* wide peak, that is, the window somewhat reduces the *frequency resolution* of the analysis. Usually the "good" effect of side-lobe suppression outweighs the "bad" effect of poorer frequency resolution; so such windows are regularly used.

The output of our **FFT** module is a "table" of frequency content ("amplitude") at each discrete frequency. We may display this as a graph, so now place a **y/t chart** module on your worksheet and configure it for two inputs. Connect **FFT** outputs 0 and 1 to **y/t chart** inputs 0 and 1, respectively. (The **y/t chart** module is "smart" enough to know that when its input comes from an **FFT** module, it should plot against *frequency,* not *time.*) This display will allow us to easily compare the spectrum produced by an unwindowed signal with that using a Hanning window. On the **y/t chart** module, set **DISPLAY WIDTH** as **BLOCK RELATION** and **1**, with **ZOOMING OPTION** as **X** and **Y DIRECTION**. For **SCALING, SETTINGS** set **DISPLAY FROM** at 0.0 and **DISPLAY TO** at 3.000 and **AUTO**.

We next want to demonstrate some *averaging*. We want to get a running average of the 10 spectra as they occur. Add a **block average** module to your worksheet and configure it for three inputs (it also gives three outputs) and **RUNNING** mode. Connect outputs 0, 1, and 2 of the **FFT** module to inputs 0, 1, and 2 of the **BLOCK AVERAGE**, respectively. Place a new **y/t chart** module with two inputs on the screen and connect inputs 0 and 1 to outputs 0 and 1 of the **BLOCK AVERAGE**, respectively. This completes the configuration of our system. Your worksheet should now look "something like" Fig. 13.3.

Running the Demonstration

Click on the **START** symbol to start the experiment and then display the two "time recorders" at the largest size that allows you to see *both* of them on the screen. You will not see *any* signal traces until the first block of data has been gathered (this will take about 11 s). Then you will see all 10.24 s of each trace. At each successive 10.24-s interval, the traces will be updated with a new block of data until the display "freezes" when the last (10th) block of data has been gathered. Wait for this frozen block and then expand to full screen the recorder with the pure sine wave as one trace. The lower trace is the *total* signal. It *contains* the sine wave, which is the upper trace, but we could never detect the presence of this sine wave "by eye." It is obscured by the other parts of the composite signal. The *frequency* spectrum (which we shall shortly display) will *clearly* show the presence and strength of this sine wave, demonstrating the utility of the frequency-spectrum analysis.

Now expand the *other* time recorder to full screen. The lower trace is the random part of the total signal and the upper trace is the low-pass filtered version of the random signal. Use **ZOOM** to expand about one-tenth of the total time to full screen width. You should see clearly how the filtering changes the frequency content of the random signal. Use the **CURSOR** to roughly estimate the "highest frequency" you can see on each trace, and use **TEXT** to write a short note on this graph, giving these two frequencies, your name, and the date. Make a hardcopy. If we had not filtered the random signal, our 100-Hz sampling rate might not have been fast enough to prevent aliasing of the random part of the signal.

Now minimize the two time-chart windows and then display the `y/t chart` that shows the windowed and unwindowed spectra, with *no* averaging, and expand it to full screen. Your display should be the "single-curve" type (shows both traces on the *same* axes) since this is the default. Switch to the "multiple-curves" display, which separates the two curves, to get clearer pictures of each. Adjust the scaling of the two traces so that they have the *same* scale and "fill the paper" as much as possible. Recall that what you are seeing is the *last* (10th) spectrum computed; the other nine have "disappeared." To see *all* these spectra, one by one, just restart the experiment and watch the display. (To get a "frozen" display of all 10 spectra, try the `WATERFALL` display later, if you wish.) Confirm distinct peaks at 1.0, 11.5, 34.5, and 42.5 Hz. The peak at 42.5 is *not* predicted as a harmonic of the triangle wave; these would be at 11.5, 34.5, 57.5, and 80.5, etc. Explain the presence of the 42.5-Hz peak and state whether it should be treated as a true part of the time-varying signal, using TEXT to write a note (include your name) on this graph, explaining your decision, then make a hard copy for a permanent record. You will note that both the windowed and unwindowed displays seem to work well here in identifying the frequency content of this signal. There *are* differences, but they are subtle, and we will shortly demonstrate the effect of windowing on frequency resolution.

Now restart the experiment and expand to full screen the chart that shows the windowed and unwindowed spectra; but now with averaging. The first block's spectrum will appear in about 11 s. Note the "jaggedness" of the spectrum contributed by the random signal; it extends from 0.0 Hz to a little beyond 10 Hz, the cutoff frequency of our low-pass filter. Averaging is particularly needed for the random portion of any frequency spectrum. Each block actually gets a different statistical sample of the random signal, so each block's spectrum will be different, even though the *statistical* parameters of the random signal do not change from block to block (so-called *stationary* random signal). By averaging several successive spectra, we, in effect, use a larger statistical sample, and the uncertainty (the jaggedness we see) is less in the average than in the individual blocks. Averaging is often useful even if the signal has strong nonrandom content, such as vibration signals from rotating machinery. Here, the machine may run at *nearly* steady speed, giving *nearly* periodic signals with distinct "spikes" in the spectrum. Actually, the speed of all real machinery exhibits some random variation, so we get a more representative spectrum by doing some averaging. Observe the averaging process as it proceeds through all 10 blocks of data; the random part of the spectrum should get somewhat "smoother" with each successive averaging operation.

Finally, we want to demonstrate the loss of frequency resolution, which is the "bad" aspect of windowing. To do this, we will change our time signal so that it has two closely spaced frequency components. When two frequencies are sufficiently close, a windowed analysis may show only a *single* broad peak rather than the two distinct peaks that are really present. Go to the `generator` module and replace the random signal with a 1.10 Hz sine wave of amplitude 2.0. We now have frequencies of 1.00 and 1.10 present in equal strength and want to see if our FFT analysis will

handle this properly. Start the experiment in the usual way and then display the `y/t`
`chart` with the unaveraged, windowed, and unwindowed spectra. Then repeat this
for the averaged spectra. Use zooming and/or scaling to get the best displays. Figure
13.4 shows some graphical results from this system.

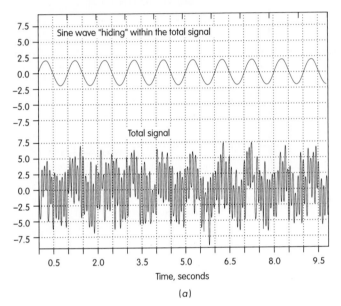

(a)

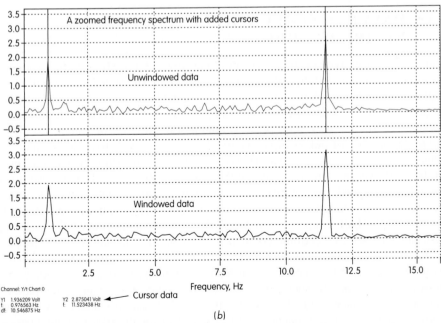

(b)

Figure 13.4
Selected results from the FFT analysis system.

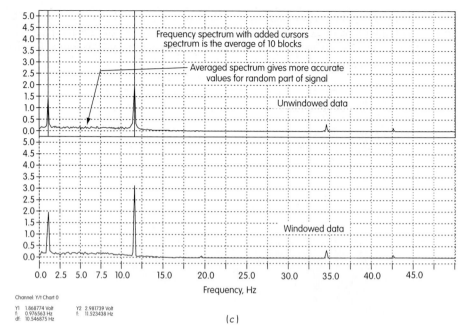

Channel: Y/t Chart 0

Y1 1.868774 Volt
f: 0.976563 Hz
df: 10.546875 Hz

Y2 2.981739 Volt
f: 11.523438 Hz

(c)

Figure 13.4
(Concluded)

13.5 DASYLAB SIMULATION EXERCISE NUMBER THREE

Our last exercise involves measurement of the speed of a rotating shaft, a common requirement in many studies of machinery operation. The speed sensor will be an inductive proximity pickup. The sensor is positioned close to a steel gear attached to the shaft, the gear having, in our example, 78 teeth. As each tooth passes the sensor, it causes a large voltage pulse, similar to a single cycle of a sine wave. If we count the number of pulses in an accurately timed interval, we can calculate the rotary speed of the shaft. Such a method of course will measure the *average* speed of the shaft over the time interval, *not* the instantaneous speed. To get closer to an instantaneous speed, we can make the time interval shorter, but we then get lower accuracy and resolution since we have fewer counts. That is, if we count for only 1 s, a 100-rpm shaft will produce $(100 \cdot 78)/60 = 130$ counts. Since pulse-counting instruments, such as electronic counter/timers, can miss the starting and stopping points of a timed count by as much as $\pm$ 1 count, our speed reading could be in error by as much as $\frac{1}{130} = 0.77$ percent. If instead we counted for 10 s, the counting error is still $\pm$ 1 count but the speed error is now 0.077 percent. Of course, the trade-off is that now our measurement is the average speed over a 10-s interval; we do not know the details of what is happening to the speed *during* the 10-s averaging period. Thus, we can trade off time resolution for accuracy of the average speed.

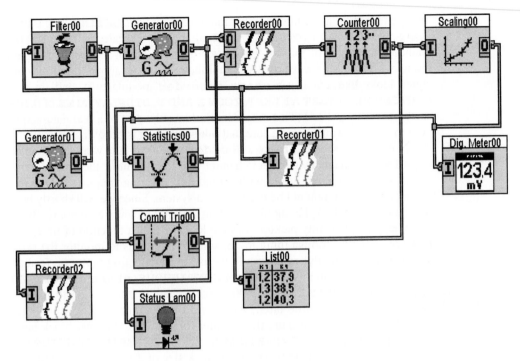

Figure 13.5
DASYLAB system for rotating-shaft speed measurement.

If our speed measurement is only one of several other measurements that we are making on our machine, we may want to use a computerized data-acquisition system to measure *all* our variables, including the shaft speed. Once we make this decision, we also can embellish our "raw" speed measurement with other useful features provided by our data-processing software, as we will see in the example below. As in the earlier two exercises, we use the DASYLAB simulation mode to demonstrate system operation. This means that we again start our DASYLAB worksheet with some means to simulate the signal coming from the proximity pickup as it measures the time-varying speed of the shaft. An easy way to do this is to use DASYLAB's frequency-modulated signal generator. Place a **generator** module on your worksheet and set it up for noise, amplitude 500, offset 160. (*Readers who do not have the software available for actual use should now use Fig. 13.5 to follow the development.*) Send the output of this generator to the input of a **filter** module set up for low pass, 0.80 Hz, 4th order, Butterworth. The output of this "noise" generator is a random number that "jumps" suddenly from one value to the next for each sample. (This generator output will shortly be used to control the frequency of another (frequency-modulated) generator set for a sine wave of 10-V amplitude.) Since real shafts cannot change their speed suddenly, we low-pass filter this "jumpy" signal to get a smoothly varying signal representing real shaft speed. To see what the commanded frequency looks like, connect a **chart recorder** module to the output of the **filter** module. Set it for **CONTINUOUS, START AT LEFT, ZOOM X AND Y, DISPLAY TIME** of 10.0 s, and scale for 0.0 to 200.0. Also, connect the output of the low-pass filter to the input of a second **generator** module, set for a sine wave of 10-V amplitude. To see what our "proximity pickup" signal looks like, connect a **chart recorder** module to it, and set it up for **CONTINUOUS, START AT LEFT, ZOOM X AND Y, DISPLAY TIME** of 0.10 s, and scale of −12.5 to 12.5. (This recorder would *not* be part of a real data-acquisition system; we just want to make sure that our simulation is behaving correctly.)

Now place a second **chart recorder** module on the worksheet and configure it for two channels. Connect channel 0 to the output of the second generator, which should be a sine wave with a randomly varying frequency. This recorder channel might or might not be used in a real system. Since we will shortly be counting the pulses and displaying the speed on a digital meter, we do not really need to see what the "proximity pickup" voltage looks like as a function of time. It *might,* however, not be a foolish thing to do, since it allows us to monitor the proximity sensor voltage directly, which could be useful in detecting faulty performance or sensor failure. Set up this recorder for **CONTINUOUS, start at left, ZOOM X AND Y,** and **DISPLAY TIME** of 100 s. Scale channel 0 for −12.5 to 12.5. (Channel 1 will be used a little later.) Also send the "proximity pickup voltage" to a **counter** module. Configure this counter for zero-crossing, per block basis, one sample per block. Go to **EXPERIMENT,** click on **EXPERIMENT SETUP** and then set the sampling rate at 1000 Hz with a block size of 2048. The sampling rate must be set fast enough to get about 10 or more samples/cycle for the highest frequency signal coming from the proximity pickup. If we do not do this, the **counter** module cannot accurately determine where the "voltage" crosses zero. With a sample rate of 1000 Hz and a block size of 2048, our counter will accumulate a count of zero-crossings over the 2.048-s period, and then repeat this over and over. Our speed measurement should thus be reported as the *average* speed over about a 2-s interval. If the time resolution is not acceptable for the purposes of the experiment, we *must* use shorter averaging time, which will reduce the accuracy unless we install a gear with more teeth. (If the teeth get *too* small, the proximity pickup can no longer distinguish individual teeth, causing error.)

The **counter** module will produce an output that is the number of zero-crossings during each block of data (every 2.048 s). This is not what we really want (which is the speed, say, in revolutions per minute, revolutions per second, or radians per second), but let us record it just to make sure everything is working correctly. Send the output of the counter module to a **list** module, configured for 15 digits and 3 decimals, display the units as "counts per block" (the options are: normal format, type of time channel "sample #," or memory per channel "9999"). Again, this module would not really be needed in a real system, but it provides some redundancy in checking for proper operation. To get a readout of rotary speed, what we really want, connect the output of the **counter** module to a **scaling** module. Configure this module for "linear function," with $a = 0.3756$ and $b = 0.0$. The number of counts per block is *proportional* to the speed but needs to be "scaled" to get an actual speed reading. Let us decide that we want to see the actual revolutions per minute. To get from counts/block to rpm, the scaling involves simply multiplying by $60/(78 \cdot 2.048) = 0.3756$. We can now send the output of the scaling module to a **DIGITAL METER**. Configure it for units of rpm, lower limit 0.0, upper limit 500, mode of "last value," option of "7 digits and 2 decimals."

In addition to the screen display of the speed on the digital meter, we might want to also graph the meter readings versus time. To do this, send the output of the scaling module to a **statistics** module, set operation "max," mode "block based," and "one value for every 1 block." Send the output of this **statistics** module to channel 1 of the 2-channel recorder set up earlier, and scale this channel for 220 to 270. Our final operation is to enable a "status lamp" that will display a screen figure and text message relating whether the speed is above or below a desired value. In a real system, this feature might or might not be needed, but we add it to give you practice with another DASYLAB function of general utility. To activate the **status display (lamp)** module, we first need to add a **combi trigger** module, which monitors the speed reading from the **scaling** module and outputs a **HI** or **LO** pulse when the speed goes above or below a set value, in our case, 250 rpm. Set up the **combi trigger** module for: "start trigger" (sample greater level, 250); "stop trigger" (sample lower level, 250) and "minimum delay hi" (1 sample), "minimum delay lo" (1 sample). Then send the output of this trigger module to the input of a **status lamp** module. Set it up for: "message on" ("Speed too high"), "message off" ("Speed too low"). This completes the setup of exercise number three; your worksheet should at this point look something like Fig. 13.5.

Running the Demonstration

You should now run the above experiment. Display simultaneously as many of the recorders, list, meter, and status lamp windows as you can comfortably fit on your screen. New data points should appear about every 2 s. Compare the status lamp message with the digital meter readings to see that the message changes correctly when the speed crosses the 250-rpm value. Channel 0 of the 2-channel recorder does *not* show an actual waveform. Why? Examine *all* the graphs and tables and verify that they are behaving as expected. If not, try to find your errors and correct them. Print a hard copy of the graph showing the output of the filter module and add a text note giving your name and the date. Figure 13.6 shows some results from this simulation. Let us now make some *changes* in the system.

1. Leave the sample rate the same (1000 Hz) but make the block size 512, rather than 2048. What is the good feature of this change? What is the bad feature of this change? Run this modified system to see that the speed readings are *very* wrong. Which module needs a number adjusted to fix this? Compute the needed numerical value, adjust the module, and run the experiment again to make sure your fix worked.

2. Now, change the sample rate to 200 Hz and leave the block size at 512. Before even running this, can you guess what problem may arise? Run this system to see that it does *not* produce good results.

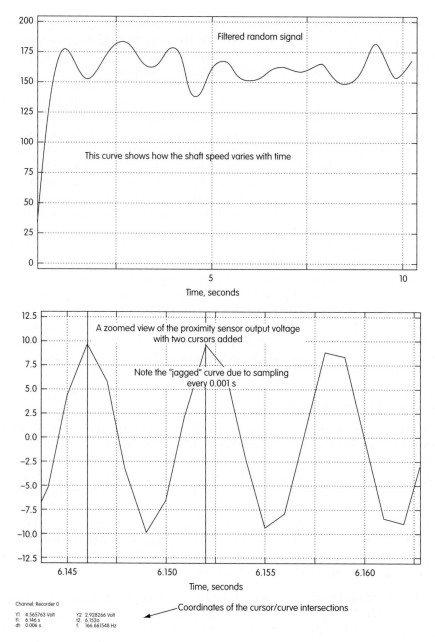

Figure 13.6
Selected results from shaft speed measurement system.

3. Your boss just decided that she wants to also get values of shaft acceleration. Modify the system to do this and check to make sure it really works.

4. Can we get the *standard deviation* of the randomly varying speed? Show how and actually do it, including a graph of the standard deviation versus time.

5. Can we get a *histogram* for the values of shaft speed? Show how and actually do it, using 20 speed readings for each histogram, that is, display a new histogram for every 20 speed readings.

13.6 A SIMPLE REAL-WORLD EXPERIMENT USING DASYLAB

Having used the simulation mode of DASYLAB to get some practice with its capabilities, we now want to show its use with a simple, but practical real-world apparatus. Our example will be an experiment that is part of a required undergraduate measurements laboratory at Ohio State's Mechanical Engineering Department. Most readers will of course not have this apparatus available to them, but our discussion should still serve the useful purpose of illustrating an actual application. We will point out only certain main features; the laboratory manual for this course leads the student through many detailed steps that we do not document here.

The apparatus (see Fig. 13.7) allows a variety of studies on the behavior of a friction brake that is applied to a rotating member driven by a $\frac{1}{4}$-hp induction motor. An air-pressure regulator allows us to set various pressures (measured with the mechanical gage), which will be applied to the pneumatic cylinder, and thus we can develop different normal forces on the friction brake. A solenoid valve, actuated manually, applies pressure suddenly, but the first needle valve can be partially closed to give a more gradual rise. A second needle valve is used as a laminar flow meter to measure the flow rate of air into the cylinder. This flow rate is related to the pressure drop across the valve, so we measure this pressure drop with a differential pressure sensor. The normal force on the brake is inferred by measuring the cylinder pressure with a sensor and then using the cylinder area to compute the force. The rotating part of the brake is attached to a $\frac{1}{4}$-hp electric induction motor, whose synchronous speed is 1800 rpm. Recall that such motors will run at a speed slightly less than 1800 rpm, depending on the load torque on the motor. We measure the speed with a DC tachometer generator. When the brake is applied, both braking surfaces (one steel and one brake-lining material) will heat up. We are interested in the temperature of the rotating brake lining material. This is a natural application for radiation temperature measurement, since this method is noncontacting; the instrument need not touch the rapidly moving surface. The friction torque on the nonrotating member of the brake is reacted into the apparatus frame through a load cell (force sensor), allowing measurement of the force and thus the torque.

Sensor signals are wired to a screw-terminal board whose flat cable connects these signals to the data-acquisition board located in the computer. Here, the DASYLAB software allows us to acquire and process the data in versatile ways to give us the results we desire, such as the friction coefficient. A few details on the sensors may be of interest. The *tachometer generator* requires no electrical power supply; it is a "self-generating" transducer that converts shaft power into electrical

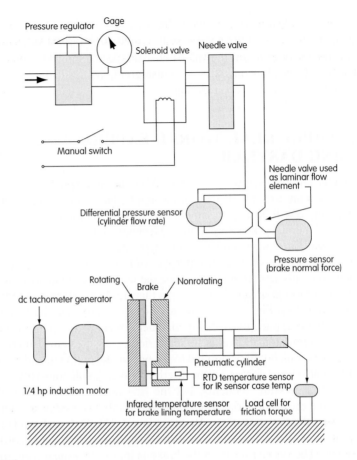

Figure 13.7
Apparatus for studies of friction brake system.

power. The power it "steals" from the shaft does not cause a measurement error because the device measures the shaft speed that exists, which is the speed that we need to know. (We do not need to know the speed that would exist if the tachometer were not present and slightly slowing the motor.) In principle, the tachometer produces a voltage linearly related to rotational speed. An actual tachometer might be a little nonlinear; we could calibrate it, using an accurate speed standard to get a nonlinear calibration curve, which we could enter into DASYLAB. More likely, we would *assume* linearity and use DASYLAB's two-point scaling module to convert tach voltages to speeds in rpm. Tachometer signals *are* generally rather noisy, so DASYLAB's averaging or filtering capabilities will be used.

The differential pressure sensor, the cylinder pressure sensor, and the load cell are all examples of MEMS (microelectromechanical systems) technology. These devices are manufactured on silicon wafers using diffused strain gage techniques, several hundred sensors being fabricated on a single wafer, leading to low cost. Our version of these sensors is the simplest and cheapest available; they have no built-in amplifiers and no adjustment circuits to set zero or sensitivity, but they do have good linearity, so a "two-point" calibration is acceptable, using the mechanical pressure gage, which is an accurate test-type unit, or dead weights for the load cell. The adjustment for zero reading and sensitivity is again easily done by our DASYLAB *software.*

One characteristic of a differential pressure sensor that is not shared with most sensors is its *common-mode rejection.* This specification is often overlooked but can be quite important. In an ideal differential sensor, if you apply the *same* pressure (no matter how large) to both sides of the diaphragm, you should get exactly *zero* output voltage. Real sensors are not this good; some are quite bad in this respect, and many users are unaware of the potential problem. It is relatively easy to do a calibration check for this feature; we just apply the same pressure to both sides, run this pressure up from zero to the highest value of interest, and plot output voltage versus input pressure. In a high-quality sensor, the voltage will be so small that we will probably not choose to correct for it in most applications. In a "cheap" sensor, the effect may be quite noticeable, and a correction may be needed. Until you run this calibration on the sensor, nobody really knows. Another common device that exhibits an analogous problem is the electronic amplifier, such as is used in the data-acquisition board in your lab computer. We often use such amplifiers in *differential mode.* Here the two input wires each carry a voltage, but the information on the wires is related to the *difference* between the two voltages. Again, an ideal amplifier puts out an amplified voltage that is exactly zero when the two input voltages are the same, but a real amplifier will not be this perfect, leading to possible errors.

The *cylinder pressure sensor* is very similar to the differential pressure sensor just described except that it has only one pressure port; the other side of the diaphragm is open to the atmosphere, making the sensor a *gage pressure* sensor. This sensor will also be calibrated using the mechanical test gage that is part of the apparatus.

The *load cell* is also a diffused strain gage type of microdevice, but the strain is now caused by a concentrated force applied externally to the "load button" of the transducer. Both pressure sensors and this load cell are excited by the same 5-V DC source, which is also used to control the solenoid air valve with a manual toggle switch. All three of the strain-gage sensors have output voltages that are the order of 50 mV full scale, so when we set up the data-acquisition board, these channels require some amplification to boost the voltages up toward the $\pm$ 10-V range of the A/D converter. These amplification settings are done through the DASYLAB software. Again because of good sensor linearity, a two-point calibration of this sensor using the zero-load point and dead weights for the full-scale load point should be adequate. Because this sensor is used to measure the friction *torque,* we also need to know the lever arm from the axis of rotation out to the force measurement point.

To measure the surface temperature of the rotating friction material, we use a *noncontacting radiation type* of sensor. This device uses a microcircuit (MEMS) thermopile, a series connection of many (typically 40 or 50) thermocouples. The measuring junctions of the thermopile are blackened to absorb the radiant energy coming from the target, while the reference junctions are thermally insulated from the measuring junctions but in good thermal contact with a heat sink that tends to stay at the same temperature as the sensor's metal housing. At thermal steady state, the measuring junctions are somewhat hotter than the reference junctions (but nowhere near the target temperature), and this temperature difference creates an output voltage, just as in an ordinary single thermocouple. In the (patented) sensor from EXERGEN,[4] the number of thermojunctions in the thermopile has been selected so that the output voltage will be nearly the same as if a *single* thermocouple were at *target* temperature, at least for a certain range of temperatures that the sensor is marketed for. This allows the user to connect the infrared sensor to any available "electronic thermometer" designed for that thermocouple (say, iron/constantan) and read out the target temperature directly. In our application, we will exceed this design range, so we cannot use a standard thermocouple readout to get temperature values. Fortunately, EXERGEN has provided a calibration table of voltage values versus temperature for a large range of temperatures, and we will simply enter this information into our DASYLAB software, which provides a facility for such lookup tables. As in ordinary thermocouples, one must account for changes in the reference junction temperature; we omit the details of how this is accomplished in this application.

As in any radiation-type temperature instrument, our sensor is affected by target emissivity. We really measure the target radiant heat flux, not the target temperature, so commercial radiation thermometers always include a means of entering the estimated emissivity of the target, which then allows the instrument to read out a target temperature. We omit the details of how we deal with the emissivity problem. Finally, our simple sensor does not use a focusing lens, and its field of view is described as "1:1." This means that at a target distance of, say, 1 in, the area whose temperature is measured is a circle of 1-in diameter. Our distance is about 0.2 in., so we measure the *average* temperature of a circle of about 0.2-in. diameter on our brake lining. If smaller target spot sizes are required, sensors with focusing lenses are available.

The last sensor to be discussed is based on the *resistance-temperature-detector* (RTD) principle. It is intended to take the temperature of the *housing* of the EXERGEN infrared noncontact sensor just explained above. This housing fits into a hole in the metal part of the friction brake, and when the brake heats up, this housing will also heat up. We contacted the EXERGEN folks, and they said this housing temperature rise was at least partially compensated internal to their sensor, and the housing heating should not be a problem in our application. Just to be on the safe side though, we decided to monitor the housing temperature, in case a correction might be needed. We found a small commercial temperature sensor made by Analog Devices[5] that covered the range of the expected housing temperatures. This device is produced by microelectronic methods and is quite inexpensive. RTDs are one of

---

[4]Exergen Corp., Watertown, MA, 800-422-3006 (www.exergen.com).

[5]Model AD22100, Analog Devices, Norwood, MA, 800-262-5643 (www.analog.com).

the most common ways of sensing temperature, though not as common as thermo-couples, the most popular device. RTD devices rely on the fact that many materials change their electrical resistance as the temperature changes. If the resistance change is large and nonlinear and the material is a semiconductor, the device is called a *thermistor.* When the resistance change is small and nearly linear (true of most metals such as nickel, platinum, and copper), then the device is called an RTD. The AD22100 uses a temperature-sensitive resistor that has a nearly linear resis-tance change with temperature. Most RTD circuits send a constant current through the temperature-sensitive resistor and use the voltage across the resistor as the output signal. The AD22100 combines the temperature-sensitive resistor, the constant-current source, three non-temperature-sensitive resistors, and an op-amp in a cubical molded plastic package about 0.2 in on a side. Three electrical leads protrude from one side, providing connections for a 5-V DC supply and the output signal. Our model is designed for a -50 to $+150°C$ range and gives an output volt-age that varies nearly linearly from 0.250 to 4.750 V. As supplied by the vendor, the error is about $\pm$ 1 percent of full scale. If one calibrates the unit, this can be reduced to about 0.2 percent. As with any contact-type temperature sensor, the accuracy depends on the sensing element being at the temperature that you wish to measure. In the AD22100, the sensitive resistor is *inside* the plastic packaging, so we cannot put it in *direct* contact with the EXERGEN's housing, while the plastic package *is* in direct contact with the ambient air, which will be much cooler than the housing. Another error source is of course the *self-heating* effect present in any RTD or thermistor.

We next briefly discuss the data-acquisition board that is installed in our personal computer: the National Instruments AT-MIO-16L-9, a 12-bit device, so it breaks the signal range into $2^{12} = 4096$ parts. For a $\pm$ 10-V full scale range, the least significant bit is 20/4096 = 0.004883 V. The amplifier has software-programmable gain, selectable from 1 to 500 V/V, so at maximum gain, the least significant bit would be 0.004883/500 = 0.000009766 V. We can thus resolve signal changes as small as 9.8 μV. (While the binary values change in increments corre-sponding to 0.000009766, the software converts to decimal values, so we do not expect to see the *displayed* signals change in increments of 0.000009766 V.) The amplifier input impedance is 1 GΩ in parallel with 50 pF of capacitance. This very high impedance means that the amplifier will not cause significant loading prob-lems with most sensors, whose impedance is usually much lower. Our highest impedance sensor is the infrared thermopile, at about 5000 Ω. The conversion time for analog to digital is about 9 μs, making the maximum sampling rate of *the board* 100,000 samples/s. Since we have 6 channels of data, our maximum sampling rate *per channel* will be about 16,000 samples/s. When the DASYLAB software is controlling the board, the allowed sampling rates may be lower, depending on the operations being performed.

The programmable-gain amplifier (available gains are 1, 10, 100, and 500 V/V) is convenient in our (and many other) systems because our sensors all have differ-ent sensitivities. We want to use as much gain as we can for each channel, so that the board's available resolution is actually employed, not wasted. DASYLAB allows us to set the desired gain for each channel and tells the amplifier when to switch from one gain to another. When the gain is switched, the amplifier goes through a transient, and we must *wait* until this transient settles before actually taking a reading. This slows down the sampling and is a drawback of programmable-gain amplifiers. This board has 16 available channels if the sensors allow the use of *single-ended* connections. When sensor signals are small (millivolts), significant electrical noise is present, and/or long sensor cables are necessary, it is better to use *differential* rather than single-ended input, but then only 8 channels are available because each channel uses 2 of the 16 input terminals. Since we need only six channels, we will use differential input. With differential input, sensors generally are "floating" (neither terminal is connected to ground), and this would cause amplifier drift and/or saturation. To avoid this, we must connect a resistor (about 100 kΩ) from the negative terminal of each channel to the board terminal called AIGND (*analog input ground*). This board is a *multifunction* device, which means that it also has digital-to-analog converters, digital inputs, counter-timer channels, and a frequency-output channel; however, we will not be using these.

Figure 13.8 shows the DASYLAB worksheet that is the *basic* setup for all parts of the experiment. Some of the possible studies require that we slightly modify this

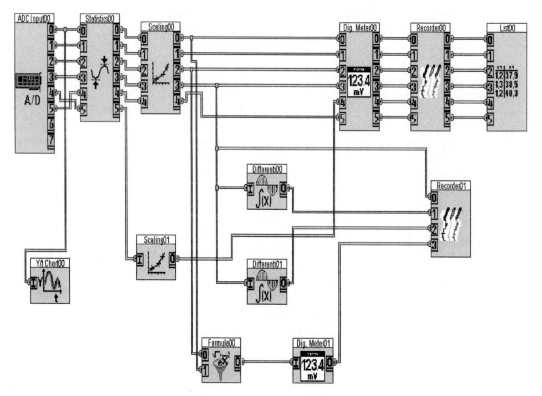

Figure 13.8
DASYLAB data acquisition/processing system for friction brake apparatus.

worksheet in order to accomplish certain goals. For example, we might need to move the y/t chart shown at the lower left from channel 0 to channel 3. The worksheet "starts" at the left with the analog-to-digital converter module **ADC Input 00**, where the analog sensor signals are sampled and digitized. Digital signals then "flow" toward the right to various processing modules that perform the operations we desire. The second module, **Statistics00**, is used to time-average each sensor signal to "smooth" any noise that might be present. If we do not do some of this averaging, then our graphs and display meters will be "jumpy" and difficult to read. The next module, **Scaling00**, is used to convert the sensor signals (which are in volts) to the proper engineering units, such as psi or rpm. All of our sensors except the infrared sensor are essentially linear, and the scaling module requires only two points on each straight-line calibration curve. The infrared sensor is nonlinear and requires a lookup table and its own separate scaling module, **Scaling01**. The **Y/tChart00** module is used as a "digital scope" to look at signal details. It is connected to the load cell signal in Fig. 13.8, but we will use it as a "roving" scope by changing its connection to whatever signal we want to see. The **Formula00** module takes the scaled friction force and normal force signals and computes from them the friction coefficient, which signal is then sent to a digital meter **Dig Meter01** for display. The friction coefficient is also passed from this digital meter to channel 3 of chart recorder module **Recorder01**. The scaled rpm signal is differentiated in **Differenti00** to get shaft acceleration, and integrated in **Differenti01** to obtain the shaft angular position. The shaft speed, acceleration, and position are sent to channels 0, 1, and 2 of **Recorder01**. Digital meter **Dig Meter00** is a multichannel display of all six channels; **Recorder00** is a graph, and **List00** is a table of all six channels.

The various sensors are assigned to the channels as follows:

Channel number	Sensor
0	load cell for friction torque
1	differential pressure sensor (flowmeter signal)
2	cylinder pressures sensor (brake normal force)
3	tachometer generator (shaft speed signal)
4	infrared temperature sensor (friction material surface temperature)
5	RTD temperature sensor (infrared sensor housing temperature)

Before starting to use *any* computerized data-acquisition system, it is always best to have a good idea of what your sensor signals look like *before* they are digitized and manipulated. The best way to do this in many cases is to observe them on an ordinary *analog* oscilloscope, which, if fast enough, shows *everything* present in the signal. Knowing what your signal looks like is a great aid in designing and using the data-acquisition system. While a fast analog scope is best, if none is available, a suitable digital scope may be used, *with care*. In the present experiment, since the signal waveforms are known to be relatively simple, we will use the data-acquisition system itself as our "scope." The main requirement here is to set the sample rate *high enough* so that we do not "miss" any important details of the waveform.

There are quite a few interesting things that we can try that do not involve the brake and illustrate important general concepts. We suggested that one should always look at all the signals on an analog scope first, but the next best thing is to use the data-acquisition system with a "high" sampling rate, which is what we will do. The noisiest and "most dynamic" signal is the tachometer voltage, so we will look only at it, as an example. The tachometer rotates at about 1750 rpm, so one revolution takes about 0.03 s. The tachometer ripple is more or less periodic with a period of 1 revolution, so if we want to see the details of the ripple, we need to sample fast enough to resolve, say, a few hundred points per revolution. We set the sample rate at 5000 Hz and the block size at 512; this makes a block last about 0.1 s, a little less than 3 revolutions. (Many of DASYLAB's operations are based on *blocks* of data; for example, we can have graphs displayed and updated for every block of data or a signal may be averaged over one or more blocks.) We can disconnect the graphing module called **Y/tchart00** from channel 0, connect it to channel 3, the tachometer signal, set up this chart for **BLOCK RELATION**, 1 and **FIXED TIME SCALE**, producing a graph of the tach voltage with 512 time points, spaced at 0.0002 s, and updated about every 0.1 s. This graph clearly shows that the ripple is a complex waveform that nearly repeats itself every revolution and has 14 peaks per revolution, suggesting that the tachometer has 14 bars in its commutator. Having discovered the detailed nature of the tachometer voltage, we are now in a good position to process this signal in the software in various ways.

For example, if we plan to differentiate the signal to get acceleration data, it is now clear that we need to do some averaging or filtering *before* invoking the differentiating module. *If we do not, the fluctuations present for a steady speed will be interpreted as a large, oscillating acceleration!* This exercise shows how we can use the data-acquisition system to study the details of a signal so that we can more intelligently plan further data processing. This should be done with *each* of the signals. If we now wanted to study brake heating for steady motor speed, past experience (or a few "dry runs") tell us that the temperature will change rather slowly, so we might want to get data only every second or so. The 5000-Hz sample rate necessary for our earlier tachometer study can thus be greatly reduced, say, to 512 Hz, and a block size of 512 used. The **Statistics00** module would then be set up to take the average of each block, so we will be getting an output that is the average of 512 samples, once each second. (There is nothing "magical" about the 512-Hz sample rate; other, slower rates could be used, as long as we get the desired averaging.)

Our graphs and digital meters will thus be updated every second, which gives easy reading and smooth graphs. We of course are *losing* any details that happen within the 1-s blocks, so this setup is good only for studies that involve rather slow changes in the variables. The **Scaling00** module would have each of its channels set up to accept the two-point calibration data for all the sensors that are essentially linear, converting sensor voltages to the appropriate engineering units. Channel 5, the infrared temperature sensor requires a nonlinear lookup table to represent its calibration, so a separate **Scaling01** module is used for this. We can use the **Formula00** module to compute the friction coefficient from the measured values of normal force (cylinder pressure times piston area), load cell force, and the known lever arm for the load cell.

A separate digital meter `Dig.Meter01` displays the friction coefficient, while `Dig.Meter00` has six small meters that show all the basic data in compact form. Note that a "meter" would normally have only an *input,* not an output signal. However, DASYLAB provides an option "copy inputs to outputs," which duplicates the input signals at individual output terminals, where they may be connected, as desired, to other modules. This feature makes the worksheet less cluttered by avoiding multiple connections and "wires." `Recorder00` also uses this option so that the same signals that are graphed against time in the recorder are also sent to the `List00` module for printing as a table. The averaged and scaled tachometer signal is sent to module `Differenti00` for differentiation (to get shaft acceleration) and `Differenti01` for integration (to get the total shaft rotation during starts and stops). (Both differentiation and integration use the module named `different`; a dialog box allows us to choose which operation is actually performed.)

The apparatus and data-acquisition and -processing system allow many different studies to be run, some of which require slight modifications to the DASYLAB worksheet. Because of the simplicity and transparency of DASYLAB's methods, these modifications are quickly and easily made, as needed. A list of these studies would include the following.

1. *Motor acceleration and deceleration studies, with and without braking.*
   Compare averaging and low-pass filtering before differentiation; try different block sizes for averaging; use a motor coast-down study, together with known motor inertia, to define the motor's friction torque versus speed curve; and estimate motor peak torque from peak acceleration and known inertia, and compare with motor rated torque.

2. *Pressure and Flow Studies.* Calibrate the pressure sensors against the mechanical test gage; calibrate the differential pressure sensor for common-mode error; study the dynamics of charging and discharging the cylinder, using various needle-valve settings; adjust the "flowmeter" needle valve to get essentially linear pressure versus flow characteristics (laminar flow); model the cylinder-charging process with a linear, constant-coefficient, differential equation, thus defining a time constant; use measured values of this time constant to get the flow resistance $(lb_f/ft^2)/(slug/s)$ of the "flowmeter"; set up DASYLAB to compute and plot the mass flow rate using this value and integrate this flow rate to get the total mass change during charging and discharging; and compare this measured result with a theoretical prediction using the perfect gas law and the initial and final cylinder pressures.

3. *Braking Studies.* Calibrate the load cell against dead weights; apply a steady normal force with the motor running at steady speed and note that the friction torque is *not* steady but exhibits a dynamic signal that appears nearly periodic (such behavior is typical of real rotating machinery, which *never* is perfectly aligned or balanced); study the dynamics of this load cell signal; how is it

affected by changing the normal force and determine what averaging or filtering is needed to get useful friction torque data to use in the friction coefficient calculation; study the effect of normal force and brake surface temperature by applying air pressures of 5 and 10 psig, alternately, for 10 s each, until a temperature of 300°F is reached; and devise a DASYLAB setup that will automatically produce *xy* plots showing two friction coefficient curves, one each for the two pressures, against temperature.

This completes our discussion of digital data acquisition using personal computers and specialized software. We have covered many basic concepts that would arise with *any* commercial data-acquisition and -processing software, and have shown some details of one particular package as it would be used in a variety of applications.

CHAPTER 14

Measurement Systems Applied to Micro- and Nanotechnology

Because of the pervasive influence of computer technology on all aspects of modern society, the manufacture of semiconductor devices, such as *dynamic random access memory* (DRAM) chips is one of the most significant industries around the world. While the design of such devices is largely the province of electrical engineers and computer scientists, their manufacture, which involves many kinds of machines and processes, requires significant contributions by mechanical engineers. (Even at the design stage for integrated-circuit chips, mechanical engineers often contribute their heat-transfer expertise since the ever-denser packaging of such devices creates difficult thermal problems which impact the layout of the circuitry, choice of materials, etc.) More recently, micro- and nanoscale devices have come to include *motion* and *force, fluid flow* and *pressure, temperature* and *heat flow,* not just voltage and current, making the contributions of mechanical engineers even more relevant. These tiny sensors and actuators are generically called MEMS (micro-electromechanical systems) and are the basis of a whole new industry. Both integrated microcircuits and MEMS involve the design, manufacture, and application of devices whose critical dimensions lie in the range of micro- and nanometers, a world with many unique problems and opportunities.[1] I wanted to include in this book a short chapter showing the significance of measurement systems in some specific industries. Almost any industry could serve this function since measurement systems are vital in most development or manufacturing efforts. I chose MEMS because of the intense current interest, the fascinating problems involved,

---

[1] J. H. Smith et al., "Intelligent Microsystems: Strategy for the Future," *Semiconductor International,* April 1998, pp. 93–98; R. H. Grace, "The Growing Presence of MEMS and MST in Automotive Applications," *Sensors,* Sept. 1999, pp. 89–96; R. Frank et al., "The Role of Semiconductor Sensors in Automotive Power Train and Engine Control," *Sensors,* Dec. 1998, pp. 48–54; J. Staley, "Platinum Thin Film and Next-Generation Micromachined Sensors," *Sensors,* April 1996, pp. 56–62; T. G. McDonald and L. Yoder, "Digital Micromirror Devices Make Projection Displays," *Laser Focus World,* Aug. 1997, pp. s5–s8.

and the fact that it is a young and growing technology rather than a well-established one.

Of the many specific topics in MEMS that one might address, I have chosen the following:

Microscale sensors

Micromotion-positioning systems

Particle measuring systems

Partial-pressure measurements in vacuum processes

Magnetic-levitation systems for wafer conveyors

Scanning-probe microscopes

Microscale sensors, one end result of microsystem manufacturing, are of course of direct interest in a book on measurements. *Positioning systems* are used at various stages of development and manufacture. They are basic to the production and use of the photomasks that define the detailed layout of the microdevices. Because of the tiny size of device features, contamination by microscopic *contaminant particles* is an ever-present problem. Detection and elimination of such particles requires specialized instrumentation. Many parts of the wafer-manufacturing process are carried out under vacuum conditions where the presence of desired or undesired gases in minute quantities is critical. *Partial-pressure analyzers* (mass spectrometers) provide this information. Moving wafers from one station to the next in the manufacturing process requires specialized robotics and conveyers. Unique *conveyer-levitation systems* have been developed that minimize the production of damaging particles. *Scanning-probe* and *atomic-force microscopes* provide vital tools for studying various effects at the micro- and nanometer level, not just for microdevice development, but for general understanding of surface phenomena, such as friction.

14.1 MICROSCALE SENSORS

We have noted several times earlier in this text that the *critical* portions of many computer-aided machines and processes are those dependent on the availability of suitable sensors. To penetrate the vast markets associated with *consumer products,* sensors for computer-aided systems must meet cost requirements much more stringent than those associated with data systems intended for engineering test work. (We should also note that sensors for consumer goods may often tolerate lower accuracy, helping them to meet the stated cost goals.) While the usual economies of scale do result in reduced costs for mass-marketed sensors, this fact alone does not always result in sufficiently low prices. Rather, a totally different sensor-manufacturing process may be needed. Since integrated-circuit manufacturing methods have been eminently successful in meeting such cost requirements in the computer and electronic field, it is natural to try to extend them to sensors. Furthermore, because sensors often require associated electronic processing (amplification,

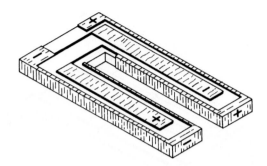

Figure 14.1
Quartz tuning-fork timing reference.

filtering, etc.), the possibility of *combining* sensing and processing functions at the microscopic (chip) level becomes very attractive.

The market for microsystems based on semiconductor processing emerged in the 1970s with bulk micromachined pressure sensors, used today as disposable blood pressure monitors in hospitals, auto engine manifold and barometric pressure measurements, and other low-cost, large-volume applications. A variety of technologies are used in manufacturing microsystems,[2] but most depend on lithography and etching from the semiconductor industry. Other techniques such as LIGA (a German acronym for *lithography, electroforming, and injection molding*) borrow only lithography from the semiconductor world. Sensing technologies include piezoresistance, piezoelectric, and capacitive. Pressure sensors often use miniature silicon diaphragms with piezoresistive strain gages diffused right into the diaphragm. Accelerometers, such as used in many air bag crash sensing systems, use miniature masses and beam springs, often with capacitive-displacement sensing. (Capacitive *actuation* is also possible; a controllable force exists between the plates of any capacitor.) Vibrating quartz-tuning-fork angular-rate sensors (replacements for spinning-wheel rate gyros) used in automotive stability augmentation systems employ piezoelectric principles both for actuation (driving one tuning fork) and sensing (measuring motion of the readout fork).

Quartz-tuning-fork technology is also used for time measurement in digital watches and clocks. Here a tiny quartz (silicon dioxide) tuning fork (Fig. 14.1) is used as an element of an electronic oscillator circuit to produce an accurate 32,768-Hz (2^{15}) signal that is easily "counted down" to 1 Hz by using divide-by-2 electronics. Several hundred such tuning forks can be manufactured "at one time" on a 1-in^2 quartz wafer by using photolithographic and chemical-milling techniques. To produce a basic *force*-sensitive device useful in force, pressure, acceleration, vacuum, and temperature sensors, the double-ended tuning fork of Fig. 14.2

---

[2]M. Madou, "Fundamentals of Microfabrication," CRC Press, Boca Raton, 1997; D. J. Elliott, "Integrated Circuit Fabrication Technology," McGraw-Hill, New York, 1982.

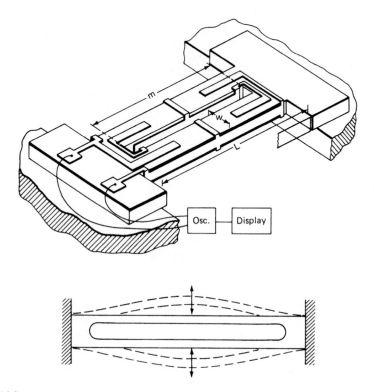

Figure 14.2
Double tuning-fork force sensor.

has been developed, whose fourth natural frequency f is related to applied axial force F by[3]

$$f = \frac{1.03\sqrt{E/\rho}\,W}{m^2}\left(1 + \frac{0.074 m^2 F}{EtW^3}\right) \tag{14.1}$$

A typical unit changes frequency from 17.5 to 19.0 kHz for a force change from 0 to 1 N.

Whereas the quartz technology briefly described above uses the piezoelectric characteristics of quartz, an alternative silicon technology employs piezoresistance or capacitance techniques to sense small strains or deflections in a micromachined silicon structure. Local introduction of dopants into the silicon produces diffused

---

[3]E. P. Eernisse and J. P. Paros, "Practical Considerations for Miniature Quartz Resonator Force Transducers," Quartzdyne, Inc., Salt Lake City, UT, 801-266-6958 (www.quartzdyne.com); Paroscientific, Inc., Redmond, WA, 425-883-8700 (www.paroscientific.com); E. P. Eernisse and R. B. Wiggins, "Tuning Fork Resonator Sensors," *Sensors,* March 1986, pp. 6–11; U.S. patents, 4,526,480, 4,535,638, and 4,550,610.

"strain gages" that can be configured into bridge circuits. When capacitance techniques are used, thin metal films deposited on silicon diaphragms form capacitors that can be used as passive deflection sensors or as electrostatic drivers in transducers based on self-excited vibration, producing a frequency output. While integrated-circuit manufacturing is mainly concerned with producing patterns over a nominally flat surface, its techniques have been modified and extended to produce the three-dimensional structures (diaphragms, beams, masses, flow channels, etc.) needed for sensors. While silicon at the wafer level is brittle and subject to catastrophic failure by cracking and chipping, reliable microdevices are achieved by careful mechanical design, using moderate stress levels, protective overload stops, etc.[4] Once immunity to overload is "designed in," the nearly perfect crystal structure of silicon results in long fatigue life. Accurate operation at low stress levels is possible because of high sensitivities and the almost complete absence of friction and material hysteresis.

Figure 14.3[5] shows an accelerometer using variable (differential) capacitance as the displacement sensor. Motion of the proof mass increases the capacitance of one capacitor and decreases that of the other. The sensing elements, single crystal silicon, $0.11 \times 0.12 \times 0.35$ in., are mounted on a substrate along with various discrete electronic components and an integrated circuit, these all being then mounted in a hermetic chip carrier. Between the base and the lid, the middle element is chemically etched to form a rigid central mass suspended by a thin, flexible membrane whose thickness sets the stiffness of this flexure, and thus the full-scale range of the accelerometer. Since the mass has a large and nearly pure translational motion, effective overrange stops are possible, giving a unit with $\pm\ 10\ g$ full-scale rating an overload capacity of 10,000 g. Gas damping is achieved by providing a number of orifices (fluid resistances) in the mass, whose motion "pumps" gas through these holes, giving the desired energy dissipation and a damping ratio of 0.7 ± 0.2.

14.2 MICRO-MOTION-POSITIONING SYSTEMS

Figure 14.4[6] gives an overview of a typical MEMS manufacturing process. The "mask set" consists of *photomasks* that contain the detailed patterns of each layer of the device, but at a larger dimensional scale. (Photomasks are made by etching the pattern into a chromium layer on a glass substrate.) The wafer is prepared for lithography by coating it with a thin layer of *photoresist,* a light-sensitive material that will receive the device pattern as projected by the photomask. In the *lithography* step, each photomask's image is projected onto a prepared wafer, with optics that reduce the dimensions to the microscopic size of the final device. The wafer is precisely moved to a new position and the imaging repeated, until the wafer is

[4]K. E. Petersen, "Silicon as a Mechanical Material," *Proc. IEEE,* vol. 70, May 1982, pp. 420–452; Special issues on sensors, *IEEE Trans. Electron. Devices,* Dec. 1979, Jan. 1982, July 1985, and June 1988; M. Dunbar, "Silicon Micromachining," *Sensors,* April 1985, pp. 10–15.

[5]Endevco, San Juan Capistrano, CA, 949-493-8181 (www.endevco.com).

[6]CaseWestern Reserve University MEMS Handbook (www.mems.cwru.edu/).

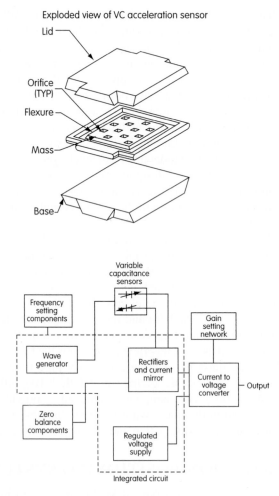

Figure 14.3
MEMS accelerometer using variable-capacitance displacement sensing.

"filled" with hundreds of copies of the photomask's pattern. The exposed photoresist is then chemically "washed away," leaving the desired pattern behind. Depending on the specific device, other operations such as film deposition, etching, or doping are sequentially applied to create the final configuration.

In this section, our focus is on the *lithography* step of the process, where the motion of the stage that carries the wafer must be precisely controlled as the multiple images of the device are transferred to the wafer. Our discussion is based on a paper[7] that details some problems and solutions in the field of micromotion control. Our example pertains to lithography and other processes where we need to position

---

[7]A. Chitayat, "Nanometer X-Y Positioning Stages for Scanning and Stepping," *J. Vac. Sci. Technol.,* B 7(t6), Nov./Dec. 1989, pp. 1412–1417; Anorad Corp., 516-231-1995 (www.anorad.com).

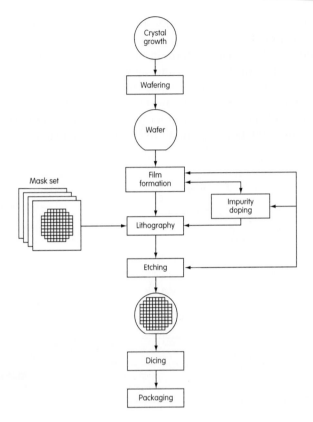

Figure 14.4
Flowchart of typical MEMS manufacturing process.

an object over a range of several centimeters (the largest wafer is about 30 cm in diameter) but the positioning must be accurate to fractions of a micrometer. This requirement for both large and small motions creates problems in choosing a proper actuator since those suitable for very small motions and high speed often are limited in the maximum travel that they can provide. For example, a typical piezoelectric actuator can position with resolution of 10 nanometers in a few milliseconds, but its total travel is only 7 micrometers. One solution to such problems, used in the paper by Chitayat is to piggyback the fine-motion actuator on top of a coarse-motion actuator,[8] giving a system that combines the best features of both.

Since the paper gives only the basic concept and none of the details, I contacted Anorad Corp. to try to get more information. As is often the case in such situations, they and the customer for whom they designed and built the system considered such details proprietary and were unwilling to provide anything beyond what appeared in the paper. At this point, I began a preliminary design study of my own, to try to

[8]A. Chitayat, op. cit., p. 1416.

come up with the simplest configuration that could meet typical specifications. After the usual iterations of trial and error, I finally arrived at the system shown in both schematic and block diagram form in Fig. 14.5. I cannot be sure whether this design is close to what Anorad Corp. came up with, but at least at the level of simulation testing, it seems to work and illustrates the main concepts. Let us now use the schematic diagram to discuss the hardware items used to implement the design concept. All our explanations will be in terms of an "analog" or continuous (rather than digital) implementation. It is quite likely that an actual system would use a

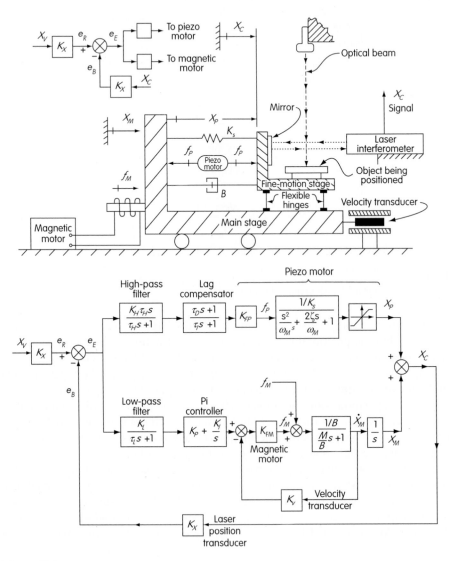

Figure 14.5
Micromotion control system used in MEMS manufacturing processes.

digital computer for at least some of the control tasks; however, such systems usually can employ quite high sampling rates, making the digital system's performance quite similar to its analog counterpart. The controlled variable in our feedback system is the absolute displacement x_C of the fine-motion stage since the wafer or other object to be positioned is mounted here. This absolute displacement is sensed with a laser interferometer, one of the few motion sensors that combines a long total travel with very fine resolution. Note that the proper operation of this *entire* system is totally dependent on the availability and proper application of the motion *sensor,* demonstrating again the importance of measurement systems in most computer-aided machines and processes.

The position error e_E is obtained by comparing the commanded position x_V with the measured position x_C, using the proportional voltages e_R and e_B. (This portion of the system might actually be digital.) A conventional, translational, brushless DC motor is used to drive the main stage. It, of course, is a strictly analog type of device. Since Anorad Corp. manufactures and markets a complete line of such motors, it is quite likely that this kind of actuator was used in the real system. Mounted "on top of" the main stage is the fine-motion stage, driven by a piezoelectric actuator, also an analog device. Because of increased recent interest in micromotion systems of various kinds, several U.S. and foreign manufacturers[9] offer rather complete lines of piezoelectric actuators. (For maximum accuracy, such actuators are available as complete closed-loop systems, using capacitive displacement sensors to overcome the piezo hysteresis. In our upcoming analysis, we assume that this "local" piezo-based feedback system has been properly designed so that we can model it with a simple and known transfer function. Note again the importance of the capacitive *sensor* to our overall system.)

We might mention at this point that our Fig. 14.5 shows only a single ("x") axis of motion. For systems with multiple axes, one can usually carry out preliminary designs considering only a single axis at a time, as we are doing. It is, however, conservative design practice to run a simulation of the complete multiaxis system once the individual axes have been "roughed out." This will reveal any interactions that might require some modifications in the preliminary single-axis designs. Translational DC motors of the type used here are available with strokes up to several feet, so selection of one for our main-stage drive, which might require only several inches of stroke, presents no problem. Piezo actuators are very limited in stroke but adequate for our application. We will assume a stroke limit of 0.002 in.

While the basic idea of piggybacking the two actuators seems intuitively feasible, its actual implementation requires some care in design. A main consideration is that the magnetic motor and main stage are not capable of following very high-speed commands and the piezoelectric stage cannot provide large motions. A solution to this incompatibility is to "decouple" the two channels using appropriate electrical filters so that each channel "sees" only signals within its capabilities. These filters are shown in the block diagram of Fig. 14.5. The piezo channel needs

---

[9]Polytec PI, Inc. (www.polytec.com); Queensgate Instruments (www.queensgate.com).

a high-pass filter since it is intended to respond to small but fast (high-frequency) components of the error signal. Note that the high-pass filter provides *no* response to steady errors, thus the magnetic motor channel must supply this capability entirely. Low-pass filtering in the magnetic motor channel prevents high-frequency signal components from reaching the motor and its electronics since the main stage drive is incapable of following these. The choice of numerical values for the filter time constants τ_H and τ_L is the main design problem for this part of the system.

Modeling of the translational magnetic motor dynamics is conventional since the behavior is essentially the same as for the well-known rotary version. Armature current is supplied by a "current" (transconductance) amplifier. This common practice suppresses the effects of armature inductance and motor back emf, giving a very simple and desirable response; the magnetic force f_M follows the amplifier voltage input instantly and proportionally. The mechanical load presented by the main-stage moving parts is modeled as inertia (mass M), viscous friction B, and an external disturbing force f_U. As is customary in servo analysis, the disturbing force is included to model any constant or time-varying forces acting on the load that we might wish to consider. The "minor feedback loop" around the motor, using a velocity transducer, another common scheme, tries to produce a velocity proportional to the command coming out of the PI (*proportional* plus *integral*) controller. This can also be thought of as a derivative mode of control which enhances stability. Again, we see that a sensor is vital to system success.

Piezoelectric actuators are much less common than magnetic motors and their modeling is rarely explained in control-system texts. One can think of these actuators as spring/mass/damper systems with an internal force generator which responds to an electrical input signal. When the actuator is attached to some useful mechanical load, the inertia, friction, and spring constant of the load simply add to what is already there in the actuator. This portion of our system is thus modeled[10] as a second-order system with a certain gain, damping ratio, and natural frequency, and driven by a force generator that produces force instantly and in proportion to an electrical input signal provided by the controller for this channel. Amplifiers for piezoactuators are somewhat specialized since much higher voltages and much smaller currents are needed than for the more common magnetic motors. For system design purposes, however, we can treat the amplifier as a simple zero-order system since it is much faster than the mechanical moving parts. Once we get to the experimental development stage of our design, we can run frequency-response tests on the subsystem from piezo-amplifier input to motion x_P output. Such test results give the actual behavior of these components, which we use to verify our earlier design assumptions and adjust them if needed. Piezoactuators usually have mechanical overload stops that limit the maximum displacement, so we include a limiter on x_P in our block diagram.

We have now described all the hardware except the controllers. Most motion-control systems use some version of the so-called PID (*proportional, integral, derivative*) control mode. In design, one generally tries the simpler versions first

[10]E. O. Doebelin, "System Dynamics," Marcel Dekker, New York, 1998, p. 307.

and only adds complexity if it is needed to meet specifications. Since we require very high steady-state accuracy in the present system, we will need to use *integral control,* which gives theoretically zero position error for a steady position command. In our system, the magnetic motor channel is quite conventional in that it uses controlled-variable derivative control ("tachometer feedback") together with a PI controller in the forward path. It appears that we *already* have "gratuitous" integral control because of the integrating effect in the load, so the inclusion of the PI controller results in *double* integral control. We decide to accept this (even though it tends to instability) because Coulomb (nonlinear) friction in the load may defeat the gratuitous integration. Our model neglects Coulomb friction, so it *does* have double integral control.

Once we decide on the *form* of control needed, the choice of numerical values for the control parameters K_P, K_I, K_V, and the loop gain is the remaining design task. For the piezo channel, integral control is neither necessary nor desirable (it would *cancel* the numerator of the high-pass filter) so we try simple lead, lag, or lead/lag controllers. It was found that a lag controller worked well when the two time constants τ_D and τ_I were properly designed.

Preliminary design of systems like that just described requires:

1. Knowledge of available hardware components
2. Knowledge of feedback-control design and analysis principles
3. Experience with similar systems
4. Access to an easy-to-use computer simulation language for dynamic systems

Using these tools I was able to come up with a preliminary design that seems to work. The most useful tool was the simulation capability, in my case the SIMULINK package used throughout this text.

Figure 14.6*a* shows the SIMULINK diagram with the final design values. I studied the response to several different types of commands **Xv**, using the **Look-Up Table** (with time as the input) to simulate typical commands and setting the two sine-wave commands to zero amplitude. Later I "disconnected" the **Look-Up Table** output from the summer and activated the two sine waves to study the sinusoidal response. For the sine-wave input, I added a "large" (beyond piezo-motion limit) 1-Hz sine wave ($0.01 \sin 6.28t$) to a small 50-Hz (too fast for the magnetic motor) wave ($0.001 \sin 314t$). This type of command is exactly what the system is intended to deal with; motion consisting of "large and slow" and "small and fast" components. Figure 14.6*b* shows the excellent response. After a starting transient with noticeable error, the system tracks the command quite accurately, as shown by the curve of error (**xv−xc**).

Disabling the sine waves and setting up the lookup table to produce small step-like and large ramplike commands, gave the results of Fig. 14.7*a.* Figure 14.7*b* shows more detail near time zero. The "steps" are actually terminated ramps with rise times of 0.001 s. If a *true* step input is applied, we get rather large overshooting and oscillating, as seen in the curve labeled "xc when xv is a step input." This can be avoided by instead commanding the terminated ramps; we get to the desired

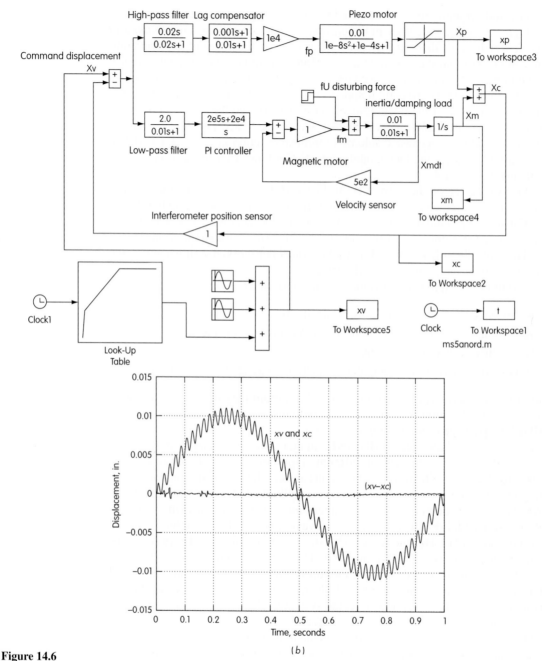

Figure 14.6
SIMULINK simulation of micromotion control system.

position just as fast, but the vibration and fatigue stressing of the oscillation is avoided. This idea of designing the *inputs* (not just the *system*) cleverly has been

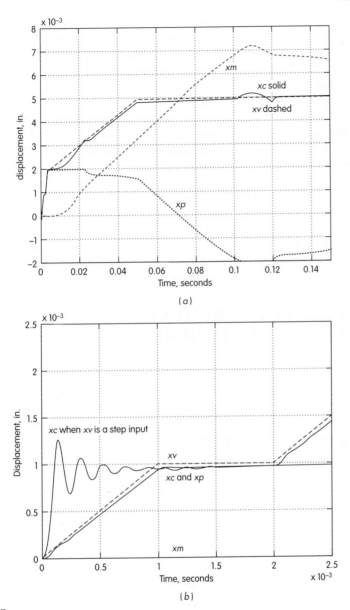

Figure 14.7
Step and ramp response of micromotion control system.

successfully implemented in even more sophisticated ways[11] for motion-control systems in general.

---

[11]Convolve, Inc., Boston, MA, 781-449-8860 (www.convolve.com).

Returning to Fig. 14.7*a,* after the two fast "step" commands, we more slowly ramp up to 0.005 in., beyond the limit of the piezomotor, so the magnetic motor must now participate. The curve **xm** shows its slower response and the *decay* of the piezo component of the displacement, as dictated by its high-pass filtered input. As **xp** decays, the magnetic motor just "makes up" this loss, keeping **xc** close to command **xv** after t = 0.05 s. However, **xp** decays *beyond* zero, requiring **xm** to go *beyond* 0.05. Between **t** = 0.05 and 0.10 s, the integral control slowly forces **xc** closer to **xv**, but some system dynamics cause a "bump" in **xc**, which however is removed, making **xc** almost exactly equal to **xv** after **t** = 0.12 s. While **xc** stays quite close to **xv** at all times, especially after **t** = 0.12, *both xm and xp continue to move.* For later times, not shown on the graph, **xm** *will* converge to 0.05 and **xp** *will* converge to 0.00. It would be nice if **xp** went to zero when **xm** *first* crossed **xv** (at about **t** = 0.07 s) so that not only would **xc** = **xv**, but *all* motion would stop at this point. However, I was unable to get this more desirable feature by an adjustment of numerical values. I could not find any references in the open literature that showed how this might be done; it may or may not be possible. I did find some references that at least dealt with the general concept of piggybacked actuators.[12]

14.3 PARTICLE INSTRUMENTS AND CLEAN-ROOM TECHNOLOGY

Many modern manufacturing processes require extremely clean operating environments, with the manufacture of electronic integrated circuits and MEMS as prime examples of great economic significance. Airborne "dirt" particles cause much of the yield loss in these processes. As miniaturization trends continue to reduce the minimum feature size used in these devices, smaller and smaller particles can cause failure, and so they become of practical interest to process designers and operators (see Fig. 14.8[13]). Particles can cause defects in many ways—by bridging over two adjacent conductors (short-circuit), by masking of the film during an etching process (open circuit), etc. In this section we briefly survey the means for reducing particle concentrations ("clean-room technology") and the measurement techniques for documenting the sizes and/or concentrations of particles that remain ("particle instruments"). Particle instruments, of course, also find important applications outside the clean-room field, such as atmospheric air quality, medical research, smoke/soot studies of combustion, cloud physics, seeding studies for laser velocimetry, etc. In addition to their measurement function, particle "instruments" are needed to *produce* particles of controllable size and concentration for use in instrument testing and calibration.[14]

[12]L. Merritt, "Space Station Freedom Stacked Position Transducers," *Motion,* May/June 1991, pp. 23–25; B. Kafai, "Positioning with Piezo," *Motion Control,* March 2000, pp. 32–35; J. Spanos and Z. Rahman, "Optical Pathlength Control on the JPL Phase B Interferometer Testbed," *JPL New Technology Rept.* NPO-19040, Jet Propulsion Lab, Pasadena, CA, August 1995.

[13]W. G. Fisher, "Particle Monitoring in Clean Room Air with the TSI 3020 Condensation Nuclear Counter," *TSI J. Particle Inst.,* vol. 2, no. 1, pp. 3–19, January–June 1987.

[14]Particle Technology Instruments, *TSI* 3000-R681, TSI Inc., St. Paul, MN.

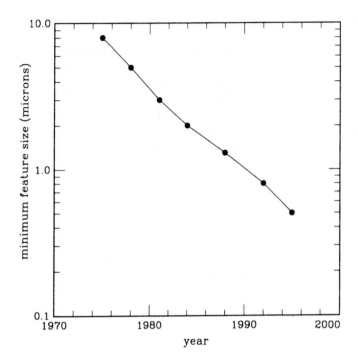

Figure 14.8
Historical reduction in size of integrated circuits.

Figure 14.9[15] relates various sources of airborne contaminants to particle size. Two essential features of clean-room technology are HEPA (high-efficiency particulate air) filters and the laminar-air-flow concept (see Fig. 14.10[16]). Even in 1950 HEPA filters which removed 99.97 percent of all particles 0.3 μm and larger were available; more recently 99.999 percent removal for 0.12-μm particles has been achieved. Since laminar (as compared to turbulent) fluid flow provides a more controllable and predictable flow path, clean rooms use a laminar air flow, from ceiling to floor, to wash away the contaminants generated by the manufacturing process, creating a clean workspace above a certain level. For lower contamination levels, higher (but still laminar) flow velocities are needed (see Fig. 14.11[17]). Figures 14.12 and 14.13 show essential clean-room design features and typical mechanical systems for implementing them.[18]

Let us turn now to methods for measuring particle size. Figure 14.14[19] lists common classes of instruments and particle-generation devices utilized in calibration. We cannot give here a complete treatment of all these devices, but we briefly

---

[15]"The Invisible Enemy," Clean Room Technology Inc., Syracuse, NY.

[16]Ibid.

[17]Ibid.; W. G. Fisher, op. cit., p. 11; R. D. Peck, "The Proposed Revision of Federal Standard 209B," *J. Environ. Sci.,* vol. 29, no. 5, pp. 42–46, 1986.

[18]"The Invisible Enemy," op. cit.

[19]*TSI* 3000-R681, op. cit.

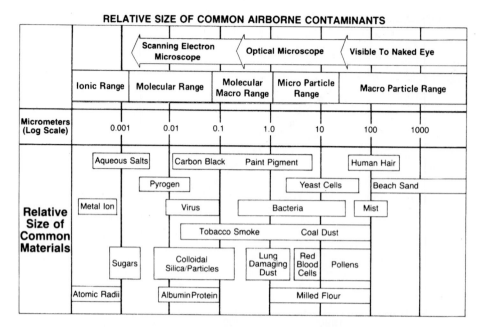

RELATIVE SIZE OF COMMON AIRBORNE CONTAMINANTS

Figure 14.9
Size range of airborne particles.

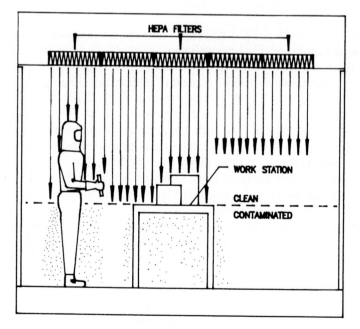

Figure 14.10
Clean room use of laminar flow principles.

Class limits for particle concentrations
(particles/ft^3) according to the proposed
revision to Federal Standard 209B (Peck, 1986).

Class	Measured particle size, μm				
	0.1	0.2	0.3	0.5	5.0
1	35	7.5	3	1	NA
10	350	75	30	10	NA
100	NA	750	300	100	NA
1000	NA	NA	NA	1000	7
10000	NA	NA	NA	10000	70
100000	NA	NA	NA	100000	700

CLEAN ROOM CLASSES AS DEFINED BY FEDERAL STANDARD 209B

Class	Max. No of particles per ft^3 of 0.5 micron and larger	Max. No. of particles per ft^3 of 5.0 micron and larger	Recommended Air Flow Velocity ft/min
100	100	0	70-110
1,000	1,000	7	20-25
10,000	10,000	65	10-15
100,000	100,000	700	4-6

Figure 14.11
Definitions of clean-room classes.

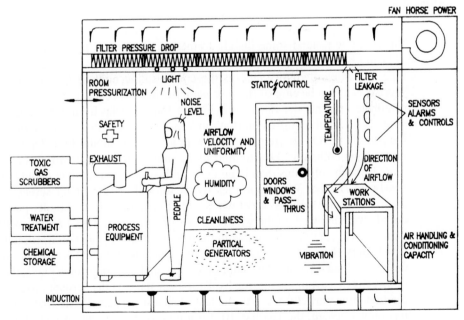

Clean Room Design Considerations

Figure 14.12
Design considerations for clean rooms.

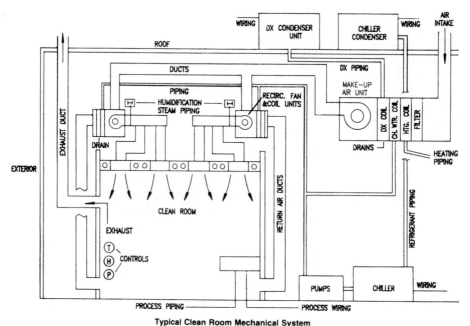

Typical Clean Room Mechanical System

Figure 14.13
Clean-room mechanical systems.

survey a few to lend some insight to the techniques and problems involved. The *diffusion battery*[20] is used to classify particles into certain size ranges so that some type of particle counter (such as the condensation nucleus type to be described shortly) can then be employed to measure particle concentration (particles per cubic centimeter) for that size range, as in Fig. 14.15*a*. Particles from 0.005 to 0.02 μm are strongly affected by diffusion (the random collisions with gas molecules) wherein the particle travels an irregular path, with its position at any time depending on its most recent collisions. Due to their lower momentum, small particles are affected more by such collisions than large ones, and particles larger than about 0.3 μm are nearly unaffected. When particles travel through a fine metal screen (typically 20-μm-diameter stainless-steel wires with 20-μm openings), smaller particles will collide with and stick to the wires because of surface attractive forces, while larger particles will continue flowing with the gas in which they are entrained. In the diffusion battery shown, the aerosol (entraining gas plus particles) is moved through the device at about 4 L/min. The stages are separated by groups of screens, the number of screens in a group increasing by 1 for each stage as the aerosol progresses from the first (fine-particle) stage to the last (coarsest particles). The first screen group has only 1 screen, the last has 10, for a total of 55 screens. An 11-stage selector

[20]Ibid.; D. Sinclair and G. S. Hoopes, "A Novel Form of Diffusion Battery," *Am. Ind. Hyg. Assoc. J.*, vol. 36, pp. 39–41, 1975.

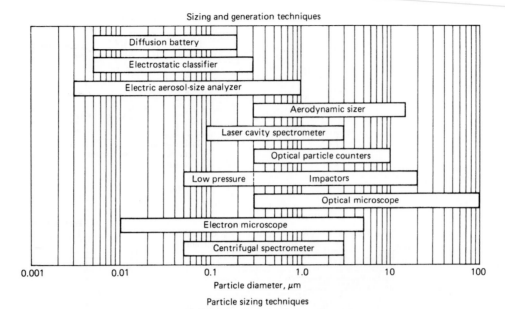

Sizing and generation techniques

Particle sizing techniques

This chart illustrates the particle diameter ranges covered by the most common aerosol sizing techniques.

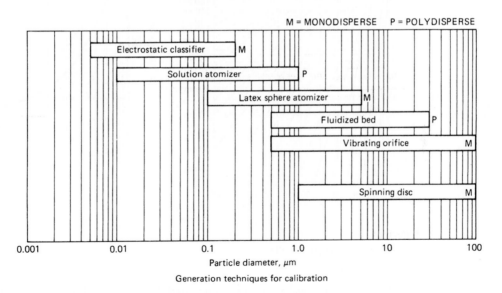

Generation techniques for calibration

Figure 14.14
Particle measurement and generation methods.

valve is used to connect a particle counter to the original, unclassified aerosol or any one of the 10 size-classified stages. By using calibration data as in Fig. 14.15*b*, counter measurements such as in Fig. 14.15*c* can be processed to yield the useful results of Fig. 14.15*d*.

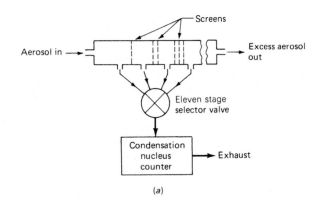

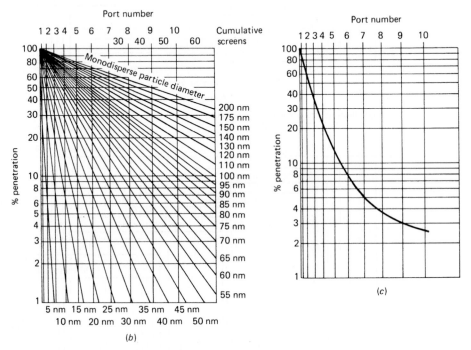

Figure 14.15
Diffusion battery for particle classification.

A commercial *condensation nucleus-particle counter* is shown in Fig. 14.16.[21] Such instruments will measure particles in the size range of 0.02 to 1.0 μm in concentrations from 10^{-2} to 10^7 particles/cm³. Rather than trying to measure the light scattered from the tiny particles themselves (a 0.1-μm particle scatters only

[21]*TSI* 3000-R681, op. cit.; J. H. Agarwal and M. Pourpriz, "A Continuous Flow CNC Capable of Counting Single Particles," TSI Inc., St. Paul, MN.

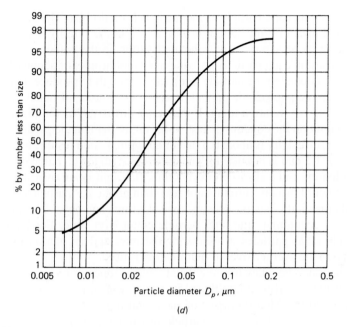

Figure 14.15
(Concluded)

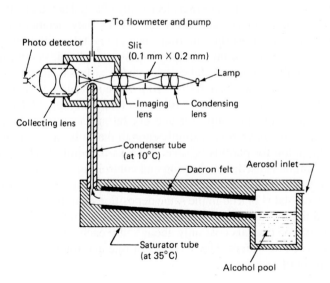

Figure 14.16
Condensation nucleus particle counter.

about twice the light that a gas molecule does, giving a very poor signal-to-noise ratio), these instruments "grow" large droplets of *m*-butyl alcohol by condensation from a vapor, using each particle as the nucleus for a droplet. The air sample to be measured is drawn into the saturator tube at a 5 cm³/s rate by a vacuum pump, causing the air to be saturated with butanol vapor. In the cooler condenser tube, vapor condenses on each particle, growing a butanol droplet of about 12-μm diameter (largely independent of particle size), large enough to produce a distinct voltage pulse from the photodiode optical detector. The choice of butanol is based on the vapor diffusivity relative to air. If the vapor diffusivity is too high, vapor will condense mainly on the condenser-tube walls, rather than on the particles. For alcohol, the vapor diffusivity is about 0.5 that of air, giving good results, whereas that of water is about 1.2 and is not usable. The optical system uses a halogen lamp (white light), two lenses, and a slit to produce a thin, flat light beam over the condenser-tube exit nozzle. For concentrations of less than about 10^3 particles/cm³, there is usually only one particle in the sensing zone at a time, and pulse-counting electronics is used to accumulate the total particle count over time. For larger concentrations, individual particles are indistinguishable, and an analog output signal proportional to the average overall illumination is used.

Japan has adopted a law limiting the allowable concentration of airborne particles smaller than 10 μm to less than 150 μg/m³ within most occupied buildings larger than 3000 m². Measurements of such total-mass concentrations are conveniently made with another type of particle instrument, the *piezobalance*[22] of Fig. 14.17. Here the aerosol stream first impinges on an impactor[23] which collects particles larger than 10 μm while smaller particles pass on to an electrostatic precipitator. In the precipitator, particles are first electrically charged and then attracted to the surface of a vibrating piezoelectric crystal, where they accumulate, gradually increasing the mass and lowering the frequency of crystal oscillation. (The crystal vibration is controlled by an intentionally unstable feedback system in a manner identical to that of the vibrating U-tube densitometer.) Mass concentrations of 0.005 to 9.999 mg/m³ for particles from 0.01 to 10 μm are measured with a resolution of 0.001 mg/m³ and an accuracy of ±10 percent of reading, ±0.01 mg/m³. For "continuous" operation, the impactor and crystal surfaces must be periodically cleaned; however, the operation is simple enough that it is easily automated when unattended operation for periods of several weeks is required. This same crystal-frequency technique is used in measuring and controlling the thickness and rate of deposition of deposited films during the manufacture of microdevice wafers.[24] Here the measuring crystal is placed in the same deposition chamber as the wafers being processed and thus accumulates on its surface a film of "identical" thickness at an "identical" rate. The crystals cost only about $5.00 each and can thus be replaced for each new batch of wafers. The accuracy of rate and thickness measurement is about 0.5 percent of reading.

---

[22]G. J. Sem and K. Tsurobayashi, "A New Mass Sensor for Respirable Dust Measurement," *Am. Ind. Hyg. Assoc. J.*, pp. 791–800, Nov. 1975; *TSI* 3000-R681, op. cit.

[23]V. A. Marple and B. H. Liu, "Characteristics of Laminar Jet Impactors," *Env. Sci. Tech.*, vol. 8, p. 648, 1974.

[24]*TM*-100 Thickness Monitor, Maxtek Inc., Torrance, CA.

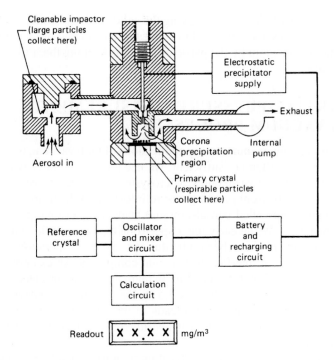

Figure 14.17
Piezobalance for particle mass concentration sensing.

Our final instrument is not a measuring device but rather an *aerosol generator,* for producing aerosols containing particles of controllable size and concentration. Of the types listed in Fig 14.14, we choose to give some details on the *vibrating-orifice monodisperse*[25] generator. (Polydisperse generators produce aerosols with a wide range of particle sizes whereas monodisperse generators are intended to give particles of mainly one size.) A cylindrical liquid jet passing through an orifice (typically of 10- to 20-μm diameter) is inherently unstable and will break up into droplets of nonuniform size. By vibrating the orifice at a fixed high frequency (10 kHz to 1 MHz), it has been found that droplets of *uniform* size are produced, one droplet for each cycle of vibration, with the droplet volume given by Q/f, where Q is the volumetric orifice flow rate and f is the vibration frequency. For example, a liquid feed rate of 0.14 cm³/min with a 60-kHz vibration frequency (a piezoelectric crystal is used) produces 60,000 droplets/s with a mean diameter of about 42 μm and standard deviation of about 1 percent for an 18-μm orifice. When *solid-*particle aerosols are wanted (assuming the solid can be dissolved in a suitable volatile liquid), droplets of solution are formed as described above and injected into a turbulent air jet to prevent coagulation. Then mixing with a high volume of air

[25]*TSI* 3000-R681, op. cit.; R. W. Vanderpool and K. L. Rabow, "Generation of Large. Solid, Monodisperse Calibration Aerosols," *TSI Quart.* vol. 10, no. 1, pp. 3–6, January–March 1984.

causes evaporation of the solvent, leaving the solid particles desired. A Krypton-85 radioactive source is sometimes used to ionize the aerosol gas, discharging any electrostatic charge picked up by the particles and thus keeping them from clinging to apparatus walls.

14.4 PARTIAL-PRESSURE MEASUREMENTS IN VACUUM PROCESSES

In both laboratory and manufacturing applications of high vacuum, measurement of the total pressure is often not sufficient; we need also to identify which gases are present and in what relative amounts. Microsystem manufacture uses many vacuum processes that benefit from partial-pressure information. Instruments that provide such information are called partial-pressure analyzers, residual-gas analyzers, or mass spectrometers. While early mass spectrometers were bulky and awkward to use, recent instruments have been miniaturized and their operation automated by providing a convenient interface to personal computers. An example unit[26] has a sensor element about 7 in. long and 3 in. in diameter, while the electronics package is about $7 \times 4 \times 5$ in. and weighs only 3.5 lb. Describing such a device as a "transducer" is now quite appropriate and leads to including a discussion of it in this edition, even though such "analytical instruments" were considered beyond the scope of earlier editions.

While these instruments can be based on various principles,[27] we present here only the most popular, the *quadrupole mass filter* (QMF).[28] We study this device not only for its importance as an instrument, but also because its design is based on a "classical" differential equation, the Mathieu equation, and we like to show whenever possible the practical utility of mathematics. In physical terms, the QMF can be thought of as an ionization gage with a mass-selective filter between the ionizer and the detector. The filter can be "tuned" to select ions in a narrow range of masses and the ion current corresponding to this selected range measured. By scanning the selected range from a low to a high value, the presence of ions of various masses can be determined and the relative strength (partial pressure) of each measured. The ion current can be plotted against the ion mass to give a mass "spectrum" for the sample. The masses of the ions can be associated with the atomic masses of known elements or compounds, thus identifying the component gases of the sample and quantifying the relative amounts of each. To get absolute (rather than relative) quantitative readings of partial pressures, the analyzer must be calibrated[29] against a (total pressure) ionization gage for each gas of interest.

---

[26]The PPT Series, MKS Instruments, Inc., Andover, MA, 800-227-8766 (www.mksinst.com).

[27]M. J. Drinkwine and D. Lichtman, "Partial Pressure Analyzers and Analysis," American Vacuum Society, New York, 1979.

[28]M. J. Drinkwine and D. L. Lichtman, op. cit.; P. H. Dawson (ed.), "Quadrupole Mass Spectrometry and Its Applications," Elsevier, New York, 1976.

[29]M. J. Drinkwine and D. L. Lichtman, op. cit., Appendix D.

Figure 14.18 shows the physical layout of the QMF. In the ion source, the gas sample is bombarded with electrons boiled off of a hot filament. When these electrons strike a gas atom or molecule, they may knock an electron out of it, leaving it with a net positive charge of 1.602×10^{-19} C (Coulomb); that is, the originally neutral atom has become a charged ion. Actually, there is some probability that more than one electron may be dislodged. Also, the electron bombardment can cause dissociation as well as ionization, that is, bombardment of nitrogen molecules (N_2) in some ionizers produces two kinds of ions, N_2^+ (mass number 28) and N^+ (mass number 14) in the approximate ratio of 7 N^+ for each 100 N_2^+. Since our analyzer filter discriminates on the basis of mass, it appears that the presence of such "fragments" (as they are called) will confuse the operation of the instrument. Actually such fragmentation is often helpful! If both carbon monoxide CO (mass number = 16 + 12 = 28) and nitrogen N_2 (mass number = 14 + 14 = 28) were present, we could not distinguish between them since they have identical masses. Since each molecule has its own known and distinctive fragmentation pattern, when we go to interpret a measured mass spectrum, these patterns allow us to resolve problems such as that just shown for nitrogen and carbon monoxide. This and other "tricks" needed to properly interpret spectra are covered in detail in the references, but we shall from here on act as if we were dealing with the ideal case where there are no fragments and the measured ions have only one electron removed.

In Fig. 14.18, the mass filter itself consists of the four metal cylindrical rods with voltages $+ \phi_0$ and $- \phi_0$ applied as shown. In the MKS instrument referenced earlier, these rods are 3.5 in. long and 0.250 in. in diameter. The energized rods create around them an electric field which will be felt by the ions as they are projected from the ion source toward the detector at the other end. To ideally accomplish the function of the filter, this electric field should be a so-called *quadrupole field.* By definition this is a field whose *xyz* components are, respectively, proportional to the *x, y,* and *z* locations in the field, that is, the field is a linear function of *x, y,* and *z*. The force in Newtons felt by a charged particle is equal to the

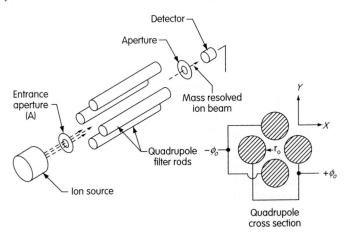

Figure 14.18
Quadrupole partial pressure analyzer (mass spectrometer).

product of the charge in Coulombs and the electric field strength in volts/meter. For a quadrupole field, the force will be directly proportional to the distance from the origin, analogous to an "omnidirectional" mechanical spring. Dawson[30] shows that to produce such a field requires that the four rods have a hyperbolic rather than cylindrical cross section. The cylindrical rods used in most *residual gas analyzers* (RGAs) are thus an approximation to the ideal geometry. Hyperbolic rods *are* used in some "high-end" analytical instruments, but in RGAs, their problems of manufacture and alignment outweigh the slight performance advantage realized. While a general quadrupole field has a nonzero z component, the QMF requires this to be zero. That is, when the ions leave the ionizer, they already have a suitable z velocity and there is no advantage to changing this, so no z force is necessary.

The potential difference $\phi(x, y, z)$ between a pair of rods is twice the potential ϕ_0 applied to a single rod. When we apply $+\phi_0$ and $-\phi_0$ voltages to a pair of rods, the potential $\phi(x, y, z)$ thus created can be shown to be

$$\phi(x, y, z) = \frac{2\phi_0(x^2 - y^2)}{2r_0^2} \tag{14.2}$$

Recall that the general relation between electric field strength E and potential ϕ is

$$E_x = -\frac{\partial \phi}{\partial x} \qquad E_y = -\frac{\partial \phi}{\partial y} \qquad E_z = -\frac{\partial \phi}{\partial z} \tag{14.3}$$

Combining these relations with Newton's law we can get the differential equations of motion for the ions. We assume that the ion's mass is m and that its charge is numerically equal to that of an electron, $q = 1.602 \times 10^{-19}$ C.

$$\frac{-2q\phi_0 x}{r_0^2} = m\frac{d^2 x}{dt^2} \tag{14.4}$$

$$\frac{2q\phi_0 y}{r_0^2} = m\frac{d^2 y}{dt^2} \tag{14.5}$$

$$0 = m\frac{d^2 z}{dt^2} \tag{14.6}$$

The z motion (along the axis of the rods) is the simplest since the acceleration is zero and thus the z velocity remains constant at whatever value it had when the ions left the ionizer. While the applied voltage ϕ_0 must be time varying in order to get the mass-filtering action we desire, let us first explore the behavior for a constant voltage. In this case, Eqs. (14.4) and (14.5) are very simple linear differential equations with constant coefficients whose solutions are well known. For the x motion, the equation is that of an undamped oscillator; thus, the ions would perform in x a sinusoidal motion of fixed amplitude and frequency, with the amplitude being determined by the initial values of the x position and velocity as the ions enter the filter. The y solution is the sum of two exponential terms, one decaying to zero and the other going toward infinity; thus, the ions would move away from the axis of the

---

[30]P. H. Dawson, op. cit., p. 10.

filter as they progressed along its length, possibly striking the filter rods if they moved far enough. While the ion mass m influences the frequency of the x motion and the time constant of the unstable y motion, the overall effect does not give the desired filtering effect.

What we want to happen is for ions with masses in a narrow range to remain close to the axis and enter the detector while ions with masses outside this range are deflected away from the axis and thus never reach the detector. It has been found that to achieve this goal, it is necessary to make the applied voltage of the form

$$\phi_0 = U - V \cos \omega t \qquad\qquad \textbf{(14.7)}$$

where U and V are constants with $V > U$. The equations now are still linear but with a time-varying coefficient, which makes the solution much more difficult. Within the general class of linear equations with time-varying coefficients, the specific equation here is called the Mathieu equation, which has been extensively studied.[31] Most applications of the Mathieu equation involve *undesirable* vibrations. However, we will shortly see that the design principle of the QMF is based on these very vibrations. While accurate numerical predictions require a careful analytical study, rough qualitative arguments may give a useful insight into the essential features. For the x axis, the alternating component of the field will cause the lighter ions (which can follow the rapid changes) to oscillate with increasing amplitude, strike the rods, and not appear at the detector. Heavier ions try to do the same, but their inertia limits the amplitude achieved and they do enter the detector. The x axis may thus be thought of as contributing a filtering effect that passes heavy ions and rejects light ones (*high-pass filter*). For the y axis, heavy ions tend to diverge due to the constant component of the field (the alternating component is too fast for them to be much affected), but the light ions can respond to those parts of the cycle that oppose the steady divergence. The light ions thus remain close enough to the axis to reach the detector, making the y-axis behave as a *low-pass filter*. With both axes acting simultaneously, only ions in a narrow band of mass values will be successful in reaching the detector, giving the desired "band-pass" filtering effect.

Careful analytical study produces the stability diagram of Fig. 14.19, which is very useful for design purposes. Any instrument operating point within the triangular stability region gives a mass-filtering effect, but the optimum point is near the apex of the triangle ($U/V = 0.168$ and $4qV/mr_0^2\,\omega^2 = 0.708$) where the range of masses that pass through the filter becomes very small and thus instrument mass resolution is very good. Practical instruments will of course work at U/V values slightly less than 0.168. Note also that the filter discriminates on the basis of q/m for the ion, *not m itself*, so that an ion which lost two electrons behaves like a "normal" one of half the mass. The "fragmentation patterns" mentioned earlier fortunately allow correct interpretation of such anomalies. To "scan" the passband of the mass filter over the desired range of masses, the field frequency ω is kept constant and the U/V ratio is kept fixed near 0.168 while U and V are individually varied.

[31]N. W. McLachlan, "Theory and Application of Mathieu Functions," Oxford University Press, New York, 1947; J. P. DenHartog, "Mechanical Vibrations," 4th ed., McGraw-Hill, New York, 1956, p. 343.

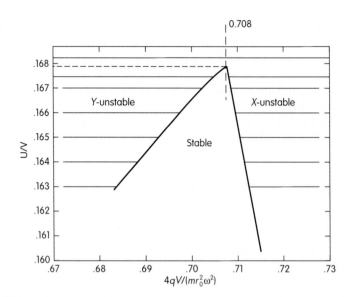

Figure 14.19
Stability diagram for quadrupole analyzer.

In addition to *analytical* results such as the stability diagram of Fig. 14.19, *numerical* solution of the differential equations is useful for showing the actual *x* and *y* motions of ions as they move through the filter. The system equations are easily simulated in SIMULINK, giving the diagram of Fig. 14.20. There we have used the numerical values:

$$r_0 = 0.005 \text{ m} \qquad V = 189 \text{ V} \qquad U/V = 0.164$$
$$\text{rod length} = 0.0889 \text{ m} \qquad \text{frequency } (\omega) = 4\pi \text{ rad}/\mu\text{s}$$

The time unit has been taken as *microseconds,* which requires that the frequency of 2.0 Mz be converted as shown. This "time scaling" also requires that the Newton's law equations have the accelerations given in $\dfrac{\text{meters}}{\text{microsecond}^2}$, which requires a division by 10^{12}. In the unit system being used, the ion mass must be given in kilograms, which requires that the mass number be multiplied by 1.675E-27. The formula relating the location (0.708) of the peak of the stability diagram to system parameters is that given on Fig. 14.19.

$$m_{\text{peak}} = \frac{4qV}{0.708 r_0^2 \omega^2} = \frac{4 \cdot 1.602 \cdot 10^{-19} \cdot 189}{0.708 \cdot 0.005^2 \cdot (4 \cdot 3.1416 \cdot 10^6)^2}$$
$$= 0.4333 \cdot 10^{-25} \text{ kg} = 25.9 \text{ amu} \tag{14.8}$$

Since we use $U/V = 0.164$ (rather than the peak 0.168), a small range of mass values will lie in the stable region near 25.9. Our simulation studies can define the actual range.

We will include in the alternating voltage component a phase angle, which is set at 0.0 radians in the simulation shown. Since ions may reach the entrance ($z = 0.0$)

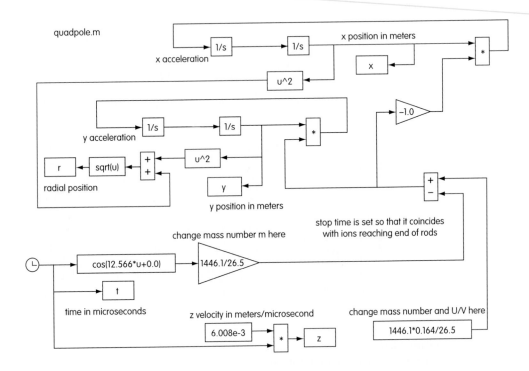

Figure 14.20
Simulation diagram for quadrupole analyzer.

of the filter at any random time relative to the state of the applied voltage, we are interested in the effect of this "phase relation" on the ion motion. By setting the phase to various values between 0 and 2π, we can study this. Ions may also enter the filter region with various x and y positions and velocities, and this will affect their motion as they proceed along. For example, if an ion has a large x or y velocity as it enters the filter, it might strike the rods and be neutralized (disabling any filtering effect) before the quadrupole field has a chance to affect its motion. Of course, the entrance aperture shown in Fig. 14.18 is intended to provide a collimated ion beam (with small radial velocity components), but such devices are never perfect. In the simulation shown, we let the initial x and y velocities be zero but take the initial displacements as 10 percent of the radius r_0. All these possible variations can be explored using the simulation, and one can thus get a good understanding of system operation.

The ions are assumed to enter the filter with a fixed and known z velocity, and the quadrupole filter is quite tolerant of variations in this quantity. It essentially determines how long the ion will "reside" within the filter and thus how many cycles of the alternating voltage it will feel. Typically, this number of cycles is the order of ten to a few hundred. One can estimate the z velocity from an energy calculation that equates the electrical energy (charge times potential difference) applied to accelerate the ion as it leaves the ionizer with the kinetic energy $mv^2/2$.

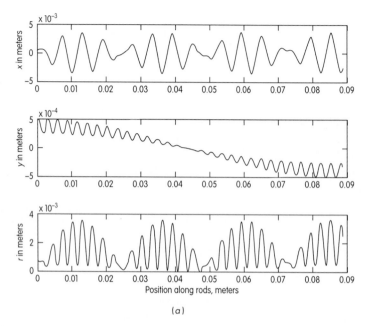

Figure 14.21
Selected results from quadrupole simulation.

We have used a typical ion energy of 10. electron volts (10. volts) (1.602E-19 coulombs) to compute the z velocity of an ion of known mass (kg). For mass number $m = 26.5$ amu, the velocity is 6008 m/s. To display how the x and y motions vary as the ion progresses along the filter z axis, we compute z in meters, using the known constant velocity and time t in microseconds. We can then ask for graphs of x and y versus z, rather than t. It is also convenient to compute x and y in millimeters and also the radial displacement, which we can compare to r_0 to see whether an ion strikes the rods and is thus lost from the beam.

Using the numerical values shown in the simulation diagram (Fig. 4.20), the graphs of Fig. 14.21a show that the x and y deflections, and thus the radial deflection r, are always less than r_0, so ions of mass 26.5 will pass through the rods and be collected for measurement at the detector. Changing only the mass m, we find (Fig. 14.21b) that $m = 25.5$ causes x instability and $m = 27.5$ (Fig. 14.21c) causes y instability; thus, ions with these masses will strike the rods and never reach the detector, confirming the mass-filtering effect of the instrument. Equation 14.8 shows that the location of the peak of the stability diagram can be changed by adjusting V, the amplitude of the oscillating component of the rod voltage; we must of course also change U so as to always maintain the desired U/V, which in our example equals 0.164. When a QMF is scanned over a range of mass numbers and the ion current is graphed against mass number we get a mass spectrum characteristic of the gases present in the vacuum system being measured. Figure 14.22[32]

[32]M. J. Drinkwine and D. L. Lichtman, op. cit., p. 83.

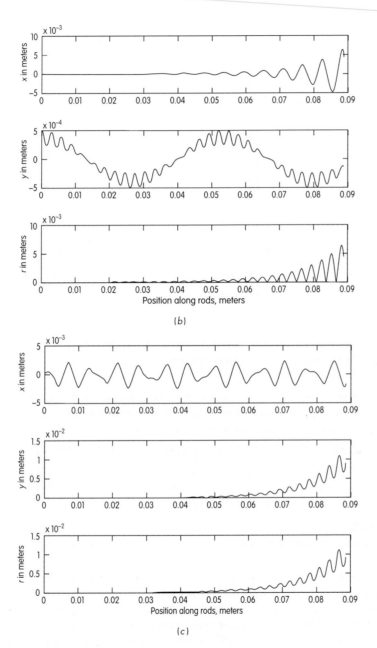

Figure 14.21
(*Concluded*)

displays an actual measured spectrum and discusses its significance. The reference by Drinkwine and Lichtman supplies many such examples, giving the reader a good insight into how such spectra are interpreted.

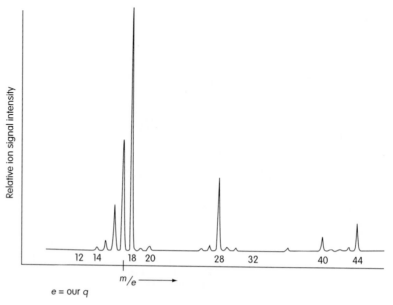

e = our q

TS-27

Type of system: UHV, stainless steel, 20 1 volume, 200 1/s DI ion pump.

Type of mass spectrometer: Extra Nuclear Quadrupole.

System pressure: 3×10^{-7} torr.

Recent system history: System rough pumped with sorption pump to 10^{-2} torr, then pumped to 3×10^{-7} torr with ion pump. The mass spectrometer filament was turned on during pump down when the pressure was 10^{-6} torr.

Spectrum features: Very large 18 peak. Also peaks at $m/e = 16, 17, 28, 40$ and 44.

Diagnosis: A typical gas phase composition midway in a pump down to UHV. H_2O (17,18) is by far the largest constituent here with CO (28), CH_4 (16), CO_2(44), Ar (40), and Ne (20) the next most prevalent gas phase species.

Figure 14.22
Typical results from quadrupole analysis of an ultra-high-vacuum (UHV) system.

The QMF can be used directly only for systems whose vacuum is sufficiently high (pressure sufficiently low) that the ion beam does not strike too many atoms or molecules of the gas being measured. For the MKS unit referenced earlier, this limit is about $1 \cdot 10^{-4}$ torr. One can however extend the useful range using an indirect approach as shown in Fig. 14.23.[33] Here the chamber can be at much higher pressures since an orifice and auxiliary turbopump create a pressure-dividing effect such that the analyzer is exposed only to its allowed pressure. Knowledge of the orifice's characteristics allows one to relate the measured pressure to the chamber pressure.

The ion detector collects the stream of ions and produces an output current related to the partial pressure of the gas component being measured. The detector is

---

[33]The PPT Series of Quadrupole Residual Gas Analyzers, MKS Instruments, Inc.

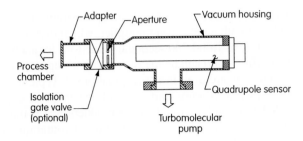

Figure 14.23
Technique to extend quadrupole use to higher pressures.

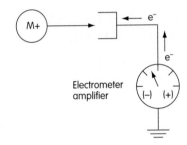

Figure 14.24
Faraday cup ion detector.

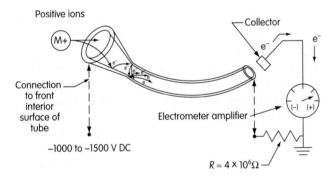

Figure 14.25
Electron multiplier ion collector.

usually a *Faraday cup* (Fig. 14.24) or an *electron multiplier* (Fig. 14.25[34]). A Faraday cup is a metal "can" that collects the ions, which take electrons from the cup, creating a current that is measured by an electrometer amplifier. One can trade off minimum detectable current (pressure) against amplifier step-response speed. A minimum current of 10^{-15} A (10^{-11} torr pressure) gives a step response of about 1 s. (10^{-13} A (10^{-9} torr) gives 0.01 second.) Amplifier response speed is related to the speed at which we can "scan" the analyzer over a range of mass numbers when we

---

[34]Ibid.

are plotting the mass spectrum of a sample. If more sensitivity is needed, the Faraday cup can be replaced with an electron multiplier, which uses secondary electron emission to produce 1000 to 1,000,000 secondary electrons for each ion entering its cone. Then, 10^{-17} A (10^{-13} torr) can be measured with a step response of 1 s.[35]

Residual gas analyzers are used in many situations.[36] Typical instruments can detect concentrations as low as a few parts per billion (ppb) and cover amu ranges of 1 to 300. In the MEMS and integrated circuit production area they started out as leak detectors but are now branching out to detection of contamination, and use as general-purpose trouble-shooting tools. Real-time control applications based on in-situ measurement of gas species is projected to be the next major application.

14.5 MAGNETIC LEVITATION SYSTEMS FOR WAFER CONVEYORS

In the wafer manufacturing process, wafers must be moved from one station to another in a clean-room environment. Conveyor systems that use moving mechanical contacts to carry electrical power to the "cart" that transports the wafers suffer from contamination of the clean room by wear particles. One solution[37] to this problem uses carts that are magnetically levitated without any physical contact to cause wear. In another of my books, a simple magnetic levitation system used to transport people is analyzed.[38] Most such systems (magnetically levitated trains, etc.) do not have the contamination problem just mentioned and use electric power transmission devices with mechanical contacts. A feedback control system is used to stabilize the levitation, and the goal is to maintain, in the face of various disturbing forces, a fixed air-gap between the relatively moving parts, using electromagnets. In the wafer-transport system that we will explain shortly, permanent magnets are cleverly combined with electromagnets in such a way that electric power requirements are drastically reduced. This allows the levitation system to be powered by a *battery* on board the transporter cart; thus, no wear-particle-producing mechanical contacts are necessary. A typical design allows operation over a full 8-h shift before the batteries need to be recharged. All these advantages were obtained by giving up the conventional requirement of a constant air gap and allowing the air gap to change in such a way that the vertical load is supported by the permanent magnets only. The electromagnets, which consume power, are used only during transient operation to

[35]P. H. Dawson, op. cit., p. 138.

[36]L. Peters, "Residual Gas Analysis: A Technology at the Crossroads," *Semiconductor International,* Oct. 1997, pp. 94–102; R. K. Waits, "Controlling Your Vacuum Process: Effective Use of a QMA," *Semiconductor International,* May 1994, pp. 79–84; "In Situ Process Gas Monitoring in PVD Systems," UTI Div. of MKS Corp.; "Mass Spectrometry Growth Seen in Benchtop, Process Areas," *R & D Magazine,* Sept. 1996, pp. 74–80; "Effective Monitoring of Wafer Fab Process Gas Environments," UTI Div. of MKS, 1986.

[37]"Toshiba Space Linear System Technical Guide," Toshiba Corp., April 1987; M. Moroshita et al., "A New Maglev System for Magnetically Levitated Carrier System," Toshiba R&D Center, Kawasaki-city, Kanagawa, 210, Japan.

[38]E. O. Doebelin, "System Modeling and Response," Wiley, New York, pp. 423–431, 1980.

guarantee stability. Fortunately, in the wafer conveyor application, a changing air gap can be tolerated, making this approach feasible.

A schematic drawing of the magnetically levitated carrier and its associated guide rail is given in Fig. 14.26. The carriers must be magnetically guided laterally

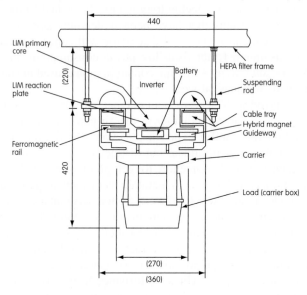

(*a*) Cross-sectional view

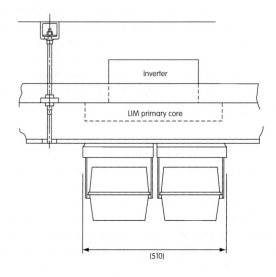

(*b*) Side view

Figure 14.26
Toshiba magnetic levitation system for wafer conveying.

and vertically, and propelled longitudinally along the guideway. All the guidance and propulsion systems are completely noncontacting so as to generate no contaminating wear particles in the clean-rooms where the system operates. A *linear induction motor* (LIM) is used for propulsion, a typical speed being about 1 m/s. While it is on a much smaller scale, this technology is closely related to the magnetically levitated and propelled trains that have been under development in Japan and Germany for many years now. Our brief discussion will concentrate entirely on the levitation system. Its unique feature is its low power consumption, allowing it to run entirely on batteries and thus avoiding the contamination caused in sliding or rolling electrical contacts by mechanical wear.

If you have "played with" a permanent magnet and a piece of iron or steel, you know that the attractive force gets larger as the air gap between magnet and steel gets smaller. If you ran careful experiments, you would find that the relation between force and air gap is not linear but distinctly curved. It is also possible to theoretically analyze this phenomenon to get predictive equations. Levitation systems used for trains are designed to maintain a constant air gap and this requires a large amount of electrical power for the electromagnets that are used. The Toshiba wafer conveyor system can tolerate a changing air gap (within limits, of course) and still provide all its necessary functions, so constant air gap is not a design requirement. A permanent-magnet system using attractive (rather than repulsive) forces can levitate a wide range of loads if we allow it to change its air gap. Unfortunately, such systems are unstable; the slightest disturbance from perfect balance between the supported load and the magnetic force causes the load to either fall away or clamp to the rail. By adding electromagnets and a feedback system to the design, stability can be guaranteed. A clever control algorithm lets the system always seek the air gap at which the applied load is exactly balanced by the permanent magnets alone, the electromagnets providing only a transient effect that produces stability. Thus, most of the time (whenever the load has become steady), the electromagnets are inactive. Since permanent magnets consume no power, the system's total power consumption is very small and can be supplied by on-board batteries, negating the need for any power-transmitting sliding or rolling contacts between the carrier and the frame.

We next want to develop a linearized model for this system since such models are usually the best for preliminary system-design purposes. Once we have the system equations, we will also write a simulation program to check out the actual performance. Figure 14.27 shows a simplified schematic diagram suitable for analysis purposes. Levitation systems (whether hybrid or pure electromagnet type) generally require sensors to measure the air gap (load position) and load velocity. These signals are required in the feedback system that provides the stability needed in practical applications. The hybrid system we are studying also needs a magnet-current sensor. The current signal itself is useful in augmenting stability, and the integral of the current turns out to be needed to allow the system to self-adjust the air gap so that all the load is supported by the permanent magnets (and none by the electromagnets) when the load is steady at any value within the design range. As is common in many electromechanical systems, op-amp circuitry is used for signal

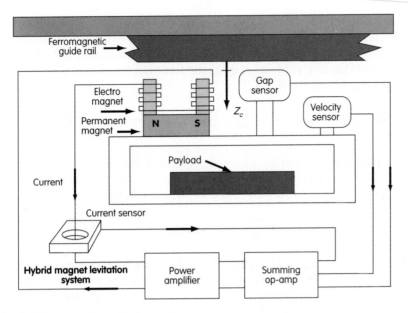

Figure 14.27
Schematic diagram of levitation system.

processing at a low power level and the op-amp output signal is boosted by a power amplifier to the higher power level needed to drive actuators.

Analysis of such systems requires basically two equations: Newton's law for the mechanical moving parts and Kirchoff's voltage loop law for the electrical circuit. These two equations are coupled because Newton's law has some electrical terms in it and Kirchoff's law has some mechanical terms in it. Magnetic circuit principles are also needed to supply certain terms that appear in these two basic equations. Let us start with Newton's law for the vertical motion z_C. Since this system has several nonlinear effects and we wish to linearize these, we take the conventional approach and assume the system is in initial equilibrium at an air gap z_{C0} at $t = 0$, when disturbances are applied and cause small perturbations of all the system variables. In our equations, symbols for all the system variables refer to these small perturbations, not the total values of these variables. Since there is no mechanical contact with the carrier and air friction is assumed negligible, the only vertical forces acting are the magnetic force f_M and the disturbing force f_U. (Recall that the steady gravity force (weight) is exactly balanced by a steady magnetic force for the initial equilibrium condition assumed; thus, this weight does not appear in our perturbation equations.) The main disturbing force in our system is caused by the addition or removal of "payload" mass (wafers, etc.). This results in a perturbation of the gravity force and a concurrent change in system mass M. The magnetic force is a nonlinear function of the instantaneous current i and air gap z_C. A theoretical analysis in the reference derives for this force a formula useful for magnet design purposes. Once hardware is available, experiments are run that give a more accurate description of the magnetic force behavior. Whether we are at the preliminary design stage (using a theoretical model) or at the experimental development stage (using a model based on measured data), our linearized model takes exactly the same form since in either case, magnetic force is a function of two independent variables, air gap and current.

$$\text{Magnetic force} \triangleq f_M(z_C, i) = f_{M0} + \frac{\partial f_M}{\partial z_C} \cdot z_C + \frac{\partial f_M}{\partial i} \cdot i \triangleq f_{M0} + K_{fz} \cdot z_C + K_{fi} \cdot i$$

$$(14.9)$$

The two partial derivatives are assumed constant in the neighborhood of the equilibrium operating point and are most accurately obtained as numerical values from experiments. Writing our Newton's law for the perturbations only, we get

$$M \frac{dz_C}{dt^2} = K_{fz} z_C + K_{fi} i + f_U$$

$$(14.10)$$

We now need to write a Kirchoff voltage loop equation for the circuit elements between the output terminals of the power amplifier. These elements are driven by the amplifier's output voltage E_{amp}. Copper-wire coils on the electromagnet plus the output resistance of the amplifier combine to give a total resistive element R. When inductive elements like our electromagnet coils have motion that changes the magnetic circuit, we need to be particularly careful in modeling these effects in our Kirchoff equation. In our system, the magnetic flux ϕ that links the coils is a nonlinear function of both current i and displacement z_C, that is, $f = f(i, z_C)$. If we moved the magnet assembly "far" (theoretically, infinitely far) from the ferromagnetic guide rail, inductive effects in the coils become insensitive to motion, but there remains a fixed inductance L_∞. The general expression for the induced voltage of a coil is $N(d\phi/dt)$, where N is the number of turns in the coil. Since our flux depends on i and z_C, we can get a linearized expression for the voltage as follows:

$$N \frac{d\phi(i, z_C)}{dt} = N \frac{\partial \phi}{\partial i} \cdot \frac{di}{dt} + N \frac{\partial \phi}{\partial z_C} \cdot \frac{dz_C}{dt}$$

$$(14.11)$$

The partial derivatives are again assumed constant for small perturbations around the equilibrium operating point. Numerical values are most accurately found from experimental measurements. Equation (14.11) has no constant term because the operating point is one of steady flux, and thus the time derivative is exactly zero. We can now write our Kirchoff voltage loop law as

$$L_\infty \frac{di}{dt} + N \frac{d\phi(i, z_C)}{dt} + Ri - E_{amp} = 0$$

$$(14.12)$$

which becomes in linearized approximation

$$\left(L_\infty + N \frac{\partial \phi}{\partial i} \right) \cdot \frac{di}{dt} + \left(N \frac{\partial \phi}{\partial z_C} \right) \cdot \frac{dz_C}{dt} + Ri - E_{amp} = 0$$

$$(14.13)$$

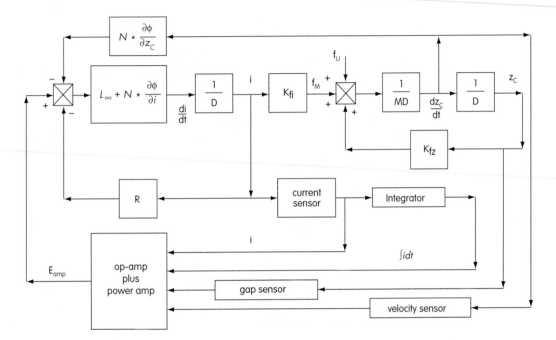

Figure 14.28
Block diagram of levitation system.

Figure 14.28 is a block diagram based on Eqs. (14.10) and (14.13); it also shows how E_{amp} is produced from the sensor and integrator signals. Each of these signals is multiplied by an adjustable constant ("C" value), and then they are all summed as:

$$E_{amp} = -C_d z_C + C_v \dot{z}_C + C_i i + C_{i\,int} \cdot \int i\,dt \qquad (14.14)$$

Equations (14.10), (14.13), and (14.14) are a complete description of the entire levitation system and can be directly used to draw a simulation diagram, as shown in Fig. 14.29. The system input is a step change in disturbing force **fU** corresponding to a suddenly added payload of 1.5 kg mass (gravity force = 14.7 N). The symbol **L** is used to represent the entire coefficient ("total inductance") of the di/dt term in Eq. (14.13), while **Npz** is used for $N\dfrac{\partial \phi}{\partial z_C}$. Some of the numerical values here are given in the Morishita et al. reference, but I had to estimate or guess at others. I adjusted my guessed values until the simulated performance agreed quite closely with an experimental step-response curve given in the reference. The step response is shown in Fig. 14.30, where we see that the current does go to zero in the steady state and the air gap is reduced by exactly the correct amount (about 1 mm) needed to allow the entire new load to be supported by the permanent magnet alone, without any drain on the battery.

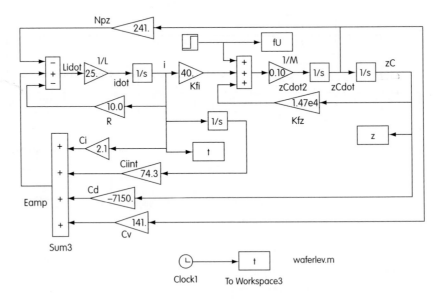

Figure 14.29
Simulation diagram of levitation system.

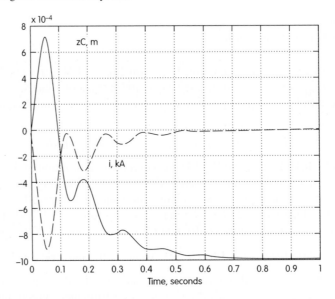

Figure 14.30
Levitation system response to adding a mass load.

While simulation allows easy study of variations in system parameters and inputs, some useful analytical results can be obtained fairly easily by combining Eqs. (14.10), (14.13), and (14.14) algebraically to get a single differential equation relating output z_C to input f_U:

$$\left\{ D^4 + \left(\frac{R - C_i}{L} \right) \cdot D^3 + \left[-\frac{C_{i\,\text{int}}}{L} - \frac{K_{fz}}{M} - \frac{K_{fi}}{ML} \cdot (-N_{pz} + C_v) \right] \cdot D^2 \right.$$

$$\left. + \frac{-K_{fi} \cdot C_d - K_{fz} \cdot (R - C_i)}{ML} \cdot D + \frac{K_{fz} \cdot C_{i\,\text{int}}}{ML} \right\} z_C$$

$$= \left[\frac{D^2}{M} + \frac{(R - C_i)}{ML} \cdot D - \frac{C_{\text{int}}}{ML} \right] f_U \qquad \qquad \textbf{(14.15)}$$

One use of this equation is to check for stability, always a potential design problem in feedback systems, using the Routh stability criterion.[39]

14.6 SCANNING-PROBE MICROSCOPES

The invention in 1981 of the *scanning-tunneling microscope* (STM) by Binnig and Rohrer at IBM Zurich (they got a Nobel physics prize for it) has led to the development of a whole family of *scanning-probe microscopes* (SPM) that have revolutionized experimental techniques for the study of all kinds of surface phenomena. These microscopes are appropriately considered in this chapter since they not only *use* microtechnology in their operation but also are *used* to study many phenomena at the micro- and nanolevel. While these instruments basically sense tiny currents or forces (rather than the light rays of conventional microscopes), the term "microscope" is apt since an *image* of microscale effects is displayed (on a TV monitor).

The basic structure of all SPMs (including the STM) is essentially the same and depends heavily on the use of piezoelectric actuators to provide controlled relative motions of a tiny sensing tip and the sample being studied. Figure 14.31*a* shows the design in simplified form. A coarse *z*-position control (perhaps a screw-type micrometer) is used to approach the sensing tip to the specimen. Once the tip is close enough, further precise positioning is done by the piezoelectric actuator, under automatic control. Since the *relative* position of tip and specimen is what is important, one could move the tip, the specimen, or both; usually the actuator is moved. Its *x* and *y* positions are usually "scanned" over the specimen surface in a prescribed pattern, like a TV raster. While the scanning is going on, the *z* position is being precisely controlled, the method depending on the type of study being done. When a scan is finished, we have a complete set of electronic *x, y,* and *z* data, so we can construct on our TV monitor an image of the measured phenomenon as it varies over the specimen surface. Figure 14.31*b*[40] shows some construction details (including that of a spring suspension for vibration isolation) of a unit intended for use in ultrahigh vacuum.

In an STM, the specimen *must* be a conductor or semiconductor, and, in addition to what is shown in Fig. 14.31*a*, a fixed bias voltage (in the range 1 mV to 4 V) is applied between the tip and the specimen. A tiny current, called the *tunneling current* (between 10 pA and 10 nA) flows between the tip and specimen, and is

[39]E. O. Doebelin, "Control System Principles and Design," Wiley, New York, 1985, pp. 187–191.

[40]Model VP@ UHV SPM, Thermomicroscopes, Sunnyvale, CA (www.thermomicro.com).

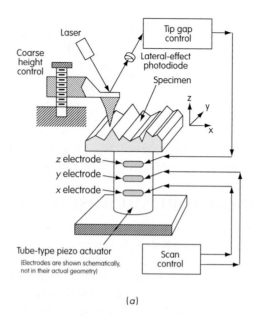

(a)

Figure 14.31

Scanning-probe microscope: (*a*) Schematic diagram. (*b*) Construction details.

measured. This current begins to flow when the tip/specimen gap is about 1 nm, that is, before actual contact. To resolve single atoms in the image, the sensing tip must be extremely sharp, and in fact, it terminates in a *single atom!* In SPM applications *other* than STM, the specimen need not be a conductor (no bias voltage is used), and instead of measuring a tunneling current, the sensing tip is mounted on a tiny cantilever beam that is used as a *force sensor.* Figure 14.32[41] shows some features of such cantilevers. They can be produced (typically 1400 per wafer) with integrated sensing tips, using **MEMS** technology. The sensing tips may be pyramidal or conical in shape, the conical form having sharper tips (curvature radius about 10 nm), which improve spatial resolution, and a greater aspect ratio (3:1 compared to 1:1) to measure steeper sidewalls and deeper trenches.

Returning now to the STM, the tunneling current depends exponentially on the tip/sample gap; a 10-percent change (0.1 nm) in an initial 1-nm gap causes the current to change by an order of magnitude.[42] This gives the STM its remarkable sensitivity. An approximate equation[43] for this relationship is

$$I = Kve^{-Bd} \tag{14.16}$$

[41]Thermomicroscopes (Park Scientific Instrument, Topometrix), Sunnyvale, CA, 800-776-1602 (www.thermomicro.com).

[42]R. Howland and L. Benator, "A Practical Guide to Scanning Probe Microscopy," Thermomicroscopes (Park Scientific, Topometrix), 1996, p. 3.

[43]"The Scanning-Probe Microscope Book," p. 9, Burleigh Instruments, Inc., Fishers, NY, 716-924-9355 (www.burleigh.com).

(b)

Figure 14.31
(Concluded)

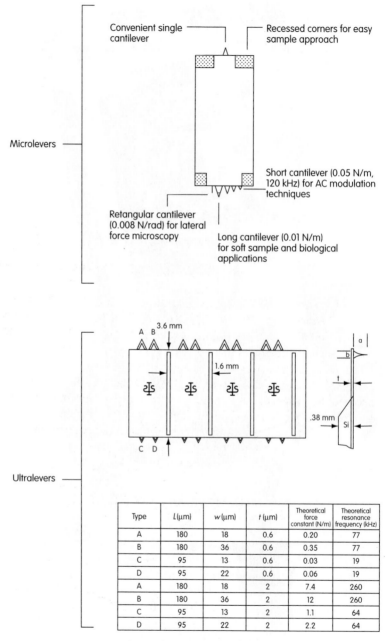

Figure 14.32
Microscale cantilevers produce by MEMS pocesses.

where K is a proportionality constant, I is the tunneling current, B is a constant proportional to the square root of the barrier height between the tip sample and tip, d is the tip-to-sample distance, and v is the bias voltage between the tip and sample.

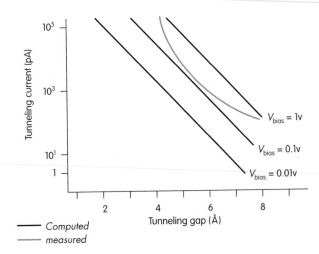

Figure 14.33
Theoretical model and measured results for current-gap relation in scanning-probe microscope.

Figure 14.33[44] illustrates this relation, including an actual measured curve for comparison to the theory. (In Fig. 14.33, and in much of the other literature, the use of the "non-SI" length unit, the angstrom (Å) where 1 Å $\triangleq$ 0.1 nm, is common.) A sample can be scanned in either of two modes: *constant height* or *constant current*. In constant-height mode, the tip travels in a horizontal plane above the sample while the current is recorded as function of the (x, y) location. This mode gives fast scans since z need not be adjusted, but it is useful only for relatively smooth surfaces. In constant-current mode, the tip's z position is adjusted to keep the current constant during the scan, which takes more time. With the current kept constant, the tip-to-sample gap will also be constant; thus, the z position accurately follows the topography of the sample surface. Because of the need for adequate sample conductivity, the STM may have to be used in high vacuum since ambient air often causes the formation of insulating oxide films on the surface. The very high sensitivity may also require a high degree of isolation[45] from building vibration and acoustic sources.

When no bias voltage is applied between the tip and sample and a cantilevered sensing tip is used to map surface topography, the technique is often called *atomic force microscopy* (AFM).[46] Here, the "atomic" forces acting between the tip and sample are important and will now be briefly discussed. The interatomic force called the van der Waals force is significant for AFM; its variation with tip-to-sample gap is shown in Fig. 14.34. When tip and sample are very close (a few angstroms apart), the force is repulsive, giving the "contact" regime shown shaded. The "noncontact" regime is for gaps of tens to hundreds of angstroms; here the

[44]R. Howland and L. Benator, op. cit.

[45]R. Wiesendanger, "Scanning Probe Microscopy and Spectroscopy," Cambridge University Press, New York, 1994, pp. 84–87.

[46]R. Howland and L. Benator, op. cit., pp. 5–13.

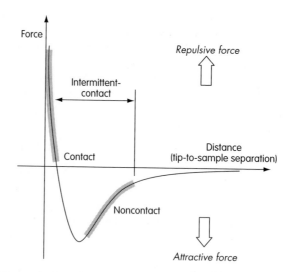

Figure 14.34
van der Waals interatomic force behavior in atomic-force microscopy.

force is attractive. AFM microscopy may use one of three modes of measurement: *contact, noncontact,* and *intermittent.*

In the contact (also called *repulsive*) mode, the cantilevered tip makes "soft" physical contact with the sample by using a cantilever with a spring constant less than the effective spring constant that holds the sample atoms together. The strength of the van der Waals force in this region means that, as the tip pushes against the sample, the cantilever *deflects* rather than forcing the tip atoms closer to the sample atoms. (If a *very* stiff cantilever is used, the sample surface will be *deformed* since the tip/sample gap will again be maintained. This phenomenon can be used to *intentionally* modify the sample surface, a process called *nanolithography,* which can also be accomplished in STM by high voltage pulses. This opens up the possibility of manipulating materials *atom by atom* to create desired properties; however, the process is at present too slow to have much practical application.) When contact mode is used in ambient air, a *capillary force* due to a thin layer of water adhering to the sample and "wicking" around the tip must also be considered. This force, about 10^{-8} N, tends to hold the tip in contact with the surface. The total force that the tip exerts on the sample is the sum of the capillary and cantilever forces, which must be balanced by the repulsive van der Waals force. The total force on the sample is typically 10^{-7} to 10^{-6} N. Contact AFM can use one of two modes: *constant height* or *constant force,* with constant force preferred for most applications. Here cantilever deflection is kept constant (using a feedback system) so that the tip follows the sample surface during the scan; the total force on the sample is also constant.

In *noncontact* AFM, a stiff cantilever is vibrated at high frequency (100 to 400 kHz) with tip/sample spacing of tens to hundreds of angstroms (see Fig. 14.34) and very low (about 10^{-12} N) sample forces. Sample topography can be measured with little or no contact or contamination from the tip, useful for soft or elastic samples or silicon wafers that must be kept "clean." Stiff cantilevers are used to prevent attractive forces from pulling the tip into contact. This, together with the low force level, means that the cantilever displacement signal is very small, requiring a sensitive "AC" type of detection. The high-frequency oscillation is provided by a separate piezoactuator attached to the base of the cantilever and operating near the resonant frequency of the cantilever, which can be modeled as a lightly damped second-order dynamic system. The resonant frequency of the cantilever depends not only on its own mass and stiffness, but also on the "stiffness" of the forces acting between tip and sample. That is, these forces are "springlike" in that they change with a changing gap, so they contribute their own stiffness to that intrinsic to the cantilever, making the resonant frequency dependent on gap. We can thus measure gap by measuring the change in resonant frequency; two methods are commonly used.[47] In one, the *amplitude* of the tip motion or the *phase angle* between the piezo input signal and the tip deflection response (both of which change when the resonant frequency changes) is measured and used in a feedback system to keep tip/sample gap constant during a scan. In the other method, direct frequency measurement, such as a frequency counter, is used.

Intermittent-contact (see Fig. 14.34) microscopy (also called *tapping mode*) is similar to the noncontact mode except that the vibrating tip is brought close enough to the sample that contact occurs at the "bottom" of travel. In imaging surface topography, it is preferred to the noncontact mode when a large scan area includes major variations in sample topography.

We have now covered what might be called the "basic" modes of SPM and now move on to a wide variety[48] of "specialty" modes. Among these are included:

1. Magnetic force microscopy (MFM)
2. Lateral force microscopy (LFM)
3. Force modulation microscopy (FMM)
4. Phase detection microscopy (PDM)
5. Electrostatic force microscopy (EFM)
6. Scanning capacitance microscopy (SCM)
7. Thermal-scanning microscopy (TSM)
8. Near-field-scanning optical microscopy (NSOM)

Most of these specialty modes require only modifications in the nature of the probe and perhaps some operating software changes; the basic instrument is largely unchanged. *Magnetic force microscopy* uses a tip coated with a thin ferromagnetic film and operates in the noncontact mode. The image contains both topographic and magnetic information, depending on the tip/sample gap; images collected at different gaps allow separation of the two effects (magnetic effects dominate at larger gaps). *Lateral force microscopy* measures twisting deflections of the cantilever caused by forces parallel to the plane of the sample surface, allowing the study of effects such as static and dynamic friction.[49] Since changes in surface slope

[47]R. Wiesendanger, op. cit., pp. 241–245.

[48]J. Leckenby, "SPM Techniques Evolve Rapidly Using Diverse Probe Materials," *R&D Magazine,* Apr. 1999, pp. 35–37.

[49]*Appl. Bull. No.* 10, Jan. 1999, CSEM Instruments, Micro Photonics, Inc., Irvine, CA, 949-461-9292 (www.microphotonics.com).

also cause lateral forces, LFM and topographic images should be collected simultaneously to separate the effects. *Force modulation microscopy* uses the contact mode but adds a high-frequency, constant-amplitude z force to the tip. Measuring the deflection caused by this force give data on the elastic properties of the sample. Since the scanner's z motion-control bandwidth is much lower than the oscillating force frequency, we can scan at constant (average) force in the usual way and use filtering to obtain both elastic and topographic data simultaneously. Force/deflection curves for slowly changing forces at a fixed (x, y) location are also used to evaluate material properties at the microlevel.[50]

Phase detection microscopy again uses a vibrating cantilever, but now the phase angle between the input force and the output motion is measured. Recall that materials have both elastic and damping properties, and that these can be measured in terms of the phase shift just mentioned. *Electrostatic force microscopy* applies a voltage between tip and sample while the cantilever hovers above the surface, allowing measurement of surface-charge carrier density. Such "voltage probing" is, for example, used to test live microprocessor chips at micro scale. *Thermal-scanning microscopy* uses a thermal probe, such as a thin film resistor,[51] integrated into a conventional contact-mode probe. Depending on the type of probe, temperature or sample thermal conductivity can be measured. *Near-field-scanning optical microscopy* uses a probe with a "light funnel" whose narrow end is about 50 nm in diameter and is scanned within 10 nm of the sample surface. The resolution limit of conventional optical microscopes is the wavelength of the light used to form the image. In NSOM, the resolution is determined by the size of the aperture, which is less than the light wavelength. There is, however, a lower usable limit to the aperture size (about 20 nm); below this there is too little light to make useful images.

Our brief discussion has presented the basic operating principles of scanning probe microscopes and listed a number of references that show how they are used to study phenomena at the micro- and nanolevel. Methods to allow more rapid scanning of larger areas are being developed.[52] A reference[53] focused on friction studies, but which also presents much useful general information, is available.

BIBLIOGRAPHY

1. Gad-el-Hak (ed.), "The MEMS Handbook," CRC, Boca Raton, 2001.

2. Lyshevski, S. E., "Nano- and Microelectromechanical Systems," CRC, Boca Raton, 2001.

3. J. Bernstein, "An Overview of MEMS Inertial Sensing Technology," *Sensors,* Feb. 2003, pp. 14–21.

---

[50]"Nanoindentation with Spherical Indenters," *Appl. Bull. No.* 11, May 1999, CSEM Instruments.

[51]M. Wendman et al., "Scanning Thermal Microscopy," Digital Instruments, Division of Veeco Inst., Foments Inc., Santo Barbara, CA, 1997, 800-873-9750 (www.di.com).

[52]S. C. Minne, S. R. Manalis, and C. F. Quate, "Bringing Scanning Probe Microscopy up to Speed," Kluwe, Boston, 1999.

[53]B. Bhushan (ed.), "Handbook of Micro/Nano Tribology," Chaps. 1, 2, 6, 7, 10, 13, and 16. Cleveland Rubber Publishing Co., Cleveland, Ohio.